Periodic Table

Representative elements: Groups 1A–2A and 3A–8A

Group 1A: Alkali metals; Group 2A: Alkaline earth metals; Group 7A: Halogens; Group 8A: Noble gases

Transition elements: the ten middle columns (Sc–Zn, Y–Cd, La–Hg, Ac–Une)

Metals | Nonmetals

Period number	1A	2A											3A	4A	5A	6A	7A	8A
1	1 **H** 1.008																	2 **He** 4.003
2	3 **Li** 6.941	4 **Be** 9.012											5 **B** 10.81	6 **C** 12.01	7 **N** 14.01	8 **O** 16.00	9 **F** 19.00	10 **Ne** 20.18
3	11 **Na** 22.99	12 **Mg** 24.31											13 **Al** 26.98	14 **Si** 28.09	15 **P** 30.97	16 **S** 32.06	17 **Cl** 35.45	18 **Ar** 39.95
4	19 **K** 39.10	20 **Ca** 40.08	21 **Sc** 44.96	22 **Ti** 47.88	23 **V** 50.94	24 **Cr** 52.00	25 **Mn** 54.94	26 **Fe** 55.85	27 **Co** 58.93	28 **Ni** 58.69	29 **Cu** 63.55	30 **Zn** 65.38	31 **Ga** 69.72	32 **Ge** 72.59	33 **As** 74.92	34 **Se** 78.96	35 **Br** 79.90	36 **Kr** 83.80
5	37 **Rb** 85.47	38 **Sr** 87.62	39 **Y** 88.91	40 **Zr** 91.22	41 **Nb** 92.91	42 **Mo** 95.94	43 **Tc** (98)	44 **Ru** 101.1	45 **Rh** 102.9	46 **Pd** 106.4	47 **Ag** 107.9	48 **Cd** 112.4	49 **In** 114.8	50 **Sn** 118.7	51 **Sb** 121.8	52 **Te** 127.6	53 **I** 126.9	54 **Xe** 131.3
6	55 **Cs** 132.9	56 **Ba** 137.3	57* **La** 138.9	72 **Hf** 178.5	73 **Ta** 180.9	74 **W** 183.9	75 **Re** 186.2	76 **Os** 190.2	77 **Ir** 192.2	78 **Pt** 195.1	79 **Au** 197.0	80 **Hg** 200.6	81 **Tl** 204.4	82 **Pb** 207.2	83 **Bi** 209.0	84 **Po** (209)	85 **At** (210)	86 **Rn** (222)
7	87 **Fr** (223)	88 **Ra** 226	89† **Ac** (227)	104 **Unq**	105 **Unp**	106 **Unh**	107 **Uns**	108 **Uno**	109 **Une**									

*lanthanides	58 **Ce** 140.1	59 **Pr** 140.9	60 **Nd** 144.2	61 **Pm** (145)	62 **Sm** 150.4	63 **Eu** 152.0	64 **Gd** 157.3	65 **Tb** 158.9	66 **Dy** 162.5	67 **Ho** 164.9	68 **Er** 167.3	69 **Tm** 168.9	70 **Yb** 173.0	71 **Lu** 175.0
†actinides	90 **Th** 232.0	91 **Pa** (231)	92 **U** 238.0	93 **Np** (237)	94 **Pu** (244)	95 **Am** (243)	96 **Cm** (247)	97 **Bk** (247)	98 **Cf** (251)	99 **Es** (252)	100 **Fm** (257)	101 **Md** (258)	102 **No** (259)	103 **Lr** (260)

Fifth Edition

Chemistry

An Introduction to General, Organic, and Biological Chemistry

Karen C. Timberlake

Los Angeles Valley College

HarperCollins*Publishers*

TO MY FAMILY, FRIENDS, AND STUDENTS

The whole art of teaching is only the art of awakening the natural curiosity of young minds.—ANATOLE FRANCE

One must learn by doing the thing; though you think you know it, you have no certainty until you try.—SOPHOCLES

Discovery consists of seeing what everybody has seen and thinking what nobody has thought.—ALBERT SZENT-GYORGI

Sponsoring Editor: Jane Piro
Development Editor: Louise Howe
Project Editor: Kristin Syverson
Art Direction: Julie Anderson
Text and Cover Design: E. Heidi Fieschko
Cover Collage and Chapter Opener Collages: E. Heidi Fieschko
Front Cover Photos: top right, Uniphoto; bottom right, E. R. Degginger; center, Comstock; bottom left, Yoav Levy/Phototake
Back Cover Photos: top right, NASA; top left, Stephen Wilkes/The Image Bank
Photo Researcher: Carol Parden
Production Administrator: Paula Keller
Compositor: York Graphic Services, Inc.
Printer and Binder: R. R. Donnelley & Sons Company
Cover Printer: The Lehigh Press, Inc.

Chemistry: An Introduction to General, Organic, and Biological Chemistry,
Fifth Edition

Library of Congress Cataloging-in-Publication Data

Timberlake, Karen.
Chemistry: an introduction to general, organic, and biological chemistry / Karen C. Timberlake.—5th ed.
p. cm.
Includes index.
ISBN 0-06-046696-0
1. Chemistry. I. Title.
QD31.2.T55 1991
540—dc20 91-37749
CIP

92 93 94 9 8 7 6 5 4 3 2

Preface

Welcome to the Fifth Edition of *Chemistry: An Introduction to General, Organic, and Biological Chemistry*. It is my hope that the reshaping of this text over five editions has resulted in a book that makes teaching and learning chemistry an enthusiastic and positive experience for both the teacher and the student. It remains my goal to assist students in their development of critical thinking and to establish a scientific framework to give students the concepts and problem-solving techniques they will need to make decisions about major issues related to the environment, medicine, and health.

Over the years I have learned that many students find chemistry a formidable subject, and thus I have made a practice of associating chemistry concepts with applications from the health sciences and related fields. This bridge between chemistry and the world of the student is designed to help students recognize the chemistry that affects their lives and future careers. I have found discussions of applications of chemistry valuable in increasing student interest, motivation, concentration, and performance in class.

New in This Edition

In response to the needs of my own students and the suggestions of teachers and reviewers, several changes have been made in this Fifth Edition.

Sample Problems

Many sample problems with complete solutions are included in every chapter. Study Checks have been added to each sample problem to give students a chance to practice the illustrated problem types. This format continually involves the student with the immediate material through active participation in the patterns of problem solving it requires.

Problem Sets

The end-of-chapter problem sets have been expanded to engage the student in solving problems with a greater range of difficulty. The more difficult problems are identified by an asterisk (*). A continuing feature is the keying of problems to the section heads and learning goals for easy reference to the chapter material. This also allows the student to select appropriate problems for any section. All of the answers to the end-of-chapter problems are given at the end of the book.

Applications

More applications to health in the form of Health Notes have been added, along with new Environmental Notes that relate chemistry to current concerns about the environment.

Appendices

Several new appendices have been included to provide additional reference material for the student. These appendices review common equalities and conversion factors, scientific (exponential) notation, percents as conversion factors, and some basic operations with the calculator including changing signs and recognizing scientific notation.

Art Program

Pedagogical changes include the use of full color to highlight major ideas. Illustrative photographs and new graphics add depth to the topics and enhance the learning experience. A larger page size allows a more attractive visual presentation. The wider margins include diagrams, tables, and photographs that illustrate the concepts.

Content

In each chapter, more difficult topics or extensions to the topics are placed at the end so they can be covered or omitted by the instructor without affecting the flow of learning throughout the chapter.

Chapter 1 incorporates the discussion of significant figures into the discussion of measurements. Chapter 2 covers temperature in conjunction with states of matter, energy, and calculations of heat energy. Chapter 3 has been reworked and includes a discussion of subshells and orbitals. Chapter 4 on bonding in compounds has been restructured to place polyatomic ions after covalent bonding. Chapter 5 covers chemical reactions and their equations and includes reactions in the atmosphere that produce smog. Also in Chapter 5, the process of writing conversion factors for the mole relationships in equations is carefully explained and then utilized in calculating quantities in reactions. Chapter 6 expands the discussion of imaging to include PET (positron emission tomography), CT (computerized tomography), and MRI (magnetic resonance imaging).

In Chapter 7 on gases, new graphics illustrate the gas relationships more clearly. All of the discussions of gas laws have been rewritten to clarify concepts. A discussion of the depletion of the ozone layer has been added. Chapter 8 covers the composition of solutions as well as their concentrations. The discussion of electrolytes and new material on solubility of salts is placed in Chapter 9 along with acids and bases and their reactions. The discussion of pH includes instructions for calculating the pH of solutions whose $[H^+]$ do not have a coefficient of 1.

Chapter 10 introduces organic chemistry by discussing the alkanes. Chapter 11 covers the unsaturated hydrocarbons and the aromatic compounds, including a new section on some chemical reactions of benzene. Chapter 12 introduces oxygen- and sulfur-containing functional groups and Chapter 13 discusses carboxylic acids and their derivatives, esters and amides. Physical properties of many classes of organic compounds are now included.

Chapter 14 begins the biomolecular section with a rewritten chapter on carbohydrates, including a new explanation of chirality of sugars and a new section on the formation of glycosidic bonds by monosaccharides. The introduction to hemiacetals and acetals now precedes Haworth structures. The formation of glycosides now precedes disaccharides. More biochemical applications have been incorporated into these chapters for classes that do not complete the biochemistry portion of the text. Chapter 15 discusses the lipid family and includes cerebrosides in glycolipids. Chapter 16 on proteins has information on sequencing of amino acids in the primary protein structure. New features include coverage of endorphins, a summary of protein structural levels, and classification of proteins by composition. Chapter 17 contains a rewritten section on classification of enzymes and enzyme inhibition. Included in this chapter is a discussion of hormones and their effects on enzyme action. The classification of enzymes, vitamins, and protein and steroid hormones is now presented in tabular form. Chapter 18, on nucleic acids, now follows enzymes and proteins to complete the biomolecules and protein synthesis. Recombinant DNA is included in the discussion of DNA replication. New features include coverage of viruses and cancer. Chapter 19 provides an exploration of metabolic pathways that produce energy in the cell and a discussion of the ways in which energy is utilized. New features include ATP energy in muscle contraction, the fermentation of pyruvic acid by yeast, and the synthesis of fatty acids and nonessential amino acids.

To Accompany the Text

Study Guide

The *Study Guide* reviews the basic concepts, provides learning drills, and gives a practice exam, all with answers, for each chapter. All sections are keyed to the learning goals in the text so students may cross-reference working tools. Students can grade their own practice exams and then check to determine whether they have mastered the material. In this way, they can identify areas of difficulty and review the material again.

Laboratory Manual

The early experiments in the *Laboratory Manual* introduce students to basic laboratory skills. Students then learn to do laboratory investigations, developing the skills of manipulating laboratory equipment, gathering and reporting data, problem solving, calculating, and drawing conclusions. In this edition, as in the past, there is an emphasis on safety in the laboratory. Hazardous chemicals and procedures that are regarded as dangerous have been omitted. For each experiment, there is a report page and a set of questions that relate the laboratory to the corresponding information in the text. Some questions require essay-type answers to promote writing skills in science.

Instructor's Manual

The *Instructor's Manual* gives chapter overviews, suggestions for lecture demonstrations, and selected solutions to worked-out problems. Experiments related to each chapter are described, and the materials required for 10 students are listed. Instructions for preparing any special solutions are included, and the laboratory skills to be demonstrated for the students are described. A complete set of sample laboratory reports from the laboratory manual completes the instructor's manual.

Test Bank

The printed **Test Bank** contains more than 500 multiple-choice, true-false, and matching questions and their answers. The test bank is also available on **Harper-Test** for the IBM-PC or Macintosh computer. This software allows you to scramble questions, add new questions, and select questions based on level of difficulty.

Transparencies

Instructors who adopt the test will receive a set of 100 two- and four-color transparency acetates providing figures and illustrations from the text.

Acknowledgments

I wish to thank my husband Bill for his invaluable assistance, repeated reviews of manuscripts and supplements, and continued support. Thanks to my son John for his cooperation and help in the preparation of this edition of the text and supplements.

The following people provided outstanding reviews that were extremely helpful with their constructive and helpful criticism and support, for which I am most grateful:

Anne Barber, Manatee Community College
Thomas M. Barnett, Johnson County Community College
Thomas Berke, Brookdale Community College
John DeKorte, Northern Arizona University
John Goodenow, Lawrence Technological University
F. H. Kruse, University of Toledo Community and Technical College
Leslie Lovett, Fairmont State College
Stanley Mehlman, State University of New York, Farmingdale
Mary Bethé Neely, Pike's Peak Community College
Roy Stein, University of Toledo Community and Technical College
Conrad Valdez, East Los Angeles College

Many thanks to the people at HarperCollins Publishers who believe in my approach to teaching chemistry and continually encourage and support me: Jane Piro, Editor; Louise Howe, Developmental Editor; Kristin Syverson, Project Editor; and Heidi Fieschko, Designer.

Writing a chemistry text is an ongoing process as students, teachers, and concepts change. I believe that teaching chemistry involves more than a transmission of chemical facts; teaching also means fostering positive attitudes toward science, encouraging students to use new thinking patterns and problem-solving techniques, and helping students to develop their reasoning powers. With this aim, I have revised *Chemistry*. I look forward to your use of this text and to hearing from both you and your students. I welcome any suggestions, criticism, or overall comments on this revision.

Karen C. Timberlake
Los Angeles Valley College
Van Nuys, CA 91401

Contents

3 Atoms and Elements 74

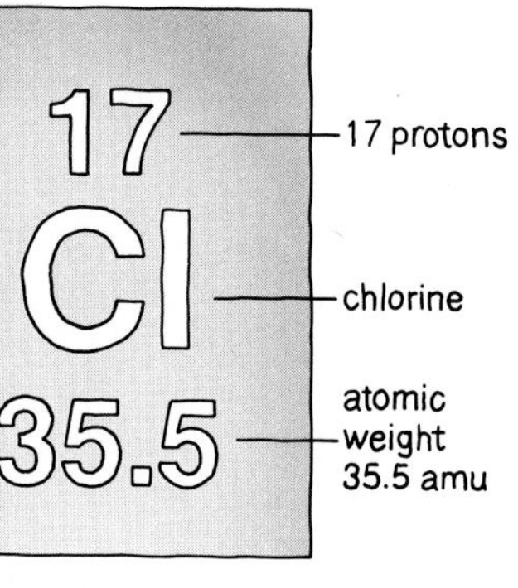

4 Compounds and Their Bonds 109

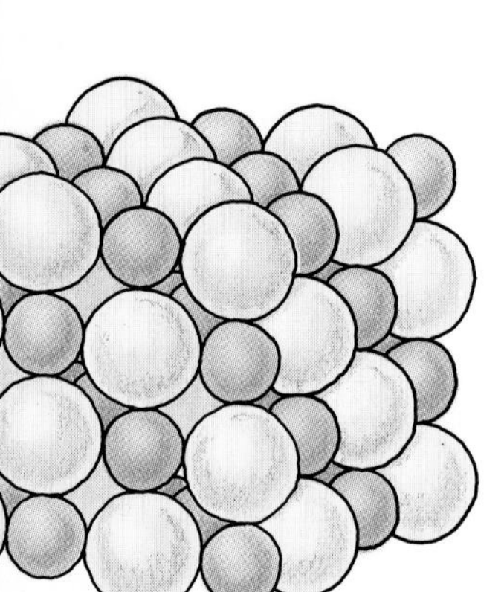

5 Chemical Quantities and Reactions 153

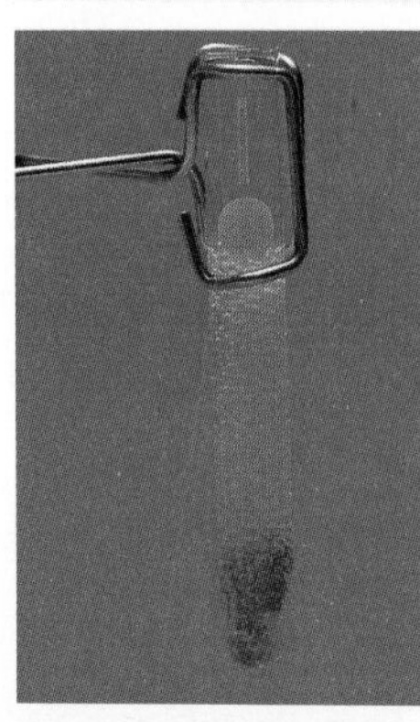

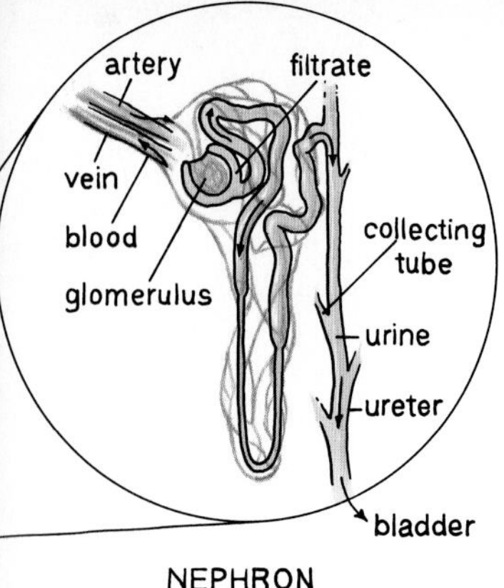

NEPHRON

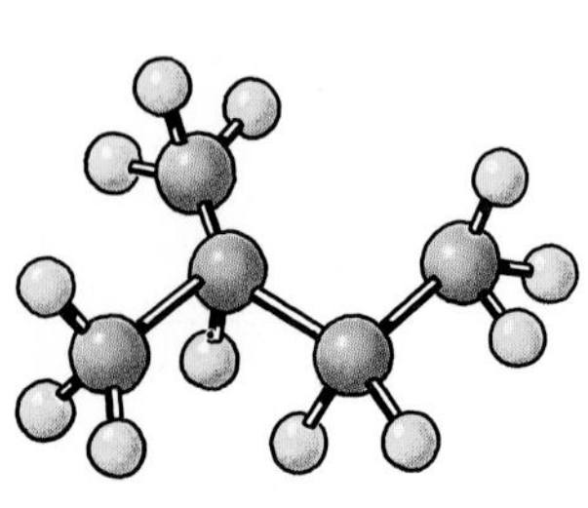

$$\begin{array}{c} CH_3 \\ | \\ CH_3CHCH_2CH_3 \end{array}$$
methylbutane

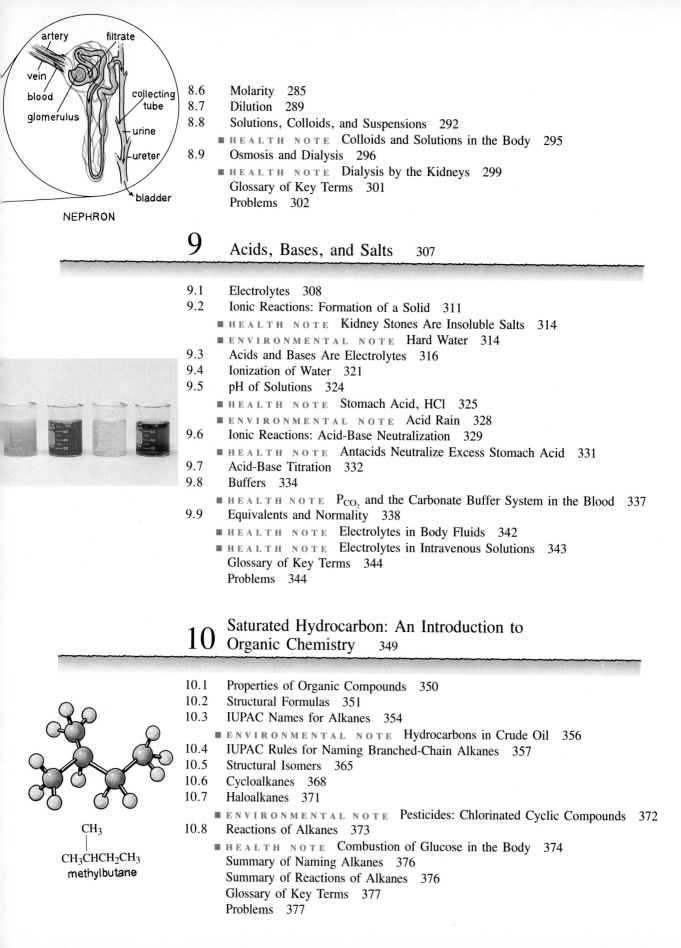

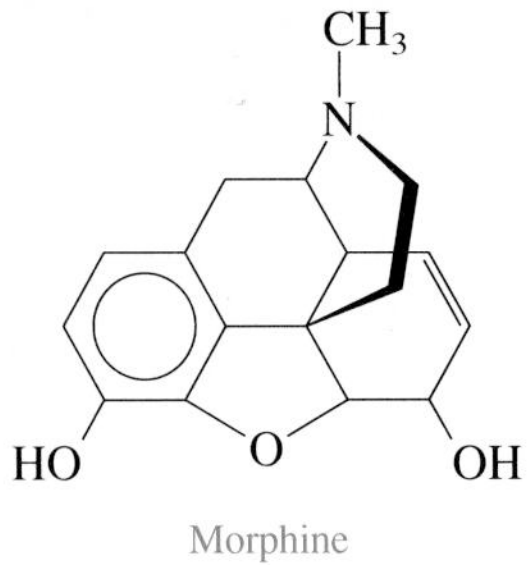

Morphine (opium)

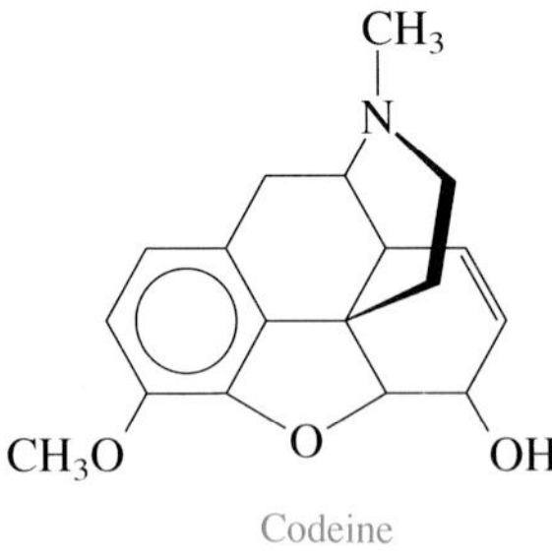

Codeine

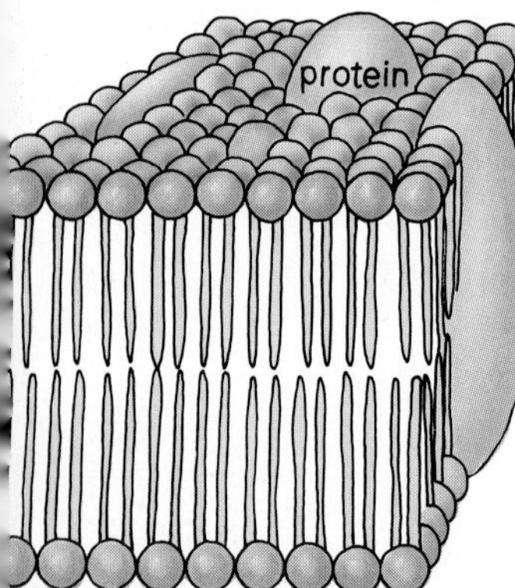

substrate

active site

enzyme

O
CH_3
HN
O
N
H

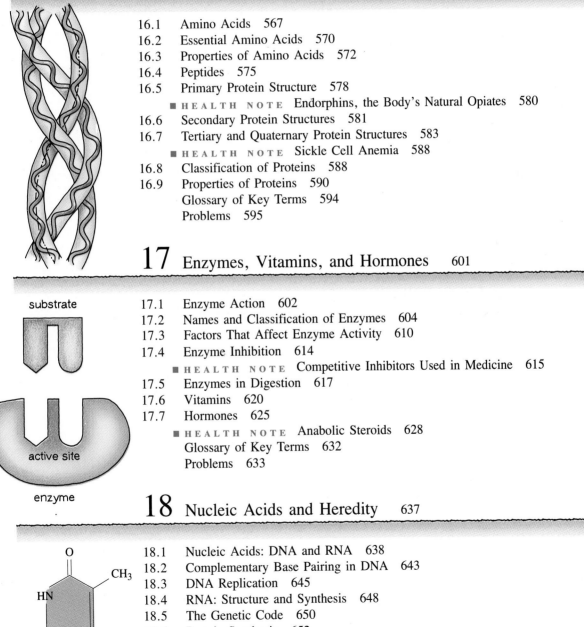

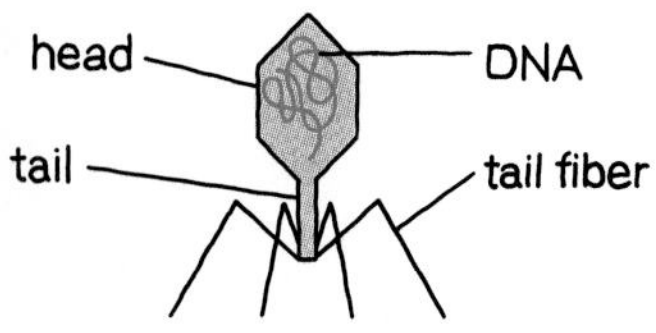

19 Metabolic Pathways and Energy Production 668

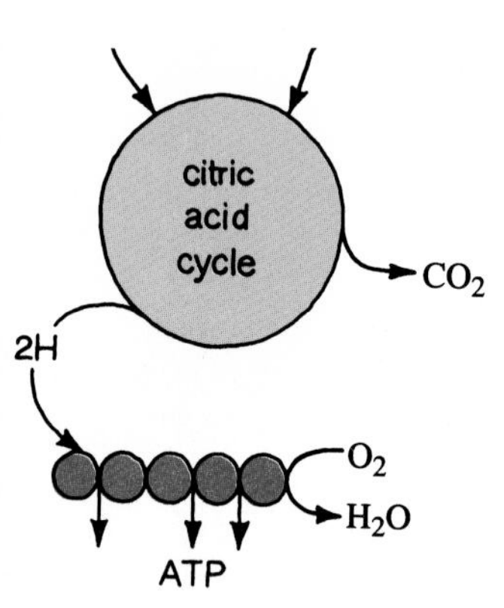

Appendices

To the Student

Here you are in chemistry, perhaps because you need a science course or just because you want to find out something about chemistry. Maybe you want to be a nurse or respiratory therapist or to enter some profession in the health sciences. If so, as you progress through this text, you will discover that chemistry is indeed exciting to learn and that it has an important relationship to the world around you. Every chapter in this Fifth Edition includes many applications of chemistry to health, medicine, and the environment. Your interest in the sciences will help you learn chemistry, and by learning chemistry you will gain a deeper understanding of physiology, medical care, and major issues of today, including pollution, global warming, the ozone layer, acid rain, nuclear energy, and recombinant DNA.

I have designed this text with you in mind. To aid your learning process, each chapter begins with a set of learning goals that tell you what to expect in the chapter and what you need to accomplish. Also at the beginning of each chapter is a section called **Scope** that relates experiences that involve the text material to your life or to specific health science areas and sets the stage for the chemical concepts discussed throughout the chapter.

As you progress through each chapter, take time to consider the learning goal for each section. To see if you have mastered the goal, do the example problems and study checks in the section. If you have difficulty with an example, study that part of the unit again before you proceed. For further self-testing, work the problems that are keyed to each section at the end of the chapter. You can check your answers by referring to the answer section in the back of the text. It is not necessary to study a chapter all the way through at one time. Instead, you may wish to cover only a few of the goals each time you study. Also, you can prepare for lecture by reading ahead in the text.

To review your knowledge of the important ideas in a chapter, read over the glossary at the end of the chapter. Study the tables and figures, which emphasize important concepts.

The study of chemistry involves some hard work, but I hope that you will find the effort rewarding when you see and understand the role of chemistry in many related fields. If you would like to share your feelings about chemistry or comments about this text, I would appreciate hearing from you.

Karen C. Timberlake
Los Angeles Valley College
Van Nuys, CA 91401

1L = 10dL
1L = 1000mL
1dL = 100mL

1

Measurements

Learning Goals

1.1 Write the names and symbols for the metric units used in measurements of length, volume, and mass.

1.2 Write the numerical value or name of a metric prefix.

1.3 Write an equality for two metric units used in measurements of length, volume, or mass.

1.4 Report answers to calculations using the correct number of significant figures.

1.5 Write a conversion factor for two units that describe the same quantity.

1.6 Set up a problem using one or more conversion factors to change from one unit to another.

1.7 Calculate the density or specific gravity of a substance, and use the density or specific gravity to calculate the mass or volume of a substance.

The metric system has been adopted by scientists and health professionals all over the world. In many countries, it is used in everyday settings as well.

SCOPE

What kinds of measurements did you make today? Perhaps you checked your weight by stepping on a scale this morning. Perhaps you measured out two eggs, two cups of flour, and a cup of milk to make your pancake batter. You may have filled the gas tank of your car with 10 gallons of gasoline and driven 12 miles to school. In each case, you were concerned with mass (weight), volume, or distance. Try to recall some of the measurements you made today.

Men and women in the health sciences use measurement every day to evaluate the health of a patient. Temperatures are taken, weights and heights are recorded, and samples of blood and urine are collected for laboratory testing. Medications are given in dosages that must be accurately measured. In the dental office, hygienists measure solutions used in fluoride treatments and in the preparation of dental materials.

1.1 Units of Measurement

Learning Goal **Write the names and symbols for the metric units used in measurements of length, volume, and mass.**

Suppose that during a recent medical examination the nurse recorded your mass (weight) as 70.0 kilograms (70.0 kg), your height as 1.78 meters (1.78 m), and your temperature as 37.0 degrees Celsius (37.0°C). The system of measurement used in these clinical evaluations is the **metric system.** Perhaps you would be more familiar with these measurements if they were stated in the American system. Then your weight would be 154 pounds (154 lb); your height, 5 feet 10 inches (5 ft 10 in.); and your temperature, 98.6 degrees Fahrenheit (98.6°F).

The metric system is used by scientists and health professionals throughout the world. It is also the common measuring system in all but a few countries in the world. In 1960, a modification of the metric system called the **International System of Units,** Système Internationale **(SI),** was adopted to provide additional uniformity. In this text, we will use metric units and introduce some of the corresponding SI units that are in use today.

Length

The metric unit **meter (m)** is used in measurements of length such as height or distance. The meter is about 39.4 in. in length, which makes it slightly longer than a yard (yd). The meter is also the SI unit for measuring length. The unit **centimeter (cm),** used to represent smaller distances, is about as wide as your little finger. For comparison, there are exactly 2.54 cm in 1 in.

$$1 \text{ m} = 39.4 \text{ in.}$$

$$2.54 \text{ cm} = 1 \text{ in.}$$

Figure 1.1 compares metric and American units of length.

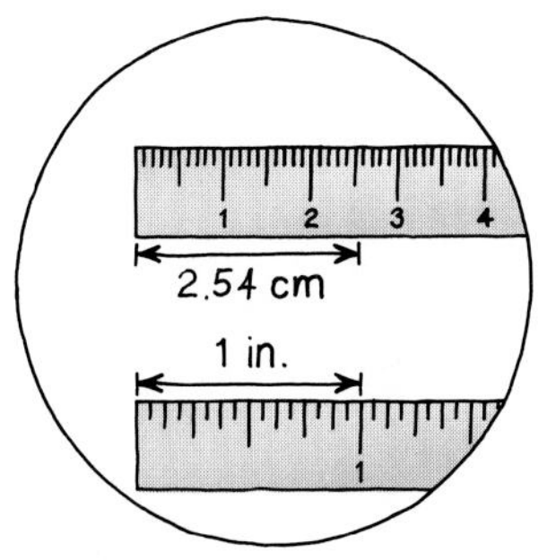

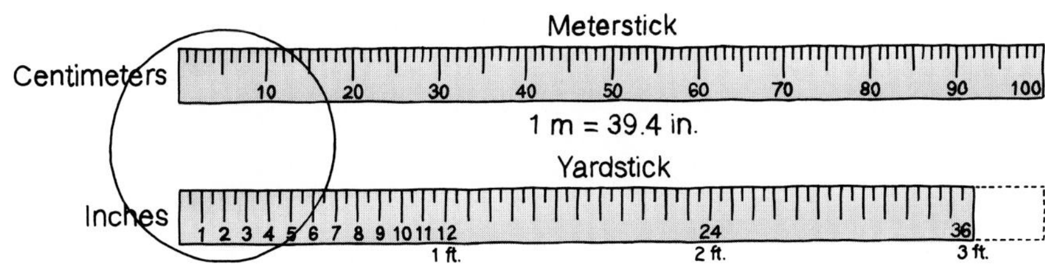

Figure 1.1
Comparison of metric and American units for measuring length.

Volume

Volume is the amount of space occupied by a substance. In chemistry and the health professions, the metric unit **liter (L)** is commonly used to measure volume. A liter is slightly larger than the quart (qt) unit of volume used in the American system. A smaller unit of volume called the **milliliter (mL)** is commonly used in hospitals and laboratories.

1 L = 1.06 qt

1 qt = 946 mL

The SI unit of volume is the cubic meter (m^3), which is equal to 1000 L. A comparison of metric and American units for volume is shown in Figure 1.2.

Mass

The **mass** of an object is a measure of the quantity of material it contains. Everything has mass. Rocks, water, people, and dogs all have mass. In the metric system, the unit for mass is the **gram (g).** A larger unit of mass, the **kilogram (kg),** is used for larger masses, such as body weight. The kilogram is also the SI unit for mass. It takes 454 g to make 1 lb, and 2.20 lb is needed to equal the mass of 1 kg. Therefore, 1 kg is larger than 1 lb.

1 lb = 454 g

2.20 lb = 1 kg

Figure 1.3 illustrates the relationship between kilograms and pounds.

Figure 1.2
Comparison of metric and American units for measuring volume.

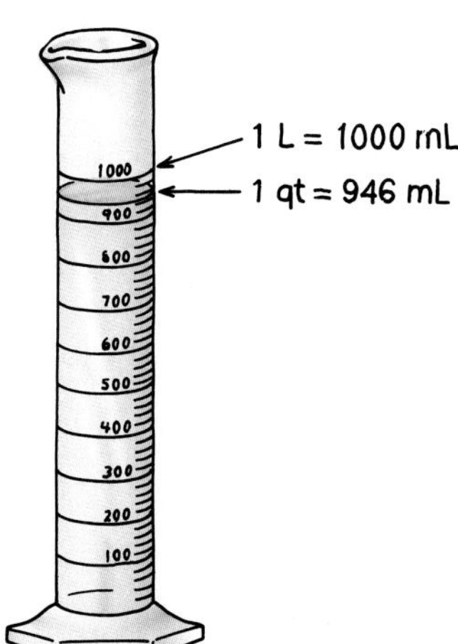

Mass and Weight

You may be more familiar with the term **weight** than with mass. However, weight and mass are not the same thing. The weight of an object depends on the pull of gravity on the object as well as on its mass. Therefore the weight of an object can vary with its location. An astronaut with a mass of 75.0 kg, for example, has a weight of 165 lb on Earth. On the moon, the gravitational pull is

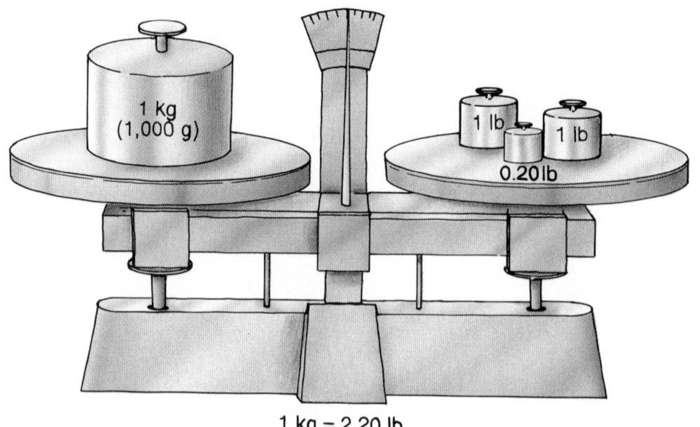

Figure 1.3
Comparison of metric and American units for measuring mass.

one-sixth that on Earth, and the astronaut has a weight of 27.5 lb. However, the mass of the astronaut remains the same as on Earth, 75.0 kg. Scientists measure mass rather than weight because mass does not vary with location.

A summary of some metric and SI units used in measurement is given in Table 1.1.

Table 1.1 Units of Measurement

Type of measurement	Metric	SI
Length	Meter (m)	Meter (m)
Volume	Liter (L)	Cubic meter (m^3)
Mass	Gram (g)	Kilogram (kg)
Time	Second (s)	Second (s)
Temperature	Celsius (°C)	Kelvin (K)

Sample Problem 1.1
Units of Measurement

Complete the following table:

Type of Measurement	*Metric Unit*	*Symbol*
Length	________	____
________	Liter	____
________	________	g

Solution:

Length	Meter	m
Volume	Liter	L
Mass	Gram	g

Study Check:
What type of measurement would use the unit centimeter?

Answer: length

1.2 Metric Prefixes

Learning Goal **Write the numerical value or name of a metric prefix.**

The National Research Council provides recommended daily allowances (RDA) of nutrients for individuals according to age and physical characteristics. Some of its recommendations for a female 23–50 years old are listed in Table 1.2.

You may notice that the gram is the basic unit of mass in all of the RDA values. However, most of the values are given in terms of smaller, more convenient units of mass, milligrams and micrograms. These smaller units of mass are indicated by using **prefixes.** Note that the prefixes *milli* and *micro* have been placed in front of the unit gram.

In the metric system, prefixes are used to indicate changes in the sizes of units by factors of 10. For example, the RDA for vitamin C is 60 milligrams. When the prefix *milli* precedes the unit gram, it makes a smaller mass unit, milligram (mg), that is equal to 0.001 g. Table 1.3 lists some of the metric prefixes, their symbols, and their decimal values. These prefix values may also be expressed in scientific notation using powers of 10, a topic reviewed in the Appendix.

Table 1.2 RDA for a 55-Kilogram Female 23–50 Years Old

Nutrient	Amount recommended
Protein	44 grams
Vitamin C	60 milligrams
Vitamin D	5 micrograms
Calcium	800 milligrams
Iron	18 milligrams
Iodine	150 micrograms
Zinc	15 milligrams

Table 1.3 Metric Prefixes[a]

Prefix	Symbol	Numerical value		Scientific notation
		Prefixes That Increase the Size of the Unit		
Mega	M	One million	1,000,000	10^6
Kilo	k	One thousand	1,000	10^3
Hecto	h	One hundred	100	10^2
Deka	da	Ten times	10	10^1
		Prefixes That Decrease the Size of the Unit		
Deci	d	One-tenth	0.1	10^{-1}
Centi	c	One-hundredth	0.01	10^{-2}
Milli	m	One-thousandth	0.001	10^{-3}
Micro	μ	One-millionth	0.000001	10^{-6}
Nano	n	One-billionth	0.000000001	10^{-9}

[a] Prefixes in boldface are used most often.

Numerical Values of Prefixes

The meaning of a metric unit that has a prefix can be written by replacing the prefix with its numerical value. For example, the prefix *kilo* in *kilometer* has the same meaning as 1000. Therefore a kilometer is a length equal to 1000 meters.

1 kilometer (1 km) = 1000 meters (1000 m)

1 kiloliter (1 kL) = 1000 liters (1000 L)

1 kilogram (1 kg) = 1000 grams (1000 g)

Similarly, the prefix *centi* has the same meaning as 0.01, and a centimeter is a length equal to 0.01 meter.

1 centimeter (1 cm) = 0.01 meter (0.01 m)

1 centiliter (1 cL) = 0.01 liter (0.01 L)

1 centigram (1 cg) = 0.01 gram (0.01 g)

Sample Problem 1.2
Prefixes

Fill in the blanks with the correct prefix or numerical value:

a. kilogram = ______ g
b. ______ liter = 0.001 L
c. centimeter = ______ m

Solution:
a. 1000 g or 10^3 g **b.** *milli*liter **c.** 0.01 m or 10^{-2} m

Study Check:
Write the numerical values for the following:
a. millimeter **b.** deciliter

Answer: **a.** 0.001 m **b.** 0.1 L

1.3 Writing Numerical Relationships

Learning Goal **Write an equality for two metric units used in measurements of length, volume, or mass.**

Length Equalities

In the health sciences, many of the objects that we measure are much smaller than a meter. An ophthalmologist may measure the diameter of the retina of the eye in centimeters (cm), and the microsurgeon may need to know the length of a nerve in micrometers (μm).

When the prefix *centi* is used with the unit meter, it indicates the unit centimeter, a length that is one-hundredth of a meter (0.01 m). A millimeter measures a length of 0.001 m. There are 1000 mm in a meter. If we compare the lengths of a millimeter and a centimeter, we find that 1 mm is 0.1 cm; there are 10 mm in 1 cm.

Some Length Equalities

1 m = 100 cm

1 m = 1000 mm

1 cm = 10 mm

Some metric units for length are compared in Figure 1.4 and examples are shown in Figure 1.5.

Figure 1.4
Comparison of some metric units used to measure length.

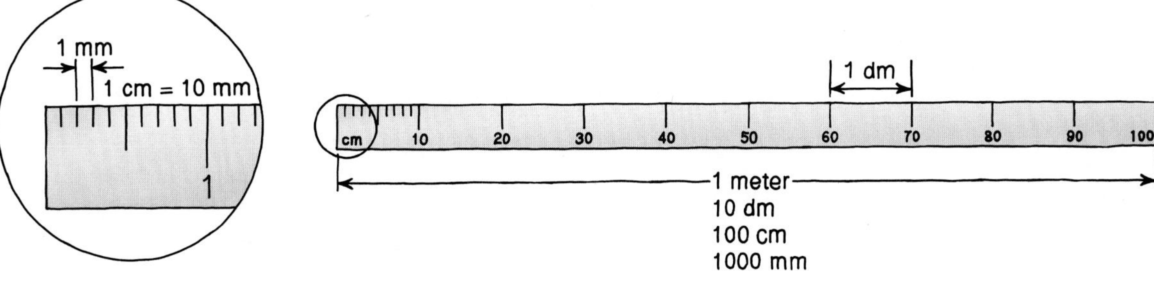

Figure 1.5
Examples of some metric measurements of length.

a dime is about 1mm thick

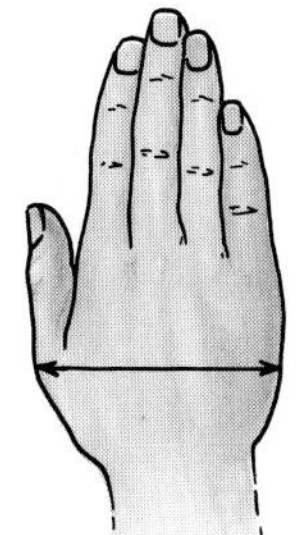

the width of the little finger is about 1cm

the width of the hand is about 1 dm

the full length of a football field is about 10 dam or 1 hm

the length of 10 football fields is about 1 km

Table 1.4 Some Typical Laboratory Test Values

Substance in blood	Typical range	
Albumin	3.5–5.0	g/dL
Ammonia	20–150	μg/dL
Calcium	8.5–10.5	mg/dL
Cholesterol	105–250	mg/dL
Iron (male)	80–160	μg/dL
Protein (total)	6.0–8.0	g/dL

Figure 1.6
Equipment used to measure liquid volumes in the hospital or in the chemistry laboratory.

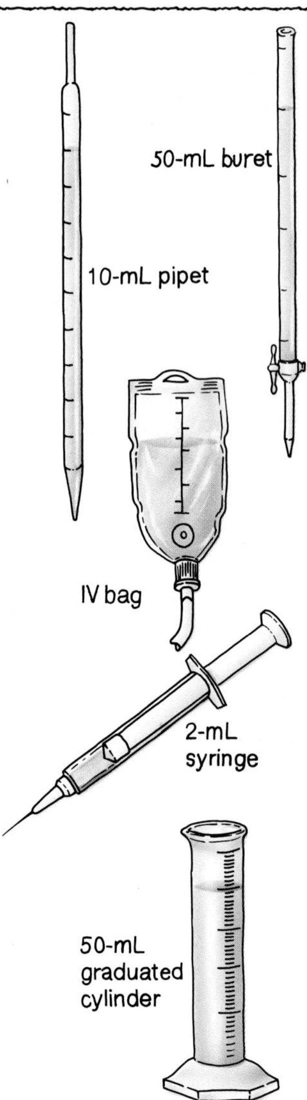

Volume Equalities

Volumes of 1 L or smaller are common in the health sciences. An intravenous solution (IV) is usually made up in a 1-L quantity. When a liter is divided into 10 equal portions, each portion is a deciliter (dL). There are 10 dL in 1 L. Laboratory results for blood work are often reported in deciliters. Notice the values listed in Table 1.4 for some substances in the blood.

Dividing a liter into 1000 parts gives the smaller volume called the milliliter. A 1-L bottle of physiological saline contains 1000 mL of the solution. Intravenous liquids are also given to patients from bottles containing 500 mL or 250 mL of a solution. Small amounts of liquids measured in a few milliliters or microliters (μL) are sometimes added to the IV solution or given as injections.

Some Volume Equalities

1 L = 10 dL

1 L = 1000 mL

1 dL = 100 mL

Typical equipment for measuring the volume of liquids in the hospital or in the chemistry laboratory is shown in Figure 1.6.

The unit **cubic centimeter** (**cm^3** or **cc**) refers to the volume of a cube whose dimensions are 1 cm on each side. A cubic centimeter has the same volume as a milliliter, and the units are often used interchangeably:

$$1\ cm^3 = 1cc = 1\ mL$$

When you see *1 cm,* you are reading about length; when you see *1 cc* or *1* cm^3 or *1 mL,* you are reading about volume. A comparison of units of volume is given in Figure 1.7.

Mass Equalities

Several units for mass are used in the health sciences. A patient's mass is recorded in kilograms, and the results of a laboratory test are reported in grams,

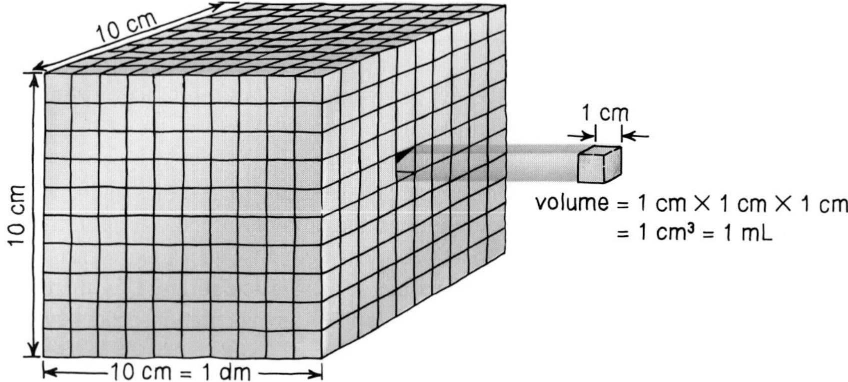

Figure 1.7
A cube measuring 10 cm on each side has a volume of 1000 cm^3, or 1 L; a cube measuring 1 cm on each side has a volume of 1 cm^3, or 1 mL.

milligrams (mg), or micrograms (μg). A kilogram is equal to 1000 g. One gram represents the same mass as 1000 mg, and 1 mg equals 1000 μg.

Some Mass Equalities

1 kg = 1000 g

1 g = 1000 mg

1 mg = 1000 μg

Sample Problem 1.3
Writing Metric Relationships

1. Place the following units in order, from smallest to largest:
   a. centimeter, kilometer, millimeter
   b. L, mL, kL, dL
   c. cg, kg, mg, g
2. Complete the following list of metric equalities:
   a. 1 L = ____ dL   c. 1 m = ____ cm
   b. 1 km = ____ m   d. 1 mg = ____ g

Solution:

1. a. millimeter, centimeter, kilometer
   b. mL, dL, L, kL
   c. mg, cg, g, kg
2. a. 10 dL   b. 1000 m   c. 100 cm   d. 0.001 g

Study Check:

1. Place the following in order, from largest to smallest:
   μg, kg, g, mg
2. Complete: a. 1 kg = ____ g   b. 1 mL = ____ L

Answer: 1. kg, g, mg, μg   2. a. 1000 g   b. 0.001 L

1.4 Significant Figures

Learning Goal **Report answers to calculations using the correct number of significant figures.**

In science, we measure such things as the length of a bacterium, the volume of a liquid, and the amount of cholesterol in a blood sample. The numbers used to report our measurements are called measured numbers. Often these measured numbers are used in calculations. It is therefore necessary to know how to use measured numbers in calculations and how to report final answers properly.

Significant Figures

In a clinical exercise, several students measured the height of a patient and recorded the following:

172.7 cm 172.6 cm 172.4 cm 172.8 cm

Although the first three digits (172) in all four measurements are identical, there are differences in the value of the last digit. Such variations occur when we reach the limit of measurement provided by our measuring device. The last digit in a measured number is obtained by estimation. This means that measurements of the same patient's height by different observers can have some variation or uncertainty in the last digit. The numbers that are reported in a measurement, including the last estimated digit, are called **significant figures.** All of the measured heights of the patient have four significant figures.

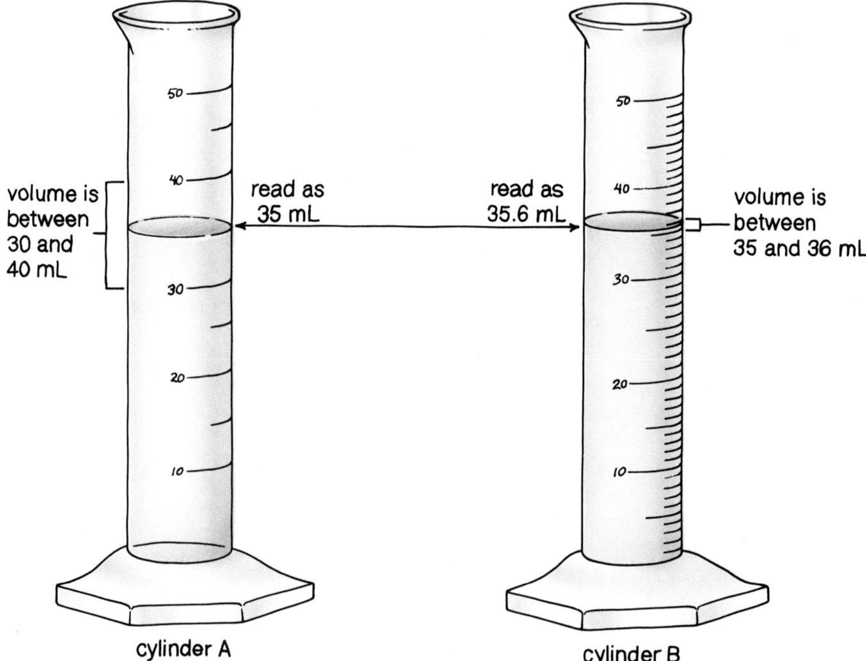

Figure 1.8
The number of units that can be used to report a volume of liquid differs with the calibrations (markings) on the graduated cylinder used. Using cylinder A, a measurement of 35 mL, which has two significant figures, can be reported. Using cylinder B, a measurement of 35.6 mL, which has three significant figures, can be reported.

When different measuring tools are used, the reported measurements have different numbers of significant figures. As shown in Figure 1.8, there are no divisions on the cylinder between 30 and 40 mL. A volume of 35 mL is recorded by estimating the mL divisions. On cylinder B, there are 1 mL divisions. The volume between 35 and 36 mL is recorded as 35.6 mL by estimating the tenths (0.1) mL divisions.

Zeros in Measured Numbers

When a zero is a part of a measured number, it counts as a significant figure. However, when a zero is used as a placeholder, it is not significant. Therefore, a zero that precedes a recorded number is not significant. Also, in very large numbers it may be difficult to tell if zeros following a number are placeholders. Zeros in large numbers are not usually significant.

Rules for Counting Significant Figures

	Examples of Measured Numbers	*Number of Significant Figures Present*
1. A number is significant if it is		
a. a nonzero digit	122.35 cm	5
	4.5 g	2
b. a zero between nonzero digits	205 mm	3
	5.082 kg	4
c. a zero that follows a number after the decimal point	5.0 cm	2
	21.40 g	4
2. A zero is not significant if it appears		
a. in front of a number	0.055 m	2
	0.0004 lb	1
b. in a large number with no decimal point	84,000 m	2
	1,065,000 kg	4

Exact Numbers

The rules for counting significant figures apply only to measured numbers. Not all numbers are measured; you can also obtain a number by counting. If you buy eight doughnuts, you have exactly eight doughnuts to eat. There is no uncertainty here because an exact number of doughnuts is counted.

Definitions that are made within the same measuring system are also exact. In the American system, 1 ft is defined as equal to exactly 12 in. In the metric system, 1 m is defined as exactly 100 cm. Since these relationships are not obtained by measurement but are counted or defined, they have no uncertainty

Table 1.5 Some Exact Numbers

	Defined numbers	
Counted Numbers	**American system**	**Metric system**
Eight doughnuts	1 ft = 12 in.	1 L = 1000 mL
Two baseballs	1 qt = 4 cups	1 m = 100 cm
Five capsules	1 lb = 16 ounces	1 kg = 1000 g

associated with them. The rules of significant figures are not applied to counted numbers or to defined numbers. Some examples of counted and defined exact numbers are listed in Table 1.5.

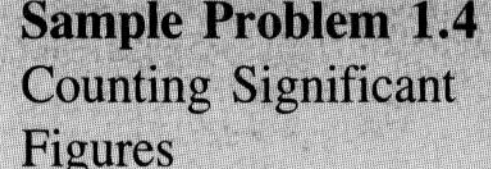

Sample Problem 1.4
Counting Significant Figures

State the number of significant figures in each of the following measured quantities:

a. 24.2 cm
b. 40.1 kg
c. 0.00025 nm
d. 450,000 km
e. 2.440 g

Solution:

a. three **b.** three **c.** two **d.** two **e.** four

Study Check:

State the number of significant figures in the following measured numbers:

a. 0.00035 g **b.** 2000 m **c.** 2.0045 L

Answer: **a.** two **b.** one **c.** five

Rounding Off Calculator Results

In mathematical calculations, the degree of uncertainty in the measurements must be indicated by the number of significant figures in the final answer. For example, dividing a mass of 7.2 g by a volume of 3.8 mL gives a calculator answer of 1.8947368 g/mL. However, each of the measurements has just two significant figures, so the calculator result must be rounded off to give a final answer with two significant figures.

$$\frac{7.2\text{ g}}{3.8\text{ mL}} \;\boxed{=}\; 1.8947368 = 1.9\text{ g/mL}$$

Calculator result — Final answer, rounded off

Rules for Rounding Off

1. If the first digit to be dropped is 4 or less, it and all following digits are simply dropped from the number.
2. If the first digit to be dropped is 5 or higher, the last retained digit of the number is increased by 1.

	Three Significant Figures	*Two Significant Figures*
Example: 8.4234 rounds off to	8.42	8.4
Example: 14.780 rounds off to	14.8	15

Sample Problem 1.5
Rounding Off

Round off each of the following numbers to give the number of significant figures indicated:

a. 35.7823 (to three figures)
b. 0.002624 (to two figures)
c. 381.268 (to four figures)
d. 0.578 (to one figure)

Solution:
a. 35.8 **b.** 0.0026 **c.** 381.3 **d.** 0.6

Study Check:
Round off each of the following numbers to three significant figures:
a. 0.12256 **b.** 214.876 **c.** 87.025

Answer: **a.** 0.123 **b.** 215 **c.** 87.0

Determining Significant Figures in Multiplication and Division

Using a calculator will permit you to solve problems much faster than you can without one. However, calculators cannot think for you. It is up to you to enter the numbers correctly, press the right keys, and adjust the calculator's result to give a proper answer. In multiplication and division, the final answer has the same number of digits as the measurement with the *fewest* significant figures (here abbreviated SF).

Multiply the following measured numbers: 24.65 by 0.67.

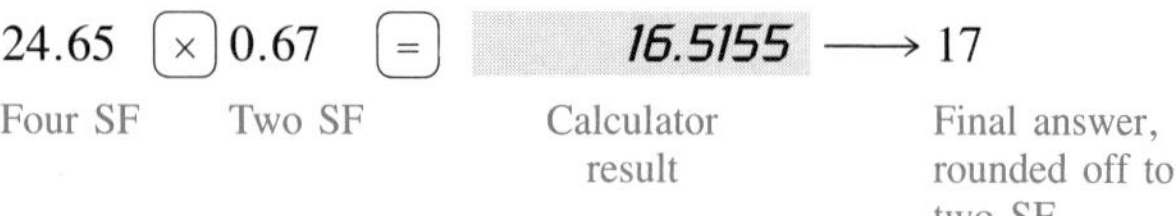

The measurement 0.67 has the least number of significant figures, two. Therefore the calculator answer is rounded off to two significant figures.

Solve the following:

$$\frac{2.85 \times 67.4}{4.39}$$

To do this problem on a calculator, enter the number and then press the operation key. In this case, we might press the keys in the following order:

2.85	⊠ 67.4	⊞ 4.39	=	43.756264	⟶ 43.8
Three SF	Three SF	Three SF		Calculator result	Final answer, rounded off to three SF

The measurements in this problem all have three significant figures. Therefore, the calculator result is rounded off to give a final answer, 43.8, that has three significant figures.

Sometimes a calculator result appears as a small whole number. You may then need to add significant zeros to obtain the final answer with the correct number of significant figures. For example, if the calculator display shows only a 4, and you need three significant numbers in your answer, the adjusted final answer is 4.00.

8.00 [÷] 2.00 [=] 4 ⟶ 4.00 (final answer with zeros added)

Three SF — Three SF

Sample Problem 1.6
Significant Figures in Multiplication and Division

Perform the following multiplication and division calculations and give the answers with the correct number of significant figures:

a. 56.8×0.37

b. $\frac{71.4}{11}$

c. $\frac{(2.075)\,(0.585)}{(8.42)\,(0.0045)}$

d. $\frac{25.0}{5.00}$

Solution:

a. Calculator answer: 21.016 ⟶ 21

b. Calculator answer: 6.4909091 ⟶ 6.5

c. Calculator answer: 32.036817 ⟶ 32

d. Calculator answer: 5 ⟶ 5.00 (must add significant zeros)

Study Check:
Solve:

a. 45.26×0.01088
b. $2.6 \div 324$
c. $\dfrac{4.0 \times 8.00}{16}$

Answer: **a.** 0.4924 **b.** 0.0080 or 8.0×10^{-3} **c.** 2.0

Determining Significant Figures in Addition and Subtraction

In addition and subtraction, the final answer is reported to the same number of decimal places as the least accurately measured number in the calculation. Consider the following examples.

Add:

2.045	(thousandths)
+ 34.1	(tenths; least accurately measured)
36.145	(calculator result)
36.1	(answer, rounded off to tenths place)

Subtract:

255	(ones)
− 175.65	(hundredths)
79.35	(calculator result)
79	(answer, rounded off to ones place)

Sample Problem 1.7
Significant Figures in Addition and Subtraction

Perform the following addition and subtraction calculations and give the answers with the correct number of significant figures:

a. 27.8 cm + 0.235 cm
b. 104.45 mL + 0.838 mL + 46 mL
c. 153.247 g − 14.82 g

Solution:
a. 28.0 cm **b.** 151 mL **c.** 138.43 g

Study Check:
Solve:

a. 82.45 mg + 1.245 mg + 0.00056 mg
b. 4.259 L − 3.8 L

Answer: **a.** 83.70 mg **b.** 0.5 L

1.5 Conversion Factors

Learning Goal **Write a conversion factor for two units that describe the same quantity.**

Many problems in chemistry and the health sciences require a change of units. You make changes in units every day. For example, suppose you spent 2.0 hours (hr) on your homework, and your instructor asked you how many minutes that was. You would answer, 120 minutes (min). Both quantities describe the same amount of time, but in different units. You knew how to change from hours to minutes because you knew a relationship (1 hr = 60 min) that involved the two units. Such a relationship is called an equality. (Recall that we used equalities earlier, in Section 1.3.) Note that each quantity has both a number and a unit.

An Equality

First Quantity		*Second Quantity*
1 hr	=	60 min
Number + Unit		Number + Unit

An equality is helpful in solving problems because we can use it to write **conversion factors,** which relate the given unit to the desired unit. One of the quantities is written as the numerator, and the other is written as the denominator. Two factors are always possible from any relationship. Be sure to include the units when you set up the conversion factor.

Conversion Factors From the Equality 1 hr = 60 min

$$\frac{\text{Numerator} \rightarrow}{\text{Denominator} \rightarrow} \frac{60\text{ min}}{1\text{ hr}} \quad \text{and} \quad \frac{1\text{ hr}}{60\text{ min}}$$

These factors are read as "60 minutes per 1 hour" and "1 hour per 60 minutes." The term *per* means "divide." Some common relationships and their corresponding conversion factors are given in Table 1.6. It is important that the equality you select to construct a conversion factor is a true, or correct, relationship.

Table 1.6 Some Common Relationships and Their Corresponding Conversion Factors

Relationship	Conversion factors
1 yard = 3 feet	$\frac{1\text{ yd}}{3\text{ ft}}$ and $\frac{3\text{ ft}}{1\text{ yd}}$
1 dollar = 100 cents	$\frac{1\text{ dollar}}{100\text{ cents}}$ and $\frac{100\text{ cents}}{1\text{ dollar}}$
1 hour = 60 minutes	$\frac{1\text{ hr}}{60\text{ min}}$ and $\frac{60\text{ min}}{1\text{ hr}}$
1 gallon = 4 quarts	$\frac{1\text{ gal}}{4\text{ qt}}$ and $\frac{4\text{ qt}}{1\text{ gal}}$

Table 1.7 Some Useful Metric Conversion Factors

Metric relationship (equality)	Conversion factors		
Length			
1 m = 1000 mm	$\frac{1\text{ m}}{1000\text{ mm}}$	and	$\frac{1000\text{ mm}}{1\text{ m}}$
1 cm = 10 mm	$\frac{1\text{ cm}}{10\text{ mm}}$	and	$\frac{10\text{ mm}}{1\text{ cm}}$
Volume			
1 L = 1000 mL	$\frac{1\text{ L}}{1000\text{ mL}}$	and	$\frac{1000\text{ mL}}{1\text{ L}}$
1 dL = 100 mL	$\frac{1\text{ dL}}{100\text{ mL}}$	and	$\frac{100\text{ mL}}{1\text{ dL}}$
Mass			
1 kg = 1000 g	$\frac{1\text{ kg}}{1000\text{ g}}$	and	$\frac{1000\text{ g}}{1\text{ kg}}$
1 g = 1000 mg	$\frac{1\text{ g}}{1000\text{ mg}}$	and	$\frac{1000\text{ mg}}{1\text{ g}}$

Figure 1.9
Metric and American units used to describe the contents of some packaged foods.

Metric–Metric Conversion Factors

In many problem-solving exercises, you will need to write conversion factors for metric–metric relationships. For example, the equality and the conversion factors for meters and centimeters are as follows:

Metric Equality	*Conversion Factors*
1 m = 100 cm	$\frac{100\text{ cm}}{1\text{ m}}$ and $\frac{1\text{ m}}{100\text{ cm}}$

Both forms are proper conversion factors for this relationship; one is just the inverse of the other. The usefulness of conversion factors is enhanced by the fact that we can turn a conversion factor over and use its inverse. Table 1.7 lists some conversion factors from the metric relationships we have discussed.

Metric–American Conversion Factors

Sometimes it is necessary to convert between American-system units and metric-system units. You might need to convert a patient's weight from pounds to kilograms. A relationship you could use is

2.20 lb = 1 kg

The corresponding conversion factors would be

$$\frac{2.20\text{ lb}}{1\text{ kg}} \quad \text{and} \quad \frac{1\text{ kg}}{2.20\text{ lb}}$$

Table 1.8 lists some useful American–metric relationships, and Figure 1.9 illustrates the different units used to describe the contents of some packaged foods.

Table 1.8 Some Metric–American Relationships and Conversion Factors

Relationship	Conversion factors		
Length			
2.54 cm = 1 in.	$\frac{2\ 54\text{ cm}}{1\text{ in.}}$	and	$\frac{1\text{ in.}}{2.54\text{ cm}}$
1 m = 39.4 in.	$\frac{1\text{ m}}{39.4\text{ in.}}$	and	$\frac{39.4\text{ in.}}{1\text{ m}}$
Volume			
946 mL = 1 qt	$\frac{946\text{ mL}}{1\text{ qt}}$	and	$\frac{1\text{ qt}}{946\text{ mL}}$
1 L = 1.06 qt	$\frac{1\text{ L}}{1.06\text{ qt}}$	and	$\frac{1.06\text{ qt}}{1\text{ L}}$
Mass			
454 g = 1 lb	$\frac{454\text{ g}}{1\text{ lb}}$	and	$\frac{1\text{ lb}}{454\text{ g}}$
1 kg = 2.20 lb	$\frac{1\text{ kg}}{2.20\text{ lb}}$	and	$\frac{2.20\text{ lb}}{1\text{ kg}}$

Sample Problem 1.8
Writing Conversion Factors for Equalities

Write conversion factors for the following equalities:

a. There are 1000 mg in 1 g.
b. One day is made up of 24 hr.
c. There are 12 eggs in 1 dozen eggs.

Solution:

Equality	*Conversion Factors*		
a. 1 g = 1000 mg	$\frac{1\text{ g}}{1000\text{ mg}}$	and	$\frac{1000\text{ mg}}{1\text{ g}}$
b. 1 day = 24 hr	$\frac{1\text{ day}}{24\text{ hr}}$	and	$\frac{24\text{ hr}}{1\text{ day}}$
c. 12 eggs = 1 dozen eggs	$\frac{12\text{ eggs}}{1\text{ dozen eggs}}$	and	$\frac{1\text{ dozen eggs}}{12\text{ eggs}}$

Study Check:
Write conversion factors for a relationship between inches and centimeters.

Answer:

Equality	*Conversion Factors*		
1 in. = 2.54 cm	$\frac{1\text{ in.}}{2.54\text{ cm}}$	and	$\frac{2.54\text{ cm}}{1\text{ in.}}$

1.6 Problem Solving Using Conversion Factors

Learning Goal **Set up a problem using one or more conversion factors to change from one unit to another.**

The process of problem solving in chemistry is much the same as the process you use when you work out everyday problems, such as changing dollars to cents, or hours to minutes. To illustrate this process, let's take a look at an everyday situation.

Suppose you decide to buy some apples. The sign at the fruit stand states that 1 lb of apples costs 38 cents. When you weigh the apples, you have 2.5 lb of apples. How much will you pay in cents for the apples? (See Figure 1.10.) A close look at the way we think about this problem might show the following:

Step 1 The quantity 2.5 lb is the given or stated quantity.

Given: 2.5 lb apples

Step 2 It is helpful to decide on a unit plan. Our unit plan for this problem is to change from pounds to cents.

Unit plan: lb ⟶ cents
Given Desired

Step 3 The relationship stated in the problem as 38 cents per pound of apples can be written in the form of conversion factors.

Relationship: 1 lb apples = 38 cents

Conversion factors: $\dfrac{\text{1 lb apples}}{\text{38 cents}}$ and $\dfrac{\text{38 cents}}{\text{1 lb apples}}$

Figure 1.10
Converting units.

Step 4 Now we can write the setup for the problem using our unit plan (step 2) and the conversion factors (step 3). First, write down the given quantity, 2.5 lb. Then multiply it by the conversion factor that has the unit lb in the denominator, because that will cancel out the given unit of lb in the numerator. The unit of cents in the numerator gives the desired unit for the answer.

Problem setup:

$$2.5\ \cancel{\text{lb}} \times \frac{38\ \text{cents}}{1\ \cancel{\text{lb}}} = 95\ \text{cents}$$

Given (stated unit) — Conversion (cancels out given) — Answer (desired unit)

Take a look at the change in the units. The unit that you want in the answer must be the one that remains when the other units have canceled out. This is a helpful way to check a problem.

$$\frac{\text{Numerator} \rightarrow}{\text{Denominator} \rightarrow} \quad \cancel{\text{lb}} \times \frac{\text{cents}}{\cancel{\text{lb}}} = \text{cents} \quad \text{(desired unit)}$$

Do the calculations on your calculator to give the following:

2.5 [×] 38 [=] 95

The answer 95 combined with the desired unit cents gives the final answer of 95 cents. With few exceptions, answers to numerical problems must contain a number and a unit.

95 cents (answer)

Sample Problem 1.9
Problem Solving Using Conversion Factors

On a recent bicycle trip, Maria averaged 35 miles per day. How many days did it take her to cover 175 miles?

Solution:

Step 1 Identify the given quantity in the question.

Given: 175 miles

Step 2 Write a unit plan.

Unit plan: miles ⟶ days

Step 3 Identify the relationships needed and write the conversion factor(s).

Relationship (from problem): 35 miles = 1 day

Conversion factors: $\frac{1 \text{ day}}{35 \text{ miles}}$ and $\frac{35 \text{ miles}}{1 \text{ day}}$

Step 4 Set up the problem starting with the given, cancel units, and carry out calculations.

Problem setup:

$$175 \cancel{\text{miles}} \times \frac{1 \text{ day}}{35 \cancel{\text{miles}}} = 5.0 \text{ days}$$

$$(175 \quad \div \quad 35 \quad = 5.0)$$

Study Check:
A recipe for shark fin soup calls for 3.5 qt of chicken broth. How many cups of broth is that? (There are 4 cups in 1 qt.)

Answer: 14 cups

Problem Solving Using Metric Factors

Table 1.7 shows several metric conversion factors for metric equalities. These can also be used in problem solving to change from one metric unit to another, as shown in Sample Problem 1.10.

Sample Problem 1.10
Problem Solving Using Metric Factors

The (RDA) for calcium for females 23–50 years old is 800 mg. How many grams of calcium is that?

Solution:

Step 1 Given: 800 mg

Step 2 Unit plan: mg $\longrightarrow$ g

Step 3 Metric relationship: 1 g = 1000 mg

Conversion factors: $\frac{1 \text{ g}}{1000 \text{ mg}}$ and $\frac{1000 \text{ mg}}{1 \text{ g}}$

Step 4 Problem setup:

$$\underset{\text{Given}}{800\ \cancel{\text{mg}}} \times \underset{\text{Metric factor}}{\frac{1\ \text{g}}{1000\ \cancel{\text{mg}}}} = \underset{\text{Answer (in grams)}}{0.8\ \text{g calcium needed every day}}$$

Study Check:
A can containing 473 mL of frozen orange juice is diluted with 1415 mL of water. How many liters of orange juice are prepared?

Answer: 1.888 L

Problem Solving Using Metric–American Factors

Conversion factors can also be made to link the metric and American systems of measurement. Figure 1.11 illustrates some common measurements in both metric and American units.

Figure 1.11
Metric and American quantities for some common items.

a 50-lb bag of potatoes has a mass of 22.7 kg
Potatoes
a 15-oz can of tomato sauce holds 425 g of sauce
TOMATO
Sauce
1 cup of coffee is 240 mL
a 12-in. pizza is 30 cm
1 qt of milk contains 946 mL of milk

Sample Problem 1.11
Problem Solving Using Metric–American Factors

The length of a newborn infant is 19.0 in. What is the length of the baby in centimeters?

Solution:

Step 1 Given: 19.0 in.

Step 2 Unit plan: in. $\longrightarrow$ cm

Step 3 Metric–American relationship: 1 in. = 2.54 cm

Conversion factors: $\frac{1 \text{ in.}}{2.54 \text{ cm}}$ and $\frac{2.54 \text{ cm}}{1 \text{ in.}}$

Step 4 Problem setup:

$$19.0 \cancel{\text{in.}} \times \frac{2.54 \text{ cm}}{1 \cancel{\text{in.}}} = 48.3 \text{ cm}$$

Study Check:
A patient weighs 154 lb. You need to prepare a medication based on the patient's mass in kilograms. What is the mass of your patient in kilograms?

Answer: 70.0 kg

Using Two or More Conversion Factors in Sequence

In many problems you will need to use two or more steps in your unit plan. Then two or more conversion factors will be required. These can be constructed from the equalities you have learned, or they may be stated in the problem. In setting up the problem, one factor follows the other. Each factor is arranged to cancel the preceding unit until you obtain the desired unit.

Sample Problem 1.12
Problem Solving Using Two Factors

A recipe for salsa requires 3.0 cups of tomato sauce. If only metric measures are available, how many milliliters of tomato sauce are needed? (There are 4 cups in 1 qt.)

Solution:
If you try to think of a relationship between cups and milliliters, you may find you can't think of one. However, you do know how to change cups to quarts, and quarts to milliliters.

Step 1 Given: 3.0 cups

Step 2 Unit plan: cups $\longrightarrow$ quarts $\longrightarrow$ milliliters

Step 3 Relationships and conversion factors:

(1) 1 qt = 4 cups $\quad \dfrac{1 \text{ qt}}{4 \text{ cups}}$ and $\dfrac{4 \text{ cups}}{1 \text{ qt}}$

(2) 1 qt = 946 mL $\quad \dfrac{1 \text{ qt}}{946 \text{ mL}}$ and $\dfrac{946 \text{ mL}}{1 \text{ qt}}$

Step 4 Problem setup:
Use factor (1) to convert from cups to quarts:

$$3.0 \cancel{\text{cups}} \times \frac{1 \text{ qt}}{4 \cancel{\text{cups}}} \times ?$$

Then use factor (2) to convert from quarts to milliliters:
The factors are arranged to cancel units.

$$\cancel{\text{cups}} \times \frac{\cancel{\text{qt}}}{\cancel{\text{cups}}} \times \frac{\text{mL}}{\cancel{\text{qt}}} = \text{mL}$$

The complete setup appears as follows:

$$\text{cups} \longrightarrow \text{qt} \longrightarrow \text{mL}$$

$$3.0 \cancel{\text{cups}} \times \frac{1 \cancel{\text{qt}}}{4 \cancel{\text{cups}}} \times \frac{946 \text{ mL}}{1 \cancel{\text{qt}}} = 710 \text{ mL}$$

Given quantity	American factor	Metric–American factor	answer (in milliliters)

The calculations are done in order on a calculator. Pay attention here to the use of the significant figures.

3.0	÷	4	×	946	=	710
Two SF		Exact		Three SF		Two SF

Study Check:
One medium bran muffin contains 4.2 g of fiber. How many ounces (oz) of fiber are obtained by eating three medium bran muffins, if 1 lb = 16 oz? (*Hint:* Number of muffins $\rightarrow$ g fiber $\rightarrow$ lb $\rightarrow$ oz.)

Answer: 0.44 oz

Using a sequence of two or more conversion factors is a very efficient way to set up and solve problems, especially if you are using a calculator. Once you have the problem set up, the calculations can be done without having to write out all of the intermediate values. This procedure is worth practicing until you understand unit cancellation and the mathematical calculations. A summary of the steps in problem solving follows.

Steps in Problem Solving

Step 1 Identify the given quantity (amount and unit).

Step 2 Write a unit plan to help you think about changing units from the given to the desired. Be sure you can supply a conversion factor for each change.

Step 3 Determine the equalities and corresponding conversion factors you will need to change from one unit to another.

Step 4 Set up the problem according to your unit plan. Arrange each conversion factor to cancel the preceding unit. Check that the units cancel to give your desired answer. Carry out the calculations and give a final answer with the correct number of significant figures and the desired unit.

Clinical Calculations Using Conversion Factors

Conversion factors are also useful in the hospital environment. For example, the dosage for a medication can be stated as a conversion factor. If you are giving an antibiotic that is available in 5-mg tablets, the dosage is written as 5 mg/1 tablet. In many hospitals, the apothecary unit of grains (gr) is still in use; there are 60 mg in 1 gr. When you do a clinical problem, you often start with a doctor's order that contains the quantity to give the patient. The medication dosage is used as a conversion factor.

Sample Problem 1.13
Clinical Calculations With Factors

Dr. Alvarez orders 0.050 g of a medication for your patient. If the dosage is 10 mg of medication per tablet, how many tablets are needed?

Solution:

Step 1 Given: 0.050 g

Step 2 Unit plan: g $\longrightarrow$ mg $\longrightarrow$ tablets

Step 3 Relationships and conversion factors:

(1) 1 g = 1000 mg $\quad \frac{1\text{ g}}{1000\text{ mg}}$ and $\frac{1000\text{ mg}}{1\text{ g}}$

(2) 1 tablet = 10 mg medication $\quad \frac{1\text{ tablet}}{10\text{ mg}}$ and $\frac{10\text{ mg}}{1\text{ tablet}}$

Step 4 Problem setup:

$$g \longrightarrow mg \longrightarrow \text{tablets}$$

$$0.050\,\cancel{g} \times \frac{1000\,\cancel{mg}}{1\,\cancel{g}} \times \frac{1\text{ tablet}}{10\,\cancel{mg}} = 5\text{ tablets}$$

Given — Metric factor — Dosage factor — Answer (in tablets)

Study Check:
An aspirin tablet contains 5 gr of aspirin. How many milligrams of aspirin are in two aspirin tablets, if 1 gr = 60 mg?

Answer: 600 mg

1.7 Density and Specific Gravity

Learning Goal **Calculate the density or specific gravity of a substance, and use the density or specific gravity to calculate the mass or volume of a substance.**

We can measure the mass of a substance on a balance, and we can determine its volume. However, the separate measurements do not tell us how tightly packed the substance might be, or whether its mass is spread out over a large volume or a small one. If we compare its mass to its volume, we can make this determination. This relationship is called the **density** of the substance. Density is defined as the mass in a unit volume of that substance.

$$\text{Density} = \frac{\text{Mass of substance}}{\text{Volume of substance}}$$

In the metric system, the densities of solids and liquids are usually expressed as grams per milliliter (g/mL) or grams per cubic centimeter (g/cm^3). The density of gases is usually stated as grams per liter (g/L). Table 1.9 gives the densities of some common substances.

Table 1.9 Densities of Some Common Substances

Solids (at 25°C)	Density (g/mL)	Liquids (at 25°C)	Density (g/mL)	Gases (at 0°C)	Density (g/gL)
Cork	0.26	Gasoline	0.66	Hydrogen	0.090
Wood (maple)	0.75	Ethyl alcohol	0.79	Helium	0.179
Ice	0.92	Olive oil	0.92	Methane	0.714
Sugar	1.59	Water (at 4°C)	1.000	Neon	0.90
Bone	1.80	Plasma (blood)	1.03	Nitrogen	1.25
Aluminum	2.70	Urine	1.003–1.030	Air (dry)	1.29
Cement	3.00	Milk	1.04	Oxygen	1.45
Diamond	3.52	Mercury	13.6	Carbon dioxide	1.96
Silver	10.5				
Lead	11.3				
Gold	19.3				

Sample Problem 1.14
Calculating Density

A 50.0-mL sample of buttermilk has a mass of 56.0 g. What is the density of the buttermilk?

Solution:

$$\text{Density} = \frac{\text{Mass}}{\text{Volume}} = \frac{56.0\text{ g}}{50.0\text{ mL}} = \frac{1.12\text{ g}}{1\text{ mL}} = 1.12\text{ g/mL}$$

Study Check:
A copper sample has a mass of 44.65 g and a volume of 5.0 cm^3. What is the density of copper in grams per cubic centimeter?

Answer: 8.9 g/cm^3

Density of Solids

The density of a solid is calculated from its mass and volume. When a solid is completely submerged in water, it displaces a volume of water equal to its own volume. In the example shown in Figure 1.12, the water level rises from 20.0 mL to 30.0 mL. This means that 10.0 mL of water is displaced and that the volume of the object is 10.0 mL.

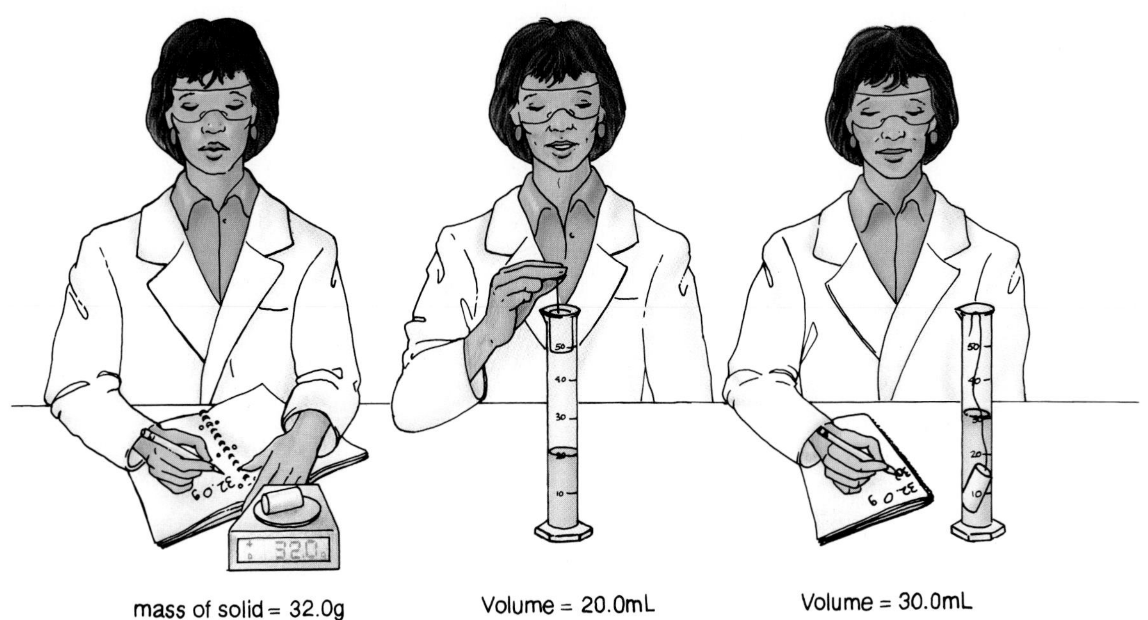

Figure 1.12
Determining the density of a solid by using volume displacement.

Sample Problem 1.15
Using Volume Displacement to Calculate Density

A lead weight used in the belt of a scuba diver has a mass of 226 g. When the weight is carefully placed in a graduated cylinder containing 200.0 mL of water, the water level rises to 220.0 mL. What is the density of the lead weight?

Solution:
Both the mass and the volume of the lead weight are needed to calculate its density. Its mass, 226 g, is given. The volume of the lead weight is equal to the volume of water displaced, which is calculated as follows:

$$\text{Volume displaced} = 220.0 \text{ mL} - 200.0 \text{ mL} = 20.0 \text{ mL}$$

$$\text{Volume of the lead weight} = 20.0 \text{ mL}$$

$$\text{Density of lead} = \frac{\text{Mass of lead}}{\text{Volume of lead}} = \frac{226 \text{ g}}{20.0 \text{ mL}} = 11.3 \text{ g/mL}$$

In the density calculation, be sure to use the volume of water the object displaces and *not* the volume of water indicated by the water level in the cylinder.

Study Check:
A total of 0.50 lb of glass marbles is added to 425 mL of water. The water level rises to indicate a volume of 528 mL. What is the density of the glass marbles?

Answer: 2.2 g/mL

Why Do Objects Sink or Float?

Whether an object sinks or floats is determined by the relative densities of the object and the substance it is immersed in. Take a look at Figure 1.13, which shows samples of lead and cork being placed in water. Since lead is denser than water, it sinks. However, the cork floats, because it is less dense than water.

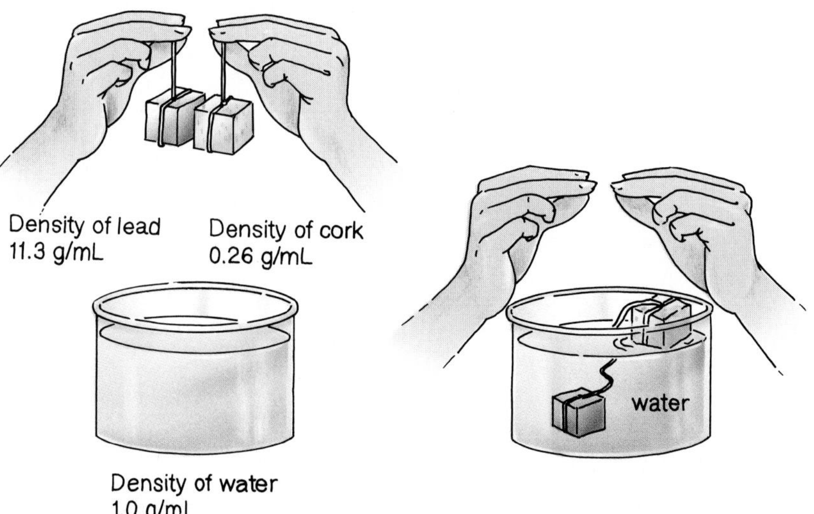

Figure 1.13
Differences in density determine whether an object will sink or float.

Sample Problem 1.16
Density and an Object's Tendency to Sink or Float

Using their densities, determine whether each of the following substances will sink or float when placed in seawater, which has a density of 1.025 g/mL.

Substance	*Density*
Gasoline	0.66 g/mL
Asphalt	1.2 g/mL
Cardboard	0.69 g/cm^3

Solution:
The gasoline and cardboard will float because they are less dense than sea water. Asphalt will sink because it is denser.

Study Check:
Using Table 1.9, explain why helium-filled balloons float in air at 25°C.

Answer: Air has a density of 1.29 g/L. Since helium has a density of 0.179 g/L, a helium-filled balloon is less dense than air and therefore floats in air.

Problem Solving Using Density as a Conversion Factor

Since density is a comparison of two units, it can be used as a conversion factor. For example, if the mass and the density of a sample are known, the volume of the sample can be calculated.

Sample Problem 1.17
Problem Solving Using Density

The cast on Malcolm's leg has a density of 2.32 g/cm^3. What is the volume of the cast if it weighs 4.25 lb?

Solution:

Step 1 Given: 4.25 lb

Step 2 Unit plan: lb $\longrightarrow$ g $\longrightarrow$ cm^3

Step 3 Relationships and conversion factors:

Metric–American: 1 lb = 454 g $\quad \dfrac{454\text{ g}}{1\text{ lb}}$ and $\dfrac{1\text{ lb}}{454\text{ g}}$

Density: 2.32 g = 1 cm^3 $\quad \dfrac{2.32\text{ g}}{1\text{ cm}^3}$ and $\dfrac{1\text{ cm}^3}{2.32\text{ g}}$

Step 4 Problem setup:

$$\underset{\text{Given}}{4.25\ \cancel{\text{lb}}} \times \underset{\text{Metric–American factor}}{\frac{454\ \cancel{\text{g}}}{1\ \cancel{\text{lb}}}} \times \underset{\text{Density as a factor}}{\frac{1\text{ cm}^3}{2.32\ \cancel{\text{g}}}} = 832\text{ cm}^3$$

Study Check:
Milk has a density of 1.04 g/mL. What is the mass of milk in a 1-qt container, if 1 qt = 946 mL?

Answer: 984 g milk

HEALTH NOTE

Determination of Percentage Body Fat

Body mass is made up of protoplasm, extracellular fluid, bone, and adipose tissue (body fat). One way to determine the amount of adipose tissue is to measure the whole-body density. After the on-land mass of the body is determined, the underwater body mass is obtained by submerging the person in water. (See Figure 1.14.) Since water helps support the body by giving it buoyancy, the apparent body mass is less in water. A higher percentage of body fat will make a person more buoyant, causing the underwater mass to be even lower. This occurs because fat has a lower density than the rest of the body.

The mass difference between the on-land mass and underwater mass is calculated as the buoyant force. This buoyant force is used to calculate the person's body volume. Several adjustments, such as subtracting the residual volume of the air trapped in the lungs and the intestine, are made. Then the mass and volume of the person are used to calculate body density. For example, suppose a 70.0-kg person has a body volume of 66.7 L.

$$\text{Body density} = \frac{\text{Body mass}}{\text{Body volume}} = \frac{70.0\ \text{kg}}{66.7\ \text{L}} = 1.05\ \text{kg/L} \quad \text{or} \quad 1.05\ \text{g/mL}$$

When the body density is calculated, it is compared to a chart that correlates the percentage of adipose tissue with body density. A person with a body density of 1.05 g/mL has 21% body fat, according to such a chart. This procedure is used by athletes in determining exercise and diet programs. ■

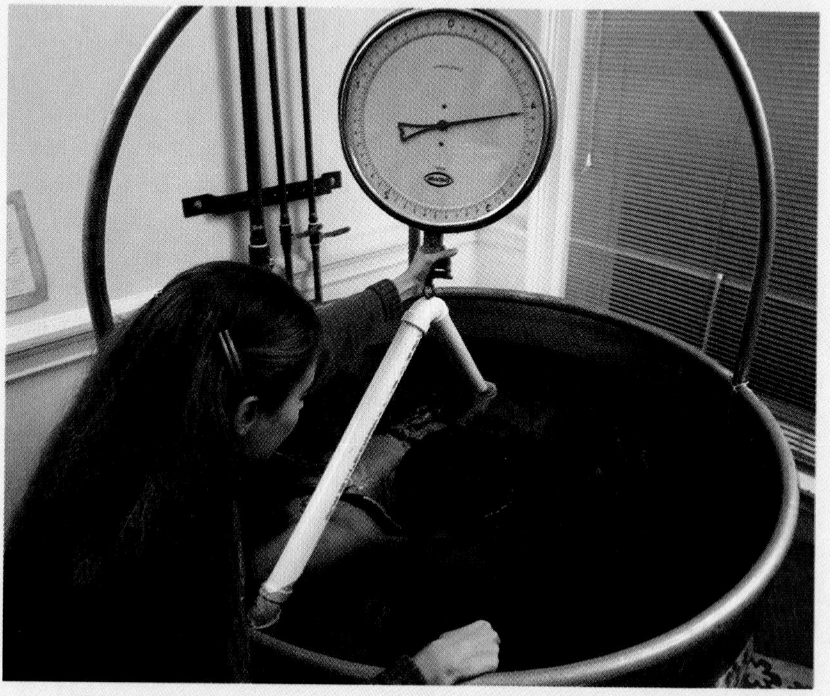

Figure 1.14
A person is submerged in a water tank to determine underwater body mass and to calculate percentage body fat.

Figure 1.15
The specific gravity of a wine is measured with a hydrometer.

Specific Gravity

Specific gravity (sp gr) is the ratio between the density of a substance and the density of water. Specific gravity is calculated by dividing the density of a sample by the density of water. However, the density units must match. For example, in this text we will use a value of 1.00 g/mL for the density of water. In American units, the density of water would be 62.4 lb/ft^3. In both cases, the specific gravity is the same, 1.00. A substance with a specific gravity of 3.00 is three times as dense as water, whereas a substance with a specific gravity of 0.50 is just half as dense as water.

$$\text{Specific gravity} = \frac{\text{Density of sample}}{\text{Density of water}}$$

In the calculations for specific gravity, all units cancel, and only a number remains. This is one of the few unitless values you will encounter in chemistry.

An instrument called a hydrometer is often used to measure the specific gravity of fluids, such as battery fluid or a sample of urine. In Figure 1.15, a hydrometer is used to measure the specific gravity of a fluid.

HEALTH NOTE

Density of Urine

The density of urine is often determined as part of a laboratory evaluation of the health of an individual. The density of urine is normally in the range of 1.003–1.030 g/mL. This is somewhat greater than the density of water because compounds such as urea are dissolved in water in the kidney to form urine. If the density of a person's urine is too low or too high, a doctor might suspect a problem with the kidneys. For example, if a urine sample shows a density of 1.001 g/mL, significantly lower than normal, malfunctioning of the kidneys is a possibility. ■

Sample Problem 1.18
Specific Gravity

What is the specific gravity of coconut oil that has a density of 0.925 g/mL?

Solution:

$$\text{sp gr of oil} = \frac{\text{Density of oil}}{\text{Density of water}} = \frac{0.925\ \cancel{\text{g/mL}}}{1.00\ \cancel{\text{g/mL}}} = 0.925 \quad \text{(no units)}$$

Study Check:
What is the specific gravity of ice if 35.0 g of ice has a volume of 38.2 mL?

Answer: 0.916

Sample Problem 1.19
Problem Solving With Specific Gravity

John took 2.0 teaspoons (tsp) of cough syrup (sp gr 1.20) for a persistent cough. If there are 5.0 mL in 1 tsp, what is the mass (in grams) of the cough syrup?

Step 1 Given: 2.0 tsp

Step 2 Unit plan: tsp ⟶ mL ⟶ g

Step 3 Relationship: 1 tsp = 5.0 mL

$$\text{Conversion factors: } \frac{1\ \text{tsp}}{5.0\ \text{mL}} \quad \text{and} \quad \frac{5.0\ \text{mL}}{1\ \text{tsp}}$$

For problem solving, it is convenient to convert the specific gravity of a sample to its density.

$$\text{Density of sample} = \text{sp gr of sample} \times \text{Density of water}$$

$$1.20 \times \frac{1.00\ \text{g}}{1\ \text{mL}} = 1.20\ \text{g/mL}$$

Relationships: 1 mL = 1.20 g

$$\text{Conversion factors: } \frac{1\ \text{mL}}{1.20\ \text{g}} \quad \text{and} \quad \frac{1.20\ \text{g}}{1\ \text{mL}}$$

Step 4 Problem setup:

$$2.0\ \cancel{\text{tsp}} \times \frac{5.0\ \cancel{\text{mL}}}{1\ \cancel{\text{tsp}}} \times \frac{1.20\ \text{g}}{1\ \cancel{\text{mL}}} = 12\ \text{g}$$

Study Check:
An ebony carving has a mass of 275 g. If ebony has a specific gravity of 1.33, what is the volume of the carving?

Answer: 207 mL

Glossary of Key Terms

centimeter (cm) A unit of length in the metric system; there are 2.54 cm in 1 in.

conversion factor A fraction in which the numerator and the denominator are quantities from an equality in measurement. For example, 1 kg equals 2.20 lb, an equality that can be written as the following conversion factors:

$$\frac{2.20\text{ lb}}{1\text{ kg}} \quad \text{and} \quad \frac{1\text{ kg}}{2.20\text{ lb}}$$

cubic centimeter (cm^3, cc) The volume of a cube that has 1-cm sides; equal to 1 mL.

density The relationship of the mass of an object to its volume expressed as grams per cubic centimeter (g/cm^3), grams per milliliter (g/mL), or grams per liter (g/L).

gram (g) The metric unit used in measurements of mass.

kilogram (kg) A metric mass of 1000 g, equal to 2.20 lb. The kilogram is the SI standard unit of mass.

liter (L) The metric unit for volume that is slightly larger than a quart.

mass A measure of the quantity of material in an object.

meter (m) The metric unit for length that is slightly longer than a yard. The meter is the SI unit of length.

metric system A system of measurement used by scientists and in most countries of the world.

milliliter (mL) A metric unit of volume equal to $\frac{1}{1000}$ L.

prefix The part of the name of a metric unit that precedes the base unit and specifies the size of the measurement. All prefixes are related on a decimal scale. Some important metric prefixes are *kilo, centi,* and *milli.*

SI units An International System of units that modifies the metric system.

significant figures The numbers recorded in a measurement.

specific gravity (sp gr) A unitless relationship between the density of a substance and the density of water:

$$\text{sp gr} = \frac{\text{Density of sample}}{\text{Density of water}}$$

weight A measure of the pull of gravity on the mass of an object.

Problems

Answers to all problems are found in the back of this book. The more difficult problems are marked with an asterisk (*).

Units of Measurement *(Goal 1.1)*

1.1 Compare the units used in measuring length in the metric system and in the American system. Do the same for volume and mass measurements.

1.2 Write the name of the metric unit and the type of measurement (mass, volume, or length) indicated for each of the following quantities:
a. 4.8 m b. 325 g c. 1.5 mL d. 480 m

Metric Prefixes *(Goal 1.2)*

1.3 Explain the use of prefixes in the metric system.

1.4 Give the symbol for each of the following metric units:
a. milligram b. deciliter c. kilometer d. kilogram

1.5 Write the name of the metric unit for each of the following symbols:
a. cm b. mm c. dL d. kg

1.6 Write the numerical value for each of the following prefixes:
a. centi b. kilo c. milli d. deci

1.7 Write the correct prefix term for each of the following numerical values:
a. 0.10 b. 10 c. 1000 d. $\frac{1}{100}$

Writing Numerical Relationships for Metric Units *(Goal 1.3)*

1.8 Place the units or prefixes in order, from smallest to largest, for each set:
a. milli, kilo, centi
b. milli, centi, micro
c. deci, milli, mega, deka
d. cg, kg, mg, g, dg
e. hm, mm, dm, m, km
f. kL, L, mL, cL

1.9 Complete the following metric relationships:
a. 1 m = ____ cm
b. 1 km = ____ m
c. 1 mm = ____ m
d. 1 L = ____ mL
e. 1 L = ____ dL
f. 1 dL = ____ L
g. 1 g = ____ kg
h. 1 g = ____ mg

Significant Figures in Measurements *(Goal 1.4)*

1.10 When is it important to count significant figures in a number?

1.11 Indicate whether each of the following statements refers to a measured number or an exact number:
a. A patient weighs 155 lb.
b. The basket holds eight apples.
c. There are 12 in. in 1 ft.
d. There were 31 students in the laboratory.
e. A laboratory test shows a blood cholesterol level of 184 mg/dL.

1.12 When are zeros counted as significant figures? When are they not significant?

1.13 State the number of significant figures in each of the following measured quantities:
a. 11.005 g
b. 0.00032 m
c. 36,000,000 km
d. 185.34 kg
e. 20.60 mL

1.14 Round off each of the following numbers to the number of significant figures indicated:
a. 1.854 (to two figures)
b. 184.2038 (to four figures)
c. 0.004738265 (to three figures)
d. 88.05785 (to two figures)
e. 1.832149 (to three figures)

1.15 For the following multiplication and division problems, give final answers rounded off to the correct number of significant figures:
a. 45.7×0.034
b. $\dfrac{34.56}{1.25}$
c. $\dfrac{(0.2465)\,(25)}{1.78}$
d. $\dfrac{2.40}{(4)\,(125)}$
e. $\dfrac{(3.5)\,(0.261)}{(8.24)\,(20.0)}$

1.16 For the following addition and subtraction problems, give final answers rounded off to the correct number of significant figures:

a. 45.48 cm + 8.057 cm
b. 23.45 g + 104.1 g + 0.025 g
c. 145.675 mL − 24.2 mL
d. 1.08 L − 0.585 L

Conversion Factors *(Goal 1.5)*

1.17 Why does an equality such as 1 m = 100 cm always have two conversion factors?

1.18 Write a numerical relationship and conversion factors for each of the following statements:

a. There are 3 ft in 1 yd.
b. One minute is 60 seconds.
c. One dollar has 4 quarters.
d. There are 4 qt in 1 gal.
e. One mile is 5280 ft.
f. There are 7 days in 1 week.

1.19 Write the numerical relationship and conversion factors for each of the following metric units:

a. centimeters and meters
b. milligrams and grams
c. centimeters and millimeters
d. liters and milliliters
e. deciliters and milliliters
f. grams and kilograms

1.20 Write the metric–American relationship and corresponding conversion factors for each of the following:

a. centimeters and inches
b. pounds and kilograms
c. quarts and milliliters
d. pounds and grams
e. liters and quarts

Problem Solving Using Conversion Factors *(Goal 1.6)*

1.21 You are explaining conversion factors and the unit factor system for solving problems to a friend. What advice would you give?

1.22 Use American conversion factors to solve the following problems:

a. How many yards are in 24 ft?
b. How many seconds are in 15 min?
c. You need 3.5 qt of oil for your car. How many gallons is that?
d. One game at the arcade requires 1 quarter. You have $3.50 in your pocket. How many games can you play?
e. You ran a total of 2.4 miles today. How far is that in feet?

1.23 Use metric conversion factors to solve the following problems:

a. A student's height is 175 cm. How tall is the student in meters?
b. A cooler has a volume of 5500 mL. What is the capacity of the cooler in liters?
c. A hummingbird has a mass of 0.055 kg. What is the mass of the hummingbird in grams?
d. The recommended daily allowance of phosphorus for an adult male is 800 mg. How many grams of phosphorus are recommended?
e. A glass of orange juice contains 0.85 dL of juice. How many milliliters of orange juice is that?
f. A package of chocolate instant pudding contains 2840 mg of sodium. How many grams of sodium is that?

1.24 Solve the following problems using one or more conversion factors:

a. A container holds 0.750 qt. How many milliliters of juice can be held in the container?

b. What is the mass in kilograms of a person who weighs 165 lb?

c. The femur, or thighbone, is the longest bone in the body. In a 6-ft person, the femur might be 19.5 in. long. What is the length of that person's femur in millimeters?

*d. A dialysis unit requires 75,000 mL of distilled water. How many gallons of water are needed? (1 gal = 4 qt)

1.25 Solve the following problems using conversion factors:

a. You need 4.0 ounces (oz) of a steroid ointment. If there are 16 oz in 1 lb, how many grams of ointment does the pharmacist need to prepare?

b. A patient receives 5.0 pints (pt) of plasma. How many milliliters of plasma were given? (1 qt = 2 pt)

c. A piece of plastic tubing measuring 560 mm in length is used in an intravenous setup. How many feet of tubing are required?

*d. A person on a diet has been losing weight at the rate of 3.5 lb per week. If the person has been on the diet for 6 weeks, how many kilograms were lost?

*e. Zippy the snail moves at the rate of 2.0 in. per hour. How many millimeters does Zippy travel in 4.0 hr?

1.26 Using conversion factors, solve the following health-related problems:

a. You have used 250 L of distilled water for a dialysis patient. How many gallons of water is that?

b. A patient needs 0.024 g of a sulfa drug. There are 8-mg tablets in stock. How many tablets should be given?

*c. The daily dose of ampicillin for the treatment of an ear infection is 115 mg/kg body weight. What is the daily dose for a 34-lb toddler?

d. An intramuscular medication is given at 5.00 mg/kg body weight. If you give 425 mg of medication to a patient, what is the patient's weight in pounds?

*e. The doctor has ordered 1.0 g tetracycline to be given every 6 hr to a patient. If your stock on hand is 500-mg tablets, how many will you need for 1 day's treatment?

*f. A doctor has ordered 325 mg of atropine IM. Atropine is available as 0.50 g/mL. How many milliliters would you need to give?

Density and Specific Gravity *(Goal 1.7)*

1.27 A letter has arrived for you that reads as follows:

Dear Cousin:

How are you? I hear you are taking chemistry now. Maybe you can help me. I have been working in our mine and recently found a brilliant stone that looks like a diamond, but I am not sure. Do you know how I could determine its density? I read in a mineralogy magazine that diamond has a density of 3.51 g/cm^3. Please describe how I can determine the density of this stone, including the tools I need to buy, the measurements I need to make, and the necessary calculations. I hope you are enjoying your chemistry class.

Sincerely, Cousin Emma

Write a letter to Cousin Emma, using complete sentences.

1.28 Calculate the density of the samples described:

a. Twenty milliliters of a salt solution that has a mass of 24 g.
b. A solid object that has a mass of 1.65 lb and a volume of 170 mL.
c. A gem that has a mass of 45.0 g. When it is placed in a graduated cylinder containing 20.0 mL of water, the level of water rises to 34.5 mL.
d. A medication, if the contents of a syringe filled to 3.00 mL have a mass of 3.85 g.
e. A 5.00-mL urine sample from a patient suffering from symptoms resembling those of diabetes mellitus. The mass of the urine sample is 5.025 g.

1.29 Use density values to solve the following problems:

a. What is the mass in grams of 150 mL of a liquid that has a density of 1.4 g/mL?
b. What is the mass of a glucose solution that fills a 500-mL IV bottle if the density of the glucose solution is 1.15 g/mL?
c. Kari, a sculptor, has prepared a mold for casting a bronze figure. The figure has a volume of 225 mL. If bronze has a density of 7.8 g/mL, how many ounces of bronze does Kari need to melt in the preparation of the bronze figure?
d. A fish tank holds 30 gal of water. Using the density of 1.0 g/mL for water, determine the number of pounds of water in the fish tank.
*e. A graduated cylinder contains 155 mL of water. A 15.0 g piece of iron (density 7.86 g/cm^3) and a 20.0 g piece of lead (density 11.3 g/cm^3) are added. What is the new water level in the cylinder?

1.30 Solve the following specific gravity problems:

a. A urine sample has a density of 1.030 g/mL. What is the specific gravity of the sample?
b. A liquid has a volume of 40.0 mL and a mass of 45.0 g. What is the specific gravity of the liquid?
c. The specific gravity of an oil is 0.85. What is its density?
*d. Butter has a specific gravity of 0.86. What is the mass in grams of 0.250 L of butter?
e. A bottle containing 325 g of cleaning solution has fallen and broken on the floor. If the solution in the bottle has a specific gravity of 0.850, what volume of solution needs to be cleaned up?
*f. Ethyl alcohol has a specific gravity of 0.79. What is the volume in quarts of 1.50 kg of alcohol?

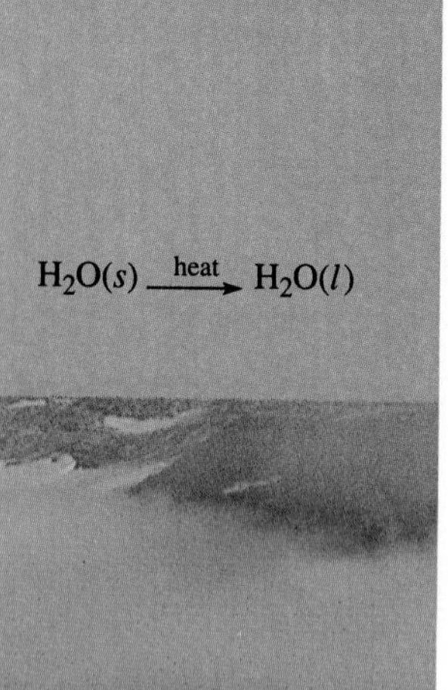

2

Energy and Matter

As the weather warms, the water molecules in ice break apart to form a liquid.

Learning Goals

2.1 Given a temperature in degrees Celsius, calculate the corresponding temperature in degrees Fahrenheit or in kelvins.
2.2 Identify the types and forms of energy in a system.
2.3 Given the mass of a water sample and the change in temperature, calculate the heat energy released or absorbed in calories or in joules.
2.4 Identify a substance as a solid, a liquid, or a gas.
2.5 Draw a heating or cooling curve and label the melting or freezing point, the boiling point, and the sections that represent solid, liquid, and gas.
2.6 Calculate the number of calories or joules absorbed or released when a given mass of water undergoes a change of state.
2.7 Calculate the energy of a food in calories, in kilocalories, or in joules.

SCOPE

Matter is all the material around us that has mass and occupies space. All matter requires energy to be set in motion. This means that energy must be used to do work by making things move. You use energy when you play tennis, run, study, work, and even sleep. You require more energy to walk 5 miles than you do to walk around the block. The burner on the stove can be used to heat the water in a teakettle. As more heat is supplied, the water gets hotter. Perhaps it even boils. The energy in your body and the energy supplied from the stove are making things move. Energy is doing work.

Matter can be a solid, a liquid, or a gas. For example, the water in an ice cube, an ice rink, or an iceberg is solid water. The water running out of a faucet or into a pool is a liquid. When water evaporates from wet clothes or boils away in a pan on the stove, it becomes a gas. Gases are usually invisible to us, but we can detect them if they have a characteristic odor or color. You know when someone opens a bottle of perfume or ammonia because you can smell the gas as it leaves the bottle and fills the room.

2.1 Temperature Measurement

Learning Goal **Given a temperature in degrees Celsius, calculate the corresponding temperature in degrees Fahrenheit or in kelvins.**

Suppose a thermometer indicated that the room temperature was 22° and your body temperature was 37.0°. You might wonder what was wrong. In these examples, the temperatures were reported in the metric units of degrees **Celsius,** °C. Temperatures of 22°C and 37.0°C are typical room and body temperatures, respectively, in the metric system. In the American system, temperature is typically reported in degrees **Fahrenheit,** °F. In our example, we would find that a temperature of 22°C is the same as 72°F, and 37.0°C is the same as 98.6°F, normal body temperature.

Celsius and Fahrenheit

Temperature is an indication of the intensity of heat in a substance. On a Celsius scale, the freezing point of water is 0°C, and the boiling point is 100°C. On a thermometer that indicates Fahrenheit temperature, water freezes at 32°F and boils at 212°F. Between the freezing and boiling points of water, there are 180 degrees on the Fahrenheit scale and 100 degrees on the Celsius scale.

$$180 \text{ Fahrenheit units} = 100 \text{ Celsius units}$$

$$\frac{180 \text{ Fahrenheit units}}{100 \text{ Celsius units}} \quad \text{or} \quad \frac{1.8 \text{ Fahrenheit units}}{1 \text{ Celsius unit}}$$

To change from a Celsius temperature to Fahrenheit, 32 degrees must be added. This adjustment is necessary because the value for the freezing point of water is 32 on the Fahrenheit scale, compared to 0 on the Celsius scale. The following equation can be used to change a Celsius temperature to its corresponding Fahrenheit temperature:

$$°F = 1.8(°C) + 32$$

Note that in the equation, Fahrenheit and Celsius temperatures are represented by °F and °C. In calculations, the temperature conversion values (1.8 and 32) are exact numbers. The new temperature is accurate to the same decimal place as the given temperature for the problem.

Sample Problem 2.1
Temperature Conversion

While traveling in Europe, Anthony checks his temperature, which is 38.2°C. What is his body temperature in degrees Fahrenheit?

Solution:
Using the temperature conversion equation, we write

$$°F = 1.8(°C) + 32$$

$$°F = 1.8(38.2) + 32$$

The multiplication of 1.8(38.2) is done first, and then 32 is added. The answer is given to the tenths place to match the accuracy of the given temperature, 38.2°C.

$$°F = 68.8 + 32$$

$$°F = 100.8°F$$

Study Check:
When making ice cream, Azita uses rock salt to chill the mixture. If the temperature of the mixture drops to −11°C, what is it in degrees Fahrenheit?

Answer: 12°F

Sample Problem 2.2
Converting to Celsius Temperature

Donyelle is going to cook a turkey at 325°F. If she uses an oven thermometer that has Celsius units, at what Celsius temperature should she set the oven?

Solution:
In this temperature problem, we need to convert from Fahrenheit to Celsius. To do this, the temperature equation is rearranged.

$$°F = 1.8(°C) + 32$$

$$°F - 32 = 1.8(°C)$$

$$\frac{(°F - 32)}{1.8} = °C$$

Entering the temperature of 325°F into the equation gives

$$°C = \frac{(325 - 32)}{1.8} = \frac{293}{1.8} = 163°C$$

Mathematically, the subtraction (325 − 32) is done first, and the result is then divided by 1.8.

Study Check:
Your patient, Mr. Lee, has a temperature of 103.6°F. What is his temperature on a Celsius thermometer?

Answer: 39.8°C

Kelvin Temperature Scale

The **Kelvin scale** is very important to scientists. It is the SI unit of temperature. On the Kelvin scale, a value of zero is assigned to the lowest possible temperature, called absolute zero. Units on the Kelvin scale are called kelvins (K); no degree symbol is used. On the Celsius temperature scale, absolute zero corresponds to −273°C. A comparison of Celsius, Fahrenheit, and Kelvin temperature scales is given in Figure 2.1. Since the size of the kelvin is the same as the size of the Celsius degree (both scales have 100 degrees between the freezing and boiling points of water), we can calculate a Kelvin temperature by adding 273 to the Celsius temperature.

$$1\ K = 1°C$$

$$K = °C + 273$$

Table 2.1 gives a comparison of some temperatures on the three temperature scales.

Table 2.1 A Comparison of Temperature Scales

Example	Kelvin	Celsius	Fahrenheit
A hot oven	505 K	232°C	450°F
Milk scalds	356 K	83°C	181°F
A desert	322 K	49°C	120°F
A high fever	313 K	40°C	104°F
Room temperature	295 K	22°C	72°F
Water freezes	273 K	0°C	32°F
A northern winter	213 K	−60°C	−76°F
Absolute zero	0 K	−273°C	−459°F

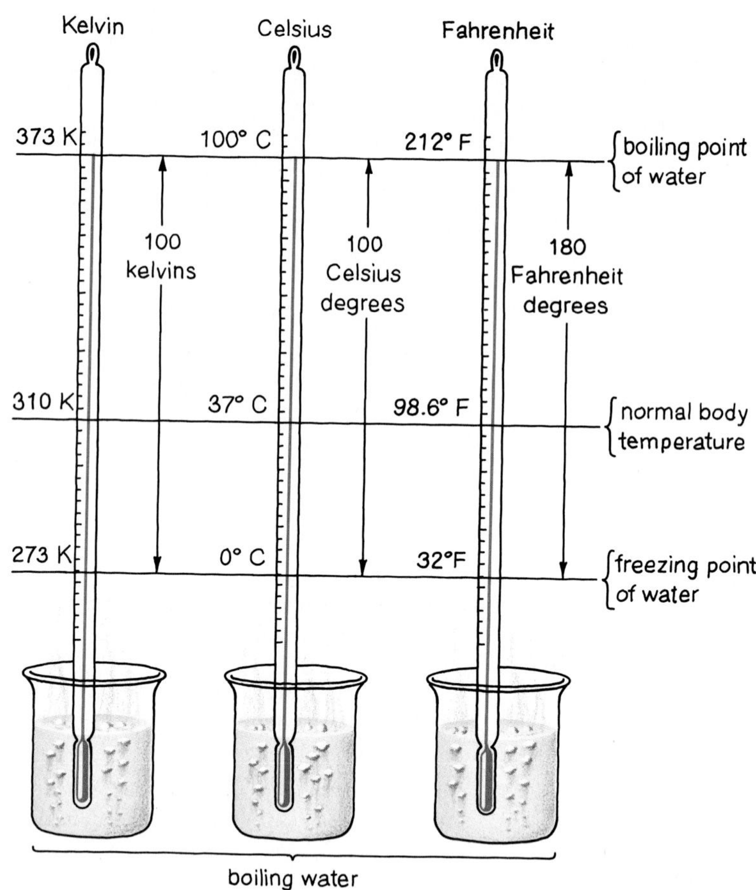

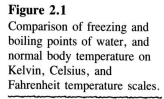

Figure 2.1
Comparison of freezing and boiling points of water, and normal body temperature on Kelvin, Celsius, and Fahrenheit temperature scales.

Sample Problem 2.3
Converting Temperature to Kelvins

What is a normal body temperature of 37°C on the Kelvin scale?

Solution:
Using the equation to convert from a Celsius temperature to kelvins, we write

$$K = °C + 273$$

$$K = 37°C + 273$$

$$= 310\ K$$

Study Check:
You are baking a pizza at 375°F. What is this temperature in kelvins? (*Hint:* Convert to Celsius temperature first.)

Answer: 464 K

Sample Problem 2.4
Converting From Kelvin to Celsius Temperature

Miriam is running a reaction cooled to 77 K by liquid nitrogen. What is the Celsius temperature of the reaction mixture?

Solution:
To find the Celsius temperature, we take the following equation and rearrange it:

$$K = {}^\circ C + 273$$

$$K - 273 = {}^\circ C$$

Using our given temperature of 77 K, we calculate the Celsius temperature for the reaction.

$${}^\circ C = 77\ K - 273 = -196^\circ C$$

Study Check:
On the planet Mercury, the average night temperature is 13 K, and the average day temperature is 683 K. What are these temperatures in Celsius degrees?

Answer: night, $-260^\circ C$; day, $410^\circ C$

2.2 Energy

Learning Goal **Identify the types and forms of energy in a system.**

When you are running, walking, dancing, or thinking, you are using energy to do work. In fact, **energy** is defined as the ability to do work. Suppose you are climbing a steep hill. While climbing that hill you are expending energy. Perhaps you become too tired to go on. We could say that you do not have sufficient energy to do any more work. Now, suppose you sit down and have lunch. In a while you will have obtained energy from the food, and you will be able to do more work and complete the climb.

Types of Energy: Potential and Kinetic

Energy may be potential energy or kinetic energy. **Potential energy** does no work; it is stored for later use. **Kinetic energy** is the energy of motion; it is doing work. A boulder resting on a mountain has potential energy because of its location. When the boulder moves and rolls down the mountain, the potential energy becomes kinetic energy. Water stored in a reservoir has potential energy. When the water goes over the dam, the stored energy becomes kinetic energy. Even the food you eat has potential energy. When you digest the food, the stored energy is converted to kinetic energy to do your work.

HEALTH NOTE

Variations in Body Temperature

Normal body temperature is considered to be 37.0°C. However, this varies throughout the day. Oral temperatures of 36.1°C are common when awakening in the morning and climb to a high of 37.2°C between 6 P.M. and 10 P.M. Elevations of temperature above 37.2°C for a person at bed rest are usually an indication of disease. However, individuals involved in prolonged exercise may also experience elevated temperatures. Body temperatures of marathon runners can range from 39°C to 41°C as heat production during exercise exceeds the body's ability to lose heat.

Changes of more than 3.5°C from the normal body temperature begin to interfere with bodily functions. Temperatures above 41.1°C can lead to convulsions, particularly in children, and cause permanent brain damage. Heatstroke (hyperpyrexia) occurs above 41.1°C. Initially, sweat production stops, and the skin becomes hot and dry. The pulse rate is elevated, and respiration becomes weak and rapid. The person generally becomes lethargic and lapses into a coma. Damage to internal organs is a major concern, and treatment must be immediate. An effective method of reducing body temperature is immersion in an ice-water bath.

In hypothermia, body temperatures can drop as low as 28.5°C. The person may appear cold and pale and have an irregular heartbeat. Unconsciousness can occur if the body temperature drops below 26.7°C. Respiration becomes slow and shallow, and oxygenation of the tissues decreases. Treatment involves providing oxygen and increasing blood volume with glucose and saline fluids. Internal temperature may be restored by injecting warm fluids (37.0°C) into the peritoneal cavity. ■

Sample Problem 2.5
Describing Potential and Kinetic Energy

State whether each of the following statements describes potential or kinetic energy:

a. the energy in gasoline
b. the energy of falling water in a waterfall
c. the energy in a candy bar
d. the energy of a stretched rubber band

Solution:

a. potential energy
b. kinetic energy
c. potential energy
d. potential energy

Forms of Energy

Energy takes several different forms. The energy obtained from a chemical compound or a piece of food is chemical energy. Light is a form of radiant energy. You are probably most familiar with the radiant energy from the sun needed for the growth of plants. The electrical energy in your home that provides electricity for radios and appliances is yet another form of energy. Mechanical energy turns the wheels of a car or the turbines of a jet engine.

Heat or thermal energy is associated with the motion of the tiny particles that make up a substance. A frozen pizza feels cold because the particles in the pizza have low kinetic energy and are moving very slowly. As the pizza is heated, the kinetic energy increases, and the pizza becomes warm. Eventually the particles in the pizza pick up enough heat to make the pizza hot and ready to eat. When you warm food in a gas or electric oven, the food gets hot because thermal energy is provided by the flame or the hot electrical coils.

Conversion of Energy

One form of energy may be converted into another. The burning of wood converts chemical energy into thermal energy and light. In an electrical power plant, the thermal energy from the burning of fossil fuels such as natural gas or coal is used to produce steam, which turns the turbines that produce electrical energy. In your home, electrical energy is converted to radiant energy when you turn on a light switch, to mechanical energy when you use a mixer or a washing machine, and to thermal energy when you use a hair dryer or a toaster (see Figure 2.2).

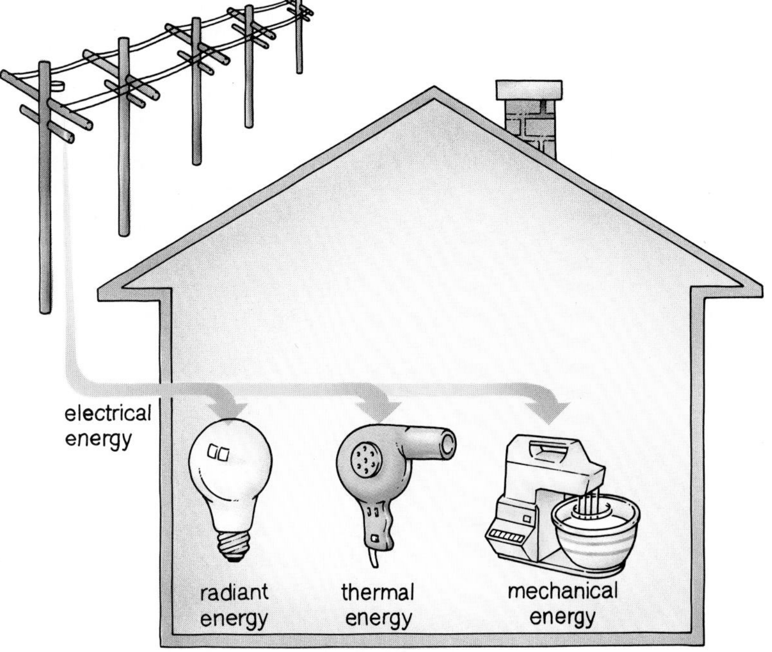

Figure 2.2
Examples of electrical energy converted into radiant energy, thermal energy, and mechanical energy.

HEALTH NOTE

Biological Work

In the cells of the body, the processes of metabolism converts potential energy stored in nutrients in our diet to kinetic energy. Kinetic energy is needed to do biological work. When you contract a muscle, you are doing mechanical work. In the movement of your legs riding a bicycle or in the involuntary movement of the heart, mechanical work is done, and kinetic energy is expended.

$$\text{Relaxed muscle} \xrightarrow{\text{Energy to do mechanical work}} \text{Contracted muscle}$$

Chemical work in the body is done for growth and maintenance of the cells. Large molecules such as proteins are built from small molecules of amino acids, a process that requires energy.

$$\text{Small molecules} \xrightarrow{\text{Energy to do chemical work}} \text{Large molecules}$$

Transport work is done when digested foodstuffs are moved from the intestinal membrane to the cells of organs and tissues. In active transport, energy is required to move certain compounds from a lower concentration to a higher concentration within the cells.

$$\text{Digestion products} \xrightarrow{\text{Energy for transport}} \text{Cellular nutrients}$$ ■

2.3 Measuring Heat Energy

Learning Goal **Given the mass of a water sample and the change in temperature, calculate the heat energy released or absorbed in calories or in joules.**

In chemistry and the health sciences, heat is commonly measured in units called **calories (cal).** One calorie is the amount of energy needed to raise the temperature of 1 g of water by 1°C. A **kilocalorie (kcal)** is used to report larger amounts of heat:

1 kcal = 1000 cal

The SI unit of energy is called the **joule (J),** pronounced "jewel." One calorie is as much energy as 4.18 J, an equality that can be written as follows:

1 cal = 4.18 J

Specific Heat

All substances absorb energy. The amount of heat a certain substance absorbs is given by its **specific heat,** which is the amount of energy needed to raise the temperature of 1 g of that substance by 1°C.

$$\text{Specific heat} = \frac{\text{calories}}{\text{gram} \times {}^\circ\text{C}} = \frac{\text{cal}}{\text{g }{}^\circ\text{C}}$$

ENVIRONMENTAL NOTE

Thermal Energy and Global Warming

Since the mid-1980s, there has been increased concern about the amount of carbon dioxide gas accumulating in Earth's atmosphere. As more fossil fuels (gasoline, coal, and natural gas) are burned, the carbon dioxide level has been increasing. The oceans absorb a certain amount of carbon dioxide, but they cannot keep up with the increase. The cutting of trees in the rain forests (deforestation) eliminates another route for the removal of carbon dioxide from the atmosphere. Many of the trees are also burned as land is cleared. It has been estimated that deforestation may account for 15–30% of the carbon dioxide placed annually in the atmosphere.

Some scientists think that the carbon dioxide in the atmosphere may be acting like the glass in a greenhouse. Sunlight can pass through to warm Earth's surface, but then the heat is kept inside, trapped by the carbon dioxide. It is predicted that more heat will be retained as the carbon dioxide level increases (see Figure 2.3).

It is not yet clear how severe the effects of global warming might be. Some scientists estimate that by around the year 2030 the atmospheric levels of carbon dioxide could double and cause the temperature of Earth's atmosphere to rise by 2–5°C. If that should happen, it could have a profound impact on Earth's climate. Ecosystems in the wetlands may be drastically affected as forest boundaries shift northward. An increase in the melting of snow and ice could raise the ocean levels by as much as 2 m.

Worldwide efforts are already being made to reduce fossil fuel use and to slow or stop deforestation. It will require the cooperation of countries throughout the world to avoid the bleak future that some scientists predict should global warming continue unchecked. ■

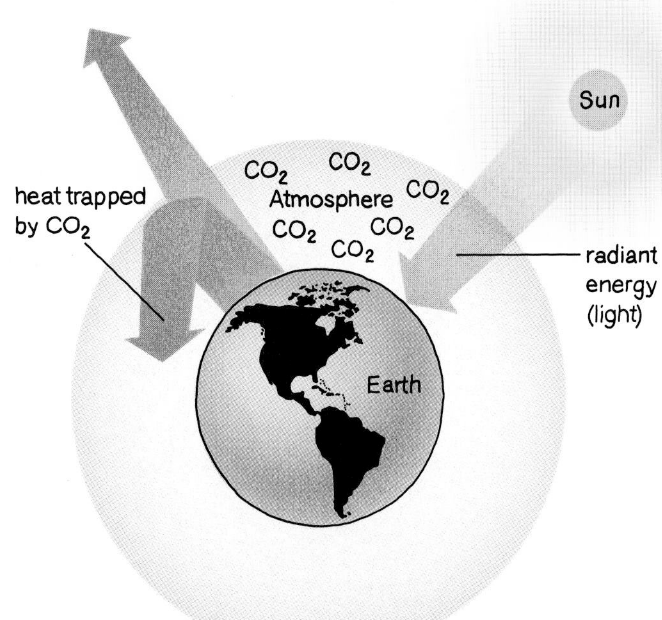

Figure 2.3
As the level of carbon dioxide (chemical formula, CO_2) increases in the atmosphere, more heat may be retained, causing an increase in Earth's temperature.

Table 2.2 Specific Heats of Some Substances

Substance	Specific heat (cal/g °C)
Water (liquid)	1.00
Ethyl alcohol	0.58
Water (ice)	0.50
Water (gas)	0.48
Wood	0.42
Aluminum	0.22
Sand	0.19
Iron	0.106
Copper	0.092
Silver	0.057
Gold	0.031

Every substance has its own specific heat. Liquid water has one of the largest specific heat values (1.00 cal/g °C), which means that water absorbs more heat than most substances. For example, on a hot day at the beach the ocean water remains cool to the touch, but the sand can burn your feet. In the body, water absorbs large amounts of heat without causing fluctuations in body temperature.

When materials with small specific heats absorb energy, they attain high temperatures rapidly. For this reason, pans used for cooking are often made of aluminum or copper. These materials heat quickly and transfer the heat to the food. The handles are made of materials (such as wood) with large specific heats so the pans can be picked up when the food is hot. Table 2.2 lists the specific heats of a variety of materials.

Calculations Using Specific Heat

The amount of heat required to heat a certain quantity of a substance to a higher temperature can be calculated using the mass of the sample, the temperature change, and the specific heat of the substance. The temperature change is often noted as ΔT (Δ is the capital Greek letter delta; read ΔT as ''delta T'').

$$\underset{\text{cal}}{\text{Heat}} = \underset{\text{g}}{\text{Mass}} \times \underset{^\circ\text{C}}{\text{Temperature change }(\Delta T)} \times \underset{\text{cal/g }^\circ\text{C}}{\text{Specific heat}}$$

Sample Problem 2.6
Calculating Heat Energy

Calculate the number of calories required to raise the temperature of 15 g of water from 24°C to 36°C.

Solution:
From Table 2.2, we find that the specific heat for water is 1.00 cal/g °C. The temperature change (ΔT) is 12°C (36°C − 24°C). Using 15 g as the mass of the water sample in the expression for heat, we can write

$$\underset{\text{Mass}}{15\ \cancel{\text{g}}} \times \underset{\Delta T}{12\cancel{^\circ\text{C}}} \times \underset{\text{Specific heat}}{\frac{1.00\ \text{cal}}{\cancel{\text{g}}\ \cancel{^\circ\text{C}}}} = \underset{\text{Heat energy}}{180\ \text{cal}}$$

Study Check:
How many calories are needed to heat 5.0 g of copper from 75°C to 125°C? Use Table 2.2 to obtain the specific heat of copper.

Answer: 23 cal

Sample Problem 2.7
Calculating Joules Released by Cooling

Some hot tea has cooled from 95°C to 23°C. Calculate the energy released in joules if the sample of tea has a mass of 50.0 g. Assume that the tea has the same specific heat as water, 1.00 cal/g °C.

Solution:
The temperature change is 72°C (95°C − 23°C). The number of calories released during cooling is calculated as

$$\underset{\text{Mass}}{50.0\ \cancel{g}} \times \underset{\Delta T}{72^\circ\cancel{C}} \times \underset{\text{Specific heat}}{\frac{1.00\ \text{cal}}{\cancel{g}\ ^\circ\cancel{C}}} = \underset{\text{Heat released}}{3600\ \text{cal}}$$

To calculate the number of joules released in the cooling process, the conversion factor for calories and joules is used:

$$3600\ \cancel{\text{cal}} \times \frac{4.18\ \text{J}}{1\ \cancel{\text{cal}}} = 15{,}000\ \text{J or } 1.5 \times 10^4\ \text{J}$$

Note that the answer contains two significant figures, and the zeros are placeholders.

Study Check:
What is the heat energy in joules released when 15.0 g of gold cools from 215°C to 35°C?

Answer: 350 J

2.4 States of Matter

Learning Goal **Identify a substance as a solid, a liquid, or a gas.**

Matter is everything around us that has mass and occupies space. All of the materials around us are composed of very small particles of matter that are in constant motion. When the particles move slowly, they are held together in rigid shapes we call **solids.** As the temperature increases, the particles within the solid move faster, eventually causing the solid to lose its rigid shape. The solid turns into a **liquid.** If more heat is added, the particles move even faster. When they are moving rapidly enough, they fly away from one another and take the form of a **gas.**

Look around you. You can probably see many examples of solids, such as tables, chairs, food, dishes—even your teeth and bones. In a solid, the particles are tightly packed together in a regular, stable arrangement. The attraction between the particles is so strong that they are held in fixed positions. However, the particles are not motionless; they vibrate slightly within the rigid structure. The result is a solid having a definite shape and a definite volume that do not depend on any container.

Figure 2.4
A liquid takes the shape of its container.

The most common liquid is water. Even your body is about 60% liquid water. The particles of a liquid have greater freedom of movement than those of a solid. In liquids such as water, the particles can slide over one another in random motion. However, the particles in a liquid are still rather strongly attracted and remain close together. A liquid maintains a definite volume, but it does not maintain a definite shape because it conforms to the shape of its container, as shown in Figure 2.4.

You can't see it, but the air you breathe is made of gases. In a gas, the particles move at high speed and continually collide with other gas particles or the walls of a container. Since they move with great speed, there is essentially no attraction between particles. This allows the particles to move far away from one another. In a gas, there are great distances between particles. Gases have no definite shape or volume; they take the shape and volume of their container, as shown in Figure 2.5. Table 2.3 compares the characteristics of the three states of matter.

Sample Problem 2.8
Identifying States of Matter

Identify each of the following as characteristic of a solid, a liquid, or a gas:

a. maintains its own volume but takes the shape of its container
b. has a definite arrangement of particles
c. has little or no attraction between particles

Solution:

a. a liquid
b. a solid
c. a gas

Study Check:
Identify each of the following substances as a solid, a liquid, or a gas:

a. a substance that has particles that are strongly attracted
b. a substance that does not have a definite volume nor a definite shape

Answer: **a.** a solid **b.** a gas

Figure 2.5
A gas takes the shape and volume of its container.

Table 2.3 Some Characteristics of Solids, Liquids, and Gases

Characteristic	Solid	Liquid	Gas
Shape	Has its own shape	Takes the shape of its container	Takes the shape of its container
Volume	Has a definite volume	Has a definite volume	Fills the volume of its container
Arrangement of particles	Definite	Random	Random
Motion	Slow	Moderate	Very fast
Attraction between particles	Strong	Moderate	None
Closeness of particles	Very close	Close	Far apart

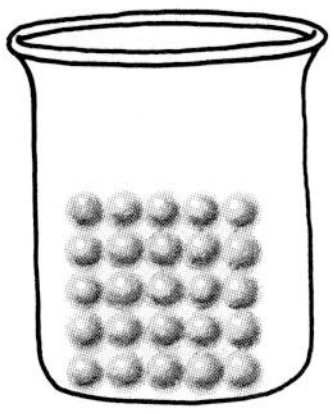

HEALTH NOTE

Homeostasis: Regulation of Body Temperature

The body is an open system through which nutrients, waste products, and heat are constantly being exchanged with the environment. The maintenance of a constant internal temperature is essential to the efficient functioning of our body cells. At low temperatures, essential metabolic reactions proceed too slowly, producing too little of the body's crucial materials. At high temperatures, the structures of the enzymes that regulate our metabolic reactions can change and the enzymes can cease to be active.

Body temperature is regulated by an elaborate series of control systems and usually does not go below 36°C (97°F) or above 40°C (104°F). In a process called homeostasis, changes in the external environment are balanced by changes in the internal environment. It is crucial to our survival that our bodies balance heat gain with heat loss. If we do not lose enough heat, our internal temperature rises; if we lose too much heat, our internal temperature drops.

The body loses heat through radiation, convection, and evaporation. During radiation, heat flows from a warmer area to a cooler one. When the external temperature increases, receptors in the skin send signals to a temperature-control center in the brain that causes the blood vessels of the skin to relax and expand. Warm blood flows to the surface of the skin, and heat radiates away. In the process of convection, the air next to the skin is heated, rises, and is replaced by cooler air. When the external temperature rises, muscular activity is usually slowed, metabolic reactions are reduced, and less internal heat is produced.

An increase in external temperature also stimulates the sweat glands in the skin, causing the production of water and some salts. As this fluid evaporates from the skin, heat is removed and the body temperature is lowered. This lowering of the skin's heat content creates the cooling effect of evaporation.

When the external environment is cold, epinephrine is released, causing an increase in metabolic rate, which increases the production of heat. The change in temperature of the skin also signals the brain to contract the blood vessels. Then less blood flows through the skin, and heat is conserved. The production of sweat also stops, lowering the amount of heat lost by evaporation. A summary of the body's temperature-regulation mechanisms is given in Figure 2.6. ■

2.5 Changes of State

Learning Goal **Draw a heating or cooling curve and label the melting or freezing point, the boiling point, and the sections that represent solid, liquid, and gas.**

Matter undergoes a **change of state** when it is converted from one state into another. You are seeing changes of state when an ice cube melts in a glass, when water boils in a teakettle, or when fog forms on a cold morning.

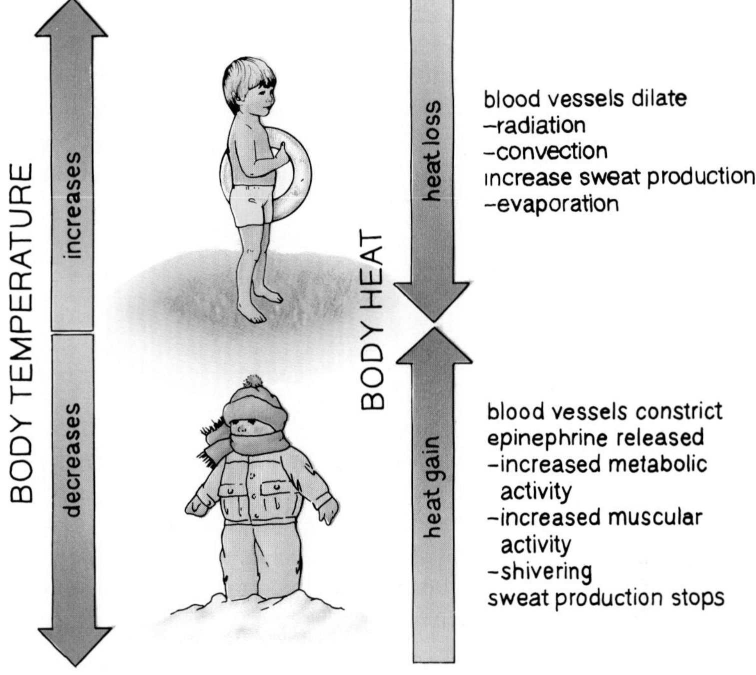

Figure 2.6
In homeostasis, external temperature changes trigger mechanisms that work to maintain a constant internal body temperature.

Melting and Freezing

Melting is a change of state from solid to liquid. A substance melts when heat is added to a solid. The motion of the particles increases until the rigid structure of the solid breaks down. The particles are free to move about in random patterns as they form a liquid. This change of state occurs at a temperature called the **melting point (mp).**

When the temperature of a liquid is lowered, the liquid loses heat. Eventually the particles are slowed enough to form a solid structure. This change of state is called **freezing,** which occurs at the **freezing point** of the liquid. Since the freezing process is the reverse of melting, the melting point and the freezing point are at the same temperature. However, each different substance has its own temperature for these processes. Ice melts at 0°C; water freezes at 0°C. Gold melts (or freezes) at 1064°C, whereas nitrogen melts (or freezes) at −210°C.

$$\text{Solid} \underset{\text{Freezing}}{\overset{\text{Melting}}{\rightleftarrows}} \text{Liquid}$$

Figure 2.7
(a) Evaporation occurs at the surface of a liquid;
(b) boiling occurs as bubbles of gas form within the liquid.

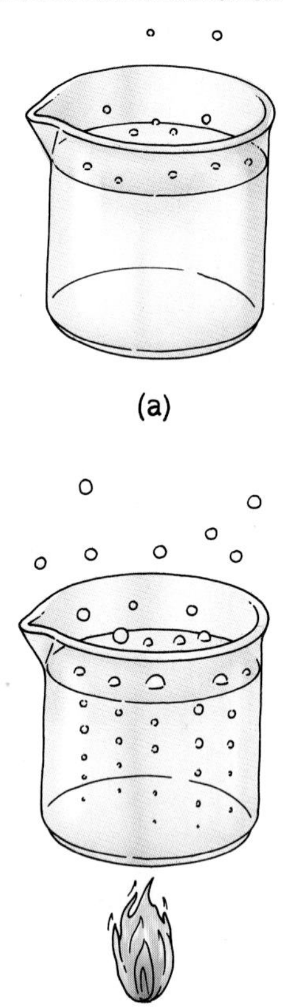

Evaporation

At the surface of a liquid, some particles gain sufficient energy to escape. They "fly away" from the liquid surface by changing into individual gas particles. The change of liquid to gas below the boiling point of a liquid occurs only at the surface of the liquid and is called **evaporation.**

Boiling and Condensation

As more heat is added to a liquid, there is an increase in the amount of liquid that can evaporate. **Boiling** in a liquid begins when gas bubbles form within the liquid as well as at the surface (Figure 2.7). The boiling of a liquid occurs at a temperature called the **boiling point (bp).**

If a gas is cooled, the particles slow down until they begin to adhere to one another and form drops of liquid. The change from gas to liquid is called **condensation** and occurs at the same temperature as boiling.

$$\text{Liquid} \underset{\text{Condensation}}{\overset{\text{Boiling}}{\rightleftharpoons}} \text{Gas}$$

Sublimation

Some particles on the surface of a solid can acquire enough energy to escape from the solid phase and go directly into the gas state. This change of state is called **sublimation.** Dry ice is solid carbon dioxide that sublimes to carbon dioxide gas, and no liquid is formed. Mothballs are substances that sublime at room temperature. They just seem to disappear. However, we can detect their presence by the odor of the gas they form. In very cold areas, snow does not melt to slush but sublimes directly into the gas state. Freeze-dried foods, such as those made for camping and hiking, are produced by sublimation (Figure 2.8). The products are frozen, placed in a vacuum chamber, and dried as the ice crystals sublime. The dried foods retain all of their nutritional value and need only water to be edible again. Freeze-dried foods are convenient on hiking trips, since they do not have to be refrigerated because bacteria cannot grow without moisture. In the reverse process, called **crystallization,** a gas changes directly into a solid.

$$\text{Solid} \underset{\text{Crystallization}}{\overset{\text{Sublimation}}{\rightleftharpoons}} \text{Gas}$$

Figure 2.9 summarizes our discussion of the various changes of state.

Figure 2.8
Some freeze-dried foods produced by sublimation.

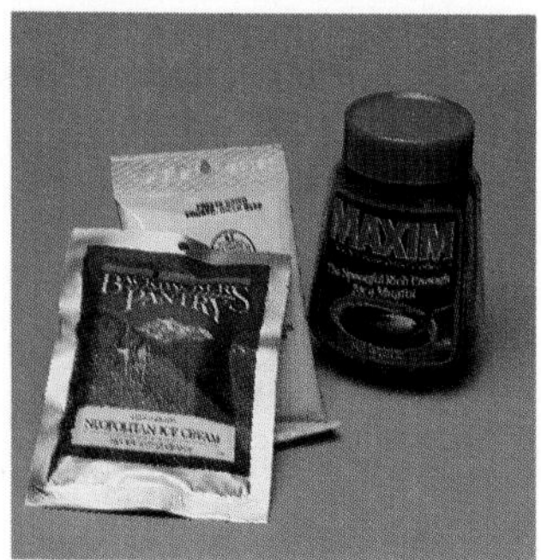

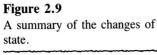

Figure 2.9
A summary of the changes of state.

Sample Problem 2.9
Identifying States and Changes of State

Indicate whether the following statements describe melting, melting point, freezing, freezing point, evaporation, boiling, boiling point, or condensation:

a. Particles of gas are formed only at the surface of a liquid.
b. A solid changes to a liquid at this temperature.
c. Gas bubbles form within a liquid.
d. On a cold day, a puddle of water changes from a liquid to a solid.

Solution:

a. evaporation
b. melting point
c. boiling
d. freezing

Heating and Cooling Curves

A diagram called a **heating curve** is used to represent the states of matter and the changes in state of a substance as heat is added. In Figure 2.10, the heating curve begins at a low temperature with a solid. As heat is added, the temperature of the solid increases until the melting point is reached. At the melting point, all of the heat is used to break apart the solid and change it to a liquid. The melting process occurs at a constant temperature, indicated by a flat line, or plateau, on the heating curve. There is no temperature change while the solid melts.

Figure 2.10
A heating curve diagrams the changes in state as heat is added to a solid.

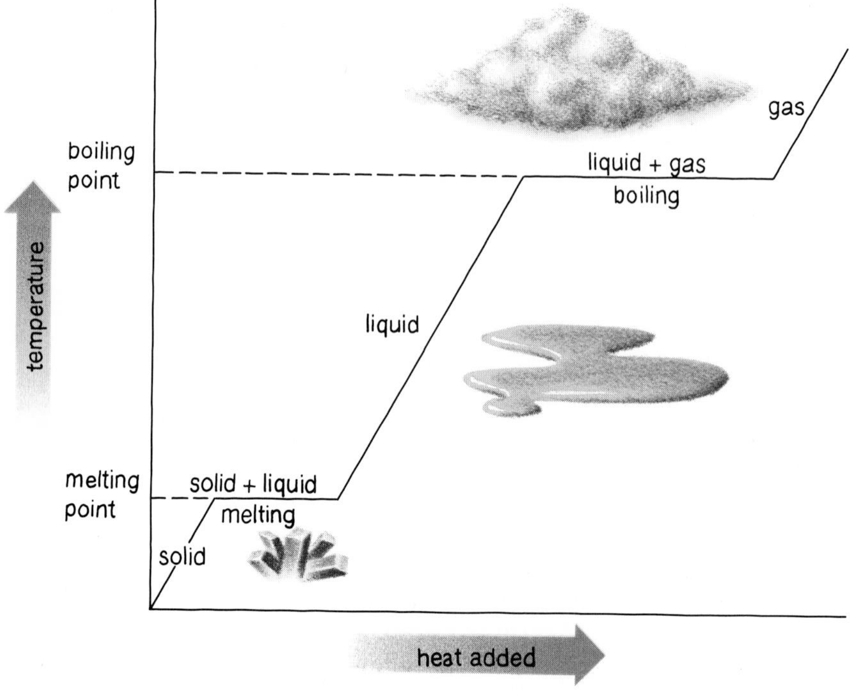

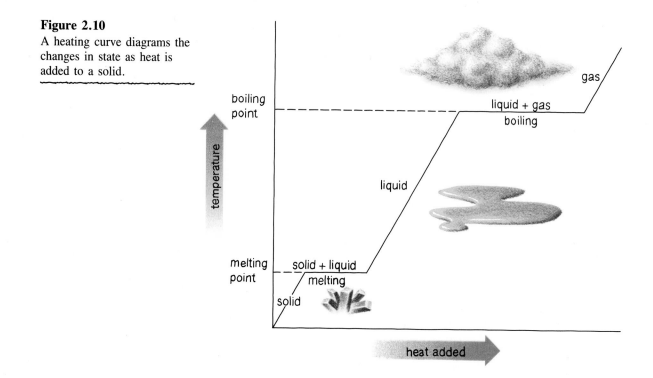

After all of the solid has become a liquid, the temperature begins to rise again. The particles in the liquid phase move faster and faster as they absorb more heat. Eventually the boiling point is reached. Another flat line on the heating curve indicates a constant temperature while the liquid changes into a gas. When all of the substance has been converted to a gas, the temperature rises again as more heat is added.

Sample Problem 2.10
Drawing a Heating Curve for Water

At sea level, water has a melting point of 0°C and a boiling point of 100°C. Draw a heating curve for water from −50°C to 150°C, labeling each section of the curve as solid, liquid, or gas, melting or boiling.

Solution:
Imagine that you are removing an ice cube from a freezer and placing it in a pan on the stove.

a. As you add heat, the temperature of the ice cube rises, which is shown as a diagonal line on the heating curve.
b. When the temperature of the ice cube reaches the melting point, 0°C, the ice cube melts. On the heating curve, draw a flat line to represent the melting process.
c. Draw another diagonal line to indicate another rise in the temperature of the water as more heat is added.
d. At its boiling point, 100°C, the water begins to boil, another change of state that is represented as a flat line on the diagram.
e. Finally, draw one more diagonal line, to represent the rise in temperature for the gaseous state as more heat is added.

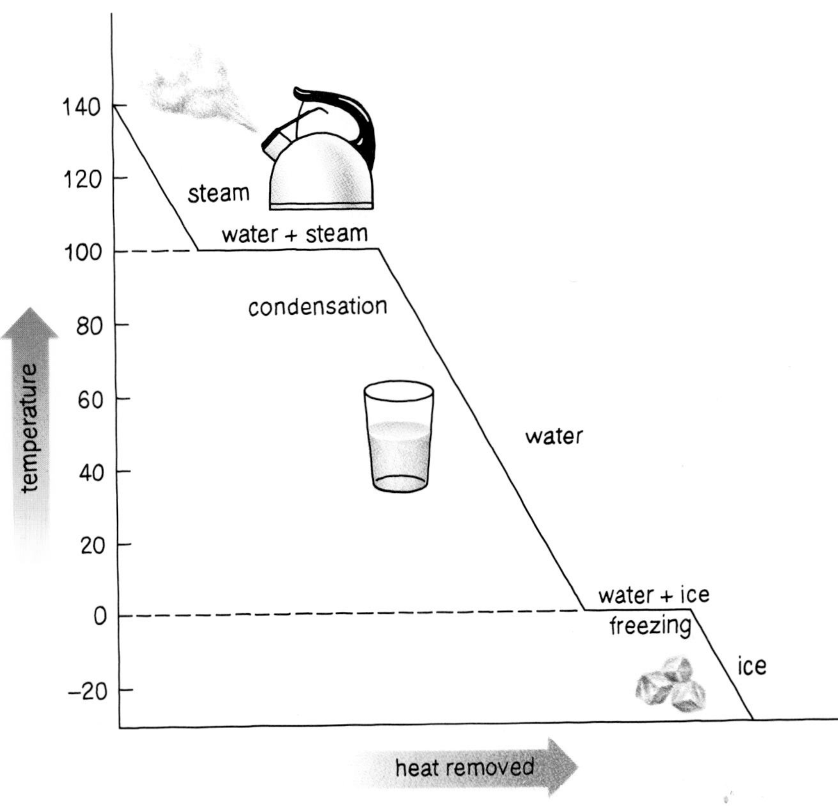

Figure 2.11
A cooling curve for water.

Cooling Curve

We can represent the cooling of a substance on a **cooling curve,** which shows a loss of heat and a drop in temperature. The cooling curve for water, shown in Figure 2.11, illustrates the process. The line on the graph begins at a high temperature for particles in the gaseous state. As heat is removed, the temperature drops. A plateau appears as the gaseous water condenses to a liquid, and another occurs when the liquid freezes.

2.6 Energy in Changes of State

Learning Goal **Calculate the number of calories or joules absorbed or released when a given mass of water undergoes a change of state.**

When a solid, such as ice, is heated, there is an increase in the motion of the particles of the solid. At the melting point, energy is used to break apart the ice structure. The amount of heat needed to melt 1 g of any solid at its melting point is called the **heat of fusion.** At 0°C, 80 cal is required to melt 1 g of ice. (We will use two significant figures for the heat of fusion of water, 80 cal/g.)

Heat of Fusion for Water at 0°C

$$\frac{80 \text{ cal}}{1 \text{ g ice}}$$

We can calculate the amount of energy needed to completely melt some ice by measuring the mass of the ice and using the heat of fusion. There is no temperature unit in the calculation for a change of state because the melting process occurs at a constant temperature.

Calculation of Heat Energy at the Melting (Freezing) Point

Heat = Mass × Heat of fusion

cal g 80 cal/g

The heat of fusion is also used when water freezes. As heat is removed, the particles of the liquid are slowed until attractive forces become so strong that solid crystals begin to form. As liquid changes to solid, heat is released. When 1 g of water freezes, 80 cal of heat is released. For this reason, water is sometimes sprayed in orchards during very cold weather. When the air temperature drops to 0°C, the water freezes and heat is released, which protects the fruit from freezing.

Sample Problem 2.11
Using the Heat of Fusion

An ice sculpture has a mass of 8.5 kg. How many kilocalories will be absorbed to completely melt the sculpture at 0°C?

Solution:
The mass in grams of the ice sculpture is calculated as follows:

$$8.5 \not{kg} \text{ ice} \times \frac{1000 \text{ g}}{1 \not{kg}} = 8500 \text{ g ice}$$

The heat in calories needed to melt the sculpture is

$$8500 \not{g} \times \frac{80 \text{ cal}}{1 \not{g}} = 680{,}000 \text{ cal}$$

Mass Heat of fusion

To calculate the number of kilocalories, we use the equality

1 kcal = 1000 cal

$$680{,}000 \not{cal} \times \frac{1 \text{ kcal}}{1000 \not{cal}} = 680 \text{ kcal or } 6.8 \times 10^2 \text{ kcal}$$

Study Check:
An ice cube tray is filled with 125 g of water and placed in a freezer. How many joules of heat are removed by the freezer compartment to form ice cubes at 0°C?

Answer: 42,000 J or 4.2×10^4 J

Heat of Vaporization

During the boiling process, water is converted from a liquid to a gas. The energy needed to convert 1 g of a liquid to a gas at its boiling point is given by the **heat of vaporization.** For water, 540 cal is needed to convert 1 g of water to a gas at a boiling point of 100°C.

Heat of Vaporization for Water at 100°C

$$\frac{540 \text{ cal}}{1 \text{ g water (liquid)}}$$

During the condensation of water at 100°C, steam is converted into liquid. The amount of energy is the same as the heat of vaporization, except that during condensation heat is released. Thus, the condensation of 1 g of steam (water vapor) releases 540 cal.

Calculation of Heat Energy at the Boiling (Condensation) Point

Heat = Mass × Heat of vaporization

cal g 540 cal/g

Sample Problem 2.12
Calculating Heat Energy for Vaporization

In a sauna, 125 g of water is converted to steam at 100°C. Calculate the energy needed in kilocalories.

Solution:
The calories needed to change 125 g of water, a liquid, to steam, a gas, at 100°C will be the product of the heat of vaporization of water, 540 cal/g, and the number of grams of water.

$$125 \not{\text{g water}} \times \frac{540 \text{ cal}}{1 \not{\text{g water}}} = 67{,}500 \text{ cal}$$

Converting calories to kilocalories,

$$67{,}500 \not{\text{cal}} \times \frac{1 \text{ kcal}}{1000 \not{\text{cal}}} = 67.5 \text{ kcal}$$

Study Check:
When steam from a pan of boiling water reaches a window, it condenses and changes into liquid. How much heat in kilojoules (kJ) is released when 25.0 g of water vapor condenses to liquid water at 100°C? (1 cal = 4.18 J)

Answer: 56.4 kJ

HEALTH NOTE

Effects of a Fever

When bacteria or viruses invade the body, they are surrounded by activated white blood cells. In the process, the hormone EP (endogenous pyrogen) is produced. It causes the brain to reset the body temperature to a higher level. Chills set in, blood vessels constrict to reduce heat loss, and body fat is broken down to produce more heat.

Some research suggests that a fever plays an important role in inhibiting the growth of bacteria. Other studies indicate that antibiotics may be more effective during a fever. A fever also triggers the production of interferon, which the body uses to fight viruses. Although a high fever can certainly be harmful, it may be that a moderate fever should run for a time before it is reduced.

Patients with high fevers can be cooled by giving them sponge baths with tepid water. The evaporation of the water from the skin lowers the body temperature. Sometimes ice packs or ice blankets are used to remove heat from the body. ■

Energy Calculations in Sequence

Many problems involve both temperature changes and changes of state. Then several types of energy calculations are needed to determine the total heat involved in the problem. These types of calculations are reviewed in Table 2.4.

Table 2.4 A Summary of Energy Calculations

Number of states	Change in state	Change in temperature	Data needed	Calculation of heat absorbed or released
One (solid, liquid, or gas)	None	Increases or decreases	Mass, temperature, specific heat	Calories = Mass × Temperature × Specific heat
Two	Solid to liquid; liquid to solid	None	Mass, heat for change	Calories = Mass × Heat of fusion
	Liquid to gas; gas to liquid	None	Mass, heat for change	Calories = Mass × Heat of vaporization

Sample Problem 2.13
Energy Calculations

Calculate the kilocalories needed to heat 40.0 g of water from 25°C to 100°C and convert it completely to steam at the boiling point, 100°C. Use the specific heat of liquid water, 1.00 cal/g °C, and the heat of vaporization for water, 540 cal/g.

Solution:
There are two steps in the heating process. Energy is needed to warm the liquid to 100°C and to convert the water to gas at the boiling point (100°C). Therefore, we need two separate energy calculations.

Step 1 Heating the water (liquid) from 25°C to 100°C:

$$\underset{\text{Mass}}{40.0\ \cancel{g}} \times \underset{\Delta T}{75\cancel{^\circ C}} \times \underset{\text{Specific heat}}{\frac{1\ \text{cal}}{\cancel{g}\ \cancel{^\circ C}}} = 3000\ \text{cal}$$

Step 2 Converting the water (liquid) to gas at 100°C:

$$\underset{\text{Mass}}{40.0\ \cancel{g}} \times \underset{\text{Heat of vaporization}}{\frac{540\ \text{cal}}{1\ \cancel{g}}} = 21{,}600\ \text{cal}$$

The total heat required is the sum of step 1 and step 2:

1. Heating water:	3,000 cal
2. Changing liquid to steam:	21,600 cal
Total heat needed:	24,600 cal

In kilocalories, the heat energy needed for the temperature change and change of state is

$$24{,}600\ \cancel{\text{cal}} \times \frac{1\ \text{kcal}}{1000\ \cancel{\text{cal}}} = 24.6\ \text{kcal}$$

Study Check:
Calculate the number of kilocalories released when 20.0 g of steam at 100°C condenses, cools to 0°C, and freezes. (*Hint:* The solution will require three energy calculations.)

Answer: 14.4 kcal

2.7 Caloric Content of Food

Learning Goal **Calculate the energy of a food in calories, in kilocalories, or in joules.**

When you are watching your food intake, the "calories" you are counting are actually kilocalories. In the field of nutrition, it is common to use the **Calorie, Cal** (with a capital "C") to mean 1000 cal, or 1 kcal. Nutritional values may also be given in kilojoules (kJ).

Nutritional Calories

1 Cal = 1000 cal

1 kcal = 1000 cal

1 kJ = 1000 J

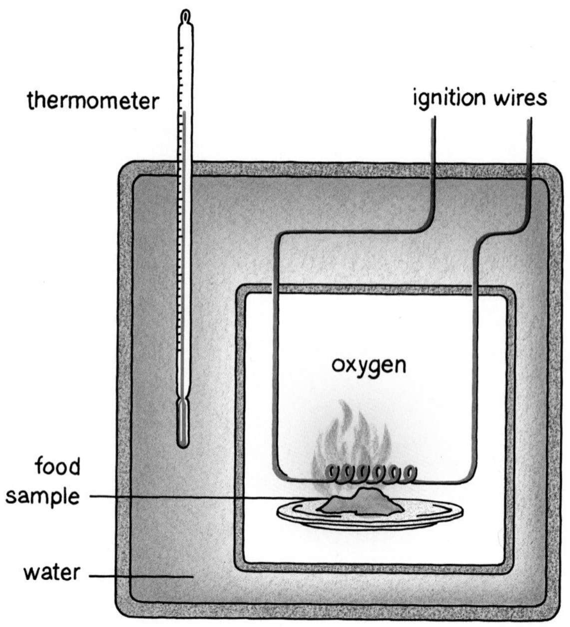

Figure 2.12
A calorimeter. The energy released by a sample of food undergoing combustion is determined by measuring the mass of the food sample, the mass of water in the calorimeter, and the change in temperature of the water.

The caloric content of a food can be determined by using an apparatus called a **calorimeter,** shown in Figure 2.12. A sample of food is placed in a container within the calorimeter, and water is added to fill a surrounding chamber. When the food sample is burned (combustion), heat is given off. The water in the outside chamber absorbs the heat, as indicated by an increase in temperature on a thermometer in the water. If we know the mass of the food sample, the temperature change, and the mass and the specific heat of water, we can calculate the caloric content of the food. The caloric values for several breakfast foods are given in Table 2.5.

Sample Problem 2.14
Calculating Food Energy

A slice of whole-wheat bread is placed in a calorimeter. The mass of the surrounding water is 1000 g, and the initial temperature of the water is 25°C. The heat released during the combustion of the bread raises the temperature of the surrounding water to 90°C. What is the heat energy in kilocalories for the slice of bread?

Solution:
The amount of heat that caused a temperature change of 65°C for 1000 g of water in the calorimeter is calculated as follows:

$$\underset{\text{Mass}}{1000\,\cancel{g}\text{ water}} \times \underset{\Delta T}{65\cancel{^\circ C}} \times \underset{\text{Specific heat}}{\frac{1\text{ cal}}{\cancel{g}\,\cancel{^\circ C}}} = 65{,}000\text{ cal}$$

To convert to kilocalories, we use the equation,

$$65{,}000\,\cancel{\text{cal}} \times \frac{1\text{ kcal}}{1000\,\cancel{\text{cal}}} = 65\text{ kcal}$$

Table 2.5 Caloric Values of Some Breakfast Foods

Item	Quantity	Energy (kcal)
Breads and Cereals		
Bagel	1 medium, plain	135
Biscuit	1, 2 in. diameter	105
Bran flakes	1 cup	105
Corn flakes	1 cup	100
Danish pastry	1, 4 in. diameter	275
Doughnut	1	125
Muffin	1, 3 in. diameter	120
Oatmeal	1 cup	130
Pancake	2, 4 in. diameter	120
Shredded wheat	1 biscuit	90
Toast	1 slice, wheat	65
Waffle	1, 7 in. diameter	205
Fruits		
Banana	1 medium	100
Cantaloupe	1/2	60
Grapefruit	1/2, pink	59
Prunes	4, uncooked	70
Fruit Juices		
Apple juice	1 cup	120
Orange juice	1 cup	110
Tomato juice	1 cup	35
Meats, Eggs, and Milk		
Bacon	2 slices	180
Eggs	2, boiled	160
	2, scrambled with milk and oil	220
Milk	1 cup, whole	160
	1 cup, lowfat	145
	1 cup, nonfat	90
Sausage	2 links, pork	125

Since the combustion of the slice of bread provided the heat to increase the temperature of the water, we can now associate that energy with the slice of bread. Thus we can say that a slice of whole-wheat bread releases 65 kcal. In nutritional terms, it has 65 Cal.

$$\text{Energy from 1 slice of bread} = 65 \text{ kcal or } 65 \text{ Cal}$$

Study Check:
When 0.20 oz of a milk-chocolate candy bar undergoes combustion in a calorimeter, the temperature of 500.0 g of water rises from 18°C to 76°C. How many kilocalories are available in 1.00 oz of milk chocolate?

Answer: 150 kcal

Figure 2.13
Nutritional information on the label of a food package lists the calories and the quantity of each foodstuff per serving of the food.

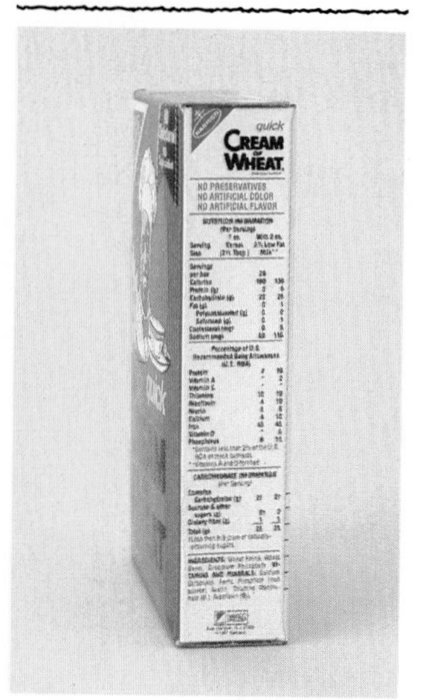

Table 2.6 Caloric Values for Major Foodstuffs

Foodstuff	Carbohydrate	Fat (lipid)	Protein
Caloric value	$\frac{4\ \text{kcal}}{1\ \text{g}}$	$\frac{9\ \text{kcal}}{1\ \text{g}}$	$\frac{4\ \text{kcal}}{1\ \text{g}}$

Caloric Food Values

The caloric value of a foodstuff is the number of kilocalories per gram of food.

$$\text{Caloric value} = \frac{\text{Kilocalories}}{\text{Grams of food}}$$

The caloric values for carbohydrates, fats, and proteins have been determined and are listed in Table 2.6.

The caloric value of a particular food is the sum of the calories from the carbohydrates, fats, and proteins in that food. The average caloric value for each major foodstuff is used to calculate the total number of kilocalories in a food.

$$\text{Kilocalories} = \text{Grams of foodstuff} \times \text{Caloric value}$$

The caloric content of many foods is listed on the package, usually in terms of the number of kilocalories in a serving, as shown in Figure 2.13. The general composition and caloric content of some selected foods are given in Table 2.7.

Table 2.7 General Composition and Caloric Content of Some Foods

Food	Protein (g)	Fat (g)	Carbohydrate (g)	Energy (kcal)
Banana, 1 medium	1	Trace	26	100
Beans, red kidney, 1 cup	15	1	42	230
Beef, lean, 3 oz	22	5	Trace	140
Carrots, raw, 1 cup	1	Trace	11	45
Catsup, 1 cup	6	1	69	290
Chicken, no skin, 3 oz	20	3	0	115
Egg, 1 large	6	6	Trace	74
Milk, whole, 3.5% fat, 1 cup	9	9	12	160
Milk, nonfat, 1 cup	9	Trace	12	90
Oatmeal, cooked, 1 cup	5	2	23	130
Oil, olive, 1 tbsp	0	14	0	125
Potato, baked	3	Trace	23	105
Salmon, 3 oz	17	5	0	120
Steak, 3 oz	20	27	0	330
Yogurt, lowfat, 1 cup	8	4	13	125

HEALTH NOTE

Steam Burns

Hot water (liquid) at 100°C will cause burns and damage to the skin. However, getting steam on the skin is even more dangerous. Let us consider 100 g of hot water at 100°C. If this water falls on a person's skin, the temperature of the water will drop to body temperature, 37°C. The heat released during cooling burns the skin. The temperature change is 63°C:

$$100\ \cancel{g} \times 63\cancel{°C} \times \frac{1\ \text{cal}}{1\cancel{g}\ \cancel{°C}} = 6300\ \text{cal heat}$$

For comparison, we can calculate the amount of heat released when 100 g of steam (gas) at 100°C hits the skin. First, the steam condenses to water (liquid) at 100°C:

$$100\ \cancel{g} \times \frac{540\ \text{cal}}{1\ \cancel{g}} = 54{,}000\ \text{cal heat}$$

Now the temperature of the 100 g of liquid drops from 100°C to 37°C, releasing still more heat—6300 cal, as we saw earlier. Using this value and the value we obtained for heat released during the condensation of the steam, we calculate the total amount of heat released as follows:

From the condensation of steam at 100°C:	54,000 cal
From the cooling of water from 100°C to 37°C:	6,300 cal
Total heat released from 100 g steam:	60,300 cal

The amount of heat released from steam is almost 10 times greater than the amount released from hot water. This tremendous release of energy is the reason for the severity of steam burns. ■

Sample Problem 2.15
Calculating Caloric Value

A 5.4-g sample of glucose is placed in a calorimeter. The mass of water in the container is 1000 g, and the initial temperature is 24°C. When the combustion of the glucose is complete, the water temperature has risen to 46°C. Calculate the caloric value (kcal/g) for the glucose.

Solution:
The amount of heat absorbed by the water is equal to the amount of heat given off by the glucose during combustion.

Heat lost (glucose) = Heat gained (water)

The heat absorbed by the water is calculated as follows:

$$\underbrace{1000\ \cancel{g}}_{\text{Mass of water}} \times \underbrace{22\cancel{°C}}_{\Delta T} \times \underbrace{\frac{1\ \text{cal}}{\cancel{g}\ \cancel{°C}}}_{\text{Specific heat}} = 22{,}000\ \text{cal}$$

To convert to kilocalories,

$$22{,}000\ \cancel{\text{cal}} \times \frac{1\ \text{kcal}}{1000\ \cancel{\text{cal}}} = 22\ \text{kcal}$$

Since 22 kcal of heat was absorbed by the water, 22 kcal must have been released during the combustion of glucose. The caloric value of glucose is calculated as

$$\frac{22 \text{ kcal}}{5.4 \text{ g glucose}} = 4.1 \text{ kcal/g glucose}$$

HEALTH NOTE

Loss and Gain of Weight

The number of kilocalories needed in your daily diet depends on your age and physical activity. Some general levels of energy needs are given in Table 2.8.

When food intake exceeds energy output, a person's body weight increases. Food intake is usually regulated by the hunger center in the hypothalamus, located in the brain. The regulation of food intake is normally proportional to the nutrient stores in body. If these nutrient stores are low, you feel hungry; if they are high, you do not feel like eating.

The regulation of food intake does not operate in obese persons. The causes of this failure may be psychological or physiological. One cause of obesity in adults may be childhood conditioning. In infancy, there is a rapid rate of formation of new fat cells as compared to that in later life. The number of fat cells produced in infancy determines the number of fat cells in the adult body. It is now believed that overfeeding children, especially infants, can lead to a lifelong struggle with obesity. In stressful situations, some people overeat in an attempt to relieve tension. Occasionally, tumors of the hypothalamus cause an excessive appetite that leads to a weight increase.

Table 2.8 Typical Energy Requirements

Age (yr)	Weight (lb)	Mass (kg)	Energy (kcal)
Child			
0–1	16	7.3	820
2–3	29	13	1360
4–6	44	20	1830
7–9	62	28	2190
10–12	83	38	2480
Young Adult			
Female 13–19	115	52	2400
Male 13–19	130	59	2980
Adult			
Female	121	55	2200
Male	143	65	3000

Table 2.9
Energy Expended by a 70-kg (154-lb) Person

Activity	Energy expended (kcal/hr)
Sleeping	60
Sitting	100
Walking	200
Swimming	500
Running	550

Weight reduction occurs when food intake is less than energy output. Many diet products contain cellulose, which has no nutritive value but provides bulk and makes you feel full. Some diet drugs depress the hunger center and must be used with caution because they excite the nervous system and can elevate blood pressure. Because muscular exercise is an important way to expend energy, an increase in daily exercise aids weight loss. Table 2.9 lists some activities and the amount of energy they require.

The body uses carbohydrates for energy before using fats and proteins. However, there are only a few hundred grams of glycogen, the form of stored carbohydrate in the body, available in the liver and muscles of the body. Glycogen provides enough energy for about half a day. In severe dieting or fasting, it is rapidly depleted. Fat stores then become the prime source of energy, and the amount of fat in the body decreases. If fat stores are depleted, as in starvation, the only energy source remaining is protein. Proteins can be converted to glucose by the liver to provide energy for the brain. However, because proteins are part of the body's structure, their depletion eventually results in death. ■

Sample Problem 2.16
Caloric Value for a Food

A 1-oz (28-g) serving of oat bran hot cereal with half a cup of whole milk contains 10 g protein, 22 g carbohydrate, and 7 g fat. Calculate the total number of kilocalories in this 1-oz serving.

Solution:
Using the caloric values for carbohydrate, fat, and protein (Table 2.6), we can calculate the total number of kilocalories:

$$\text{Carbohydrate:}\quad 22\,\cancel{g} \times \frac{4\text{ kcal}}{1\,\cancel{g}} = 88\text{ kcal}$$

$$\text{Fat:}\quad 7\,\cancel{g} \times \frac{9\text{ kcal}}{1\,\cancel{g}} = 63\text{ kcal}$$

$$\text{Protein:}\quad 10\,\cancel{g} \times \frac{4\text{ kcal}}{1\,\cancel{g}} = 40\text{ kcal}$$

$$= 191\text{ kcal}$$

Rounding off the answer for significant figures,

Total energy in 1 oz of oat bran hot cereal = 190 kcal

Study Check:
How many kilocalories are in a piece of chocolate cake that contains 3 g protein, 35 g carbohydrate, and 11 g fat?

Answer: 250 kcal

Glossary of Key Terms

boiling The conversion of a liquid to a gas that occurs when a gas forms within the liquid.

boiling point (bp) The temperature at which boiling occurs.

calorie (cal) The amount of heat energy that raises the temperature of 1 g of water 1°C.

Calorie (Cal) The dietary unit of energy, equal to 1000 cal, or 1 kcal.

calorimeter An instrument used to measure the amount of heat released by a food sample that undergoes combustion.

Celsius scale A temperature scale on which the freezing point of water is 0°C and the boiling point is 100°C.

change of state The transformation of one state to another; for example, solid to liquid, liquid to gas.

condensation The formation of a liquid from a gas when the gas is cooled to the condensation point (same temperature as boiling).

cooling curve A graph that indicates the changes of state for a substance as heat is removed.

crystallization The formation of a solid from a liquid by cooling; this occurs at the same temperature as melting.

energy The ability to do work.

evaporation The formation of a gas (vapor) by the escape of high-energy particles from the surface of a liquid.

Fahrenheit scale A temperature scale used in the United States on which water freezes at 32°F and water boils at 212°F.

freezing A change of state from liquid to solid.

freezing point (fp) The temperature at which a liquid freezes to form a solid.

gas A state of matter characterized by no definite shape or volume. Particles in a gas move rapidly and exhibit little or no attraction to one another.

heat The energy associated with the motion of particles in a substance.

heat of fusion The energy required to melt or freeze 1 g of a substance at its melting (freezing) point. For water, at 0°C, 80 cal is needed to melt 1 g of ice; 80 cal is released when 1 g of water freezes.

heat of vaporization The energy required to vaporize (condense) 1 g of a substance at its boiling (condensation) point. For water, at 100°C, 1 g of liquid requires 540 cal to vaporize; 1 g of steam gives off 540 cal when it condenses.

heating curve A graph that represents the transitions from one state to the next as heat is added to a substance.

joule (J) The SI unit of heat energy; 4.18 J = 1 cal.

Kelvin scale A temperature scale (SI) that assigns absolute zero to the lowest possible temperature. One kelvin (K) also represents a unit of temperature on the Kelvin scale: K = °C + 273.

kinetic energy A type of energy that is actively doing work; energy of motion.

kilocalorie An amount of heat energy equal to 1000 calories.

liquid A state of matter that takes the shape of its container but has its own volume. A liquid has moderate attractions between particles.

matter Anything that has mass and occupies space.

melting The conversion of a solid to a liquid.

melting point (mp) The temperature at which a solid become a liquid (melts).

potential energy An inactive type of energy that is stored for use in the future.

solid A state of matter that has its own shape and volume, with little motion—only vibrations—and strong attractions between particles.

specific heat A quantity of heat that raises the temperature of 1 g of a substance by 1°C.

sublimation The change of state in which a solid is transformed directly into a gas.

temperature An indication of the intensity of heat in a substance.

Problems

Temperature Measurement (*Goal 2.1*)

2.1 Your friend who is visiting from France just took her temperature. When she reads 99.8°, she becomes concerned that she is quite ill. How would you explain this temperature problem to your friend? Be sure to compare Celsius and Fahrenheit temperature scales and convert her temperature to Celsius, which is the temperature scale she knows in France.

2.2 Solve the following temperature conversions:

a. 37.0°C = ____°F	d. 65.3°F = ____°C
b. 25°C = ____°F	e. 110°F = ____°C
c. 155°C = ____°F	f. −25°F = ____°C

2.3 Solve the following temperature conversions:

a. 62°C = ____K	d. 545 K = ____°C
b. −27°C = ____K	e. 224 K = ____°C
c. 72°F = ____K	f. 875 K = ____°F

2.4 a. A patient has a fever of 106°F. What does this read on a Celsius thermometer?
b. A 4-year-old child has a temperature of 38.7°C. Since high fevers cause convulsions in children, it is recommended that phenobarbital be given if the temperature exceeds 101.0°F. Should phenobarbital be given now?
c. Hot compresses are being prepared for the patient in room 32B. The water is heated to 145°F. What is the temperature of the hot water in degrees Celsius?

***2.5** A young woman recovered from extreme hypothermia, during which her temperature had dropped to 20.6°C. What was her temperature on (a) the Fahrenheit scale and (b) the Kelvin scale?

Energy (*Goal 2.2*)

2.6 Discuss the changes in the potential and kinetic energy of a roller coaster ride as the roller coaster car climbs up a hill and goes down the other side.

2.7 Indicate whether each statement describes potential or kinetic energy:
a. water at the top of a waterfall
b. water falling in a waterfall
c. the energy in a lump of coal
d. a skier at the top of a hill
e. the energy you will obtain from your food
f. a glacier on the side of a mountain
g. an earthquake
h. a car speeding down the freeway
i. a ski jumper going down the ramp

2.8 In each of the following, energy is changed from one form to another. Name each form of energy.
a. using a hair dryer
b. using a solar-powered calculator
c. burning gasoline in a car engine
d. sunlight falling on a solar water heater
e. turning on a light switch

Measuring Heat Energy *(Goal 2.3)*

2.9 Why are the handles of cooking pans made out of wood when the pans are made of aluminum, iron, or copper?

2.10 What information is necessary to calculate the amount of heat energy required to heat a substance?

2.11 Determine the amount of heat needed in the following:
a. the number of calories to heat 25 g of water from 22°C to 44°C
b. the number of joules to heat 15 g of water from 13°C to 85°C
*c. the number of calories to heat 10.0 g of silver from 15°C to 237°C (specific heat of silver, 0.057 cal/g °C)

2.12 An electric power plant releases heat into a nearby stream. The temperature of 5.0 kg of stream water increases from 22°C to 28°C. How many kilocalories of heat were absorbed by the water?

2.13 Calculate the amount of heat released in each of the following situations:
a. the number of calories released when 85 g of water cools from 95°C to 21°C
b. the number of joules released when 0.50 kg of water cools from 65°C to 13°C
*c. the number of kilocalories lost when 250 g of sand cools from 45°C to 28°C (specific heat of sand, 0.19 cal/g °C)

States of Matter *(Goal 2.4)*

2.14 Indicate whether each of the following statements describes a gas, a liquid, or a solid.
a. This substance has no definite volume nor shape.
b. The particles of this substance are not attracted to one another.
c. The particles of this substance are held in a definite structure.
d. The particles of this substance are very far apart.
e. This substance occupies the entire volume of the container.
f. This substance has a definite volume but takes the shape of the container.

Changes of State *(Goal 2.5)*

2.15 Identify the following changes of state as melting, freezing, evaporation, boiling, condensation, or sublimation:
a. In a liquid, the particles at the surface change to a gas.
b. A substance changes from a liquid to a solid.
c. Bubbles of gas form within a liquid and escape.

d. A solid changes to a gas, but no liquid forms.
e. The water in the clouds changes to rain.
f. A breakdown of the solid structure of a substance occurs.

2.16 a. How does the production of sweat during heavy exercise cool the body?
b. Why do clothes dry more quickly on a hot summer day than on a cold winter day?
*c. For sports injuries and in some localized surgeries, a spray such as ethyl chloride may be used to numb an area of the skin. Explain how a substance that evaporates quickly can numb the skin.

2.17 Draw a heating curve for a sample of ice that is heated from −20°C to 140°C. The melting point of water is 0°C; the boiling point is 100°C. Indicate on the graph the portion of the curve that corresponds to each of the following:
a. solid c. liquid e. gas
b. melting point d. boiling point

2.18 The melting point of benzene is 5°C, and its boiling point is 80°C. Sketch a heating curve for benzene from 0°C to 100°C.
a. What is the state of benzene at 15°C?
b. What happens on the curve at 5°C?
c. What is the state of benzene at 60°C? At 90°C?
d. At what temperature will both liquid and gas be present?

2.19 What happens to a sample of steam, initially at 110°C, that is cooled to −10°C? Sketch the cooling curve.

Energy in Changes of State *(Goal 2.6)*

2.20 Using the heat of fusion, calculate the amount of heat needed in the following problems:
a. the number of calories needed to melt 115.0 g of ice
b. the number of joules needed to melt 75.0 g of ice
c. the number of calories needed to melt a 50.0-g ice cube at 0°C, and then to heat the water from 0°C to 65°C

2.21 A bag of ice was placed on a burn on a patient's hand. The ice bag contained 220 g of ice at 0°C. When the ice bag was removed, all of the ice inside had melted, and the liquid water inside had a temperature of 21°C. How many kilocalories of heat were absorbed by the ice bag?

2.22 Using the heat of vaporization, calculate heat energy in the following problems:
a. How many kilocalories are needed to completely vaporize (boil) 50.0 g of water at 100°C?
b. How many calories are released when 20.0 g of steam (gas at 100°C) condenses to liquid (at 100°C) and then cools to 25°C?
*c. How many kilocalories are required to melt 80.0 g of ice at 0°C, warm the liquid to 100°C, and then convert the liquid to steam at 100°C?

***2.23** An ice cube tray holds 4.0 cups of water. If tap water has a temperature of 25°C, how many calories must be removed to cool and freeze the water at 0°C? (1 qt = 4 cups; 1 qt = 946 mL; density of H_2O = 1.00 g/mL)

***2.24** A 115-g sample of steam at 100°C escapes from a volcano. It condenses, cools, and finally falls as snow at 0°C. How many kilocalories of heat are released?

***2.25** Water is sprayed on the ground of an orchard when temperatures are near freezing.

a. How would the water protect the fruit from freezing?
b. How many kilocalories of heat are released if 5.0 kg of water at 15°C is sprayed on the ground, cools, and freezes at 0°C?

Caloric Content of Food *(Goal 2.7)*

2.26 The combustion of the following foods has been determined by calorimetry. Use the data given to calculate the number of kilocalories for each food:

Food	Water in Calorimeter (g)	Temperature (°C)	
		Initial	Final
a. celery (1 stalk)	500	25	35
b. waffle (7 in. diameter)	5000	20	62
c. popcorn (1 cup, no oil)	1000	25	50

2.27 Use the caloric values for carbohydrates (4 kcal/g), fats (9 kcal/g), and proteins (4 kcal/g) to complete the following table:

Food	Carbohydrate (g)	Fat (g)	Protein (g)	Energy (kcal)
a. orange, 1	16	0	1	—
b. apple, 1	—	0	0	72
c. Danish pastry, 1	30	15	5	—
d. avocado, 1	13	—	5	405

2.28 One cup of clam chowder contains 9 g protein, 12 g fat, and 16 g carbohydrate. How many kilocalories are in the clam chowder?

2.29 A diet calls for 220 g carbohydrate, 65 g lipid (fat), and 85 g of protein in one day. How many kilocalories per day does this diet provide?

2.30 A high-protein diet includes 70 g carbohydrate, 150 g protein, and 5 g fat. How many kilocalories does this diet provide?

2.31 A typical diet in the United States provides 15% of the calories from protein, 45% from carbohydrates, and 40% from fats. Calculate the total number of grams of protein, carbohydrate, and fat to be included each day in diets having the following caloric requirements:

a. 1000 kcal
b. 1800 kcal
c. 2600 kcal

***2.32** You have just eaten a quarter-pound hamburger with cheese, french fries, and a chocolate shake. With the following nutritional information, determine how many hours you will need to run to burn off the kilocalories in your meal. See Table 2.9. (Assume you are a 70-kg person.)

Item	**Protein**	**Fat**	**Carbohydrate**
quarter-pounder with cheese	31 g	29 g	34 g
french fries	3 g	11 g	26 g
chocolate shake	11 g	9 g	60 g

${}^{88}_{38}Sr$

${}^{23}_{11}Na$

${}^{63}_{29}Cu$

${}^{13}_{56}Ba$

3

Atoms and Elements

Fireworks display vibrant colors that result when electrons in atoms of various elements are excited by heat. Atoms of strontium, sodium, copper and barium produce the colors red, yellow, green and blue.

Learning Goals

3.1 Given the name of an element, write its correct symbol; from the symbol, write the correct name.

3.2 Describe the electrical charge, mass (amu), and location in an atom for a proton, a neutron, and an electron.

3.3 Given the atomic number and the mass number of an atom, state the number of protons, neutrons, and electrons.

3.4 Use the periodic table to identify the group and the period of an element, and whether it is a metal or a nonmetal.

3.5 Given the name or symbol of one of the first 20 elements in the periodic table, write the electron arrangement.

3.6 Use the electron arrangement of an element to state its group number and to explain periodic law.

3.7 Write the electron configuration using subshell notation.

SCOPE

All of the matter that surrounds us is composed of primary substances called **elements.** There are different kinds of elements. The elements calcium and phosphorus build your teeth and bones. The hemoglobin that carries oxygen in your blood contains the element iron. The elements carbon, hydrogen, oxygen, and nitrogen derived from the digestion of food are used by the cells of your body to build proteins. Today, there are 109 different elements. Of these, 88 occur naturally and are found in different combinations, providing the great number of compounds that make up our world. The rest of the elements have been produced artificially and are not found in nature.

You have probably seen the element aluminum. Imagine that you are tearing a piece of aluminum foil into smaller and smaller pieces. Now imagine that you have a piece so small that you can no longer break it down further. Then you would have an atom of aluminum, the smallest particle of an element that still retains the characteristics of that element.

3.1 Elements and Symbols

Learning Goal **Given the name of an element, write its correct symbol; from the symbol, write the correct name.**

Elements are primary substances from which all other things are built. They cannot be broken down into simpler substances. Many of the elements were named for planets, mythological figures, minerals, colors, geographic locations, and famous people. Some sources of the names of elements are listed in Table 3.1.

Table 3.1 Some Elements and Their Names

Element	Source of Name
Uranium	The planet Uranus
Titanium	Titans (mythology)
Chlorine	*Chloros,* ''greenish yellow'' (Greek)
Iodine	*Ioeides,* ''violet'' (Greek)
Magnesium	Magnesia, a mineral
Californium	California
Curium	Marie and Pierre Curie

Chemical Symbols

Chemical symbols are abbreviations of the names of the elements. They are usually formed from one or two letters of the name. A few symbols are derived from the ancient Latin or Greek names for elements. For example, the Latin word for sodium is *natrium,* which means ''alkali metal''; the symbol is Na. The symbol for potassium, K, comes from the Latin name *kalium,* ''potash.''

Table 3.2 lists the names and symbols of some common elements. Learning the names and symbols of these elements will greatly help your learning of chemistry. A complete list of all the elements (both naturally occurring and artificially produced) and their symbols appears on the inside front cover of this text. Figure 3.1 shows samples of some common elements and objects made of those elements.

Table 3.2 Names and Symbols of Some Common Elements

Name[a]	Symbol	Name[a]	Symbol
Aluminum	Al	Lead (*plumbum*)	Pb
Argon	Ar	Lithium	Li
Barium	Ba	Magnesium	Mg
Boron	B	Mercury (*hydrargyrum*)	Hg
Bromine	Br	Neon	Ne
Cadmium	Cd	Nickel	Ni
Calcium	Ca	Nitrogen	N
Carbon	C	Oxygen	O
Chlorine	Cl	Phosphorus	P
Cobalt	Co	Potassium (*kalium*)	K
Copper (*cuprum*)	Cu	Silicon	Si
Fluorine	F	Silver (*argentum*)	Ag
Gold (*aurum*)	Au	Sodium (*natrium*)	Na
Helium	He	Strontium	Sr
Hydrogen	H	Sulfur	S
Iodine	I	Tin (*stannum*)	Sn
Iron (*ferrum*)	Fe	Zinc	Zn

[a]Names given in parentheses are ancient Latin or Greek words from which the symbols are derived.

Figure 3.1
Some common elements and objects made of them: carbon (C) as graphite in a pencil, and in a diamond; copper (Cu) in a penny; silicon (Si) in a computer chip; nickel (Ni) in a nickel; silver (Ag) in a pair of earrings; aluminum (Al) in aluminum foil.

HEALTH NOTE

Latin Names for Elements in Clinical Usage

In medicine, the Latin names for sodium and potassium are often used. The condition in which there is too much sodium in the body is called hypernatremia, and a low sodium level is called hyponatremia. In the case of potassium, both the modern name and the Latin name are used. For example, a high potassium level may be called hyperpotassemia or hyperkalemia; a below-normal potassium level may be called hypopotassemia or hypokalemia. ■

A new system of naming was proposed in 1976 for elements 104 and higher. Each digit in the number is named as follows:

0 nil	1 un	2 bi	3 tri	4 quad
5 pent	6 hex	7 sept	8 oct	9 en

Thus the name for element 104 would be *unnilquadium* (1, *un*; 0, *nil*; 4, *quad*; the ending *ium* is added). Its symbol has three letters, Unq. The name for element 105 would be unnilpentium (Unp).

Element	*Name*	*Symbol*
104	unnilquadium	Unq
105	unnilpentium	Unp
106	unnilhexium	Unh

Sample Problem 3.1
Writing Chemical Symbols

Write the chemical symbol for each of the following elements:

a. carbon **b.** nitrogen **c.** chlorine **d.** copper

Solution:

a. C **b.** N **c.** Cl **d.** Cu

Sample Problem 3.2
Naming Chemical Elements

Give the name of the element that corresponds to each of the following chemical symbols:

a. Zn **b.** K **c.** H **d.** Fe

Solution:

a. zinc **b.** potassium **c.** hydrogen **d.** iron

HEALTH NOTE

Elements Essential to Health

Several elements are essential for the well-being and survival of the human body. Some examples and the amounts present in a 60-kg person are listed in Table 3.3. ■

Table 3.3 Elements Essential to Health

Element	Symbol	Amount in a 60-kg Person	Where Found
Oxygen	O	39 kg	Water, carbohydrates, fats, proteins
Carbon	C	11 kg	Carbohydrates, fats, proteins
Hydrogen	H	6 kg	Water, carbohydrates, fats, proteins
Nitrogen	N	2 kg	Proteins, DNA, RNA
Calcium	Ca	1 kg	Bones, teeth
Phosphorus	P	0.6 kg	Bones, teeth, DNA, RNA
Potassium	K	0.2 kg	Inside cells (important in conduction of nerve impulses)
Sulfur	S	0.2 kg	Some amino acids
Sodium	Na	0.1 kg	Body fluids (important in nerve conduction and fluid balance)
Magnesium	Mg	0.1 kg	Bone (important in enzyme function)
Chlorine	Cl	0.1 kg	Outside cells (major electrolyte)

3.2 The Atom

Learning Goal **Describe the electrical charge, mass (amu), and location in the atom for a proton, a neutron, and an electron.**

Atoms are the smallest particles of an element that retain the characteristics of that element. Billions of atoms are packed together to build you and all the matter around you. The paper in this book contains atoms of carbon, hydrogen, and oxygen. The ink on this paper, even the dot over the letter *i,* contains huge numbers of atoms. There are as many atoms in that dot as there are seconds in 10 billion years.

The concept of the atom is relatively recent. Although the Greek philosophers in A.D. 500 reasoned that matter must contain minute particles they called *atomos,* the idea of atoms did not become a scientific theory until 1808. Then John Dalton (1766–1844) developed an atomic theory that proposed that atoms were responsible for the combinations of elements found in compounds.

Atomic Theory

1. All matter is made up of tiny particles called atoms.
2. All atoms of a given element are similar to one another and different from atoms of other elements.
3. Atoms of two or more different elements combine to form compounds. A particular compound is always made up of the same kinds of atoms and always has the same number of each kind of atom.
4. A chemical reaction involves the rearrangement, separation, or combination of atoms. Atoms are never created nor destroyed during a chemical reaction.

Atoms are the building blocks of everything we see around us, yet we cannot see an atom or even a billion atoms with the naked eye. However, when billions and billions of atoms are packed together, the characteristics of each atom are added to those of the next until we can see the characteristics we associate with the element. For example, a small piece of the shiny, copper-colored element we call copper consists of many, many copper atoms. Through a special kind of microscope called a tunneling microscope, we can now see images of individual atoms, such as the atoms of carbon in graphite shown in Figure 3.2.

Subatomic Particles

By the early part of the twentieth century, growing evidence indicated that the atom was not a solid sphere, as Dalton had imagined. New experiments showed that atoms were composed of even smaller bits of matter called **subatomic particles.** Much of the chemistry of an element depends upon the subatomic particles that are the building blocks of the atoms.

There are three subatomic particles of interest to us, the proton, neutron, and electron. Two of these carry electrical charges. The **proton** has a positive charge (+), and the **electron** carries a negative charge (−). The **neutron** has no electrical charge; it is neutral.

Figure 3.2
Graphite, a form of carbon, magnified millions of times by a scanning tunneling microscope. This instrument generates an image of the atomic structure. The round yellow objects are atoms.

Figure 3.3
Attraction and repulsion of electrical charges.

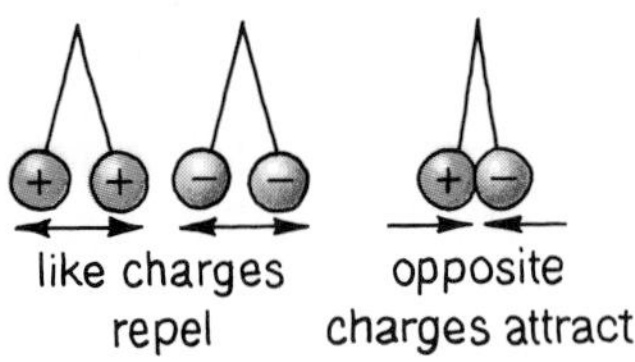

Like charges repel; they push away from each other. When you brush your hair on a dry day, electrical charges that are alike build up on the brush and in your hair; as a result your hair flies away from the brush. Opposite or unlike charges attract. The crackle of clothes taken from the clothes dryer indicates the presence of electrical charges. The clinginess of the clothing is due to the attraction of opposite, unlike charges, as shown in Figure 3.3.

Nucleus of the Atom

The protons and neutrons in an atom are tightly packed in a tiny space at the center of the atom called the **nucleus.** The rest of the atom is extremely large, but it consists mostly of empty space and the electrons. If we imagine a football stadium, the nucleus would be about the size of a golf ball in the center. The rest of the large volume would be mostly empty space occupied only by the fast-moving electrons. (See Figure 3.4.)

Mass

All of the subatomic particles are extremely small compared to the things you see around you. Because the mass of a subatomic particle is so minute, chemists find it convenient to use a very small unit of mass called an **atomic mass unit (amu).** An atomic mass unit is defined as one-twelfth of the mass of a carbon-12 atom, a standard to which the mass of every other atom is compared. (As we shall see later, a carbon-12 atom has six protons and six neutrons in its nucleus which gives it a mass of 12 amu.) A proton has a mass of 1 amu; so also does a neutron. Since the electron is so light, its mass is usually ignored in atomic mass calculations. Table 3.4 summarizes some information about the subatomic particles in an atom.

Table 3.4 Subatomic Particles in the Atom

Subatomic Particle	Symbol	Electrical Charge	Approximate Mass	Location in Atom
Proton	p or p^+	1+	1 amu	Inside nucleus
Neutron	n or n^0	None	1 amu	Inside nucleus
Electron	e^-	1−	0	Outside nucleus

Figure 3.4
Arrangement of the subatomic particles in an atom. The protons and neutrons are located in the nucleus at the center of the atom; the electrons are located outside the nucleus.

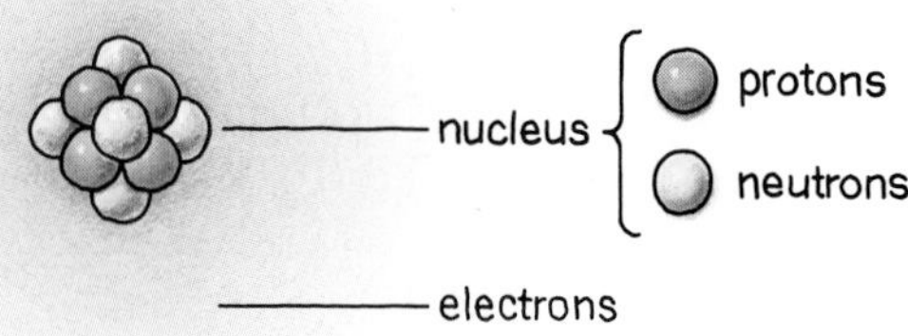

Sample Problem 3.3
Identifying Subatomic Particles

Complete the following table for subatomic particles:

Name	*Symbol*	*Mass (amu)*	*Charge*	*Location in Atom*
Electron	______	______	______	______
______	______	1	0	______

Solution:

Electron	e^-	0	1−	Outside nucleus
Neutron	n^0	1	0	Nucleus

Study Check:
Give the symbol, mass, electrical charge, and location of a proton in the atom.

Answer: The symbol for a proton is p or p^+. It has a mass of 1 amu, carries a positive charge of 1+, and is found in the nucleus of an atom.

3.3 Atomic Number and Mass Number

Learning Goal **Given the atomic number and the mass number of an atom, state the number of protons, neutrons, and electrons.**

All of the atoms of the same element always have the same number of protons. This feature distinguishes atoms of one element from atoms of all the other elements. An **atomic number,** which is equal to the number of protons in the nucleus of an atom, is used to identify each element.

Atomic number = Number of protons in an atom

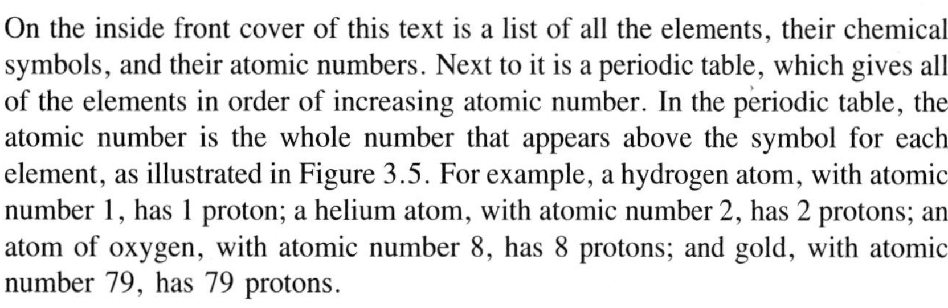
On the inside front cover of this text is a list of all the elements, their chemical symbols, and their atomic numbers. Next to it is a periodic table, which gives all of the elements in order of increasing atomic number. In the periodic table, the atomic number is the whole number that appears above the symbol for each element, as illustrated in Figure 3.5. For example, a hydrogen atom, with atomic number 1, has 1 proton; a helium atom, with atomic number 2, has 2 protons; an atom of oxygen, with atomic number 8, has 8 protons; and gold, with atomic number 79, has 79 protons.

Figure 3.5
Atomic numbers appear above the symbols of the elements listed in the periodic table. The atomic number indicates the number of protons in an atom of that element.

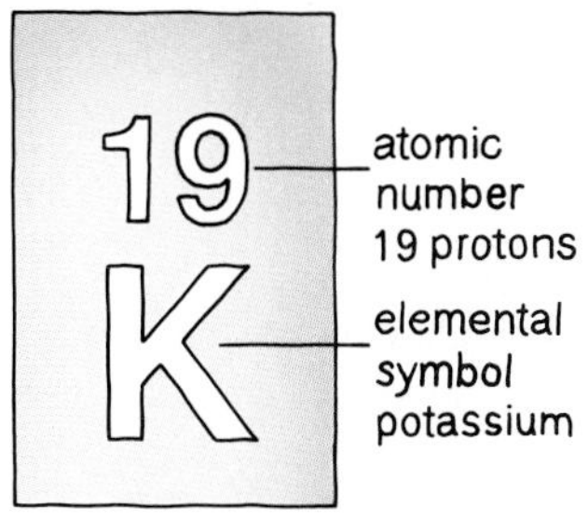

Atoms Are Neutral

An atom is electrically neutral. That means that the number of protons in an atom is equal to the number of electrons. This electrical balance gives an atom an overall charge of zero. Thus, in every atom, the atomic number also gives the number of electrons. Table 3.5 lists the atomic number and corresponding number of protons and electrons for some atoms.

Table 3.5 Number of Protons and Electrons in Some Atoms

Element	Symbol	Atomic Number	Number of Protons	Number of Electrons	Overall Charge
Hydrogen	H	1	1	1	0
Helium	He	2	2	2	0
Oxygen	O	8	8	8	0
Sodium	Na	11	11	11	0
Iron	Fe	26	26	26	0
Gold	Au	79	79	79	0

Sample Problem 3.4
Using Atomic Number to Find the Number of Protons and Electrons

Using the periodic table, state the atomic number, number of protons, and number of electrons for an atom of each of the following elements:

a. nitrogen
b. magnesium
c. bromine

Solution:

a. atomic number 7; 7 protons and 7 electrons
b. atomic number 12; 12 protons and 12 electrons
c. atomic number 35; 35 protons and 35 electrons

Study Check:
Consider an atom that has 26 electrons.

a. How many protons are in its nucleus?
b. What is its atomic number?
c. What is its name, and what is its symbol?

Answer: **a.** 26 **b.** 26 **c.** iron, Fe

Mass Number

The protons and neutrons in an atom provide essentially all of its mass. The **mass number** of an atom is equal to the sum of the number of protons and the number of neutrons in the nucleus. Since we are counting the subatomic particles in the nucleus, the mass number is always a whole number.

Mass number = Total number of protons and neutrons in one atom of an element

For example, an atom of potassium with 19 protons and 20 neutrons has a mass number of 39. By knowing the atomic number and the number of neutrons, you can calculate the mass number of an atom.

Table 3.6 Composition of Some Atoms of Different Elements

Element	Symbol	Atomic Number	Mass Number	Number of Protons	Number of Neutrons	Number of Electrons
Hydrogen	H	1	1	1	0	1
Nitrogen	N	7	14	7	7	7
Phosphorus	P	15	31	15	16	15
Chlorine	Cl	17	37	17	20	17
Iron	Fe	26	56	26	30	26

Element:	K	O	Al	Fe
Protons:	19	8	13	26
Neutrons:	20	8	14	30
Mass number (total):	39	16	27	56

Table 3.6 illustrates the relationship between atomic number, mass number, and the number of protons, neutrons, and electrons in some atoms of different elements.

Sample Problem 3.5
Calculating Mass Number

Calculate the mass number of an atom by using the information given:

a. 5 protons and 6 neutrons
b. 18 protons and 22 neutrons
c. atomic number 48, and 64 neutrons

Solution:

a. $5 + 6 =$ mass number $= 11$
b. $18 + 22 =$ mass number $= 40$
c. $48 + 64 =$ mass number $= 112$

Study Check:
What is the mass number of a silver atom that has 60 neutrons?

Answer: Since the atomic number of silver is 47, it has 47 protons. The mass number, the sum of 47 protons and 60 neutrons, is 107.

Sample Problem 3.6
Calculating Numbers of Protons and Neutrons

For an atom of phosphorus that has a mass number of 31, determine the following:

a. the number of protons
b. the number of neutrons
c. the number of electrons

Solution:

a. On the periodic table, the atomic number of phosphorus is 15. A phosphorus atom has 15 protons.

b. The number of neutrons in this atom is found by subtracting the atomic number from the mass number. The number of neutrons is 16.

Mass number	−	Atomic number	=	Number of neutrons
31	−	15	=	16

c. Since an atom is neutral, there is an electrical balance of protons and electrons. Since the number of electrons is equal to the number of protons, the phosphorus atom has 15 electrons.

Study Check:
How many neutrons are in the nucleus of a bromine atom that has a mass number of 80?

Answer: 45 neutrons

Isotopes

We have seen that all atoms of the same element have the same number of protons and electrons. However, the atoms are not completely identical because they can have different numbers of neutrons. **Isotopes** are atoms of the same element that have different numbers of neutrons. For example, all atoms of the element magnesium (Mg) have 12 protons. However, some magnesium atoms have 12 neutrons, others have 13 neutrons, and still others have 14 neutrons. The differences in numbers of neutrons for these magnesium atoms cause their mass numbers to be different, but not their chemical behavior. The three isotopes of magnesium have the same atomic number, but different mass numbers.

To distinguish between the different isotopes of an element, we can write an **isotope symbol** that indicates the mass number and the atomic number of the atom.

Symbol for an Isotope of Magnesium

Mass number → 24
Symbol of element → Mg
Atomic number → 12

$$^{24}_{12}\mathbf{Mg}$$

Magnesium has three naturally occurring isotopes, as Table 3.7 shows. Table 3.8 lists nuclear symbols for the isotopes of some selected elements.

Table 3.7 Isotopes of Magnesium

	$12e^-$ (12p^+, 12n^0)	$12e^-$ (12p^+, 13n^0)	$12e^-$ (12p^+, 14n^0)
Nuclear symbol	$^{24}_{12}Mg$	$^{25}_{12}Mg$	$^{26}_{12}Mg$
Number of protons	12	12	12
Number of electrons	12	12	12
Mass number	24	25	26
Number of neutrons	12	13	14

Table 3.8 Isotopes of Some Common Elements

Element	Isotope Symbols[a]			
Lithium	$^{6}_{3}Li$	$^{7}_{3}Li$		
Carbon	$^{12}_{6}C$	$^{13}_{6}C$	$^{14}_{6}C$	
Oxygen	$^{16}_{8}O$	$^{17}_{8}O$	$^{18}_{8}O$	
Sulfur	$^{32}_{16}S$	$^{33}_{16}S$	$^{34}_{16}S$	$^{36}_{16}S$
Chlorine	$^{35}_{17}Cl$	$^{37}_{17}Cl$		
Copper	$^{63}_{29}Cu$	$^{65}_{29}Cu$		

Sample Problem 3.7
Identifying Protons and Neutrons in Isotopes

State the number of protons and neutrons in the following isotopes of **neon** (Ne).

a. $^{20}_{10}Ne$ **b.** $^{21}_{10}Ne$ **c.** $^{22}_{10}Ne$

Solution:

The atomic number of Ne is 10; each isotope has 10 protons. The number of neutrons in each isotope is found by subtracting the atomic number (10) from each mass number.

a. 10 protons; 10 neutrons (20 − 10)
b. 10 protons; 11 neutrons (21 − 10)
c. 10 protons; 12 neutrons (22 − 10)

Study Check:
Write a symbol for the following isotopes:
a. a nitrogen atom with 8 neutrons
b. an atom with 20 protons and 22 neutrons
c. an atom with mass number 27 and 14 neutrons

Answer:
a. $^{15}_{7}N$ **b.** $^{42}_{20}Ca$ **c.** $^{27}_{13}Al$

Atomic Weight

In laboratory work, a scientist generally uses samples that contain many atoms of an element. Among those atoms are all of the various isotopes with their different masses. To obtain a convenient mass to work with, chemists use the mass of an "average atom" of each element. This average atom has an **atomic weight,** which is the weighted average mass of all of the naturally occurring isotopes of that element.

To calculate an atomic weight, the percent abundance of each isotope must be known. For example, in chlorine (Cl), 75.5 percent of the atoms have a mass number of 35; the other 24.5 percent have a mass number of 37. Using these values and the mass number of each isotope (which is quite close to each isotope's mass in amu), we can calculate the atomic weight for the element.

Determination of the Atomic Weight of Chlorine

Isotope	*Amount In Sample*		*Mass Number*		*Contribution to Average Atom*
$^{35}_{17}Cl$	$\frac{75.5}{100}$	×	35	=	26.4 amu
$^{37}_{17}Cl$	$\frac{24.5}{100}$	×	37	=	9.1 amu
Atomic weight Cl				=	35.5 amu

This calculation illustrates the way in which isotopes determine the overall atomic weight for an element. Most elements consist of several isotopes, and this is one reason atomic weights are seldom whole numbers.

On the periodic table, the atomic weight is given below the symbol of each element; the mass number is not given. The element chlorine is shown in Figure 3.6 as it appears in the periodic table.

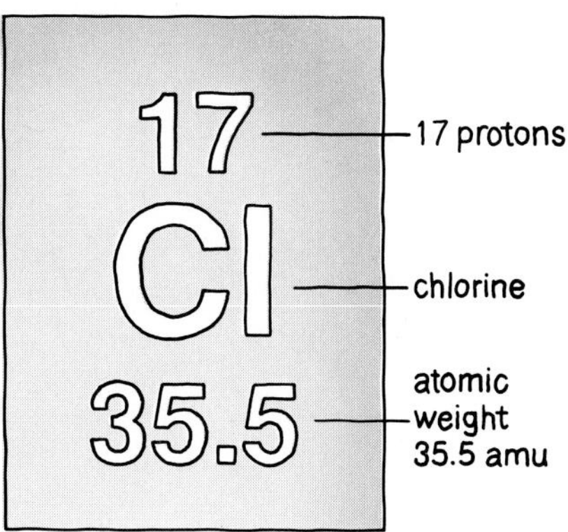

Figure 3.6
Information that is given for chlorine in the periodic table.

Sample Problem 3.8
Identifying Atomic Weight

Using the periodic table, write the atomic weight for each of the following elements:

a. hydrogen **b.** iron **c.** sulfur **d.** potassium

Solution:

a. 1.0 amu **b.** 55.8 amu **c.** 32.1 amu **d.** 39.1 amu

3.4 The Periodic Table

Learning Goal **Use the periodic table to identify the group and the period of an element, and whether it is a metal or a nonmetal.**

By the late 1800s, scientists began to find similarities in the behavior of the elements that were known at the time. A Russian chemist, Dmitri Mendeleev, suggested that the elements could be arranged in groups that showed similar chemical and physical properties. Today, we have the modern **periodic table,** which arranges the elements by increasing atomic number in such a way that similar properties repeat at periodic intervals.

Periods and Groups

A single horizontal row in the periodic table is called a **period.** Each row is counted from the top of the table down as Period 1, Period 2, and so on. The first period contains only the elements hydrogen (H) and helium (He). The second period, which is the second row of elements, contains lithium (Li), beryllium (Be), boron (B), carbon (C), nitrogen (N), oxygen (O), fluorine (F), and neon (Ne). The third period begins with sodium (Na) and ends at argon (Ar). There are seven periods in the periodic table.

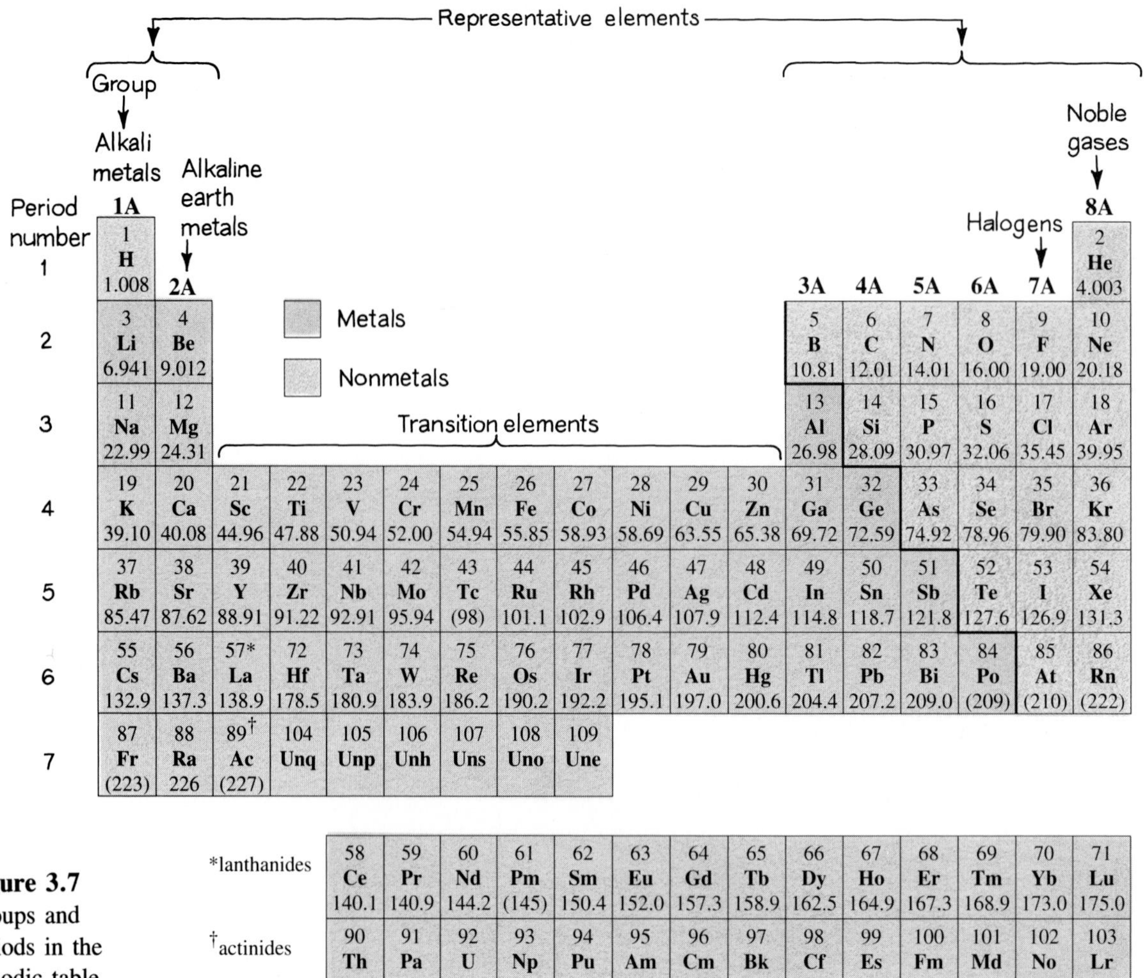

Figure 3.7 Groups and periods in the periodic table.

A **group** is a vertical column of elements that have similar physical and chemical properties. The groups are identified by numbers that go across the top of the periodic table. The elements in the periodic table are divided into sections: the A groups, which are the **representative elements** (1A–8A), and the **transition elements,** as shown in Figure 3.7.

Sample Problem 3.9
Groups and Periods in the Periodic Table

State whether each set represents elements in a group, a period, or neither:

a. F, Cl, Br, I
b. Na, Al, P
c. K, Al, O

Solution:

a. The elements F, Cl, Br, and I are part of a group of elements; they all appear in the vertical column 7A.

b. The elements Na, Al, and P all appear in the third row or third period in the periodic table.
c. Neither. The elements K, Al, and O are not part of the same group and they do not belong to the same period.

Study Check:
a. What elements are found in Period 2?
b. What elements are found in Group 2A?

Answer:
a. Period 2: Li, Be, B, C, N, O, F, Ne
b. Group 2A: Be, Mg, Ca, Sr, Ba, Ra

Sample Problem 3.10
Stating Group and Period Number

Identify the period number and the group number for the following elements:
a. calcium **b.** tin

Solution:
The period is found by counting down the horizontal rows of the elements on the periodic table, and the group number is found at the top of the vertical column that contains that element.
a. Calcium (Ca) is in Period 4 and Group 2A.
b. Tin (Sn) is in Period 5 and Group 4A.

Study Check:
Give the symbols of the elements that are represented by the following period and group numbers:
a. Period 3, Group 5A
b. Period 6, Group 8A

Answer: **a.** P **b.** Rn

Classification of Groups

Several groups in the periodic table have special names, as Figure 3.7 shows. Group 1A elements, lithium (Li), sodium (Na), potassium (K), rubidium (Rb), cesium (Cs), and francium (Fr) are part of a family of elements known as the **alkali metals.** The elements within this group exhibit similar properties (see Figure 3.8). The alkali metals are soft, shiny metals that are good conductors of heat and electricity and have relatively low melting points. They react vigorously with water and form white products when they combine with oxygen.

Figure 3.8
Lithium (Li), sodium (Na), and potassium (K), some alkali metals from Group 1A.

Group 2A elements beryllium (Be), magnesium (Mg), calcium (Ca), strontium (Sr), barium (Ba), and radium (Ra) are called the **alkaline earth metals.** They are also shiny metals like those in Group 1A, but they are not quite so reactive.

The **halogens** are found on the right side of the periodic table in Group 7A. They include the elements fluorine (F), chlorine (Cl), bromine (Br), and iodine (I), as shown in Figure 3.9. On the periodic table, each halogen is listed as a single symbol. However, halogens exist as combinations of two atoms joined to form diatomic molecules. For example, a sample of chlorine consists of Cl_2 molecules. The other halogens exist as F_2, Br_2 and I_2 molecules.

Group 8A contains the **noble gases,** helium (He), neon (Ne), argon (Ar), krypton (Kr), xenon (Xe), and radon (Rn). They are quite unreactive and are seldom found in combination with other elements.

Figure 3.9
Chlorine (Cl_2), bromine (Br_2), and iodine (I_2), some halogens from Group 7A.

Metals and Nonmetals

Another feature of the periodic table is the heavy zigzag line that separates the elements into **metals** and the **nonmetals.** The metals are those elements on the left of the line except for hydrogen, and the nonmetals are the elements on the right.

In general, most **metals** are shiny solids. They can be shaped into wires (ductile), or hammered into a flat shape (malleable). Metals are often good conductors of heat and electricity. They usually melt at higher temperatures than nonmetals. All of the metals are solids at room temperature, except for mercury (Hg), which is a liquid. Some typical metals are sodium (Na), magnesium (Mg), copper (Cu), gold (Au), silver (Ag), iron (Fe), and tin (Sn).

Nonmetals are not very shiny, malleable, or ductile, and they are often poor conductors of heat and electricity. They typically have low melting points and low densities. You may have heard of nonmetals such as hydrogen (H), carbon (C), nitrogen (N), oxygen (O), chlorine (Cl), and sulfur (S). Table 3.9 compares some characteristics of silver, a metal, with those of sulfur, a nonmetal.

HEALTH NOTE

Calcium and Strontium

Calcium (Ca) and strontium (Sr) are two elements of the alkaline earth group, Group 2A. The chemical behavior of strontium is so similar to that of calcium that when strontium is ingested, it replaces some of the calcium in the bones and teeth. This similarity in behavior caused great concern among scientists during nuclear testing, since radioactive strontium is a product of certain nuclear reactions. If the radioactive strontium were to drift to cattle grazing lands, it could become a part of cow's milk and eventually find its way to the bones of young children. Once it is there, the effects of the radioactivity are detrimental to proper growth and development. This is what happened in Chernobyl in 1986. ■

Table 3.9 Some Characteristics of Silver, a Metal, and Sulfur, a Nonmetal

Silver	Sulfur
A shiny metal	A dull, yellow nonmetal
Extremely ductile	Brittle
Can be hammered into sheets (malleable)	Shatters when hammered
Good conductor of heat and electricity	Poor conductor, good insulator
Used in coins, jewelry, tableware	Used in gunpowder, rubber, fungicides
Density 10.5 g/mL	Density 3.1 g/mL
Melting point 962°C	Melting point 113°C

Sample Problem 3.11
Classifying Elements as Metals and Nonmetals

Using a periodic table, classify the following elements as metals or nonmetals:

a. Na **b.** Si **c.** Cl **d.** Cu

Solution:

a. Metal; sodium is located to the left of the heavy zigzag line that separates metals and nonmetals.
b. Nonmetal; silicon is on the right side of the zigzag line.
c. Nonmetal; chlorine is on the right side of the zigzag line.
d. Metal; copper is located to the left of the zigzag line.

HEALTH NOTE

Some Important Trace Elements in the Body

Some metals and nonmetals known as trace elements are essential to the proper functioning of the body. Although they are required in very small amounts, their absence can disrupt major biological processes and cause illness. The trace elements listed in Table 3.10 are present in the body combined with other elements. ■

Table 3.10 Some Important Trace Elements in the Body

Element	Adult RDA[a]	Biological Function	Deficiency Symptoms	Dietary Sources
Iron (Fe)	10 mg (males) 18 mg (females)	Formation of hemoglobin; enzymes	Dry skin, spoon nails, decreased hemoglobin count, anemia	Beef, kidneys, liver; egg yolk, oysters, spinach, beans, apricots, raisins, whole-wheat bread
Copper (Cu)	2.0–5.0 mg	Necessary in many enzyme systems; growth; aids formation of red blood cells and collagen	Uncommon; anemia, decreased white cell count, and bone demineralization	Nuts, organ meats, whole-wheat grains, shellfish, eggs, poultry, leafy green vegetables
Zinc (Zn)	15 mg	Amino acid metabolism; enzyme systems, energy production; collagen	Retarded growth and bone formation; skin inflammation, loss of taste and smell, poor healing	Wheat germ, shellfish, milk, lima beans, fish, eggs, whole grains, turkey (dark meat), cheddar cheese
Manganese (Mn)	2.5–5.0 mg	Necessary for some enzyme systems; collagen formation; bone formation; central nervous system; fat and carbohydrate metabolism; blood clotting	Abnormal skeletal growth; impairment of central nervous system	Cereals, peas, beans, lettuce, wheat bran, meat, poultry, fish
Iodine (I)	150 μg	Necessary for activity of thyroid gland	Hypothyroidism; goiter; cretinism	Iodized table salt, seafood
Fluorine (F)	1.5–4.0 mg	Necessary for solid teeth formation and retention of calcium in bones with aging	Dental cavities	Tea, fish, milk, eggs, water in some areas, supplementary drops, toothpaste

[a]RDA, recommended daily allowance

3.5 Electron Arrangement in the Atom

Learning Goal **Given the name or symbol of one of the first 20 elements in the periodic table, write the electron arrangement.**

The chemical behavior of an element is primarily determined by the way the electrons are arranged about the nucleus. Every electron has a specific amount of energy. Electrons of similar energy are grouped in an energy level called a **shell.** The shells closest to the nucleus contain electrons with the lowest energies, whereas shells farther away from the nucleus contain electrons with higher energies.

Each electron shell can hold a different number of electrons. As shown in Table 3.11, shell 1, the lowest energy level, can hold up to 2 electrons; shell 2 can hold up to 8 electrons; shell 3 can take 18 electrons; and shell 4 has room for 32 electrons.

Table 3.11 Capacity of Some Electron Shells

Electron Shell	Maximum Number of Electrons
1	2
2	8
3	18
4	32

There are additional energy levels or shells in an atom, but they are beyond our consideration in this text. In the atoms of the elements known today, electrons can occupy energy levels as high as shell 7.

Electron Arrangements for the First Twenty Elements

The **electron arrangement** of an atom gives the number of electrons in each shell. We might imagine electron shells as floors in a hotel. The ground floor fills first, then the second floor, and so on. In Figure 3.10, two electrons are placed in shell 1, the lowest energy level. The next eight electrons go into shell 2. Both shell 1 and shell 2 are now filled. Shell 3 initially takes eight electrons, it then stops filling for a while, even though it is capable of holding more electrons. This break in filling is due to an overlapping in the energy values of shell 3 and shell 4. At this point, shell 4 takes the next two electrons.

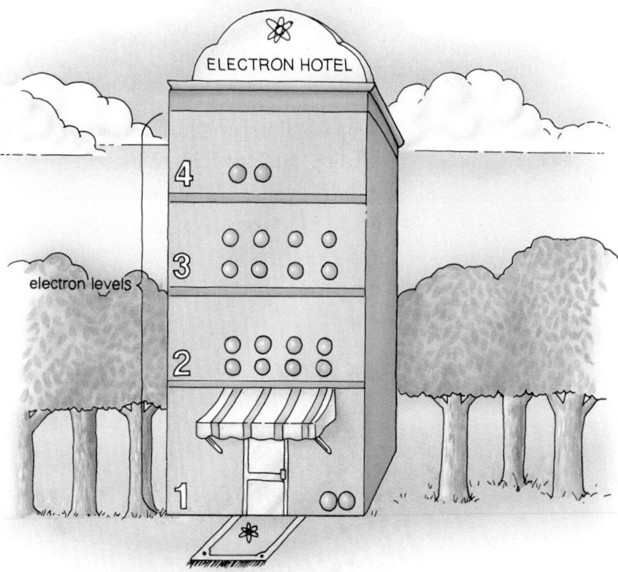

Figure 3.10
Electron occupancy for the first 20 electrons.

The electron arrangements for the first 20 elements can be written by placing electrons in shells beginning with the lowest energy. The single electron of hydrogen and the two electrons of helium can be placed in shell 1. When we wish to draw a simple diagram of the electron arrangement, we indicate the nucleus of the atom and draw curved lines to represent each of the occupied shells. In the illustrations that follow, one isotope has been chosen for each element shown. The electron configurations for hydrogen and helium would appear as follows:

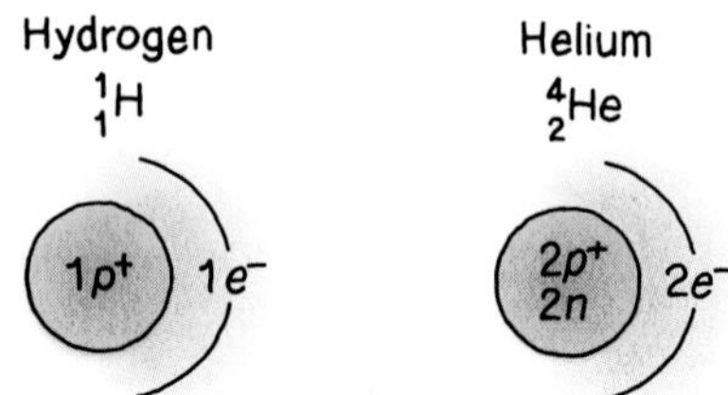

The elements of the second period (lithium, Li, to neon, Ne) have enough electrons to fill the first shell and begin filling the second shell. For example, lithium has 3 electrons. Two of those electrons complete shell 1. The remaining electron goes into the second shell. As we go across the period, more and more electrons enter the second shell. For example, an atom of carbon, with a total of 6 electrons, fills shell 1 with 2 electrons; and 4 remaining electrons enter the second shell. The last element in Period 2 is neon. The 10 electrons in an atom of neon completely fill the first and second shells.

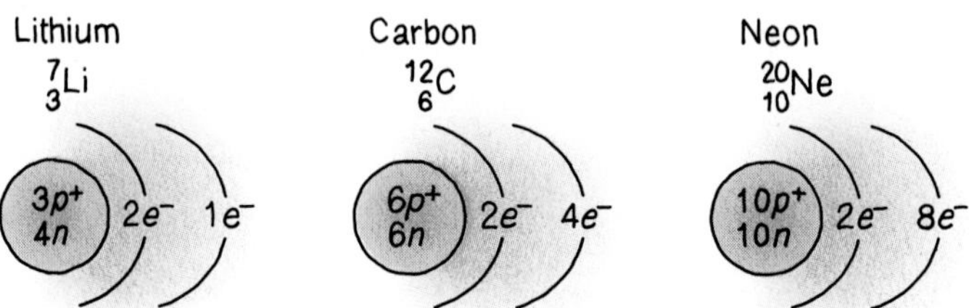

In an atom of sodium, atomic number 11, the first and second electron shells are filled and the last electron enters the third shell. The rest of the elements in the third period continue to add to the third shell. For example, a sulfur (s) atom with 16 electrons has 2 electrons in the first shell, 8 electrons in the second shell, and 6 electrons in the third shell. At the end of the period, we find that argon has 8 electrons in the third shell.

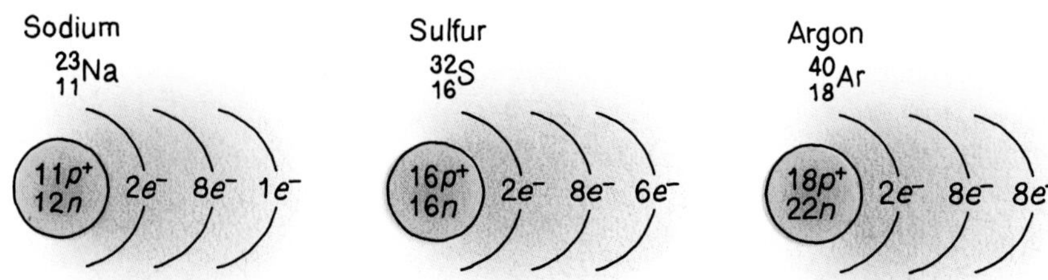

At this point, the third shell with eight electrons stops filling for a while. The remaining electrons in potassium and sodium actually begin to fill the fourth shell. The third shell will continue filling in elements with higher atomic numbers. The electron arrangements for the first 20 elements are summarized in Table 3.12. Although there are many more elements in the periodic table, we will not consider the electron arrangements of elements beyond atomic number 20.

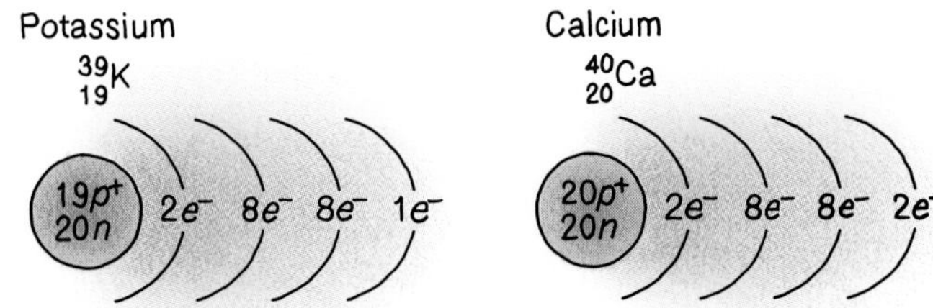

Table 3.12 Electron Arrangements for the First Twenty Elements

Element	Symbol	Atomic Number	Number of Electrons in Shell 1	2	3	4	Element	Symbol	Atomic Number	Number of Electrons in Shell 1	2	3	4
Hydrogen	H	1	1				Sodium	Na	11	2	8	1	
Helium	He	2	2				Magnesium	Mg	12	2	8	2	
Lithium	Li	3	2	1			Aluminum	Al	13	2	8	3	
Beryllium	Be	4	2	2			Silicon	Si	14	2	8	4	
Boron	B	5	2	3			Phosphorus	P	15	2	8	5	
Carbon	C	6	2	4			Sulfur	S	16	2	8	6	
Nitrogen	N	7	2	5			Chlorine	Cl	17	2	8	7	
Oxygen	O	8	2	6			Argon	Ar	18	2	8	8	
Fluorine	F	9	2	7			Potassium	K	19	2	8	8	1
Neon	Ne	10	2	8			Calcium	Ca	20	2	8	8	2

Sample Problem 3.12
Writing Electron Arrangements

Write the electron arrangement for each of the following:

a. oxygen
b. chlorine

Solution:

a. Oxygen has an atomic number of 8. Therefore, there are 8 electrons in the electron arrangement:

$$2\ e^- \quad 6\ e^-$$

b. An atom of chlorine has 17 protons and 17 electrons. The electrons are arranged as follows:

$$2\ e^- \quad 8\ e^- \quad 7\ e^-$$

Study Check:
What element has the following electron arrangement?

$2\ e^-$ $8\ e^-$ $2\ e^-$

Answer: magnesium

Energy-Level Changes

Whenever possible, electrons occupy the lowest energy levels. However, if a source of energy such as heat or light is available, an electron can absorb a certain amount of energy and jump to a higher energy level. Electrons in this higher energy state, however, are unstable. They drop back to one of the lower, more stable energy levels, releasing some of their energy, as illustrated in Figure 3.11. Neon lights are an example of how electrons that gain energy can emit energy as light (see Figure 3.12).

The amount of energy released when electrons drop to lower energy levels can be large or small. High-energy emissions include x-rays and gamma rays. Low-energy emissions include infrared rays (heat), radio waves, and microwaves. Figure 3.13 illustrates some forms of energy emitted when electrons drop to lower energy levels in atoms.

Most energy cannot be seen. Only the energies in the visible region are detected by our eyes. For example, when sunlight passes through a prism or a raindrop, it is separated into the colors known as the visible spectrum, which we see in rainbows.

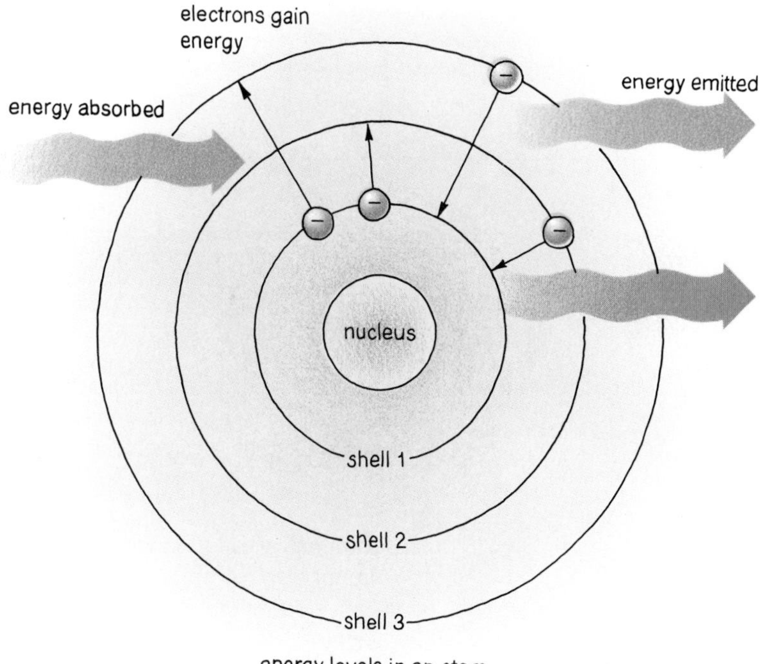

Figure 3.11
Electrons gain energy and jump to higher energy levels. When they drop back to lower energy levels, energy is emitted.

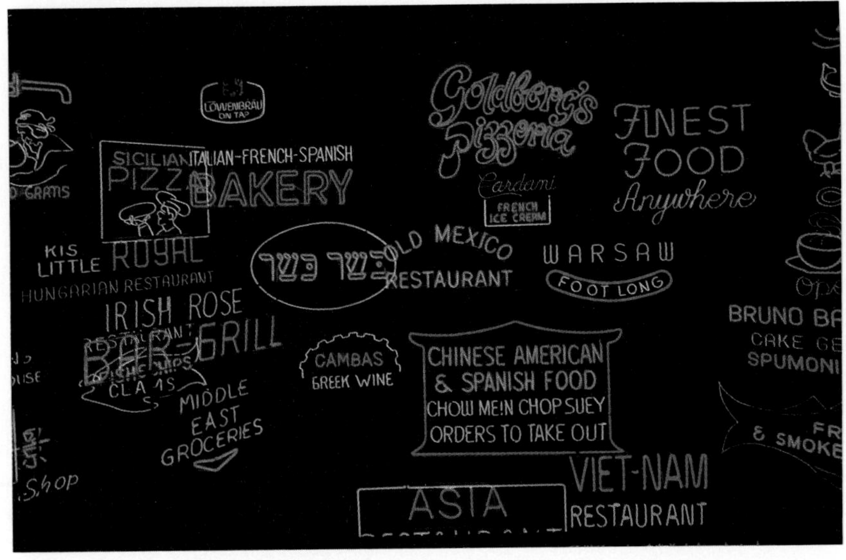

Figure 3.12
When an electric current flows through a gas-filled tube, electrons jump to higher energy levels and fall again, emitting light of different colors.

Figure 3.13
Types of energy emitted as electrons drop from higher to lower energy levels.

Energy	Type of Radiation	Wavelength (nm)
high energy		short wavelength
	cosmic rays	10^{-5}
	gamma rays	10^{-4}
	x-rays	10^{-1}
	ultraviolet rays	10^{2}
	visible light	10^{3}
	infrared rays	10^{4}
	microwaves	10^{7}
	TV	10^{9}
	radio waves	10^{11}
		10^{12}
low energy		long wavelength

400 nm violet
blue
500 nm green
yellow
600 nm orange
red
700 nm

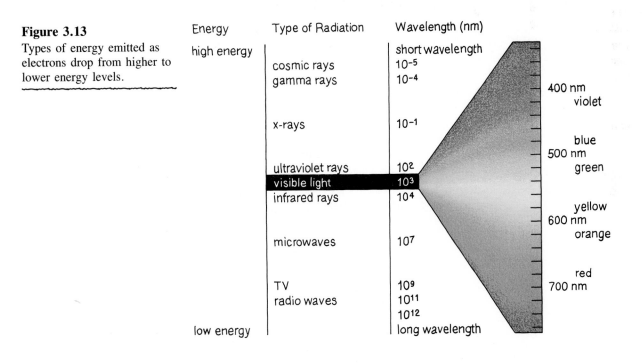

3.6 Periodic Law

Learning Goal **Use the electron arrangement of an element to state its group number and to explain periodic law.**

We have seen that the groups in the periodic table contain elements having similar properties. This repetition of physical and chemical properties with increasing atomic number is known as the **periodic law.** If we now observe the

ENVIRONMENTAL NOTE

Biological Reactions to Sunlight

Our everyday life depends on sunlight, but exposure to sunlight can have damaging effects on living cells, and too much exposure can even cause their death. The list of damaging effects of sunlight includes sunburn; wrinkling; premature aging of the skin; changes in the DNA of the cells, which can lead to skin cancers and melanomas; inflammation of the eyes; and, perhaps, cataracts. Some drugs, like the acne medications Accutane and Retinin A, as well as antibiotics, diuretics, sulfonamides, and estrogens, make the skin extremely photosensitive and can cause undesirable changes in its reaction to sunlight. Using a sunscreen is now recommended by doctors to prevent the adverse effects of sun exposure.

High-energy radiation is the most damaging biologically. Most of the radiation in this range is absorbed in the epidermis of the skin. The degree to which radiation is absorbed depends on the thickness of the epidermis, the hydration of the skin, the amount of coloring pigments and proteins of the skin, and the arrangement of the blood vessels. In light-skinned people, 85–90% of the radiation is absorbed by the epidermis, with the rest reaching the dermis layer. In dark-skinned people, 90–95% of the radiation is absorbed by the epidermis, with a smaller percentage reaching the dermis.

However, medicine does take advantage of the beneficial effect of sunlight. Phototherapy can be used to treat certain skin conditions, including psoriasis, eczema, and dermatitis. In the treatment of psoriasis, for example, oral drugs are given to make the skin more photosensitive; exposure to ultraviolet (UV) radiation follows. Low-energy radiation is used to break down bilirubin in neonatal jaundice. Sunlight is also a factor in stimulating the immune system. (See Figure 3.14.) ■

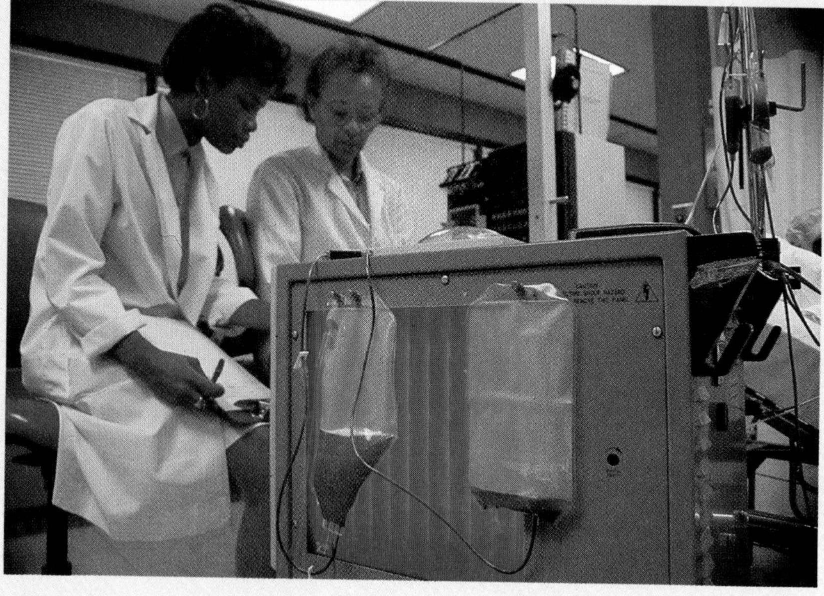

Figure 3.14
In cutaneous T-cell lymphoma, an abnormal increase in T cells causes painful ulceration of the skin. The skin is treated by photopheresis, in which the patient receives a photosensitive chemical, and then blood is removed from the body and exposed to ultraviolet light. The blood is returned to the patient, and the treated T cells stimulate the immune system to respond to the cancer cells.

Table 3.13 A Comparison of Electron Arrangements of Some Group 1A Elements

Element	Atomic Number	Number of Electrons in Shell				
		1	2	3	4	
Lithium	3	2	1			One electron in each outer shell
Sodium	11	2	8	1		
Potassium	19	2	8	8	1	

electron arrangements for the elements in a group, we find another similarity. All the elements in a group (also called family) of elements have the same number of electrons in their outermost shells.

For example, the elements lithium, sodium, and potassium are part of Group 1A. All of these elements have one electron in their outer shells, as seen in Table 3.13. The outer shell is the shell with the highest energy that contains one or more electrons. Thus, the similarity of chemical and physical properties among elements in a group can now be attributed to their having the same number of electrons in each of their outermost shells.

Group Number

The **group numbers** 1A–8A appear at the top of the periodic table. Each group number is equal to the number of electrons in the outer shell of the elements in that column. All elements in Group 1A have one electron in their outer shells, elements in Group 2A have two electrons in their outer shells, elements in Group 3A have three electrons in their outer shells, and so on. We are most interested in the number of electrons in the outer shells because these electrons have the greatest effect on the way an atom forms compounds. Table 3.14 shows the electron arrangement by group for the first 20 elements.

Sample Problem 3.13
Using Group Numbers

Using the periodic table, write the group number and the number of electrons in the outer electron level of the following elements:

a. sodium **b.** sulfur **c.** aluminum

Solution:

a. Sodium is in Group 1A; sodium has one electron in the outer electron level.

b. Sulfur is in Group 6A; sulfur has six electrons in the outer electron level.

c. Aluminum is in Group 3A; aluminum has three electrons in the outer electron level.

Table 3.14 Electron Arrangements, by Group, for the First Twenty Elements

		Number of Electrons in Shell			
Group Number	**Element**	**1**	**2**	**3**	**4**
1A	Hydrogen	1			
	Lithium	2	1		
	Sodium	2	8	1	
	Potassium	2	8	8	1
2A	Beryllium	2	2		
	Magnesium	2	8	2	
	Calcium	2	8	8	2
3A	Boron	2	3		
	Aluminum	2	8	3	
4A	Carbon	2	4		
	Silicon	2	8	4	
5A	Nitrogen	2	5		
	Phosphorus	2	8	5	
6A	Oxygen	2	6		
	Sulfur	2	8	6	
7A	Fluorine	2	7		
	Chlorine	2	8	7	
8A	Helium	2			
	Neon	2	8		
	Argon	2	8	8	

3.7 Subshells and Orbitals

Learning Goal **Write the electron configuration using subshell notation.**

The electrons in the electron shells or main energy levels can be described in yet more detail. Within each shell, the electrons with identical energy are grouped as **subshells.** The different types of subshells are identified by the letters *s, p, d,* and *f*. The *s* subshell is lowest in energy, followed by the *p* subshell, then the *d* subshell, and finally the highest energy subshell, the *f* subshell.

Order of Increasing Energy for Subshells

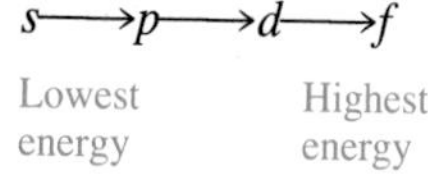

The number of subshells in each shell is equal to the numerical value of that shell. As shown in Figure 3.15, shell 1 has only one subshell, 1*s*. Shell 2 has two subshells, 2*s* and 2*p*. The 2*s* subshell is lower in energy than the 2*p*. Shell 3 has three subshells, 3*s*, 3*p*, and 3*d*. The fourth shell (shell 4) consists of four subshells, 4*s*, 4*p*, 4*d*, and 4*f*. Note that some of the subshells in the third and fourth shells are so close in energy that they overlap. The 4*s* subshell is lower in energy than the 3*d* subshell. In electron arrangements with subshells, the 4*s* is filled before the 3*d*.

Number of Electrons in Subshells

Each type of subshell holds a specific number of electrons. Any *s* subshell holds just 2 electrons. A *p* subshell can hold up to 6 electrons, a *d* subshell takes up to 10 electrons, and an *f* subshell holds a maximum of 14 electrons. The number of electrons in the subshells adds up to give the number of electrons in each electron shell, as Table 3.15 shows.

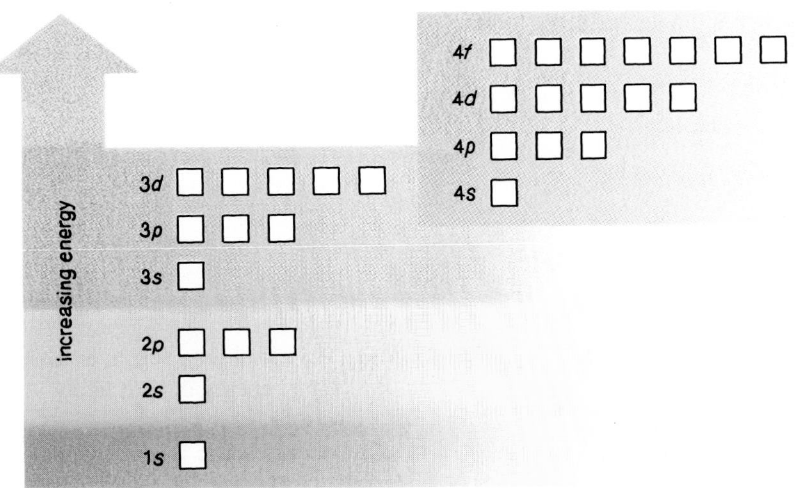

Figure 3.15 The energy of the electron subshells increases as their distance from the nucleus increases. Electrons fill lowest available subshells first.

Table 3.15 Number of Electrons Within the Subshells

Shell	Subshell	Number of Electrons	Shell Capacity
1	1*s*	2	2
2	2*s*	2	8
	2*p*	6	
3	3*s*	2	18
	3*p*	6	
	3*d*	10	
4	4*s*	2	32
	4*p*	6	
	4*d*	10	
	4*f*	14	

Sample Problem 3.14
Subshells

Describe the third electron shell in terms of the following:

a. maximum number of electrons
b. number and designation of subshells
c. number of electrons in each subshell

Solution:

a. The maximum number of electrons for the third shell is 18.
b. The third shell has three subshells: 3*s*, 3*p*, and 3*d*.
c. The 3*s* subshell can accommodate 2 electrons, the 3*p* subshell can hold up to 6 electrons, and the 3*d* subshell will take up to 10 electrons.

Study Check:
State the number of electrons that would fill the following:

a. the second shell
b. the 3*p* subshell
c. the 5*s* subshell

Answer:
a. 8 **b.** 6 **c.** 2

Orbitals

An **orbital** is described as a region in space around the nucleus in which an electron of a certain energy is most likely to be found. An orbital can hold only one or two electrons.

There is a special shape for orbitals in each type of subshell. An *s* orbital is spherical, with the nucleus at the center. You might think of a 1*s* orbital as analogous to a golf ball. That means that the electrons of the 1*s* subshell are most likely to be in a spherical space. The 2*s* orbital might be like a baseball, which is bigger than a golf ball but has the same shape. A 3*s* orbital is spherical too, but perhaps it is like a basketball, fitting over the other two. Each of the *s* orbitals has a maximum of two electrons, and there is just one *s* orbital for every *s* subshell.

A *p* subshell consists of three *p* orbitals, which have shapes like dumbbells. Recall that a *p* subshell can hold six electrons. If each *p* orbital holds just two electrons, then three *p* orbitals are needed to build the *p* subshell. The three dumbbell-shaped *p* orbitals are arranged in three different directions (x, y, z axes) around the nucleus. As shown in Figure 3.16, all *p* orbitals have the same shape; only their size increases as the energy level increases.

Electron Arrangements Using Subshells

In an electron arrangement with subshells the number of electrons in each subshell is shown as a superscript. For example, the electron arrangement for a fluorine atom with nine electrons shows two electrons in the 1*s* subshell, two

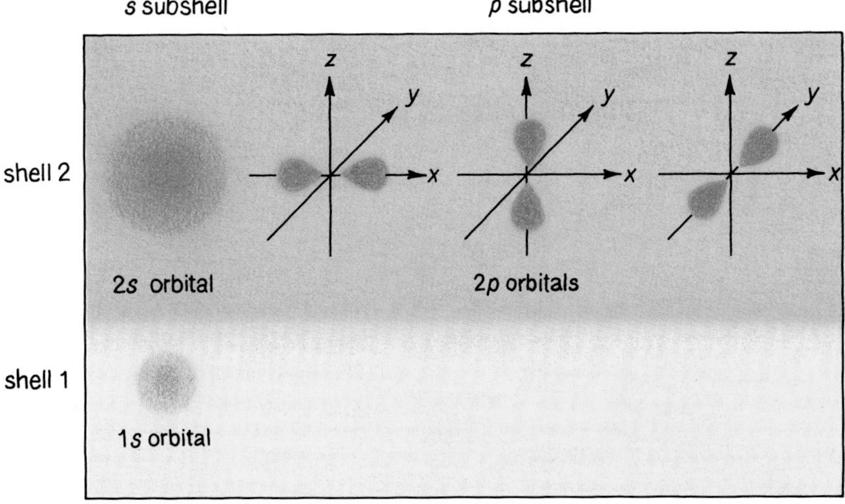

Figure 3.16
Shapes of *s* and *p* orbitals in shells 1 and 2.

electrons in the 2*s* subshell, and five electrons in the 2*p* subshell. The shorthand method of writing this electron arrangement for fluorine is as follows:

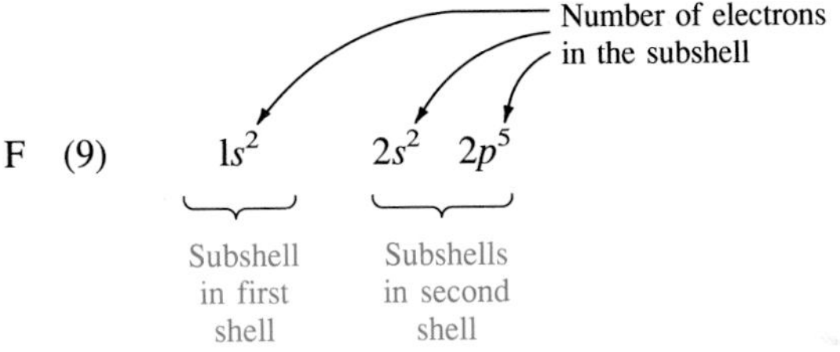

Examples of electron arrangements for atoms of oxygen, sodium, phosphorus, chlorine, and calcium follow.

Element (Atomic Number)	*Electron Arrangement with Subshells*
O (8)	$1s^2 2s^2 2p^4$
Na (11)	$1s^2 2s^2 2p^6 3s^1$
P (15)	$1s^2 2s^2 2p^6 3s^2 3p^3$
Cl (17)	$1s^2 2s^2 2p^6 3s^2 3p^5$
Ca (20)	$1s^2 2s^2 2p^6 3s^2 3p^6 4s^2$

Sample Problem 3.15
Writing Electron Arrangements with Subshells

Write the electron arrangement with subshells for argon.

Solution:
An atom of argon has 18 electrons. The order of filling the subshells begins with the 1*s*, followed by the 2*s* and 2*p*. This completes the first and second shells, with a total of 10 electrons.

$$1s^2\ 2s^2\ 2p^6$$

There are still 8 electrons to account for. The next subshells to fill are the 3*s* and 3*p*, giving us a total of 18 electrons and the correct electron arrangement for argon.

$$1s^2\ 2s^2\ 2p^6\ 3s^2\ 3p^6$$

Study Check:
Write the electron arrangement with subshells for sulfur.

Answer: $1s^2\ 2s^2\ 2p^6\ 3s^2\ 3p^4$

Glossary of Key Terms

alkali metals Elements of Group 1A except hydrogen that are soft, shiny metals with one outer shell electron.

alkaline earth metals Group 2A elements having two electrons in their outer shells.

atom The smallest particle of an element that retains the characteristics of the element.

atomic mass unit (amu) A small mass unit used to measure the mass of very small particles such as atoms and subatomic particles; 1 amu is equal to one-twelfth the mass of a carbon-12 atom.

atomic number A number that is equal to the number of protons in an atom.

atomic weight The weighted average mass of all the naturally occurring isotopes of an element.

chemical symbol An abbreviation that represents the name of an element.

electron A negatively charged subatomic particle having a very small mass that is usually ignored in calculations; its symbol is e^-.

electron arrangement An organization of electrons within the atom by increasing energy shells.

element A primary substance that cannot be separated into any simpler substances.

group A vertical column in the periodic table that contains elements having similar physical and chemical properties.

group number A number that appears at the top of each vertical column (group) in the periodic table and indicates the number of electrons in the outermost shell.

halogen Group 7A elements of fluorine, chlorine, bromine, and iodine.

isotope An atom that differs only in mass number from another atom of the same element. Isotopes have the same atomic number (number of protons) but different numbers of neutrons.

isotope symbol An abbreviation used to indicate the mass number and atomic number of an isotope.

mass number The total number of neutrons and protons in the nucleus of an atom.

metal An element that is shiny, malleable and a good conductor of heat and electricity. The metals are located to the left of the zigzag line in the periodic table.

neutron A neutral subatomic particle having a mass of 1 amu and found in the nucleus of an atom; its symbol is n or n°.

noble gas An element in Group 8A of the periodic table, generally unreactive and seldom found in combination with other elements.

nonmetal An element with little or no luster that is a poor conductor of heat and electricity. The nonmetals are located to the right of the zigzag line in the periodic table.

nucleus The compact, very dense center of an atom, containing the protons and neutrons of the atom.

orbital A region in space in which an electron is most likely to be found.

period A horizontal row of elements in the periodic table.

periodic law The repetition of similar chemical and physical properties with increasing atomic number due to the reappearance of the same number of electrons in the outermost shells of atoms of those elements.

periodic table An arrangement of elements by increasing atomic number such that elements having similar chemical behavior are grouped in vertical columns.

proton A positively charged subatomic particle having a mass of 1 amu and found in the nucleus of an atom; its symbol is p or p^+.

representative element An element found in Groups 1A to 8A of the periodic table.

shell An energy level containing electrons of similar energies.

subatomic particle A particle within an atom; protons, neutrons, and electrons are subatomic particles.

subshell A group of electrons having identical energies.

transition element An element located between groups 2A and 3A on the periodic table.

Problems

Elements and Symbols *(Goal 3.1)*

3.1 Write the symbols for the following elements:

a. copper	f. barium	k. sulfur
b. silicon	g. lead	l. aluminum
c. potassium	h. neon	m. unnilquadium
d. cobalt	i. oxygen	n. unnilhexium
e. iron	j. lithium	o. hydrogen

3.2 Write the correct name of the element for each symbol:

a. C	f. F	k. He
b. Cl	g. Ar	l. Ni
c. I	h. Zn	m. Hg
d. P	i. Mg	n. Ca
e. Ag	j. Na	o. Unp

The Atom *(Goal 3.2)*

3.3 Use *proton, neutron,* and/or *electron* to identify the subatomic particle or particles that each of the following statements describes:

a. has the smallest mass
b. carries a positive charge
c. is located outside the nucleus
d. is electrically neutral
e. carries a negative charge
f. has a mass about the same as a proton
g. is located in the nucleus
h. is found in the largest part of the atom

Atomic Number and Mass Number *(Goal 3.3)*

3.4 State whether you would use the atomic number or the mass number or both to obtain the following information:

a. number of protons in an atom
b. number of neutrons in an atom
c. number of particles in the nucleus
d. number of electrons in an atom

3.5 Write the atomic number, symbol, and mass number for the following neutral atoms:

a. an atom with 15 protons and 16 neutrons
b. an atom with 35 protons and 45 neutrons
c. an atom with 11 electrons and 12 neutrons
d. an atom with 26 electrons and 30 neutrons
e. an oxygen atom with 10 neutrons

3.6 Calculate the number of neutrons in each of the following atoms:

a. an atom with atomic number 17 and mass number 37
b. an atom with 20 electrons and mass number 40
c. a calcium atom with mass number 44
d. an atom of copper with mass number 65

3.7 Complete the following table for the indicated atoms:

Atomic Number	Mass Number	Protons	Neutrons	Electrons	Name	Symbol
a. ___	27	___	___	___	___	Al
b. 12	___	___	12	___	___	___
c. ___	___	6	7	___	___	___
d. ___	___	16	15	___	___	___
e. ___	34	___	___	16	___	___
f. 20	___	___	22	___	___	___

3.8 State the number of protons, neutrons, and electrons in the following atoms:

a. ${}^{27}_{13}Al$ b. ${}^{52}_{24}Cr$ c. ${}^{34}_{16}S$ d. ${}^{56}_{26}Fe$
e. ${}^{2}_{1}H$ f. ${}^{70}_{30}Zn$

3.9 Write an isotope symbol for the following:

a. an atom having mass number 44 and atomic number 20
b. an atom having 28 protons and 31 neutrons
c. an atom having mass number 24 and 13 neutrons
d. an atom having 35 electrons and 45 neutrons
e. a chlorine atom having 18 neutrons
f. a silver atom having 62 neutrons

3.10 There are four isotopes of sulfur (mass numbers 32, 33, 34, and 36).

a. Write the isotope symbols for each of these atoms.
b. How are these isotopes alike?
c. How are they different?
d. Why is the atomic weight of sulfur not a whole number?

The Periodic Table *(Goal 3.4)*

3.11 Using the periodic table, write the atomic weight (to the tenths place) of each of the following elements:

a. oxygen c. iron e. magnesium g. sodium
b. nitrogen d. hydrogen f. chlorine h. phosphorus

3.12 How does a group of elements in the periodic table differ from a period of elements?

3.13 Indicate whether each of the following statements describes a group or a period in the periodic table:

a. contains the elements C, N, and O
b. begins with helium
c. is a vertical column of elements
d. ends with neon
e. contains Na, K, and Rb
f. begins with atomic number 3

3.14 Give the symbol of the element that fits each of the following period and group number combinations:

a. Period 2, Group 1A
b. Period 1, Group 1A
c. Period 3, Group 6A
d. Period 4, Group 2A
e. Period 4, Group 8A
f. Period 5, Group 7A
g. Period 6, Group 2A

3.15 Use the periodic table to identify each of the following sets of elements as part of a group or a period of elements. If the set does *not* represent a group or a period, write *none*.

a. B, Al, Ga
b. Na, Mg, Al
c. Cl, Br, I
d. Si, P, S
e. Cu, Ag, Au
f. O, S

3.16 a. How do metal and nonmetals differ in physical properties?
b. Where are metals and nonmetals located in the periodic table?

3.17 Classify the following elements as metals or nonmetals.

a. phosphorus
b. magnesium
c. silver
d. fluorine
e. nickel
f. sulfur
g. silicon
h. nitrogen
i. aluminum
j. sodium

3.18 The following are trace elements that have been found to be crucial to the biochemical and physiological processes in the body. Indicate whether each is a metal or a nonmetal.

a. zinc
b. cobalt
c. manganese (Mn)
d. iodine
e. copper
f. selenium (Se)
g. nickel
h. iron

Electron Arrangement in the Atom *(Goal 3.5)*

3.19 Write the electron arrangement for each of the following elements: (*Example*: sodium 2, 8, 1.)

a. carbon
b. argon
c. aluminum
d. sulfur
e. potassium
f. phosphorus
g. nitrogen
h. neon

3.20 Write the electron arrangement for each of the following atoms:

a. an atom with 13 protons and 14 neutrons
b. an atom with mass number 40 and atomic number 18
c. $^{18}_{8}O$

3.21 Identify the elements that have the following electron arrangements:

Energy Level:	*1*	*2*	*3*
a.	$2\ e^-$	$1\ e^-$	
b.	$2\ e^-$	$8\ e^-$	$2\ e^-$
c.	$1\ e^-$		
d.	$2\ e^-$	$8\ e^-$	$7\ e^-$
e.	$2\ e^-$	$6\ e^-$	

3.22 Use the word *emit* or the word *absorb* to complete the following statements:
a. Electrons can jump to higher energy levels when they ____ a specific amount of energy.
b. When electrons drop to lower energy levels, they ____ a certain amount of energy.

Periodic Law (Goal 3.6)

3.23 The elements boron and aluminum are in the same group in the periodic table.
a. Write the electron arrangements for B and Al.
b. How many electrons are in the outer energy level for each?
c. What is their group number?

3.24 Write the number of electrons in the outer energy level and the group number for each of the following elements. (*Example*: fluorine, $7e^-$; Group 7A.)
a. magnesium c. oxygen e. lithium g. silicon
b. chlorine d. nitrogen f. neon h. argon

3.25 Would you expect Mg, Ca, and Sr to display similar physical and chemical behavior? Explain your answer.

Subshells and Orbitals (Goal 3.7)

***3.26** Indicate the maximum number of electrons that can occupy the following:
a. $2p$ orbital
b. $2p$ subshell
c. shell 3
d. $3s$ orbital
e. $3s$ subshell

***3.27** Using subshell notation, write the electron arrangement for each of the following elements:
a. magnesium b. phosphorus c. argon
d. sulfur e. chlorine f. potassium

***3.28** Identify the elements that have the following electron arrangements for subshells:
a. $1s^1$
b. $1s^2\ 2s^2\ 2p^3$
c. $1s^2\ 2s^2\ 2p^6\ 3s^1$
d. $1s^2\ 2s^2\ 2p^5$
e. $1s^2\ 2s^2\ 2p^6\ 3s^2\ 3p^4$
f. $1s^2\ 2s^2\ 2p^6\ 3s^2\ 3p^6\ 4s^2$

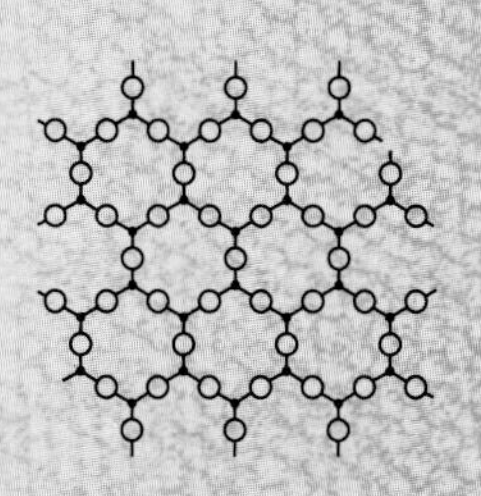

4

Compounds and Their Bonds

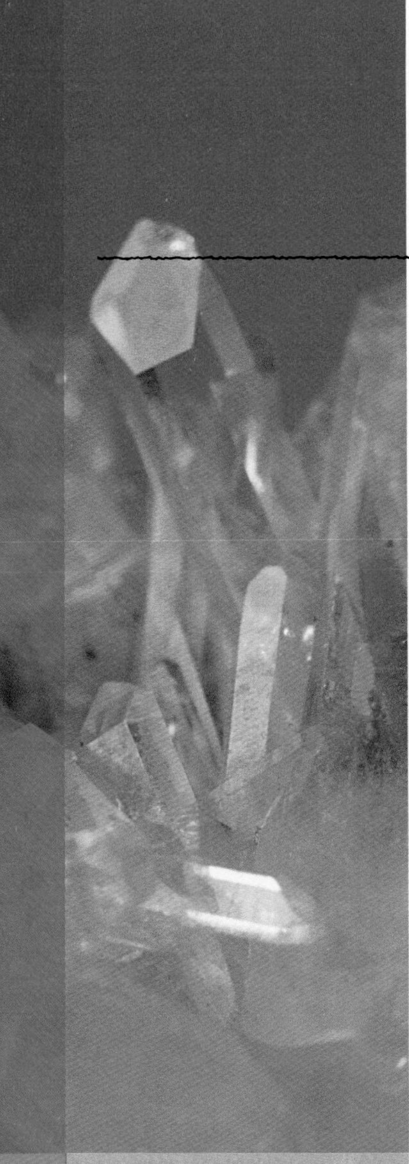

The elements silicon and oxygen combine to form a compound we know as quartz, which is seen in crystals and some semi-precious stones.

Learning Goals

4.1 Using the periodic table, write the electron-dot structures for the first 20 elements.

4.2 Using the octet rule, determine whether an element will form a compound to acquire a noble gas arrangement.

4.3 Write the formulas of positive and negative ions of some of the representative elements.

4.4 Given the formula of an ionic compound, write the correct name; given the name, write the correct formula.

4.5 Given the formula of a compound containing a metal and a nonmetal, write the correct name; given the name, write the correct formula.

4.6 Using electron-dot structures, diagram the bonds in a covalent compound.

4.7 Using electronegativity differences, classify a bond as covalent, polar, or ionic.

4.8 Given the formula of a covalent compound, write its correct name; given the name of a covalent compound, write its formula.

4.9 Write the formula of a polyatomic ion; write the correct name or formula of a compound containing a polyatomic ion.

SCOPE

Most of the things you see and use every day are made of compounds, in which atoms of one element are combined with atoms of other elements. Although there are 109 elements known today, there are millions of different compounds because of the many different ways in which atoms may combine. In ionic compounds such as table salt (sodium chloride, NaCl), electrons are transferred from one atom to another to form positive and negative ions. Strong attractive forces called ionic bonds hold the ions together. In covalent compounds such as water (H_2O), electrons are shared by the atoms, an attraction called a covalent bond.

The substances necessary for life include ionic and covalent compounds. For example, the human body is about 60–65% water, a covalent compound made from the elements hydrogen and oxygen. Some other compounds necessary for life are carbohydrates, fats, and proteins. All are made of the elements carbon, hydrogen, and oxygen; proteins contain nitrogen and sulfur also. These compounds, obtained from your diet, are needed to build and repair cells, provide energy, and direct metabolic processes. Ionic compounds such as sodium chloride and potassium chloride (KCl) are needed for the proper functioning of the heart muscle and nerve conduction.

4.1 Valence Electrons

Learning Goal **Using the periodic table, write the electron-dot structures for the first 20 elements.**

For the atoms of the representative elements in the periodic table, the electrons in the outer shell play an important role in determining their chemical properties. These influential electrons, called **valence electrons,** are located in the valence shell, which is the highest or outermost energy level of an atom. For example, all of the elements in Group 1A, such as potassium, have one valence electron. Recall that in an atom of potassium, the electrons are arranged in the first four energy levels as 2, 8, 8, 1. Since the fourth shell is the outermost shell for potassium, the electron in the fourth shell is a valence electron. In an electron-dot structure, this valence electron is placed next to the symbol for the element, as shown in Figure 4.1.

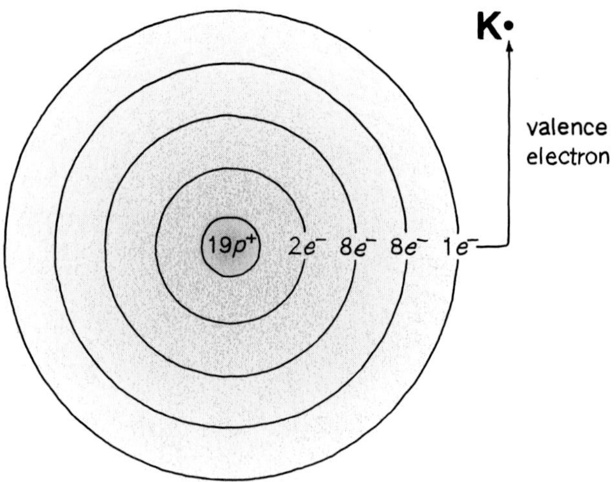

Figure 4.1
The electron arrangement for potassium and its electron-dot structure.

Table 4.1 Electron Arrangement

Element	Atomic number	Group number	Number of electrons in shell 1	2	3	4	Number of valence electrons
Li	3	1A	2	[1]			1
Ca	20	2A	2	8	8	[2]	2
Al	13	3A	2	8	[3]		3
C	6	4A	2	[4]			4
N	7	5A	2	[5]			5
S	16	6A	2	8	[6]		6
Cl	17	7A	2	8	[7]		7

The elements in Group 2A have two valence electrons; similarly, the elements in Group 3A have three valence electrons, elements in Group 5A have five valence electrons, and elements in Group 7A have seven valence electrons. If we compare the electron arrangement and the group number of an element, we find that the number of valence electrons in an atom is equal to the group number for that element. Some examples of the relationship between electron arrangement, group number, and number of valence electrons are shown in Table 4.1.

Sample Problem 4.1
Counting Valence Electrons

Using the electron arrangement, state the number of valence electrons in atoms of each of the following elements:

a. O **b.** Na

Solution:

a. The element oxygen has six valence electrons, as shown in its electron arrangement.

2 [6]

b. The element sodium has one valence electron, since there is one electron in its highest energy level, shell 3.

2 8 [1]

Electron-Dot Structures

An **electron-dot structure** is a convenient way to represent the valence shell. Valence electrons are shown as dots placed on the sides, top, or bottom of the symbol for the element. It does not matter on which of the four sides you place the dots. However, one to four valence electrons are usually arranged as single dots. When there are more than four electrons, the electrons begin to pair up. Any of the following would be an acceptable electron-dot structure for magnesium, which has two valence electrons:

Possible Electron-Dot Structures for the Two Valence Electrons in Magnesium

Mg·   Mg   ·Mg   ·Mg·   Mg·

Electron-dot structures for the first 20 elements are given in Table 4.2. Note that although helium has just two valence electrons, it appears in Group 8A with the rest of the noble gases because shell 1, its outermost shell, is filled.

Table 4.2 Electron-Dot Structures for the First Twenty Elements

	Group Number							
	1A	**2A**	**3A**	**4A**	**5A**	**6A**	**7A**	**8A**
Number of valence electrons	**1 e^-**	**2 e^-**	**3 e^-**	**4 e^-**	**5 e^-**	**6 e^-**	**7 e^-**	**8 e^-**
	H·							He:
	Li·	Be·	·B·	·C·	·N·	·O:	·F:	:Ne:
	Na·	Mg·	·Al·	·Si·	·P·	·S:	·Cl:	:Ar:
	K·	Ca·						

Sample Problem 4.2
Writing Electron-Dot Structures

Write the electron-dot structure for each of the following elements:

a. chlorine **b.** aluminum **c.** argon

Solution:

a. Since the group number for chlorine is 7A, we can state that chlorine has seven valence electrons.

·Cl:

b. Aluminum, in Group 3A, has three valence electrons.

·Al·

c. Argon, in Group 8A, has eight valence electrons.

:Ar:

4.2 The Octet Rule

Learning Goal **Using the octet rule, determine whether an element will form a compound to acquire a noble gas arrangement.**

The noble gases do not combine readily with other elements. One explanation for their lack of reactivity is that their eight valence electrons are in a particularly stable arrangement. All of the noble gases have an octet of eight valence electrons, except for helium, which is stable with two electrons in its first shell. The electron arrangements and electron-dot structures for some noble gases are shown in Table 4.3.

Table 4.3 Electron Arrangement for Some Noble Gases

	Number of electrons in shell				
Noble gas	**1**	**2**	**3**	**4**	**Electron-dot structure**
He	[2]				He :
Ne	2	[8]			:N̈ë:
Ar	2	8	[8]		:Är̈:
Kr	2	8	18	[8]	:K̈r̈:

Except for the noble gases, most elements tend to combine with other elements to form a **compound.** In most of the compounds formed by representative elements (Groups 1A–7A), each atom in the compound has acquired a noble gas arrangement. The **octet rule** indicates that the atoms of the elements in a compound lose, gain, or share valence electrons in order to produce a stable, noble gas arrangement of eight electrons. Thus, the octet rule provides a basic key to understanding the ways in which atoms bond together in compounds.

Sample Problem 4.3
Octet Rule

Write the electron-dot structures for neon (Ne), sodium (Na), and fluorine (F). Which atoms have octets and which do not? Which atoms will most likely form compounds? Why?

Solution:

:N̈ë: Na · ·F̈̈:

Neon, a noble gas, already has an octet of eight valence electrons and is not likely to form compounds. Sodium, however, has just one valence electron, and fluorine has seven. They would be expected to form compounds so that each of them will have an octet of valence electrons.

4.3 Ions

Learning Goal **Write the formulas of positive and negative ions of some of the representative elements.**

Positive Ions

In ionic compounds, metals lose their valence electrons in order to attain a noble gas arrangement. By losing electrons, they form **ions,** which are atoms that have a different number of electrons than protons. For example, when a sodium atom loses its outer electron, the remaining electrons have a noble gas arrangement, as shown in Figure 4.2.

By losing an electron, sodium now has 10 electrons instead of 11. Since there are still 11 protons in its nucleus, the atom is no longer neutral. It has become a sodium ion and has an electrical charge, called an **ionic charge,** of 1+. In the symbol for the ion, the ionic charge is written in the upper right-hand corner, as in Na^+.

Comparing the Charges in the Sodium Atom and the Sodium Ion

	Na atom	*Na^+ ion*
Charge from electrons	11−	10−
Charge from protons	11+	11+
Overall ionic charge	0	1+

In general, metals in ionic compounds have lost their valence electrons to form positively charged ions. Positive ions are also called **cations** (pronounced *cat'-ions*). Another metal, magnesium, in Group 2A, attains a noble gas arrangement by losing two valence electrons to form a positive ion with a 2+ ionic charge. (See Figure 4.3.)

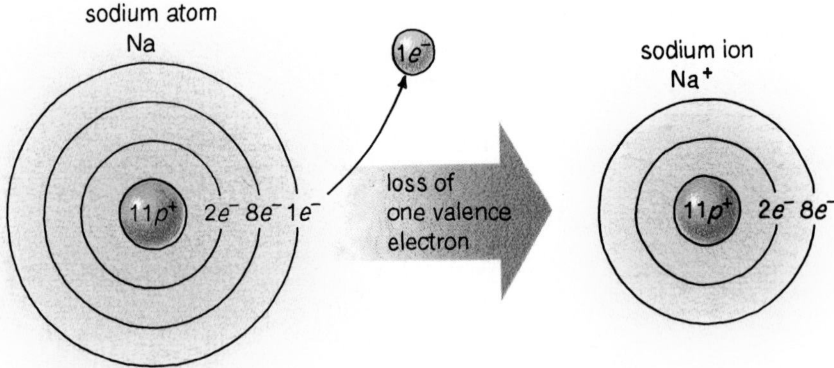

Figure 4.2
The loss of a valence electron changes a sodium atom to a sodium ion, which has a stable electron arrangement.

HEALTH NOTE

Some Uses for Noble Gases

Noble gases may be used when it is necessary to have a substance that is unreactive. Scuba divers normally use a pressurized mixture of nitrogen and oxygen gases for breathing under water. However, when the air mixture is used at depths where pressure is high, the nitrogen gas is absorbed into the blood, where it can cause mental disorientation. To avoid this problem, a breathing mixture of oxygen and helium may be substituted. The diver still obtains the necessary oxygen, but the unreactive helium that dissolves in the blood does not cause mental disorientation. However, its lower density does change the vibrations of the vocal cords, and the diver will sound like Donald Duck.

Helium is also used to fill blimps and balloons. When dirigibles were first designed, they were filled with hydrogen, a very light gas. However, when they came in contact with any type of spark or heating source, they exploded violently because of the extreme reactivity of hydrogen gas with oxygen present in the air. Today blimps are filled with unreactive helium gas, which presents no danger of explosion.

Lighting tubes are generally filled with a noble gas such as neon or argon. So, while the electrically heated filaments that produce the light get very hot, the surrounding noble gases do not react with the hot filament. If heated in air, the elements that constitute the filament would soon burn up. ■

Comparing the Charges in the Magnesium Atom and the Magnesium Ion

	Mg atom	*Mg^{2+} ion*
Charge from electrons	12−	10−
Charge from protons	12+	12+
Overall ionic charge	0	2+

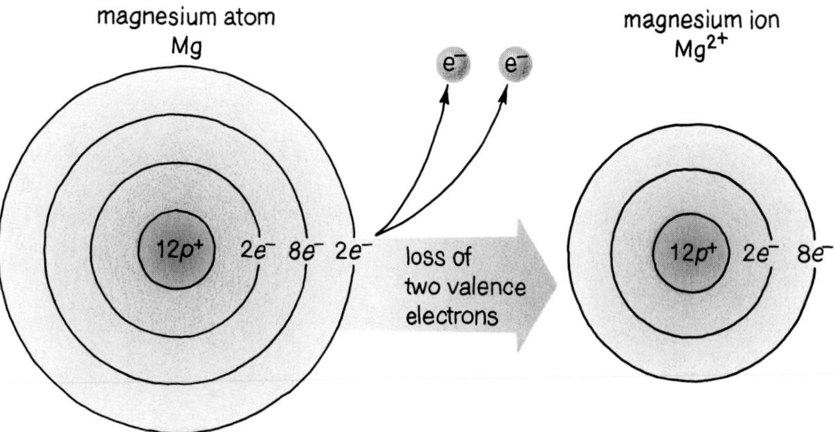

Figure 4.3
A magnesium ion has a stable electron arrangement after it loses two valence electrons.

Metals That Form Two Positive Ions

We have seen that the metals in Group 1A and Group 2A form only one type of positive ion. Therefore, we are able to predict their ionic charges from the periodic table. For example, sodium always forms Na^+, and magnesium always forms Mg^{2+}. Aluminum, in Group 3A, always forms the same ion, Al^{3+}. The other metals in the periodic table, such as the transition metals, also form positive ions. However, we are not able to predict their ionic charges because they can form more than one type of positive ion.

For example, iron, a transition metal, forms Fe^{2+} and Fe^{3+}. In some compounds, iron loses two electrons, producing the Fe^{2+} ion. In other compounds, iron loses three electrons and forms the Fe^{3+} ion. (The reasons for these different ionic charges are more complex than the octet rule and will not be considered in this text.)

When two different ions are possible for the same element, a naming system is needed that will differentiate between the ions. Using the name *iron chloride,* for example, does not specify whether the Fe^{2+} or the Fe^{3+} ion is the positive ion in the compound. Therefore, for metals that form two or more positive ions, a Roman numeral placed after the name of the metal in parentheses indicates the charge for that particular ion.

Ionic charge shown as superscript:	Fe^{2+}	Fe^{3+}
Ionic charge indicated by Roman numeral:	Iron(II)	Iron(III)

An older system names Fe^{2+} as the *ferrous ion* and Fe^{3+} as the *ferric ion*. The ending *ous* is added to the root of the Latin name, in this case *ferrum,* to identify the ion with the lower ionic charge; the *ic* ending is used to indicate the ion with the higher ionic charge.

Ferrous	Indicates the Fe^{2+} ion (the lower charge)
Ferric	Indicates the Fe^{3+} ion (the higher charge)

Typically, the transition metals form at least two positive ions. However, zinc and silver form only one type of positive ion. Their ionic charges are fixed,

Table 4.4 Some Metals That Form Two Positive Ions

Element	Possible ions	Systematic name	Older name
Iron	Fe^{2+}	Iron(II)	Ferrous
	Fe^{3+}	Iron(III)	Ferric
Copper	Cu^+	Copper(I)	Cuprous
	Cu^{2+}	Copper(II)	Cupric
Gold	Au^+	Gold(I)	Aurous
	Au^{3+}	Gold(III)	Auric
Tin	Sn^{2+}	Tin(II)	Stannous
	Sn^{4+}	Tin(IV)	Stannic
Lead	Pb^{2+}	Lead(II)	Plumbous
	Pb^{4+}	Lead(IV)	Plumbic

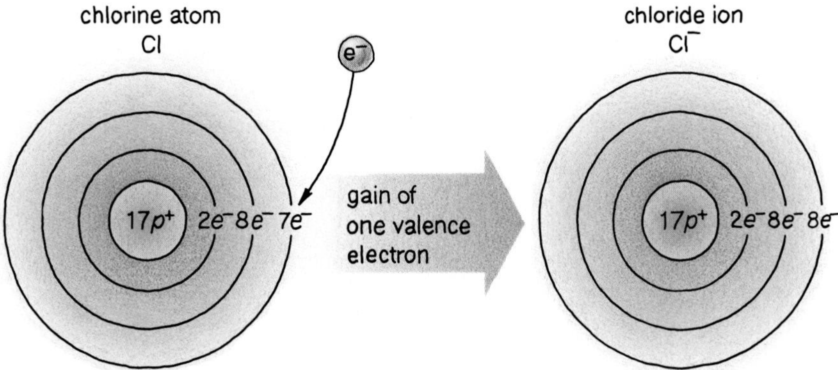

Figure 4.4
When a chlorine atom gains one valence electron, it becomes a chloride ion and has a stable electron arrangement.

so their elemental names are sufficient for naming. The zinc ion is always Zn^{2+}, and the silver ion is Ag^+.

Zinc	Zn^{2+}
Silver	Ag^+

Table 4.4 lists the common ions of some metals that produce more than one positive ion.

Negative Ions

A nonmetal forms a negative ion when it gains one or more valence electrons in the valence shell. For example, when an atom of chlorine with seven valence electrons obtains one more electron, it completes a noble gas arrangement, as shown in Figure 4.4. The resulting particle is a chloride ion having a negative ionic charge (Cl^-). An ion with a negative ionic charge is also called an **anion** (pronounced *an'-ion*).

Comparing the Charges in the Chlorine Atom and the Chloride Ion

	Cl atom	*Cl^- ion*
Charge from electrons	17−	18−
Charge from protons	17+	17+
Overall ionic charge	0	1−

Sample Problem 4.4
Calculating Ionic Charge

Write the symbol and the ionic charge for each of the following ions:
a. a nitrogen ion that has 7 protons and 10 electrons
b. a calcium ion that has 20 protons and 18 electrons
c. a lithium ion that has 3 protons and 2 electrons

Solution:
a. 7 p^+ and 10 e^- = N^{3-}
b. 20 p^+ and 18 e^- = Ca^{2+}
c. 3 p^+ and 2 e^- = Li^+

Study Check:
An iron ion is written Fe^{2+}. How many protons and electrons does it have?

Answer: 26 p^+(atomic number 26); 24 e^-

Sample Problem 4.5
Writing Ions

Consider the elements aluminum and oxygen.

a. Identify each as a metal or a nonmetal.
b. State the number of valence electrons for each.
c. State the number of electrons that must be lost or gained for each to acquire an octet.
d. Write the symbol of each resulting ion, including its ionic charge.

Solution:

	Aluminum	*Oxygen*
a.	metal	nonmetal
b.	3 valence electrons	6 valence electrons
c.	loses 3 e^-	gains 2 e^-
d.	Al^{3+}	O^{2-}

Ionic Charges From Group A Numbers

Group numbers can be used to determine the ionic charges of the ions in that group. We have seen that metals lose electrons to form positive ions. The elements in Groups 1A, 2A, and 3A lose the same number of electrons (1, 2, and 3, respectively) to produce ions with the same ionic charge. Group 1A metals form ions with 1+charges, Group 2A metals form ions with 2+ charges, and Group 3A metals form ions with 3+ charges.

When the nonmetals from Groups 5A, 6A, and 7A form ions, they acquire negative charges. Group 5A nonmetals form ions with 3− charges, Group 6A nonmetals form ions with 2− charges, and Group 7A nonmetals form ions with 1− charges. The elements of Group 4A are not discussed here because they do not typically form simple ions. Table 4.5 lists the ionic charges for ions within groups of representative elements, and Figure 4.5 shows how some of these ions are positioned in the periodic table. Note that some transition metal ions are included, even though their ionic charges are not easily obtained from group numbers and have to be memorized.

Table 4.5 Ionic Charges for Representative Elements

Group number	Number of electrons	Electron change to give an octet	Ionic charge	Examples
Metals				
1A	1	Lose 1	1+	Li^+, Na^+, K^+
2A	2	Lose 2	2+	Mg^{2+}, Ca^{2+}
3A	3	Lose 3	3+	Al^{3+}
Nonmetals				
5A	5	Gain 3	3−	N^{3-}, P^{3-}
6A	6	Gain 2	2−	O^{2-}, S^{2-}
7A	7	Gain 1	1−	F^-, Cl^-, Br^-, I^-

1A	2A											3A	4A	5A	6A	7A	8A
Li^+	Be^{2+}													N^{3-}	O^{2-}	F^-	
Na^+	Mg^{2+}											Al^{3+}		P^{3-}	S^{2-}	Cl^-	
K^+	Ca^{2+}						Fe^{2+} Fe^{3+}			Cu^+ Cu^{2+}	Zn^{2+}						
Rb^+	Sr^{2+}									Ag^+			Sn^{2+} Sn^{4+}				
Cs^+	Ba^{2+}												Pb^{2+} Pb^{4+}				

Metals

Nonmetals

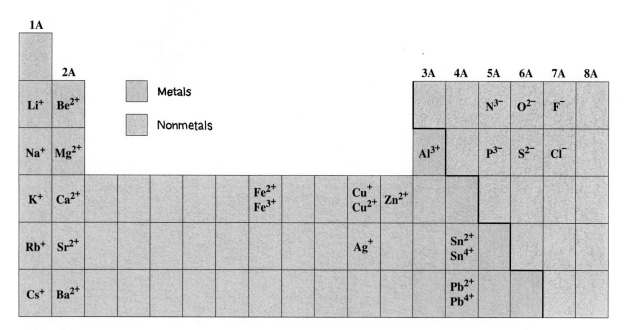

Figure 4.5
Some typical ions formed by metals and nonmetals.

Sample Problem 4.6
Using Group Number to Determine Ionic Charge

Using the group numbers of the elements, write the formula for an ion for each of the following:

a. Al **b.** Br

Solution:

a. The element aluminum (Al) is found in Group 3A of the periodic table. Since atoms in Group 3A have three valence electrons, an aluminum atom would lose three electrons to form an ion with a 3+ ionic charge.

$$Al \xrightarrow{\text{Loses } 3\,e^-} Al^{3+}$$

b. The element bromine (Br) is found in Group 7A. Since bromine has seven valence electrons, it would gain one electron to have an octet.

$$Br \xrightarrow{\text{Gains } 1\,e^-} Br^-$$

Study Check:
Write the formula for an ion of nitrogen.

Answer: N^{3-}

HEALTH NOTE

Some Important Ions in the Body

There are a number of ions in body fluids that have important physiological and metabolic functions. Some of them are listed in Table 4.6. ■

Table 4.6 Ions in the Body

Ion	Occurrence	Function	Source	Result of too little	Result of too much
Na^+	Principal cation outside the cell	Regulation and control of body fluids	Salt, seafood, meat	Hyponatremia, anxiety, diarrhea, circulatory failure, decrease in body fluid	Hypernatremia, little urine, thirst, edema
K^+	Principal cation inside the cell	Regulation of body fluids and cellular functions	Bananas, orange juice, skim milk, prunes, meat	Hypokalemia (hypopotassemia), lethargy, muscle weakness, failure of neurological impulses	Hyperkalemia (hyperpotassemia), irritability, nausea, little urine, cardiac arrest
Ca^{2+}	Cation outside the cell; 90% of calcium in the body in bone as $Ca_3(PO_4)_2$ or $CaCO_3$	Major cation of bone; muscle smoothant	Milk, cheese, butter, meat, some vegetables	Hypocalcemia, tingling finger-tips, muscle cramps, tetany	Hypercalcemia, relaxed muscles, kidney stones, deep bone pain, nausea
Mg^{2+}	Cation outside the cell; 70% of magnesium in the body in bone structure	Essential for certain enzymes, muscles, and nerve control	Widely distributed (part of chlorophyll of all green plants), nuts, grains	Disorientation, hypertension, tremors, slow pulse	Drowsiness
Cl^-	Principal anion outside the cell	Gastric juice, regulation of body fluids	Salt, seafood, meat	Same as for Na^+	Same as for Na^+

4.4 Ionic Compounds

Learning Goal **Given the formula of an ionic compound, write the correct name; given the name, write the correct formula.**

Ionic compounds consist of positive and negative ions. The ions are held together by strong electrical attractions between the opposite charges, called **ionic bonds.** Consider the ionic compound sodium chloride, NaCl. Its **formula,** NaCl,

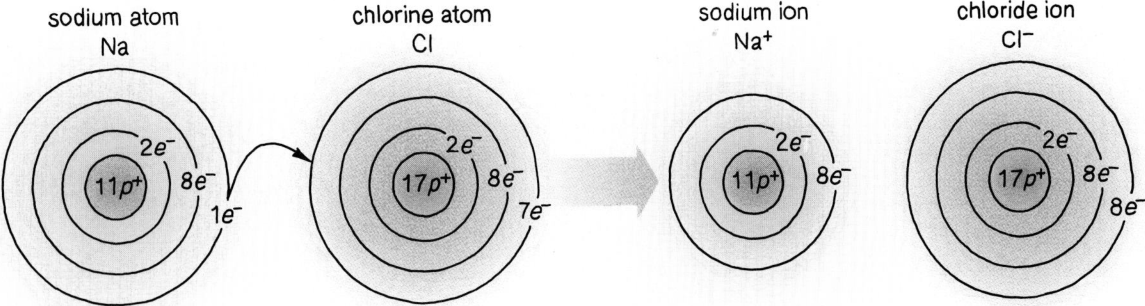

Figure 4.6
In the formation of sodium chloride, NaCl, the valence electron of sodium is transferred to complete the outer shell of a chlorine atom, forming an Na^+ cation and a Cl^- anion. The resulting ionic compound is held together by ionic bonds between the positive and negative ions.

indicates that it consists of Na^+ ions and Cl^- ions. Figure 4.6 illustrates how NaCl is formed through the transfer of an electron from the metal sodium to the nonmetal chlorine.

Sodium atom	Chlorine atom		Sodium ion	Chloride ion	Sodium chloride, an ionic compound
$Na\cdot$	$\cdot\ddot{\underset{..}{Cl}}:$	$\longrightarrow$	Na^+	$[:\ddot{\underset{..}{Cl}}:]^-$	$=$ NaCl

Charge Balance in Ionic Compounds

The formula of an ionic compound indicates the number and kinds of ions that make up the ionic compound. The sum of the ionic charges in the formula is always zero. For example, the NaCl formula indicates that there is one sodium ion, Na^+, for every chloride ion, Cl^-, in the compound. Note that the ionic charges of the ions do not appear in the formula of the ionic compound.

Charge Balance for Ions in the Ionic Compound NaCl

Na^+	+	Cl^-	=	NaCl
1+	+	1−	=	0
Total positive charge		Total negative charge		Overall charge of formula

The physical and chemical properties of an ionic compound such as NaCl are very different from those of the original elements. For example, the original elements of NaCl were sodium, a soft, shiny metal, and chlorine, a yellow-green poisonous gas. Yet, as positive and negative ions, they form table salt, NaCl, a white, crystalline substance that is common in our diet. Strong ionic bonds resulting from the electrical attraction of Na^+ ions and Cl^- ions are responsible for the high melting point of NaCl, 800°C. To account for these properties, scientists diagram the solid form of NaCl as an alternating pattern of Na^+ and Cl^- ions packed together in a lattice structure, as shown in Figure 4.7.

Subscripts in Formulas

Consider another ionic compound, one containing ions of magnesium and chlorine. To achieve noble gas configuration, a magnesium atom loses its two va-

Figure 4.7
(a) Crystals of the ionic compound sodium chloride, and (b) their structure. Note how Na^+ and Cl^- ions are packed together in the solid NaCl crystal.

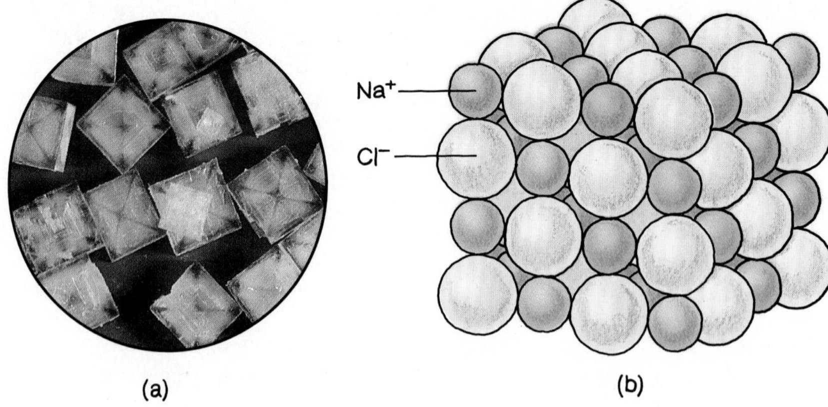

lence electrons, and two chlorine atoms each gain one electron. When more than one ion of an element is required to complete the electron transfer, that number is shown below the line as a subscript following the symbol of that element. For the ionic compound formed between magnesium and chlorine, $MgCl_2$, the 2 below the line is the subscript indicating that there are two chlorine atoms in the formula.

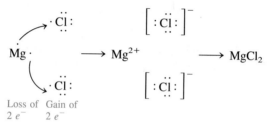

$$\text{Mg}\cdot + 2\,\cdot\ddot{\underset{..}{\text{Cl}}}: \longrightarrow \text{Mg}^{2+}\ 2\left[:\ddot{\underset{..}{\text{Cl}}}:\right]^- \longrightarrow \text{MgCl}_2$$

Loss of $2\,e^-$ Gain of $2\,e^-$

Ionic charge balance: $2+ \ + 2(1-) = 0$

Sample Problem 4.7
Diagramming an Ionic Compound

Diagram the formation of the ionic compound aluminum fluoride, AlF_3.

Solution:
In their electron-dot structures, aluminum has three valence electrons and fluorine has seven. The aluminum loses its three valence electrons, and each fluorine atom gains an electron, to give an ion with a noble gas arrangement in the ionic compound AlF_3.

$$\cdot\dot{\text{Al}}\cdot + 3\,\cdot\ddot{\underset{..}{\text{F}}}: \longrightarrow \text{Al}^{3+}\ 3\left[:\ddot{\underset{..}{\text{F}}}:\right]^- \longrightarrow \text{AlF}_3$$

Loss of $3\,e^-$ Gain of $3\,e^-$

Ionic charge balance: $3+ \ + \ 3(1-) \ = \ 0$

Study Check:
Diagram the formation of lithium sulfide, Li_2S.

Answer:

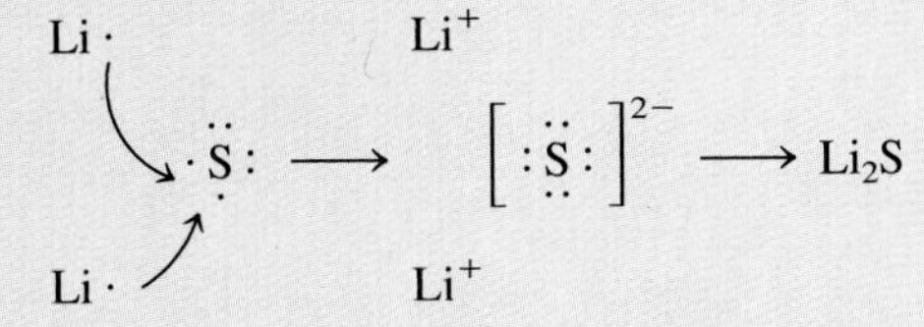

$$\mathrm{Li}\cdot + \cdot\ddot{\underset{\cdot}{S}}: + \mathrm{Li}\cdot \longrightarrow \mathrm{Li}^+ \left[:\ddot{\underset{\cdot\cdot}{S}}:\right]^{2-} \mathrm{Li}^+ \longrightarrow \mathrm{Li_2S}$$

Loss of 2 e^- Gain of 2 e^-

$2(1+) + 2- = 0$

Writing Ionic Formulas From Ionic Charges

We have seen that the formula of an ionic compound represents the number of positive ions and negative ions in that compound that give an overall charge of zero. Thus, we can now write a formula directly from the ionic charges of the positive and negative ions. Suppose we wish to write the formula of the ionic compound containing Na^+ and S^{2-} ions. To balance the ionic charge of the S^{2-} ion, we will need to place two Na^+ ions in the formula. This gives the formula Na_2S, which has an overall charge of zero.

Na^+ Na^+	+	S^{2-}	=	Na_2S
Positive charge		Negative charge		Net charge of formula
$2 \times (1+)$	+	$2-$	=	0

Sample Problem 4.8
Writing Formulas From Ionic Charges

Use ionic charge balance to write the formula for the ionic compound containing Na^+ and N^{3-}.

Solution:
Determine the number of each ion needed for charge balance. The charge for nitrogen (3−) is balanced by three Na^+ ions. Writing the positive ion first gives the formula Na_3N.

$$\begin{matrix} Na^+ & & \\ Na^+ & N^{3-} \longrightarrow & Na_3N \\ Na^+ & & \end{matrix}$$

Charge balance: $3(1+) + 3- = 0$

Study Check:
Write the formula of a compound containing Al^{3+} and O^{2-} ions.

Answer:
Charge balance is achieved using two Al^{3+} ions, and three O^{2-} ions.

$$\begin{matrix} & O^{2-} & \\ Al^{3+} & & \\ & O^{2-} & \longrightarrow Al_2O_3 \\ Al^{3+} & & \\ & O^{2-} & \end{matrix}$$

$$2\,(3+) + 3\,(2-) = 0$$

4.5 Naming Ionic Compounds

Learning Goal **Given the formula of a compound containing a metal and a nonmetal, write the correct name; given the name, write the correct formula.**

In Section 4.4, we saw that an ionic compound consists of the positive ion of a metal and the negative ion of a nonmetal. The metal ion is named by its elemental name. The nonmetal ion is named by replacing the ending of its elemental name by **ide.** Table 4.7 lists the names of some important metal and nonmetal ions.

Table 4.7 Formulas and Names of Some Common Ions

Group number	Formula of ion	Name of ion	Group number	Formula of ion	Name of ion
Metals			*Nonmetals*		
1A	Li^+	Lithium	5A	N^{3-}	Nitride
	Na^+	Sodium		P^{3-}	Phosphide
	K^+	Potassium			
2A	Mg^{2+}	Magnesium	6A	O^{2-}	Oxide
	Ca^{2+}	Calcium		S^{2-}	Sulfide
	Ba^{2+}	Barium			
3A	Al^{3+}	Aluminum	7A	F^-	Fluoride
				Cl^-	Chloride
				Br^-	Bromide
				I^-	Iodide
Transition metals					
	Fe^{2+}	Iron(II)			
	Fe^{3+}	Iron(III)			
	Cu^+	Copper(I)			
	Cu^{2+}	Copper(II)			
	Ag^+	Silver			
	Zn^{2+}	Zinc			

Sample Problem 4.9
Naming Ions

Write the name of each of the following ions:
a. Al^{3+} **b.** S^{2-} **c.** Fe^{3+}

Solution:

a. Aluminum ion; a metal ion uses the elemental name.

b. Sulfide ion; a nonmetal changes the ending of the name of the element to *ide*.

c. Iron(III) ion; a transition metal ion uses the name of the element and shows the ionic charge.

Study Check:
Write the formula of each of the following ions:
a. nitride ion **b.** calcium ion

Answer:
a. N^{3-} **b.** Ca^{2+}

Naming Ionic Compounds Containing Two Elements

In the name of an ionic compound made up of two elements, the positive metal ion is named first, and the negative nonmetallic ion is named next. Subscripts are never mentioned; they are understood as a result of the charge balance of the ions in the compound.

Formula of ionic compound:	$MgCl_2$	
Ions:	Mg^{2+}	Cl^-
	Metal	Nonmetal
Name of ionic compound:	Magnesium chloride	

Table 4.8 gives formulas and names of some ionic compounds.

Table 4.8 Formulas and Names of Some Ionic Compounds

	Ion		
Formula	**Metal**	**Nonmetal**	**Name**
MgS	Mg^{2+}	S^{2-}	Magnesium sulfide
K_2O	K^+	O^{2-}	Potassium oxide
$CaCl_2$	Ca^{2+}	Cl^-	Calcium chloride
Na_3N	Na^+	N^{3-}	Sodium nitride
Al_2O_3	Al^{3+}	O^{2-}	Aluminum oxide

Sample Problem 4.10
Naming Ionic Compounds

Write the name of each of the following ionic compounds:
a. Na_2O **b.** Al_2S_3

Solution:

Compound	*Ions and Names*		*Name of Compound*
a. Na_2O	Na^+	O^{2-}	
	Sodium	Oxide	Sodium oxide
b. Al_2S_3	Al^{3+}	S^{2-}	
	Aluminum	Sulfide	Aluminum sulfide

Study Check:
Name the compound Ca_3P_2.

Answer: Calcium phosphide

Naming Compounds That Include Transition Metals

As we saw earlier, when a metal forms more than one type of positive ion, the ionic charge of that metal in a particular formula must be included as a Roman numeral when the metal is named. The selection of the correct Roman numeral depends upon the calculation of the ionic charge of the metal in the formula of a specific ionic compound. For example, we know that in the formula $CuCl_2$ the positive charge of the copper ion must balance the negative charge of two chloride ions. Since we know that chloride ions each have a 1− charge, there must be a total negative charge of 2−. Balancing the 2− by the positive charge gives a charge of 2+ for Cu (a Cu^{2+} ion):

$CuCl_2$

Cu charge	+	Cl^- charge	= 0
(?)	+	2(1−)	= 0
(2+)	+	2−	= 0

Since copper forms ions with two different charges, Cu^+ and Cu^{2+}, we need to use the Roman numeral system and place II after *copper* when naming the compound:

Copper(II) chloride

Table 4.9 lists names of some ionic compounds of metals that form more than one type of positive ion. A summary of how ionic compounds are named follows.

Table 4.9 Some Ionic Compounds of Metals That Form Two Kinds of Positive Ions

Compound	Systematic name	Older name
$FeCl_2$	Iron(II) chloride	Ferrous chloride
$FeCl_3$	Iron(III) chloride	Ferric chloride
Cu_2S	Copper(I) sulfide	Cuprous sulfide
$CuCl_2$	Copper(II) chloride	Cupric chloride
$SnCl_2$	Tin(II) chloride	Stannous chloride
$PbBr_4$	Lead(IV) bromide	Plumbic bromide

Rules for Naming Ionic Compounds

1. For compounds in which the metal forms only one type of positive ion (Groups 1A and 2A; Al, Zn, Ag):
   a. Name the metal by its element name.
   b. Name the nonmetal by its element name but change the ending of its name to *ide*.
2. For compounds in which the metal forms more than one type of positive ion:
   a. Assign the known ionic charge to the negative ion.
   b. Multiply the charge of the negative ion by its subscript to find the total negative charge.
   c. State the total positive charge required to balance the negative charge.
   d. Assign the positive charge to the metal ion. If there are two or more positive ions, first divide the total positive charge by the number of metal ions.
   e. Name the compound using a Roman numeral after the name of the metal ion.

Sample Problem 4.11
Naming Ionic Compounds

Write the name for each of the following ionic compounds:

a. FeO
b. $AlBr_3$
c. Cu_2S

Solution:

a. Since iron forms Fe^{2+} and Fe^{3+}, it is necessary to determine the ionic charge of Fe in FeO:

$$\text{Fe charge} + O^{2-}\text{ charge} = 0$$
$$? \quad + (2-) \quad = 0$$
$$(2+) \quad + (2-) \quad = 0$$

The charge on Fe is therefore 2+. So we place the Roman numeral II in parentheses after *iron:*

Iron(II) oxide

b. Since aluminum is in Group 3A and forms only Al^{3+}, the name of the metal is sufficient. A Roman numeral is not needed.

$AlBr_3$ Aluminum bromide

c. As a transition metal, copper forms more than one positive ion. Balancing the charge in Cu_2S, we determine the ionic charge of the copper ion:

$$2(\text{Cu charge}) + S^{2-}\text{ charge} = 0$$

$$2(?) \quad + (2-) \quad = 0$$

$$2(1+) \quad + (2-) \quad = 0$$

The charge on Cu is therefore 1+. So we place the Roman numeral I in parentheses after copper:

Copper(I) sulfide

Although the total positive charge is 2+, it is divided between two copper ions; therefore, each copper ion in the formula is indicated as Cu^+.

Study Check:
Write the name of the compound whose formula is $AuCl_3$.

Answer: gold(III) chloride

Writing Formulas From the Name of an Ionic Compound

The formula of an ionic compound can be written from its name, since the first term in the name describes the metal ion and the second term specifies the nonmetal ion. Subscripts are added as needed to balance the charge. The steps for writing a formula from the name of an ionic compound follow.

Steps for Writing a Formula From the Name of an Ionic Compound

Step 1 Write the positive ion from the first term in the name, and the negative ion from the second term.

Step 2 Balance the ionic charges to give an overall charge of zero.

Step 3 Write the formula, placing the symbol for the metal first and the symbol for the nonmetal second. Use appropriate subscripts for charge balance.

Sample Problem 4.12
Writing Formulas for Ionic Compounds

Write the correct formula for iron(III) chloride.

Solution:

Step 1 Write the names of the metal ion and the nonmetal ion. Since the Roman numeral (III) indicates the ionic charge of the particular iron ion, we can then write Fe^{3+} for the positive ion. The chloride ion has an ionic charge of 1−.

Iron(III)	Chloride
Fe^{3+}	Cl^-

Step 2 Balance the charges.

Ions needed to balance charge:	Fe^{3+}	Cl^-
		Cl^-
		Cl^-
Total charge:	$(3+) + 3(1-) = 0$	

Step 3 Write the formula of the compound, metal first, using subscripts when needed.

$FeCl_3$

4.6 Covalent Bonds

Learning Goal **Using electron-dot structures, diagram the bonds in a covalent compound.**

When atoms of two nonmetals combine, covalent compounds result. The atoms, which are similar or even identical, are held together by sharing electrons. This type of bond is called a **covalent bond.**

The simplest covalent compound is a hydrogen molecule, H_2. A hydrogen atom attains a noble gas arrangement when its first shell is filled with two electrons. To accomplish this, two hydrogen atoms share their single electrons and form a covalent bond:

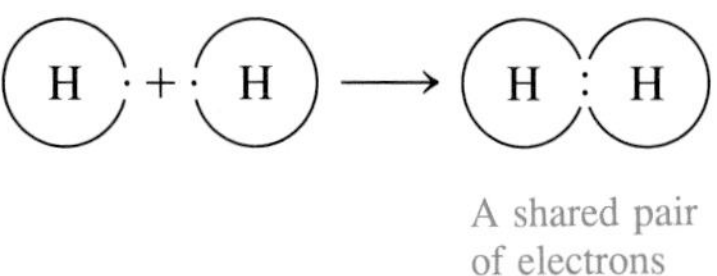

A shared pair of electrons

The covalent bond representing a shared pair of electrons can also be written as a line between the two atoms. Atoms that are bonded through covalent bonds produce a **molecule** of that compound.

H : H	=	H—H	=	H_2
A shared pair of electrons		A covalent bond		A hydrogen molecule

In the molecules of most covalent compounds, each of the atoms acquires an octet of eight electrons. For example, a fluorine molecule, F_2, consists of two fluorine atoms. Each atom needs one more valence electron for a noble gas arrangement. An octet for each is achieved when the fluorine atoms share a pair of electrons in their outer shells.

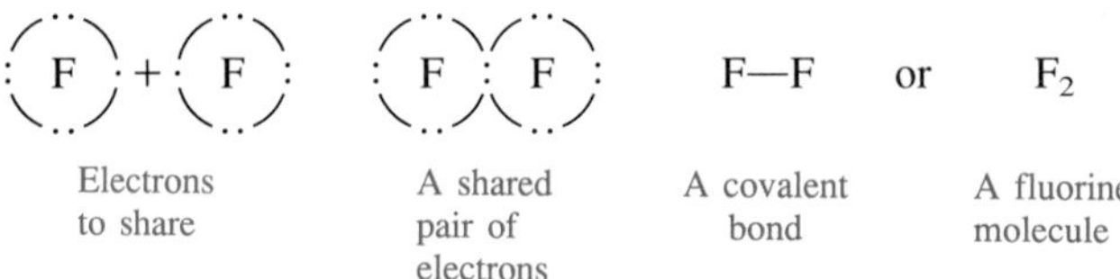

Hydrogen (H_2) and fluorine (F_2) are examples of several nonmetal elements whose natural state is diatomic; that is, they consist of molecules of two atoms. The elements that exist as diatomic molecules are listed in Table 4.10.

The number of electrons that an atom shares and the number of covalent bonds it forms are equal to the number of electrons needed to acquire a noble gas arrangement, usually an octet. For example, carbon has four valence electrons. Since carbon needs to acquire four more electrons for an octet, it forms four covalent bonds by sharing its four valence electrons. Table 4.11 relates the group numbers of some elements to the number of covalent bonds the elements typically form.

Table 4.10 Elements That Exist as Diatomic, Covalent Molecules

Element	Diatomic molecule	Name	Element	Diatomic molecule	Name
H	H_2	Hydrogen	F	F_2	Fluorine
N	N_2	Nitrogen	Cl	Cl_2	Chlorine
O	O_2	Oxygen	Br	Br_2	Bromine
			I	I_2	Iodine

Table 4.11 Covalent Bonds Required by Various Nonmetals

Group number	Nonmetal	Number of valence electrons	Number of electrons shared to complete an octet	Number of covalent bonds
1A	H	1	1	1
4A	C, Si	4	4	4
5A	N, P	5	3	3
6A	O, S	6	2	2
7A	F, Cl, Br, I	7	1	1

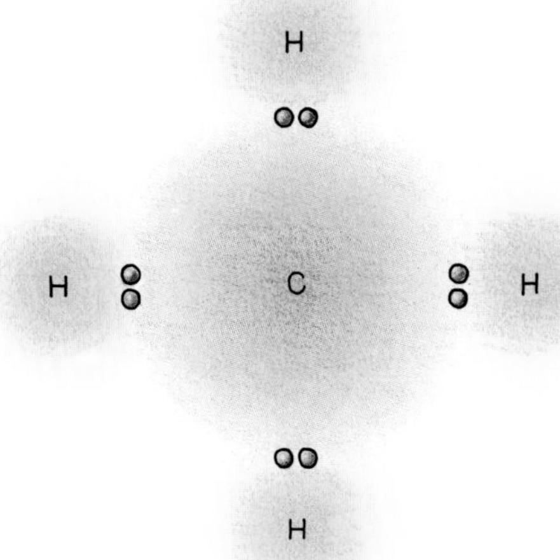

Figure 4.8
Diagram of a carbon atom sharing electrons with four atoms of hydrogen to form four covalent bonds. The carbon atom is stable with eight electrons in the outer shell, and each hydrogen is stable with a filled outer shell of two electrons.

Consider a compound of carbon and hydrogen. To attain an octet, carbon shares four electrons, and hydrogen shares one electron. In a molecule, a carbon atom can form four covalent bonds with four hydrogen atoms. (See Figure 4.8.) The electron-dot structure for the molecule is written with the carbon atom in the center and the hydrogen atoms on the sides. Each valence electron in carbon is paired with a hydrogen electron.

Table 4.12 gives the electron-dot structures for several covalent molecules.

Table 4.12 Electron-Dot Structures for Some Covalent Compounds

H_2	CH_4	NH_3	SCl_2	Cl_2
Structures using electron dots only				
H:H	H H:C:H H	H:N:H H	:S:Cl: :Cl:	:Cl:Cl:
Structures using bonds and electron dots				
H—H	H | H—C—H | H	H—N—H | H	:S—Cl: | :Cl:	:Cl—Cl:

Sample Problem 4.13
Writing Electron-Dot Structures for Covalent Compounds

Draw an electron-dot structure for water, H_2O.

Solution:
Oxygen, which has six valence electrons, shares two electrons to form two covalent bonds. Two atoms of hydrogen, each having one valence electron, will form two covalent bonds, one for each atom:

Electron-dot structure for H_2O

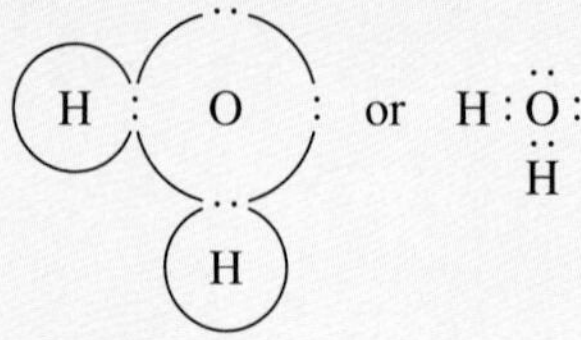

Study Check:
Write the electron-dot structure for NH_3.

Answer:

H : N : H or H:N:H
H

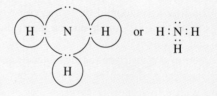

4.7 Bond Polarity

Learning Goal **Using electronegativity differences, classify a bond as covalent, polar, or ionic.**

In covalent bonds, as we have seen, electrons may be shared between atoms that are identical, as in Cl_2, or between atoms that are different, as in HCl. When electrons are shared between identical nonmetal atoms, they are shared equally. However, when they are shared between atoms of different nonmetals, they are shared unequally because one of the atoms in the covalent bond has a stronger attraction for the pair of electrons than the other. The ability of an atom to attract the shared electrons is called its electronegativity, and it is indicated by a number called an **electronegativity value.** The element with the greater electronegativity value pulls the shared electron closer to its nucleus.

In the table of electronegativities shown in Figure 4.9, fluorine, the most electronegative element, is assigned a value of 4.0. By contrast, the metals in Groups 1A and 2A have relatively low electronegativities. As we read from left to right across any period of the periodic table, we see that the electronegativity values increase, which means that the atoms of those elements to the right have a greater attraction for electrons.

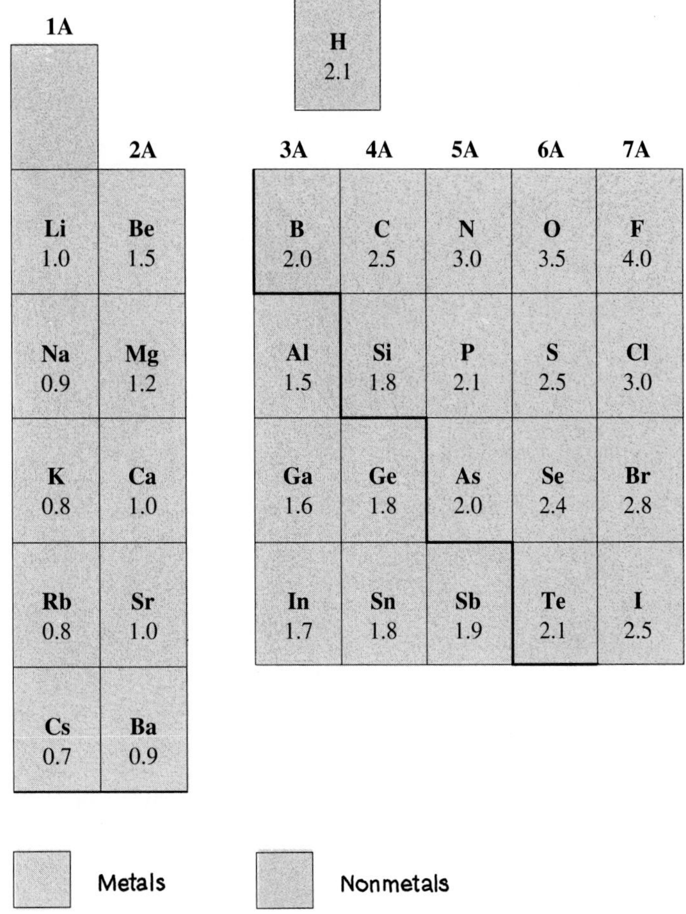

Figure 4.9
Electronegativity values for representative elements in the periodic table.

The **polarity** of a bond is a result of the difference in the electronegativity values of the two atoms in a bond. When the atoms are identical in a covalent bond, they have the same attraction for the shared pairs of electrons and the same electronegativity values. Therefore, the electronegativity difference is zero.

When electrons are shared between nonmetal atoms having different electronegativity values, they are pulled closer to the atom with the higher electronegativity to form a **polar bond.** For example, in HCl the shared pair of electrons is pulled closer to the more electronegative chlorine atom, as shown in Figure 4.10.

In a polar bond, the more electronegative atom may be shown with a partial negative charge, indicated by the lowercase Greek letter delta with a negative sign, δ^-. The atom with the lower electronegativity is farther from the electron pair and is shown with a partial positive charge, indicated by the Greek letter delta with a positive sign, δ^+.

Covalent bonds with electronegativity differences up to 1.6 are considered to be polar. When the difference in electronegativity is greater than 1.6, the differences in electron attraction are so great that electrons are transferred from one atom to the other. Ions form and the bonding is ionic. Table 4.13 gives some examples of how bond type can be predicted from electronegativity differences.

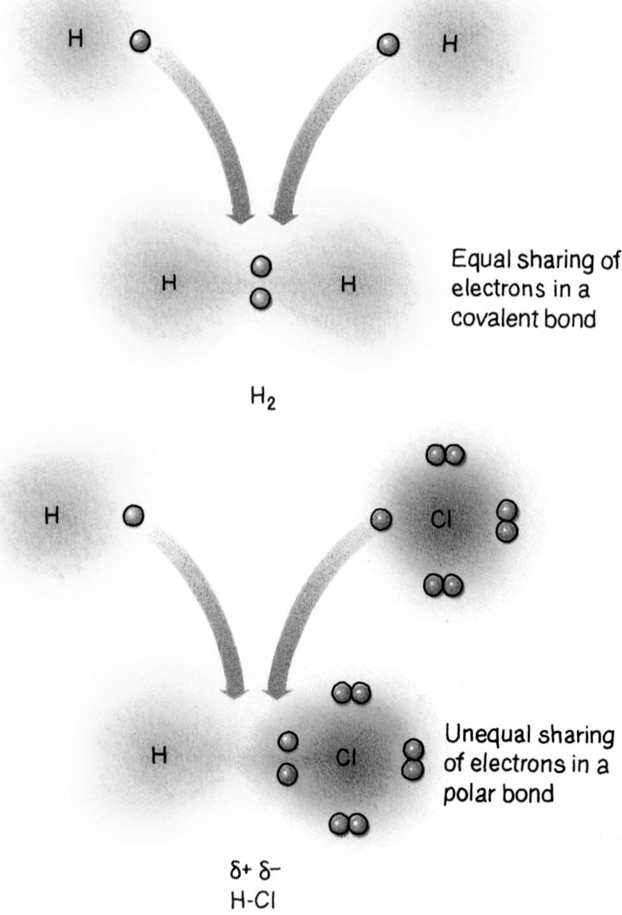

Figure 4.10
Equal and unequal sharing of electrons in the covalent bond of H_2 and the polar bond of HCl.

Table 4.13 Predicting Bond Type From Electronegativity Differences

Molecule		Type of electron sharing	Electronegativity difference[a]	Bond type
H_2	H—H	Shared equally	2.1 − 2.1 = 0	Covalent
Cl_2	Cl—Cl	Shared equally	3.0 − 3.0 = 0	Covalent
HI	$\overset{\delta+}{\text{H}}$—$\overset{\delta-}{\text{I}}$	Shared unequally	2.5 − 2.1 = 0.4	Polar
HCl	$\overset{\delta+}{\text{H}}$—$\overset{\delta-}{\text{Cl}}$	Shared unequally	3.0 − 2.1 = 0.9	Polar
NaCl	Na^+ Cl^-	Electron transfer	3.0 − 0.9 = 2.1	Ionic
MgO	Mg^{2+} O^{2-}	Electron transfer	3.5 − 1.2 = 2.3	Ionic

[a]Values are taken from Figure 4.9.

Sample Problem 4.14
Identifying Bond Polarity

Indicate whether bonds between the following atoms would be covalent, polar, or ionic. If polar, indicate the partial charges with δ^+ and δ^- notation.

a. carbon and chlorine
b. fluorine and fluorine
c. aluminum and oxygen

Solution:

a. The bond between the two nonmetals C and Cl has an electronegativity difference calculated as 3.0 − 2.5 or 0.5. The covalent bond is polar. Since the chlorine atom has the greater electronegativity (3.0), it is designated as partially negative, δ^-, in the polar bond.

$$\overset{\delta+}{\text{C}}\text{—}\overset{\delta-}{\text{Cl}}$$

b. The bond F—F occurs between identical atoms. It is a nonpolar covalent bond, as indicated by an electronegativity difference of zero (4.0 − 4.0 = 0).

c. The bond between the metal Al and the nonmetal O has an electronegativity difference of 2.0 (3.5 − 1.5 = 2.0). Since this difference is greater than 1.6, the bond is an ionic bond.

A Review of Bonding

As we have seen, the types of bonds that hold atoms together range from the covalent bonds, in which electrons are shared equally, all the way to the ionic bond, in which electrons are completely transferred from one atom to another. This variation in bonding is continuous; there is no definite point where one type of bond stops and the next starts. However, for purposes of discussion, we can state some general rules (summarized in Figure 4.11) that may be used to predict the type of bond formed between two elements. Examples of the different bonding types are given in Table 4.14.

Rules for Predicting Bond Type

1. A bond between two atoms of the same element, or between atoms having the same electronegativity value is a nonpolar covalent bond; we refer to it just as covalent, for simplicity.
2. A covalent bond between atoms of two different nonmetals having different electronegativities is polar covalent; we refer to it just as polar, for simplicity.
3. A bond between a metal and a nonmetal is usually ionic.

Figure 4.11
A summary of electronegativity difference and bond type.

Typical Elements	Two Nonmetals (identical)	Two Nonmetals (different)	Metal and Nonmetal
Electron Bonding	shared equally	δ+ : δ− shared unequally	+ −
Electronegativity Difference	0	increases → 1.6	over 1.6
Bond Type	covalent	polar increases →	ionic
Typical Compound	H_2 Cl_2 Br_2	H_2O CCl_4 HBr	Li_2O NaCl $BaCl_2$

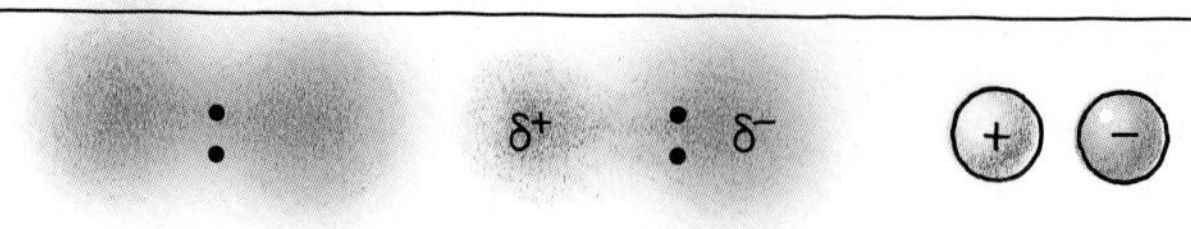

Table 4.14 Examples of Types of Bonding

Elements to bond	Electron-dot structure	Type of elements	Type of bonding	Molecule or ionic unit	Formula
F and F	:F̤̈· :F̤̈·	Two nonmetals (the same element)	Covalent	F—F	F_2
P and Cl	·P̤̈· :C̤̈l·	Two nonmetals (different elements)	Polar	Cl—P—Cl with Cl bonded below P	PCl_3
Na and F	Na· ·F̤̈:	Metal, nonmetal	Ionic	Na^+ F^-	NaF
Ca and O	Ċa· :Ö·	Metal, nonmetal	Ionic	Ca^{2+} O^{2-}	CaO

Sample Problem 4.15
Predicting Bond Polarity

Predict the type of bond between each of the following pairs of elements:
a. Ca and Cl **b.** P and S **c.** Br and Br

Solution:

a. A bond between calcium, a metal, and chlorine, a nonmetal, would be ionic.
b. A bond between phosphorus and sulfur, two different nonmetals, is polar.
c. A bond between atoms of the same element would be covalent.

Multiple Covalent Bonds

Up to now, we have looked at covalent bonding in molecules having only single bonds. In many covalent compounds, atoms share two or three pairs of electrons to complete their electron octets. A **double bond** is the sharing of two pairs of electrons, and in a **triple bond,** three pairs of electrons are shared.

Double and triple bonds are formed when single covalent bonds fail to complete the octets of all the atoms in the molecule. For example, in the electron-dot structure for the covalent compound N_2, an octet is achieved when each nitrogen atom shares three electrons. Thus, three covalent bonds, or a triple bond, will form.

Octets

$$\cdot\ddot{N}\cdot \quad \cdot\ddot{N}\cdot \longrightarrow \ddot{N}\vdots\vdots\ddot{N} = N \equiv N$$

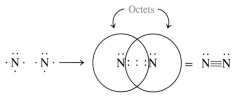

Three shared pairs Triple bond

We can use the following steps to write electron-dot structures:

Steps for Writing Electron-Dot Structures

Step 1 Determine the total number of valence electrons in all of the atoms.

Step 2 Place single bonds between each set of atoms. Each single bond uses two of the available valence electrons.

Step 3 Subtract the electrons used to bond the atoms, and arrange the remaining valence electrons to give each atom an octet. Hydrogen needs only a single bond, or two electrons.

Step 4 If octets of all the atoms cannot be completed using the remaining electrons, rearrange some of the electrons so that another pair or two are shared between two of the atoms as a double or triple bond. Check that all of the atoms have octets.

Sample Problem 4.16
Drawing Electron-Dot Structures Having Multiple Bonds

Draw the electron-dot structure of the compound CO_2. (Carbon is the central atom.)

O C O

Solution:
Oxygen has six valence electrons, and carbon has four. Each oxygen atom needs two electrons, and the carbon atom needs four electrons.

Step 1 Using group numbers, calculate the total number of valence electrons available:

O C O
$6\ e^- + 4\ e^- + 6\ e^- = 16$ valence electrons

Step 2 Connect the atoms by single bonds:

O : C : O

Uses $2\,e^-$ Uses $2\,e^-$

Step 3 Arrange the remaining valence electrons to satisfy octets:

$:\ddot{O}:\underset{\cdot\cdot}{\ddot{C}}:\ddot{O}:$

Not octets

Step 4 Octets cannot be completed using the remaining 12 electrons. Therefore, double bonds are needed instead of single bonds. Rearrange the electron dots, placing them between atoms to form double bonds:

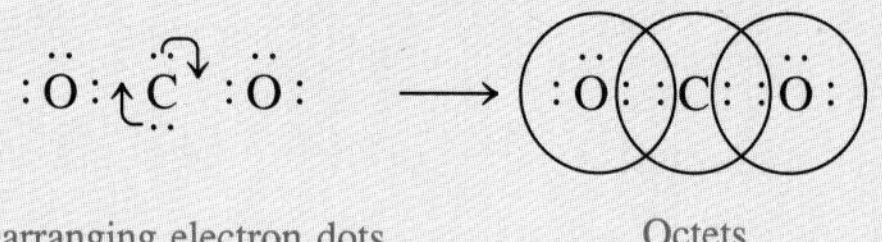

Study Check:
Determine the number of valence electrons and arrange them in the electron-dot structure of the HCN molecule.

Answer:

Number of valence electrons: $1 + 4 + 5 = 10\ e^-$

A triple bond gives octets to carbon and nitrogen.

H : C ⋮ ⋮ N :

4.8 Naming Covalent Compounds

Learning Goal **Given the formula of a covalent compound, write its correct name; given the name of a covalent compound, write its formula.**

In the name of a covalent compound, the first nonmetal in the formula is named by its elemental name; the second nonmetal is named by its elemental name with the ending changed to *ide*. All subscripts indicating two or more atoms of an element are expressed as prefixes and attached in front of each name. Table 4.15 lists some prefixes used in naming covalent compounds.

Unlike the names of ionic compounds, the names of covalent compounds need prefixes because several different compounds can be formed from the same two nonmetals. For example, carbon and oxygen can form two different compounds, carbon monoxide, CO, and carbon dioxide, CO_2. Nitrogen and oxygen

Table 4.15
Prefixes Used in Naming Covalent Compounds

Number of atoms	Prefix
1	Mono
2	Di
3	Tri
4	Tetra
5	Penta
6	Hexa
7	Hepta
8	Octa

also form several different covalent molecules. We could not distinguish between the following compounds by using the name *nitrogen oxide*. Therefore, prefixes must be used.

Some Covalent Compounds Formed by Nitrogen and Oxygen

NO Nitrogen monoxide
N_2O Dinitrogen monoxide
N_2O_3 Dinitrogen trioxide
N_2O_4 Dinitrogen tetroxide
N_2O_5 Dinitrogen pentoxide

In the name of a covalent compound, the prefix *mono* is understood for the first element and is usually omitted. When the vowels *o* and *o* or *a* and *o* appear together, the first vowel is omitted. Table 4.16 lists the formulas, names, and commercial uses of some other covalent compounds.

Table 4.16 Some Common Covalent Compounds

Formula	Name	Commercial uses
CS_2	Carbon disulfide	Manufacture of rayon
CO_2	Carbon dioxide	Carbonation of beverages; fire extinguishers, propellant in aerosols, dry ice
SiO_2	Silicon dioxide	Manufacture of glass
NCl_3	Nitrogen trichloride	Bleaching of flour in some countries (prohibited in U.S.)
SO_2	Sulfur dioxide	Preserving fruits, vegetables; disinfectant in breweries, bleaching textiles
SO_3	Sulfur trioxide	Manufacture of explosives
SF_6	Sulfur hexafluoride	Electrical circuits
ClO_2	Chlorine dioxide	Bleaching pulp (for making paper), flour, leather
ClF_3	Chlorine trifluoride	Rocket propellant

Sample Problem 4.17
Naming Covalent Compounds

Name each of the following covalent compounds:
a. NCl_3 **b.** P_4O_6

Solution:

a. In NCl_3, the first nonmetal is nitrogen. We could use the term *mononitrogen,* but the prefix *mono* is usually dropped from the name of the first element. The second nonmetal is chlorine, which is named *chloride*. The subscript indicating three chlorine atoms is shown as the prefix *tri* in front of the name of the second element, trichloride.

NCl_3
Nitrogen trichloride

b. In P_4O_6, the first nonmetal is phosphorus. Since there are four P atoms, it is named *tetraphosphorus*. The second nonmetal, consisting of six oxygen atoms, is named *hexoxide*. (When the vowels *a* and *o* appear together, as would happen in *hexa* plus *oxide*, the ending of the prefix is dropped.) Putting the names for the two nonmetals together, we have

P_4O_6

Tetraphosphorus hexoxide

Study Check:
Using high temperatures and pressures, scientists have recently combined some of the noble gases with other elements. One of the compounds produced was XeF_6. How would you name this compound?

Answer: xenon hexafluoride

Sample Problem 4.18
Writing Formulas From Names of Covalent Compounds

Write the formulas of the following covalent compounds:

a. sulfur dichloride
b. dinitrogen pentoxide

Solution:

a. The first nonmetal is sulfur. Since there is no prefix given, we can assume that there is one atom of sulfur. The second nonmetal in the formula is chlorine. The prefix *di* indicates that there are two atoms of chlorine, which means that the subscript 2 appears with chlorine in the formula.

Sulfur dichloride

SCl_2

b. The first nonmetal in the name is nitrogen. The prefix *di* means that there are two atoms of nitrogen and therefore the subscript 2 is needed. The second nonmetal is oxygen. The prefix *pent(a)* means "five"—there are five atoms of oxygen so the subscript 5 is needed.

Dinitrogen pentoxide

N_2O_5

Study Check:
What is the formula of iodine pentafluoride?

Answer: IF_5

4.9 Polyatomic Ions

Learning Goal **Write the formula of a polyatomic ion; write the correct name or formula of a compound containing a polyatomic ion.**

There are ionic compounds that contain three or more elements. Such compounds contain a **polyatomic ion,** which is a group of atoms that has an overall electrical charge. Most polyatomic ions consist of a nonmetal such as phosphorus, sulfur, carbon, or nitrogen covalently bonded to one or more oxygen atoms. These oxygen-containing polyatomic ions have negative charges of 1−, 2−, or 3− because one, two, or three electrons were added to the atoms in the group to complete their octets. Only one of the common polyatomic ions, NH_4^+, is positively charged.

Naming Polyatomic Ions

The name of the most common form of the oxygen-containing polyatomic ion ends in *ate*. The *ite* ending is used for the name of a related ion that has one fewer oxygen atom. Recognizing these endings will help you identify polyatomic ions in the name of a compound. The hydroxide ion (OH^-) and cyanide ion (CN^-) are exceptions to this naming pattern. There is no easy way to learn polyatomic ions. You will need to memorize the number of oxygen atoms and the charge associated with each ion, as shown in Table 4.17. By learning the formulas and the names of the more common forms shown in the boxes, you can derive the related ions. For example, by learning that sulfate is written SO_4^{2-}, we can write the formula of sulfite, which has one fewer oxygen atom, as SO_3^{2-}. Or the formula of hydrogen carbonate can be written by placing a hydrogen in front of the formula for carbonate (CO_3^{2-}) and decreasing the charge from 2− to 1− to give HCO_3^-.

Writing Formulas For Compounds Containing Polyatomic Ions

No polyatomic ion exists by itself. Like any ion, a polyatomic ion must be associated with ions of opposite charge. The bonding between polyatomic ions and other ions is one of electrical attraction and is thus ionic. For example, the compound sodium sulfate consists of sodium ions (Na^+) and sulfate ions (SO_4^{2-}) held together by ionic bonds. Figure 4.12 diagrams the formation of Na_2SO_4.

Writing correct formulas for compounds containing polyatomic ions follows the same rules of charge balance that we used for writing the formulas of ionic compounds. The total negative and positive charges must equal zero. For example, consider the formula for a compound containing calcium ions and carbonate

Table 4.17 Names and Formulas of Some Common Polyatomic Ions

Nonmetal	Formula of ion	Name of ion
Hydrogen	OH^-	Hydroxide
Nitrogen	NH_4^+	Ammonium
	NO_3^-	Nitrate
	NO_2^-	Nitrite
Chlorine	ClO_3^-	Chlorate
	ClO_2^-	Chlorite
Carbon	CO_3^{2-}	Carbonate
	HCO_3^-	Hydrogen carbonate (or bicarbonate)
	CN^-	Cyanide
Sulfur	SO_4^{2-}	Sulfate
	HSO_4^-	Hydrogen sulfate (or bisulfate)
	SO_3^{2-}	Sulfite
	HSO_3^-	Hydrogen sulfite (or bisulfite)
Phosphorus	PO_4^{3-}	Phosphate
	HPO_4^{2-}	Hydrogen phosphate
	$H_2PO_4^-$	Dihydrogen phosphate
	PO_3^{3-}	Phosphite

ions. The *ate* ending of carbonate indicates that it is a polyatomic ion. The ions are written as

Ca^{2+}	CO_3^{2-}
Calcium ion	Carbonate ion

Ionic charge: $(2+) + (2-) = 0$

Since one ion of each balances the charge, the formula can be written as

$CaCO_3$

Calcium carbonate

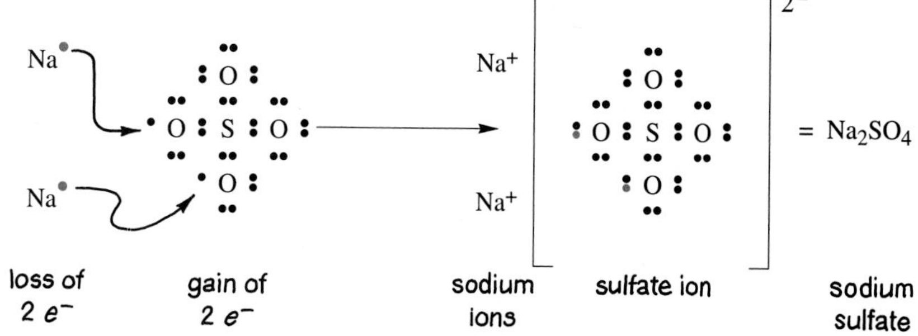

Figure 4.12
Diagram of the formation of the ionic compound sodium sulfate, which contains a polyatomic ion.

When more than one polyatomic ion is needed for charge balance, parentheses are used to enclose the formula of the ion. A subscript is written outside the closing parenthesis. Consider the formula for magnesium nitrate. The ions in this compound are the magnesium ion and the nitrate ion, a polyatomic ion.

Mg^{2+}	NO_3^-
Magnesium ion	Nitrate ion

To balance the positive charge of 2+, two nitrate ions are needed. The formula, including the parentheses around the bicarbonate ion, is as follows:

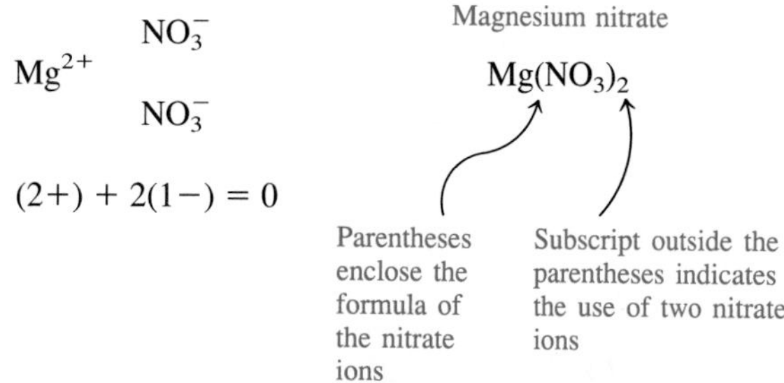

Sample Problem 4.19
Writing Formulas Having Polyatomic Ions

Write the formula for a compound containing aluminum ions and bicarbonate ions.

Solution:
The positive ion in the compound is the aluminum ion, Al^{3+}, and the negative ion is the bicarbonate ion, HCO_3^-, which is a polyatomic ion.

Al^{3+}	HCO_3^-
Aluminum ion	Bicarbonate ion

Three HCO_3^- ions are required to balance the Al^{3+} charge.

Al^{3+} HCO_3^-
HCO_3^-
HCO_3^-

$$(3+) + 3 \times (1-) = 0$$

The formula for the compound is written by enclosing the formula of the bicarbonate ion, HCO_3^-, in parentheses and writing the subscript 3 just outside the last parenthesis:

$$Al(HCO_3)_3$$

Study Check:
Write the formula for a compound containing ammonium ions and phosphate ions.

Answer: $(NH_4)_3PO_4$

Naming Compounds Containing Polyatomic Ions

In naming ionic compounds containing polyatomic ions, we write the positive ion, usually a metal, first, and then we write the name of the polyatomic ion. It is important that you learn to recognize the polyatomic ion in the formula and name it correctly. As with other ionic compounds, no prefixes are used.

Na_2SO_4	$FePO_4$	$Al_2(CO_3)_3$
Na_2 $\boxed{SO_4}$	Fe $\boxed{PO_4}$	$Al_2(\boxed{CO_3})_3$
Sodium sulfate	Iron(III) phosphate	Aluminum carbonate

Table 4.18 lists the formulas and names of some ionic compounds that include polyatomic ions, and also gives their uses in medicine and industry.

Table 4.18 Some Compounds That Contain Polyatomic Ions

Formula	Name	Use
$BaSO_4$	Barium sulfate	Radiopaque medium
$CaCO_3$	Calcium carbonate	Antacid, calcium supplement
$Ca_3(PO_4)_2$	Calcium phosphate	Calcium replenisher
$CaSO_3$	Calcium sulfite	Preservative in cider and fruit juices
$CaSO_4$	Calcium sulfate	Plaster casts
$AgNO_3$	Silver nitrate	Topical anti-infective
$NaHCO_3$	Sodium bicarbonate	Antacid
$Zn_3(PO_4)_2$	Zinc phosphate	Dental cements
$FePO_4$	Iron(III) phosphate	Food and bread enrichment
K_2CO_3	Potassium carbonate	Alkalizer, diuretic
$Al_2(SO_4)_3$	Aluminum sulfate	Antiperspirant, Anti-infective
$AlPO_4$	Aluminum phosphate	Antacid
$MgSO_4$	Magnesium sulfate	Cathartic, Epsom salts

HEALTH NOTE

Polyatomic Ions in Bone and Teeth

Bone structure consists of two parts: a solid mineral material, and a second phase made up primarily of collagen proteins. The mineral substance is a compound called hydroxyapatite, a solid formed from calcium ions, phosphate ions, and hydroxide ions. This material is deposited in the web of collagen to form a very durable bone material.

$Ca_{10}(PO_4)_6(OH)_2$

Hydroxyapatite

In most individuals, bone material is continuously being absorbed and re-formed. After age 40, more bone material may be lost than formed, a condition called osteoporosis. Bone mass reduction occurs at a faster rate in women than in men, and at different rates in different parts of the body skeleton. The reduction in bone mass can be as much as 50% over a period of 30 to 40 years. It is recommended that persons over 35, especially women, include a daily calcium supplement in their diet. ■

Sample Problem 4.20
Naming Compounds Containing Polyatomic Ions

Name the following ionic compounds:

a. $CaSO_4$
b. $Cu(NO_2)_2$

Solution:

	Compound Formula	Ions Present	Name of Compound
a.	$CaSO_4$	Ca^{2+}, calcium	calcium sulfate
	Ca[SO_4]	SO_4^{2-}, sulfate	
b.	$Cu(NO_2)_2$	Cu^{2+}, copper(II)	copper(II) nitrite
	Cu([NO_2])$_2$	NO_2^-, nitrite One fewer oxygen than nitrate, NO_3^-	

Study Check:
What is the name of $Ca_3(PO_4)_2$?

Answer: calcium phosphate

Summary of Naming Compounds

Throughout this chapter we have examined strategies for naming ionic and covalent compounds. Now we can summarize the rules. In general, compounds having two elements are named by stating the first element, followed by the second element with an *ide* ending. If the first element is a metal, the compound is ionic;

if the first element is a nonmetal, the compound is covalent. For ionic compounds, it is necessary to determine whether the metal can form more than one type of positive ion; if so, a Roman numeral following the name of the metal indicates the particular ionic charge. In naming covalent compounds having two elements, prefixes are necessary to indicate the number of atoms of each nonmetal as shown in that particular formula. Ionic compounds having three or more elements include some type of polyatomic ion. They are named by ionic rules, but have an *ate* or *ite* ending when the polyatomic ion has a negative charge. Table 4.19 summarizes some naming rules for elements, ionic compounds, and covalent compounds.

Table 4.19 Rules for Naming Elements, Ionic Compounds, and Covalent Compounds

Type	Formula feature	Naming procedure
Element	Symbol of element; may be diatomic. *Examples:* Ca Cl_2	Use name of element. *Examples:* Calcium Chlorine
Ionic compound (two elements)	Symbol of metal followed by symbol of nonmetal; subscripts used for charge balance. *Examples:* Na_2O Fe_2S_3	Use element name for metal; Roman numeral required if two or more positive ions possible. For nonmetal use element name with *ide* ending. *Examples:* Sodium oxide Iron(III) sulfide
Ionic compound (three elements)	Usually symbol of metal followed by a polyatomic ion composed of nonmetals; parentheses may enclose polyatomic ion for charge balance. *Examples:* $Mg(NO_3)_2$ $CuSO_4$ $(NH_4)_2CO_3$	Use element name for metal, with Roman numeral if needed, followed by name of polyatomic ion. *Examples:* Magnesium nitrate Copper(II) sulfate Ammonium carbonate
Covalent compound (two elements)	Symbols of two nonmetals; subscripts show number of atoms in a molecule. *Examples:* N_2O_3 CCl_4	Place prefixes before each element name if there are two or more atoms in the formula; add *ide* ending to second element. *Examples:* Dinitrogen trioxide Carbon tetrachloride

Glossary of Key Terms

anion A negatively charged ion such as Cl^-, O^{2-}, or SO_4^{2-}.

cation A positively charged ion such as Na^+, Mg^{2+}, or Al^{3+}.

compound A combination of atoms in which noble gas arrangements are attained through electron transfer or electron sharing.

covalent bond A sharing of valence electrons by atoms.

double bond A sharing of two pairs of electrons by two atoms.

electron-dot structure The representation of an atom that shows each valence electron as a dot above, below, or on the sides of the symbol of the element.

electronegativity value A number that indicates the relative ability of an element to attract electrons.

formula The group of symbols that represent the atoms or ions of a unit of a compound.

ion An atom having an electrical charge because of a loss or gain of electrons.

ionic bond The attraction between oppositely charged ions that results from the transfer of electrons.

ionic charge The difference between the number of protons (positive) and the number of electrons (negative), written in the upper right corner of the symbol for the element.

ionic compound A compound of positive and negative ions held together by ionic bonds.

molecule The smallest unit of two or more atoms held by covalent bonds that is representative of a covalent compound.

octet rule Elements in Groups 1A–7A react with other elements by forming ionic or covalent bonds that produce a noble gas arrangement, usually eight electrons in the outer shell.

polar bond A covalent bond in which the electrons are shared unequally, resulting in a partially negative end (δ^-) and a partially positive end (δ^-).

polarity A measure of the unequal sharing of electrons, indicated by the difference in electronegativity values.

polyatomic ion A group of covalently bonded nonmetal atoms that has an overall electrical charge.

triple bond A sharing of three pairs of electrons by two atoms.

valence electrons The electrons present in the outermost shell of an atom, which are largely responsible for the chemical behavior of the element.

Problems

Valence Electrons *(Goal 4.1)*

4.1 Write the electron configuration and the number of valence electrons in an atom of each of the following elements:

a. N e. S
b. O f. Na
c. Ar g. Al
d. K h. Cl

4.2 Write the group number and the electron-dot structure for each element:

a. sulfur f. carbon
b. nitrogen g. oxygen
c. calcium h. fluorine
d. sodium i. lithium
e. potassium j. chlorine

4.3 State the group number for each of the following electron-dot structures:

a. $\dot{\mathrm{X}}\cdot$ b. $\cdot\underset{\cdot}{\ddot{\mathrm{X}}}\cdot$ c. $:\underset{\cdot\cdot}{\ddot{\mathrm{X}}}\cdot$ d. $\mathrm{X}\cdot$ e. $\cdot\dot{\mathrm{X}}\cdot$

The Octet Rule *(Goal 4.2)*

4.4 Write the electron arrangement for each of the following elements. Indicate whether the element would lose electrons, gain electrons, or be stable as written.
a. neon d. argon
b. oxygen e. phosphorus
c. lithium

4.5 State the number of electrons lost or gained when the following elements form ions:
a. Mg f. O
b. P g. Group 2A
c. Group 7A h. F
d. Na i. Li
e. Al j. N

Ions *(Goal 4.3)*

4.6 Write the atoms or ions that have the following numbers of protons and electrons:
a. 9 protons, 10 electrons
b. 12 protons, 10 electrons
c. 15 protons, 18 electrons
d. 10 protons, 10 electrons
e. 19 protons, 18 electrons
f. 20 protons, 18 electrons

4.7 Write the electron arrangement for each of the following ions:
a. Na^+ b. S^{2-} c. Ca^{2+} d. Cl^- e. N^{3-}

4.8 Write ions for the following:
a. chlorine f. fluorine
b. magnesium g. calcium
c. potassium h. sulfur
d. oxygen i. sodium
e. aluminum j. lithium

4.9 Give the name of each of the following ions:
a. Na^+ e. O^{2-}
b. N^{3-} f. Al^{3+}
c. F^- g. Ca^{2+}
d. Cl^- h. S^{2-}

4.10 Write the names of the following ions (include the Roman numeral when necessary):
a. Fe^{2+} e. Ag^+
b. Cu^{2+} f. Cu^+
c. Fe^{3+} g. Zn^{2+}
d. Pb^{4+} h. Sn^{2+}

Ionic Compounds *(Goal 4.4)*

4.11 Use electron-dot structures to diagram the formation of the following ionic compounds:

a. KCl b. $MgCl_2$ c. Na_3N d. MgS e. $AlCl_3$

4.12 Write the correct ionic formula for compounds formed between the following ions:

a. Na^+ and O^{2-}
b. Al^{3+} and Cl^-
c. Ba^{2+} and Cl^-
d. Mg^{2+} and O^{2-}
e. Zn^{2+} and Cl^-
f. Al^{3+} and S^{2-}
g. Li^+ and S^{2-}
h. K^+ and I^-
i. Fe^{3+} and Cl^-
j. Cu^{2+} and S^{2-}
k. Cu^+ and O^{2-}

4.13 Write the correct formula for ionic compounds formed by the following metals and nonmetals:

a. sodium and sulfur
b. aluminum and oxygen
c. calcium and chlorine
d. barium and bromine
e. lithium and chlorine
f. potassium and nitrogen

Naming Ionic Compounds *(Goal 4.5)*

4.14 Write names for the following ionic compounds:

a. Al_2O_3
b. $CaCl_2$
c. Na_2O
d. Mg_3N_2
e Na_2S
f. K_3P
g. MgO
h. LiBr
i. $FeCl_2$
j. $SnCl_4$

4.15 Write the formula (including charge) for each of the following:

a. potassium ion
b. copper(II) ion
c. iron(III) iron
d. ferrous ion
e. silver ion
f. zinc ion

4.16 Complete the following table by writing the formula and name of each compound formed from the given ions:

	Cl^-	O^{2-}	N^{3-}
K^+			
Cu^{2+}			
Mg^{2+}			
Fe^{3+}			

4.17 Indicate the valence of the metal ion in each of the following ionic compounds:
a. $SnCl_2$ d. Ag_2O
b. FeO e. Fe_2O_3
c. Cu_2S

4.18 Write names for the following ionic compounds:
a. $FeCl_2$ e. Ag_3P
b. CuO f. Na_2S
c. Fe_2S_3 g. ZnF_2
d. $CuCl$ h. AlP

4.19 Write formulas for the following ionic compounds:
a. magnesium chloride f. iron(III) oxide
b. sodium sulfide g. barium fluoride
c. copper(I) oxide h. aluminum chloride
d. zinc phosphide i. silver sulfide
e. barium nitride j. copper(II) chloride

Covalent Bonds *(Goal 4.6)*

4.20 Write the electron-dot structure for each of the following covalent molecules:
a. Br_2 e. NCl_3
b. H_2S f. CCl_4
c. HF g. H_2O
d. OF_2 h. SiF_4

4.21 Write the formulas of all the nonmetals that form diatomic molecules in their elemental state.

4.22 The following covalent compounds have double or triple bonds. Write the electron-dot structure for each. (The order of atoms is indicated in parentheses.)
a. SO_2 (O S O) c. N_2
b. CH_2O $\left(\begin{matrix} & O & \\ H & C & H \end{matrix}\right)$ d. N_2Cl_2 (Cl N N Cl)

Bond Polarity *(Goal 4.7)*

4.23 Place the symbols for partially positive (δ^+) and partially negative ($\delta-$) above the appropriate atoms in the following covalent bonds:
a. H—F b. C—Cl c. N—O d. O—H e. O—S

4.24 Identify the bonding between the following pairs of elements as covalent, polar, or ionic. If no bond forms, write *none*.
a. Cl and Cl e. S and Cl i. H and H
b. Ne and O f. Na and Cl j. N and H
c. F and F g. O and F
d. Mg and F h. K and S

Naming Covalent Compounds *(Goal 4.8)*

4.25 Name the following covalent compounds:

a. H_2
b. CBr_4
c. SCl_2
d. HF
e. NI_3
f. CS_2
g. P_2O_5
h. Cl_2O

4.26 Write the formula of each of the following covalent compounds:

a. carbon tetrachloride
b. carbon monoxide
c. phosphorus trichloride
d. dinitrogen tetroxide
e. oxygen difluoride
f. hydrogen monochloride
g. dinitrogen monoxide
h. chlorine

Polyatomic Ions *(Goal 4.9)*

4.27 Name the following polyatomic ions:

a. SO_4^{2-}
b. CO_3^{2-}
c. PO_4^{3-}
d. NO_3^-
e. OH^-
f. SO_3^{2-}

4.28 Write the formulas for the following polyatomic ions:

a. bicarbonate
b. ammonium
c. phosphate
d. hydrogen sulfate
e. nitrite
f. sulfite
g. hydroxide
h. phosphite

4.29 Circle the polyatomic ion in each of the following formulas and write the correct name of the compound:

a. Na_2CO_3
b. NH_4Cl
c. Li_3PO_4
d. $Cu(NO_2)_2$
e. $FeSO_3$
f. KOH
g. $NaNO_3$
h. $CuCO_3$
i. $NaHCO_3$
j. $BaSO_4$

4.30 Complete the following table by writing the correct formula and the name of each compound formed by the given ions:

	OH^-	NO_2^-	CO_3^{2-}	HSO_4^-	PO_4^{3-}
Li^+					
Cu^{2+}					
Ba^{2+}		$Ba(NO_2)_2$ barium nitrite			
NH_4^+					
Al^{3+}					
Pb^{4+}					

4.31 Write the correct formula for each of the following compounds:

a. barium hydroxide
b. sodium sulfate
c. iron(II) nitrate
d. zinc phosphate
e. silicon tetrachloride
f. aluminum chloride
g. ammonium oxide
h. magnesium bicarbonate
i. sodium nitrite
j. copper(I) sulfate

4.32 Name the following compounds:

a. Al_2O_3
b. K_2CO_3
c. OF_2
d. Na_3PO_4
e. $(NH_4)_2SO_4$
f. N_2
g. NBr_3
h. $Fe(NO_3)_3$
i. MgO
j. $CuCl_2$

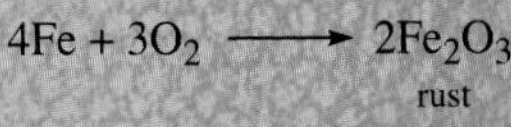

5

Chemical Quantities and Reactions

When exposed to air and water, iron becomes coated with rust.

Learning Goals

5.1 Given the chemical formula of a compound, determine the formula weight.

5.2 Given the chemical formula of a substance, calculate the molar mass.

5.3 Given the number of moles of a substance, calculate the mass in grams; given the mass, calculate the number of moles.

5.4 Classify a change in a substance as a chemical change or a physical change.

5.5 Given a chemical equation, state its meaning in words; given a word description of a reaction, write a chemical equation for it.

5.6 Write a balanced equation for a chemical reaction.

5.7 Use a balanced equation to write conversion factors for the molar relationships between reactants and products.

5.8 Given a balanced equation and the quantity of one of the compounds in the equation, calculate the amount in moles or grams of another compound in the equation.

SCOPE

You are now ready to learn something about how a chemist measures certain quantities of atoms, molecules, and ions. In Chapter 4, we learned that a formula indicates the elements that make up a compound. Now we will see how a compound undergoes a chemical reaction to form another type of compound. By using chemical equations, the scientist calculates how much reactant to use and how much product it will make. In this way, researchers develop new drugs, cosmetics, fabrics, and dyes.

Understanding chemical reactions and equations is important in our everyday lives. The carburetor or fuel-injection system of an engine is adjusted to allow the correct amounts of fuel and oxygen from the air to mix so the engine will run properly. In medicine, the correct dosages of medications must be given to ensure the proper reactions within the body. In chemistry, the amount of each reactant is also carefully measured to obtain the correct products in the desired amounts.

5.1 Formula Weight

Learning Goal **Given the chemical formula of a compound, determine the formula weight.**

In Chapter 4, we learned that atoms combine to form covalent and ionic compounds. The number and kinds of atoms in these compounds are indicated in the formula for the compound.

Sample Problem 5.1
Counting the Atoms in a Formula

State the number and kinds of atoms in each of the following formulas:

a. aspirin, $C_9H_8O_4$
b. caffeine, $C_8H_{10}N_4O_2$
c. $Ca_3(PO_4)_2$

Solution:

a. $C_9H_8O_4$: 9 carbon atoms, 8 hydrogen atoms, and 4 oxygen atoms
b. $C_8H_{10}N_4O_2$: 8 carbon atoms, 10 hydrogen atoms, 4 nitrogen atoms, and 2 oxygen atoms
c. $Ca_3(PO_4)_2$: 3 calcium atoms, 2 phosphorus atoms, and 8 oxygen atoms

Study Check:
A molecule of the antibiotic streptomycin contains 21 carbon atoms, 39 hydrogen atoms, 7 nitrogen atoms, and 12 oxygen atoms. What is the formula for streptomycin?

Answer: $C_{21}H_{39}N_7O_{12}$

To determine the **formula weight** of a compound, we use the periodic table to find the atomic weights of all the elements in the compound.

For example, the formula for carbon dioxide is CO_2. Its formula weight is the sum of the weights of one carbon atom and two oxygen atoms. Since atomic weights are expressed in atomic mass units (amu), the formula weight is also expressed in amu. Using the atomic weights of one carbon atom and two oxygen atoms, we calculate a formula weight of 44.0 amu for CO_2, as follows:

Number of Atoms in the Formula	*Atomic Weight*	*Total Weight for Each Element*
1 carbon atom	× 12.0 amu	= 12.0 amu
2 oxygen atoms	× 16.0 amu	= 32.0 amu
Formula weight of CO_2		= 44.0 amu

Sample Problem 5.2
Calculating Formula Weight

Calculate the formula weight of $CaCO_3$, a compound used as an antacid and as a means of replenishing calcium in the diet.

Solution:
In the formula $CaCO_3$, the subscript 3 indicates that there are three atoms of oxygen. The symbols for calcium and carbon do not have subscripts, so there is only one atom of calcium and one atom of carbon. The atomic weight of each element is obtained from the periodic table, and the formula weight is calculated as follows:

Number of atoms in the formula		*Atomic weight*		*Total weight for each element*
1 $\cancel{\text{atom}}$ Ca	×	$\frac{40.1 \text{ amu}}{1 \cancel{\text{atom}} \text{ Ca}}$	=	40.1 amu
1 $\cancel{\text{atom}}$ C	×	$\frac{12.0 \text{ amu}}{1 \cancel{\text{atom}} \text{ C}}$	=	12.0 amu
3 $\cancel{\text{atoms}}$ O	×	$\frac{16.0 \text{ amu}}{1 \cancel{\text{atom}} \text{ O}}$	=	48.0 amu
Formula weight of $CaCO_3$			=	100.1 amu

Study Check:
Calculate the formula weight of aspartame, $C_{14}H_{18}N_2O_5$, an artificial sweetener.

Answer:

$$\overset{\text{C}}{(14 \times 12.0)} + \overset{\text{H}}{(18 \times 1.0)} + \overset{\text{N}}{(2 \times 14.0)} + \overset{\text{O}}{(5 \times 16.0)} = 294.0 \text{ amu}$$

Sample Problem 5.3
Calculating Formula Weight With Polyatomic Ions

The compound aluminum nitrate, $Al(NO_3)_3$, is an active ingredient in antiperspirants. What is its formula weight?

Solution:
When a formula contains a polyatomic ion, we must consider the subscript outside of the parentheses. In the formula $Al(NO_3)_3$, we have one atom of aluminum, three atoms of nitrogen, and nine atoms of oxygen.

Number of atoms in the formula		*Atomic weight*		*Total weight for each element*
1 ~~atom~~ Al	×	$\frac{27.0 \text{ amu}}{1 \text{ atom Al}}$	=	27.0 amu
3 ~~atoms~~ N	×	$\frac{14.0 \text{ amu}}{1 \text{ atom N}}$	=	42.0 amu
9 ~~atoms~~ O	×	$\frac{16.0 \text{ amu}}{1 \text{ atom O}}$	=	144.0 amu
Formula weight of $Al(NO_3)_3$			=	213.0 amu

Study Check:
Calcium bisulfite, $Ca(HSO_3)_2$, is used in papermaking and as a disinfectant to wash brewery casks. Calculate the formula weight (to the tenths place) of calcium bisulfite.

Answer:

$$\underset{Ca}{(1 \times 40.1)} + \underset{H}{(2 \times 1.0)} + \underset{S}{(2 \times 32.1)} + \underset{O}{(6 \times 16.0)} = 202.3 \text{ amu}$$

5.2 The Mole

Learning Goal **Given the chemical formula of a substance, calculate the molar mass.**

At the store, you buy eggs by the dozen, not by the egg. In an office, pencils are ordered by the gross and paper by the ream. A restaurant orders soda by the case. For each of these terms, *dozen, gross, ream,* and *case,* we know the number of items it represents. For example, when you buy a dozen doughnuts, you know you will get 12 doughnuts in the box. (See Table 5.1 and Figure 5.1.) In chemistry, where we count large numbers of very small particles, it is convenient to use a unit called a **mole,** which contains 6.02×10^{23} items such as atoms, molecules, or ions. This number is called **Avogadro's number,** after Amedeo Avogadro, an eighteenth-century Italian physicist. Avogadro's number is a very large number. It looks like this when written out:

Avogadro's Number

$$602{,}000{,}000{,}000{,}000{,}000{,}000{,}000 = 6.02 \times 10^{23}$$

Table 5.1 Number of Items in Some Typical Collections

Collection	Number of items
1 trio of singers	3 singers
1 dozen eggs	12 eggs
1 case of soda	24 cans of soda
1 gross of pencils	144 pencils
1 ream of paper	500 sheets of paper
1 mole of atoms	6.02×10^{23} atoms
1 mole of molecules	6.02×10^{23} molecules

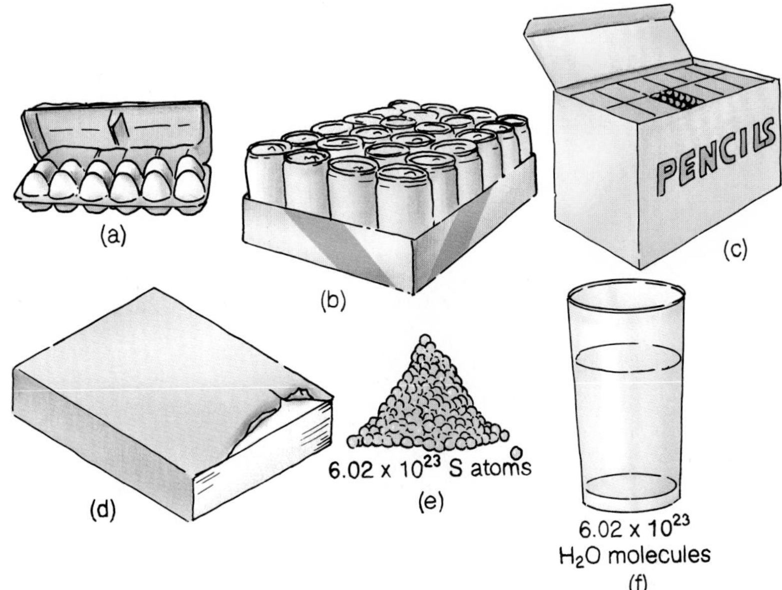

Figure 5.1
Some common collections and the number of items in them. (a) A dozen eggs is 12 eggs. (b) A case of soda contains 24 cans of soda. (c) A gross of pencils is 144 pencils. (d) A ream of paper contains 500 sheets. (e) A mole of sulfur contains 6.02×10^{23} sulfur atoms. (f) A mole of water contains 6.02×10^{23} water molecules.

A mole of any element contains the same number of atoms as a mole of any other element. For example, 1 mole of carbon, 1 mole of aluminum, and 1 mole of sulfur each contain 6.02×10^{23} atoms.

One mole of a compound also contains Avogadro's number of particles. One mole of a covalent compound contains 6.02×10^{23} molecules, and 1 mole of an ionic compound consists of 6.02×10^{23} formula units. A formula unit is that group of positive and negative ions in an ionic compound represented by its formula. Table 5.2 gives some examples of 1-mole quantities.

Molar Mass

It is highly unlikely that you will ever weigh a single atom or molecule. They are much too small to weigh out, even on the most accurate balance. Instead, we work with samples containing many atoms or molecules whose mass we can weigh in grams on a laboratory balance. We have seen that 1 mole of any substance contains the same number of particles. However, the mass of a mole of an element or a compound depends upon the mass of its particles. The **molar mass** of a substance is the mass in grams that is equal numerically to the sub-

Table 5.2 Comparison of One-Mole Samples of Various Substances

Substance	Number of particles	Type of particle
Elements		
1 mole carbon	6.02×10^{23}	Carbon atom
1 mole aluminum	6.02×10^{23}	Aluminum atom
1 mole sulfur	6.02×10^{23}	Sulfur atom
Compounds		
1 mole NaCl	6.02×10^{23}	NaCl formula unit
1 mole water (H_2O)	6.02×10^{23}	H_2O molecule
1 mole sucrose ($C_{12}H_{22}O_{11}$)	6.02×10^{23}	Sucrose molecule
1 mole vitamin C ($C_6H_8O_6$)	6.02×10^{23}	Vitamin C (ascorbic acid) molecule

stance's atomic or formula weight. For example, we can look at the periodic table and find the atomic weight of carbon, 12.0 amu. We say that the mass of 1 mole of carbon is 12.0 g. This molar mass of carbon also contains 6.02×10^{23} particles.

1 mole C = 12.0 g C
1 mole C = 6.02×10^{23} C atoms

Earlier, we saw that the compound carbon dioxide (CO_2) has a formula weight of 44.0 amu. Now we can say that 1 mole of carbon dioxide has a mass of 44.0 g and contains 6.02×10^{23} carbon dioxide molecules.

1 mole CO_2 = 44.0 g CO_2
1 mole CO_2 = 6.02×10^{23} CO_2 molecules

See Figure 5.3 for more examples. Using the periodic table, we can calculate the molar mass from the formula weight of any substance. (See Table 5.3.)

Figure 5.2
Clockwise, starting at upper left: one mole samples of ferric oxide, salt (sodium chloride), sulfur, and carbon. Note: Fe_2O_3 = 159.7 g; NaCl = 58.5 g; S = 32 g; C = 12 g.

Table 5.3 One-Mole Samples: Their Molar Mass and Number of Particles

Substance	Formula weight[a] (amu)	Molar mass (g)	Number of particles
Elements			
Carbon (C)	12.0	12.0	6.02×10^{23} C atoms
Sodium (Na)	23.0	23.0	6.02×10^{23} Na atoms
Iron (Fe)	55.8	55.8	6.02×10^{23} Fe atoms
Compounds			
NaF (preventative for dental caries)	42.0	42.0	6.02×10^{23} NaF formula units
$CaCO_3$ (antacid)	100.1	100.1	6.02×10^{23} $CaCO_3$ formula units
$C_6H_{12}O_6$ (glucose)	180.0	180.0	6.02×10^{23} glucose molecules
$C_8H_{10}N_4O_2$ (caffeine)	194.0	194.0	6.02×10^{23} caffeine molecules

[a]Atomic weights are taken to the tenths (0.1) place.

Sample Problem 5.4
Expressing Formula Weight as Molar Mass

Calculate the molar mass of each of the following compounds:

a. NaBr, a sedative **b.** $FeSO_4$, an iron supplement
c. $Al(OH)_3$, an antacid

Solution:
From the periodic table, we can obtain the atomic weight of each element in the formula and add them all to give a formula weight. Expressing the formula weight in grams gives the molar mass.

a. Formula weight of NaBr = 23.0 amu + 79.9 amu
= 102.9 amu
Molar mass of NaBr = 102.9 g

b. Formula weight of $FeSO_4$ = 55.8 amu + 32.1 amu + (4 × 16.0 amu)
= 151.9 amu
Molar mass of $FeSO_4$ = 151.9 g

c. Formula weight of $Al(OH)_3$ = 27.0 amu + (3 × 16.0 amu) + (3 × 1.0 amu)
= 78.0 amu
Molar mass of $Al(OH)_3$ = 78.0 g

Study Check:
Calculate the molar mass of carmine, $C_{22}H_{20}O_{13}$, a red pigment used in paints, inks, and stains in microbiology.

Answer:

$$\underset{C}{(22 \times 12.0\ \text{amu})} + \underset{H}{(20 \times 1.0\ \text{amu})} + \underset{O}{(13 \times 16.0\ \text{amu})} = 492.0\ \text{amu}$$

Molar mass = 492.0 g

5.3 Calculations Using Molar Mass

Learning Goal **Given the number of moles of a substance, calculate the mass in grams; given the mass, calculate the number of moles.**

The molar mass of elements and compounds is one of the most useful numerical values in chemistry. It is needed in laboratory work and provides conversion factors to change from moles to grams of a substance, or from grams to moles. To do these calculations, we need to write the molar mass of a substance as a conversion factor. Since the molar mass relates the unit gram and the unit mole, we can write

$$\text{Molar mass} = \frac{\text{Mass in grams}}{1\ \text{mole}} = \frac{\text{g}}{\text{mole}}$$

For example, 1 mole of magnesium has a mass of 24.3 g.

1 mole Mg = 24.3 g Mg

For this equality, two conversion factors for the molar mass of magnesium can be written.

$$\frac{24.3 \text{ g Mg}}{1 \text{ mole Mg}} \quad \text{and} \quad \frac{1 \text{ mole Mg}}{24.3 \text{ g Mg}}$$

One mole of the compound H_2O has a mass of 18.0 g: $(2 \times 1.0) + (1 \times 16.0)$.

1 mole H_2O = 18.0 g H_2O

The conversion factors derived from the molar mass of H_2O are written as

$$\frac{18.0 \text{ g } H_2O}{1 \text{ mole } H_2O} \quad \text{and} \quad \frac{1 \text{ mole } H_2O}{18.0 \text{ g } H_2O}$$

These conversion factors derived from molar mass are used in problems that call for calculations of the grams or moles of a substance. Remember that we must first calculate the molar mass of the substance in order to write its conversion factors. Then we can apply the system learned in Chapter 1 to set up problems with the appropriate units.

When the number of grams of an element or a compound is given, the number of moles can be calculated. Be sure to obtain the molar mass first.

Sample Problem 5.5
Calculating the Mass of an Element From Moles

The metal silver is used in the manufacture of tableware, mirrors, jewelry, and dental alloys. If the design for a piece of jewelry requires 1.50 moles of silver, how many grams of silver would be needed?

Solution:
Using the periodic table, we see that the atomic weight of silver (Ag) is 107.9 amu. Expressing the atomic weight of silver as its molar mass, we can write:

1 mole Ag = 107.9 g Ag

The conversion factors from this molar mass are

$$\frac{107.9 \text{ g Ag}}{1 \text{ mole Ag}} \quad \text{and} \quad \frac{1 \text{ mole Ag}}{107.9 \text{ g Ag}}$$

The problem can be set up by writing the given amount of 1.50 moles of silver followed by the conversion factor of molar mass that cancels the mole unit. For this problem, we will use the first of the two conversion factors.

$$1.50 \, \cancel{\text{mole Ag}} \times \frac{107.9 \text{ g Ag}}{1 \, \cancel{\text{mole Ag}}} = 162 \text{ g Ag}$$

Given — Molar mass conversion factor — Answer (rounded to three significant figures)

Study Check:
Calculate the number of grams present in 0.0550 mole of gold (Au).

Answer: 10.8 g Au

Sample Problem 5.6
Calculating the Mass of a Compound From Moles

Camphor has a formula of $C_{10}H_{16}O$. Calculate the number of grams in 3.50 moles of camphor.

Solution:
Since camphor is a compound, we must first calculate its formula weight, which is 152.0 amu.

$$\begin{aligned} 10 \text{ atoms C} \times 12.0 \text{ amu} &= 120.0 \text{ amu} \\ 16 \text{ atoms H} \times 1.0 \text{ amu} &= 16.0 \text{ amu} \\ 1 \text{ atom O} \times 16.0 \text{ amu} &= \underline{16.0 \text{ amu}} \\ & \quad 152.0 \text{ amu} \end{aligned}$$

Expressing the formula weight as the molar mass of camphor gives

$$1 \text{ mole } C_{10}H_{16}O = 152.0 \text{ g } C_{10}H_{16}O$$

The molar mass of camphor can now be used to write two possible molar conversion factors:

$$\frac{152.0 \text{ g camphor}}{1 \text{ mole camphor}} \quad \text{and} \quad \frac{1 \text{ mole camphor}}{152.0 \text{ g camphor}}$$

The problem setup uses the first conversion factor:

$$3.50 \, \cancel{\text{moles camphor}} \times \frac{152.0 \text{ g camphor}}{1 \, \cancel{\text{mole camphor}}} = 532 \text{ g camphor}$$

Given — Molar mass conversion factor — Answer

Study Check:
Calculate the number of grams present in 0.65 mole of ammonium carbonate, $(NH_4)_2CO_3$, a substance found in baking powder.

Answer: 62 g

Sample Problem 5.7
Calculating Moles From a Given Mass of an Element

The total bone mass of a typical 70.0-kg person contains 115 g of phosphorus. How many moles of phosphorus are in the bone mass?

Solution:
Using the periodic table, we find phosphorus has an atomic weight of 31.0 amu. Therefore, 1 mole of phosphorus has a mass of 31.0 g.

$$1 \text{ mole P} = 31.0 \text{ g P}$$

The conversion factors derived from this molar mass would be:

$$\frac{31.0 \text{ g P}}{1 \text{ mole P}} \quad \text{and} \quad \frac{1 \text{ mole P}}{31.0 \text{ g P}}$$

The problem setup begins with the given quantity of 115 g phosphorus. A conversion factor from molar mass cancels the unit of gram.

$$\underset{\text{Given}}{115 \cancel{\text{g P}}} \times \underset{\text{Molar mass conversion factor}}{\frac{1 \text{ mole P}}{31.0 \cancel{\text{g P}}}} = \underset{\text{Answer}}{3.71 \text{ moles P}}$$

Study Check:
Calculate the number of moles in 255 g of iron (Fe).

Answer: 4.57 moles Fe

Sample Problem 5.8
Calculating Moles From a Given Mass of a Compound

Calcium iodate, $Ca(IO_3)_2$, is added to table salt to make iodized salt. Iodized salt provides sufficient iodine in a person's diet to prevent goiter, a condition that occurs in the thyroid when iodine levels are too low. If 655 g of calcium iodate are used in the manufacture of iodized salt, how many moles are used?

Solution:
The formula weight of $Ca(IO_3)_2$ is 389.9 amu.

$$\begin{aligned} 1 \text{ Ca} \times 40.1 \text{ amu} &= 40.1 \text{ amu} \\ 2 \text{ I} \times 126.9 \text{ amu} &= 253.8 \text{ amu} \\ 6 \text{ O} \times 16.0 \text{ amu} &= \underline{96.0 \text{ amu}} \\ & 389.9 \text{ amu} \end{aligned}$$

In grams, the molar mass of $Ca(IO_3)_2$ is 389.9 g.

$$1 \text{ mole } Ca(IO_3)_2 = 389.9 \text{ g } Ca(IO_3)_2$$

Written in the form of conversion factors, this molar mass gives

$$\frac{389.9 \text{ g } Ca(IO_3)_2}{1 \text{ mole } Ca(IO_3)_2} \quad \text{and} \quad \frac{1 \text{ mole } Ca(IO_3)_2}{389.9 \text{ g } Ca(IO_3)_2}$$

The problem setup begins with the stated amount of 655 g $Ca(IO_3)_2$ and the conversion factor that cancels grams.

$$655 \cancel{\text{g } Ca(IO_3)_2} \times \frac{1 \text{ mole } Ca(IO_3)_2}{389.9 \cancel{\text{g } Ca(IO_3)_2}} = 1.68 \text{ moles } Ca(IO_3)_2$$

Given — Molar mass conversion factor — Answer (rounded)

Study Check:
If, in one day, a person takes eight tablets of an antacid containing 500 mg $CaCO_3$ per tablet, how many moles of $CaCO_3$ have been ingested?

Answer:

$$8 \cancel{\text{tablets}} \times \frac{500 \cancel{\text{mg } CaCO_3}}{1 \cancel{\text{tablet}}} \times \frac{1 \cancel{\text{g}}}{1000 \cancel{\text{mg}}} \times \frac{1 \text{ mole } CaCO_3}{100.1 \cancel{\text{g } CaCO_3}} = 0.04 \text{ mole } CaCO_3$$

5.4 Chemical Change

Learning Goal **Classify a change in a substance as a chemical change or a physical change.**

There are many chemical changes that take place around us and within us. You have probably noticed that iron forms a red-orange compound called rust when it is exposed to the sulfur and water in air. Perhaps you have cleaned a dish or a piece of jewelry made of silver because the metal had changed from a shiny, silver color to a dull, gray-black. You may have dropped an antacid tablet in a glass of water and noticed fizzing and bubbling as a gas was released.

As a result of a chemical change, new substances with new characteristics are formed. Iron (Fe) reacts with oxygen and silver (Ag) reacts with sulfur to produce new substances, rust (Fe_2O_3) and tarnish (Ag_2S), that have different properties than the reactants. The bubbling of the antacid is the result of a gas, carbon dioxide, that forms when the tablet dissolves in water. In each example, a **chemical change** occurs because atoms of the initial substances are reorganized into new substances with different compositions; a **chemical reaction** has taken place. Therefore, the rusting of iron, the tarnishing of silver, and the fizzing of an antacid are all chemical reactions as evidenced by chemical changes.

Sometimes a change in appearance is a physical change. You have probably seen ice cubes melt in a glass, and water in a teakettle escape as steam when the liquid water boils. The ice cubes in the glass and the water in the teakettle have gone through a physical change. No new substances are formed. The molecules of water in the ice cubes, the liquid, and the vapor are all identical (H_2O). When physical properties change without the formation of new substances, a **physical change** has taken place. Let us look at some everyday examples of physical and chemical changes in Table 5.4 and Figure 5.3.

Table 5.4 Some Chemical and Physical Changes

Chemical changes	Physical changes
Rusting nail	Melting ice
Bleaching a stain	Boiling water
Burning a log	Tearing paper
Tarnishing silver	Sawing a board in half
Fermenting grapes	Breaking a glass

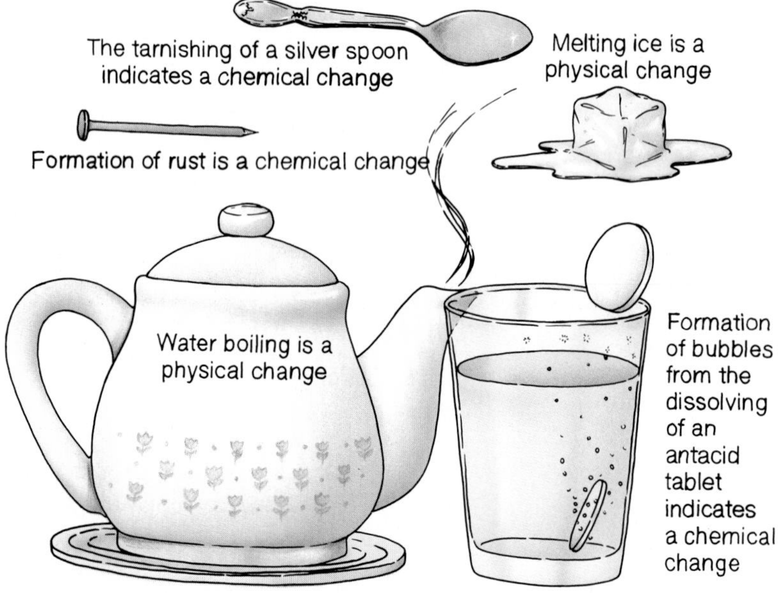
The tarnishing of a silver spoon indicates a chemical change
Melting ice is a physical change
Formation of rust is a chemical change
Water boiling is a physical change
Formation of bubbles from the dissolving of an antacid tablet indicates a chemical change

Figure 5.3
Examples of chemical change include iron rusting, silver tarnishing, and the fizzing (bubbling) of an antacid tablet in water. Examples of physical change include water (ice) melting and water (liquid) boiling.

Sample Problem 5.9
Classifying Chemical and Physical Changes

Identify each of the following as a physical or a chemical change:

a. freezing water
b. burning a match
c. breaking a chocolate bar
d. digesting a chocolate bar

Solution:

a. Physical. Freezing water involves only a change from liquid water to ice. No change has occurred in the composition of the water itself.

b. Chemical. Burning a match causes the formation of new substances that were not present prior to striking the match.

c. Physical. Breaking a chocolate bar does not affect the composition of the substances in it.

d. Chemical. The digestion of the chocolate bar converts its components into new substances for utilization in the body.

Study Check:

Identify each of the following as a physical or a chemical change:

a. chopping a carrot

b. developing a Polaroid picture

c. inflating a balloon

Answer:

a. physical **b.** chemical **c.** physical

5.5 Chemical Equations

Learning Goal **Given a chemical equation, state its meaning in words; given a word description of a reaction, write a chemical equation for it.**

Whenever you put together a bicycle, build a model airplane, prepare a new recipe, mix a medication, or clean a patient's teeth, you must follow a set of directions. These directions tell you what materials to use, how much you need, and the amount of product you will obtain. The chemist also has a set of directions called a **chemical equation** for every chemical reaction.

As an analogy, we might imagine that we are going to make some pancakes. In the recipe, we see that we need some milk, some eggs, and pancake mix. These ingredients are called **reactants** by the chemist, and the pancakes we make will be the **products.** We can write an equation with the reactants on the left and an arrow that points toward the products on the right. The pancakes are the new substances that result from the reaction:

Writing an Equation

$$\text{Reactants} \xrightarrow{\text{React to form}} \text{Products}$$

$$\text{Milk} + \text{Eggs} + \text{Pancake mix} \longrightarrow \text{Pancakes}$$

However, our recipe and the equation do not give us enough information; they are not yet complete. We need to know the amounts of each reactant to use before we can make the pancakes. This quantity is indicated by a number placed

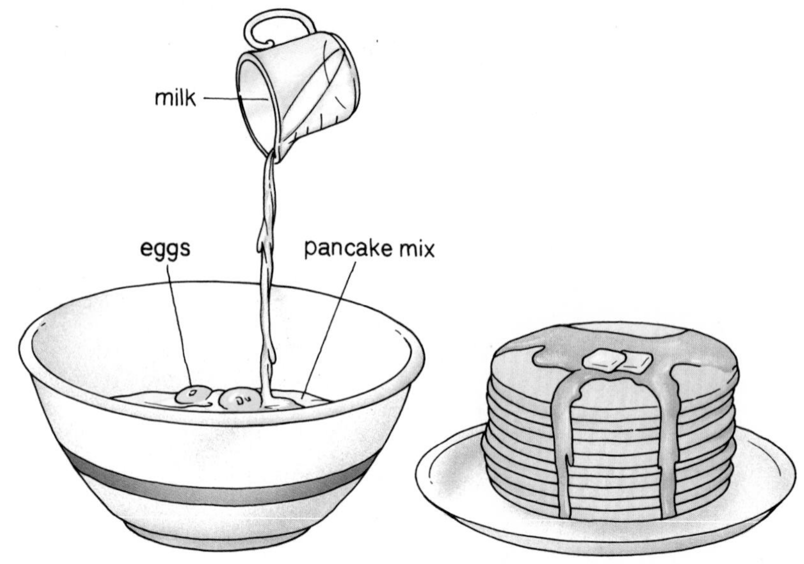

Figure 5.4
In preparing pancakes, the proper amounts of milk, eggs, and pancake mix are mixed together. After the ingredients undergo a change, the pancakes are ready to eat.

in front of each reactant. In an equation, this number is called a **coefficient.** Our equation for making pancakes will be complete after we write the following:

1 cup milk + 2 eggs + 2 cups pancake mix ⟶ 16 pancakes

A chemical equation is like a recipe because it describes the reactants and products of the reaction. Figure 5.4 shows the reactants and the products of our cooking experiment.

In a chemical equation, chemical formulas are used to represent the reactants and the products. For example, the gas methane is used as a fuel for stoves and heating systems. Its reaction with oxygen is described by the following equation, written in symbols and in words.

$$\underset{\text{Methane}}{CH_4} + 2O_2 \longrightarrow CO_2 + 2H_2O$$ Methane reacts with oxygen to produce carbon dioxide and water

A diagram of the reaction is shown in Figure 5.5. In the equation, the numbers that appear in front of the formulas for oxygen (O_2) and water (H_2O) are the

Figure 5.5
In a gas burner, there is a chemical reaction in which one molecule of methane (CH_4) reacts with two molecules of oxygen (O_2) to form one molecule of carbon dioxide (CO_2) and two molecules of water (H_2O).

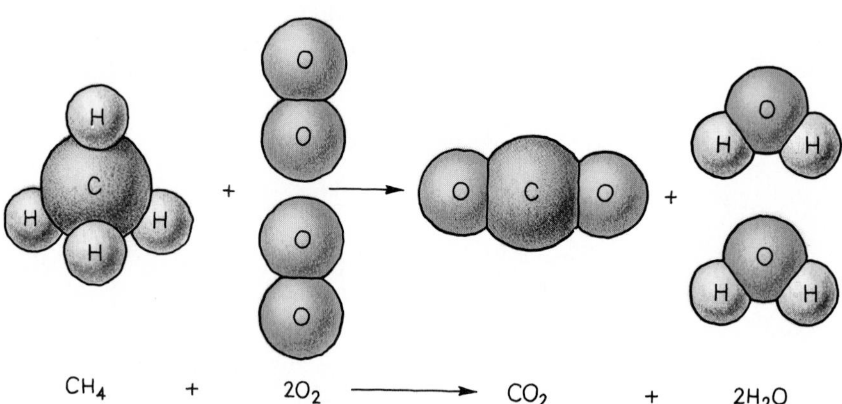

coefficients. They indicate that two molecules of oxygen are needed to react with each molecule of methane to produce one molecule of carbon dioxide and two molecules of water.

Sample Problem 5.10
Describing a Chemical Equation

State the meaning in words of the following chemical equations:

a. $C_3H_8 + 5O_2 \longrightarrow 3CO_2 + 4H_2O$
Propane

b. $4NH_3 + 3O_2 \longrightarrow 2N_2 + 6H_2O$
Ammonia

Solution:

a. One molecule of propane reacts with five molecules of oxygen to form three molecules of carbon dioxide and four molecules of water.

b. Four molecules of ammonia and three molecules of oxygen react to form two molecules of nitrogen and six molecules of water.

Study Check:
Three molecules of oxygen react with one molecule of carbon disulfide to form one molecule of carbon dioxide and two molecules of sulfur dioxide. Write an equation for the reaction.

Answer:

$$3O_2 + CS_2 \longrightarrow CO_2 + 2SO_2$$

Some Symbols Used in Equations

In a chemical equation, several symbols may appear that tell you something about the physical state of the reactants and products. The following list gives some of these symbols.

Symbol	*Meaning*
(*s*)	Solid
(*l*)	Liquid
(*g*)	Gas
(*aq*)	Dissolved in water

Here are examples of equations that include several of the foregoing symbols for physical state:

$$2P(s) + 3H_2(g) \longrightarrow 2PH_3(g)$$
$$2Na(s) + 2H_2O(l) \longrightarrow 2NaOH(aq) + H_2(g)$$
$$Mg(s) + 2HCl(aq) \longrightarrow MgCl_2(aq) + H_2(g)$$

(a)

(b)

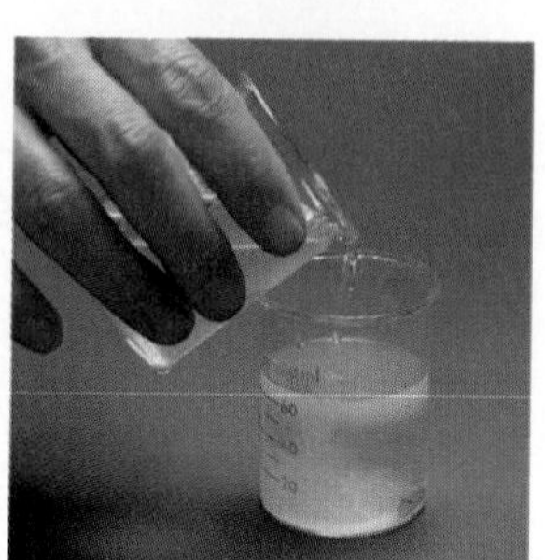

Figure 5.6
Some chemical reactions. The balanced equations for these reactions are
(a) $Zn + CuSO_4 \rightarrow Cu + ZnSO_4$;
(b) $Na_2SO_4 + BaCl_2 \rightarrow BaSO_4 + 2NaCl$;
(c) $2Mg + O_2 \rightarrow 2MgO$; and
(d) $Zn + 2HCl \rightarrow ZnCl_2 + H_2$.

5.6 Balancing Chemical Equations

Learning Goal **Write a balanced equation for a chemical reaction.**

Using the **law of conservation of matter,** we can say that atoms are neither created nor destroyed during a chemical reaction; the atoms of the reactants are just rearranged to form the products. No new atoms enter the reaction, nor do any of the original atoms disappear. Therefore, the chemical equation that represents the reaction must be balanced by showing the same number of atoms for each kind of element on both sides of the equation.

Consider the reaction in which hydrogen reacts with oxygen to form water. In this reaction, two molecules of hydrogen (H_2) react with one molecule of oxygen (O_2) to form two molecules of water. The four hydrogen atoms in the reactants are balanced by four hydrogen atoms in the products; two oxygen atoms in the reactants are balanced by two oxygen atoms in the products.

```
                        H   H
H—H   O                  \ /
    + |   ⟶            H  O  H
H—H   O                  \ /
                          O
```

$$2H_2 + O_2 \longrightarrow 2H_2O$$

We say that the equation for this reaction is balanced. Examples of several other reactions and their balanced equations are shown in Figure 5.6.

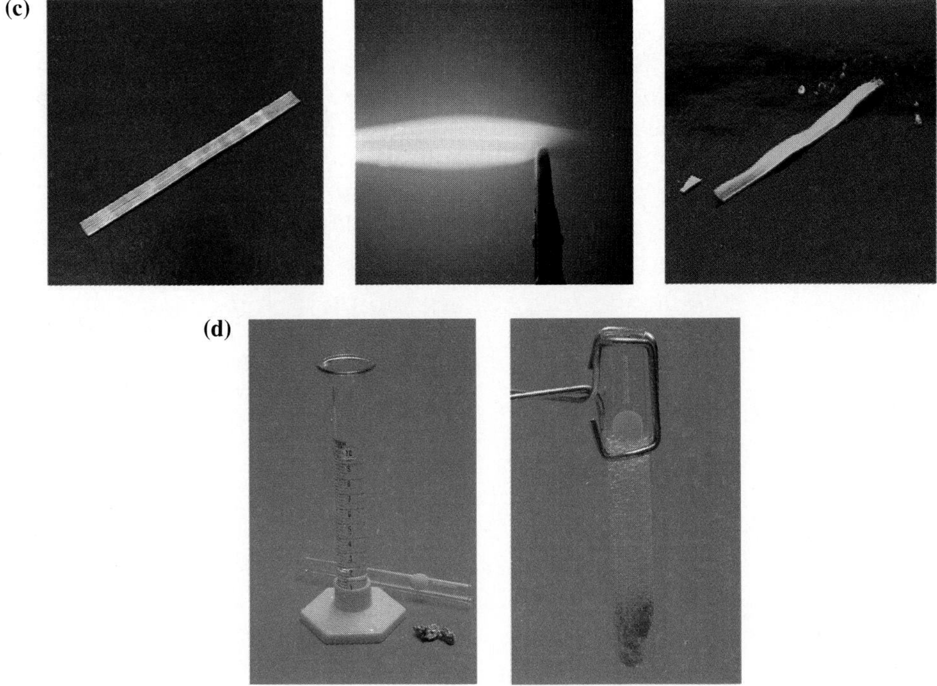

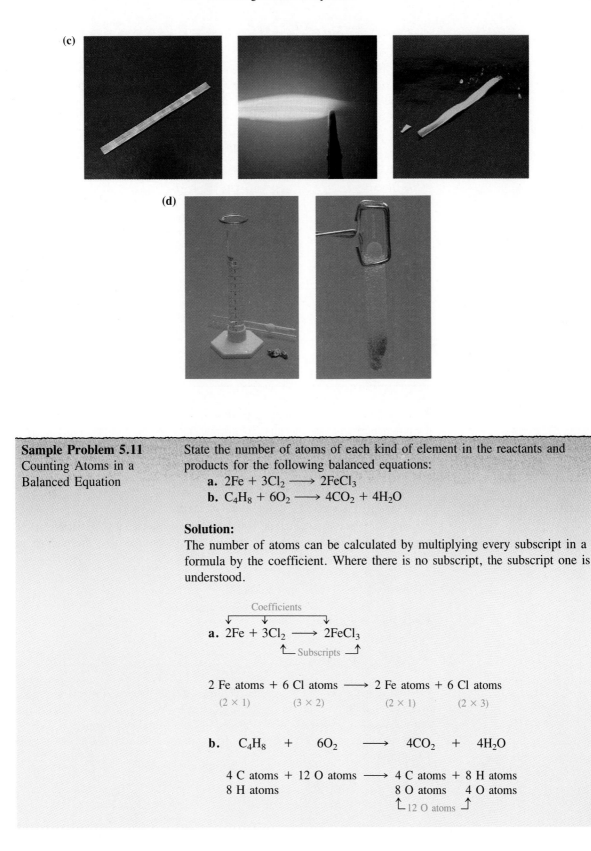

Sample Problem 5.11
Counting Atoms in a Balanced Equation

State the number of atoms of each kind of element in the reactants and products for the following balanced equations:

a. $2Fe + 3Cl_2 \longrightarrow 2FeCl_3$

b. $C_4H_8 + 6O_2 \longrightarrow 4CO_2 + 4H_2O$

Solution:

The number of atoms can be calculated by multiplying every subscript in a formula by the coefficient. Where there is no subscript, the subscript one is understood.

Coefficients

a. $2Fe + 3Cl_2 \longrightarrow 2FeCl_3$

Subscripts

2 Fe atoms + 6 Cl atoms ⟶ 2 Fe atoms + 6 Cl atoms
(2 × 1) (3 × 2) (2 × 1) (2 × 3)

b. $C_4H_8 + 6O_2 \longrightarrow 4CO_2 + 4H_2O$

4 C atoms + 12 O atoms ⟶ 4 C atoms + 8 H atoms
8 H atoms 8 O atoms 4 O atoms
12 O atoms

Study Check:
One of the components of acid rain is nitric acid, produced when the nitrogen dioxide found in smog reacts with the water and oxygen in the air. State the number and kinds of atoms on both sides of the equation for the reaction of nitrogen dioxide in air.

$$4NO_2 + 2H_2O + O_2 \longrightarrow \underset{\text{Nitric acid}}{4HNO_3}$$

Answer:
There are 4 N atoms, 4 H atoms, and 12 O atoms in the reactants and also in the products.

Balancing a Chemical Equation

We have seen that in a chemical equation the same number of atoms must be in the products and in the reactants. To balance an equation, numbers (coefficients) are placed in front of each formula to equalize the number of each kind of atom on both sides of the equation. Consider the following equation:

$$H_2 + Cl_2 \longrightarrow HCl$$

We can determine if an equation is balanced by setting up a score sheet for the reaction, listing the number of atoms on the reactant side and the number of atoms on the product side.

$$H_2 + Cl_2 \longrightarrow HCl \quad \text{(not balanced)}$$

Reactants	***Products***
2 H	1 H
2 Cl	1 Cl

We find that the hydrogen and the chlorine are not balanced; there are more hydrogen and chlorine atoms on the reactant side. We might start balancing the chlorine first. A coefficient of 2 is placed in front of the HCl formula. (*Note:* A coefficient is never placed in between the symbols of the formula nor written as a subscript.)

$$H_2 + Cl_2 \longrightarrow 2HCl \quad \text{(Balanced)}$$

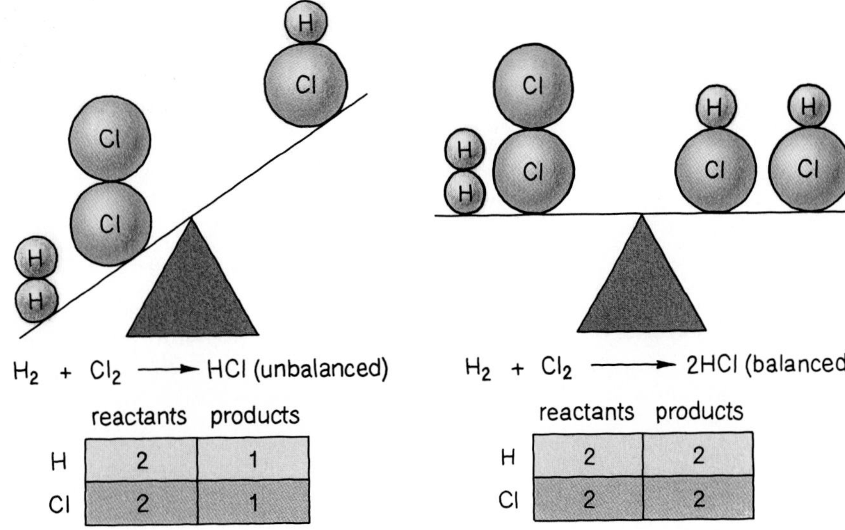

Figure 5.7
Unbalanced and balanced equations.

Checking the hydrogen, we find that it is now balanced; the coefficient 2 placed in front of the HCl formula also balances the hydrogen.

Reactants		***Products***
2 H	=	2 H
2 Cl	=	2 Cl

Figure 5.7 illustrates the unbalanced and the balanced equation.

Some hints for balancing equations follow. Remember, they are only hints. Balancing equations in this way is largely a matter of trial and error.

Balancing Chemical Equations

Step 1 Count the number of atoms for each element or ion on the reactant side and then on the product side.

Step 2 Determine which atoms need to be balanced.

Step 3 Pick one element at a time to balance. The most likely starting place is a metal, or the elements in a formula that has subscripts. Hydrogen, oxygen, and polyatomic ions are usually balanced last.

Step 4 Start balancing one of the elements by placing a coefficient in front of the formula containing that elements. (*Note:* No changes can be made in any of the subscripts of the formulas when you balance an equation.)

Step 5 Check to see if the equation is completely balanced. Sometimes balancing one element will undo the balance of another. Then you must return to the other element and rebalance. If there are any fractions used as coefficients, multiply all the coefficients by 2 if the fraction is ½, and by 3 if it is ⅓. The final ratio of coefficients should be small whole numbers that are not divisible by a whole number other than 1.

Sample Problem 5.12
Balancing Equations Using Coefficients

Use coefficients to balance the following equations:

a. $C + SO_2 \longrightarrow CS_2 + CO$

b. $Ca + H_3PO_4 \longrightarrow Ca_3(PO_4)_2 + H_2$

Solution:

a. We start by checking the number of atoms of each element on the reactant and product sides.

$$C + SO_2 \longrightarrow CS_2 + CO$$

Reactants	***Products***
1 C	2 C
1 S	2 S
2 O	1 O

Generally, it is convenient to begin balancing the atoms in formulas that have subscripts. The sulfur can be balanced by placing the coefficient 2 in front of SO_2.

$$C + 2SO_2 \longrightarrow CS_2 + CO$$

This gives four oxygen atoms on the reactant side, which we can balance by placing the coefficient 4 in front of CO on the product side:

$$C + 2SO_2 \longrightarrow CS_2 + 4CO$$

Now we can balance the carbon atoms, making sure that we count all the carbon atoms on the product side. One carbon atom in CS_2 and four carbon atoms in 4CO give a total of five carbon atoms on the product side. Carbon is balanced by placing the coefficient 5 in front of the symbol for carbon on the reactant side:

$$5C + 2SO_2 \longrightarrow CS_2 + 4CO$$

Our final check for a correctly balanced equation shows that the atoms in the reactants and in the products are equal.

Reactants		***Products***
5 C	=	5 C
2 S	=	2 S
4 O	=	4 O

b. $Ca + H_3PO_4 \longrightarrow Ca_3(PO_4)_2 + H_2$
Counting the number of atoms on the reactant side and on the product side, we find that the equation is unbalanced. When reactions involve the same polyatomic group on both sides of the equation, the group can be counted as a unit.

Reactants	***Products***
1 Ca	3 Ca
3 H	2 H
1 PO_4	2 PO_4

We might begin balancing with one of the elements in the formula that has several subscripts, $Ca_3(PO_4)_2$. Balancing calcium first requires the coefficient 3 in front of the symbol for calcium.

$$3Ca + H_3PO_4 \longrightarrow Ca_3(PO_4)_2 + H_2$$

To balance the phosphate (PO_4) group, we use the coefficient 2 in front of H_3PO_4.

$$3Ca + 2H_3PO_4 \longrightarrow Ca_3(PO_4)_2 + H_2$$

The placement of the coefficient 2 now gives six hydrogen atoms. To balance hydrogen, we place the coefficient 3 in front of H_2.

$$3Ca + 2H_3PO_4 \longrightarrow Ca_3(PO_4)_2 + 3H_2$$

Our final check of the atoms on both sides of the equation shows that the equation is completely balanced.

Reactants		***Products***
3 Ca	=	3 Ca
6 H	=	6 H
2 PO_4	=	2 PO_4

Study Check:
Balance the following equation:

$$C_2H_6 + O_2 \longrightarrow CO_2 + H_2O$$

Answer:
Upon initial balancing, you may have the fraction 3½ in front of O_2.

$$C_2H_6 + 3\frac{1}{2}O_2 \longrightarrow 2CO_2 + 3H_2O$$

To obtain small whole numbers, multiply all of the coefficients in the entire equation by 2.

$$\underset{(1 \times 2)}{2C_2H_6} + \underset{(3\frac{1}{2} \times 2)}{7O_2} \longrightarrow \underset{(2 \times 2)}{4CO_2} + \underset{(3 \times 2)}{6H_2O}$$

Energy in Chemical Reactions

Typically, a chemical reaction involves a change in energy. For example, when the fuel methane, CH_4, is burned in a gas stove or a heating system, it produces heat in addition to its products, carbon dioxide and water. When a reaction such as the combustion of methane produces heat, it is said to be an **exothermic reaction.** Other reactions need energy or heat to occur; they are called **endothermic reactions.** For example, the formation of hydrogen iodide requires energy along with the reactants hydrogen and iodine. We refer to the energy produced or required in a reaction as the **heat of reaction.**

Exothermic Reactions

$$\underset{\text{Methane}}{CH_4} + 2O_2 \longrightarrow CO_2 + 2H_2O + 211 \text{ kcal}$$

$$2Mg + O_2 \longrightarrow 2MgO + 287 \text{ kcal}$$

Endothermic Reactions

$$H_2 + I_2 + 12 \text{ kcal} \longrightarrow 2HI$$
$$PCl_5 + 22 \text{ kcal} \longrightarrow PCl_3 + Cl_2$$

Sample Problem 5.13
Energy in Equations

Two gases from the air, N_2 and O_2, react in the hot engine of a car to give a smog-forming pollutant called nitrogen oxide (NO). The reaction requires 43 kcal of energy. Write a balanced equation for the reaction, including the energy.

Solution:
First, we set up the reaction in terms of the reactants and products.

$$N_2 + O_2 \longrightarrow NO$$

Then we balance the equation by placing the coefficient 2 in front of NO.

$$N_2 + O_2 \longrightarrow 2NO$$

Since energy is required, the reaction is endothermic, which means that the energy belongs with the reactants.

$$N_2 + O_2 + 43 \text{ kcal} \longrightarrow 2NO$$

Study Check:
When 0.0435 mole of ethanol (C_2H_5OH) is burned in a calorimeter, the temperature of 2.00 kg of water rises 7.1 °C. Using the specific heat of water of 1.00 cal/g °C, calculate the heat released (kcal/mole), and place it in the following equation for the combustion of ethanol:

$$C_2H_5OH + 3O_2 \longrightarrow 2CO_2 + 3H_2O$$

Answer:

$$\text{Heat} = \text{Mass} \times \Delta T \times \frac{1.00 \text{ cal}}{\text{g °C}}$$

$$\text{kcal} = 2000 \cancel{\text{g } H_2O} \times 7.1\text{°C} \times \frac{1 \cancel{\text{cal}}}{\cancel{\text{g °C}}} \times \frac{1 \text{ kcal}}{1000 \cancel{\text{cal}}} = 14.2 \text{ kcal}$$

The energy released by 1 mole of ethanol is

$$\frac{14.2 \text{ kcal}}{0.0435 \text{ mole}} = 326 \text{ kcal/mole } C_2H_5OH$$

Since the reaction is exothermic, the energy released by the combustion of 1 mole of ethanol is included with the products.

$$C_2H_5OH + 3O_2 \longrightarrow 2CO_2 + 3H_2O + 326 \text{ kcal}$$

5.7 Mole Relationships in Chemical Equations

Learning Goal **Use a balanced equation to write conversion factors for the molar relationships between reactants and products.**

We have examined reactions and balanced equations in terms of the numbers of atoms and molecules taking part in a chemical reaction. However, when we do experiments in the laboratory, or prepare medications in the hospital, or follow our recipe for pancakes, we work with large samples containing billions of atoms and molecules. The law of conservation of matter tells us that the total amount of

HEALTH NOTE

Smog Production and Health Concerns

There are two types of smog. One, called photochemical smog, requires sunlight to initiate reactions that produce pollutants such as nitrogen oxides and ozone. The other type of smog, called industrial or London smog, occurs in industrial areas where coal containing high amounts of sulfur is burned and sulfur dioxide is given off.

Photochemical smog is most prevalent in cities where people are dependent on the use of cars for transportation. On a typical day in Los Angeles, for example, nitrogen monoxide (NO) emissions from car exhaust increase as traffic increases on the roads. In the air, NO is converted to nitrogen dioxide (NO_2). As the day warms, the interaction of sunlight converts the NO_2 to NO and oxygen atoms, which react with oxygen and form ozone.

Let's take a closer look at some of the reactions that produce these pollutants in the air. The nitrogen oxide found in automobile exhaust is formed when N_2 and O_2 react at high temperatures in car and truck engines.

$$N_2 + O_2 \xrightarrow{\text{Heat}} 2NO$$

The NO reacts with oxygen in the air to produce NO_2, a reddish brown gas that is irritating to the eyes and damaging to the respiratory tract. (See Figure 5.8.)

$$2NO + O_2 \longrightarrow 2NO_2$$

As we have seen, when NO_2 is exposed to sunlight it is converted into NO and oxygen atoms. Oxygen atoms are very reactive because they do not have an octet of electrons. When they react with oxygen molecules present in the atmosphere, ozone is formed.

$$NO_2 \xrightarrow{\text{Sunlight}} NO + \underset{\text{Oxygen atom}}{O}$$

$$O + O_2 \longrightarrow \underset{\text{Ozone}}{O_3}$$

In the upper atmosphere (the stratosphere), ozone is beneficial because it protects us from the harmful ultraviolet radiation that comes from the sun. However, in the lower atmosphere, ozone with its acrid smell irritates the eyes and respiratory tract, where it causes coughing, decreased lung function, and fatigue. It also causes deterioration of fabrics, cracks rubber, and damages trees and vegetable crops.

Industrial smog is prevalent in areas where coal with a high sulfur content is burned to produce electricity. During combustion, the sulfur is converted to sulfur dioxide:

$$S + O_2 \longrightarrow SO_2$$

The SO_2 is damaging to plants, suppressing growth, and it is corrosive to metals such as steel. SO_2 is also damaging to humans and can cause lung impairment and respiratory difficulties. The SO_2 in the air reacts with more oxygen to form SO_3, which combines with water to form sulfuric acid, an extremely corrosive acid:

$$2SO_2 + O_2 \longrightarrow 2SO_3$$
$$SO_3 + H_2O \longrightarrow \underset{\text{Sulfuric acid}}{H_2SO_4}$$

The presence of sulfuric acid in rivers and lakes causes an increase in the acidity of the water, reducing the ability of animals and plants to survive. ■

Figure 5.8
Nitrogen dioxide, a reddish brown gas that irritates the respiratory tract, is one of the pollutants in photochemical smog.

matter before and after reaction must be equal. It also means that the total mass of the reactants must be conserved. Let us see what this means in terms of our pancake recipe analogy.

If we weighed the reactant ingredients, we would find that their total mass was equal to the mass of the product pancakes. (Any water loss is included in the mass of the pancakes.) None of the mass of the ingredients is lost; neither the milk, the eggs, nor the pancake mix disappears. They only change their properties as they react to form pancakes. The mass of each ingredient becomes part of the total mass of the pancakes. Mass has been conserved. (See Figure 5.9.)

$$\text{1 cup milk} + \text{2 eggs} + \text{2 cups pancake mix} \longrightarrow \text{16 pancakes}$$

$$250\text{ g} + 150\text{ g} + 300\text{ g} = 700\text{ g}$$
$$\text{Mass of the reactants} = \text{Mass of the products}$$
$$700\text{ g} = 700\text{ g}$$

The mass of the reactants and products in a chemical equation can be compared in a similar way. Consider the reaction in which hydrogen and chlorine react to give hydrogen chloride:

$$H_2 + Cl_2 \longrightarrow 2HCl$$

Figure 5.9
Mass is conserved in a recipe. The pancakes produced are equal in mass to the ingredients.

Up to now, we have interpreted such an equation only in terms of individual particles. We can say that there are two hydrogen atoms and two chlorine atoms on the reactant side and on the product side. Since there is a relationship between the number of atoms or molecules in a mole, we can interpret an equation in terms of moles. The coefficients in an equation can also be read as the number of moles of each reactant and product in the reaction.

Equation coefficients = Moles of reactants and products

If we calculate the mass of all the reactants and the mass of all the products, we find that they are equal. Such a result is expected according to the law of conservation of matter. The mole interpretation and the mass of the reactants and the products in this equation are calculated as follows:

	H_2	+ Cl_2	$\longrightarrow$	2HCl
Moles:	1 mole H_2	+ 1 mole Cl_2	produces	2 moles HCl

	Reactants		*Product*
	H_2	Cl_2	2HCl
Molar mass:	2.0 g/mole	71.0 g/mole	36.5 g/mole
Moles:	× 1 mole	× 1 mole	× 2 moles
Mass:	2.0 g H_2	71.0 g Cl_2	73.0 g HCl
	73.0 g reactants		73.0 g product

In Figure 5.10, the conservation of matter is illustrated for the reaction between silver and sulfur.

2Ag + S	$\longrightarrow$	Ag_2S
2 moles Ag + 1 mole S	=	1 mole Ag_2S
2 × 107.9 g + 32.1 g	=	247.9 g
247.9 g reactants	=	247.9 g product

Sample Problem 5.14
Mole Relationships in Equations

Demonstrate the law of conservation of matter by calculating the mass of the reactants and the mass of the product in the following equation:

$$2Na + Cl_2 \longrightarrow 2NaCl$$

Solution:
Multiplying each of the formula weights by the number of moles (indicated by the coefficients) gives the mass of the reactants and the product.

Reactants				*Product*
2Na	+	Cl_2	$\longrightarrow$	2NaCl
23.0 g/mole		71.0 g/mole		58.5 g/mole
× 2 moles Na		× 1 mole Cl_2		× 2 moles NaCl
46.0 g	+	71.0 g	=	117.0 g
117.0 g reactants			=	117.0 g product

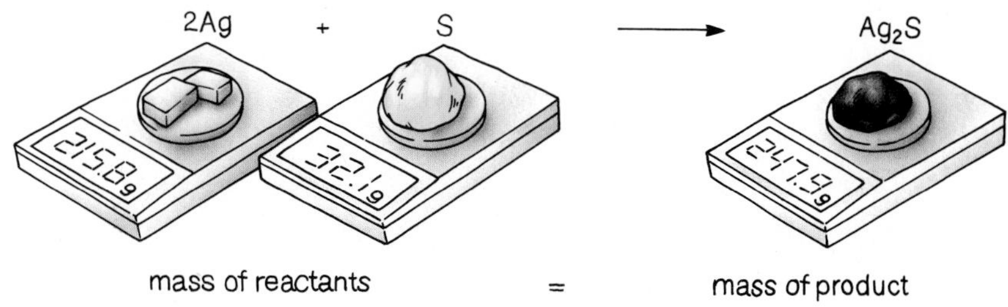

Figure 5.10
Matter is conserved when silver and sulfur form silver sulfide. Reactants and products are shown on laboratory balances to illustrate the law of conservation of matter.

Conversion Factors From Balanced Equations

A chemical equation is also a source of several conversion factors. Since we can interpret the coefficients in terms of mole relationships between the reactants and products, we can write mole factors for any two compounds in the equation. Let's consider the following reaction, in which nitrogen reacts with hydrogen to produce ammonia (NH_3):

$$N_2 + 3H_2 \longrightarrow 2NH_3$$

We can read this equation as "1 mole of N_2 and 3 moles of H_2 react to form 2 moles of NH_3." By considering two substances at a time, we can find the following relationships:

$\mathbf{N_2} + \mathbf{3H_2} \rightarrow 2NH_3$	1 mole N_2 reacts with 3 moles H_2
$\mathbf{N_2} + 3H_2 \rightarrow \mathbf{2NH_3}$	1 mole N_2 produces 2 moles NH_3
$N_2 + \mathbf{3H_2} \rightarrow \mathbf{2NH_3}$	3 moles H_2 produce 2 moles NH_3

From the mole relationships indicated by the coefficients in the equation, we can write six conversion factors:

$$N_2 \text{ and } H_2\text{:} \quad \frac{1 \text{ mole } N_2}{3 \text{ moles } H_2} \quad \text{and} \quad \frac{3 \text{ moles } H_2}{1 \text{ mole } N_2}$$

$$N_2 \text{ and } NH_3\text{:} \quad \frac{1 \text{ mole } N_2}{2 \text{ moles } NH_3} \quad \text{and} \quad \frac{2 \text{ moles } NH_3}{1 \text{ mole } N_2}$$

$$H_2 \text{ and } NH_3\text{:} \quad \frac{3 \text{ moles } H_2}{2 \text{ moles } NH_3} \quad \text{and} \quad \frac{2 \text{ moles } NH_3}{3 \text{ moles } H_2}$$

Sample Problem 5.15
Writing Conversion Factors From Coefficients

State the mole conversion factors for the substances in the following balanced equation:

$$2Al + 3Cl_2 \longrightarrow 2AlCl_3$$

Solution:
There are six conversion factors that can be derived from the coefficients in this equation.

Al and Cl_2: $\frac{2 \text{ moles Al}}{3 \text{ moles } Cl_2}$ and $\frac{3 \text{ moles } Cl_2}{2 \text{ moles Al}}$

Al and $AlCl_3$: $\frac{2 \text{ moles Al}}{2 \text{ moles } AlCl_3}$ and $\frac{2 \text{ moles } AlCl_3}{2 \text{ moles Al}}$

Cl_2 and $AlCl_3$: $\frac{3 \text{ moles } Cl_2}{2 \text{ moles } AlCl_3}$ and $\frac{2 \text{ moles } AlCl_3}{3 \text{ moles } Cl_2}$

Study Check:
Consider the following equation:

$$4Na + O_2 \longrightarrow 2Na_2O$$

Write the conversion factors that relate (a) moles of Na to moles of Na_2O; (b) moles of O_2 to moles of Na.

Answer:

a. $\frac{2 \text{ moles } Na_2O}{4 \text{ moles Na}}$ and $\frac{4 \text{ moles Na}}{2 \text{ moles } Na_2O}$

b. $\frac{4 \text{ moles Na}}{1 \text{ moles } O_2}$ and $\frac{1 \text{ mole } O_2}{4 \text{ moles Na}}$

5.8 Calculations With Equations

Learning Goal **Given a balanced equation and the quantity of one of the compounds in the equation, calculate the amount in moles or grams of another compound in the equation.**

Mole-to-Mole Conversions

Using mole conversion factors, we can now answer some questions about the amounts of reactants and products in a reaction. If you are given the number of moles for one substance, you can use a mole conversion factor to find the number of moles of any of the other substances in the reaction.

$$\text{Moles A} \xrightarrow{\text{Conversion factor}} \text{Moles B}$$

Sample Problem 5.16
Using Mole Relationships to Calculate Moles

Consider the reaction of aluminum and chlorine in the formation of aluminum chloride

$$2Al + 3Cl_2 \longrightarrow 2AlCl_3$$

How many moles of chlorine react with 6 moles of aluminum?

Solution:
In the reaction, we are interested in the amount of chlorine that combines with 6 moles of aluminum.

$$6 \text{ moles Al} \times \text{Conversion factor} = ? \text{ moles } Cl_2$$

The coefficients in the equation indicate that 2 moles of Al react with 3 moles of Cl_2. This mole relationship gives the following conversion factors

$$\frac{2 \text{ moles Al}}{3 \text{ moles } Cl_2} \quad \text{and} \quad \frac{3 \text{ moles } Cl_2}{2 \text{ moles Al}}$$

Now we can set up the problem with the given of 6 moles Al followed by the conversion factor that cancels the unit of moles Al.

$$\underset{\text{Given}}{6 \cancel{\text{ moles Al}}} \times \underset{\substack{\text{Conversion factor} \\ \text{(from coefficients)}}}{\frac{3 \text{ moles } Cl_2}{2 \cancel{\text{ moles Al}}}} = \underset{\text{Answer}}{9 \text{ moles } Cl_2}$$

The answer tells us that 9 moles of Cl_2 are needed to completely react with 6 moles of Al.

Study Check:
In the following reaction, how many moles of iron are produced when 10 moles of Fe_2O_3 react?

$$Fe_2O_3 + 3H_2 \longrightarrow 2Fe + 3H_2O$$

Answer:

$$10 \cancel{\text{ moles } Fe_2O_3} \times \underset{\text{Conversion factor}}{\frac{2 \text{ moles Fe}}{1 \cancel{\text{ mole } Fe_2O_3}}} = 20 \text{ moles Fe}$$

The answer tells us that 20 moles of iron can be produced when 10 moles of Fe_2O_3 react.

Calculating Mass in Chemical Reactions

In the laboratory, you will be measuring quantities of reactants or products in grams, not moles. However, the mole relationships are still the key to solving problems that give the amounts of reactants or products in grams. The conversion of moles to grams, or grams to moles, can be accomplished by including the molar mass of the given compounds in the problem-solving setup.

Sample Problem 5.17
Calculating Grams Using Mole Relationships

Propane is used as a fuel for camp stoves and some specially equipped automobiles. It undergoes combustion in air to form carbon dioxide and water according to the following equation:

$$\underset{\text{Propane}}{C_3H_8} + 5O_2 \longrightarrow 3CO_2 + 4H_2O$$

How many grams of carbon dioxide are produced when 2.00 moles of propane react? (You may assume that there is sufficient oxygen for the reaction.)

Solution:
From the equation, we see that 3 moles of CO_2 are produced for every 1 mole of propane that reacts. This mole relationship gives two conversion factors for carbon dioxide and propane:

$$\frac{3 \text{ moles } CO_2}{1 \text{ mole } C_3H_8} \quad \text{and} \quad \frac{1 \text{ mole } C_3H_8}{3 \text{ moles } CO_2}$$

First we need to calculate the number of moles of CO_2 produced when 2.00 moles of propane react.

$$2.00 \cancel{\text{moles } C_3H_8} \times \frac{3 \text{ moles } CO_2}{1 \cancel{\text{mole } C_3H_8}} = 6.00 \text{ moles } CO_2$$

This answer gives us moles of CO_2. We can change the moles of CO_2 to grams by using the molar mass as a conversion factor. We calculate the molar mass of CO_2 as 44.0 g/mole (12.0 + 2 × 16.0).

$$\underset{\text{Moles } CO_2}{6.00 \cancel{\text{moles } CO_2}} \times \underset{\text{Molar mass}}{\frac{44.0 \text{ g } CO_2}{1 \cancel{\text{mole } CO_2}}} = \underset{\text{Grams } CO_2}{264 \text{ g } CO_2}$$

These calculations may be combined as a series of conversion factors. Our unit plan is as follows:

$$\text{Moles } C_3H_8 \xrightarrow[\text{coefficients}]{\text{Factor from}} \text{Moles } CO_2 \xrightarrow{\text{Molar mass}} \text{Grams } CO_2$$

$$2.00\ \cancel{\text{moles } C_3H_8} \times \frac{3\ \cancel{\text{moles } CO_2}}{1\ \cancel{\text{mole } C_3H_8}} \times \frac{44.0\text{ g } CO_2}{1\ \cancel{\text{mole } CO_2}} = 264\text{ g } CO_2$$

Study Check:
Using the same equation, find the number of moles of water that can be produced from 55 g of C_3H_8.

$$C_3H_8 + 5O_2 \longrightarrow 3CO_2 + 4H_2O$$

Propane

Answer:

$$55\ \cancel{\text{g } C_3H_8} \times \underbrace{\frac{1\ \cancel{\text{mole } C_3H_8}}{44.0\ \cancel{\text{g } C_3H_8}}}_{\text{Molar mass of } C_3H_8} \times \underbrace{\frac{4\text{ moles } H_2O}{1\ \cancel{\text{mole } C_3H_8}}}_{\text{Factor from coefficients}} = 5.0\text{ moles } H_2O$$

A similar problem may give the mass in grams of one compound and ask for the mass in grams of another compound. However, we cannot go directly from grams of one substance to grams of another. First, we must change the grams of the initial substance to moles using its molar mass. Then we can calculate the moles of the desired substance using the mole relationship found in the equation. Finally, the moles of the desired compound can be changed into grams using its molar mass.

This process of converting grams to moles of one substance and moles to grams of a related substance in an equation is the key in calculating the quantities of reaction compounds. Consider that we need to convert from compound A to compound B.

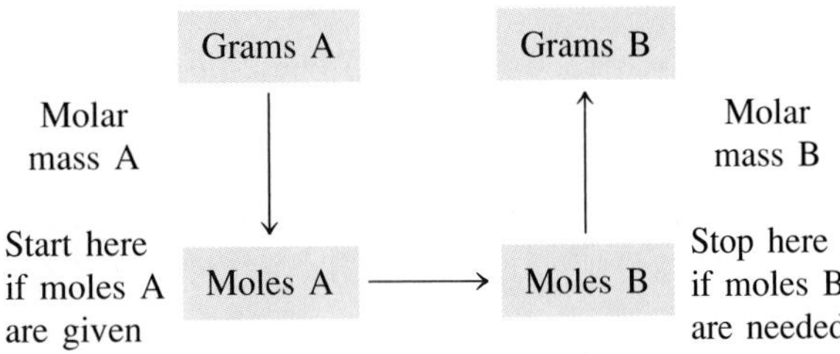

Sample Problem 5.18
Mass Relationships in Balanced Equations

Hydrogen and oxygen react to form water according to the following equation:

$$2H_2 + O_2 \longrightarrow 2H_2O$$

How many grams of H_2O can be produced from 48 g O_2?

Solution

Step 1 Change grams O_2 to moles O_2.
To use the mole relationships in the equation, we need to change 48 g O_2 to moles of O_2 using its molar mass of 32.0 g/mole.

$$\text{Formula weight } O_2 = 2 \text{ atoms O} \times 16.0 \text{ amu/atom O} = 32.0 \text{ amu}$$
$$\text{Molar mass } O_2 = 32.0 \text{ g/mole}$$

$$48\,\cancel{\text{g } O_2} \times \frac{1 \text{ mole } O_2}{32.0\,\cancel{\text{g } O_2}} = 1.5 \text{ moles } O_2$$

Step 2 Change moles O_2 to moles H_2O.
The conversion factors for the mole relationship between H_2O and O_2 are set up using the coefficients in the equation.

$$\frac{2 \text{ moles } H_2O}{1 \text{ mole } O_2} \quad \text{and} \quad \frac{1 \text{ mole } O_2}{2 \text{ moles } H_2O}$$

We can now change from moles O_2 to moles H_2O using the conversion factor that cancels moles O_2.

$$1.5\,\cancel{\text{moles } O_2} \times \frac{2 \text{ moles } H_2O}{1\,\cancel{\text{mole } O_2}} = 3.0 \text{ moles } H_2O$$

Step 3 Change moles H_2O to grams H_2O.
Note that we know the number of moles of H_2O produced, we can calculate grams H_2O using its molar mass of 18.0 g/mole.

$$\text{Formula weight } H_2O = (2 \text{ H} \times 1.0 \text{ amu}) + (1 \text{ O} \times 16.0 \text{ amu}) = 18.0 \text{ amu}$$
$$\text{Molar mass } H_2O = 18.0 \text{ g/mole}$$

$$3.0\,\cancel{\text{moles } H_2O} \times \frac{18.0 \text{ g } H_2O}{1\,\cancel{\text{mole } H_2O}} = 54 \text{ g } H_2O$$

It is also possible to place all of these conversion factors in one sequence leading to the desired unit of grams H_2O. The unit plan would be:

$$\text{Grams } O_2 \longrightarrow \text{Moles } O_2 \longrightarrow \text{Moles } H_2O \longrightarrow \text{Grams } H_2O$$

Note that the unit plan must include the mole relationship of oxygen (stated substance) and water (desired substance). Using the same conversion factors, we solve the problem with the following continuous setup:

$$48\ \text{g O}_2 \times \underset{\text{Molar mass O}_2}{\frac{1\ \text{mole O}_2}{32.0\ \text{g O}_2}} \times \underset{\text{Factor from coefficients}}{\frac{2\ \text{moles H}_2\text{O}}{1\ \text{mole O}_2}} \times \underset{\text{Molar mass H}_2\text{O}}{\frac{18.0\ \text{g H}_2\text{O}}{1\ \text{mole H}_2\text{O}}} = 54\ \text{g H}_2\text{O}$$

Study Check:
During combustion, ethane (C_2H_6) reacts with oxygen (O_2) to produce carbon dioxide (CO_2) and water (H_2O). How many grams of oxygen are needed to react with 150 g of ethane?

Answer:
First, we need to write the equation with the reactants and products.

$$C_2H_6 + O_2 \longrightarrow CO_2 + H_2O$$

Balancing the equation gives

$$2C_2H_6 + 7O_2 \longrightarrow 4CO_2 + 6H_2O$$

The problem is then set up with conversion factors of molar mass for the reactants and their mole relationship from the coefficients.

$$150\ \text{g C}_2\text{H}_6 \times \underset{\text{Molar mass C}_2\text{H}_6}{\frac{1\ \text{mole C}_2\text{H}_6}{30.0\ \text{g C}_2\text{H}_6}} \times \underset{\text{Factor from coefficients}}{\frac{7\ \text{moles O}_2}{2\ \text{moles C}_2\text{H}_6}} \times \underset{\text{Molar mass O}_2}{\frac{32.0\ \text{g O}_2}{1\ \text{mole O}_2}} = 560\ \text{g O}_2$$

The steps in problem solving using balanced equations can be summarized as follows:

Problem Solving with Equations

Step 1 Convert the mass in grams of the given substance to moles using its molar mass.

Step 2 Convert from moles of the given substance to moles of the desired substance using conversion factors derived from the coefficients of the balanced equation.

Step 3 If necessary, change moles of the desired substance to grams using its molar mass.

Sample Problem 5.19
Calculating Grams Using Molar Mass and Coefficient Factors

The air pollutant SO_3 is formed when sulfur released into the air reacts with oxygen.

$$2S + 3O_2 \longrightarrow 2SO_3$$

How many grams of sulfur trioxide are produced when 45.5 g of sulfur react with oxygen in the air?

Solution:
The problem states the number of grams of sulfur that are to be converted to grams of sulfur trioxide. Recall that it is necessary to go through the mole relationship indicated in the equation for the given and desired substances. To do this, we write the following unit plan:

$$\text{Grams S} \longrightarrow \text{Moles S} \longrightarrow \text{Moles } SO_3 \longrightarrow \text{Grams } SO_3$$

To calculate these changes, we need the following conversion factors:

$$\text{Grams S} \xrightarrow[\text{mass S}]{\text{Molar}} \text{Moles S} \xrightarrow[\text{coefficients}]{\text{Factor from}} \text{Moles } SO_3 \xrightarrow[\text{mass } SO_3]{\text{Molar}} \text{Grams } SO_3$$

From the formula weights, we can determine that the molar mass of S is 32.1 g/mole and the molar mass of SO_3 is 80.1 g/mole ($32.1 + 3 \times 16.0$). The conversion factors obtained from the coefficients of S and SO_3 in the equation are

$$\frac{2 \text{ moles } SO_3}{2 \text{ moles S}} \quad \text{and} \quad \frac{2 \text{ moles S}}{2 \text{ moles } SO_3}$$

We can now set up the problem following the unit plan:

$$45.5 \cancel{\text{g S}} \times \underbrace{\frac{1 \cancel{\text{mole S}}}{32.1 \cancel{\text{g S}}}}_{\text{Molar mass S}} \times \underbrace{\frac{2 \cancel{\text{moles } SO_3}}{2 \cancel{\text{moles S}}}}_{\text{Factor from coefficients}} \times \underbrace{\frac{80.1 \text{ g } SO_3}{1 \cancel{\text{mole } SO_3}}}_{\text{Molar mass } SO_3} = 114 \text{ g } SO_3$$

Our answer tells us that 114 g of sulfur trioxide would be produced from 45.5 g of sulfur.

Study Check:
Using the same equation, calculate the mass in grams of sulfur that reacts with 75.5 g of oxygen.

Answer: 50.5 g S

Glossary of Key Terms

Avogadro's number The number of items such as atoms, molecules, or ions in a mole, equal to 6.02×10^{23}.

chemical change The formation of a new substance that has a different composition and different properties than the initial substance.

chemical equation A shorthand way to represent a chemical reaction using chemical formulas to indicate the reactants and products.

chemical reaction The process by which a chemical change takes place.

coefficient A whole number placed in front of a formula in an equation to balance the number of atoms or moles of atoms of each element in the equation.

endothermic reaction A reaction that requires heat.

exothermic reaction A reaction that gives off heat.

formula weight The sum of the atomic weights of all the atoms in a formula.

heat of reaction The heat energy required or produced by a chemical reaction.

law of conservation of matter A law that states that atoms are neither created nor destroyed in a chemical reaction but are only rearranged.

molar mass The mass of 1 mole of an element or compound, equal to the atomic or formula weight expressed in grams.

mole A collection containing 6.02×10^{23} items such as atoms, molecules, or ions.

physical change A change in which the physical properties change but the composition of the substance does not.

product A substance produced in a chemical reaction.

reactant An initial substance that undergoes change in a chemical reaction.

Problems

Formula Weight *(Goal 5.1)*

5.1 State the total number of atoms of each element in each of the following formulas:

a. Al_2O_3 (absorbant and abrasive)
b. $Al(OH)_3$ (gastric antacid)
c. $Al_2(SO_4)_3$ (antiperspirant)
d. $(NH_4)_2CO_3$ (baking powder)
e. $Mg(OH)_2$ (antacid, laxative)
f. $C_{14}H_{29}NO_4S$ (penicillin, an antibiotic)

5.2 Calculate the formula weight of each of the following compounds:

a. KCl (restores electrolytes)
b. Li_2CO_3 (antidepressant)
c. $C_7H_5NO_3S$ (saccharin, a noncaloric sweetener)
d. $(NH_4)_2SO_4$ (fertilizer)
e. C_2H_6O (ethanol)
f. $C_{22}H_{24}N_2O_8$ (tetracycline, an antibiotic)

The Mole *(Goal 5.2)*

5.3 Using the periodic table, state the molar mass of each of the following elements:

a. Na e. Fe
b. Cl f. Mg
c. Pb g. H
d. Cu h. C

5.4 Calculate the molar mass of each of the following compounds:
a. C_3H_6 (cyclopropane, an inhaled anesthetic)
b. $C_{20}H_{18}O_4$ (cyclocumarol, an anticoagulant)
c. NH_4NO_3 (a diuretic)
d. $Al(OH)_3$ (an antacid)
e. $NaHCO_3$ (baking soda)
f. $C_{29}H_{50}O_2$ (vitamin E)

Calculations Using Molar Mass *(Goal 5.3)*

5.5 Calculate the mass in grams of each of the following:
a. 2.0 moles of calcium
b. 0.12 mole of sulfur
c. 10.0 moles of aluminum
d. 4.5 moles of carbon
e. 0.50 mole of tin
f. 0.0085 mole of copper
g. 1.0 mole of potassium
h. 2.5 moles of phosphorus

5.6 Calculate the mass in grams of each of the following:
a. 0.50 mole of NH_3
b. 2.0 moles Na_2O
c. 0.40 mole $Ca(NO_3)_2$
d. 5.0 moles HCl
e. 7.5 moles C_2H_6O
f. 0.10 mole $(NH_4)_3PO_4$

5.7 The compound $MgSO_4$ is called Epsom salts. How many grams will you need to prepare a bath with 5.0 moles of Epsom salts?

5.8 In a bottle of soda, there is 0.25 mole of CO_2. How many grams of CO_2 are in the bottle?

5.9 Cyclopropane, C_3H_6, is an anesthetic given by inhalation. How many grams are in a 0.25-mole sample of cyclopropane?

5.10 The sedative Demerol has the formula $C_{15}H_{22}ClNO_2$. How many grams are present in 0.025 mole of Demerol?

5.11 Calculate the number of moles in each of the following:
a. 50.0 g of silver
b. 40.0 g of copper
c. 10.0 g of carbon
d. 150 g of iron
e. 400 g of sulfur
f. 1.50 g of silicon

5.12 A nickel has a mass of 5.10 g. If it is 100% pure nickel, how many moles of nickel does it contain?

5.13 A gold nugget weighs 35.0 g. How many moles of gold are in the nugget?

***5.14** You have collected 20.0 lb of aluminum cans. How many moles of aluminum do you have?

5.15 Calculate the number of moles in each of the following samples:
a. 1.00 g of H_2
b. 160 g of CH_4
c. 222 g of $CaCl_2$
d. 100 g of SO_2
e. 36.0 g of H_2O
f. 0.200 g of $Cu(NO_3)_2$

5.16 A can of Drāno contains 480 g of NaOH. How many moles of NaOH are in the can of Drāno?

***5.17** The human body contains about 60.0% water, H_2O. If a person weights 70.0 kg, how many moles of water are present in that person's body?

***5.18** An ethanol blood level of 400 mg alcohol per 100 mL of blood can cause coma and may be fatal. At this level, how many moles of ethanol (C_2H_6O) are present in 100 mL of blood?

Chemical Change *(Goal 5.4)*

5.19 Identify each of the following as a chemical or a physical change:

a. chewing a gumdrop
b. ignition of fuel in the space shuttle
c. drying clothes
d. neutralizing stomach acid with an antacid tablet
e. formation of snowflakes
f. using oxygen in body cells
g. formation of green leaves in plants

Chemical Equations *(Goal 5.5)*

5.20 State the meaning of the following equations in terms of atoms, molecules, or formula units:

a. $2NO + O_2 \longrightarrow 2NO_2$
b. $2H_2S + 3O_2 \rightarrow 2SO_2 + 2H_2O$
c. $C_2H_6O + 3O_2 \rightarrow 2CO_2 + 3H_2O$
Ethanol
*d. $K_2SO_4 + 2AgNO_3 \rightarrow Ag_2SO_4 + 2KNO_3$

5.21 Write an equation for each of the following reactions using correct formulas for the elements and compounds:

a. Four molecules of ammonia (NH_3) and three molecules of oxygen react to form two molecules of nitrogen and six molecules of water.
b. Four iron atoms and three oxygen molecules react to form two formula units of iron(III) oxide.
c. Two molecules of propanol (C_3H_8O) and nine molecules of oxygen react to form six molecules of carbon dioxide and eight molecules of H_2O.

5.22 State the number of atoms of each elements in the reactants and in the products for each of the following equations:

a. $2Na + Cl_2 \longrightarrow 2NaCl$
b. $N_2H_4 + 2H_2O_2 \longrightarrow N_2 + 4H_2O$
c. $P_4O_{10} + 6H_2O \longrightarrow 4H_3PO_4$
d. $C_5H_{12} + 8O_2 \longrightarrow 5CO_2 + 6H_2O$

Balancing Chemical Equations *(Goal 5.6)*

5.23 Balance the following equations:
a. $N_2 + O_2 \longrightarrow NO$
b. $HgO \longrightarrow Hg + O_2$
c. $Fe + O_2 \longrightarrow Fe_2O_3$
d. $Na + Cl_2 \longrightarrow NaCl$
e. $Cu_2O + O_2 \longrightarrow CuO$

5.24 Balance the following equations:
a. $Al + Cl_2 \longrightarrow AlCl_3$
b. $P_4 + O_2 \longrightarrow P_4O_{10}$
c. $C_3H_8 + O_2 \longrightarrow CO_2 + H_2O$
d. $Sb_2S_3 + HCl \longrightarrow SbCl_3 + H_2S$
e. $Fe_2O_3 + C \longrightarrow Fe + CO$

5.25 Balance the following equations:
a. $Mg + AgNO_3 \longrightarrow Mg(NO_3)_2 + Ag$
b. $CuCO_3 \longrightarrow CuO + CO_2$
c. $Al + CuSO_4 \longrightarrow Cu + Al_2(SO_4)_3$
d. $Pb(NO_3)_2 + NaCl \longrightarrow PbCl_2 + NaNO_3$
e. $Zn + H_2SO_4 \longrightarrow ZnSO_4 + H_2$
f. $Al_2(SO_4)_3 + KOH \longrightarrow Al(OH)_3 + K_2SO_4$
g. $K_2SO_4 + BaCl_2 \longrightarrow BaSO_4 + KCl$
h. $CaCO_3 \longrightarrow CaO + CO_2$

5.26 Indicate whether the following reactions are exothermic or endothermic:
a. lighting your Bunsen burner in the laboratory
$CH_4(g) + 2O_2(g) \longrightarrow CO_2(g) + 2H_2O(g) + 213\text{ kcal}$
Methane
b. dehydrating limestone
$Ca(OH)_2 + 15.6\text{ kcal} \rightarrow CaO + H_2O$
c. formation of aluminum oxide and iron from aluminum and iron(III) oxide
$2Al + Fe_2O_3 \longrightarrow Al_2O_3 + 2Fe + 204\text{ kcal}$

Mole Relationships in Chemical Equations *(Goal 5.7)*

5.27 Give an interpretation of the following equations in terms of mole relationships:
a. $NaCl + AgNO_3 \longrightarrow AgCl + NaNO_3$
b. $4Al + 3O_2 \longrightarrow 2Al_2O_3$

5.28 Demonstrate the law of conservation of matter by calculating the total mass of the reactants and the total mass of the products in each of the following balanced equations:
a. $N_2 + O_2 \longrightarrow 2NO$
b. $CaCO_3 \longrightarrow CaO + CO_2$
c. $2SO_2 + O_2 \longrightarrow 2SO_3$
d. $4Al + 3O_2 \longrightarrow 2Al_2O_3$

5.29 Write all of the mole factors for the equations listed in Problem 5.27.

Calculations With Equations *(Goal 5.8)*

5.30 Copper metal reacts with sulfur to form copper(I) sulfide, as shown in the following equation:

$2Cu + S \longrightarrow Cu_2S$

a. How many moles of S are needed to react with 2.0 moles of Cu?
b. How many moles of Cu_2S can be produced from 5.0 moles of S?
c. How many grams of Cu_2S can be produced when 2.5 moles Cu react?

5.31 Nitrogen gas reacts with hydrogen gas to produce ammonia, as shown in the following equation:

$N_2(g) + 3H_2(g) \longrightarrow 2NH_3(g)$

a. How many moles of H_2 are needed to react with 1.0 mole of N_2?
b. How many grams of H_2 are needed to react with 2.5 moles of N_2?
c. How many grams of NH_3 will be produced when 12 g of H_2 react?

5.32 Ammonia and oxygen react to form nitrogen and water.

$4NH_3 + 3O_2 \longrightarrow 2N_2 + 6H_2O$

Ammonia

a. How many moles of O_2 are needed to react with 8.0 moles of NH_3?
b. How many grams of N_2 will be produced when 170 g of NH_3 react?
c. How many grams of O_2 must react to produce 90.0 g H_2O?
*d. How many pounds of water are formed when 34 g of ammonia undergo reaction?

***5.33** At a winery, the glucose in grapes undergoes fermentation to produce ethyl alcohol and carbon dioxide, as shown in the following equation:

$C_6H_{12}O_6 \longrightarrow 2C_2H_6O + 2CO_2$

a. How many moles of CO_2 are produced when 500.0 g of glucose undergo fermentation?
b. How many grams of ethanol would be formed from the reaction of 0.240 kg of glucose?

***5.34** Gasohol is a fuel containing ethyl alcohol (C_2H_6O) that burns in oxygen (O_2) to given carbon dioxide (CO_2) and water (H_2O).

a. State the reactants and products for this reaction in the form of an equation.
b. Balance the equation.
c. How many moles of O_2 are needed to react completely with 4.0 moles of ethyl alcohol?
d. How many grams of H_2O will be produced when 24.0 g of C_2H_6O react?
e. If a car produces 88 g of CO_2, how many grams of O_2 are used up in the reaction?

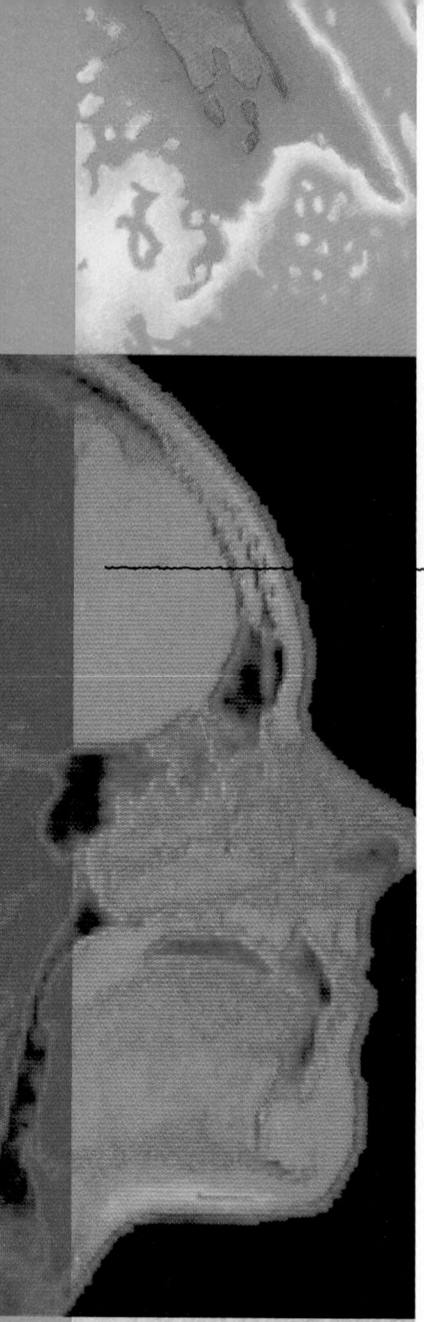

A PET (CT) scan of the brain is created by a scanning technique using positron emitting isotopes (x-rays).

6

Nuclear Radiation

Learning Goals

6.1 Describe three types of ionizing radiation and the protection required for each.

6.2 Write a nuclear equation for the radioactive decay of a nucleus.

6.3 Describe the methods and units used to detect and measure radiation.

6.4 Describe the use of radioisotopes in nuclear medicine.

6.5 Given the half-life of a radioisotope, calculate the amount that is active after one or more half-lives.

6.6 Write a nuclear equation for the production of a radioisotope from a nonradioactive isotope.

6.7 Describe the nuclear processes of fission and fusion.

SCOPE

Radiation is a tool of nuclear medicine that can be detected and measured. Using radioactive isotopes, a radiologist determines the size and shape of an organ, locates a tumor, and measures the metabolic activity of the cells in the body. Radioactive isotopes are used to measure the amounts of substances such as drugs and hormones in blood and urine samples.

The amount of radiation received by a patient is carefully monitored, since radiation can cause damage in the cells of the body. Medically, radiation is used to treat cancer and diseases of the blood or bone. The radiation penetrates the abnormal cells, inhibiting their growth. Nearby normal cells are also affected by the radiation, but they have a greater capacity to recover than do the abnormal cells.

6.1 Radioactivity

Learning Goal **Describe three types of ionizing radiation and the protection required for each.**

Most naturally occurring isotopes of elements up to atomic number 19 have stable nuclei. Elements that have higher atomic numbers (20 to 83) consist of a mixture of isotopes, some of which may have unstable nuclei. These unstable isotopes, called **radioactive** isotopes, or **radioisotopes,** undergo changes in their nuclei to become more stable. These radioactive nuclei emit high-energy radiation such as alpha (α) particles, beta (β) particles, or gamma (γ) rays. Elements of atomic number 84 and higher consist only of radioactive isotopes. So many protons and neutrons are crowded together in their nuclei that the strong repulsions between the protons make those nuclei unstable.

In Chapter 3, we wrote symbols to distinguish among the different isotopes of an element. Recall that an atom's mass number is equal to the sum of the protons and neutrons in the nucleus, and its atomic number is equal to the number of protons. In the symbol for an isotope, the mass number is written in the upper left corner of the symbol, and the atomic number is written in the lower left corner. For example, a radioactive isotope of iodine used in the diagnosis and treatment of thyroid conditions has a mass number of 131 and an atomic number of 53. We can write its symbol as

Mass number (protons and neutrons)
Element
Atomic number (protons)

$$^{131}_{53}I$$

This isotope is also called iodine-131 or I-131. (Note that radioactive isotopes are named by writing the mass number after the element's name or symbol.) When necessary, we can obtain the atomic number from the periodic table. Table 6.1 compares some stable, nonradioactive isotopes with some radioactive isotopes.

Table 6.1 Stable and Radioactive Isotopes of Some Elements

Magnesium	Iodine	Uranium
	Stable Isotopes	
$^{24}_{12}Mg$	$^{127}_{53}I$	None
Magnesium-24	Iodine-127	
	Radioactive Isotopes	
$^{23}_{12}Mg$	$^{125}_{53}I$	$^{235}_{92}U$
Magnesium-23	Iodine-125	Uranium-235
$^{27}_{12}Mg$	$^{131}_{53}I$	$^{238}_{92}U$
Magnesium-27	Iodine-131	Uranium-238

Types of Radiation

We cannot feel, taste, or smell radiation, but high-energy radiation is capable of creating havoc within the cells of our bodies. Different forms of radiation may be emitted from an unstable nucleus when a change takes place among its protons and neutrons. Energy is released, and a new, more stable nucleus is formed. One type of radiation consists of alpha particles. An **alpha particle** contains two protons and two neutrons, which gives it a mass number of 4 and an atomic number of 2. Because it has two protons, an alpha particle has a charge of 2+. That makes it identical to a helium nucleus. It is shown as the Greek letter alpha (α) or as the symbol for helium.

α or $^{4}_{2}He$

Alpha particle

Another type of radiation occurs when a radioisotope emits **beta particles.** A beta particle, which is identical to an electron, has a charge of 1− and a mass number of 0. It is represented by the Greek letter beta ($\boldsymbol{\beta}$) or by the symbol for the electron (e^-). When symbol $^{0}_{-1}e$ is used for a beta particle, the charge of 1− (shown as a subscript −1) is used in place of an atomic number.

β or $^{0}_{-1}e$

Beta particle

Beta particles are produced by unstable nuclei when neutrons are changed into protons. These high-energy electrons do not come from the orbitals of the atoms, and they do not exist in the nucleus until there is a transformation of neutrons within the nucleus.

$$^{1}_{0}n \longrightarrow {}^{1}_{1}H + {}^{0}_{-1}e$$

Neutron in the nucleus	New proton remains in the nucleus	Electron formed and emitted as a beta particle

Gamma rays are high-energy radiation released as an unstable nucleus undergoes a rearrangement to give a more stable, lower-energy nucleus. A gamma ray is shown as the Greek letter gamma (γ). Since gamma rays are energy only, there is no mass or charge associated with their symbol.

γ
Gamma ray

Table 6.2 summarizes the types of radiation we will use in nuclear equations.

Table 6.2 Some Common Forms of Nuclear Radiation

Type of radiation	Symbol	Mass number	Atomic number	Charge
Alpha particle	α, ${}^{4}_{2}\text{He}$	4	2	2+
Beta particle	β, ${}^{0}_{-1}e$	0	0	1−
Gamma ray	γ	0	0	0
Proton	${}^{1}_{1}\text{H}$	1	1	1+
Neutron	${}^{1}_{0}\text{n}$	1	0	0
Positron	β^+, ${}^{0}_{1}e$	0	1	1+

Sample Problem 6.1
Writing Formulas for Radiation Particles

Write the nuclear symbol for an alpha particle.

Solution:
The alpha particle contains two protons and two neutrons. It has a mass number of 4 and an atomic number of 2.

α or ${}^{4}_{2}\text{He}$

Study Check:
What is the symbol used for beta radiation?

Answer: β or ${}^{0}_{-1}e$

Ionizing Radiation

When high-energy radiation strikes molecules in its path, electrons may be knocked away. The result of this **ionizing radiation** is the formation of unstable ions or radicals. A radical is a particle that has an unpaired electron. For example, when radiation passes through the human body, it may interact with water molecules, removing electrons and producing H_2O^+ ions or it may produce radicals:

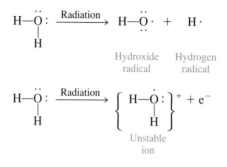

$$H{-}\ddot{\underset{|}{\text{O}}}{:} \quad \underset{H}{} \xrightarrow{\text{Radiation}} H{-}\ddot{\underset{\cdot\cdot}{\text{O}}}\cdot \; + \; H\cdot$$

Hydroxide radical — Hydrogen radical

$$H{-}\ddot{\text{O}}{:} \xrightarrow{\text{Radiation}} \left\{ H{-}\dot{\text{O}}{:} \right\}^{+} + e^{-}$$

Unstable ion

When ionizing radiation strikes the cells of the body, the unstable ions or radicals that form can cause undesirable chemical reactions. The cells in the body most sensitive to radiation are the ones undergoing rapid division—those of the bone marrow, skin, reproductive organs, and intestinal lining, as well as all cells of growing children. Damaged cells may lose their ability to produce necessary materials. For example, if radiation damages cells of the bone marrow, red blood cells may no longer be produced. If sperm cells or ova or the cells of a fetus are damaged, birth defects may result. In contrast, cells of the nerves, muscles, liver, and adult bones are much less sensitive to radiation because they undergo little or no cellular division.

Cancer cells are another example of rapidly dividing cells. Since cancer cells are highly sensitive to radiation, large doses of radiation are used to destroy them. The surrounding normal tissue, dividing at a slower rate, shows a greater resistance to radiation and suffers less damage. In addition, normal tissue is able to repair itself more readily than cancerous tissue. However, this repair is not always complete. Long-range effects of ionizing radiation include a shortened life span, malignant tumors, leukemia, anemia, and genetic mutations.

Radiation Protection

Since many cells in the body are sensitive to radiation, it is important that the radiologist, doctor, and nurse working with radioactive isotopes use proper radiation protection. Proper **shielding** is necessary to prevent exposure. Alpha particles are the heaviest of the radiation particles; they travel only a few centimeters in the air before they collide with air molecules, acquire electrons, and become helium atoms. A piece of paper, clothing, or the skin can be used as protection against alpha particles. Lab coats and gloves will provide sufficient shielding. However, if ingested or inhaled, alpha particles can bring about serious internal damage; because of their size they cause much ionization in a short distance.

Beta particles have a very small mass and move much faster and farther than alpha particles, traveling as much as several meters through air. They can pass through paper and penetrate as far as 4–5 mm into the skin. External exposure to beta particles can burn the surface of the skin, but they are stopped before they can reach the internal organs. Heavy clothing such as lab coats and gloves are needed to protect the skin from beta particles. (See Figure 6.1.)

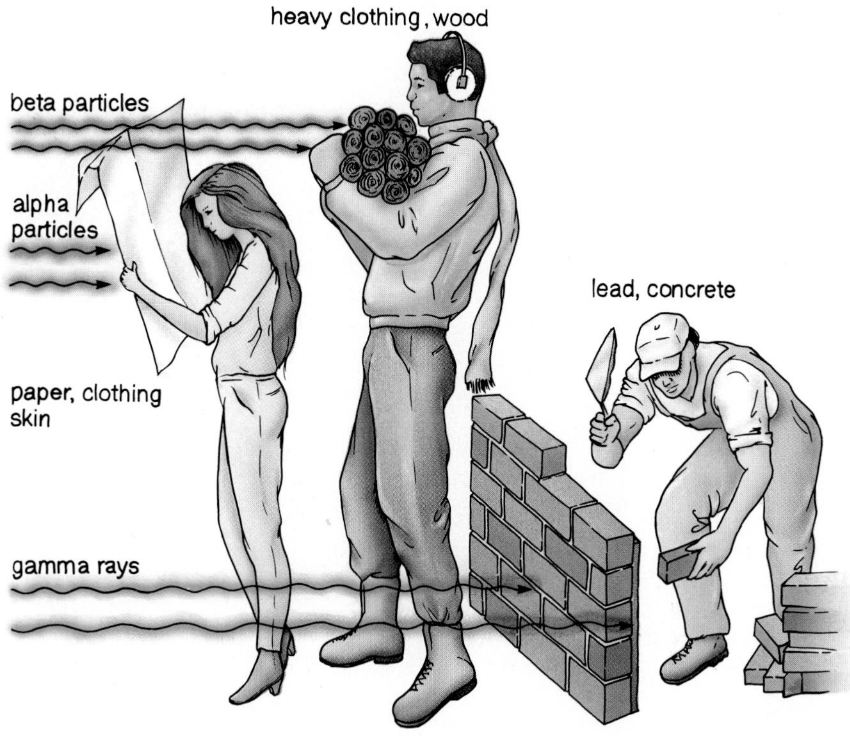

Figure 6.1
Shielding material needed to protect a person from alpha, beta, and gamma radiation.

Gamma rays travel great distances through the air and pass through many materials, including body tissues. Only very dense shielding, such as lead or concrete, will stop them. Since they can penetrate so deeply, exposure to gamma rays can be extremely hazardous. Even the syringe used to give an injection of a gamma-emitting radioisotope is placed inside a special lead-glass cover.

When preparing radioactive materials, the radiologist wears special gloves and works behind leaded windows. Long tongs are used within the work area to pick up vials of radioactive material, keeping them away from the hands and body. (See Figure 6.2.) Table 6.3 summarizes the shielding materials required for the various types of radiation.

Try to keep the time you must spend in a radioactive area to a minimum. A certain amount of radiation is emitted every minute. Remaining in a radioactive area twice as long exposes a person to twice as much radiation.

Keep your distance! The greater the distance from the radioactive source, the lower the intensity of radiation received. If you double your distance from the radiation source, the intensity of radiation drops to $(\frac{1}{2})^2$ or one-fourth of its previous value, as Figure 6.3 shows. This is one of the reasons dentists and x-ray technicians leave the room and stand behind a shield or lead-lined wall to take your x-rays. They are exposed to radiation every day and must minimize the amount of radiation they receive.

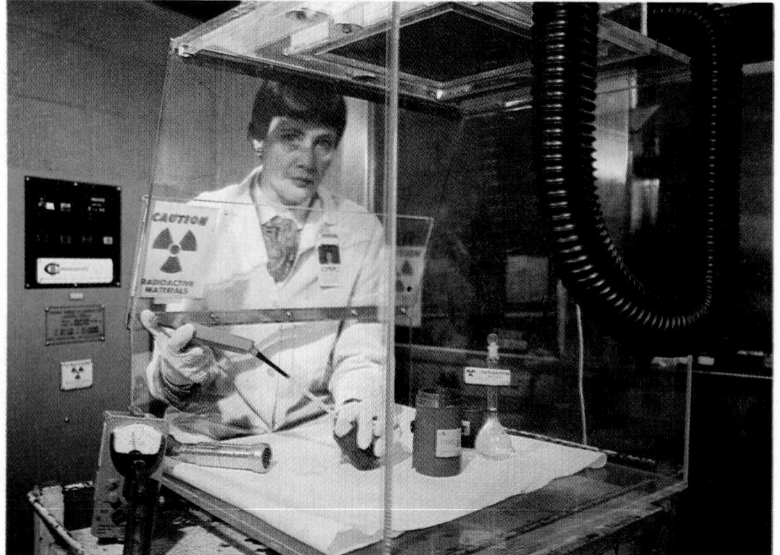

Figure 6.2
A technician in a radiation laboratory wears special protective clothing and stands behind a lead shield when preparing radioisotopes.

Table 6.3 Types of Radiation and Shielding Required

		Distance Particle Travels		
Type	**Symbol**	**Through air**	**Into tissue**	**Shielding**
Alpha	α	2–4 cm	0.05 mm	Paper, clothing
Beta	β	200–300 cm	4–5 mm	Heavy clothing, lab coats, gloves
Gamma	γ	500 m	50 cm	Lead, concrete

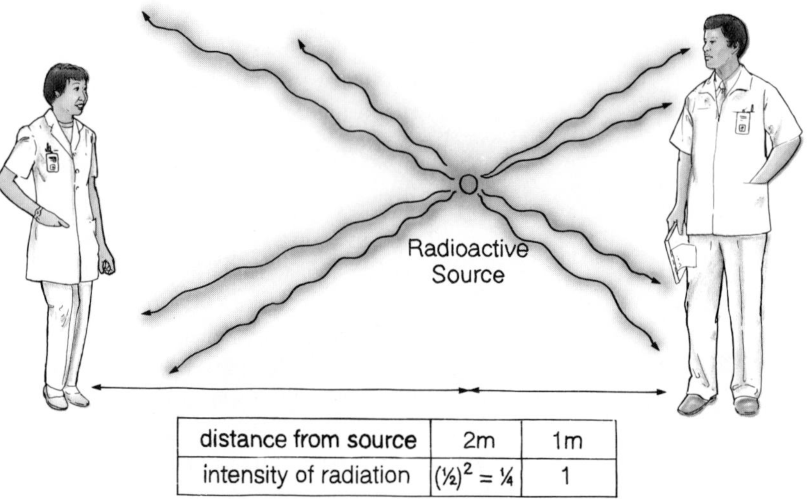

distance from source	2m	1m
intensity of radiation	$(½)^2 = ¼$	1

Figure 6.3
The intensity of radiation decreases as the distance from a radioactive source increases.

Sample Problem 6.2
Radiation Protection

How does the type of shielding for alpha radiation differ from that used for gamma radiation?

Solution:
Alpha radiation is stopped by paper and clothing. However, lead or concrete is needed for protection from gamma radiation.

Study Check:
Besides shielding, what other methods help reduce exposure to radiation?

Answer:
Limiting the time one spends near a radioactive source and staying as far away as possible will reduce exposure to radiation.

6.2 Nuclear Equations

Learning Goal **Write a nuclear equation for the radioactive decay of a nucleus.**

When a nucleus spontaneously breaks down by emitting radiation, the process is called **radioactive decay.** It can be shown as a **nuclear equation** using the symbols for the radioactive nucleus, the new nucleus, and the radiation emitted.

$$\text{Radioactive nucleus} \longrightarrow \text{New nucleus} + \text{Radiation } (\alpha, \beta, \gamma)$$

A nuclear equation is balanced when the sum of the mass numbers and the sum of the atomic numbers of the particles and atoms on one side of the equation are equal to their counterparts on the other side.

Alpha Emitters

Alpha emitters are radioisotopes that decay by emitting alpha particles. For example, uranium-238 decays to thorium-234 by emitting an alpha particle, as shown in Figure 6.4.

Mass numbers
Symbols
Atomic numbers

$$^{238}_{92}\text{U} \longrightarrow {}^{234}_{90}\text{Th} + {}^{4}_{2}\text{He}$$

Unstable nucleus — New nucleus — Alpha particle

The alpha particle emitted contains 2 protons, which gives the new nucleus 2 fewer protons, or 90 protons. That means that the new nucleus has an atomic number of 90 and is therefore thorium (Th). Since the alpha particle has a mass number of 4, the mass number of the thorium isotope is 234, 4 less than that of the original uranium nucleus.

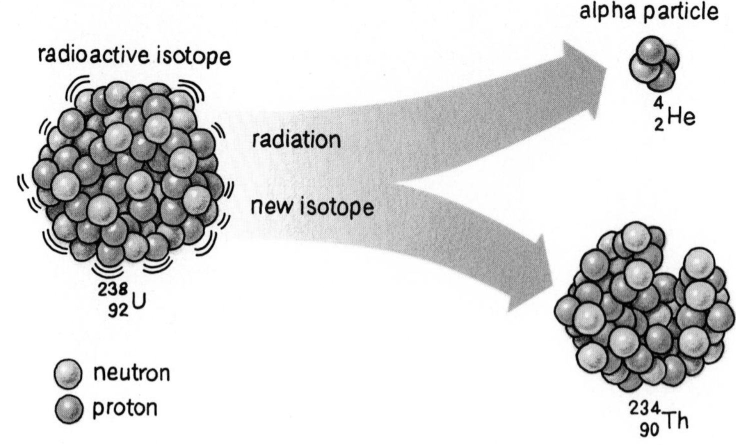

Figure 6.4
An unstable uranium-238 nucleus undergoes radioactive decay by emitting an alpha particle.

Completing a Nuclear Equation

In another example of radioactive decay, radium-226 emits an alpha particle to form a new isotope whose mass number, atomic number, and identity we must determine. We would first write an incomplete nuclear equation:

Mass numbers Symbols Atomic numbers	$^{226}_{88}Ra$	$\longrightarrow$	?	$+\ ^{4}_{2}He$
	Radioisotope		New isotope	Alpha particle

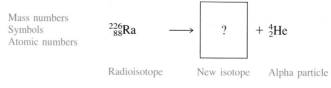

To complete the equation, we must make the mass number of the reactant equal the combined mass number of the products. Therefore, the sum of the mass number of the alpha particle and the mass number of the new isotope must equal 226, the mass number of radium. We can calculate the mass number of the new isotope by subtracting:

$226 - 4 = 222$ (mass number of new isotope)

The atomic number of radium (88) must equal the sum of the atomic numbers of the alpha particle and the new isotope. Therefore, we obtain the atomic number of the new isotope by the following calculation:

$88 - 2 = 86$ (atomic number of new isotope)

In the periodic table, the element that has an atomic number of 86 is radon (Rn). We complete the nuclear equation by writing Rn, the symbol for radon:

$^{226}_{88}Ra$	$\longrightarrow$	$^{222}_{86}Rn$	$+\ ^{4}_{2}He$
Radioisotope		New isotope	Alpha particle

ENVIRONMENTAL NOTE

Radon in Our Homes

The presence of radon has become a much publicized environmental and health issue because of radiation danger. Radioisotopes that produce radon, such as radium-226 and uranium-238, are naturally present in many types of rocks and soils. Radium-226 emits an alpha particle and is converted into radon gas, which diffuses out of the rocks and soil.

$$^{226}_{88}\text{Ra} \longrightarrow ^{222}_{86}\text{Rn} + ^{4}_{2}\text{He}$$

As uranium-238 decays, it also forms radium-226, which in turn produces radon. Uranium-238 has been found in particularly high levels in an area between Pennsylvania and New England.

Outdoors, radon gas poses little danger, since it dissipates in the air. However, if the source of radon is under a house or building, the gas can enter the house through cracks in the foundation or other openings. Then the radon is inhaled by those living or working there. Inside the lungs, radon emits alpha particles to form polonium-218, which is known to cause cancer when present in the lungs.

$$^{222}_{86}\text{Rn} \longrightarrow ^{218}_{84}\text{Po} + ^{4}_{2}\text{He}$$

Some researchers have estimated that 10% of all lung cancer deaths in the United States are due to radon. The Environmental Protection Agency (EPA) recommends that the maximum level of radon not exceed 4 picocuries (pCi) per liter of air in a home. One (1) picocurie (pCi) is equal to 10^{-12} curies (Ci). In California, 1% of all the houses surveyed exceeded the EPA's recommended maximum radon level. ■

Sample Problem 6.3
Writing an Equation for Alpha Decay

Smoke detectors, now required in many homes and apartments, contain an alpha emitter such as americium-241. The alpha particles emitted ionize air molecules, producing a constant stream of electrical current. However, when smoke particles enter the detector, they interfere with the formation of ions in the air, and the electric current is interrupted. This causes the alarm to sound and warns the occupants of the danger of fire. Complete the following nuclear equation for the decay of americium-241:

$$^{241}_{95}\text{Am} \longrightarrow \boxed{\;?\;} + ^{4}_{2}\text{He}$$

Solution:
The mass number of the new nucleus is 237, obtained by subtracting the mass number of the alpha particle (4) from the mass number of americium (241).

Mass number of Am	−	Mass number of alpha particle	=	Mass number of new nucleus
241	−	4	=	237

The atomic number of the new nucleus is obtained by subtracting the atomic number of the alpha particle (2) from the atomic number of americium (95).

Atomic number of Am	−	Atomic number of alpha particle	=	Atomic number of new nucleus
95	−	2	=	93

The element whose atomic number is 93 is neptunium, Np. The completed nuclear equation for the reaction is written as follows:

$$^{241}_{95}Am \longrightarrow ^{237}_{93}Np + ^{4}_{2}He$$

Study Check:
Write a balanced nuclear equation for the alpha emitter polonium-214.

Answer:

$$^{214}_{84}Po \longrightarrow ^{210}_{82}Pb + ^{4}_{2}He$$

Beta Emitters

A beta emitter is a radioisotope that decays by emitting beta particles. To form a beta particle, the unstable nucleus converts a neutron into a proton. The newly formed proton adds to the number of protons already in the nucleus and increases the atomic number by 1. However, the mass number of the newly formed nucleus stays the same. For example, carbon-14 decays by emitting a beta particle and forming a nitrogen isotope. (See Figure 6.5.)

In the nuclear equation of a beta emitter, the mass number of the radioisotope and the mass number of the new nucleus are the same, and the atomic

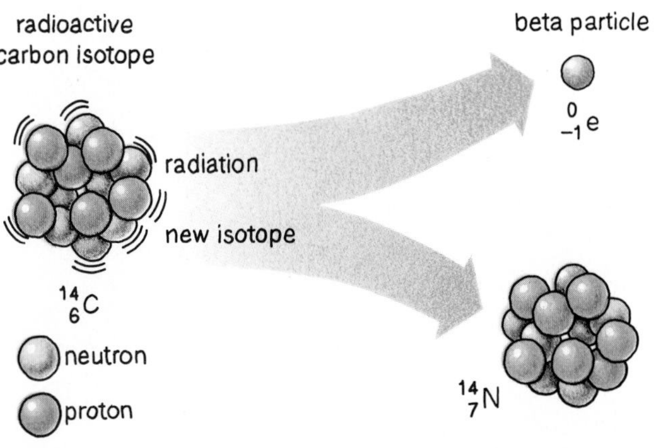

Figure 6.5
A carbon-14 nucleus undergoes beta decay when a neutron is converted to a beta particle and a proton.

number of the new nucleus increases by 1, indicating a change of one element into another. This is the nuclear equation for the beta decay of carbon-14:

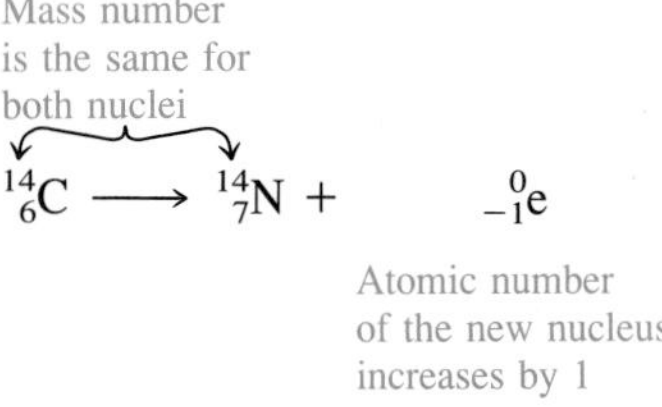
Mass number
is the same for
both nuclei
$^{14}_{6}C \longrightarrow ^{14}_{7}N + \quad ^{0}_{-1}e$
Atomic number
of the new nucleus
increases by 1

Sample Problem 6.4
Writing an Equation for Beta Decay

Cobalt-60, a radioisotope used in the treatment of cancer, decays by emitting a beta particle. Write the nuclear equation for its decay.

Solution:
The radioisotope is a cobalt isotope that has a mass number of 60. Looking at the periodic table, we find that the atomic number of cobalt is 27. The products of the nuclear decay are a beta particle and a new isotope:

$$^{60}_{27}Co \longrightarrow \boxed{\quad ? \quad} + ^{0}_{-1}e$$

Since the mass number does not change in beta emission, we can assign a mass number of 60 to the new isotope. The atomic number of the new nucleus will be 1 more than that of cobalt, or 28.

$$\text{Atomic number of new isotope} = \text{Atomic number of Co} + 1$$
$$28 = 27 + 1$$

Since the element that has an atomic number of 28 is nickel (Ni), we can write the equation for the beta decay of the cobalt radioisotope as

$$^{60}_{27}Co \longrightarrow ^{60}_{28}Ni + ^{0}_{-1}e$$

Study Check:
Iodine-131, a beta emitter, is used to check thyroid function and to treat hyperthyroidism. Write its nuclear equation.

Answer:

$$^{131}_{53}I \longrightarrow ^{131}_{54}Xe + ^{0}_{-1}e$$

Gamma Emitters

There are very few pure gamma emitters, although gamma radiation accompanies most alpha and beta radiation. In radiology, one of the most commonly used

HEALTH NOTE

Beta Emitters in Medicine

The radioactive isotopes of several biologically important elements are beta emitters. When a radiologist wants to treat a malignancy within the body, a beta emitter may be used. The short range of penetration into the tissue by beta particles is advantageous for certain conditions. For example, some malignant tumors increase the fluid within the body tissues. A compound containing phosphorus-32, a beta emitter, is injected into the body cavity where the tumor is located. The beta particles travel only a few mm through the tissue, so only the malignancy and any tissue within that range are affected. The growth of the tumor is slowed or stopped, and the production of fluid decreases. Phosphorus-32 is also used to treat leukemia, polycythemia vera (excessive production of red blood cells), and lymphomas.

$$^{32}_{15}P \longrightarrow {}^{32}_{16}O + {}^{0}_{-1}e$$

Another beta emitter, iron-59, is used in blood tests to determine the level of iron in the blood, and the rate of production of red blood cells by the bone marrow.

$$^{59}_{26}Fe \longrightarrow {}^{59}_{27}Co + {}^{0}_{-1}e$$ ■

gamma emitters is an unstable form of technetium (Tc). This high-energy, excited state is sometimes called the metastable (m) state; metastable technetium may be written as technetium-99, Tc-99m, or ^{99m}Tc. By emitting energy in the form of gamma rays, the unstable nucleus becomes more stable. Figure 6.6 summarizes the changes in the nucleus for alpha, beta, and gamma radiation.

$$^{99m}_{43}Tc \longrightarrow {}^{99}_{43}Tc + \gamma$$

6.3 Detecting and Measuring Radiation

Learning Goal **Describe the methods and units used to detect and measure radiation.**

One of the most common instruments for detecting beta and gamma radiation is the Geiger–Müller counter, or Geiger counter, shown in Figure 6.7. It consists of a metal tube filled with a gas such as argon. Wires connect the metal tube to a battery. When radiation enters a window on the end of the tube, it produces ions in the gas; the ions produce an electrical current. Each burst of current is amplified to give a click and a readout on a meter.

$$Ar + Radiation \longrightarrow Ar^{+} + e^{-}$$

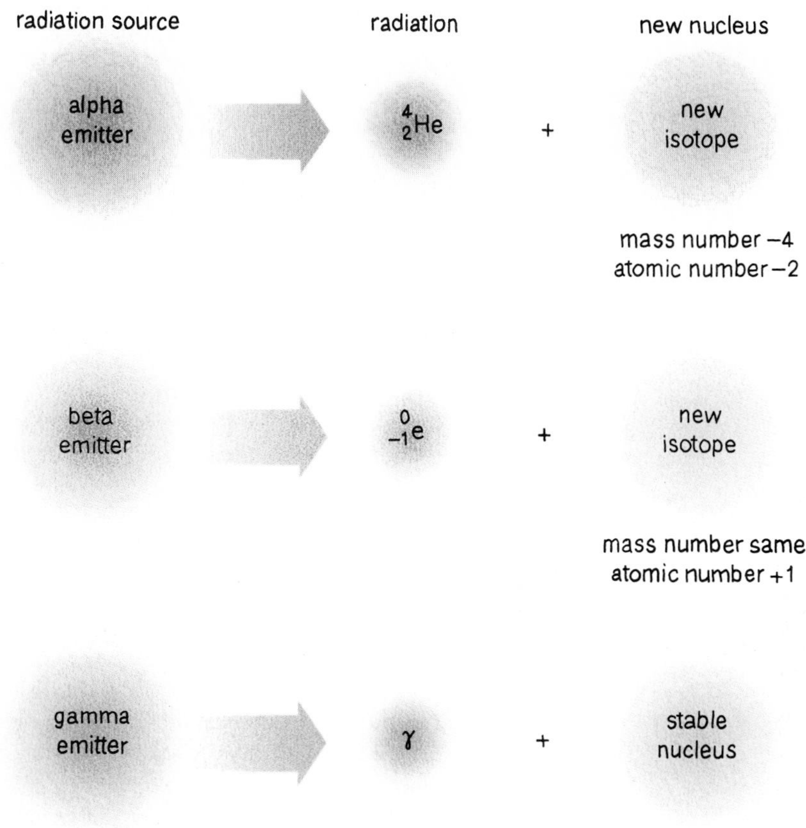

Figure 6.6
Changes in the nucleus for alpha, beta, and gamma radiation.

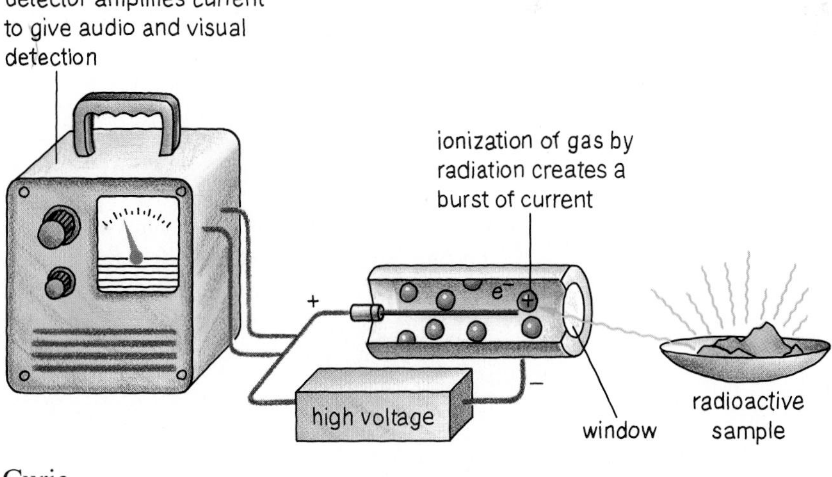

Figure 6.7
Ionization of gas in a Geiger counter.

Curie

When a radiology laboratory obtains a radioisotope, the activity of the sample is measured in terms of the number of disintegrations or nuclear transformations produced by the sample per second. The **curie (Ci)** is the unit used to express nuclear disintegration. One curie is equal to 3.7×10^{10} disintegrations per second (s), the number of atoms that decay in 1.0 g of radium in 1 s.

$$1 \text{ curie (Ci)} = 3.7 \times 10^{10} \text{ disintegrations/s}$$

The curie was named for Marie Curie, a Polish scientist, who along with Pierre Curie discovered the radioactive elements radium and polonium.

Rad

The **rad** (for radiation absorbed dose) is a unit that measures the amount of radiation absorbed by a gram of material such as body tissue. One rad is the absorption of 10^{-5} J of energy per gram of tissue. (Recall that joules and calories are units of energy, and that 1 cal = 4.18 J.)

$$1 \text{ rad} = 10^{-5} \text{ J/g}$$

Rem

The **rem** (short for radiation equivalent in humans) is a unit that measures the biological damage caused by the various kinds of radiation. The rem considers that the biological effects of alpha, beta, and gamma radiation on tissue are not the same. If the heavy, positively charged alpha particles reach the tissues, they cause more ionization and therefore more damage than do beta particles or gamma rays. To determine the rem dose, the absorbed dose in rads is multiplied by a factor called RBE (for radiation biological effectiveness) that adjusts the biological damage that would result for a particular form of radiation.

$$\text{Rem} = \text{Rad} \times \text{RBE}$$

Rem	=	Rad	×	RBE
Effect of radiation in humans		Dose of radiation absorbed		Biological effect factor

For beta and gamma radiation, and x-rays, the RBE is about 1, so the biological damage in rems is about equal to the absorbed radiation (rad). For high-energy protons and neutrons, the RBE factor is about 10, and for alpha particles, it is 20. Thus, the biological damage in rems for alpha particles would be the absorbed dose (rad) multiplied by 20. Table 6.4 summarizes some common units of radiation measurement.

Table 6.4 Some Units of Radiation Measurement

Measurement	Unit	Meaning
Activity	Curie (Ci)	3.7×10^{10} disintegrations/s
Absorbed dose	Rad	10^{-5} J/g
Biological damage in humans	Rem	Rad × RBE

Sample Problem 6.5
Measuring Activity

In a treatment for leukemia, phosphorus-32, which has an activity of 2 millicuries (mCi), is used. If phosphorus-32 is a beta emitter, how many beta particles are emitted in 1 s?

Solution:
We can calculate the number of beta particles from a radioisotope's activity. Since 1 Ci is 3.7×10^{10} disintegrations/s, there must be 3.7×10^{10} beta particles produced in a second.

$$2\ \cancel{\text{mCi}} \times \frac{1\ \cancel{\text{Ci}}}{1000\ \cancel{\text{mCi}}} \times \frac{3.7 \times 10^{10}\ \beta\ \text{particles}}{\cancel{\text{s}}\ \cancel{\text{Ci}}} \times 1\ \cancel{\text{s}} = 7.4 \times 10^{7}\ \text{beta particles}$$

Study Check:
An iodine-131 source has an activity of 0.25 Ci. How many radioactive atoms will disintegrate in 1 min?

Answer:
5.6×10^{11} iodine-131 atoms

Background Radiation

We are all exposed to low levels of radiation every day. Naturally occurring radioactive isotopes are part of the atoms of wood, brick, and concrete in our homes and the buildings where we work and go to school. This radioactivity, called background radiation, is present in the soil, in the food we eat, in the water we drink, and in the air we breathe. For example, one of the naturally occurring isotopes of potassium, potassium-40 is radioactive. It is found in the body because it is always present in any potassium-containing food. Other naturally occurring radioisotopes in air and food are carbon-14, radon-222, strontium-90, and iodine-131.

The atomosphere is another natural source of radiation. We are constantly exposed to radiation (cosmic rays) produced in space by the sun. At higher elevations, the amount of radiation from outer space is greater because there are fewer air molecules to absorb the radiation. People living at high altitudes or flying in an airplane receive more radiation from cosmic rays. For example, a person living in Denver receives about twice the cosmic radiation as a person living in Los Angeles.

You may also receive some radiation from nuclear testing, although such testing has decreased in recent years. If nuclear testing occurs, radioactive isotopes enter the atmosphere and are carried around the world by wind currents. They reach the earth during rainstorms and other weather phenomena. A person living close to a nuclear power plant normally does not receive much additional radiation, perhaps 0.1 millirem (mrem) in 1 year. (One rem equals 1000 mrem.) However, in the accident at the Chernobyl nuclear power plant in 1986, it is estimated that people in a nearby town received as much as 1 rem/hr.

In addition to naturally occuring radiation from construction materials in our homes, we receive radiation from television. In the medical clinic, dental and

HEALTH NOTE

Maximum Permissible Dose

Any person working with radiation, such as a radiologist, a radiation technician, or a nurse whose patient has received a radioisotope, must wear some type of detection apparatus to measure exposure to radiation. A standard for occupational exposure is the maximum permissible dose (MPD). If this dose is not exceeded, the probability of injury is minimized. The detection apparatus is usually a badge, ring, or pin containing a small piece of photographic film. The window that covers the film absorbs light and beta rays, but gamma rays penetrate the covering and expose the film. The darker the film, the more exposure to gamma radiation the person has had. These badges are checked periodically to prevent exposure to more than the maximum permissible dose. The maximum permissible dose for occupational exposure is 5 rem (5000 mrem) per year. ■

chest x-rays also add to our radiation exposure. Table 6.5 lists some common sources of radiation. The average person in the United States receives about 0.170 rem or 170 mrem of radiation annually.

Radiation Sickness

The larger the dose of radiation received at one time, the greater the effect on the body. Exposure to radiation under 25 rem usually cannot be detected. Whole-body exposure of 100 rem produces a temporary decrease in the number of white blood cells. If the exposure to radiation is 100 rem or higher, the person suffers

Table 6.5 Average Annual Radiation Received by a Person in the United States

Source	Dose (mrem)
Natural	
The ground	15
Air, water, food	30
Cosmic rays	40
Wood, concrete, brick	50
Medical	
Chest x-ray	50
Dental x-ray	20
Upper gastrointestinal tract x-ray	200
Other	
Television	2
Air travel	1
Global fallout	2
Cigarette smoking	35

Table 6.6
Lethal Doses of Radiation for Some Life-Forms

Life-Form	LD_{50} (rem)
Insect	100,000
Bacterium	50,000
Rat	800
Human	500
Dog	300

the symptoms of radiation sickness: nausea, vomiting, fatigue, and a reduction in white cell count. A whole-body dosage greater than 300 rem can lower the white cell count to zero. The patient suffers diarrhea, hair loss, and infection.

Lethal Dose

Exposure to radiation of about 500 rem is expected to cause death in 50% of the people receiving that dose. This amount of radiation is called the lethal dose for one-half the population, or the LD_{50}. The LD_{50} varies for different life-forms, as Table 6.6 shows. Radiation dosages of about 600 rem would be fatal to all humans within a few weeks.

6.4 Medical Applications of Radioisotopes

Learning Goal **Describe the use of radioisotopes in nuclear medicine.**

Suppose a radiologist wants to determine the condition of an organ in the body. How is this done? In some cases, the patient is given a radioisotope that is known to concentrate in that organ. The cells in the body cannot differentiate between a nonradioactive atom and a radioactive one. All atoms of an element, including any radioactive isotopes, have the same electron arrangement and the same chemistry in the body. The difference is that radioactive atoms can be detected because they emit radiation as they move to the same organs as the nonradioactive atoms of an element. Some radioisotopes used in nuclear medicine are listed in Table 6.7.

Table 6.7 Some Radioisotopes Used in Nuclear Medicine

Element	Radioisotope	Medical use
Chromium	^{51}Cr	Spleen imaging, blood volume
Technetium	^{99m}Tc	Brain, lung, liver, spleen, bone, and bone marrow scans
Gallium	^{67}Ga	Treatment of lymphomas
Phosphorus	^{32}P	Treatment of leukemia, polycythemia vera, and lymphomas; detection of brain and breast tumors
Sodium	^{24}Na	Study of vascular disease, extracellular volume, and blood volume determination
Strontium	^{85}Sr	Bone imaging for the diagnosis of bone damage and bone disease
Iodine	^{125}I	Thyroid imaging; study of plasma volume and fat absorption
Iodine	^{131}I	Study of thyroid; treatment of thyroid conditions such as hyperthyroidism

HEALTH NOTE

Visualization of the Brain and Skull

Normally, radioactive isotopes do not enter brain cells because there is a protective blood–brain barrier. However, if a blood vessel has broken, or if there is a brain tumor drawing on the blood supply, radioisotopic uptake does occur within the brain and appears on a CATscan. (See Figure 6.8). ■

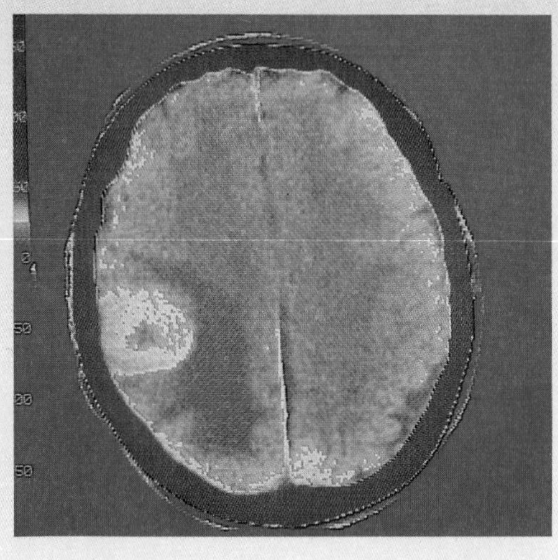

Figure 6.8
CAT scan showing brain with melanoma tumor.

After a patient receives a radioisotope, the radiologist determines the level and location of radioactivity emitted by the radioisotope. An apparatus called a scanner is used to produce an image of the organ. (See Figure 6.9.) The scanner moves slowly across the patient's body above the region where the organ containing the radioisotope is located. The gamma rays emitted from the radioisotope in the organ can be used to expose a photographic plate, producing a **scan** of the organ. On a scan, an area of decreased or increased radiation can indicate such conditions as a disease of the organ, a tumor, a blood clot, or edema.

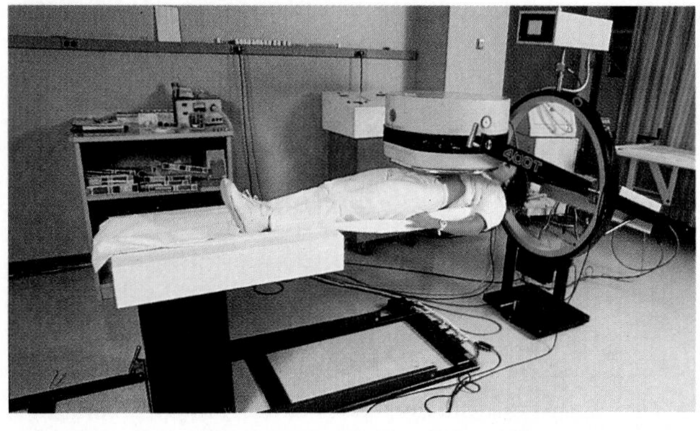

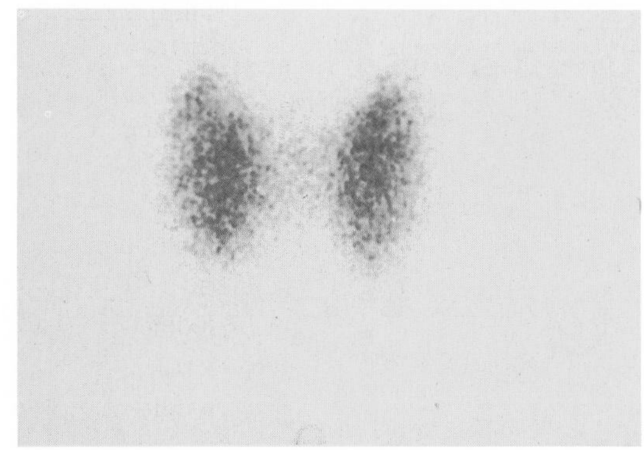

Figure 6.9
A scanner is used to detect radiation from radioisotopes in the organs of the chest (left) and thyroid (right).

HEALTH NOTE

Use of Radioactive Iodine in Testing Thyroid Function

A common method of determining thyroid function is the use of radioactive iodine uptake (RAIU). Taken orally, the radioisotope iodine-131 mixes with the iodine already present in the thyroid. Twenty-four hours later, the amount of iodine taken up by the thyroid is determined. A detection tube held up to the area of the thyroid gland detects the radiation coming from the iodine-131 that has located there.

The iodine uptake is directly proportional to the activity of the thyroid. A patient with a hyperactive thyroid will have a higher than normal level of radioactive iodine, whereas a patient with a hypoactive thyroid will record low values.

If the patient has hyperthyroidism, treatment is begun to lower the activity of the thyroid. One treatment involves giving the patient a therapeutic dosage of radioactive iodine, which has a higher radiation count than the diagnostic dose. The radioactive iodine goes to the thyroid where its radiation destroys some of the thyroid cells. The thyroid produces less thyroid hormone, bringing the hyperthyroid condition under control. ■

Positron Emission Tomography (PET)

Radioisotopes that emit a positron are used in an imaging method called positron emission tomography (PET). A positron is a particle emitted from the nucleus that has the same mass as an electron but has a positive charge.

β^+ or ${}^{0}_{1}e$

Positron

Carbon-11 is an example of a radioisotope that emits a positron when it decays.

$${}^{11}_{6}\text{C} \longrightarrow {}^{0}_{1}e + {}^{11}_{5}\text{B}$$

Positron

A positron is produced when a proton changes to a neutron. The new nucleus retains the same mass number but has a lower atomic number. The positron exists for only a moment before it collides with an electron. Some of a positron's mass is converted into bursts of gamma energy, which are detected

$${}^{0}_{1}e + {}^{0}_{-1}e \rightarrow 2\,\gamma$$

positron | electron in an atom | gamma rays produced

Medically, positron emitters such as carbon-11, oxygen-15, and nitrogen-13 are used to diagnose conditions involving blood flow, metabolism, and, particularly, the functioning of the brain. Glucose labelled with carbon-11 is used to detect damage in the brain from epilepsy, stroke, and Parkinson's disease. After the

HEALTH NOTE

Radiation Doses in Diagnostic and Therapeutic Procedures

We can compare the levels of radiation exposure commonly used during diagnostic and therapeutic procedures in nuclear medicine. In diagnostic procedures, the radiologist minimizes radiation damage by using only enough radioisotope to evaluate the condition of an organ or tissue. (See Table 6.8.)

The doses used in **radiation therapy** are much greater than those used for diagnostic procedures. For example, a therapeutic dose would be used to destroy the cells in a malignant tumor. Although there will be some damage to surrounding tissue, the healthy cells are more resistant to radiation and can repair themselves. (See Table 6.9.) ■

Table 6.8
Radiation Doses Used for Diagnostic Procedures

Organ	Dose (rem)
Liver	0.3
Thyroid	50.0
Lung	2.0

Table 6.9
Radiation Doses Used for Therapeutic Procedures

Condition	Dose (rem)
Lymphoma	4500
Skin cancer	5000–6000
Lung cancer	6000
Brain tumor	6000–7000

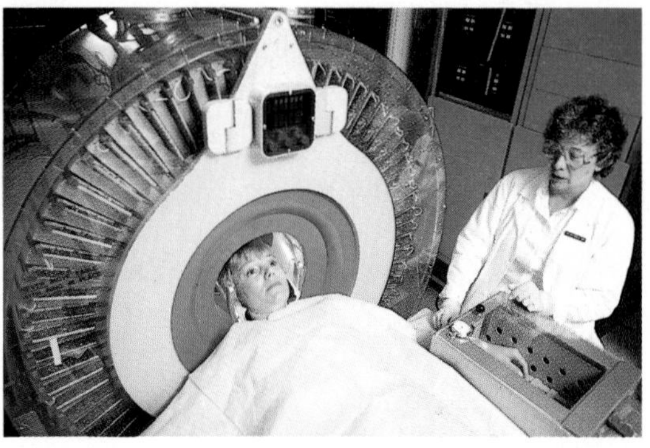

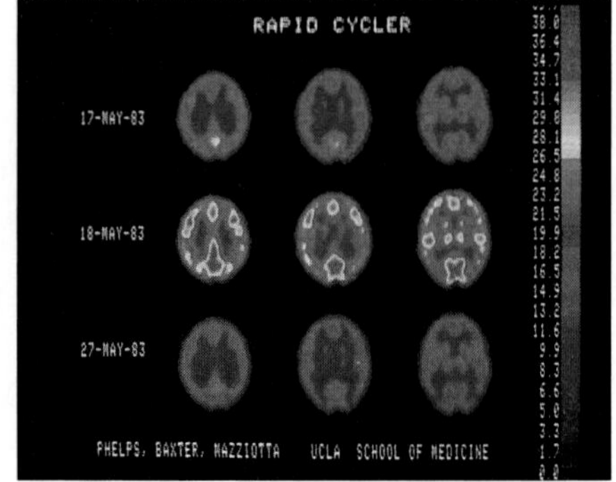

Figure 6.10 A PET scan of the brain.

radioisotope is injected, the gamma rays from the positrons emitted are detected by computerized equipment to create a three-dimensional image of the organ. (See Figure 6.10.)

Computerized Tomography (CT)

Another imaging method used to detect changes within the body is computerized tomography (CT). A computer monitors the degree of absorption of 30,000 x-ray beams directed at the brain at successive layers. The differences in absorption based upon the densities of the tissues and fluids in the brain provide a series of images of the brain. This technique is successful in the identification of brain hemorrhages, tumors, and atrophy.

Figure 6.11
An MRI scan of a heart.

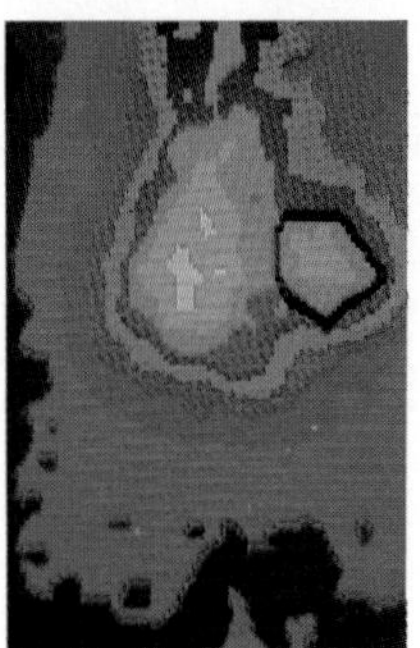

Magnetic Resonance Imaging (MRI)

In another imaging technique called magnetic resonance imaging (MRI), no radiation is used. The atomic nuclei in atoms such as hydrogen emit small amounts of energy when they are excited by a strong magnetic field. Such low-energy changes are harmless to the patient, but they can be detected by a computer to create an image on a television monitor. (See Figure 6.11.) Magnetic resonance imaging is used in the detection of multiple sclerosis, abnormalities of the spine and brain, tumors, and birth malformations. In some cases, MRI is replacing the use of x-rays and other scanning techniques, including the CT scanner.

6.5 Half-Life of a Radioisotope

Learning Goal **Given the half-life of a radioisotope, calculate the amount that is active after one or more half-lives.**

The time it takes for one-half of a radioactive sample to decay is called its **half-life.** For example, iodine-131, a radioactive isotope of iodine used in the diagnosis and treatment of thyroid disorders, has a half-life of 8 days. If we began with a sample containing 1000 atoms of iodine-131, there would be 500 atoms remaining after 8 days. In another 8 days, we would have 250 atoms left. Thus, in each half-life of 8 days, one-half of the sample has disintegrated.

1000 atoms I-131 → 8 days → 500 atoms I-131 → 8 days → 250 atoms I-131

500 I-131 atoms decay 250 I-131 atoms decay

This information can be summarized as follows:

Time elapsed	0	8 days	16 days	24 days
Number of half-lives elapsed	0	1	2	3
Quantity of I-131 remaining	1000 atoms	500 atoms	250 atoms	125 atoms

A decay curve is a diagram of the decay of a radioactive isotope. Figure 6.12 shows such a curve for the iodine-131 we have discussed.

Naturally occurring isotopes of the elements usually have long half-lives, as shown in Table 6.10. They disintegrate slowly and produce radiation over a long period of time, even hundreds or millions of years. In contrast, many of the radioisotopes used in nuclear medicine have much shorter half-lives. They disintegrate rapidly and produce almost all their radiation in a short period of time.

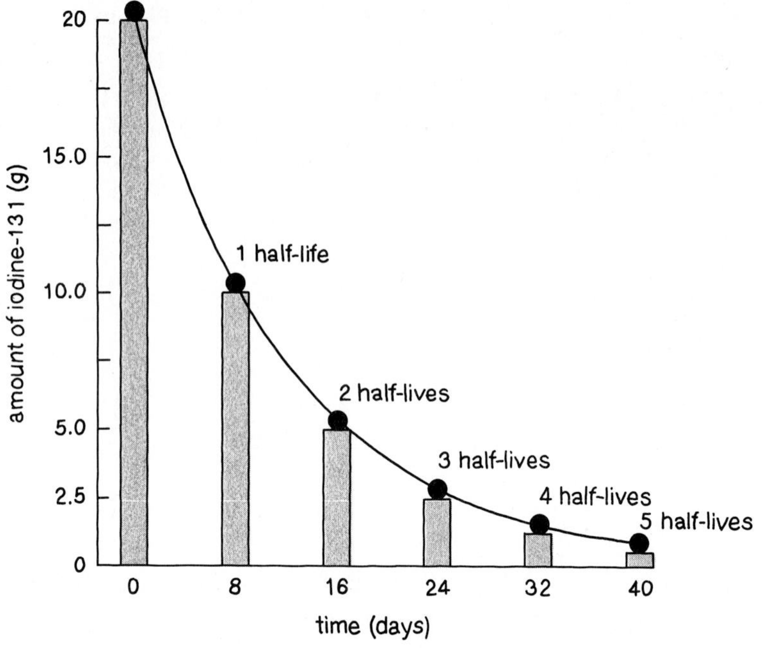

Figure 6.12
Decay curve for iodine-131. One-half of the sample decays with each half-life. Iodine-131 has a half-life of 8 days.

Table 6.10 Half-Lives of Some Radioisotopes

Element	Radioisotope	Half-life	Type of radiation
Naturally Occurring Radioisotopes			
Carbon	^{14}C	5730 yr	β
Potassium	^{40}K	1.3×10^9 yr	β, γ
Radium	^{226}Ra	1600 yr	α, γ
Uranium	^{238}U	4.5×10^9 yr	α, γ
Some Medical Radioisotopes			
Carbon	^{11}C	20 min	β^+
Chromium	^{51}Cr	28 days	γ
Iodine	^{131}I	8 days	β, γ
Iodine	^{125}I	60 days	γ
Iron	^{59}Fe	46 days	β, γ
Phosphorus	^{32}P	14 days	β
Oxygen	^{15}O	2 min	β^+
Potassium	^{42}K	12 hr	β, γ
Sodium	^{24}Na	15 hr	β, γ
Strontium	^{85}Sr	64 days	γ
Technetium	^{99m}Tc	6.0 hr	γ

For example, technetium-99m emits half of its radiation in the first 6 hr. This means that a small amount of the radioisotope given to a patient is essentially gone within two days. The decay products of technetium-99m are totally eliminated by the body.

Sample Problem 6.6
Calculating Quantity of Radioisotopes

Nitrogen-13, which has a half-life of 10 min, is used to image organs in the body. For the diagnostic procedure, the patient receives an injection of a compound containing the radioisotope. Originally, the nitrogen-13 has an activity of 40 microcuries (μCi). If the procedure requires 30 min, what is the remaining activity of the radioisotope?

Solution:

$$\text{Number of half-lives} = 30\ \cancel{\text{min}} \times \frac{1\ \text{half-life}}{10\ \cancel{\text{min}}} = 3$$

The activity of the radioisotope after 3 half-lives is

40 μCi → 10 min → 20 μCI → 10 min → 10 μCi → 10 min → 5 μCi

Another way to calculate the activity of radioactive nitrogen-13 left in the sample is to construct a chart to show the number of half-lives, elapsed time, and the amount of radioactive isotope that is left in the sample.

Time elapsed	0	10 min	20 min	30 min
Number of half-lives elapsed	0	1	2	3
Activity of N-13 remaining	40 μCi	20 μCi	10 μCi	5 μCi

Study Check:
Iron-59, used in the determination of bone marrow function, has a half-life of 46 days. If the laboratory receives a sample of 8.0 g of iron-59, how many grams are still active after 184 days?

Answer:

0.50 g

ENVIRONMENTAL NOTE

Dating Ancient Objects

A technique known as radiological dating is used by geologists, archaeologists, and historians as a way to determine the age of ancient objects. The age of an object derived from plants or animals (such as wood, fiber, natural pigments, bone, and cotton or woolen clothing) is determined by measuring the amount of carbon-14, a naturally occurring radioactive form of carbon. In 1960, Willard Libby received the Nobel Prize for the work he did developing carbon-14 dating techniques during the 1940s. Carbon-14 is produced in the upper atmosphere by the bombardment of high-energy neutrons from cosmic rays.

$$ {}^{1}_{0}n \quad + \quad {}^{14}_{7}\text{N} \quad \longrightarrow $$

Neutron from cosmic rays — Nitrogen in atmosphere

$$ {}^{14}_{6}\text{C} \quad + \quad {}^{1}_{1}\text{H} $$

Radioactive carbon-14 — Proton

The carbon-14 reacts with oxygen to form radioactive carbon dioxide, $^{14}CO_2$. Since carbon dioxide is continuously absorbed by living plants during the process of photosynthesis, some carbon-14 will be taken into the plant. After the plant dies, no more carbon-14 is taken up, and the amount of carbon-14 contained in the plant decreases as it undergoes radioactive decay emitting β particles.

$$ {}^{14}_{6}\text{C} \longrightarrow {}^{14}_{7}\text{N} + {}^{0}_{-1}e $$

Scientists use the half-life of carbon-14 (5730 years) to calculate the amount of time that has passed since the plant died, a process called **carbon dating.** The smaller the amount of carbon-14 remaining in the sample, the greater the number of half-lives that have passed. Thus, the approximate age of the sample can be determined. For example, a wooden beam found in an ancient Indian dwelling might have one-half of the carbon-14 found in living plants today. Thus, the dwelling was probably constructed about 5730 years ago, one half-life of carbon-14. (See Figure 6.13.) This technique is useful for dating samples that have ages up to 30,000 years. For older objects, carbon dating is not reliable.

A radiological dating method used for determining the age of rocks is based on the radioisotope uranium-238, which decays through a series of reactions to lead-206. The uranium-238 isotope has a very long half-life, about 4×10^9 (4 billion) years. Measurements of the amounts of uranium-238 and lead-206 enable geologists to determine the age of rock samples. The older rocks will have a higher percentage of lead-206 because more of the uranium-238 has decayed. The age of rocks brought back from the moon by the *Apollo* missions was determined using uranium-238. They were found to be about 4×10^9 years old, approximately the same age calculated for Earth. ■

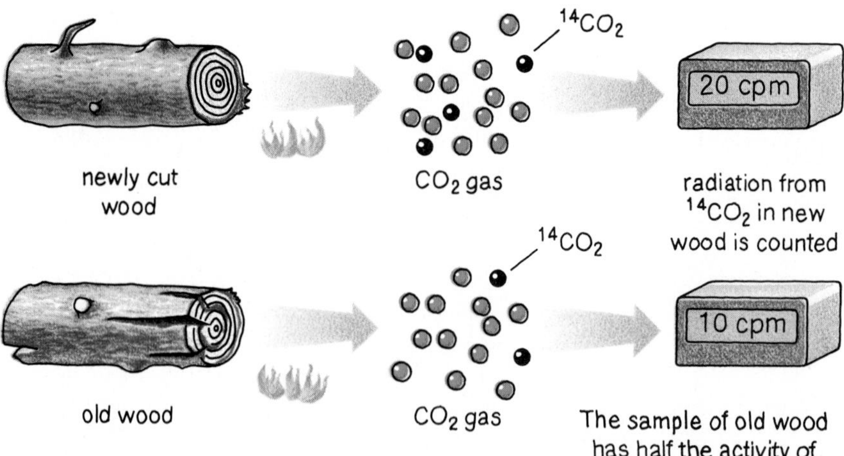

Figure 6.13
The age of an ancient boat, wood from a prehistoric Indian village, and the like can be determined by carbon-14 dating.

Sample Problem 6.7
Dating Using Half-Lives

In Los Angeles, the remains of ancient animals have been unearthed at the La Brea tar pits. Suppose a bone sample from the tar pits is subjected to the carbon-14 dating method. If the sample shows about two half-lives have passed, about when did the animal live in the tar pits?

Solution:
We can calculate the age of the bone sample by using the half-life of carbon-14 (5730 years).

$$2\ \cancel{\text{half-lives}} \times \frac{5730\ \text{years}}{1\ \cancel{\text{half-life}}} = 11{,}000\ \text{years}$$

We would estimate that the animal lived in the tar pits about 11,000 years ago, or about 9000 B.C.

6.6 Producing Radioisotopes

Learning Goal **Write a nuclear equation for the production of a radioisotope from a nonradioactive isotope.**

Today, more than 1500 radioisotopes are produced by converting stable, nonradioactive isotopes into radioactive ones. To do this, a stable atom is bombarded by fast-moving alpha particles, protons, or neutrons. When one of these particles is absorbed by a stable nucleus, the nucleus becomes unstable and the atom is now a radioactive isotope.

When a nonradioactive isotope such as boron-10 is bombarded by an alpha particle, it is converted to nitrogen-13, a radioisotope, as shown in Figure 6.14. In this bombardment reaction, a neutron is emitted.

$$^{4}_{2}\text{He} + {}^{10}_{5}\text{B} \longrightarrow {}^{13}_{7}\text{N} + {}^{1}_{0}n$$

Bombarding particle | Stable isotope | Radioactive isotope | Neutron

The process of changing one element into another is called **transmutation.**

All of the known elements that have atomic numbers greater than 92 have been produced by bombardment; none of these elements occurs naturally. Most have been produced in only small amounts and exist for such a short time that it is difficult to study their properties. An example is element 105, unnilpentium, which is produced when californium-249 is bombarded with nitrogen-15.

$$^{249}_{98}\text{Cf} + {}^{15}_{7}\text{N} \longrightarrow {}^{260}_{105}\text{Unp} + 4\ {}^{1}_{0}n$$

Figure 6.14
Transmutation: the formation of a radioactive isotope by nuclear bombardment.

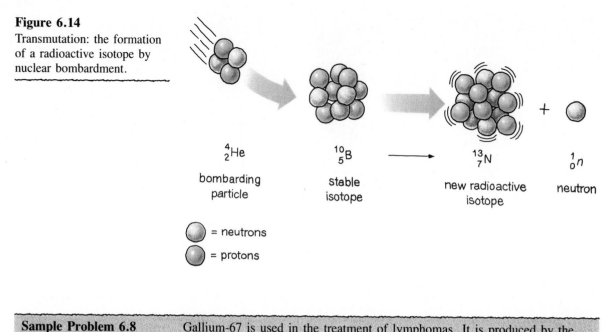

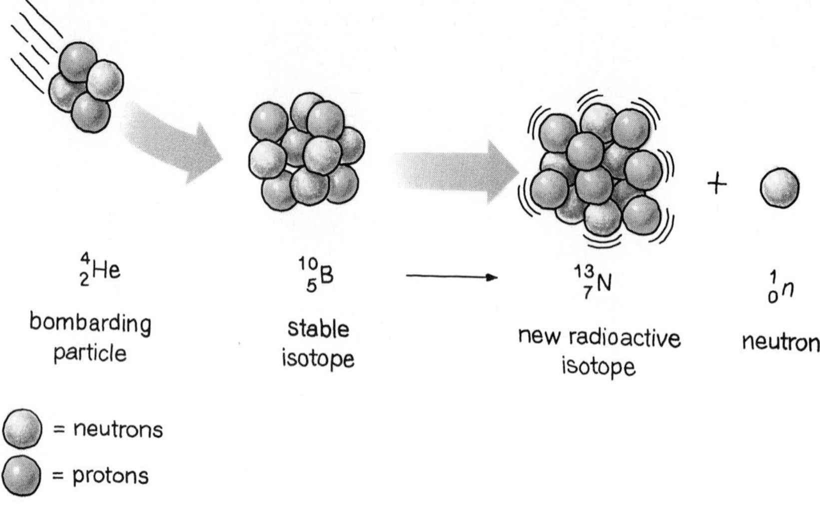

Sample Problem 6.8
Writing Equations for Transmutations

Gallium-67 is used in the treatment of lymphomas. It is produced by the bombardment of zinc-66 by a proton. Write the equation for this nuclear bombardment.

Solution:
The reactants in this nuclear equation are zinc-66 and a proton. The radioactive isotope formed is gallium-67.

$$^{66}_{30}Zn + ^{1}_{1}H \longrightarrow ^{67}_{31}Ga$$

Proton — New radioisotope

Study Check:
Write the equation for the bombardment of aluminum-27 by an alpha particle to produce the radioactive isotope phosphorus-30 and one neutron.

Answer:

$$^{27}_{13}Al + ^{4}_{2}He \longrightarrow ^{30}_{15}P + ^{1}_{0}n$$

Sample Problem 6.9
Completing a Nuclear Equation for a Bombardment Reaction

Write the missing symbol in the following bombardment reaction:

$$^{58}_{28}Ni + ^{1}_{1}H \longrightarrow \boxed{?} + ^{4}_{2}He$$

Proton

Solution:
The sum of the mass numbers for nickel and hydrogen is 59. Therefore, the mass number of the new isotope must be 59 minus 4, or 55. The sum

of the atomic numbers is 29. The atomic number of the new isotope is 29 minus 2, or 27. The element that has atomic number 27 is cobalt (Co).

$$^{58}_{28}\text{Ni} + ^{1}_{1}\text{H} \longrightarrow \boxed{^{55}_{27}\text{Co}} + ^{4}_{2}\text{He}$$

Proton New isotope

Study Check:
Complete the following bombardment equation:

$$^{249}_{98}\text{Cf} + ^{10}_{5}\text{B} \longrightarrow ^{257}_{103}\text{Lr} + ?$$

Answer:

$2\ ^{1}_{0}n$

HEALTH NOTE

Figure 6.15
Hospital generator for the production of technetium-99m.

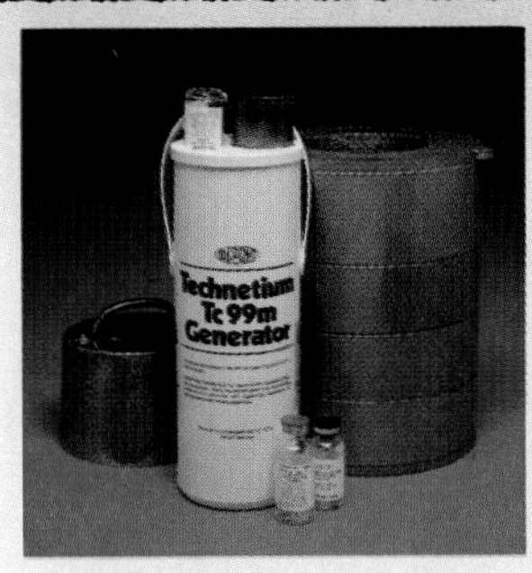

Producing Radioisotopes for Nuclear Medicine

Technetium-99m is a radioisotope used in nuclear medicine for several diagnostic procedures, including the detection of brain tumors and the examination of the liver and spleen. The source of technetium-99m is molybdenum-99, which is produced in a nuclear reactor by neutron bombardment of molybdenum-98.

$$^{98}_{42}\text{Mo} + ^{1}_{0}n \longrightarrow ^{99}_{42}\text{Mo}$$

Many radiology laboratories have a small generator containing the radioactive molybdenum-99, which decays to give the technetium-99m radioisotope. (See Figure 6.15.)

$$^{99}_{42}\text{Mo} \longrightarrow ^{99m}_{43}\text{Tc} + ^{0}_{-1}e$$

The technetium-99m radioisotope has a half-life of 6 hours and decays by emitting gamma rays. Gamma emission is most desirable for diagnostic work because the gamma rays pass through the body to the detection equipment.

$$^{99m}_{43}\text{Tc} \longrightarrow ^{99}_{43}\text{Tc} + \gamma$$ ■

6.7 Nuclear Fission and Fusion

Learning Goal **Describe the nuclear processes of fission and fusion.**

Nuclear Fission

In the 1930s, scientists bombarding uranium-235 with neutrons discovered that two medium-weight nuclei were produced along with a great amount of energy.

This was the discovery of a new kind of nuclear reaction called nuclear **fission.** The energy generated by nuclear fission, splitting the atom, was called atomic energy. When uranium-235 absorbs a neutron, it breaks apart to form two smaller nuclei, several neutrons, and a great amount of energy. (See Figure 6.16.) A typical equation for nuclear fission is

$$^{235}_{92}\text{U} + {}^{1}_{0}n \longrightarrow {}^{139}_{56}\text{Ba} + {}^{94}_{36}\text{Kr} + 3\ {}^{1}_{0}n + \text{Energy}$$

If we could weigh these products with great accuracy, we would find that their total mass is slightly less than the mass of the starting materials. The missing mass has been converted into energy, consistent with the famous equation derived by Albert Einstein:

$$E = mc^2$$

E is the energy released, m is the mass lost, and c is the speed of light, 3×10^{10} cm/s. Even though the mass loss is very small, when it is multiplied by the speed of light squared the result is a large value for the energy released. The fission of 1 g of uranium-235 would produce about as much energy as 3 tons of coal.

Chain Reaction

Fission begins when a neutron collides with the nucleus of a uranium atom. The resulting nucleus is unstable and splits into smaller nuclei. This fission process also releases several neutrons and large amounts of gamma radiation and energy. The neutrons emitted have high energies and bombard more uranium-235 nuclei. As fission continues, there is a rapid increase in the number of high-energy neutrons capable of splitting more uranium atoms, a process called a **chain**

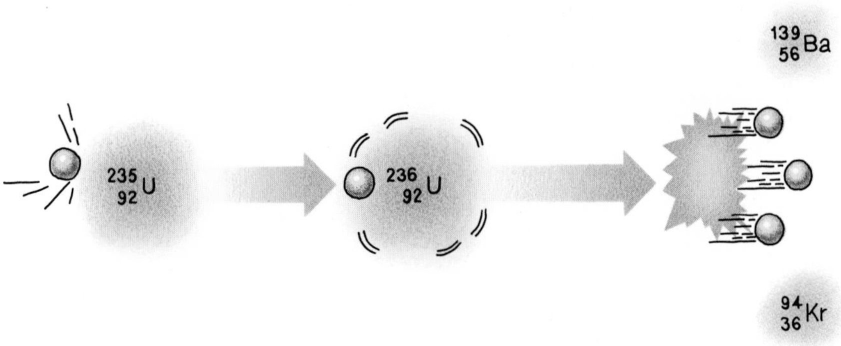

Figure 6.16
Diagram of nuclear fission. After absorbing a fast-moving neutron, a $^{235}_{92}$U nucleus undergoes fission to produce two smaller nuclei, three neutrons, and a great amount of energy.

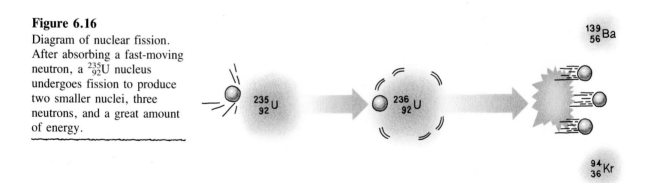

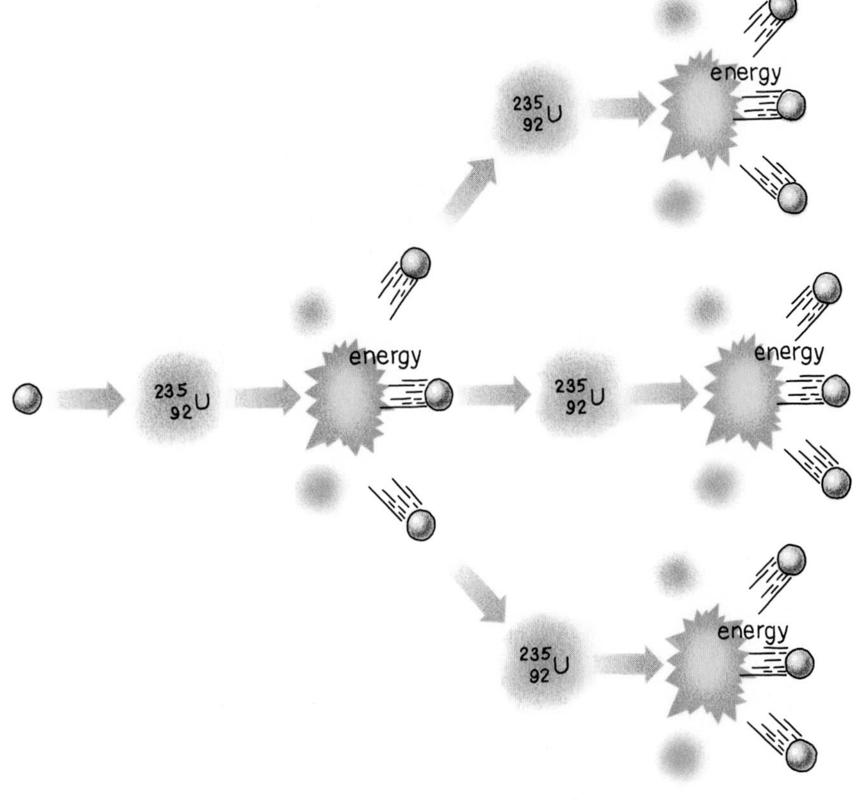

Figure 6.17
Nuclear chain reaction by fission of uranium-235.

reaction, shown in Figure 6.17. To sustain a nuclear chain reaction, sufficient quantities of uranium-235 must be brought together to provide a critical mass in which almost all the neutrons immediately collide with more uranium-235 nuclei. So much heat and energy build up that an atomic explosion occurs.

Nuclear Power Plants

In a nuclear power plant, the quantity of uranium-235 is held below a critical mass, so it cannot sustain a chain reaction. The fission reactions are slowed by placing control rods among the samples of uranium. These rods absorb some of the fast-moving neutrons. In this way, less fission occurs, and there is a slower, controlled production of energy. The heat from the controlled fission is used to produce steam. The steam drives a generator, which produces electricity as shown in Figure 6.18. Approximately 10% of the electrical energy produced in the United States is generated in nuclear power plants.

Although nuclear power plants help meet some of our energy needs, there are major problems. One of the most serious problems is the production of radioactive by-products that have very long half-lives. It is essential that these waste products be stored safely for a very long time in a place where they do not

Figure 6.18
Generating electricity at Indian Point, a nuclear power plant in Buchanan, New York.

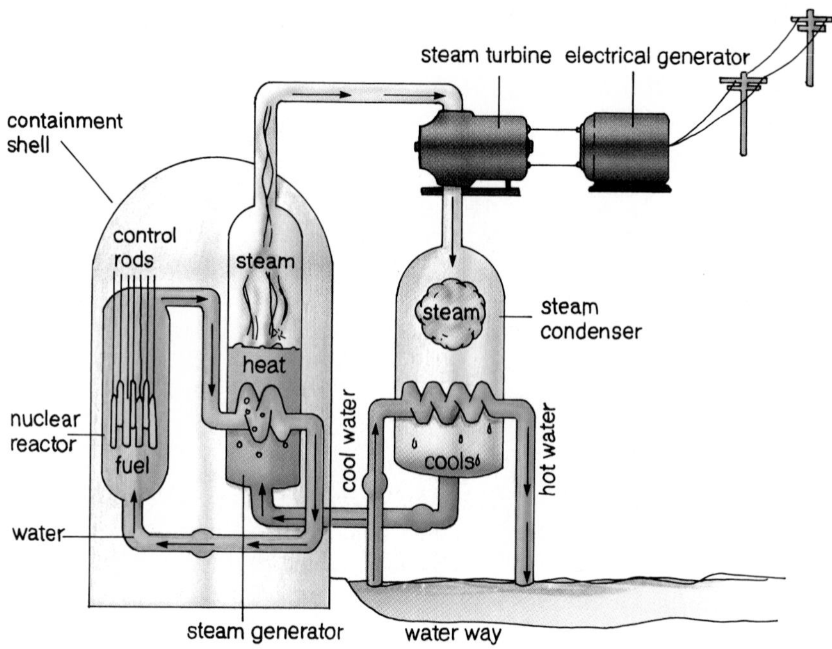

contaminate the environment. Early in 1990, the Environmental Protection Agency gave its approval to the storing of radioactive hazardous wastes in chambers 2150 ft underground. It will be a matter of time before we know whether this type of storage is a good idea.

Nuclear Fusion

In **fusion,** two small nuclei combine to form a larger nucleus. Mass is lost, and a tremendous amount of energy is released, even more than the energy released from nuclear fission. However, a very high temperature (100,000,000°C) is required to overcome the repulsion of the hydrogen nuclei and cause them to undergo fusion. Fusion reactions occur continuously in the sun, providing us

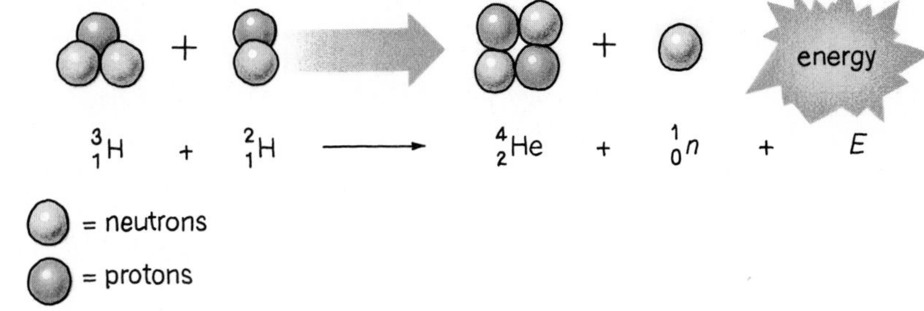

Figure 6.19
The fusion process: Energy is released when small masses combine to form a larger mass.

with heat and light. The fusion reaction shown in Figure 6.19 involves the combination of two isotopes of hydrogen.

$$\underset{\text{Tritium}}{^{3}_{1}H} + \underset{\text{Deuterium}}{^{2}_{1}H} \longrightarrow {}^{4}_{2}He + {}^{1}_{0}n + \text{Energy}$$

The fusion reaction has tremendous potential as a possible source for future energy needs. One of the advantages of fusion as an energy source is that the deuterium isotope of hydrogen ($^{2}_{1}H$) is plentiful in the oceans. Although scientists expect some radioactive waste from fusion reactors, the amount is expected to be much less than that from fission, and the waste products should have shorter half-lives. However, fusion is still in the experimental stage because the extremely high temperatures needed have been difficult to reach and even more difficult to maintain. Research groups around the world are attempting to develop the technology needed to make the fusion reaction a reality in our lifetimes.

Sample Problem 6.10
Identifying Fission and Fusion

Classify the following as pertaining to nuclear fission, nuclear fusion, or both:

a. Small nuclei combine to form larger nuclei.
b. Large amounts of energy are released.
c. Very high temperatures are needed for reaction.

Solution:

a. fusion **b.** both fusion and fission **c.** fusion

Study Check:
Would you expect the following reaction to be an example of a fission or fusion reaction?

$$^{2}_{1}H + {}^{1}_{1}H \longrightarrow {}^{3}_{2}He$$

Answer:
fusion

Glossary of Key Terms

alpha particle (α) A nuclear particle containing two protons and two neutrons; its symbol is ^4_2He or α.

beta particle (β) A particle identical to an electron that forms in the nucleus as a result of nuclear change and is emitted with high energy; its symbol is β or $^{0}_{-1}e$.

carbon dating A technique used to date ancient specimens that contain carbon. The age is determined by the amount of carbon-14 that remains in the sample.

chain reaction A fission reaction that will continue once it has been initiated with high-energy neutrons bombarding a heavy nucleus such as uranium-238.

curie (Ci) A unit of radiation equal to 3.7×10^{10} disintegrations/s.

fission A process in which heavy nuclei are split into smaller atoms, releasing neutrons and large amounts of energy.

fusion A reaction in which large amounts of energy are released when small nuclei combine to form larger nuclei.

gamma ray High-energy radiation emitted as a nucleus becomes more stable; its symbol is γ.

half-life The time it takes for one-half of a radioactive sample to decay.

ionizing radiation High-energy radiation, such as alpha particles, beta particles, and gamma rays, that produces unstable ions and radicals when it reacts with molecules such as water in the body. The resulting unstable products cause biological damage and may produce cancers.

nuclear equation An equation that represents a nuclear transformation in which the sum of atomic numbers and mass numbers must be equal for reactants and products.

rad Radiation absorbed dose. A measure of the amount of radiation absorbed by the body.

radiation therapy The use of high doses of radiation to destroy harmful tissues in the body.

radioactive An isotope with an unstable nucleus.

radioactive decay The breakdown of a radioactive nucleus accompanied by the release of radiation.

radioisotope A radioactive form of an element.

rem Radiation equivalent in humans. A measure of the biological damage caused by the various kinds of radiation.

scan The image of an organ in the body created by radioactive isotopes that have accumulated in that organ.

shielding Materials necessary to provide protection from dangerous radiation sources.

transmutation The formation of a radioactive isotope by bombarding a stable nucleus with fast-moving particles.

Problems

Radioactivity *(Goal 6.1)*

6.1 Naturally occurring potassium consists of three isotopes: nonradioactive potassium-39, radioactive potassium-40, and radioactive potassium-41. In what ways are the isotopes similar, and in what ways do they differ?

6.2 Supply the missing information in the following table:

Medical Use	Nuclear Symbol	Mass Number	Number of Protons	Number of Neutrons
Spleen imaging	$^{51}_{24}Cr$	______	______	______
Malignancies	______	60	27	______
Blood volume	______	______	26	33
Hyperthyroidism	$^{131}_{53}I$	______	______	______
Leukemia treatment	______	32	______	17

6.3 Write a symbol for the following:
a. alpha particle
b. neutron
c. beta particle
d. proton
e. gamma ray

6.4 Identify the symbol for X in each of the following nuclear particles:
a. ${}^{0}_{-1}X$ b. ${}^{4}_{2}X$ c. ${}^{1}_{0}X$ d. ${}^{0}_{0}X$ e. ${}^{1}_{1}X$

6.5 Why does beta radiation penetrate farther than alpha radiation in solid material?

6.6 How does ionizing radiation cause damage to the cells of the body?

6.7 Why are cancer cells more sensitive to radiation than nerve cells?

6.8 As a nurse in an oncology unit, you sometimes give an injection of a radioisotope. What are three ways you can minimize your exposure to radiation?

6.9 What is the purpose of placing a lead apron on a patient who is receiving routine dental x-rays?

Nuclear Equations *(Goal 6.2)*

6.10 How can you check that a nuclear equation is correctly balanced?

6.11 Write a balanced nuclear equation for the alpha decay of the following radioactive isotopes:
a. ${}^{208}_{84}Po$ b. ${}^{232}_{90}Th$ c. ${}^{251}_{102}No$ d. ${}^{220}_{86}Rn$

6.12 The following radioisotopes are alpha emitters. Write the nuclear equation for each nuclear reaction.
a. curium-243 b. einsteinium-252 c. californium-251

6.13 Write a balanced nuclear equation for the beta decay of each of the following radioactive isotopes:
a. ${}^{25}_{11}Na$ b. ${}^{20}_{8}O$ c. ${}^{92}_{38}Sr$ d. ${}^{42}_{19}K$

6.14 Iron has four stable isotopes whose mass numbers are 54, 56, 57, and 58. Iron-59 and iron-60 are beta emitters. Write equations for their nuclear reactions.

6.15 Write the nuclear symbol for the missing particle or isotope in the following nuclear equations:

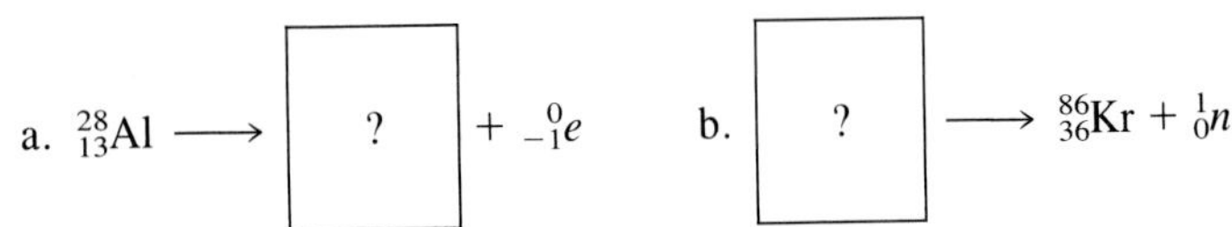

c. $^{66}_{29}Cu \longrightarrow ^{66}_{30}Zn + \boxed{?}$

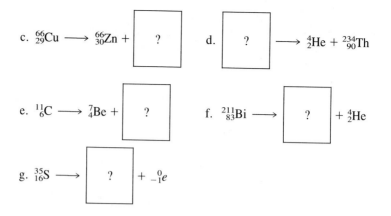

d. $\boxed{?} \longrightarrow ^{4}_{2}He + ^{234}_{90}Th$

e. $^{11}_{6}C \longrightarrow ^{7}_{4}Be + \boxed{?}$

f. $^{211}_{83}Bi \longrightarrow \boxed{?} + ^{4}_{2}He$

g. $^{35}_{16}S \longrightarrow \boxed{?} + ^{0}_{-1}e$

Detecting and Measuring Radiation *(Goal 6.3)*

6.16 How does a Geiger counter detect radiation?

6.17 State the name of the unit that describes each of the following radiation measurements:
a. the activity of a sample
b. the absorbed dose of radiation by tissue
c. the biological effect of radiation

6.18 A sample of iodine-131 has an activity of 3.0 Ci. How many disintegrations occur in the sample within 20 s?

6.19 The recommended dosage of iodine-131 is 4.20 μCi/kg body weight. How many microcuries of iodine-131 are needed for a 70.0-kg patient with hyperthyroidism?

6.20 The dosage of technetium-99m for a lung scan is 20 μCi/kg body weight. How many millicuries should be given to a 50.0-kg patient? (1 mCi = 1000 μCi)

6.21 a. A patient receives 50 mrads in a chest x-ray. What would be the absorbed dose in millirems?
b. Suppose a person absorbed 50 mrads of alpha radiation. What would that be in millirems? How does it compare to the dosage in part (a)?

6.22 Why would an airline pilot be exposed to more background radiation than the airline reservationist who works at the ticket counter?

6.23 In radiation therapy, a patient receives high doses of radiation. What kinds of symptoms of radiation sickness might be exhibited by the patient?

Medical Applications of Radioisotopes *(Goal 6.4)*

6.24 Bone and bony structures consist primarily of $Ca_3(PO_4)_2$. Why are the radioisotopes of calcium-47, phosphorus-32, and strontium-85 used in the diagnosis and treatment of bone lesions and bone tumors?

6.25 A patient with polycythemia vera (excessive production of red blood cells) receives radioative phosphorus-32. Why would this treatment reduce the production of red blood cells in the bone marrow of the patient?

6.26 Treatment with iodine-131 decreases the amount of hormone produced by the thyroid gland. Why?

***6.27** What method is used to obtain a scan in each of the following techniques?
a. radioisotope
b. computerized tomography (CT scans)
c. magnetic resonance imaging (MRI scans)
d. positron emission tomography (PET scans)

Half-Life of a Radioisotope *(Goal 6.5)*

6.28 Why do radioisotopes used for diagnosis in nuclear medicine have short half-lives?

6.29 Technetium-99m is an ideal radioisotope for scanning organs because it has a half-life of 6.0 hr and is a pure gamma emitter. Suppose that 80.0 mg had been prepared in the technetium generator this morning. How many milligrams would remain after:
a. one half-life
b. two half-lives
c. 18 hr
d. 24 hr

6.30 A sample of sodium-24 with an activity of 12 mCi is used to study the rate of blood flow in the circulatory system. If sodium-24 has a half-life of 15 hr, what is the activity of the sodium after 2½ days?

6.31 Strontium-85, used for bone scans, has a half-life of 64 days. How long will it take for the radiation level of strontium-85 to drop to one-fourth of its original level? To one-eighth?

***6.32** A nurse was accidentally exposed to potassium-42 while doing some brain scans for possible brain tumors. The error was not discovered until 36 hr later, when the activity of the potassium-42 sample was 2.0 μCi/g. If potassium-42 has a half-life of 12 hr, what was the activity of the sample at the time the nurse was exposed?

***6.33** Fluorine-18, which has a half-life of 110 min, is used in PET scans. If a 100-mg quantity of fluorine-18 is shipped from the radioisotope supplier at 8:00 A.M. and arrives at the radiology laboratory at 1:30 P.M., how many milligrams are still active?

6.34 A wooden object from the site of a Mayan temple has a carbon-14 activity of 10 counts per minute (cpm) compared to a reference piece of wood cut today, which has an activity of 40 cpm. If the half-life for carbon-14 is 5730 years, what is the age of the ancient wood object?

Producing Radioisotopes *(Goal 6.6)*

6.35 Complete the following nuclear equations for bombardment reactions by writing the nuclear symbol for the unknown component:

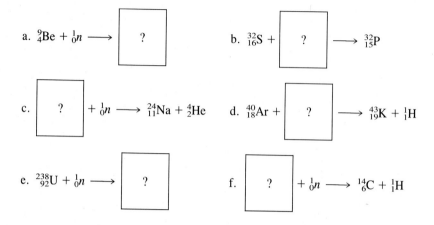

a. $^{9}_{4}Be + ^{1}_{0}n \longrightarrow$ [?]

b. $^{32}_{16}S +$ [?] $\longrightarrow ^{32}_{15}P$

c. [?] $+ ^{1}_{0}n \longrightarrow ^{24}_{11}Na + ^{4}_{2}He$

d. $^{40}_{18}Ar +$ [?] $\longrightarrow ^{43}_{19}K + ^{1}_{1}H$

e. $^{238}_{92}U + ^{1}_{0}n \longrightarrow$ [?]

f. [?] $+ ^{1}_{0}n \longrightarrow ^{14}_{6}C + ^{1}_{1}H$

***6.36** When californium-249 is bombarded by oxygen-18, a new element 106 and four neutrons are produced. Write the balanced nuclear equation for the transmutation reaction.

Nuclear Fission and Fusion *(Goal 6.7)*

6.37 How does a chain reaction occur in nuclear fission?

6.38 Complete the following fission reaction:

$^{235}_{92}U + ^{1}_{0}n \longrightarrow ^{131}_{50}Sn +$ [?] $+ 2\ ^{1}_{0}n +$ Energy

6.39 Indicate whether each of the following is characteristic of the fission process, the fusion process, or both.

a. Neutrons bombard a nucleus.
b. This process occurs in the sun.
c. A large nucleus splits into smaller nuclei.
d. Small nuclei combine to form larger nuclei.
e. Very high temperatures are required to initiate the reaction.
f. Less radioactive waste results.
g. Hydrogen and helium nuclei are the reactants.
h. Large amounts of energy are released when the reaction occurs.
i. This is used in some electrical power plants to produce electricity.

$PV = nRT$
(The Ideal Gas Law)

7

Gases

he nitrogen in air breathed nder pressure at depths of 100 to 00 feet can be harmful to ivers; thus the oxygen in uba tanks is mixed with elium.

Learning Goals

7.1 Describe the properties of a gas according to the kinetic theory of gases.

7.2 Describe the units of measurement used for pressure and change from one unit to another.

7.3 Use the pressure–volume relationship (Boyle's law) to determine the new pressure or volume of a certain amount of gas at a constant temperature.

7.4 Use the temperature–volume relationship (Charles' law) to determine the new temperature or volume of a certain amount of gas at a constant pressure.

7.5 Use the temperature-pressure relationship (Gay-Lussac's law) to determine the new temperature or pressure of a certain amount of gas at a constant volume.

7.6 Use the combined gas law to find the new pressure, volume, or temperature of a gas when changes in two of these properties are given.

7.7 Describe the relationship between the amount of a gas and its volume and use this relationship in calculations.

7.8 Use partial pressures to calculate the total pressure of a mixture of gases.

7.9 Use the ideal gas law to solve for pressure, volume, temperature, or amount of a gas when given values for the other properties.

SCOPE

We all live at the bottom of a sea of gases called the atmosphere. The most important of these gases is oxygen, which constitutes about 21% of the atmosphere. Without oxygen life on this planet would be impossible, because oxygen is vital to all life processes of plants and animals. Ozone (O_3), formed in the upper atompshere by the interaction of oxygen with ultraviolet light, absorbs some of the harmful ultraviolet radiation from the sun before it can strike the earth's surface. The other gases in the atmosphere include nitrogen (78% of the atmosphere) and argon, carbon dioxide (CO_2), and water vapor (the chief components of the remaining 1%). Carbon dioxide gas, a product of human cellular metabolism, is used by plants in a process called photosynthesis, which produces the oxygen that is essential to respiration in humans.

The atmosphere has become a dumping ground for other gases, such as methane, chlorofluorohydrocarbons (CFCs), sulfur dioxide, and nitric oxides. The chemical reactions of these gases with sunlight and oxygen in the air are contributing to air pollution, ozone depletion, global warming, and acid rain. Such chemical changes can seriously affect our health and the way all of us live. A knowledge of the behavior of gases and some of the laws that govern gas behavior can help us understand the nature of matter and allow us to make decisions concerning the important environmental and health issues we face today.

7.1 Properties of Gases

Learning Goal **Describe the properties of a gas according to the kinetic theory of gases.**

In Chapter 2, we saw that the behavior of gases is quite different from that of liquids and solids. Gas particles are far apart, whereas particles of both liquids and solids are held close together because of strong attractive forces. This means that a gas has no definite shape or volume and will completely fill any container. Because there are great distances between its particles, a gas is less dense than a solid or liquid and can be compressed. A model for the behavior of a gas, called the **kinetic theory,** helps us understand gas behavior.

Kinetic Theory of Gases

1. A gas is composed of very small particles (molecules or atoms).
2. The particles of a gas are very far apart. Thus, a gas is mostly empty space, and we assume that the volume of a container of gas is the same as the volume of the gas.
3. Gas particles move rapidly, colliding with other gas particles and with the walls of the container.
4. Gas particles do not attract or repel one another.
5. The kinetic energy of gas particles is related to the temperature of the gas; particle motion increases when the temperature increases.

The kinetic theory helps explain some of the characteristics of gases. For example, we can quickly smell perfume from a bottle that is opened on the other side of a room, because its particles move rapidly in all directions. They move faster at higher temperatures, and more slowly at lower temperatures. Sometimes tires and gas-filled containers explode when temperatures are too high. From the kinetic theory, we know that gas particles move faster when heated, hit the walls of a container with more force, and cause a buildup of pressure inside a container.

Sample Problem 7.1
Kinetic Theory of Gases

How are the following characteristics considered in the kinetic theory of gases?

a. attraction between gas particles
b. velocity of gas particles

Solution:

a. It is assumed that there are no attractions between the particles in a gas.
b. The particles of a gas move rapidly.

Study Check:
What effect does increasing the temperature have on the pressure of a gas?

Answer:
Since increasing the temperature increases the motion of the gas particles, they move faster and hit the walls of the container with more force, increasing the pressure of the gas in the container.

When we study a gas, we describe, measure, and relate four properties: pressure (*P*), volume (*V*), temperature (*T*), and the amount of gas involved, usually expressed as the number of moles (*n*).

Pressure (*P*)

The pressure of a gas is the result of a force that is created when gas particles hit the walls of a container. In measuring the pressure of a gas, typical units used are atmosphere (atm) and millimeters of mercury (mm Hg); the latter is also called torr. Other pressure units include inches or centimeters of mercury, and kilopascals.

Volume (*V*)

Since a gas completely fills its container, the volume of the gas is equal to the volume of the container. The volume of a gas in a 10-mL vial is 10 mL; the volume of oxygen in a 25-L tank is 25 L. The units for the volume of a gas are usually milliliters (mL) or liters (L).

Temperature (T)

All calculations with gases use the Kelvin temperature scale. If a gas were to reach a temperature of absolute zero (0 K), its particles would have no energy or motion. In Chapter 2 you learned that the relationship between degrees Celsius and kelvins is

$$K = °C + 273$$

Amount of Gas (n)

When you add air to car or bicycle tires, you increase the quantity of gas and therefore increase the pressure in the tires. In gas laws, the amount of gas in a container is usually stated in moles (n).

A summary of the four properties of a gas is given in Table 7.1.

Table 7.1 Properties of a Gas

Property	**Units of measurement**
Pressure (P)	Atmosphere (atm); torr or millimeter of mercury (mm Hg)
Volume (V)	Liter (L); milliliter (mL)
Temperature (T)	Kelvin (K)
Amount (n)	Mole

Sample Problem 7.2
Identifying Gas Properties

Use the word *pressure, volume, temperature,* or *amount* to indicate the property of a gas described in each of the following statements or measurements:

a. 800 torr
b. 295 K
c. 4.0 L
d. the space occupied by a gas

Solution:

a. pressure
b. temperature
c. volume
d. volume

Study Check:
When air is added to a bicycle tire, the number of moles of air increases. What property of a gas is described?

Answer: amount

7.2 Gas Pressure

Learning Goal **Describe the units of measurement used for pressure and change from one unit to another.**

When water boils in a pan covered with a lid, the collisions of the molecules of steam (gas) lift up the lid. The gas molecules exert **pressure,** which is defined as a force acting on a certain area.

$$\text{Pressure } (P) = \frac{\text{Force}}{\text{Area}}$$

The air that covers the surface of the earth, the atmosphere, contains vast numbers of gas particles. Because the air particles have mass, they are pulled toward the earth by gravity, where they exert an **atmospheric pressure.**

The atmospheric pressure can be measured by using a **barometer,** as shown in Figure 7.1. A long glass tube is closed on one end and filled with mercury. Then its open end is placed in a dish of mercury. The weight of the mercury in the tube begins to push it out of the tube. However, the pressure of the atmosphere against the mercury in the dish will push the mercury back up the tube.

Eventually, the mercury in the tube reaches a level where its weight is equal to the atmospheric pressure. At a pressure of 1 atmosphere (atm), the mercury column would be 760 mm high. We say that the atmospheric pressure is 760 mm Hg (millimeters of mercury), which is the same as a standard **atmosphere (atm).**

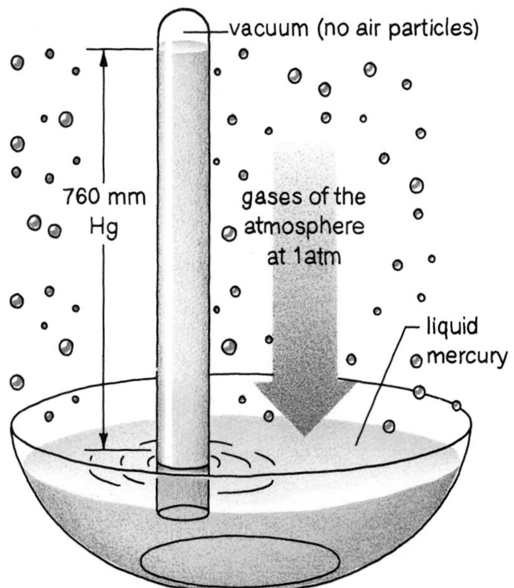

Figure 7.1
A barometer: A column of mercury 760 mm high is supported by the pressure exerted by the gases in the atmosphere at sea level (P = 1 atm).

This pressure is often expressed as 760 **torr,** a pressure unit named to honor Evangelista Torricelli, the inventor of the barometer. Because they are equal, units of torr and mm Hg are used interchangeably.

1 mm Hg = 1 torr
1 atm = 760 mm Hg = 760 torr

In SI units, pressure is measured in pascals (Pa); 1 atm is equal to 101,325 Pa. Since a pascal is a very small unit, it is likely that pressures would be reported in kilopascals. However, the unit is not yet in common use.

1 atm = 101,325 Pa = 101.3 kPa

If you have a barometer in your home, it probably gives pressure in inches of mercury. One atmosphere is also equal to the pressure of a column of mercury that is 29.9 in. high. Weather reports are given in inches of mercury. A lowering of pressure often indicates rain or snow, whereas an increase in pressure (referred to as a high-pressure system) usually brings dry and sunny weather.

The American equivalent of 1 atm is 14.7 pounds per square inch (lb/in.^{2}). This is the measurement you see on a pressure gauge when you check the air pressure in the tires of a car or bicycle. Table 7.2 summarizes the various units used in the measurement of pressure.

Table 7.2 Units for Measuring Pressure

Unit	Abbreviation	Unit equivalent to 1 atm
Atmosphere	atm	1 atm
Millimeters of Hg	mm Hg	760 mm Hg
Torr	torr	760 torr
Inches of Hg	in. Hg	29.9 in. Hg
Pounds per square inch	lb/in.^{2} (psi)	14.7 lb/in.^{2}
Pascal	Pa	101,325 Pa

Atmospheric Pressure and Altitude

At sea level, where there are more air particles, the atmospheric pressure is about 1 atm. In Los Angeles, atmospheric pressure can fluctuate between 730 and 760 mm Hg, depending on the weather. At higher altitudes, there are fewer air particles and the atmospheric pressure is less than 1 atm. In Denver (1.6 km), a typical atmospheric pressure might be 630 mm Hg, and if you were climbing Mount Everest (9.3 km) it might get as low as 270 mm Hg.

Deep-sea divers must be concerned about increasing pressures on their ears and lungs when they dive below the surface of the ocean. Since water is denser than air, the pressure on a diver increases rapidly as the diver descends. At a depth of 33 ft below the surface of the ocean, there is an additional atmosphere of pressure exerted by the water on a diver, for a total of 2 atm. At 100 ft down, there is a total of 4 atm on a diver. The air tanks a diver carries continuously adjusts the pressure of the breathing mixture to match the increase of pressure on the body.

Sample Problem 7.3
Units of Pressure Measurement

A sample of neon gas has a pressure of 0.50 atm. Give the pressure of the neon in:

a. millimeters of Hg **b.** inches of Hg

Solution:

a. The equality 1 atm = 760 mm Hg can be written as conversion factors:

$$\frac{760 \text{ mm Hg}}{1 \text{ atm}} \quad \text{or} \quad \frac{1 \text{ atm}}{760 \text{ mm Hg}}$$

Using the appropriate conversion factor, the problem is set up as

$$0.50 \not{\text{atm}} \times \frac{760 \text{ mm Hg}}{1 \not{\text{atm}}} = 380 \text{ mm Hg}$$

b. One atm is equal to 29.9 in. Hg. Using this equality as a conversion factor in the problem setup, we obtain

$$0.50 \not{\text{atm}} \times \frac{29.9 \text{ in. Hg}}{1 \not{\text{atm}}} = 15 \text{ in. Hg}$$

Study Check:
What is the pressure in atmospheres for a gas that has a pressure of 655 torr?

Answer: 0.862 atm

Vapor Pressure

When liquids evaporate, gas vapor forms at the surface of the liquid and escapes. If the liquid is in an open container, it can all eventually evaporate. In a closed container, the vapor accumulates and creates pressure called **vapor pressure.** Each liquid exerts its own vapor pressure at a given temperature. As temperature increases, more vapor forms, and vapor pressure increases. Table 7.3 lists the vapor pressure of water at various temperatures.

Vapor Pressure and Boiling Point

A liquid reaches its boiling point when its vapor pressure becomes equal to the atmospheric pressure. As boiling occurs, bubbles of the gas form within the

Table 7.3 Vapor Pressure of Water

Temperature (°C)	Vapor pressure (mm Hg)
0	5
10	9
20	18
30	32
37	47[a]
40	55
50	93
60	149
70	234
80	355
90	528
100	760

[a]At body temperature.

liquid and quickly rise to the surface. For example, at an atmospheric pressure of 760 mm Hg, water will boil at 100°C, the temperature at which its vapor pressure reaches 760 mm Hg. (See Figure 7.2.)

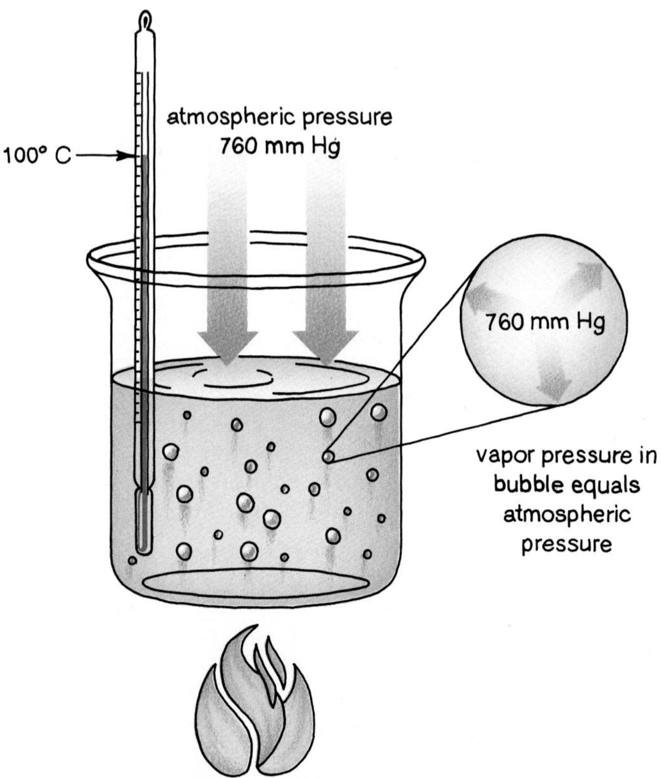

Figure 7.2
Vapor pressure and boiling point for water at 1 atm pressure (760 mm Hg).

At higher altitudes, atmospheric pressures are lower and the boiling point of water is less than 100°C. Earlier, we saw that the typical atmospheric pressure in Denver is 630 mm Hg. This means that water in Denver needs a vapor pressure of 630 mm Hg to boil. Since water has a vapor pressure of 630 mm Hg at 95°C, water boils at 95°C in Denver. Water boils at even lower temperatures at still lower atmospheric pressures, as Table 7.4 shows.

Table 7.4 Altitude, Atmospheric Pressure, and the Boiling Point of Water

Location	Altitude (km)	Atmospheric pressure (mm Hg)	Boiling point (°C)
Sea level	0	760	100
Los Angeles	0.09	752	99
Las Vegas	0.70	700	98
Denver	1.60	630	95
Mount Whitney	4.00	467	87
Mount Everest	9.30	270	70

People who live at high altitudes often use pressure cookers to obtain higher temperatures when preparing food. When the pressure exerted is greater than 1 atm, a temperature higher than 100°C is needed before water will boil. Laboratories and hospitals use devices called autoclaves to sterilize laboratory and surgical equipment. An autoclave, like a pressure cooker, is a closed container that increases the pressure above the water so it will boil at higher temperatures. Table 7.5 shows how the boiling point of water increases as pressure increases.

Table 7.5 Pressure and the Boiling Point of Water

Pressure (mm Hg)	Boiling point (°C)
760	100
800	100.4
1075	110
1520 (2 atm)	120
2026	130
7600 (10 atm)	180

7.3 Pressure and Volume (Boyle's Law)

Learning Goal **Use the pressure–volume relationship (Boyle's law) to determine the new pressure or volume of a certain amount of gas at a constant temperature.**

Imagine that you can see air particles hitting the walls inside a bicycle pump. What happens to the pressure inside the pump as we push down on the handle?

As the air is compressed, the air particles are crowded together. In the smaller volume, more collisions occur, and the air pressure increases.

When a change in one property (in this case, volume) causes a change in another property (in this case, pressure), those properties are said to be related to each other. Furthermore, when one change causes a change in the opposite direction, such as an increase in pressure causing a decrease in volume, the properties are said to be inversely related and have an **inverse relationship.** The relationship between the pressure and volume of a gas is known as **Boyle's law.** The law states that the volume (V) of a sample of gas changes inversely with the pressure (P) of the gas as long as there has been no change in the temperature (T) or amount of gas (n).

Sample Problem 7.4
Relationship of Pressure and Volume

Complete the information for the pressure and volume of a gas sample at constant temperature.

	Pressure	*Volume*
a.	increases	______
b.	______	increases

Solution:

a. Pressure and volume are inversely related. When the pressure of a gas increases, the volume decreases.

b. If the volume of the gas sample increases, the pressure of the gas decreases.

Study Check:

If the pressure of a gas decreases, what happens to its volume if no change has occurred in the temperature or in the amount of the gas?

Answer:

The volume increases.

In mathematical terms, the initial pressure (P_i) multiplied by the initial volume (V_i) of a gas gives a constant (C) value.

$$P_iV_i = \text{Constant } (C)$$

If we change the volume or pressure of the gas without any change occurring in the temperature or in the amount of the gas, the new final presssure and volume will also equal the same constant.

$$P_fV_f = C$$

Since the constant has the same value under our conditions, we can set the initial and final PV conditions of Boyle's Law equal to each other.

$$P_iV_i = P_fV_f \quad \text{(no change in number of moles and temperature)}$$

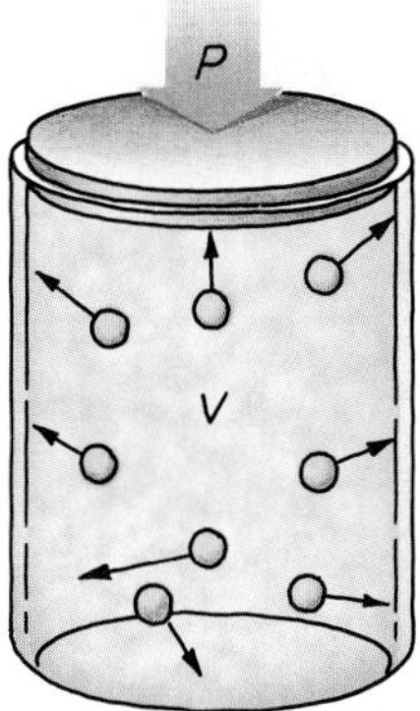

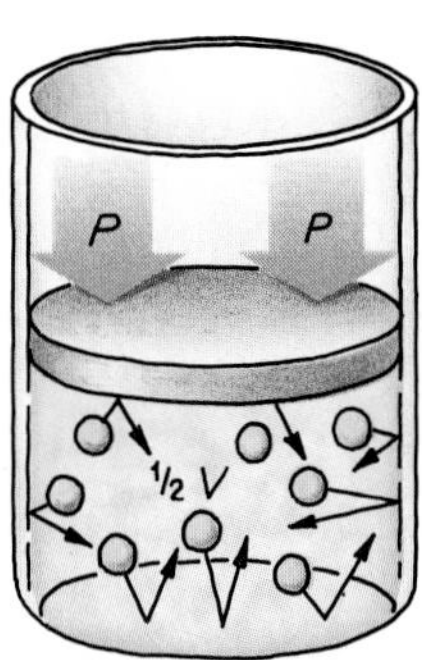

Figure 7.3
When the volume of a gas decreases, its pressure increases as long as there is no change in the temperature or the amount of the gas.

Sample Problem 7.5
Calculating Pressure When Volume Changes

A sample of hydrogen gas (H_2) has a volume of 4.0 L and a pressure of 1.0 atm. What is the new pressure if the volume is decreased to 2.0 L?

Solution:
In calculations with gas laws, it is helpful to organize the data in a table of initial and final conditions.

Initial Conditions	*Final Conditions*	*Change*
$P_i = 1.0$ atm	$P_f = ?$	P must increase
$V_i = 4.0$ L	$V_f = 2.0$ L	V decreases

In this problem, we want to know the final pressure (P_f) for the change in volume. Figure 7.3 illustrates the change.

The problem setup begins with the initial pressure of 1.0 atm followed by a volume factor that is written from the initial and final volumes.

$$\text{New pressure} = \text{Initial pressure} \times \text{Volume factor}$$

We can write two possible volume factors from the given information. One factor will decrease pressure, and the other will increase pressure.

Volume Factors

$$\frac{2.0 \text{ L}}{4.0 \text{ L}} \quad \text{and} \quad \frac{4.0 \text{ L}}{2.0 \text{ L}}$$

This factor will decrease pressure — This factor will increase pressure

The choice of factor depends on the way pressure will change as a result of the volume change. In this problem the volume has decreased. Since pressure and volume are inversely related, the pressure must increase. Therefore, we will use the volume factor that increases pressure.

$$\underset{\text{New pressure}}{P_f} = \underset{\text{Initial pressure}}{1.0 \text{ atm}} \times \underset{\text{Volume factor that increases pressure}}{\frac{4.0 \cancel{\text{L}}}{2.0 \cancel{\text{L}}}} = 2.0 \text{ atm}$$

The pressure of the hydrogen gas increases from 1.0 atm to 2.0 atm. This is the change expected, since the volume decreased.

Alternatively, the pressure–volume equation can be solved for the final pressure by dividing both sides by V_f.

$$\frac{P_iV_i}{V_f} = \frac{P_f\cancel{V_f}}{\cancel{V_f}}$$

$$P_f = \frac{P_iV_i}{V_f}$$

Substituting the values listed in the table of information into the equation for P_f gives

$$P_f = \frac{(1.0 \text{ atm})(4.0 \cancel{\text{L}})}{(2.0 \cancel{\text{L}})} = 2.0 \text{ atm}$$

Study Check:
A sample of helium gas has a volume of 250 mL at 800 torr. If the volume is changed to 500 mL, what is the new pressure, assuming no change in temperature or number of moles?

Answer: 400 torr

Sample Problem 7.6
Calculating Volume When Pressure Changes

In the hospital respiratory unit, the gauge on a 10.0-L tank of compressed oxygen reads 4500 mm Hg. How many liters of oxygen can be delivered from the tank at a pressure of 750 mm Hg?

Solution:
Placing our information in a table gives the following:

Initial Conditions	*Final Conditions*	*Change*
$P_i = 4500$ mm Hg	$P_f = 750$ mm Hg	P decreases
$V_i = 10.0$ L	$V_f = ?$	V must increase

According to Boyle's law, a decrease in the pressure will cause an increase in the volume. The initial volume (V_i) must be multiplied by a pressure factor that gives a greater volume.

$$V_f = V_i \times \frac{P_i}{P_f}$$

$$V_f = 10.0\ \text{L} \times \frac{4500\ \cancel{\text{torr}}}{750\ \cancel{\text{torr}}} = 60\ \text{L}$$

New volume — Initial volume — Pressure factor that increases volume

Study Check:
A sample of methane gas (CH_4) has a volume of 125 mL at 0.600 atm pressure and 25°C. How many milliliters will it occupy at a pressure of 1.50 atm and 25°C?

Answer: 50.0 mL

HEALTH NOTE

Pressure–Volume Relationship in Breathing

The importance of Boyle's law becomes more apparent when you consider the mechanics of breathing. Our lungs are elastic, balloon-like structures contained within an airtight chamber called the thoracic cavity. The diaphragm, a muscle, forms the flexible floor of the cavity.

Inspiration
The process of taking a breath of air begins when the diaphragm flattens, and the rib cage expands, causing an increase in the volume of the thoracic cavity. The elasticity of the lungs allows them to expand when the thoracic cavity expands. According to Boyle's law, the pressure inside the lungs will decrease when their volume increases. This causes the pressure inside the lungs to fall below the pressure of the atmosphere. This difference in pressures produces a **pressure gradient** between the lungs and the atmosphere. In a pressure gradient, molecules flow from an area of greater pressure to an area of lower pressure, a process called **diffusion.** Thus, we inhale as air flows into the lungs (*inspiration*), until the pressure within the lungs becomes equal to the pressure of the atmosphere.

Expiration
Expiration or the exhalation phase of breathing, occurs when the diaphragm relaxes and moves back up into the thoracic cavity to its resting position. This reduces the volume of the thoracic cavity, which squeezes the lungs and decreases their volume. Now the pressure in the lungs is greater than the pressure of the atmosphere, so air flows out of the lungs. Thus, breathing is a process in which pressure gradients are continuously created between the lungs and the environment as a result of the changes in the volume and pressure. (See Figure 7.4.) ■

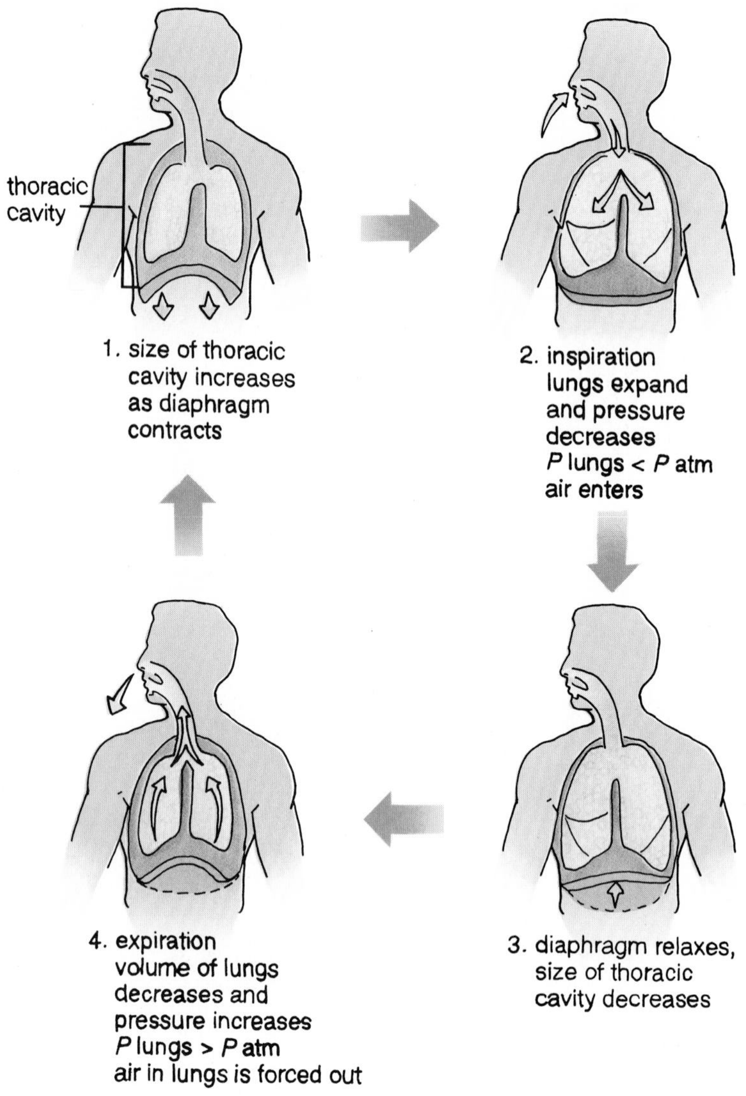
thoracic
cavity
1. size of thoracic
cavity increases
as diaphragm
contracts
2. inspiration
lungs expand
and pressure
decreases
P lungs < P atm
air enters
3. diaphragm relaxes,
size of thoracic
cavity decreases
4. expiration
volume of lungs
decreases and
pressure increases
P lungs > P atm
air in lungs is forced out

Figure 7.4
The mechanics of breathing are based on a pressure–volume relationship.

7.4 Temperature and Volume (Charles' Law)

Learning Goal **Use the temperature–volume relationship (Charles' law) to determine the new temperature or volume of a certain amount of gas at a constant pressure.**

In preparing a hot-air balloon for a flight, the air in the balloon is heated with a small propane heater. As the air warms, its volume expands. The decrease in density allows the balloon to rise. (See Figure 7.5.)

To study the effect of changing temperature on the volume of a gas, we must not change the pressure or the amount of the gas. Suppose we increase the Kelvin temperature of a gas sample. The kinetic theory shows that the activity (kinetic

Figure 7.5
As the air inside a hot-air balloon is heated, its volume expands and its density decreases. The hot-air balloon rises and floats in the air.

energy) of the gas will also increase. To keep pressure constant, the volume of the container must increase. (See Figure 7.6.) By contrast, if the temperature of the gas is lowered, the volume of the container must be reduced in order to maintain the same pressure.

Charles' law states that the temperature in kelvins and the volume of a gas are directly related when there is no change in pressure or the amount of gas. A **direct relationship** is one in which the related properties increase or decrease together. For two different conditions, the relationship of volume (V) to temperature (T) is constant as long as pressure (P) and number of moles (n) do not change.

$$\frac{V}{T} = \text{Constant } (C)$$

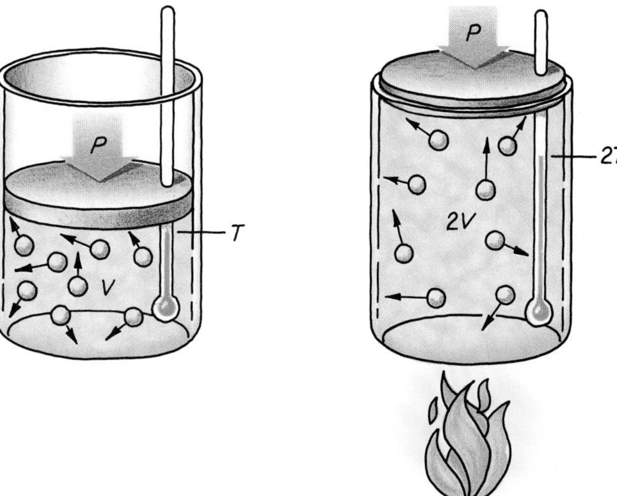

Figure 7.6
The Kelvin temperature of a gas is directly related to the volume of the gas when there is no change in pressure or amount.

For initial and final conditions, we can write

$$\frac{V_i}{T_i} = \frac{V_f}{T_f} \quad \text{(no change in pressure or number of moles)}$$

Remember that all temperatures used in gas law calculations must be in kelvins (K).

Sample Problem 7.7
Relationship of Volume and Temperature

Complete the information for the volume and temperature of a gas when the pressure is kept constant.

	Volume	*Temperature*
a.	_______	decreases
b.	increases	_______

Solution:

a. If the temperature of a gas decreases, its volume also decreases.
b. When the volume of a gas increases, the temperature also increases.

Study Check:
If the volume of a sample of gas decreases, how must the temperature change if the pressure is to remain constant?

Answer:
Temperature must decrease.

Sample Problem 7.8
Calculating Volume When Temperature Changes

A sample of neon gas at 760 torr has a volume of 5.0 L and a temperature of 17°C. Find the new volume of the gas after the temperature has been increased to 47°C at 760 torr.

Solution:
In this problem, temperature is increased while pressure remains constant. Since the temperatures are given in degrees Celsius, we must first change them to kelvins.

$$T_i = 17°\text{C} + 273 = 290 \text{ K}$$

$$T_f = 47°\text{C} + 273 = 320 \text{ K}$$

Placing the information for the gas in a table gives the following:

Initial Conditions	*Final Conditions*	*Change*
$T_i = 290$ K	$T_f = 320$ K	T increases
$V_i = 5.0$ L	$V_f = ?$	V must increase

According to the temperature–volume relationship, an increase in temperature will increase the volume. The new volume is calculated by multiplying the initial volume by a temperature factor that gives a greater volume.

$$V_f = V_i \times \text{Temperature factor}$$

$$V_f = 5.0\text{ L} \times \frac{320\,\cancel{\text{K}}}{290\,\cancel{\text{K}}} = 5.5\text{ L}$$

New volume — Initial volume — Temperature factor that increases volume

The new volume of 5.5 L is larger than the initial volume of 5.0 L, as a result of a temperature increase.

Study Check:
A mountain climber inhales 500.0 mL of air at a temperature of −10°C. What volume will the air occupy in the lungs if the climber's body temperature is 37°C?

Answer: 589 mL

7.5 Temperature and Pressure (Gay-Lussac's Law)

Learning Goal **Use the temperature–pressure relationship (Gay-Lussac's law) to determine the new temperature or pressure of a certain amount of gas at a constant volume**

If we could watch the molecules of a gas as the temperature rises, we would notice that they move faster and hit the sides of the container more often and with greater force. If we keep the volume of the container the same, we would observe an increase in the pressure. A temperature–pressure relationship, also known as **Gay-Lussac's law,** states that the pressure of a gas is directly related to its Kelvin temperature. This means that an increase in temperature increases the pressure of a gas, and a decrease in temperature decreases the pressure of the gas, as long as the volume and number of moles of the gas stay the same. (See Figure 7.7.)

The ratio of pressure (P) to temperature (T) is the same under all conditions as long as volume (V) and amount of gas (n) do not change.

$$\frac{P_i}{T_i} = \frac{P_f}{T_f}$$

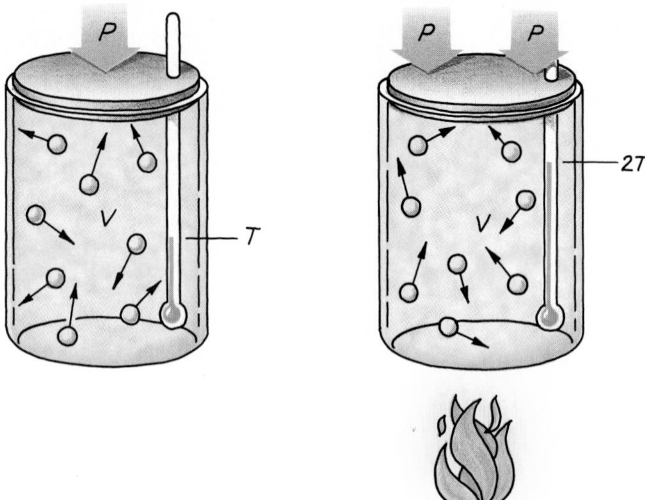

Figure 7.7
The Kelvin temperature of a gas is directly related to the pressure of the gas when there is no change in volume or amount.

Sample Problem 7.9
Relationship of Temperature and Pressure

Complete the information for the temperature and pressure of a gas when volume is kept constant.

	Temperature	*Pressure*
a.	increases	_______
b.	_______	decreases

Solution:

a. If the temperature increases, the pressure of the gas also increases.
b. If the pressure decreases, the temperature of the gas also decreases.

Study Check:
If the temperature in a tire decreases, how will pressure change?

Answer:
Pressure will decrease.

Sample Problem 7.10
Calculating Pressure When Temperature Changes

Aerosol cans can be dangerous if they are heated because they can explode. Suppose a can of hair spray with a pressure of 4.0 atm at a room temperature of 27°C is thrown into a fire. If the temperature of the gas inside the aerosol can reaches 402°C, what will be its pressure? The aerosol can may explode if the pressure inside exceeds 8.0 atm. Would you expect the can to explode?

Solution:
Since the temperatures are given in degrees Celsius, we must first change them to kelvins.

$$T_i = 27°\text{C} + 273 = 300 \text{ K}$$

$$T_f = 402°\text{C} + 273 = 675 \text{ K}$$

Placing the information for the gas in a table gives the following:

Initial Conditions	*Final Conditions*	*Change*
$P_i = 4.0$ atm	$P_f = ?$	P must increase
$T_i = 300$ K	$T_f = 675$ K	T increases

Since the temperature of the gas has increased, the final pressure must increase. In the calculation, we use a temperature factor that gives a greater pressure.

$$P_f = 4.0 \text{ atm} \times \frac{675 \cancel{\text{K}}}{300 \cancel{\text{K}}} = 9.0 \text{ atm}$$

New pressure — Initial pressure — Temperature factor that increases pressure

Since the calculated pressure exceeds 8.0 atm, we might expect the can to explode.

Study Check:
In a storage area where the temperature has reached 55°C, the pressure of oxygen gas in a 15.0-L steel cylinder is 965 torr. To what temperature would the gas have to be cooled in order to reduce the pressure to 850 torr?

Answer: 16°C

7.6 The Combined Gas Law of Pressure, Volume, and Temperature Relationships

Learning Goal **Use the combined gas law to find the new pressure, volume, or temperature of a gas when changes in two of these properties are given.**

All of the pressure–volume–temperature relationships for gases that we have studied may be combined into a single relationship called the **combined gas law.** This expression is useful for studying the effect of changes in two of these variables on the third as long as the amount of gas (number of moles) remains constant.

$$\frac{P_iV_i}{T_i} = \frac{P_fV_f}{T_f} \quad \text{(no change in amount of gas)}$$

By remembering the combined gas law, we can derive the gas laws we have studied by omitting those properties that do not change. Table 7.6 summarizes the pressure–volume–temperature relationships of gases.

Table 7.6 Summary of Gas Laws

Combined gas law	Properties held constant	Properties that change	Relationship
$\frac{P_iV_i}{T_i} = \frac{P_fV_f}{T_f}$	n, T	P, V	Boyle's law: $P_iV_i = P_fV_f$
$\frac{P_iV_i}{T_i} = \frac{P_fV_f}{T_f}$	n, P	V, T	Charles' law: $\frac{V_i}{T_i} = \frac{V_f}{T_f}$
$\frac{P_iV_i}{T_i} = \frac{P_fV_f}{T_f}$	n, V	P, T	Gay-Lussac's law: $\frac{P_i}{T_i} = \frac{P_f}{T_f}$

Sample Problem 7.11
Using the Combined Gas Law

A 25.0-mL bubble is released from a diver's air tank at a pressure of 4.0 atm and a temperature of 11°C. What is the volume of the bubble when it reaches the ocean surface, where the pressure is 1.0 atm and the temperature is 18°C?

Solution:
Since the temperatures are given in degrees Celsius, we must first change them to kelvins.

$$T_i = 11°\text{C} + 273 = 284 \text{ K}$$

$$T_f = 18°\text{C} + 273 = 291 \text{ K}$$

Placing the information in a table gives the following:

Initial Conditions	*Final Conditions*	*Change*
$P_i = 4.0$ atm	$P_f = 1.0$ atm	P decreases
$V_i = 25.0$ mL	$V_f = ?$	
$T_i = 284$ K	$T_f = 291$ K	T increases

To set up the calculation, we will first consider the effect of the pressure change, and then the effect of the temperature change, on the volume. Since the effect of the pressure change will increase volume (Boyle's law), we multiply the initial volume by a pressure factor that gives a greater volume.

$$V_f = 25.0 \text{ mL} \times \frac{4.0 \text{ atm}}{1.0 \text{ atm}} \times ?$$

New volume   Initial volume   × Pressure factor that increases volume

Now we can consider the effect of the temperature increase, which would cause another increase in volume (Charles' law). We multiply again, this time by a temperature factor that increases the volume.

$$V_f = 25.0 \text{ mL} \times \frac{4.0 \cancel{\text{atm}}}{1.0 \cancel{\text{atm}}} \times \frac{291 \cancel{\text{K}}}{284 \cancel{\text{K}}} = 102 \text{ mL}$$

New volume = Initial volume × Pressure factor that increases volume × Temperature factor that increases volume

The pressure decrease and temperature increase combine in their effect, which is to increase the volume.

Study Check:
A weather balloon is filled with 15.0 L of helium at a temperature of 25°C and a pressure of 685 mm Hg. What is the pressure of the helium in the balloon in the upper atmosphere when the temperature is −35°C and the volume becomes 34.0 L?

Answer: 241 mm Hg

7.7 Volume and Moles (Avogadro's Law)

Learning Goal **Describe the relationship between the amount of a gas and its volume and use this relationship in calculations.**

In our study of the gas laws, we have looked at changes in properties for a specified amount (n) of gas. Now we will consider how the properties of a gas change when there is a change in number of moles or grams. For example, when you blow up a balloon, its volume increases because you add more air molecules. If a basketball gets a hole in it, and some of the air leaks out, its volume decreases. In 1811, Amedeo Avogadro stated that the volume of a gas is directly related to the number of moles of a gas when temperature and pressure are not changed. We refer to this statement as **Avogadro's law.** If the moles of a gas are doubled, then the volume will double as long as we do not change the pressure or the temperature. (See Figure 7.8.) For two conditions, we can write

$$\frac{V_i}{n_i} = \frac{V_f}{n_f}$$

Figure 7.8
If the amount (number of moles) of a gas is doubled, the volume of the gas doubles when there is no change in pressure or temperature.

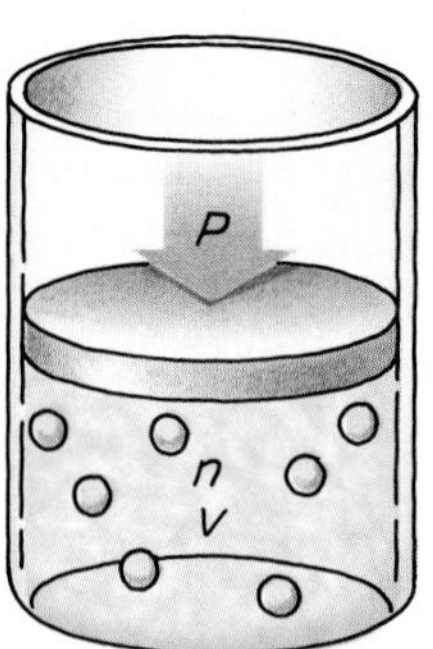

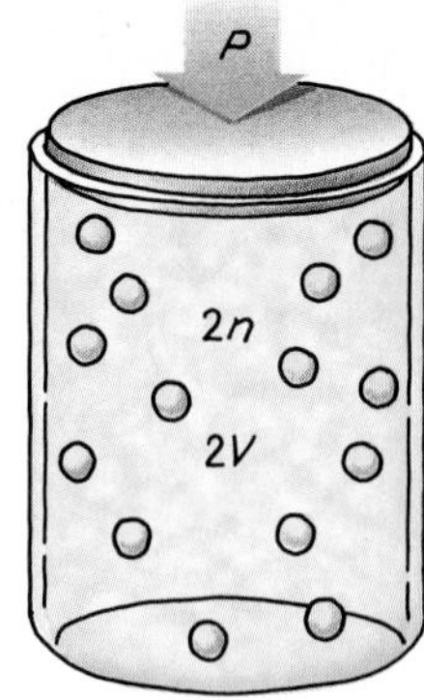

Sample Problem 7.12
Calculating Volume for a Change in Moles

A balloon with a volume of 220 mL is filled with 2.0 moles of helium. To what volume will the balloon expand if 3.0 moles of helium are added, to give a total of 5.0 moles of helium, and the pressure and temperature do not change?

Solution:
A data table for our given information can be set up as follows:

Initial Conditions	*Final Conditions*	*Change*
$V_i = 220$ mL	$V_f = ?$	V must increase
$n_i = 2.0$ moles	$n_f = 5.0$ moles	n increases

Since the amount of gas (number of moles) has increased, the volume of the balloon must increase according to Avogadro's law.

$$V_f = 220 \text{ mL} \times \frac{5.0 \cancel{\text{moles}}}{2.0 \cancel{\text{moles}}} = 550 \text{ mL}$$

New volume — Initial volume — Mole factor that increases volume

Study Check:
At a certain temperature and pressure, 8.00 g of oxygen has a volume of 5.00 L. What is the volume after 4.00 g of oxygen is added to the balloon?

Answer: 7.50 L

Standard Temperature and Pressure (STP)

If two different scientists have two different gas samples, each at a different temperature, volume, and pressure, they cannot tell if they have the same amount of gas. However, if they adjust the temperature and pressure to match, they could

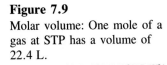

Figure 7.9
Molar volume: One mole of a gas at STP has a volume of 22.4 L.

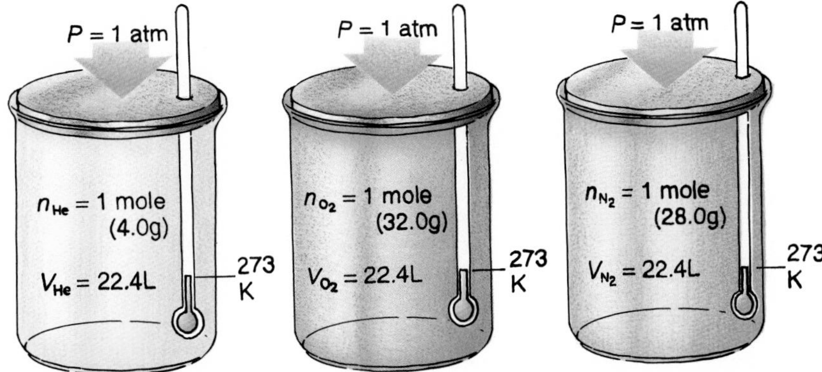

compare the two gas samples. For this purposes, scientists use the standard conditions of temperature and pressure **(STP)**, which are 0°C (273 K) and 1 atm (760 mm Hg). Then, according to Avogadro's law, a mole of any gas at STP will occupy the same volume.

Molar Volume

Suppose we have three containers, one filled with 1 mole of oxygen gas (O_2) another filled with 1 mole of nitrogen gas (N_2) and one filled with 1 mole of helium gas. At STP conditions (1 atm and 273 K), the gases would have identical volumes of 22.4 L. (See Figure 7.9.) Thus, the volume of 1 mole of any gas at STP is 22.4 L, a value called the **molar volume** of a gas.

Molar Volume

1 mole of gas (STP) = 22.4 L

As long as a gas is at STP conditions (273 K and 1 atm), the molar volume can be used to convert between the number of moles of a gas and its volume.

Sample Problem 7.13
Calculations Using Molar Volume

What is the volume of 64.0 g O_2 gas at STP?

Solution:
The mass of the gas is changed to moles by its molar mass, which is 32.0 g/mole. Since the gas is at STP, we can use the molar volume to convert from moles to liters of gas.

$$64.0\ \cancel{\text{g O}_2} \times \underbrace{\frac{1\ \cancel{\text{mole O}_2}}{32.0\ \cancel{\text{g O}_2}}}_{\text{Molar mass of O}_2} \times \underbrace{\frac{22.4\ \text{L O}_2}{1\ \cancel{\text{mole O}_2}}}_{\text{Molar volume of a gas}} = 44.8\ \text{L O}_2$$

Study Check:
How many grams of nitrogen (N_2) gas are in 5.6 L of the gas at STP?

Answer: 7.0 g N_2

7.8 Partial Pressures (Dalton's law)

Learning Goal **Use partial pressures to calculate the total pressure of a mixture of gases.**

Many gases are composed not of a single gas but of a mixture of gases. For example, the air you breathe is a mixture of mostly oxygen and nitrogen gases. However, we can still use the gas laws we have studied, because we assume that particles of all gases behave in the same way. Therefore, the total pressure of the gases in a mixture is a result of the collisions of the gas particles regardless of what type of gas they are.

In a mixture of gases, each gas exerts the pressure it would exert if it were the only gas in the container. This is called its **partial pressure.** John Dalton determined that the total pressure of a gas mixture is the sum of the partial pressures of the gases in the mixture. We call this statement **Dalton's law.** (See Figure 7.10.)

$$P_{total} = P_1 + P_2 + P_3 + \ldots$$

Total pressure of gas mixture — Sum of partial pressures of each gas in mixture

Sample Problem 7.14
Calculating the Total Pressure of a Gas Mixture

A 10-L gas tank contains propane (C_3H_8) gas at a pressure of 300 torr. Another 10-L gas tank contains methane (CH_4) gas at a pressure of 500 torr. In preparing a gas fuel mixture, the gases from both tanks are combined in 10-L container at the same temperature. What is the pressure of the gas mixture?

Solution:
Using Dalton's law of partial pressure, we find that the total pressure of the gas mixture is the sum of the partial pressures of the gases in the mixture.

$$\begin{aligned} P_{total} &= P_{propane} + P_{methane} \\ &= 300 \text{ torr} + 500 \text{ torr} \\ &= 800 \text{ torr} \end{aligned}$$

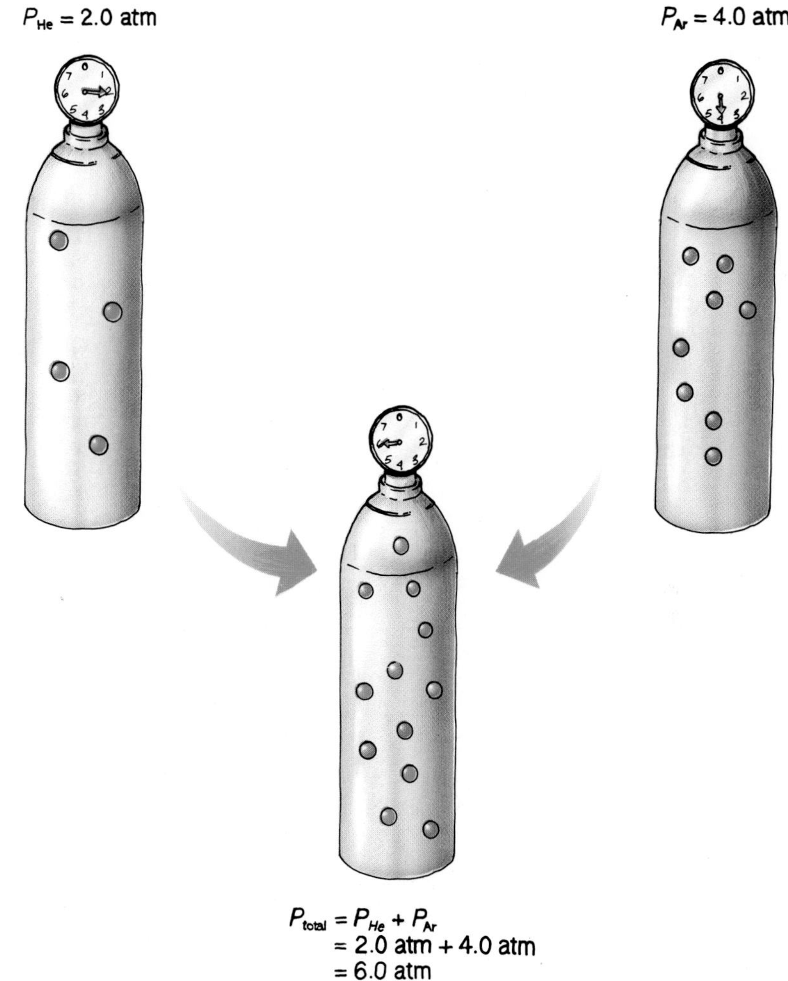

Figure 7.10
The total pressure of a gas mixture, which depends on the total number of gas particles present, is the sum of the partial pressure of the individual gases.

Therefore, when both propane and methane are placed in the same container, the total pressure of the mixture is 800 torr.

Study Check:
A gas mixture consists of helium with a partial pressure of 315 mm Hg, nitrogen with a partial pressure of 204 mm Hg, and argon with a partial pressure of 422 mm Hg. What is the total pressure in atmospheres?

Answer: 1.24 atm

Air Is a Gas Mixture

The air you breathe is a mixture of gases. What we call the atmospheric pressure is actually the sum of the partial pressures of the gases in the air. Table 7.7 lists the partial pressures of the gases in air on a typical day.

Table 7.7 Composition of Air

Gas	Pressure (torr)	Percentage
Nitrogen, N_2	594.0	78
Oxygen, O_2	160.0	21
Carbon dioxide, CO_2	0.3	1
Water vapor, H_2O	5.7	
Total air	760.0	100

Sample Problem 7.15
Partial Pressure of a Gas in a Mixture

A mixture of oxygen and helium is prepared for a scuba diver who is going to descend 200 ft below the ocean surface. At that depth, the diver breathes a gas mixture that has a pressure of 7.0 atm. If the partial pressure of the oxygen at that depth is 1.5 atm, what is the partial pressure of the helium?

Solution:
From Dalton's law of partial pressures, we know that the total pressure is equal to the sum of the partial pressures:

$$P_{\text{total}} = P_{O_2} + P_{\text{He}}$$

To solve for the partial pressure of helium (P_{He}), we rearrange the expression to give the following:

$$P_{\text{He}} = P_{\text{total}} - P_{O_2}$$
$$P_{\text{He}} = 7.0 \text{ atm} - 1.5 \text{ atm}$$
$$= 5.5 \text{ atm}$$

Thus, in the gas mixture that the diver breathes, the partial pressure of the helium is 5.5 atm.

Study Check:
An anesthetic consists of a mixture of cyclopropane gas, C_3H_6, and oxygen gas, O_2. If the mixture has a total pressure of 825 torr, and the partial pressure of the cyclopropane is 73 torr, what is the partial pressure of the oxygen in the anesthetic?

Answer: 752 torr

Amount of Gas Dissolved in a Liquid

The amount of gas that dissolves in a liquid depends on the pressure of the gas. **Henry's law** states that the amount of gas that will dissolve in a liquid at a particular temperature is directly related to the pressure of that gas above the liquid. (See Figure 7.11.)

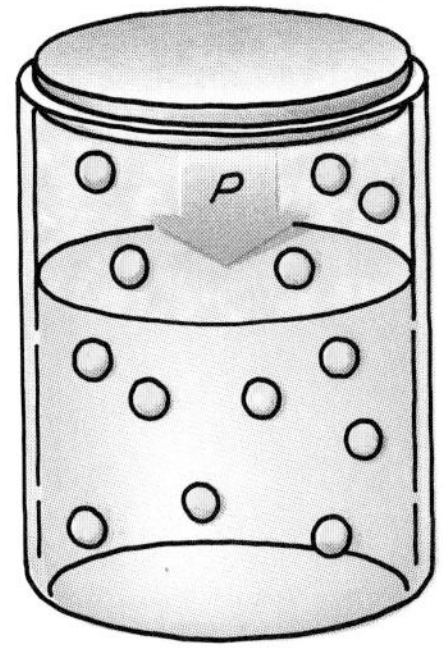

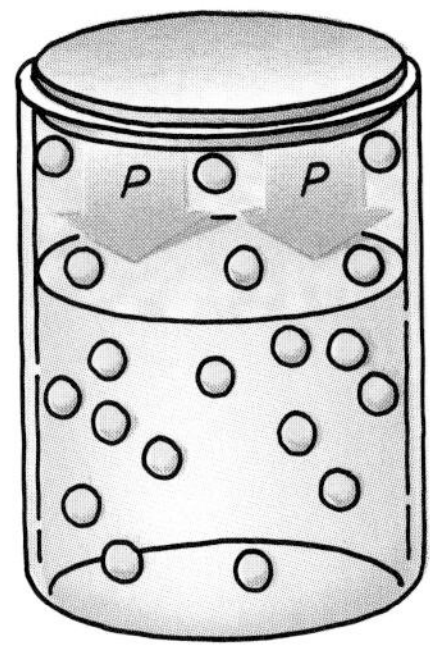

Figure 7.11
Henry's Law: The amount of gas dissolved in a liquid is related to the pressure of that gas above the liquid. When the pressure of the gas increases, the number of gas molecules that enter the liquid increases.

A can of soda contains dissolved carbon dioxide gas. When it is opened, much bubbling occurs. Sometimes the amount of escaping gas is so great that the liquid sprays out of the container. (See Figure 7.12.) When the can was originally filled and sealed, the pressure of the CO_2 gas in it was high, to make sure that sufficient $CO_2(g)$ dissolved in the liquid. When the can is opened, the CO_2 pressure drops, less CO_2 is soluble, and bubbles of gas rapidly escape from the beverage.

Blood Gases

Our cells continuously use oxygen and produce carbon dioxide. Both gases are transported in the bloodstream between the cells and the lungs. They diffuse in and out of the lungs through the membranes of the alveoli, the tiny air sacs at the ends of the airways in the lungs. An exchange of gases occurs in which oxygen from the air diffuses into the lungs and into the blood, while carbon dioxide produced in the cells is carried to the lungs to be exhaled. In Table 7.8, partial pressures are given for the gases in air that we inhale (inspired air), air in the alveoli, and the air that we exhale (expired air). The partial pressure of water vapor increases within the lungs, since the vapor pressure of water is 47 torr at body temperature.

bubbles form as CO_2 escapes from solution when pressure of gas is reduced

Figure 7.12
When a can of soda is opened, the CO_2 gas in the liquid rapidly escapes.

Table 7.8 Partial Pressures of Gases During Breathing

	Partial pressure (torr)		
Gas	**Inspired air**	**Alveolar air**	**Expired air**
Nitrogen, N_2	594.0	573	569
Oxygen, O_2	160.0	100	116
Carbon dioxide, CO_2	0.3	40	28
Water vapor, H_2O	5.7	47	47
Total	760.0	760	760

HEALTH NOTE

Hyperbaric Chambers

A burn patient may undergo treatment for burns and infections in a hyperbaric chamber, a device in which pressures can be obtained that are two to three times greater than atmospheric pressure. A greater oxygen pressure increases the level of dissolved oxygen in the blood and tissues, where it fights bacterial infections. The hyperbaric chamber may also be used during surgery, to help counteract carbon monoxide (CO) poisoning, and to treat some cancers. (See Figure 7.13.)

The blood is normally capable of dissolving up to 95% of the oxygen. Thus, if the partial pressure of the oxygen is 2280 torr (3 atm), 95% of that or 2160 torr of oxygen can dissolve in the blood where it saturates the tissues. In the case of carbon monoxide poisoning, this oxygen can replace the carbon monoxide that has attached to the hemoglobin.

A patient undergoing treatment in a hyperbaric chamber must also undergo decompression (reduction of pressure) at a rate that slowly reduces the concentration of dissolved oxygen in the blood. If decompression is too rapid, the oxygen dissolved in the blood may form gas bubbles in the circulatory system.

If divers do not decompress slowly, they suffer a similar condition called the bends. While below the surface of the ocean, divers breathe air at higher pressures. At such higher pressures, nitrogen gas will dissolve in their blood. If they ascend to the surface too quickly, the dissolved nitrogen forms bubbles in the blood that can produce life-threatening blood clots. The gas bubbles can also appear in the joints and tissues of the body and be quite painful. A diver suffering from the bends is placed immediately in a decompression chamber where pressure is first increased and then slowly decreased. The dissolved nitrogen can then diffuse through the lungs until atmospheric pressure is reached. ■

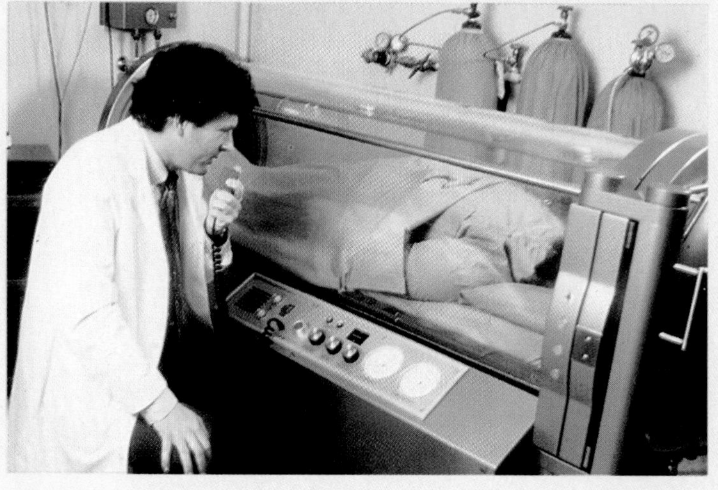

Figure 7.13
A hyperbaric chamber exposes a patient to oxygen at a pressure that is two to three times atmospheric pressure.

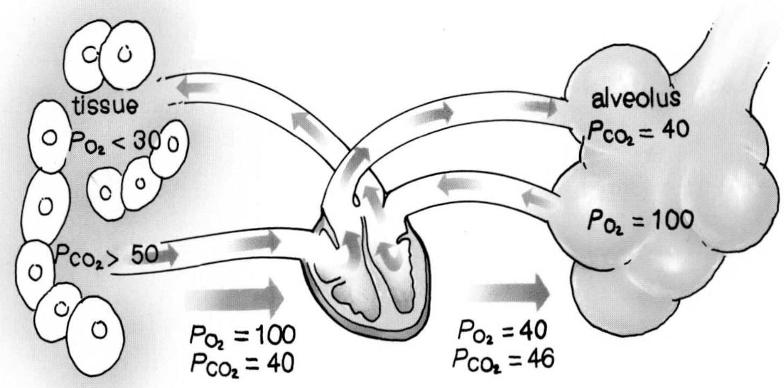

Figure 7.14
The diffusion of oxygen and carbon dioxide across the membranes of the alveoli and tissues during the exchange of blood gases.

Table 7.9 Partial Pressures of Oxygen and Carbon Dioxide in Blood and Tissues

	Partial pressure (torr)		
Gas	**Oxygenated blood**	**Deoxygenated blood**	**Tissues**
O_2	100	40	30 or less
CO_2	40	46	50 or greater

Oxygen normally has a partial pressure of 100 torr in the alveoli. Since the partial pressure of oxygen in venous blood is 40 torr, oxygen diffuses from the alveoli into the bloodstream. Most of the oxygen combines with hemoglobin, which carries it to the tissues of the body. There the partial pressure of oxygen can be very low, less than 30 torr, causing oxygen to diffuse from the blood into the tissues.

As oxygen is used in the cells of the body during metabolic processes, carbon dioxide is produced, so the partial pressure of CO_2 may be as high as 50 torr or more. Carbon dioxide diffuses from the tissues into the bloodstream and is carried to the lungs. There it diffuses out of the blood, where CO_2 has a partial pressure of 46 torr, into the alveoli, where the CO_2 is at 40 torr and is exhaled. (See Figure 7.14.) Table 7.9 gives the partial pressures of blood gases in the tissues and in oxygenated and deoxygenated blood.

7.9 The Ideal Gas Law

Learning Goal **Use the ideal gas law to solve for pressure, volume, temperature, or amount of a gas when given values for the other properties.**

The four properties used in the measurement of a gas—pressure (P), volume (V), temperature (T) and amount of a gas (n)—are found in the **ideal gas law** expression, which is written as follows:

Ideal Gas Law

$$PV = nRT$$

The ideal gas law is a useful expression when you are given the measurements for any three of the four properties of a gas.

Property of a Gas	*Unit*
Pressure (P)	Atmosphere (atm)
Volume (V)	Liter (L)
Amount (n)	Mole (n)
Temperature (T)	Kelvin (K)

In the ideal gas law, the R value, called the **universal gas constant,** is the combination of the constants from the gas laws relating volume, pressure, temperature, and amount of a gas. The R value can be determined by placing the volume (22.4 L) of 1.00 mole at standard temperature (273 K) and pressure (1.00 atm) in the ideal gas equation and solving the equation for R.

$$\frac{PV}{nT} = \frac{\cancel{n}R\cancel{T}}{\cancel{n}\cancel{T}}$$

$$\frac{PV}{nT} = R$$

Universal gas constant

At STP,

$$\frac{(1.00 \text{ atm})(22.4 \text{ L})}{(1.00 \text{ mole})(273 \text{ K})} = 0.0821 \frac{\text{L} \cdot \text{atm}}{\text{mole} \cdot \text{K}}$$

So the universal gas constant has a value of 0.0821 L · atm/mole · K.

Sample Problem 7.16
Calculations Using the Ideal Gas Law

Nitrous oxide, N_2O, is an anesthetic known as "laughing gas." What is the pressure in atmospheres of 0.35 mole of N_2O gas at 22°C in a 5.00-L tank?

Solution:
Organizing the given measurements gives

$P = ?$

$V = 5.00$ L

$n = 0.35$ mole

$T = 22°C$ (295 K)

To calculate the pressure of the N_2O gas, we can rearrange the ideal gas law to solve for P:

$$PV = nRT$$

$$\frac{P\cancel{V}}{\cancel{V}} = \frac{nRT}{V}$$

$$P = \frac{nRT}{V}$$

After checking that the measured units match the units of the gas constant R, we can substitute the values into the expression and solve for pressure. All units cancel except for the pressure in atmospheres.

$$P = \frac{(0.35\ \cancel{\text{mole}})(0.0821\ \cancel{\text{L}} \cdot \text{atm}/\cancel{\text{mole}} \cdot \cancel{\text{K}})(295\ \cancel{\text{K}})}{(5.00\ \cancel{\text{L}})} = 1.7\text{ atm}$$

Study Check:
Chlorine gas, Cl_2, is used to purify water in swimming pools. How many grams of chlorine are in a 7.00-L tank of gas at a pressure of 760.0 mm Hg and 24°C? (*Hint:* Solve for the number of moles of Cl_2, and change the units of other properties to match the units of the gas constant R.)

Answer: 20.4 g Cl_2

ENVIRONMENTAL NOTE

Ozone Depletion in the Atmosphere

From 10 to 30 miles above the earth, ozone is continuously produced by ultraviolet (UV) radiation that strikes oxygen molecules, first producing oxygen atoms.

$$O_2 \xrightarrow{\text{UV light}} 2O$$ Oxygen atoms are produced from O_2

The oxygen atoms are highly reactive because they have an unpaired electron.

$$:\ddot{\underset{\cdot}{O}}\cdot \longleftarrow$$ Unpaired electron makes oxygen atom highly reactive

$$O + O_2 \longrightarrow O_3$$ Oxygen atoms combine with O_2 to form O_3, ozone

The ozone in the upper atmosphere acts as a shield for plants and animals on the earth by absorbing a portion of the ultraviolet rays of the sun. The ultraviolet radiation that does get through can produce sunburn, cataracts, and skin cancers.

Normally, there is a balance between the formation and decomposition of ozone in the atmosphere. When an ozone molecule absorbs UV energy, it splits into an oxygen molecule and an oxygen atom.

$$\underset{\text{Ozone}}{O_3} \xrightarrow{\text{UV light}} O_2 + :\ddot{\underset{\cdot}{O}}\cdot$$

However, since the early 1970s scientists have become concerned that certain compounds entering the atmosphere are threatening the formation of the ozone layer. These compounds are the chlorofluorocarbons, (CFCs), which have been used as propellants for hair sprays, paints, and deodorants, as well as refrigerants in air conditioners. Although they appear to be unreactive in the lower atmosphere, it has become evident that in the upper atmosphere CFCs decompose in the presence of ultraviolet light to produce chlorine atoms. These chlorine atoms have unpaired electrons and are highly reactive.

$$\underset{\text{Freon, a chlorofluorocarbon}}{CFCl_3} \xrightarrow{\text{UV light}} CFCl_2 + \underset{\text{A highly reactive radical with an unpaired electron}}{:\ddot{\underset{..}{Cl}}\cdot}$$

The chlorine atoms break down ozone molecules, and at the same time another reaction forms more of the unstable chlorine atoms. It has been estimated that one chlorine atom can destroy as many as 100,000 ozone molecules.

$$:\ddot{\underset{..}{Cl}}\cdot + O_3 \longrightarrow ClO + O_2$$

$$ClO + :\ddot{\underset{\cdot}{O}}\cdot \longrightarrow :\ddot{\underset{..}{Cl}}\cdot + O_2$$

It is difficult to predict the overall effect of compounds such as CFCs on our atmosphere. Since 1985, researchers have observed the thinning of the ozone layer over the South Pole. In some areas as much as 50% of the ozone has been depleted, and at certain times of the year an ozone hole appears. (See Figure 7.15.) Such phenomena had never occurred before in the history of ozone observations. The concern of scientists is that a thinning of the ozone layer over more populated areas would increase the ultraviolet radiation received by plants and animals, with adverse effects. By the early 1980s, CFCs were banned by the United States for use in spray containers. By 1987, 24 nations had agreed to cut back on CFC use. Some states have banned the use of car air conditioners that use CFCs. Chemical companies are working on substitutes for CFCs, but so far they appear to be costly. Ultimately, the consumer will probably pay more for new products, but that may be the price for retaining the life-protecting ozone layer. ■

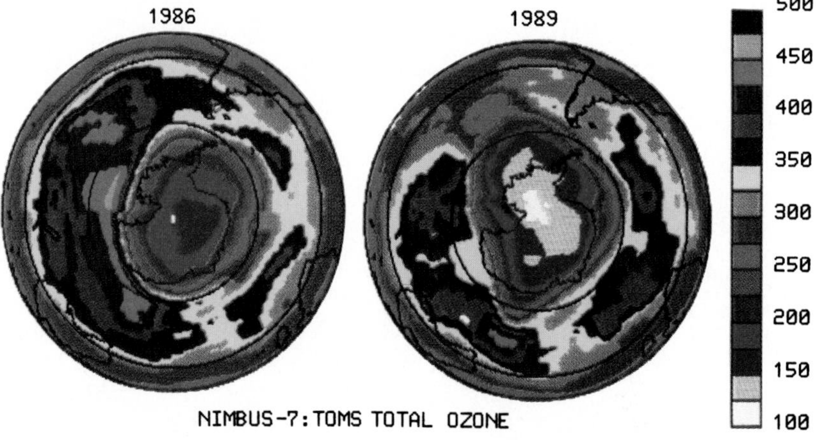

Figure 7.15
At certain times of the year, an ozone hole appears in an area over the South Pole where depletion of the ozone layer is most severe.

Glossary of Key Terms

atmosphere (atm) The pressure exerted by a column of mercury 760 mm high.

atmospheric pressure The pressure exerted by the atmosphere.

Avogadro's law A law that states that the volume of gas is directly related to the number of moles of gas in the sample when pressure and temperature do not change.

barometer An instrument used to measure atmospheric pressure.

Boyle's law A gas law stating that the pressure of a gas is inversely related to the volume, when temperature and the moles of gas do not change; that is, if volume decreases, pressure increases.

Charles' law A gas law stating that the volume of a gas changes directly with a change in Kelvin temperature, when pressure and the moles of gas do not change.

combined gas law A relationship that combines the gas laws relating pressure, volume, and temperature.

$$\frac{P_iV_i}{T_i} = \frac{P_fV_f}{T_f}$$

Dalton's Law The total pressure exerted by a mixture of gases in a container is the sum of the partial pressures that each gas would exert alone.

diffusion The movement of gas particles from a more concentrated area (high pressure) to a less concentrated area (low pressure).

direct relationship A relationship in which two properties increase or decrease together.

Gay-Lussac's law A gas law stating that the pressure of a gas changes directly with a change in temperature when the number of moles of a gas and its volume are held constant.

Henry's Law The amount of gas dissolved in a liquid at a certain temperature is directly related to the pressure of the gas above the liquid.

ideal gas law A law that combines the four measured properties of a gas in the equation $PV = nRT$.

inverse relationship A relationship in which two properties change in opposite directions.

kinetic theory A model used to explain the behavior of gases.

molar volume A volume of 22.4 L occupied by 1 mole of a gas at STP conditions of 0°C (273 K) and 1 atm.

partial pressure The pressure exerted by a single gas in a gas mixture.

pressure The force exerted by gas particles that hit the walls of a container.

pressure gradient A difference in pressure causing gas particles to move from the area of high pressure to the area of low pressure.

STP Standard conditions of 0°C (273 K) temperature and 1 atm pressure used for the comparison of gases.

torr A unit of pressure equal to 1 mm Hg; 760 torr = 1 atm.

Universal gas constant (*R*) A numerical value that relates the properties P, V, n, and T in the ideal gas law, $PV = nRT$.

vapor pressure The pressure exerted by the particles of vapor above a liquid.

Problems

Properties of a Gas *(Goal 7.1)*

7.1 What does the kinetic theory say about the following:

a. particles of a gas
b. motion of gas particles
c. distance between gas particles
d. attraction between gas particles
e. effect of temperature on gases
f. velocity of gas particles

7.2 What are the four properties used to describe a gas?

7.3 Use the word *pressure, volume, temperature,* or *amount* to indicate the property of a gas that is described in each of the following statements or mreasurements:
a. 350 K
b. 1200 torr
c. 4.00 moles helium
d. determines the kinetic energy of the gas particles
e. space occupied by a gas
f. 10.0 L

Gas Pressure *(Goal 7.2)*

7.4 Which of the following statement(s) describes the pressure of a gas?
a. the force of the gas particles on the walls of the container
b. the number of gas particles in a container
c. the volume of the container
d. 3.00 atm
e. 750 torr

7.5 An oxygen tank contains oxygen (O_2) at a pressure of 2.00 atm. What is the pressure in the tank in terms of the following units?
a. torr
b. $lb/in.^2$
c. mm Hg

7.6 The helium gas in a balloon has a pressure of 735 mm Hg. What is the pressure in the balloon in terms of the following units?
a. atm
b. torr
c. in. Hg

7.7 Indicate whether each of the following statements describes vapor pressure, atmospheric pressure, or boiling point.
a. the temperature at which bubbles of vapor appear within the liquid
b. the pressure exerted by a gas above the surface of its liquid
c. the pressure exerted on the earth by the particles in the air
d. the temperature at which the vapor pressure of a liquid becomes equal to atmospheric pressure

7.8 In which pair would boiling occur?

	Atmospheric Pressure	Vapor Pressure
a.	760 mm Hg	700 mm Hg
b.	640 torr	640 torr
c.	1.2 atm	912 mm Hg
d.	1020 mm Hg	760 mm Hg

7.9 Give an explanation for the following observations:
a. Water boils at 87°C on the top of Mount Whitney.
b. Food cooks more quickly in a pressure cooker than in an open pan.
c. Water used to sterilize surgical equipment is heated to 120°C at 2.0 atm in an autoclave.

Pressure and Volume (Boyle's Law) *(Goal 7.3)*

7.10 Complete the following using the pressure–volume relationship:

	Pressure	Volume
a.	______	decreases
b.	decrease	______
c.	______	increases

7.11 Solve for the new pressure when there is no change in temperature or amount of gas.

a. A 10.0-L balloon contains helium gas at a pressure of 655 torr. What is the new pressure when the volume expands to 20.0 L?

b. The air in a 50.0-L tank has a pressure of 725 torr. What is the new pressure when the air is placed in a smaller tank that has a volume of 30.0 L?

c. A sample of nitrogen (N_2) gas has a volume of 425 mL at a pressure of 1.50 atm. If the volume is decreased to 215 mL, what is the new pressure?

7.12 Calculate the new volume for the following changes in pressure:

a. Cyclopropane, C_3H_6, is a general anesthetic. A 5.0-L sample has a pressure of 5.0 atm. What is the volume of the anesthetic given to a patient at a pressure of 1.0 atm?

b. The volume of air in a person's lungs is 615 mL at a pressure of 760 mm Hg. Inhalation occurs as the pressure in the lungs drops to 752 mm Hg. To what volume did the lungs expand?

c. An emergency tank of oxygen holds 20.0 L of oxygen (O_2) gas at a pressure of 15 atm. What volume of O_2 can be given to a patient if the gas is released at a pressure of 1.0 atm?

7.13 Use the word *inspiration* or *expiration* to describe the part of the breathing cycle that occurs as a result of each of the following:

a. The diaphragm flattens out (contracts).

b. The diaphragm relaxes, moving up into the thoracic cavity.

c. The volume of the lungs decreases.

d. The pressure within the lungs is less than that of the atmosphere.

Temperature and Volume (Charles' Law) *(Goal 7.4)*

7.14 State the change that occurs for volume or temperature when P and n are not changed:

	Volume	Temperature
a.	______	decreases
b.	increases	______
c.	decreases	______

7.15 A balloon contains 3150 mL of helium gas at 75°C. What is the new volume of the gas when the temperature is changed to the following, if n and P are not changed?

a. 55°C

b. 682 K

c. −25°C

7.16 A gas has a volume of 4.00 L at 0°C. What final temperature in degrees Celsius is needed to cause the volume of the gas to change to the following, if n and P are not changed?

a. 10.0 L
b. 1200 mL
c. 2.50 L

Temperature and Pressure (Gay-Lussac's Law) *(Goal 7.5)*

7.17 State the expected change for temperature and pressure according to Gay-Lussac's law, with no change in n and V:

	Pressure	Temperature
a.	______	increases
b.	decreases	______
c.	______	decreases

7.18 Using the temperature–pressure relationship, solve for the new pressure when the following temperature changes occur, with n and V constant:

a. A gas sample has a pressure of 1200 torr at 155°C. What is the final pressure of the gas after the temperature has dropped to 0°C?
b. An aerosol can has a pressure of 1.40 atm at 12°C. What is the final pressure in the aerosol can if it is used in a room where the temperature is 35°C?
*c. A fire extinguisher has a pressure of 150 lb/in.^{2} at 25°C. What is the pressure in atmospheres if the fire extinguisher is used at a temperature of 75°C? (1 atm = 14.7 lb/in.^{2})

7.19 Solve for the new temperature in degrees Celsius when pressure is changed:

a. A 10.0-L container of helium gas has a pressure of 250 torr at 0°C. To what temperature does the sample need to be heated to obtain a pressure of 1500 torr?
*b. A sample of hydrogen (H_2) gas at 127°C has a pressure of 2.00 atm. At what temperature will the pressure of the H_2 decrease to 0.25 atm?

The Combined Gas Law *(Goal 7.6)*

7.20 Solve the following problems for changes in pressure, volume, and temperature using the combined gas law:

a. A 2460-mL sample of a gas at 285°C and 620 torr is compressed to a volume of 820 mL and 150°C. What is the new pressure of the gas?
b. A 1.50-L sample of carbon dioxide (CO_2) has a pressure of 2.00 atm and a temperature of 15°C. What is the pressure when the volume of the sample is increased to 5.00 L, and the temperature is increased to 125°C with no change in moles of gas?
*c. A weather balloon has a volume of 750 L when filled with helium at 8°C and a pressure of 380 torr. What is the new volume of the balloon in the atmosphere, where the pressure is 0.20 atm and the temperature is −45°C, with no change in number of moles of gas?
d. A 100-mL bubble of hot gases at 250°C and 2.4 atm escapes from an active volcano. What is the new volume of the bubble outside the volcano, where the temperature is −15°C and the pressure is 0.80 atm?

Volume and Moles (Avogadro's Law) *(Goal 7.7)*

7.21 Solve the following problems when no changes occur in pressure and temperature:

a. A sample of 4.00 moles of argon has a volume of 10.0 L. A small leak causes half of the molecules to escape. What is the new volume of the gas?
b. A balloon containing 1.00 mole of helium has a volume of 440 mL. What is the new volume after 2.00 more moles of helium are added to the balloon at the same pressure and temperature?
c. A 1500-mL sample of SO_2 contains 6.00 moles SO_2. How many moles of SO_2 are lost when the volume is reduced to 750 mL at the same pressure and temperature?

7.22 Use the molar volume of a gas to solve the following:

a. How many moles of O_2 are present in 44.8 L of O_2 gas at STP?
b. How many moles of CO_2 are present in 4.00 L of CO_2 gas at STP?
c. How many grams of neon are contained in 11.2 L of gas at STP?
d. How many moles of H_2 are in 1600 mL of H_2 gas at STP?

7.23 Use molar volume to solve the following problems:

a. What volume (L) is occupied by 2.5 moles of N_2 at STP?
b. What volume (mL) is occupied by 0.420 mole helium at STP?
c. What is the volume (L) of 6.40 g of O_2 at STP?
d. What volume (mL) is occupied by 50.0 g neon at STP?

Partial Pressures (Dalton's Law) *(Goal 7.8)*

7.24 A steel cylinder contains a mixture of nitrogen gas (N_2) at 400 torr, oxygen gas (O_2) at 115 torr, and helium gas at 225 torr. What is the total pressure of the gas mixture?

7.25 What is the total pressure in atmospheres of a gas mixture containing argon gas at 0.25 atm, helium gas at 350 mm Hg, and nitrogen gas at 360 torr?

7.26 A gas mixture exerts a pressure of 2550 torr. The mixture contains oxygen, nitrogen, and helium. If the partial pressure of oxygen is 425 torr, and the partial pressure of helium is 320 torr, what is the partial pressure of nitrogen in the mixture?

***7.27** Nitrogen is prepared and a 250 mL sample is collected over water at 30°C and a total pressure of 745 mm Hg.

a. Using the vapor pressure of water in Table 7.3, what is the partial pressure of the nitrogen?
b. What is the volume of the nitrogen at STP?

***7.28** A gas mixture with a total pressure of 2400 torr is used by a scuba diver. If the mixture contains 2.0 moles of helium and 6.0 moles of oxygen, what is the partial pressure of each gas in the sample?

The Ideal Gas Law *(Goal 7.9)*

7.29 What is the pressure, in atmospheres, of 2.0 moles of helium gas placed in a 10.0-L container at 27°C?

7.30 How many liters are occupied by 5.0 moles of methane gas, CH_4, at a temperature of 0°C and 2.00 atm?

7.31 A steel cylinder of oxygen has a volume of 20.0 L at 22°C, and the oxygen has a pressure of 35 atm. How many moles of oxygen are in the container?

***7.32** A gas cylinder contains 64 g oxygen (O_2) at a temperature of 25°C and a pressure of 850 torr. What is the volume in liters of the oxygen gas in the cylinder?

***7.33** A cylinder contains 1250 mL of krypton at a temperature of 35°C and a pressure of 475 mm Hg. How many moles of krypton are in the cylinder?

***7.34** A sample of ammonia gas, NH_3, is collected in a container that has a volume of 525 mL at a temperature of 15°C and a pressure of 455 torr. How many grams of NH_3 are in the container?

$Cl(s) \longrightarrow Na^+(aq) + Cl^-(aq)$

8

Solutions

Learning Goals

8.1 Identify the solute and the solvent in a solution.

8.2 Describe the polar nature of water.

8.3 Describe the formation of solutions for polar and nonpolar solutes.

8.4 Identify the factors that affect solubility and the rate of formation of a solution.

8.5 Calculate the percent concentration of a solution; use percent concentration to calculate the amount of solute or solution.

8.6 Calculate the molarity of a solution; solve problems using molarity as the conversion factor to find the number of moles of solute or the volume of the solution.

8.7 Calculate the final volume or new concentration of a solution obtained by dilution.

8.8 From its properties, identify a mixture as a solution, a colloid, or a suspension.

8.9 Describe the processes of osmosis and dialysis.

ie desalinization of sea water ill provide an alternative urce of fresh water as the rld population grows.

SCOPE

Solutions are everywhere around us. Each consists of a mixture of at least one substance dissolved in another. The air we breathe is a solution of oxygen and nitrogen gases. Carbon dioxide gas dissolved in water makes the soda in our carbonated drinks. When we make solutions of coffee or tea, we use hot water to dissolve substances from coffee beans or tea leaves. The ocean is also a solution, consisting of many salts such as sodium chloride dissolved in water. In the hospital, the antiseptic tincture of iodine is a solution of iodine dissolved in alcohol.

Another type of mixture, called a colloid, contains large particles. Mayonnaise, whipped cream, and gelatin are all colloids. In a third type of mixture, called a suspension, the particles are so large they settle out. You may have been given suspensions such as liquid penicillin or calamine lotion, which had to be shaken before use.

In the processes of osmosis and dialysis, we will see how water, essential nutrients, and waste products enter and leave the cells of the body. In osmosis, water flows in and out of the cells of the body. In dialysis, small particles in solution as well as water diffuse through semipermeable membranes.

8.1 Solutes and Solvents

Learning Goal **Identify the solute and the solvent in a solution.**

A **solution** is a type of mixture in which one substance called the **solute** is dispersed uniformly in another substance called the **solvent.** Since the solute and the solvent do not react with each other, they can be mixed in any proportion. A little salt dissolved in water tastes slightly salty. When more salt is dissolved, the water tastes very salty. Usually, the solute (in this case, salt) is the substance present in the smaller amount, whereas the solvent (in this case, water) is the larger amount. The salt solution forms when the particles of the solute (salt) become evenly dispersed among the molecules of the solvent (water). (See Figure 8.1.)

Figure 8.1
A solution of copper sulfate ($CuSO_4$) is formed when particles of solute become evenly dispersed among the particles of the solvent.

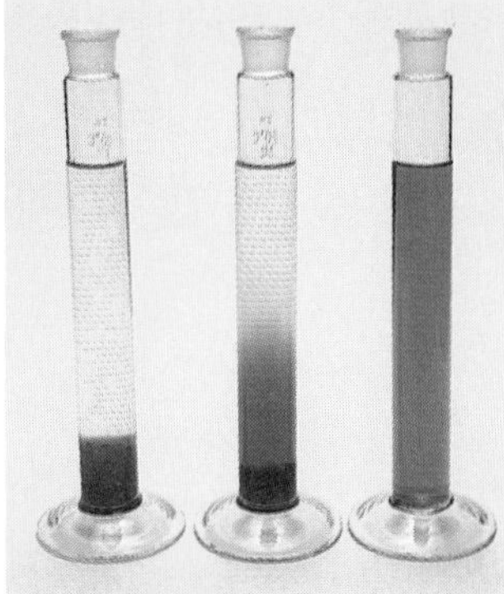

Types of Solutes and Solvents

Solutes and solvents may be solids, liquids, or gases. The solution that forms has the same physical state as the solvent. When sugar is dissolved in a glass of water, a liquid sugar solution forms. Sugar is the solute, and water is the solvent. The carbonated solutions of soda water and soft drinks are prepared by dissolving CO_2 gas in water. The CO_2 gas is the solute, and water is the solvent. Table 8.1 lists some solutes and solvents and their solutions.

Table 8.1 Some Examples of Solutions

Type	Example	Solute	Solvent
Gas solutions			
Gas in a gas	Air	Oxygen (gas)	Nitrogen (gas)
Liquid solutions			
Gas in a liquid	Soda water	Carbon dioxide (gas)	Water (liquid)
	Household ammonia	Ammonia (gas)	Water (liquid)
Liquid in a liquid	Vinegar	Acetic acid (liquid)	Water (liquid)
Solid in a liquid	Seawater	Sodium chloride (solid)	Water (liquid)
	Tincture of iodine	Iodine (solid)	Alcohol (liquid)
Solid solutions			
Liquid in a solid	Dental amalgam	Mercury (liquid)	Silver (solid)
Solid in a solid	Brass	Zinc (solid)	Copper (solid)
	Steel	Carbon (solid)	Iron (solid)

Sample Problem 8.1
Identifying a Solute and a Solvent

Identify the solute and the solvent in each of the following solutions:

a. 1.0 g of sugar dissolved in 100 g of water
b. 50 mL of water mixed with 20 mL isopropyl alcohol (rubbing alcohol)

Solution:

a. Sugar is the smaller quantity that is dissolving. It is the solute; water is the solvent.
b. Since both water and isopropyl alcohol are liquids, the one with the smaller volume, isopropyl alcohol, is the solute. Water is the solvent.

Study Check:
A tincture of iodine is prepared with 0.10 g I_2 and 10.0 mL ethyl alcohol. What is the solute, and what is the solvent?

Answer:
Iodine is the solute, and ethyl alcohol is the solvent.

8.2 Water: An Important Solvent

Learning Goal **Describe the polar nature of water.**

As the most common solvent in nature, water is called the universal solvent. In a water molecule, H_2O, an oxygen atom shares electrons with two hydrogen atoms. Since the oxygen atom has a greater attraction for the shared electron pairs, the oxygen atom has a partial negative (δ^-) charge, and the hydrogen atoms have partial positive (δ^+) charges. The water molecule contains polar bonds as well as two unshared pairs of electrons, so the water molecule is polar. Water is therefore a polar solvent.

$\mathrm{H{:}\ddot{\underset{..}{O}}{:}}$ H	$\mathrm{H^{\delta+}{-}\ddot{O}{:}^{2\delta-}}$ $\mathrm{H^{\delta+}}$
Electron-dot structure for water	Polarity of the water molecule

Hydrogen Bonding in Water

The electrical attractions between polar water molecules are called **hydrogen bonds.** Hydrogen bonds form between the oxygen atom of one water molecule and the hydrogen atoms of other water molecules. (See Figure 8.2.) Because more energy is required to break apart the hydrogen bonds, liquid water has an unusually high boiling point.

Surface Tension

Imagine that you have filled a glass up to the rim with water. Carefully adding a few more drops of water or dropping in some pennies does not cause the water to overflow. Instead the water seems to adhere to itself, forming a dome that rises slightly above the rim of the glass. This effect is the result of the polarity of water. Throughout the liquid in the glass, water molecules are attracted in all directions by surrounding water molecules. However, the water molecules on the surface are pulled like a skin toward the rest of the water in the glass. As a result, the water molecules on the surface become more tightly packed, a feature called

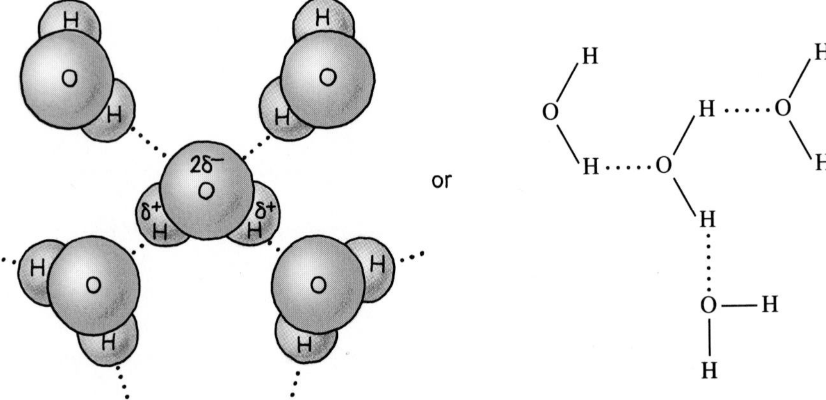

Figure 8.2
Hydrogen bonding occurs between the oxygen atom of one water molecule and the hydrogen atom of other water molecules.

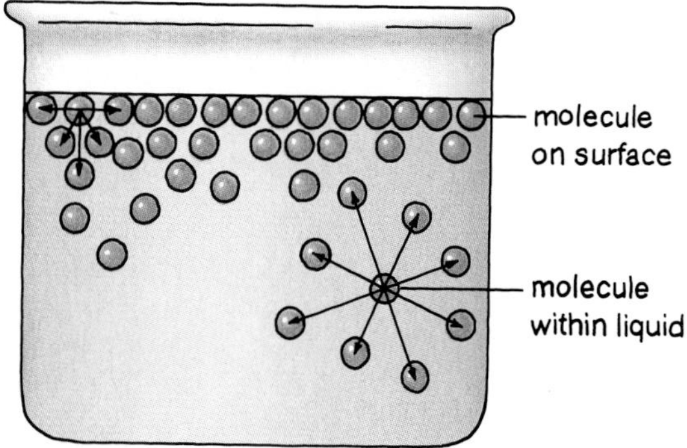

Figure 8.3
Water molecules within the liquid are attracted in all directions by surrounding water molecules. On the surface, they are attracted only to the side and downward, causing surface tension.

surface tension. Because of surface tension, a needle floats on top of water, certain water bugs can travel across the surface of a pond or lake, and drops of water are spherical. (See Figure 8.3.)

HEALTH NOTE

Water in the Body

The average adult contains about 60% water by weight, and the average infant about 75%. About 60% of the body's water is contained within the cells as intracellular fluids; the other 40% makes up extracellular fluids, which include the interstitial fluid in tissues, and the plasma in the blood. These external fluids carry nutrients and waste materials between the cells and the circulatory system. (See Figure 8.4.)

Every day you lose between 1500 and 3000 mL of water from the kidneys as urine, from the skin as perspiration, from the lungs as you exhale, and from the gastrointestinal tract. Serious dehydration can occur in an adult if there is a 10% net loss in total body fluid, and a 20% loss of fluid can be fatal. An infant suffers severe dehydration with a 5–10% loss in body fluid.

Water lost is continually replaced by the liquids and foods in the diet and from metabolic processes that produce water in the cells of the body. (See Figure 8.5.) Table 8.2 lists the amounts of water contained in some foods. ■

8.3 Formation of Solutions

Learning Goal **Describe the formation of solutions for polar and nonpolar solutes.**

In solutions of gases, the particles mix easily because they are moving rapidly, are far apart, and are not attracted to or repelled by the other gas particles. However, to form solutions of liquids, or solids dissolved in liquids, there must be an attraction between the solute and the solvent. Then the solute and solvent will separate into individual ions or molecules and form a solution. If no attrac-

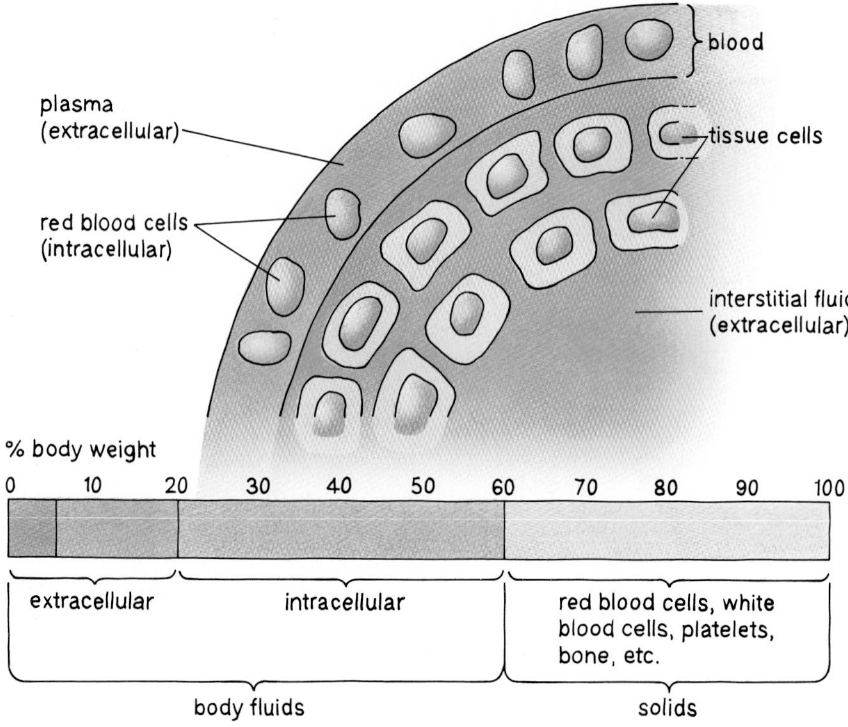

Figure 8.4
Distribution of fluids in the body. The adult human is about 60% water by weight, which goes into intracellular and extracellular fluids.

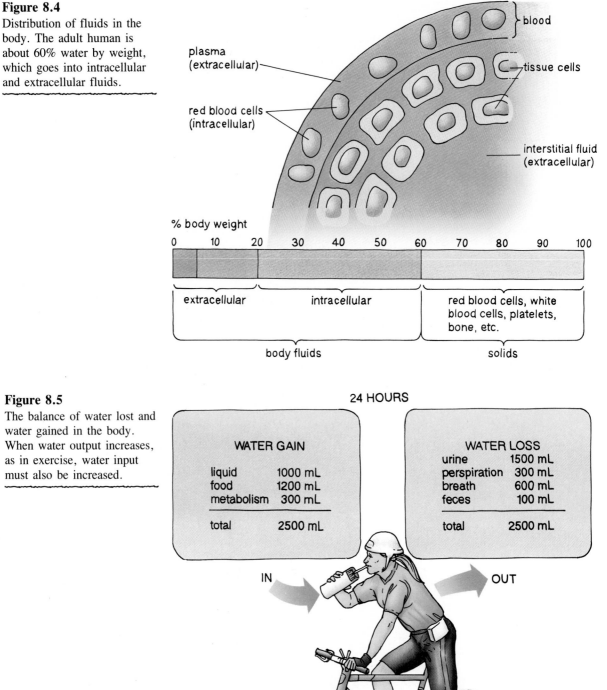

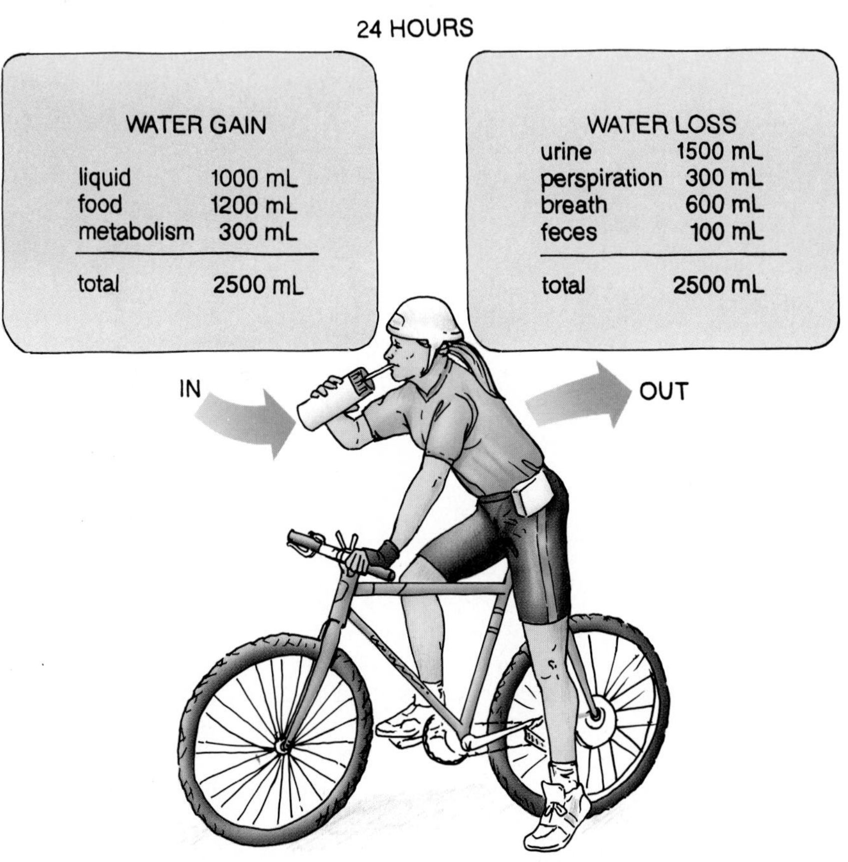

24 HOURS
WATER GAIN
liquid 1000 mL
food 1200 mL
metabolism 300 mL
total 2500 mL
WATER LOSS
urine 1500 mL
perspiration 300 mL
breath 600 mL
feces 100 mL
total 2500 mL
IN
OUT

Figure 8.5
The balance of water lost and water gained in the body. When water output increases, as in exercise, water input must also be increased.

Table 8.2 Percentage of Water in Some Foods

Food	Water (%)	Food	Water (%)
Vegetables		***Meats/Fish***	
Carrot	88	Chicken, cooked	71
Celery	94	Hamburger, broiled	60
Cucumber	96	Salmon	71
Tomato	94	***Grains***	
Fruits		Cake	34
Apple	85	French bread	31
Banana	76	Noodles, cooked	70
Cantaloupe	91	***Milk Products***	
Grapefruit	89	Cottage cheese	78
Orange	86	Milk, whole	87
Strawberry	90	Yogurt	88
Watermelon	93		

tion between solute and solvent occurs, the particles of the solute stay together and will not mix with the solvent.

Dissolving Ionic Solutes

The formation of a solution depends on similar polarities of the solute and solvent particles. An ionic compound, such as sodium chloride (NaCl), dissolves in water because water is a polar solvent. As the water molecules bombard the surface of the salt, as shown in Figure 8.6, the Cl^- ions are attracted to the partially positive hydrogen atoms of the water molecules. The Na^+ ions are attracted to the partially negative oxygen atoms. These attractions pull the ions into the water, where they undergo hydration: They become surrounded by several water molecules. A sodium chloride solution has formed.

Sample Problem 8.2
Describing the Solution Process of a Salt

Complete the following statement describing the formation of a KCl solution:
The salt KCl consists of positive ______ ions and negative ______ ions. The ______ end of a water molecule will be attracted to the K^+ ions, pulling the ions into solution. The ______ end of a water molecule will be attracted to the Cl^- ions. The dissolved potassium and chloride ions are surrounded by ______ molecules, a process called ______.

Solution:
The salt KCl consists of positive *potassium* ions and negative *chloride* ions. The *negative oxygen* end of a water molecule will be attracted to the K^+ ions, pulling the ions into solution. The *positive hydrogen* end of a water molecule will be attracted to the Cl^- ions. The dissolved potassium and chloride ions are surrounded by *water* molecules, a process called *hydration*.

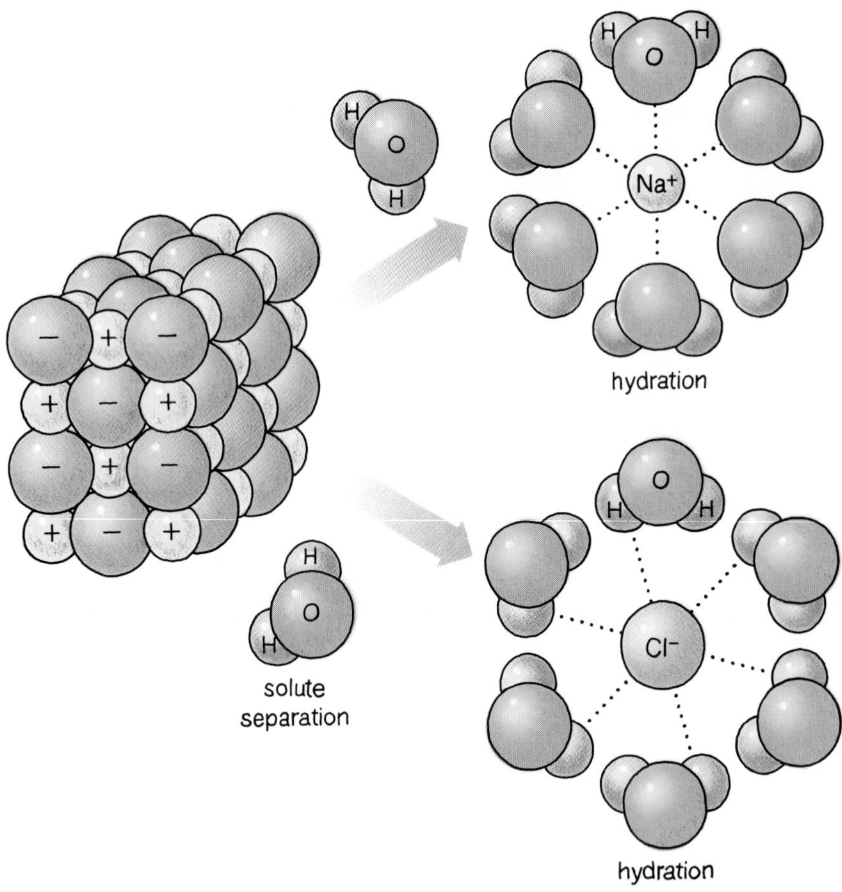

Figure 8.6
The dissolving of NaCl, an ionic solute, by water. Polar water molecules attract Na^+ and Cl^-; the ions then break away from the solute and move into solution, where they are hydrated.

Dissolving Polar Solutes

Water is also a solvent for solutes that are polar (not ionic), such as table sugar which is called sucrose ($C_{12}H_{22}O_{11}$). In sugar, there are many O—H groups that attract water molecules. In forming a solution, the sugar disperses in water as sucrose molecules. (See Figure 8.7.)

Dissolving Nonpolar Solutes

Nonpolar compounds, such as iodine (I_2), oil, and grease, do not dissolve in water because they are not attracted by the polar water molecules. Instead, nonpolar solutes form solutions in nonpolar solvents, such as pentane (C_5H_{12}). These examples illustrate a rule of thumb, "like dissolves like," which we can use to predict the formation of a solution.

Like Dissolves Like

> Polar (or ionic) solutes form solutions with polar solvents; nonpolar solutes form solutions with nonpolar solvents.

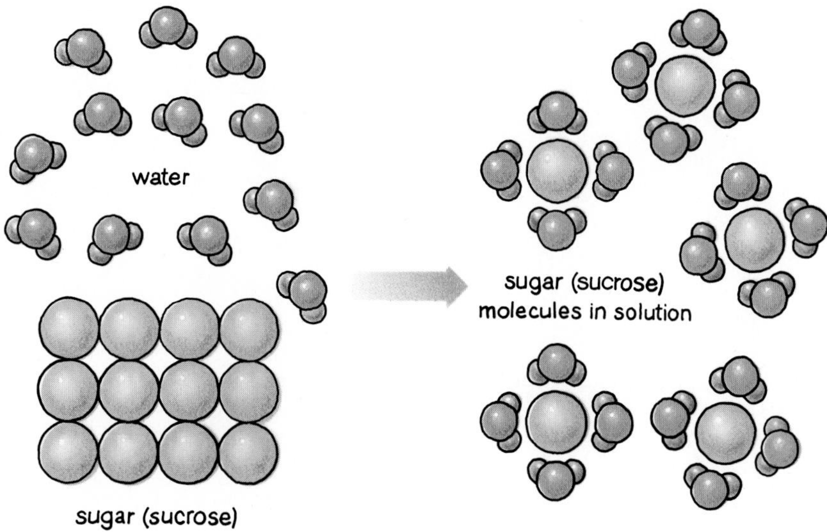

Figure 8.7
The formation of a sugar solution by polar sucrose molecules ($C_{12}H_{22}O_{11}$) and polar water molecules.

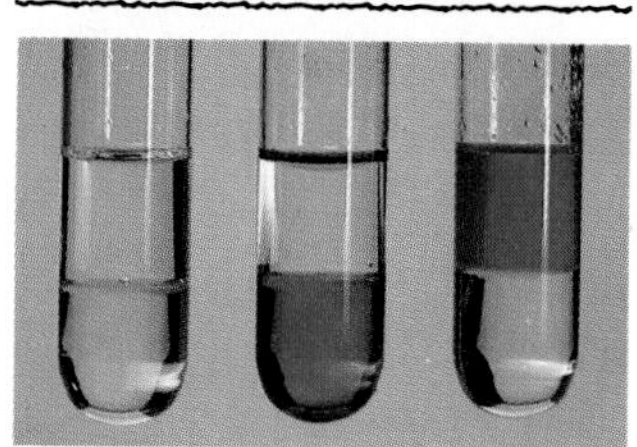

Figure 8.8
Like dissolves like: A cylinder contains a nonpolar solvent (CH_2Cl_2) as the lower layer, with water, a polar solvent, as the upper layer. The nonpolar I_2 dissolves in the nonpolar solvent, and the polar $NiNO_3$ dissolves in the polar solvent.

Table 8.3 lists examples of solutes and solvents that will form solutions and those that will not. Figure 8.8 illustrates some polar and nonpolar solutions.

Table 8.3 Formation of Polar and Nonpolar Solutions

	Solvent	
Solute	**Water (polar)**	**Pentane (nonpolar)**
NaCl (ionic)	Yes	No
Oil (nonpolar)	No	Yes
I_2 (nonpolar)	No	Yes
HCl (polar)	Yes	No
Sucrose (polar)	Yes	No

Sample Problem 8.3
Predicting Solution Formation

Predict whether a solution will form with the following solutes and solvents:

a. the salt KCl in water
b. the salt KCl in hexane, a nonpolar solvent
c. oil, a nonpolar solute, in water

Solution:

a. Yes. The salt KCl consists of ions. Ionic compounds are soluble in polar solvents, such as water.
b. No. An ionic compound will not dissolve in a solvent that is not polar.
c. No. A nonpolar solute will not be soluble in water, which is polar.

Study Check:
Why will oil, a nonpolar solute, dissolve in hexane, a nonpolar solvent?

Answer:
Like dissolves like; both the solute and the solvent are nonpolar substances.

HEALTH NOTE

Fat-Soluble and Water-Soluble Vitamins

Vitamins are classified as water-soluble or fat-soluble. The water-soluble vitamins are soluble in body fluids, because their molecules include ionic or polar groups. They include vitamin C (ascorbic acid) and the B vitamins such as B_1 (thiamine), B_2 (riboflavin), and B_3 (niacin). Excesses of water-soluble vitamins are filtered out by the kidneys and excreted in the urine.

The fat-soluble vitamins A, D, E, and K are soluble in nonpolar solvents, including the body fat (adipose tissue), where they are stored. Their molecules are nonpolar in nature and excesses are not eliminated through the urine.

For vitamins A and D, cases of hypervitaminosis have been reported when large amounts of the vitamins are consumed. Symptoms of hypervitaminosis for vitamin A include nausea, abdominal pain, pain in the joints, and chronic headache. Hypervitaminosis for vitamin D can lead to the formation of calcium salts, which can cause kidney damage and high blood pressure. ■

8.4 Solubility and Saturation

Learning Goal **Identify the factors that affect solubility and the rate of formation of a solution.**

There is often a limit to the amount of solute that can dissolve in a solvent. The **solubility** of a substance is usually defined as the maximum number of grams of that substance that can dissolve in 100 g of solvent at a certain temperature. For example, at 20°C the solubility of sugar is 204 g in 100 g of water. Adding more sugar to the sugar solution will not result in more dissolved solute, since the maximum amount of sugar is already in solution at that temperature.

Effect of Temperature

Most solids become more soluble in water as the temperature increases. For example, if you add sugar to ice tea, a layer of undissolved sugar forms on the bottom of the glass when the maximum solubility is reached. However, if the tea is heated, more sugar dissolves, since the solubility of sugar is greater in hot water. Figure 8.9 illustrates the effect of higher temperatures on the solubility of some solids.

Gases are less soluble in liquids at higher temperatures. As the temperature increases, increasing numbers of gas molecules escape from the solution. At high temperatures, bottles containing carbonated solutions may burst, as more gas molecules leave the solution and increase the gas pressure inside the bottle. Biologists have found that increased temperatures in rivers and lakes cause the amount of oxygen available to decrease until the warm water can no longer support a biological community. This is the reason electrical generating plants are required to have their own ponds of water to use with their cooling towers. The solubilities of some gases at increasing temperatures are shown in Figure 8.10.

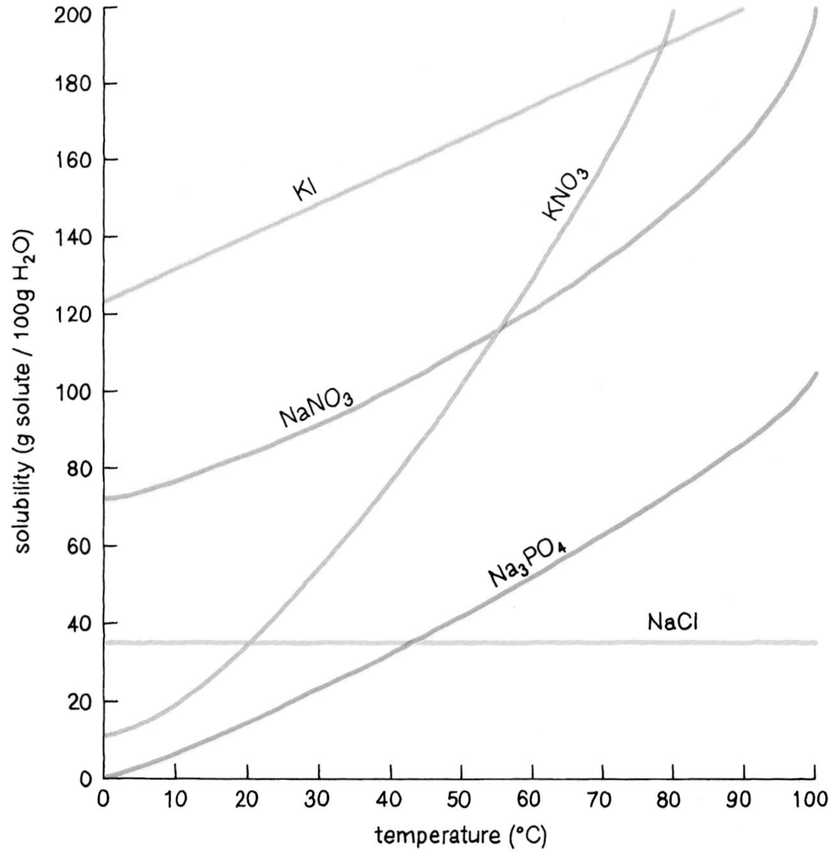

Figure 8.9
Many solids are more soluble in water as the temperature increases.

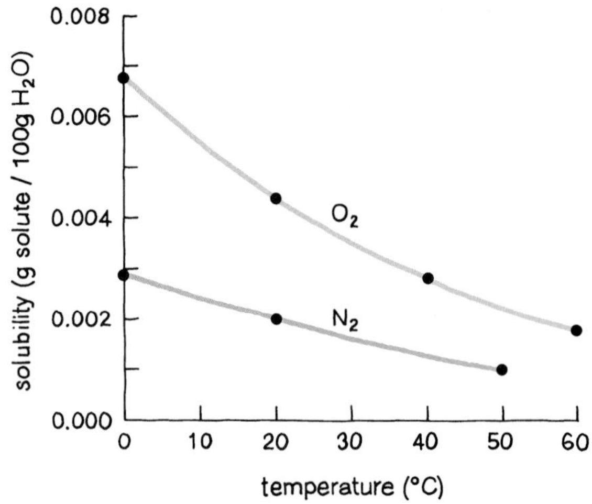

Figure 8.10
The solubility of gases decreases as the temperature increases.

Sample Problem 8.4
Factors Affecting Solubility

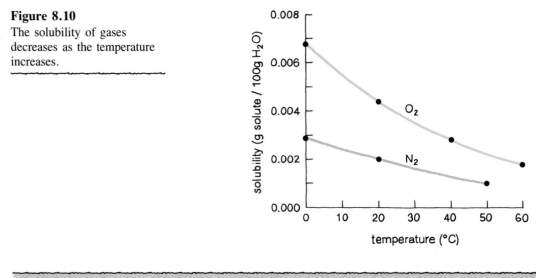

Indicate whether the solubility of the solute will increase or decrease in each of the following situations:

a. using 80°C water instead of 25°C water to dissolve sugar
b. increasing the temperature of the water in a lake (O_2 gas is the solute we are interested in)

Solution:

a. Using warmer water to dissolve sugar will increase its solubility.
b. The solubility of the dissolved O_2 in the lake will decrease because gases are less soluble at higher temperatures.

Saturated Solutions

A solution that contains the maximum amount of dissolved solute at a certain temperature is called a **saturated solution.** If we try to add more solute, undissolved crystals of solute will form on the bottom of the container. However, when we add more solute to a solution and it all dissolves, we call the solution an **unsaturated solution.** It does not contain the maximum amount of solute that can be dissolved. Therefore, more solute can still be added to an unsaturated solution. For example, the solubility of NaCl at 20°C is 36 g/100 g water. If 100 g of water at 20°C contains 15 g of NaCl, the solution is unsaturated. The NaCl solution becomes saturated only after an additional 21 g of NaCl is added. (See Figure 8.11.)

Sample Problem 8.5
Saturated Solutions

AT 70°C, the solubility of KCl is 50 g/100 g water. In the laboratory, a student mixes 75 g of KCl with 100 g of water at a temperature of 70°C.

a. How much of the KCl will dissolve?
b. Is the solution saturated?
c. What is the mass in grams of the undissolved KCl that forms on the bottom of the container?

Solution:

a. A total of 50 g of the KCl will dissolve because that is its solubility and, therefore, the maximum amount of KCl allowed in solution at 70°C.

b. Yes, the solution is saturated.

c. The mass of the undissolved KCl is 25 g.

Study Check:

At 30°C, the solubility of $CuSO_4$ is 25 g/100 g water. How many grams of $CuSO_4$ are needed to make a saturated $CuSO_4$ solution with 300 g of water at 30°C?

Answer: 75 g $CuSO_4$ (3 × 25)

HEALTH NOTE

Gout: A Problem of Saturation in Body Fluids

Gout is a disease that affects adults, primarily men, over the age of 40. Attacks of gout occur when the concentration of uric acid in the plasma exceeds its solubility, and undissolved uric acid is deposited in the joints, causing severe pain. At body temperature, 37°C, the solubility of uric acid is 7 mg/100 mL plasma. Typically, uric acid concentrations range from 2 mg to 6 mg/100 mL plasma. If the uric acid concentration exceeds 7 mg/100 mL plasma, solid uric acid crystals can appear in the cartilage, tendons, and soft tissues, as well as in the tissues of the kidneys, where they can cause renal damage.

Foods in the diet that contribute to high levels of uric acid are often rich in protein and include certain meats, sardines, mushrooms, and asparagus. Certain enzyme defects can also cause an overproduction of uric acid. Treatment for gout includes a reduction in protein intake and the restriction of certain foods. Drugs such as colchicine may be used to reduce uric acid production or to increase the excretion of uric acid. ■

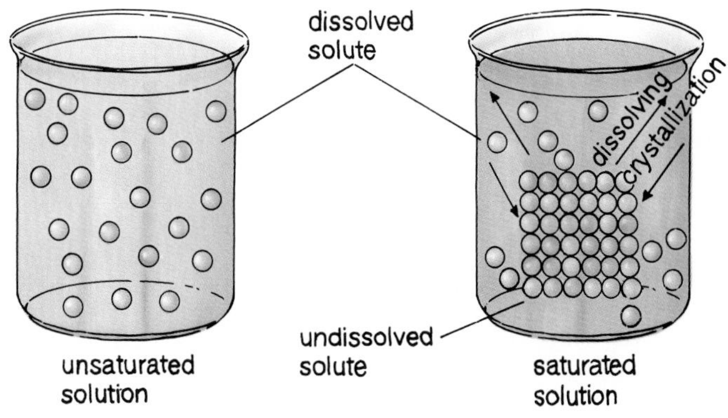

Figure 8.11
In an unsaturated solution, all solute dissolves. In a saturated solution, the maximum amount of solute for a given temperature has dissolved.

Rate of Solution

The time it takes to form a solution depends on how fast the solute leaves the solid and spreads throughout the solvent. One way to form a solution faster is to increase the surface area of the solute. By breaking the solute into smaller particles, more surface area can come in contact with the solvent. The solution then forms more rapidly. It is important to note that we are not changing the solubility of the solute—only the time it takes to form that solution. For example, in a glass of water, the sugar in a spoonful of powdered sugar dissolves faster than the sugar in a sugar cube. The more compact sugar cube has a smaller surface area, so there are relatively fewer molecules on the surface to interact with water. Eventually the final composition of the solutions will be identical, however.

Another way to increase the rate of solution is by stirring or agitating the solution. If the solute particles are dispersed more quickly, more solvent can interact with the remaining undissolved solute.

A higher temperature also increases the rate of solution by increasing the rate at which the solvent bombards the solute and moves the solute into the solution.

The rate of solution increases when:

1. The solute is crushed.
2. The solution is stirred.
3. The solution is heated.

Sample Problem 8.6
Rate of Solution Formation

You are going to dissolve some sugar cubes in a glass of water. Indicate whether each of the following conditions will increase or decrease the rate of solution for the sugar cubes.

a. using hot water
b. crushing the sugar cubes
c. placing the glass of water and sugar cubes in the refrigerator

Solution:

a. An increase in the temperature of the water increases the rate of solution.
b. Increasing the surface area of the sugar that comes in contact with the water molecules will increase the rate of solution.
c. Cooling the solution will slow the movement of the solute and the solvent to decrease the rate of solution.

8.5 Percent Concentration

Learning Goal **Calculate the percent concentration of a solution; use percent concentration to calculate the amount of solute or solution.**

Another way to describe a solution is to indicate the amount of solute that is dissolved in a certain amount of the solution. This quantity is called the **concentration** of a solution.

$$\text{Concentration} = \frac{\text{Amount of solute}}{\text{Amount of solution}}$$

Weight/Weight Percent

The term **percent concentration** means ''parts of solute per 100 parts of solution.'' A weight (or mass) percent (w/w) indicates the number of grams of solute in a given mass of solution.

$$\text{Weight percent} = \frac{\text{Grams of solute}}{\text{Grams of solution}} \times 100$$

Note that the mass of solution is composed of both solute and solvent:

$$\text{Grams of solution} = \text{Grams of solute} + \text{Grams of solvent}$$

For example, 10 g of KCl and 40 g of water would make 50 g of solution with a 20% (w/w) concentration.

$$\text{Weight percent} = \frac{10 \text{ g KCl}}{50 \text{ g solution}} \times 100 = 20\%$$

Volume/Volume Percent

Since the volumes of liquids or gases can be easily measured, their solution concentrations may be stated in volume percent (v/v). The volume percent gives the volume of the solute in a given volume of solution.

$$\text{Volume percent} = \frac{\text{Volume of solute}}{\text{Volume of solution}} \times 100$$

For example, a solution containing 25 mL of alcohol in 100 mL of solution is a 25% (v/v) solution.

$$\text{Volume percent} = \frac{25 \text{ mL alcohol}}{100 \text{ mL solution}} \times 100 = 25\%$$

Weight/Volume Percent

In the health sciences, a weight/volume percent (w/v) is commonly used. Grams are used for the amount of solute, and milliliters for the amount of solution. In this percent, the units (g/mL) do not actually cancel. However, the weight/volume percent is generally used for dilute solutions, so that 100 mL of solution has essentially the same mass as 100 g of water.

$$\text{Weight/volume percent} = \frac{\text{Grams of solute}}{\text{Milliliters of solution}} \times 100$$

Suppose we mix 25 g of glucose and enough water to prepare 500 mL of glucose solution. The weight/volume percent concentration is calculated as

$$\text{Weight/volume percent} = \frac{25 \text{ g glucose}}{500 \text{ mL solution}} \times 100 = 5\%$$

Sample Problem 8.7
Calculating Percent Concentration

Calculate the weight/volume percent (w/v) concentration of a solution prepared by mixing 50 g KI with enough water to make 250 mL of KI solution.

Solution:
Placing grams of solute and milliliters of solution in the expression for weight/volume percent concentration gives

$$\text{Weight/volume percent} = \frac{\overset{\text{Mass of solute}}{50 \text{ g KI}}}{\underset{\text{Volume of solution}}{250 \text{ mL solution}}} \times 100 = 20\% \text{ KI}$$

The laboratory preparation of this solution is shown in Figure 8.12.

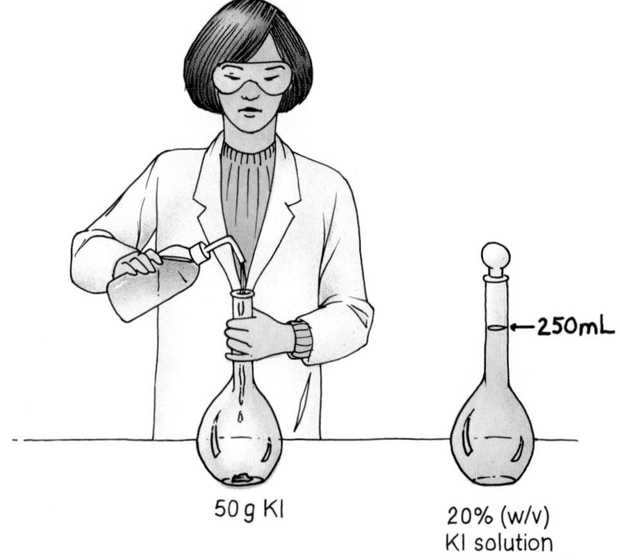

Figure 8.12
A 20% (w/v) KI solution is prepared by (1) measuring 50 g of KI, (2) placing the KI in a 250-mL volumetric flask, (3) adding some water to dissolve the KI, and (4) filling the flask with water to the 250-mL mark.

Study Check:
What is the volume percent (v/v) of a solution prepared by adding 12 mL of bromine (Br_2) to enough carbon tetrachloride to give 250 mL of solution?

Answer: 4.8% (v/v)

Calculations Using Percent Concentration

To use percent concentration in a calculation, it is important to interpret the percent concentration as units of solute and solution. Some examples of percent concentrations and their meanings are given in Table 8.4.

Table 8.4 Interpreting Percent Concentrations

Percent concentration of solution	Meaning in numbers	Meaning in words
10% (w/w) KCl	$\frac{10 \text{ g KCl}}{100 \text{ g solution}}$	There are 10 g of KCl in every 100 g of KCl solution
5% (w/v) glucose	$\frac{5 \text{ g glucose}}{100 \text{ mL solution}}$	There are 5 g of glucose in every 100 mL of glucose solution
15% (v/v) acetone	$\frac{15 \text{ mL acetone}}{100 \text{ mL solution}}$	There are 15 mL of acetone in every 100 mL of acetone solution

To solve solution problems, we can use a percent concentration as a conversion factor. For example, we have seen that a 5% (w/v) glucose solution contains 5 g of glucose in every 100 mL of solution. This relationship can be written as the following percent conversion factors:

$$\frac{5 \text{ g glucose}}{100 \text{ mL solution}} \quad \text{and} \quad \frac{100 \text{ mL solution}}{5 \text{ g glucose}}$$

Sample Problem 8.8
Using Weight/Volume Percent to Calculate Grams of Solute

A laboratory technician must prepare 400 mL of a 5% (w/v) glucose solution. How many grams of glucose must be measured out?

Solution:
Using the 5% (w/v) concentration as a conversion factor, we can calculate the mass of glucose needed for the solution:

$$\underset{\text{Given}}{400\ \cancel{\text{mL}}} \times \underset{\text{Percent conversion factor}}{\frac{5\text{ g glucose}}{100\ \cancel{\text{mL}}\text{ solution}}} = 20\text{ g glucose}$$

Study Check:
How many grams of NaCl are needed to prepare 3.0 L of a 1% (w/v) NaCl solution?

Answer: 30 g NaCl

Sample Problem 8.9
Using Weight/Volume Percent to Calculate Volume Needed

A nurse is to give a patient 50 g of lipid using a 10% (w/v) lipid solution. How many milliliters should be given?

Solution:
The 10% (w/v) concentration can be expressed as the following percent conversion factors:

$$\frac{10\text{ g lipid}}{100\text{ mL}} \quad\text{and}\quad \frac{100\text{ mL}}{10\text{ g lipid}}$$

Using the factor that cancels grams of lipid (g lipid), we calculate the volume required as

$$\underset{\text{Solute needed}}{50\ \cancel{\text{g lipid}}} \times \underset{\text{Percent conversion factor}}{\frac{100\text{ mL}}{10\ \cancel{\text{g lipid}}}} = \underset{\text{Volume of 10\% w/v lipid}}{500\text{ mL lipid solution}}$$

The nurse must give 500 mL of the 10% (w/v) lipid solution so that the patient receives 50 g lipid.

Study Check:
An antifreeze mixture is made with ethylene glycol and water. How many liters of a 15% (v/v) antifreeze solution can be prepared from 60.0 mL of ethylene glycol?

Answer: 0.40 L

8.6 Molarity

Learning Goal **Calculate the molarity of a solution; solve problems using molarity as the conversion factor to find the number of moles of solute or the volume of the solution.**

The molarity of a solution is an important concentration for chemists. In the expression for **molarity (*M*),** the amount of solute is given in moles, and the volume of the solution is given in liters.

$$\text{Molarity } (M) = \frac{\text{Moles of solute}}{\text{Liters of solution}}$$

For example, if we dissolve 6.0 moles of NaOH in enough water to prepare 3.0 L of solution, the resulting NaOH solution has a molarity of 2.0 *M*.

$$M = \frac{\text{Moles of solute}}{\text{Liters of solution}} = \frac{6.0 \text{ moles NaOH}}{3.0 \text{ L solution}} = 2.0\ M \text{ NaOH}$$

Sample Problem 8.10
Calculating the Molarity of a Solution

What is the molarity (*M*) of each of the following solutions?

a. 2.0 moles calcium chloride ($CaCl_2$) in 500 mL of solution
b. 60.0 g NaOH in 0.250 L of solution

Solution:

a. Since the calculation of molarity requires liters, we first change the given volume from milliliters to liters.

$$500 \cancel{\text{mL}} \times \frac{1 \text{ L}}{1000 \cancel{\text{mL}}} = 0.5 \text{ L solution}$$

Using the number of moles of solute and the volume in liters of the solution, we calculate the molarity as

$$\text{Molarity } (M) = \frac{\text{Moles } CaCl_2}{\text{liters of solution}} = \frac{2.0 \text{ moles } CaCl_2}{0.5 \text{ L solution}} = 4\ M$$

b. Since molarity requires moles of solute, we must change the grams of NaOH to moles of NaOH by using the formula weight of NaOH (40.0).

$$60.0 \cancel{\text{g NaOH}} \times \frac{1 \text{ mole NaOH}}{40.0 \cancel{\text{g NaOH}}} = 1.50 \text{ moles NaOH}$$

Amount of NaOH given in grams   Molar mass of NaOH

Now we use the moles of NaOH and the volume of the solution in liters to calculate the molarity of the solution.

$$M = \frac{1.50 \text{ moles NaOH}}{0.250 \text{ L}} = 6.00\ M$$

Study Check:
What is the molarity of a solution that contains 75 g of KNO_3 in 350 mL of solution?

Answer: 2.1 *M*

Calculations Using Molarity

In calculations, it is helpful to restate the molarity of a solution in units of moles per liter. Some examples of molar solutions are given in Table 8.5.

Table 8.5 Some Examples of Molar Solutions

Molarity	In units	Meaning
1 *M* HCl	$\frac{1 \text{ mole HCl}}{1 \text{ L solution}}$	1 mole HCl per liter of solution
2.0 *M* $CaCl_2$	$\frac{2.0 \text{ moles } CaCl_2}{1 \text{ L solution}}$	2.0 moles $CaCl_2$ per liter of solution
3.5 *M* KCl	$\frac{3.5 \text{ moles KCl}}{1 \text{ L solution}}$	3.5 moles KCl per liter of solution
6.0 *M* Na_2CO_3	$\frac{6.0 \text{ moles } Na_2CO_3}{1 \text{ L solution}}$	6.0 moles Na_2CO_3 per liter of solution

When we need to calculate moles of a solute or volume of a solution, the units of molarity can be stated as conversion factors. For example, a 4.0 *M* KCl solution can be written in the form of the following molarity conversion factors:

$$\frac{4.0 \text{ moles KCl}}{1 \text{ L solution}} \quad \text{and} \quad \frac{1 \text{ L solution}}{4.0 \text{ moles KCl}}$$

Sample Problem 8.11
Using Molarity to Calculate Moles of Solute

How many moles of NaCl are present in 4.0 L of a 2.0 *M* NaCl solution?

Solution:
Using the molarity (2.0 *M*) of the solution as a conversion factor, we calculate the number of moles of NaCl:

$$4.0\,\cancel{L} \times \frac{2.0 \text{ moles NaCl}}{1\,\cancel{L}\text{ solution}} = 8.0 \text{ moles NaCl}$$

Given volume — Molarity conversion factor

Study Check:
How many moles of HCl are present in 750 mL of a 6.0 *M* HCl solution?

Answer: 4.5 moles HCl

Often, we must prepare a molar solution by weighing out the proper number of grams of a substance. To do this, we convert the number of moles needed in the solution to grams of solute using the molar mass of the substance.

Sample Problem 8.12
Using Molarity and Molar Mass to Calculate Grams of Solute for a Solution

How many grams of KCl would you need to weigh out in preparing 0.250 L of a 2.00 *M* KCl solution?

Solution:
The number of moles of KCl can be determined by using the given volume of solution and the molarity as a conversion factor.

$$0.250\,\cancel{\text{L solution}} \times \frac{2.00 \text{ moles KCl}}{1\,\cancel{\text{L solution}}} = 0.500 \text{ mole KCl}$$

Given volume — Molarity conversion factor

After we determine the moles of solute needed, the number of grams of KCl is calculated using the molar mass of KCl (74.6 g/mole):

$$0.500\,\cancel{\text{mole KCl}} \times \frac{74.6 \text{ g KCl}}{1\,\cancel{\text{mole KCl}}} = 37.3 \text{ g KCl}$$

Moles KCl — Molar mass KCl

An illustration of how this solution is prepared in the laboratory is given in Figure 8.13.

Study Check:
How many grams of baking soda, $NaHCO_3$, are in 325 mL of 4.50 *M* $NaHCO_3$?

Answer: 123 g $NaHCO_3$

Figure 8.13
A 2.00 *M* KCl solution is obtained when 0.500 mole KCl is placed in a 250-mL volumetric flask, and enough water is added to make 250 mL (0.250 L) of KCl solution.

The volume of the solution can be calculated if the number of moles of solute and the molarity of the solution are given. The following sample problem shows how this type of calculation is carried out.

Sample Problem 8.13
Calculating Volume of a Molar Solution

What volume, in liters, of 1.5 *M* HCl solution is needed to provide 6.0 moles of HCl?

Solution:
Using the molarity as a conversion factor, we can set up the following calculation:

$$6.0 \text{ \sout{moles HCl}} \times \frac{1 \text{ L solution}}{1.5 \text{ \sout{moles HCl}}} = 4.0 \text{ L HCl}$$

Given — Molarity conversion factor

Study Check:
How many milliliters of 2.0 *M* NaOH solution will provide 20.0 g NaOH?

Answer: 250 mL

8.7 Dilution

Learning Goal **Calculate the final volume or new concentration of a solution obtained by dilution.**

In the process of **dilution,** a solution is prepared by the addition of solvent to a more concentrated solution. For example, you might prepare some orange juice for breakfast by adding several cans of water to the orange juice concentrate. (See Figure 8.14.)

When more solvent is added to a solution, the volume increases, causing a decrease in the concentration. However, the amount of solute does not change. It is the same in the diluted solution as in the original sample.

$$\underset{\text{Decreases}}{\text{Percent concentration}} = \frac{\overset{\text{Stays the same}}{\text{grams of solute}}}{\underset{\text{Increases when water is added}}{\text{volume (mL)}}}$$

For example, in 20 mL of a 10% (w/v) $CoCl_2$ solution, there are 2 g $CoCl_2$.

$$20 \cancel{\text{mL solution}} \times \frac{10 \text{ g } CoCl_2}{100 \cancel{\text{mL solution}}} = 2 \text{ g } CoCl_2$$

Figure 8.14
Dilution of orange juice at breakfast time.

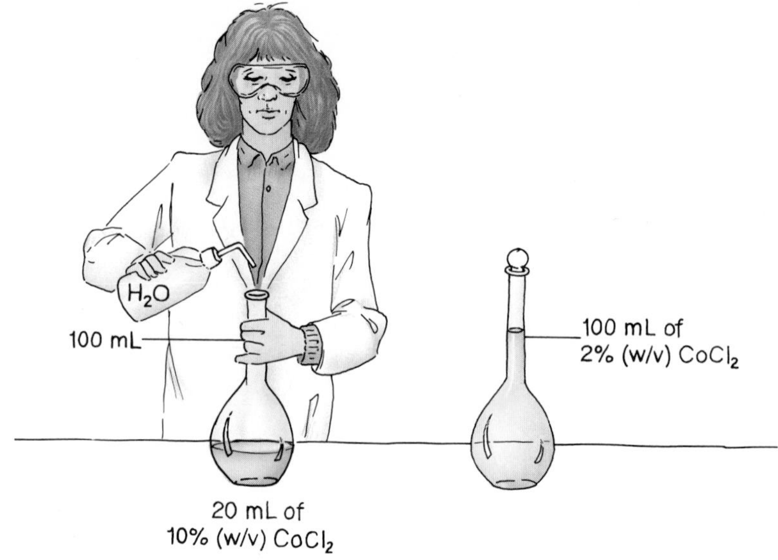

Figure 8.15
A 20-mL quantity of a 10% $CoCl_2$ solution is placed in a volumetric flask, and water is added up to the 100 mL mark. The diluted solution is 2% (w/v) $CoCl_2$.

When water is added, the concentration of the solution is lowered. Suppose we add 80 mL water. The 2 g of $CoCl_2$ from the initial solution are now in 100 mL of solution. (See Figure 8.15.)

Initial volume	+ Water added	= New volume
20 mL	+ 80 mL	= 100 mL

The relationship of the volumes of the initial and the new diluted solutions can be stated in the form of a dilution factor:

$$\text{Dilution factor} = \frac{20\ \cancel{\text{mL}}}{100\ \cancel{\text{mL}}} = \frac{1}{5} \quad \text{or} \quad 1:5$$

$$= \frac{\text{Initial volume}}{\text{New volume}}$$

The percent concentration of the new (diluted) solution can be calculated using the dilution factor. Since the percent concentration must decrease, we multiply the initial concentration by the dilution factor that gives a lower concentration.

$$10\% \times \frac{1}{5} = 2\%$$

10%	$\times \frac{1}{5} =$	2%
Initial concentration	Dilution factor	New concentration of diluted solution

We can summarize the dilution quantities as follows:

	Initial Solution	*Diluted Solution*	
Mass of solute	2 g	2 g	(no change)
Volume of solution	20 mL	100 mL	(increases)
Percent (w/v)	10%	2%	(decreases)

Sample Problem 8.14
Calculating Percent Concentration of a Diluted Solution

What will be the new concentration of a solution if 1500 mL of water is added to 500 mL of a 20% (w/v) solution of KCl?

Solution:
The final volume of the diluted solution will be

$$\underset{\text{Initial volume}}{500 \text{ mL}} + \underset{\text{Water added}}{1500 \text{ mL}} = \underset{\text{New volume}}{2000 \text{ mL}}$$

The dilution factor is

$$\frac{\text{Initial volume}}{\text{New volume}} = \frac{500 \cancel{\text{mL}}}{2000 \cancel{\text{mL}}} = \frac{1}{4} \quad \text{or} \quad 1:4$$

The concentration of the new solution after dilution is

$$\underset{\text{Initial concentration}}{20\%} \times \underset{\text{Dilution factor that decreases concentration}}{\frac{1}{4}} = \underset{\text{Concentration of new solution}}{5\%}$$

Study Check:
How much water must be added to 100 mL of a 20% NaOH solution to prepare a 2% NaOH solution?

Answer: 900 mL

Dilution of Molar Solutions

The calculations are similar for the dilution of a molar solution since the amount of solute (moles) is the same in both initial and diluted solutions.

Sample Problem 8.15
Diluting a Molar Solution

To what volume must 300 mL of a 6 *M* HCl solution be diluted to prepare a 2 *M* HCl solution?

Solution:
We must use the initial and final concentrations to calculate a dilution factor:

$$\text{Dilution factor} = \frac{2 \cancel{M}}{6 \cancel{M}} = \frac{1}{3} \quad \text{or} \quad 1:3$$

Since volume must increase, the dilution factor that increases volume is used.

$$300 \text{ mL} \times \frac{3}{1} = 900 \text{ mL}$$

Initial volume	Dilution factor that increases volume	New volume

In the dilution of a 6 M solution to 2 M, the volume must be increased by adding 600 mL of water to give a final volume of 900 mL.

Study Check:
You have 250 mL of a 1.0 M HCl solution. How many milliliters of water should you add to dilute the solution to a 0.50 M HCl solution?

Answer: 250 mL

8.8 Solutions, Colloids, and Suspensions

Learning Goal **From its properties, identify a mixture as a solution, a colloid, or a suspension.**

The solute particles in a solution play an important role in determining the properties of that solution. In most of the solutions we have discussed so far, the solute is dissolved as small, single particles that are uniformly dispersed throughout the solvent to give a homogeneous solution. The particles are so small that they go through filters and through semipermeable membranes, a type of membrane that allows only certain kinds of particles to pass through. Solutions do not settle out even after they have been stored for long periods of time. When you observe a solution, such as salt water, you cannot visually distinguish the solute from the solvent. The solution appears transparent even when a light shines through it because the particles are too small to scatter the light. Although solutions are transparent, they don't have to be colorless.

Colloids

The particles in colloidal dispersions, or **colloids,** are larger than solution particles. They can be large molecules, such as proteins, or groups of molecules or ions. Colloids are homogeneous mixtures that do not separate or settle out. Colloidal particles are small enough to pass through filters, but too large to pass through semipermeable membranes.

Figure 8.16
The Tyndall effect: When light is reflected by colloids, it becomes visible.

When a beam of light shines through a colloid, it produces the **Tyndall effect,** in which the large particles reflect the light and make the light beam visible, as shown in Figure 8.16. The Tyndall effect causes a light beam to be visible in fog because the water droplets in the air reflect the light. When there is smoke or smog in the air, the blues and greens of sunlight are reflected more, so that we observe the oranges and reds of a beautiful sunset.

Types of Colloids

Colloids are classified by the type of substance and the dispersing medium. There are four types of colloids; aerosols, foams, emulsions, and sols. *Aerosols* consist of liquid or solid particles dispersed in a gas. *Foams* are dispersions of gases in liquids or solids. An *emulsion* is a liquid dispersed in another liquid or a solid. For example, when milk is not homogenized, the suspended particles of cream separate. The process of homogenization breaks the cream particles into colloids that remain dispersed within the solution. In a *sol,* particles of a solid are dispersed in a liquid or a solid to give a more rigid solution. Table 8.6 gives several examples of colloids.

Table 8.6 Colloids

Examples	Type	Substance dispersed	Dispersing medium
Fog, clouds, sprays	Aerosol	Liquid	Gas
Dust, smoke	Aerosol	Solid	Gas
Shaving cream, whipped cream, soapsuds	Foam	Gas	Liquid
Styrofoam, marshmallows	Foam	Gas	Solid
Mayonnaise, butter, homogenized milk, hand lotions	Emulsion	Liquid	Liquid
Cheese, butter	Emulsion	Liquid	Solid
Blood plasma, paints (latex), gelatin	Sol	Solid	Liquid
Cement, pearls	Sol	Solid	Solid

Suspensions

A **suspension** is a heterogeneous, nonuniform mixture that is very different from solutions or colloids. The particles of a suspension are so large that they can often be seen with the naked eye. They are trapped by either filters or membranes.

The weight of suspended particles causes them to settle out soon after mixing. If you stir muddy water, it mixes but then quickly separates as the suspended particles settle to the bottom and leave clear water at the top. You can find suspensions among the medications in a hospital or in your medicine cabinet. These include Kaopectate, calamine lotion, antacid mixtures, and liquid penicillin. It is important to "shake well before using" in order to suspend all the particles before giving a medication that is a suspension.

Water-treatment plants make use of the properties of suspensions to purify water. Coagulants such as aluminum sulfate or ferric sulfate are added to untreated water. These chemicals react with small, suspended particles in the water and form larger particles called floc which are suspensions. The water is sent through a filtering system that traps the large particles and lets the clean water go through. The treated water is chlorinated and sent into the water system of a city.

Table 8.7 compares the properties of solutions, colloids, and suspensions, and Figure 8.17 illustrates some of these properties.

Table 8.7 Comparison of Solutions, Colloids, and Suspensions

Type of mixture	Type of particle	Effect of light	Settling	Separation
Solution	Small particles such as single atoms, ions, or molecules	Transparent	Particles do not settle	Particles cannot be separated by filters or semipermeable membranes
Colloid	Larger molecules or groups of molecules or ions	Tyndall effect occurs	Particles do not settle	Particles can be separated by semipermeable membranes but not by filters
Suspension	Very large particles that may be visible	Opaque (not transparent)	Particles settle rapidly	Particles can be separated by filters

Sample Problem 8.16
Classifying Types of Mixtures

Classify the solution in each of the following descriptions as a solution, colloid, or suspension:

a. a homogeneous mixture in which a beam of light is visible
b. a mixture that settles rapidly upon standing
c. a solution whose solute particles pass through both filters and membranes

Solution:

a. colloid
b. suspension
c. solution

Figure 8.17
Properties of different types of solutions: (a) Suspended particles settle out; (b) only suspended particles are removed by filters; (c) only solution particles move through a semipermeable membrane and become separated from colloidal and suspended particles.

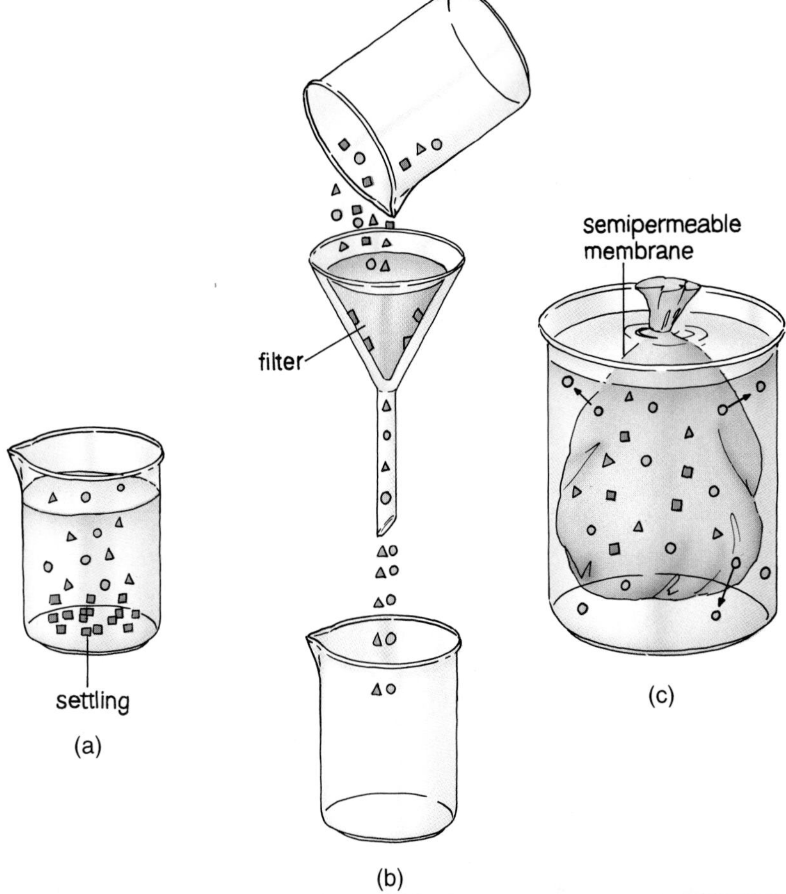

HEALTH NOTE

Colloids and Solutions in the Body

Colloids in the body are separated from solutions by semipermeable membranes. For example, the intestinal lining allows solution particles to pass into the blood and lymph circulatory systems. However, the large colloid molecules from foods are too large to pass through the membrane, and they remain in the intestinal tract. Digestion breaks down large colloidal particles, such as starch and protein, into smaller particles, such as glucose and amino acids, that can pass through the intestinal membrane and enter the circulatory system. Certain foods, such as bran, a fiber, cannot be broken down by human digestive processes, and they move through the intestine intact.

The cell membranes also separate solutions from colloids. For example, large proteins, such as enzymes, are formed inside cells. Since they are colloids, they remain inside the cell. However, many of the nutrients that must be obtained by cells, such as oxygen, amino acids, electrolytes, glucose, and minerals, are solutions that can pass through cellular membranes. Waste products, such as urea and carbon dioxide, also form solutions and pass out of the cell to be excreted. ■

8.9 Osmosis and Dialysis

Learning Goal **Describe the processes of osmosis and dialysis.**

Osmosis is the movement of a solvent, usually water, through a semipermeable membrane. For example, we might place some water on one side of a semipermeable membrane and a sucrose (sugar) solution on the other side as shown in Figure 8.18. In osmosis, water flows into the sucrose solution, but sucrose cannot pass through the semipermeable membrane. As water moves into the compartment containing the sugar, the volume increases. The increased weight of sucrose solution exerts a downward pressure called **osmotic pressure.** The osmotic pressure of the sucrose pushes water back through the semipermeable membrane preventing any further increase in volume.

Osmotic pressure depends on the number of particles in the solution. The greater the number of particles dissolved, the higher the osmotic pressure. In the body, blood, tissue fluids, lymph, and plasma all exert osmotic pressure. **Physiological solutions** that are used to replace body fluids, such as 0.9% NaCl (physiological saline) or 5% glucose solutions, do not usually have the same kinds of particles as body fluids, but they exert the same osmotic pressures.

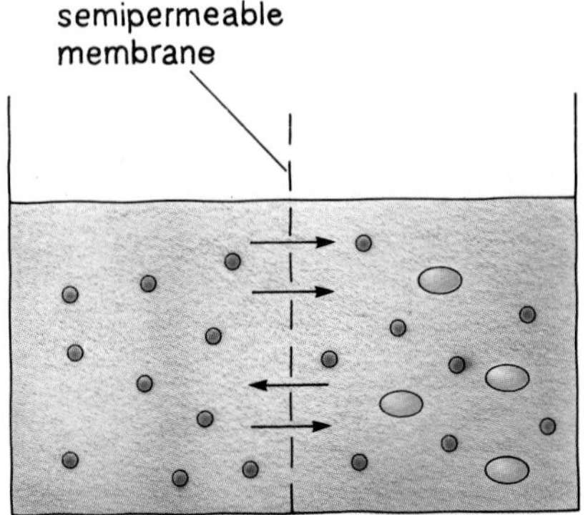

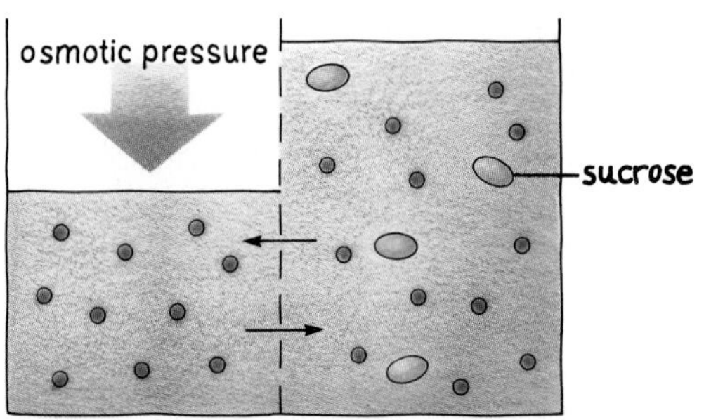

Figure 8.18
Osmosis, the diffusion of water through a semipermeable membrane into the compartment with the greater solute concentration.

Sample Problem 8.17
Osmotic Pressure

A 2% sucrose solution and an 8% sucrose solution are separated by a semipermeable membrane.

a. Which sucrose solution exerts the greater osmotic pressure?
b. In what direction does water flow initially?
c. Which side will increase in volume?

Solution:

a. The 8% sucrose solution has the higher solute concentration, more solute particles, and the greater osmotic pressure.
b. Initially, water will flow out of the 2% solution into the more concentrated 8% solution.
c. Since water is entering the 8% solution, its volume will increase.

Isotonic Solutions

Fluids that are placed intravenously in the body, such as 0.9% NaCl or 5% glucose, are called **isotonic solutions.** *Iso* means "equal to," and *tonic* refers to the osmotic pressure of the cell. Therefore, an isotonic solution has the same osmotic pressure as the body fluids and will not alter the volume of the cells. If a red blood cell is placed in an isotonic solution, it retains its normal volume because there is an equal flow of water into and out of the cell (see Figure 8.19a).

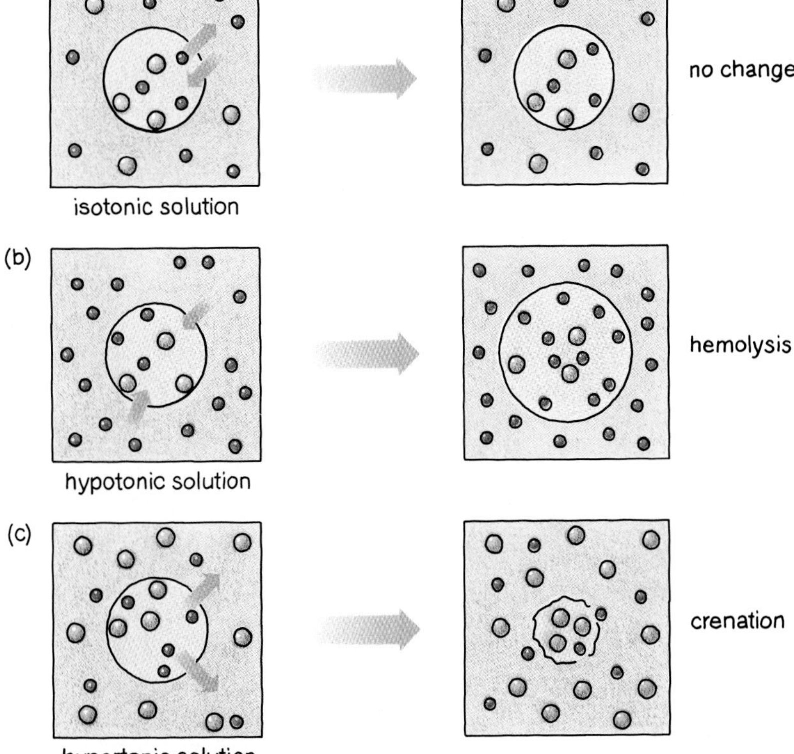

Figure 8.19
Effects of (a) isotonic, (b) hypotonic, and (c) hypertonic solutions on red blood cells. In an isotonic solution, a red blood cell retains its normal volume. In a hypotonic solution, the red blood cell swells (hemolysis) because water flows into the cell. In a hypertonic solution, water leaves the red blood cell, causing it to shrink (crenation).

Hypotonic and Hypertonic Solutions

If a red blood cell is surrounded by a solution that is not isotonic, the difference in osmotic pressure inside and outside the cell causes a change in the volume of the cell. Pure water is a **hypotonic solution** because its osmotic pressure (zero) is less than the osmotic pressure of the cells (*hypo* means "lower than"). Since the solution in a red blood cell exerts osmotic pressure, water flows into the cell, as shown in Figure 8.19b. The increase in fluid volume causes the cell to swell, and possibly burst, a process called **hemolysis.** A similar process occurs when you place dehydrated food, such as raisins or dried fruit, in water. The water enters the cells and the food becomes plump and smooth.

Suppose we place red blood cells in a 10% NaCl solution. Since red blood cells have an osmotic pressure equal to the osmotic pressure of a 0.9% NaCl solution, the 10% NaCl solution has a much higher osmotic pressure; it is a **hypertonic solution** (*hyper* means "greater than"). In this case, the difference in osmotic pressure causes water to leave the red blood cells. Fluid volume is lost and the cell shrinks, a process called **crenation** (Figure 8.19c). A similar process occurs when making pickles, in which a hypertonic brine (a salt solution) causes the cucumbers to shrivel as they lose water.

Sample Problem 8.18
Isotonic, Hypotonic, and Hypertonic Solutions

Indicate whether each of the following solutions is isotonic, hypotonic, or hypertonic, and whether red blood cells placed in each solution will undergo hemolysis, crenation, or no change:

a. a 5% glucose solution
b. a 0.2% NaCl solution
c. a 10% glucose solution

Solution:

a. A 5% glucose solution is an isotonic solution because it has the same osmotic pressure as body fluids. Red blood cells will not undergo any change in an isotonic solution.
b. A 0.2% NaCl solution is a hypotonic solution and will cause red blood cells to absorb water and undergo hemolysis.
c. A 10% glucose solution is hypertonic compared to body fluids. In this solution, red blood cells will lose water and undergo crenation.

Dialysis

Dialysis is a process that is similar to osmosis. In dialysis, a semipermeable membrane, called a dialyzing membrane, permits small molecules and ions in solutions as well as water molecules to pass through, but it retains larger particles, such as colloids. Dialysis is a way to separate solution particles from colloids.

Suppose we fill a cellophane bag with a solution of NaCl, glucose, starch, and protein and place it in pure water. Cellophane is a dialyzing membrane, and

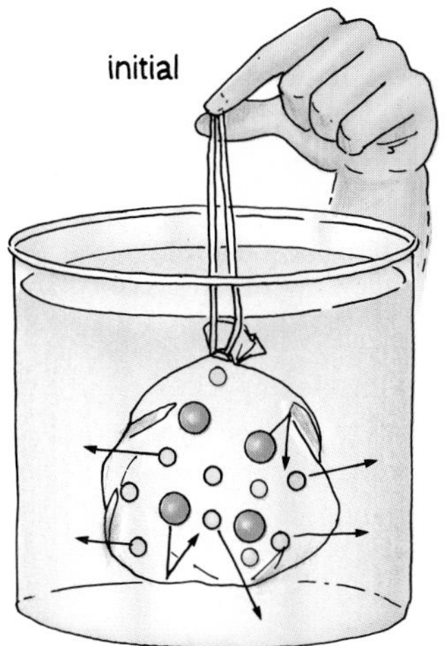

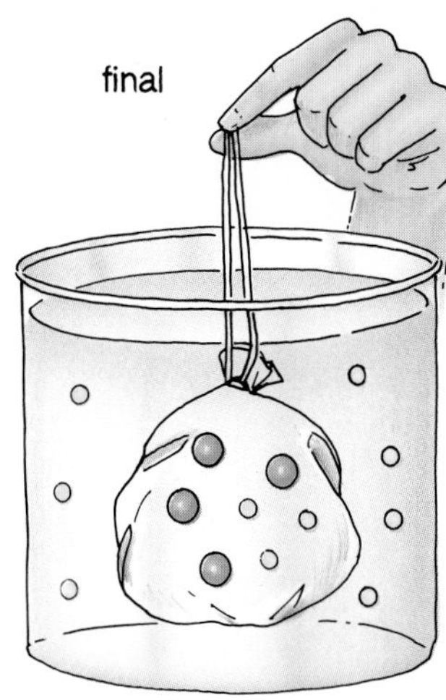

Figure 8.20
Dialysis, the separation of solutions from colloids in water.

the solution particles of sodium ions, chloride ions, and small glucose molecules will pass through it into the surrounding water. However, the colloids starch and protein remain inside. Water molecules will flow by osmosis into the colloids within the cellophane bag. Eventually, the total concentrations of sodium ions, chloride ions, and glucose inside and outside the dialysis bag become equal. To remove more NaCl or glucose, the cellophane bag must be placed in a fresh sample of pure water. (See Figure 8.20.)

Sample Problem 8.19
Dialysis

A KCl solution and starch, a colloid, are placed in a dialysis bag. How will the concentration of each change when the dialysis bag is placed in pure water?

Solution:
The ions K^+ and Cl^- will dialyze. Their concentrations will decrease inside the dialysis bag but increase outside the bag until the same concentrations of K^+ and Cl^- are on both sides of the bag. The colloidal starch particles remain inside. Water flowing into the dialysis bag by osmosis will dilute the contents and lower the concentration of the solutes that remain.

HEALTH NOTE

Dialysis by the Kidneys and the Artificial Kidney

The fluids of the body undergo dialysis by the membranes of the kidneys, which remove waste materials, excess salts, and water. In an adult, each kidney contains about 2 million nephrons. (See Figure 8.21.) At the top of each nephron, there is a network of arterial capillaries called the glomerulus.

As blood flows into the glomerulus, small particles, such as amino acids, glucose, urea, water, and certain ions, will move through the capillary membranes into the nephron. As this solution moves through the nephron, substances still of value to the body (such as amino acids, glucose, certain ions, and 99% of the water) are reabsorbed. The major waste product, urea, forms urine in the bladder to be excreted.

Hemodialysis

If the kidneys fail to dialyze waste products, increased levels of urea can become life-threatening in a relatively short time. A person with kidney failure can use an artificial kidney, which cleanses the blood by **hemodialysis,** as shown in Figure 8.22.

A typical artificial kidney machine contains a large tank filled with about 100 L of distilled water. In the center of this dialyzing bath (dialysate), there is a dialyzing coil or membrane made of cellulose tubing. As the patient's blood flows through the dialyzing coil, the highly concentrated waste products dialyze out of the blood. No blood is lost because the membrane is not permeable to large particles such as red blood cells.

Dialysis patients do not produce much urine. As a result, they retain large amounts of water between dialysis treatments, which produces a strain on the heart. The intake of fluids for a dialysis patient may be restricted to as little as a few teaspoons of water a day. In the dialysis procedure, the pressure of the blood is increased as it circulates through the dialyzing coil so water can be squeezed out of the blood. For some dialysis patients, 2–10 L of water may be removed during one treatment. Dialysis patients have from two to three treatments a week, each treatment requiring about 5–7 hr. Some of the newer treatments require less time. For many patients, dialysis is done at home with a home dialysis unit. ■

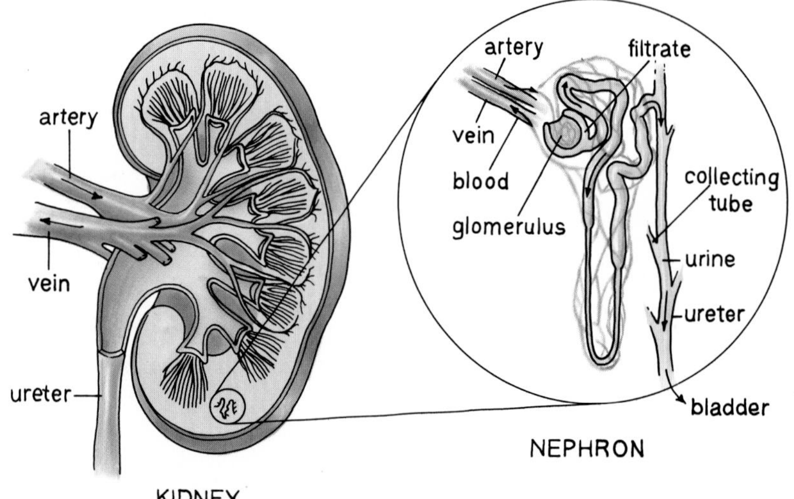

Figure 8.21
The nephron, the working unit of the kidney. Water and small particles dialyze out of the blood across the membrane of the glomerulus. Most of the reusable substances, including water, glucose, and amino acids, are reabsorbed from the filtrate, leaving urea and other waste products to be excreted in the urine.

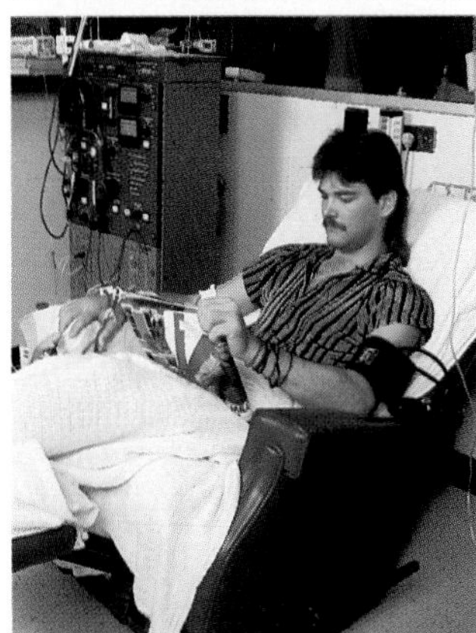

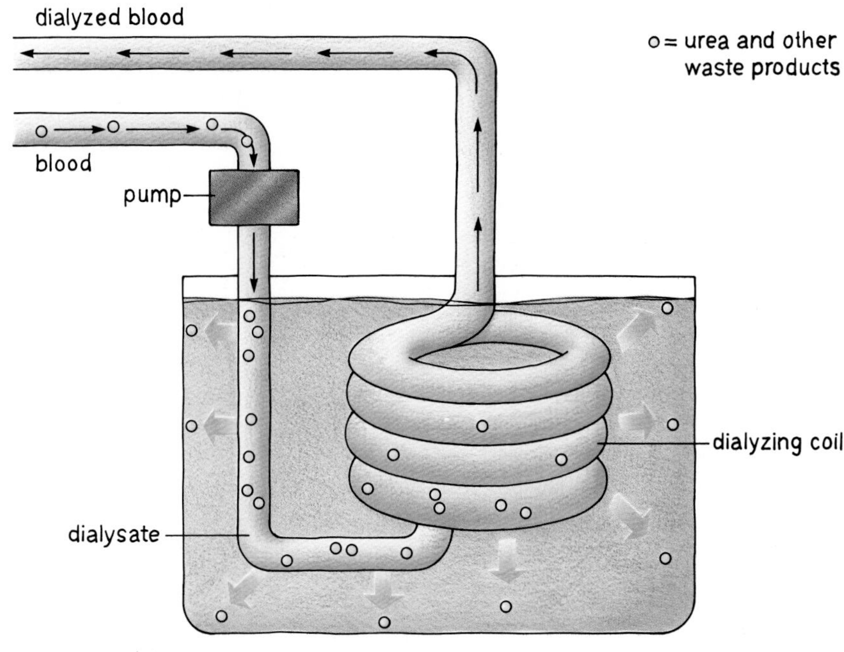

Figure 8.22
Hemodialysis, the dialysis of the blood by an artificial kidney. The initial dialysate consists of water and electrolytes. As blood flows through the dialyzing coil, urea and other waste products dialyze out of the coil and into the dialysate.

Glossary of Key Terms

colloid A mixture having particles that are moderately large. Colloids scatter light in the Tyndall effect and pass through filters but will not pass through semipermeable membranes.

concentration A measure of the amount of solute that is dissolved in a specified amount of solution.

crenation The shriveling of a cell due to water leaving the cell when the cell is placed in a hypertonic solution.

dialysis A process in which water and small solution particles move through a semipermeable membrane.

dilution The decrease in concentration of a solution by the addition of more solvent.

hemodialysis A mechanical cleansing of the blood by an artificial kidney using the principle of dialysis.

hemolysis A swelling and/or bursting of red blood cells in a hypotonic solution due to an increase in fluid volume.

hydrogen bond The attraction between a partially positive hydrogen atom in one water molecule and the partially negative oxygen atom of another.

hypertonic solution A solution that has a higher osmotic pressure than the red blood cells of the body have.

hypotonic solution A solution that has a lower osmotic pressure than the red blood cells of the body have.

isotonic solution A solution that has the same osmotic pressure as that of the red blood cells of the body.

molarity (*M*) The number of moles of solute in 1 L of solution.

osmosis The flow of a solvent, usually water through a semipermeable membrane into a solution of higher concentration.

osmotic pressure The pressure created by the weight of a column of fluid that prevents the flow of more water into the liquid.

percent concentration The amount of solute in 100 parts of solution defined as weight/weight, weight/volume, or volume/volume.
physiological solution A solution that exerts the same osmotic pressure as body fluids.
saturated solution A solution containing the maximum amount of solute that can dissolve at a given temperature. Any additional solute will remain undissolved in the container.
semipermeable membrane A membrane that permits the passage of certain substances while blocking or retaining others.
solubility The maximum amount of solute that can dissolve in 100 g of solvent, usually water, at a given temperature.
solute The component in a solution that changes state upon dissolving; if no change in state occurs, it is the component present in smaller quantity.
solution A homogeneous mixture in which the solute is made up of small, single particles (ions or molecules) that can pass through filters and semipermeable membranes.
solvent The substance in which the solute dissolves; usually the component present in greatest amount.
surface tension A characteristic of water in which the hydrogen bonding of water molecules pulls its surface molecules together so tightly that they form a ''skin.''
suspension A mixture in which the particles are large enough and heavy enough to settle out and/or be retained by both filters and semipermeable membranes.
Tyndall effect The opaque path of a light beam through a colloid as light is reflected by the colloidal particles.
unsaturated solution A solution that contains less solute than its solubility level.

Problems

Solutes and Solvents *(Goal 8.1)*

8.1 Identify the solute and the solvent in the following solutions:
a. 10 g NaCl and 100 g H_2O
b. 50 mL ethanol, $C_2H_5OH(l)$, and 10 mL H_2O
c. 2.0 L oxygen (O_2), 8.0 L nitrogen (N_2)
d. 100 g silver and 40 g mercury
e. 100 mL H_2O and 5.0 g sugar

Water: An Important Solvent *(Goal 8.2)*

8.2 Match the following statements with the terms *hydrogen bonding, surface tension,* or *polar solvent:*
a. the pulling together of water molecules at the surface to form a thick layer than will support the weight of small bugs
b. attractive forces between water molecules
c. behavior of water as a solvent

Formation of Solutions *(Goal 8.3)*

8.3 Match the following statements with the terms *oxygen atom, hydrogen atom,* or *hydration*.
a. the portion of the water molecule attracted to the positive ions of a salt
b. the surrounding of ions in a solution by several water molecules
c. the portion of the water molecule attracted to negative ions in a salt

8.4 State whether each of the following solutes will be soluble in water or in hexane, a nonpolar solvent:

a. KCl, ionic
b. sugar (sucrose), polar
c. benzene, nonpolar
d. I_2, nonpolar
e. vegetable oil, nonpolar
f. KNO_3, a salt

Solubility and Saturation *(Goal 8.4)*

8.5 Using the solubilities of KBr and KI given in the accompanying table, state whether each solution would be saturated or unsaturated:

Solubility (g/100 g H_2O)		
T (°C)	**KBr**	**KI**
20	65	145
40	80	160
60	90	175
80	100	190
100	110	210

a. 75 g KBr in 100 g H_2O at 40°C
b. 160 g KI in 100 g H_2O at 40°C
c. 100 g KBr in 100 g H_2O at 100°C
*d. 100 g KBr in 200 g H_2O at 80°C
e. 50 g KI in 100 g H_2O at 20°C

8.6 State whether each of the following refers to a saturated or unsaturated solution:

a. A solid piece of solute added to a solution does not change in size.
b. A sugar cube dissolves completely when added to a cup of coffee.
c. All of the solute dissolves.
d. A layer of sugar forms on the bottom of a glass of ice tea.

8.7 Tap water containing dissolved chlorine gas cannot be used in a fish tank until the chlorine is removed. Why can tap water that has been warmed to room temperature for at least a day, or boiled and cooled, be used to fill a fish tank?

***8.8** A solution containing 100 g of KBr in 100 g of H_2O at 100°C is cooled to 20°C. Using the solubility table in Problem 8.5 for KBr and KI, answer the following questions:

a. Is the solution saturated or unsaturated at 100°C?
b. Is the solution saturated or unsaturated at 20°C?
c. At what temperature will 350 g of KI form a saturated solution with 200 g of water?

8.9 Indicate whether the following preparations will increase or decrease the rate of solution of KCl in water:

a. stirring the mixture
b. using large chunks of KCl
c. placing the beaker of KCl and water in ice
d. heating the beaker containing the KCl and water
e. crushing the KCl into a fine powder before mixing it with the water

Percent Concentration *(Goal 8.5)*

8.10 What is the difference between a 5% (w/w) glucose solution and a 5% (w/v) glucose solution?

8.11 Calculate the percent concentration of the following solutions:
a. (w/v) 2.0 g sucrose in 100 mL of solution
b. (w/w) 20.0 g KCl in 150 g of solution
c. (w/v) 75.0 g of Na_2SO_4 in 0.50 L of solution
d. (v/v) 3.0 mL of acetone in 40 mL of solution
*e. (w/v) 0.300 kg of glucose in 5.00 L of solution
f. (w/w) 40.0 g of $CaCl_2$ in 250 g of solution

8.12 Write two conversion factors for solutions having the following labels:
a. 10% (w/w) NaOH
b. 2.5% (w/v) NaCl
c. 15% (v/v) CH_3OH
d. 5.0% (w/v) $C_6H_{12}O_6$ (glucose)
e. 20% (w/w) K_2CO_3

8.13 Calculate the number of grams of solute needed to prepare the following solutions:
a. 50.0 mL of a 5.0% (w/v) KCl solution
b. 225 g of a 10.0% (w/w) NaCl solution
*c. 1.25 L of a 4.0% (w/v) NH_4Cl solution
d. 75 g of a 40% (w/w) NaOH solution

8.14 How many milliliters of solute are needed to prepare the following solutions?
a. 250 mL of a 10.0% (v/v) acetic acid solution
b. 400 mL of a 15% (v/v) isopropyl alcohol solution
*c. 1.5 L of a 7.5% (v/v) methyl alcohol solution

8.15 A patient receives 100 mL of 20% (w/v) mannitol solution every hour.
a. How many grams of mannitol are given in 1 hr?
b. How many grams of mannitol does the patient receive in one day?

***8.16** A patient receives all of his/her nutrition from fluids. Every 12 hr, 750 mL of a solution that is 4% (w/v) amino acids from protein and 25% (w/v) dextrose (glucose) is given, along with 500 mL of a 10% (w/v) lipid (fat).
a. In one day, how many grams of amino acids, dextrose, and lipid are given to the patient?
b. How many kilocalories does the patient obtain in one day?

8.17 How many grams of solution are needed to provide the following amounts of solute?
a. 10.0 g HCl from a 1.0% (w/w) solution
b. 5.0 g $LiNO_3$ from a 25% (w/w) solution
c. 40.0 g KOH from a 10.0% (w/w) solution

8.18 Calculate the number of milliliters needed to provide the following amounts of solute:
a. 150 g of glucose from a 5% (w/v) glucose solution
b. 400 g of NaOH from a 10% (w/v) NaOH solution

c. 50 g of NaCl from a 2.0% (w/v) NaCl solution
d. 2.0 g of KBr from a 8.0% (w/v) solution

8.19 A patient needs 100 g of glucose in the next 12 hr. How many liters of a 5% (w/v) glucose solution should be given?

***8.20** An 80-proof brandy is 40% (v/v) ethyl alcohol. (The numerical term *proof* is twice the percent concentration of alcohol in the beverage.) How many milliliters of alcohol are present in 750 mL of brandy?

Molarity *(Goal 8.6)*

8.21 Calculate the molarity (M) of the following solutions:
a. 4.0 moles of KOH in 2.0 L of solution
b. 2.0 moles of glucose in 4.0 L of solution
c. 5.0 moles of NaOH in 500 mL of solution
d. 0.10 mole of NaCl in 40 mL of solution

8.22 Calculate the molarity (M) of the following solutions:
a. 36.5 g HCl in 1.0 L of solution
*b. 8.0 g of NaOH in 200 mL of solution
*c. 320 g of glucose, $C_6H_{12}O_6$, in 500 mL of solution

8.23 Write two conversion factors for the following molar solutions:
a. 6.00 M HCl
b. 0.250 M $NaHCO_3$
c. 1.0 M H_2SO_4
d. 0.50 M KBr

8.24 Calculate the number of moles in the following:
a. 1.0 L of a 3.0 M NaCl solution
b. 5.0 L of a 2.0 M $CaCl_2$ solution
c. 200 mL of a 4.0 M glucose solution
d. 50 mL of a 10 M sucrose solution

8.25 Calculate the number of grams of solute needed to prepare each of the following solutions:
a. 1.0 L of a 1.0 M NaOH solution
b. 4.0 L of a 2.0 M KCl solution
*c. 500 mL of a 1.0 M NaCl solution
d. 200 mL of a 6 M NaOH solution
e. 750 mL of a 1.0 M glucose ($C_6H_{12}O_6$) solution

8.26 What volume in liters provides the following amounts of solute?
a. 2.0 moles of NaOH from a 2.0 M NaOH solution
b. 10 moles of NaCl from a 1 M NaCl solution
c. 80.0 g of NaOH from a 1 M NaOH solution
d. 1.0 mole glucose from a 2.0 M glucose solution

Dilution *(Goal 8.7)*

8.27 Calculate the final volume needed to prepare the following diluted solutions:
a. a 5% (w/v) KCl solution from 10.0 mL of 20% (w/v) KCl
b. a 10% (w/v) mannitol solution from 50.0 mL of 30% (w/v) mannitol
c. a 5% (w/v) NaCl solution from 450 mL of 10% (w/v) NaCl

d. a 3.0 *M* HCl solution from 40.0 mL of 6.0 *M* HCl
e. a 2.0 *M* NaOH solution from 5.0 mL of 12 *M* NaOH

8.28 Calculate the new concentration for each solution prepared by the following procedures:
a. diluting 10.0 mL of a 15% (w/v) NaCl solution to 50.0 mL
*b. adding 150 mL of water to 50 mL of 20% (w/v) KOH
c. diluting 20.0 mL of 4.0 *M* $CaCl_2$ to a new volume of 80.0 mL
*d. adding 180 mL of water to 20.0 mL of 10.0 *M* sucrose solution

Solutions, Colloids, and Suspensions *(Goal 8.8)*

8.29 Identify the following as characteristic of a solution, colloid, or suspension:
a. The solution is clear and cannot be separated by a semipermeable membrane.
b. The mixture appears cloudy when a beam of light is passed through it.
c. Particles of this mixture remain inside a semipermeable membrane.
d. The mixture settles out upon standing.
e. The particles can be separated by filters.
f. A beam of light is not visible in this solution.
g. The particles are very large and visible.

Osmosis and Dialysis *(Goal 8.9)*

8.30 A 10% (w/v) starch solution is separated from pure water by an osmotic membrane.
a. In which direction will water flow initially?
b. Which compartment will increase in volume?
c. Does the starch or the water exert the greatest osmotic pressure?

8.31 A 0.1% (w/v) albumin solution and a 2% (w/v) albumin solution are separated by a semipermeable membrane. (Albumins are proteins that are colloids.)
a. In which direction will water flow initially?
b. Which solution will increase in volume?
c. Which solution has the higher osmotic pressure?

8.32 Consider the following liquids or solutions:

H_2O	1% (w/v) glucose
5% (w/v) glucose	10% (w/v) NaCl
0.90% (w/v) NaCl	

Compared to a red blood cell, which of these solutions

a. is isotonic?	c. is hypotonic?	e. causes hemolysis?
b. is hypertonic?	d. causes crenation?	f. causes no change?

8.33 Each of the following mixtures is placed in a dialyzing bag that is immersed in distilled water. Which substances will dialyze?
a. NaCl, starch, and amino acids (solution)
b. albumin (colloidal protein), KCl, and glucose
c. urea (solution) and NaCl

***8.34** A patient on dialysis has a high level of urea, a high level of sodium, and a low level of potassium in the blood. Why is the dialyzing solution prepared with a high level of potassium but no sodium or urea?

9

Acids, Bases, and Salts

Stalactites are downward growing formations of calcium carbonate caused by a solution of Ca^{2+} and HCO_3^- dripping through the ceiling of a limestone cave.

Learning Goals

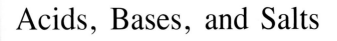

9.1 Write an equation to show the dissociation of electrolytes in water.

9.2 Predict the formation of a solid when two solutions of electrolytes are mixed.

9.3 Write an equation to show the formation of an acidic solution or a basic solution.

9.4 Write an equation to show the ionization of water; write the ion-product expression for water. Use the ion product to calculate $[H^+]$ and $[OH^-]$ in solution.

9.5 Use the pH of a solution to classify the solution as acidic, neutral, or basic; calculate the pH, $[H^+]$, and $[OH^-]$ of the solution.

9.6 Write a balanced equation for the neutralization reaction of an acid and a base.

9.7 Calculate the concentration of an acidic solution from the volume and the molarity of the basic solution used to neutralize the acid.

9.8 Identify the components of a buffer and the role of each component in maintaining the pH of a solution.

9.9 Use equivalent weight and normality to determine the number of equivalents of an acid, base, or salt.

SCOPE

Water is the solvent for most solutions, including the fluids in biological systems. In the human body, ions such as K^+ Na^+, Cl^-, H^+, and HCO_3^-, along with molecules such as glucose and urea, all form solutions with water. Significant changes in the concentrations of these substances in the body fluids can indicate illness or injury. Therefore, their measurement can be a valuable diagnostic tool.

Why do lemons, grapefruits, and vinegar taste sour? They all contain acidic solutions. We use acids to digest our food, etch glass, and curdle milk to produce cottage cheese or yogurt. Sometimes we take antacids such as milk of magnesia to offset the effects of too much acid. Solutions that neutralize acids are called basic solutions.

The pH of a solution describes its acidity. The pH of body fluids, including blood and urine, is regulated primarily by the lungs and the kidneys. Major changes in the pH of body fluids can severely affect biological activities within the cells.

9.1 Electrolytes

Learning Goal **Write an equation to show the dissociation of electrolytes in water.**

Solutes that produce ions when they dissolve in water are called **electrolytes.** As the ions move through the solution, they conduct an electrical current. When the solution contains only ions and no molecules of the solute, the compounds is called a **strong electrolyte.** For example, a sodium chloride (NaCl) solution is a clear liquid and looks like water. However, we know from its salty taste that it contains dissolved sodium ions and chloride ions. To show the formation of the NaCl solution, we can write an ionic equation for the separation of the ions. Because NaCl is an ionic solute, it is also a strong electrolyte and therefore separates completely into ions when it dissolves. The solid form of the salt is written as NaCl(*s*). The formula H_2O above the reaction arrow indicates that the solute is dissolved in the solvent water. The dissolved ions in solution are written as $Na^+(aq)$ and $Cl^-(aq)$ or, more simply, as Na^+ and Cl^- where (*aq*) is understood. (See Figure 9.1.)

$$\text{NaCl}(s) \xrightarrow{H_2O} \text{Na}^+(aq) + \text{Cl}^-(aq)$$

100% dissociation — Ions in solution

Other ionic compounds dissolve in a similar way. In all ionic equations, the dissolved ions must show an electrical balance; that is, the total positive charge must equal the total negative charge. For example, magnesium chloride, $MgCl_2$, dissolves in water to give one magnesium ion and two chloride ions in solution.

$$\text{MgCl}_2(s) \xrightarrow{H_2O} \text{Mg}^{2+}(aq) + 2\text{Cl}^-(aq)$$

Charge balance for ions in solution

Figure 9.1
An aqueous solution of the strong electrolyte sodium chloride (NaCl) contains sodium ions (Na^+) and chloride (Cl^-) ions.

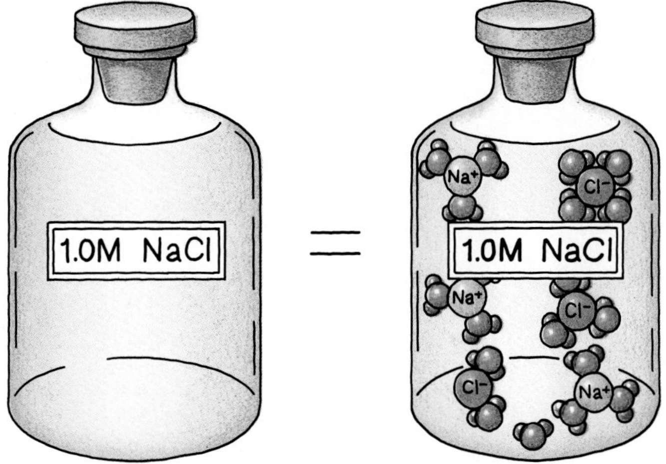

When writing equations for dissolving ionic compounds that include polyatomic ions, the polyatomic ion remains a unit in solution. For example, the ionic compound Na_3PO_4 dissolves in water to give three Na^+ ions and one PO_4^{3-} ion.

$$Na_3PO_4(s) \xrightarrow{H_2O} 3Na^+(aq) + PO_4^{3-}(aq)$$

Polyatomic ion in solution

Sample Problem 9.1
Writing Equations for Ionic Compounds Dissolving in Water

Write the equation for the formation of an aqueous solution of each of the following ionic compounds:

a. KCl
b. Na_2SO_4

Solution:

a. The KCl formula indicates one potassium ion, K^+, for every chloride ion, Cl^-. The equation for the formation of its solution in water is

$$KCl(s) \xrightarrow{H_2O} K^+(aq) + Cl^-(aq)$$

b. The Na_2SO_4 formula indicates that two sodium ions, Na^+, are present in solution for every sulfate ion, SO_4^{2-}. In the equation, the sulfate ion is written as a unit.

$$Na_2SO_4(s) \xrightarrow{H_2O} 2Na^+(aq) + SO_4^{2-}(aq)$$

Some compounds are **weak electrolytes:** They dissolve in water and ionize slightly to give a small number of ions in solution. For example, when HF(*g*) dissolves in water, it dissolves as HF molecules, but then a few molecules ionize to give a small number of H^+ and F^- ions in solution. Thus, a solution of a weak electrolyte contains mostly the molecular form of the compound but also a few ions.

$$HF \xrightarrow{H_2O} H^+ + F^-$$

As the number of H^+ and F^- ions build up in solution, some of the ions recombine to give HF molecules. This reverse reaction eventually becomes equal to the rate of the forward reaction. Then the concentrations of HF, H^+, and F^- no longer change. There is a balance between molecules and ions, which is indicated by a double arrow ($\rightleftharpoons$). The ionization of HF may be written as follows:

$$HF \underset{\text{equals recombination}}{\overset{H_2O \text{ Ionization}}{\rightleftharpoons}} H^+ + F^-$$

or

$$HF \overset{H_2O}{\rightleftharpoons} H^+ + F^-$$

Most covalent compounds that dissolve in water are **nonelectrolytes,** because they dissolve as molecules, produce no ions, and their solutions do not conduct electricity. For example, glucose ($C_6H_{12}O_6$) is a nonelectrolyte that reacts with water to produce a solution of sugar molecules.

$$\underset{\text{Glucose}}{C_6H_{12}O_6(s)} \xrightarrow{H_2O} \underset{\text{Solution of glucose molecules}}{C_6H_{12}O_6(aq)}$$

Sample Problem 9.2
Identifying Electrolytes and Nonelectrolytes

For each of the following, indicate whether their aqueous solutions contain ions, molecules, or a mixture of molecules and ions:

a. Na_2SO_4
b. urea, $CH_4N_2O(s)$, a nonelectrolyte

Solution:

a. A solution of Na_2SO_4, a salt, contains only ions of the dissolved salt, $Na^+(aq)$ and $SO_4^{2-}(aq)$.
b. A nonelectrolyte such as urea dissolves in water as urea molecules only.

Study Check:
Boric acid, H_3BO_3, is a weak electrolyte. Would you expect a boric acid solution to contain ions, molecules, or both?

Answer: Mostly molecules of boric acid, and a few ions

Figure 9.2
In solution, a strong electrolyte produces ions and conducts an electrical current; a weak electrolyte produces only a few ions and conducts a weak electrical current. No ions exist in a solution of a nonelectrolyte; only molecules are present, and there is no electrical current.

Testing for Electrolytes

The presence of electrolytes in solution can be determined by using an apparatus consisting of two electrodes attached by wires to a light bulb. The light bulb glows when current flows through it, and that only happens when there are ions in the solution. A strong electrolyte produces many ions, which makes the light bulb glow brightly. When the electrodes are placed in a solution of a weak electrolyte, which produces only a few ions, the light bulb glows dimly. When the electrodes are placed in a nonelectrolyte solution, the light bulb will not glow because there are no ions to conduct an electrical current. (See Figure 9.2.)

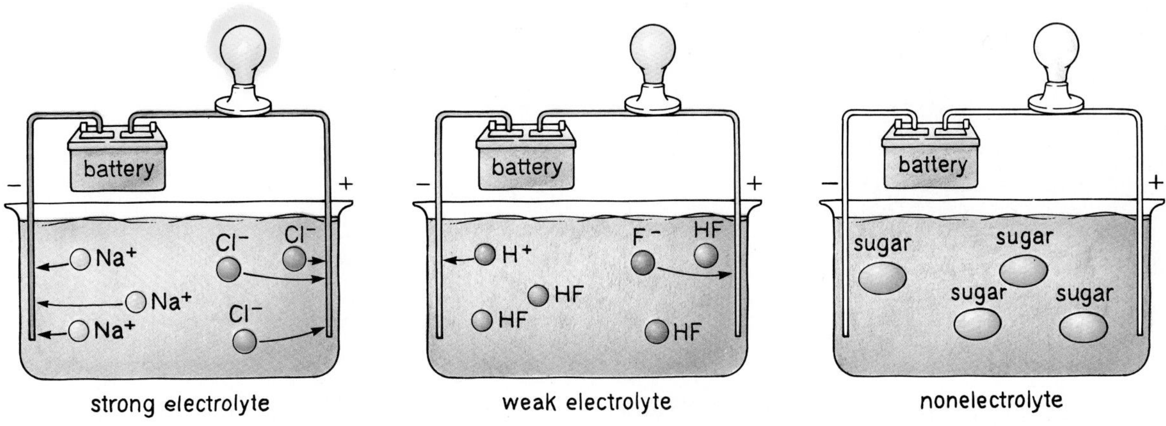

9.2 Ionic Reactions: Formation of a Solid

Learning Goal **Predict the formation of a solid when two solutions of electrolytes are mixed.**

An ionic compound is called a **soluble salt** when it dissolves in water. However, there are some ionic compounds that are **insoluble salts;** they do not dissolve in water. They remain in their solid crystalline form, as Figure 9.3 shows (see also Figure 9.4). To determine which ionic compounds are soluble in water, we need to learn the following **solubility rules:**

Figure 9.3
Insoluble salts in water: (a) $Ni(OH)_2$, (b) CdS, (c) $PbCrO_4$, and (d) FeS.

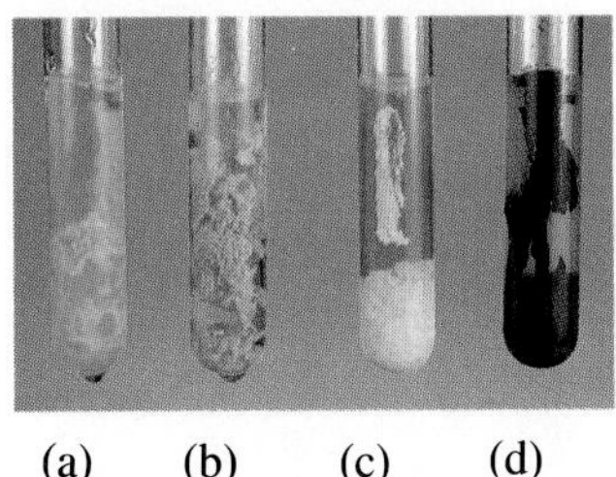
(a) (b) (c) (d)

Solubility Rules for Ionic Compounds

1. An ionic compound is soluble in water if it contains one of the following ions: Li^+, Na^+, K^+, NH_4^+, or NO_3^-.
2. Most chloride (Cl^-) compounds are soluble. However, AgCl, Hg_2Cl_2, and $PbCl_2$ are not soluble.

Figure 9.4
The insoluble salt $BaSO_4$ is used as an opaque medium to enhance x-rays of the gastrointestinal tract. $BaSO_4$ is so insoluble that it does not dissolve in gastric fluids. If Ba^{2+} were produced in solution, it would be toxic to the patient.

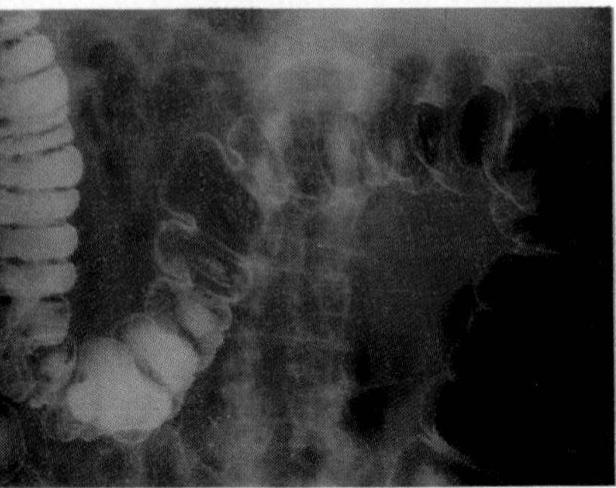

3. Most sulfate (SO_4^{2-}) compounds are soluble. However, $BaSO_4$, $PbSO_4$, and $CaSO_4$ are not soluble.
4. Most other compounds, including hydroxides (OH^-), carbonates (CO_3^{2-}), sulfides (S^{2-}), and phosphates (PO_4^{3-}) have limited solubility in water and are considered insoluble.

We can use these solubility rules to predict whether an ionic solid would be expected to be soluble and form ions in water. Table 9.1 illustrates the use of these rules.

Table 9.1 Using Solubility Rules

Ionic compound	Solubility in water	Reasoning
K_2S	Soluble	Contains K^+ (rule 1)
$Ca(NO_3)_2$	Soluble	Contains NO_3^- (rule 1)
$PbCl_2$	Insoluble	Is an insoluble chloride (rule 2)
$MgSO_4$	Soluble	Is a sulfate (usually soluble; rule 3)
$NaOH$	Soluble	Contains Na^+ (rule 1)
$AlPO_4$	Insoluble	Has no soluble ions (rule 4)

Sample Problem 9.3
Predicting the Solubility of Salts in Water

Predict whether the following salts will be soluble in water:

a. $MgCl_2$
b. $CaCO_3$
c. Na_2S

Solution:

a. Most salts containing Cl^- ions are soluble (rule 2).
b. Carbonate salts are insoluble (rule 4).
c. Soluble; salts with Na^+ dissolve in water (rule 1).

Study Check:
Predict the solubility of the salts $BaCl_2$ and $BaSO_4$.

Answer: $BaCl_2$ is soluble; $BaSO_4$ is insoluble.

Predicting the Formation of a Solid in Solution

Sometimes, when we mix two solutions, a solid called a precipitate forms and separates from the solution. This type of reaction occurs when an ion from one solution comes in contact with an ion from the second solution to form crystals of an insoluble salt. Such a process is called **precipitation.** Suppose we mix solutions of $AgNO_3$ and NaCl, and a white solid forms. How do we decide on the formula of the precipitate? For the moment, we can think of the new mixture as containing four ions: Ag^+, NO_3^-, Na^+, and Cl^-.

Ag^+	Na^+
NO_3^-	Cl^-

If we write all of the combinations of salts possible with the positive and negative ions, and use the solubility rules, we can identify the ion combinations as soluble or insoluble.

Reactants		*New Combinations*	
$AgNO_3$	NaCl	$NaNO_3$	AgCl
Soluble	Soluble	Soluble	Insoluble

Since AgCl is an insoluble salt, we would expect that the Ag^+ and Cl^- ions will form solid AgCl, as shown in Figure 9.5, and leave the Na^+ and NO_3^- ions in solution.

$$Ag^+(aq) + Cl^-(aq) \longrightarrow AgCl(s)$$

The formation of the insoluble salt AgCl(*s*) can be written as an equation using the formulas of the initial reactants:

$$AgNO_3(aq) + NaCl(aq) \longrightarrow AgCl(s) + NaNO_3(aq)$$

To predict the formation of a solid from mixtures of two solutions, we need to consider the possible new combinations of positive and negative ions in the reactants. If the new ion combinations represent only soluble salts, then no solid forms and no precipitation is observed. If any one of the new combinations of ions represents an insoluble salt, then a solid precipitate will form. For example, the result of mixing solutions of Na_2S and $CuCl_2$ is illustrated in Figure 9.6. The Cu^{2+} and S^{2-} ions combine to form a black precipitate of CuS, leaving the Na^+ and Cl^- ions in solution.

Figure 9.5 (left)
The precipitation of AgCl occurs when solutions of $AgNO_3$ and NaCl are mixed. The soluble salt of Na^+ and NO_3^- remains in solution.

Figure 9.6 (right)
Mixing solutions of Na_2S and $CuCl_2$ gives a black precipitate of CuS and leaves Na^+ and Cl^- in solution.

HEALTH NOTE

Kidney Stones Are Insoluble Salts

The kidneys continuously excrete substances that have low solubilities. But if the urine becomes saturated with these substances, they crystallize and form kidney stones. One type, renal calculus, formed in the urinary tract is composed primarily of a calcium phosphate known as hydroxyapatite, $Ca_5(PO_4)_3OH$. The excessive ingestion of mineral salts and a deficiency of water available to form urine can lead to the saturation of the urine with minerals and consequently to the kidney stones. Typically, one stone is formed every 2–3 years, and the tendency to form the stones is strongly familial.

When a kidney stone passes through the urinary tract, it can cause considerable pain and discomfort, necessitating the use of painkillers and surgery. Patients prone to kidney stones are advised to drink six to eight glasses of water every day in order to prevent saturation levels of minerals in the urine. ■

Sample Problem 9.4
Predicting the Formation of a Precipitate

Indicate the precipitate (if any) that forms when the following solutions are mixed:

a. $K_2CO_3(aq)$ and $BaCl_2(aq)$
b. $Na_2S(aq)$ and $KI(aq)$

Solution:

a. The new combinations are KCl, a soluble salt, and $BaCO_3$, an insoluble salt, which forms a precipitate.
b. The new combinations of K_2S and NaI are both soluble salts; no precipitate forms.

Study Check:
When solutions of $CdCl_2$ and Na_2S are mixed, a yellow precipitate forms. Predict the formula of the insoluble salt.

Answer: $CdS(s)$

ENVIRONMENTAL NOTE

Hard Water

Water in different parts of the world contains varying amounts of minerals. Of these minerals, the ions Ca^{2+}, Fe^{3+}, and Mg^{2+} form insoluble substances when combined with soap, leaving a "scum" or "bathtub ring" in a container. Water containing these ions is called hard water. Greater amounts of soap are needed to produce a soapy solution in hard water because some of the soap is converted to the insoluble form (scum).

$$Ca^{2+} + 2\ Soap^- \longrightarrow Ca(soap)_2(s)$$

Water that does not contain Ca^{2+}, Fe^{3+}, or Mg^{2+} is called soft water. No scum forms with soap, and less soap is required to form a soapy solution. In the past, phosphates were added to detergents to help soften water by forming insoluble precipitates with Ca^{2+}, Fe^{3+}, or Mg^{2+} ions. However, the phosphates have been removed from many detergents because of detrimental environmental effects. Phosphates accelerate the growth of algae and plants in rivers and streams, causing a decrease in available oxygen and endangering the survival of fish and other aquatic animals.

Hard water can be softened in other ways. In distillation, the hard-water sample is boiled, and the water vapor formed is condensed through a cooling apparatus to give soft water, leaving the minerals containing Ca^{2+}, Fe^{3+}, and Mg^{2+} behind.

Several types of *ion-exchangers* are used for commercial or home water softening. In ion-exchangers, the offending ions are replaced by ions that do not affect water hardness. This is accomplished by passing the hard water through a column packed with a substance containing sodium ions. As the hard-water ions move through the column, they are replaced by sodium ions. The soft water leaving the ion-exchanger now contains sodium ions instead of hard-water ions. This type of ion-exchanger should be avoided by people who must maintain a low-sodium diet. (See Figure 9.7.) ■

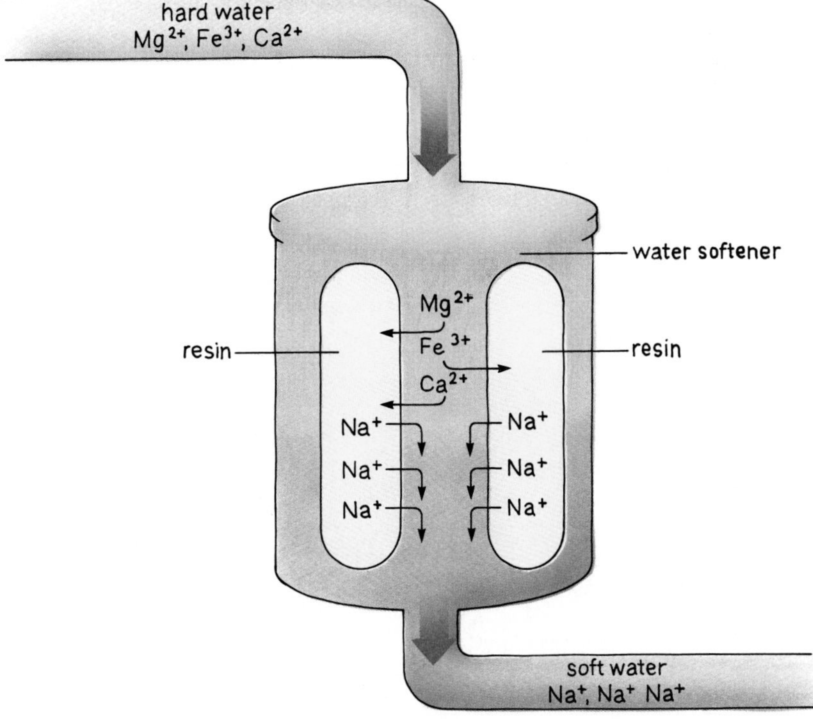

Figure 9.7
Softening water by means of an ion-exchange resin in which Mg^{2+}, Fe^{3+}, and Ca^{2+} ions are exchanged for Na^+ ions.

9.3 Acids and Bases

Learning Goal **Write an equation to show the formation of an acidic solution or a basic solution.**

Acids are electrolytes that produce hydrogen ions (H^+) in water. For example, hydrogen chloride gas, a covalent (molecular) compound, reacts with water to give hydrochloric acid, HCl(*aq*), a strong acid that exists as H^+ ions and Cl^- ions. We can write the ionization of HCl(*g*) in water as

$$HCl(g) \xrightarrow{H_2O} H^+(aq) + Cl^-(aq)$$

Acids exist as molecular compounds until they dissolve in water to produce hydrogen ions. Actually, the hydrogen ion does not exist by itself in water. A proton is transferred from the HCl molecule to a water molecule to form a **hydronium ion,** H_3O^+, and a chloride ion, Cl^-. The formation of ions from molecular compounds is called an ionization reaction.

$$H{-}\ddot{\underset{..}{Cl}}{:} + {:}\ddot{O}{-}H \longrightarrow \left[H{-}\ddot{O}{-}H\right]^+ + {:}\ddot{\underset{..}{Cl}}{:}^-$$
(with H bonded below O in water and in the hydronium ion)

Polar bond

Hydrogen chloride — Water — Hydronium ion

$$HCl(g) + H_2O \longrightarrow H_3O^+(aq) + Cl^-(aq)$$

We will use the H^+ symbol for convenience for acids, but remember that it is the H_3O^+ ion that is present in acidic solutions. (See Figure 9.8.)

Properties of Acids

There are several properties that are characteristic of acids:

1. Acids produce hydrogen ions (H^+) in solution.
2. Acids have a sour taste.

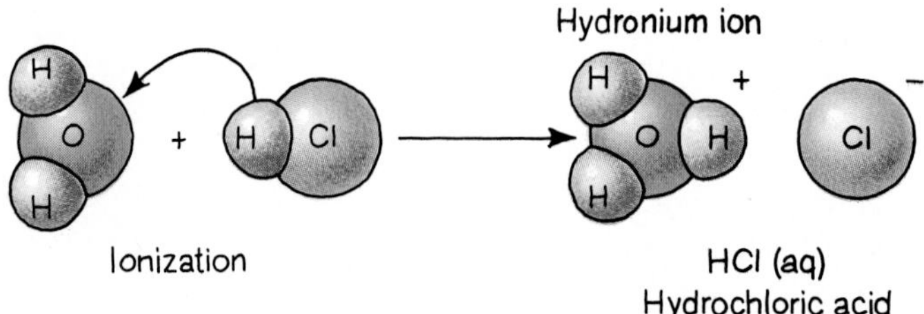

Figure 9.8
Hydrogen chloride gas ionizes in water to produce hydrochloric acid, which exists as hydronium ions $(H_3O)^+$ and chloride ions (Cl^-).

3. Acids turn blue litmus (a vegetable dye) red.
4. Acids are electrolytes.
5. Acids react with bases.

Strengths of Acids

Only a few acids will conduct a strong electrical current. These **strong acids** ionize completely to produce many hydrogen ions in solution. The most common strong acids are hydrochloric acid (HCl), sulfuric acid (H_2SO_4), and nitric acid (HNO_3). As strong acids, they are completely (100%) ionized.

Strong Acids

$$HCl \xrightarrow{H_2O} H^+(aq) + Cl^-(aq)$$
Hydrochloric acid

$$H_2SO_4 \xrightarrow{H_2O} 2H^+(aq) + SO_4^{2-}(aq)$$
Sulfuric acid

$$HNO_3 \xrightarrow{H_2O} H^+(aq) + NO_3^-(aq)$$
Nitric acid

It is important to know when you are using a strong acid in the laboratory. A strong acid can severely burn the skin and damage the eyes. If you spill an acid on your skin or get some in your eyes, be sure to flood the area immediately with water to lower the concentration of hydrogen ions and lessen any damage.

In a **weak acid,** only a few of the dissolved molecules ionize to give hydrogen ions. Even at high concentrations, weak acids contain only a small number of hydrogen ions, so they are not nearly as damaging as strong acids are to the skin and eyes. In fact, you may use some weak acids in the home, as Figure 9.9 shows. Citric acid is found in lemon or grapefruit juice, and vinegar is a 5% acetic acid solution. In carbonated soft drinks, CO_2 dissolves to form carbonic acid, H_2CO_3.

Figure 9.9
Some common household items that contain weak acids include vinegar, lemons, carbonated beverages, and vitamin C.

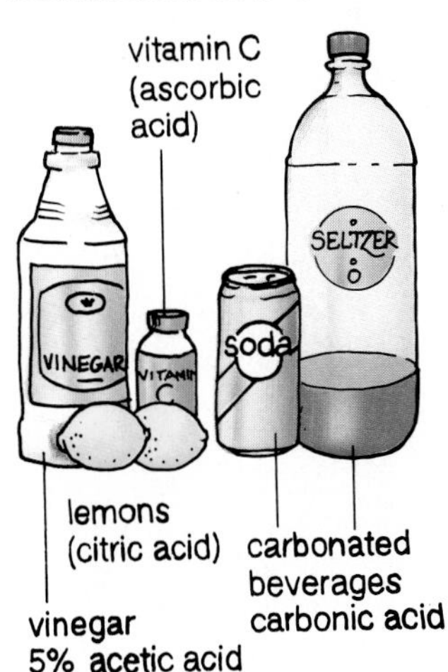

In equations for the ionization of weak acids, a double arrow shows the presence of both molecules and ions of the acid:

$$HC_2H_3O_2 \overset{H_2O}{\rightleftharpoons} H^+(aq) + C_2H_3O_2^-(aq)$$
Acetic acid

$$H_2CO_3 \overset{H_2O}{\rightleftharpoons} H^+(aq) + HCO_3^-(aq)$$
Carbonic acid

Figure 9.10 compares the ionization of strong and weak acids in solution.

Figure 9.10
A strong acid ionizes completely in solution; a weak acid consists mostly of molecules and ionizes slightly in solution to give a small number of ions.

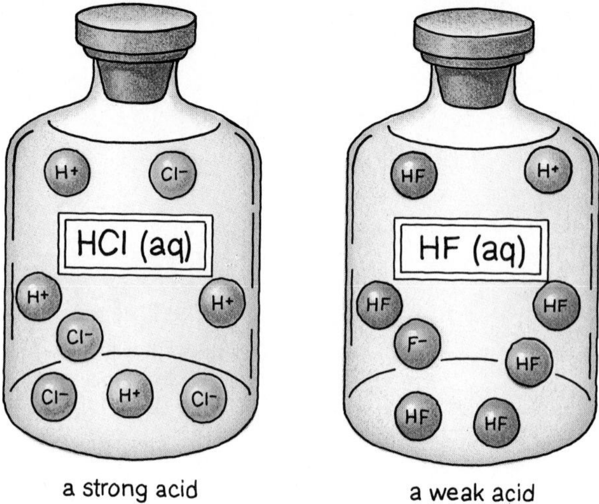

Sample Problem 9.5
Writing Equations for the Ionization of Acids

Write an equation for the ionization in water of nitric acid (HNO_3) and of formic acid ($HCHO_2$), an irritant found in ant stings.

Solution:
Nitric acid is a strong acid and ionizes completely to form hydrogen ions (H^+) and nitrate ions (NO_3^-):

$$HNO_3 \xrightarrow{H_2O} H^+(aq) + NO_3^-(aq)$$

Formic acid is a weak acid. As a weak acid, it consists mostly of the molecular form of the acid, with a few ions in solution. For weak acids, the first hydrogen in the formula is the only hydrogen that ionizes in water.

$$HCHO_2 \xrightleftharpoons{H_2O} H^+(aq) + CHO_2^-(aq)$$

Study Check:
Hypochlorous acid, HOCl, is a weak acid that forms when chlorine reacts with water in swimming pools and hot tubs. Write an equation for its ionization in water.

Answer:

$$HOCl \xrightleftharpoons{H_2O} H^+(aq) + OCl^-(aq)$$

Figure 9.11
Some common bases include detergents, antacids, drain openers, and oven and window cleaners.

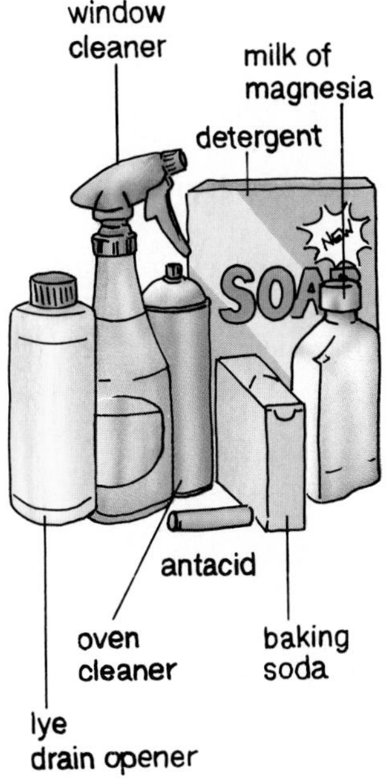

Bases

A typical **base** is an ionic compound that separates into a metal ion and a hydroxide ion (OH^-) when dissolved in water. For example, sodium hydroxide (NaOH) is a base that dissociates in water to give Na^+ ions and OH^- ions.

$$NaOH(s) \xrightarrow{H_2O} Na^+(aq) + OH^-(aq)$$

Sodium hydroxide — Sodium ion — Hydroxide ion

Properties of Bases

Some common bases illustrated in Figure 9.11 include detergents, antacids, drain openers, and oven and window cleaners. There are several properties that are associated with bases:

1. Bases produce hydroxide ions (OH^-) in solution.
2. Bases have a bitter taste.
3. Bases have a slippery, soapy feel.
4. Bases turn red litmus (a vegetable dye) blue.
5. Bases are electrolytes.
6. Bases react with acids.

Sample Problem 9.6
Identifying Acids and Bases

Identify each of the following solutions as an acid or a base.

a. contains hydroxide ions
b. tastes sour
c. turns blue litmus red
d. feels slippery or soapy

Solution:

a. A base contains hydroxide ions.
b. An acid tastes sour.
c. An acid turns blue litmus red.
d. A base feels slippery or soapy.

Strengths of Bases

In water, a **strong base** completely separates into ions, one of which is the hydroxide ion (OH^-). Strong bases consist of the hydroxides of elements in Groups 1A and 2A. Of these, only NaOH and KOH are commonly used in the chemistry laboratory.

$$NaOH(s) \xrightarrow{H_2O} Na^+(aq) + OH^-(aq)$$

Sodium hydroxide

$$KOH(s) \xrightarrow{H_2O} K^+(aq) + OH^-(aq)$$

Potassium hydroxide

At high concentrations, hydroxide ions cause damage to the skin and eyes. Strong bases, such as NaOH (also known as lye), are used in household products to dissolve grease in ovens and to clean drains. The use of such products in the home should be carefully supervised.

Most hydroxides are not very soluble in water, although they do dissociate 100% into ions. One such hydroxide is magnesium hydroxide, $Mg(OH)_2$, an antacid known as milk of magnesia that is ingested to counteract the effects of too much acid in the stomach.

$$Mg(OH)_2(s) \xrightarrow{H_2O} Mg^{2+}(aq) + 2OH^-(aq)$$

Sample Problem 9.7
Dissociation of a Base in Water

Write an equation for the dissociation of LiOH in water.

Solution:
LiOH is a hydroxide of lithium, a Group 1A element. It completely separates in water to give a basic solution of lithium ions (Li^+) and hydroxide ions (OH^-).

$$LiOH(s) \xrightarrow{H_2O} Li^+(aq) + OH^-(aq)$$

Study Check:
Calcium hydroxide is used in some antacids to counteract excess stomach acid. Write an equation for the formation of a calcium hydroxide solution.

Answer:

$$Ca(OH)_2(s) \xrightarrow{H_2O} Ca^{2+}(aq) + 2OH^-(aq)$$

Weak bases produce only a few hydroxide ions when they react with water. The most common weak base is an aqueous solution of ammonia (NH_3), found in several household window cleaners. Ammonia reacts with water by accepting a proton. The resulting hydroxide ion makes the solution basic. The double arrow in the equation indicates that only a few of the dissolved ammonia molecules ionize at any given time.

$$NH_3(g) + H_2O(l) \rightleftharpoons NH_4^+(aq) + OH^-(aq)$$

Ammonia Water Ammonium ion Hydroxide ion

9.4 Ionization of Water

Learning Goal **Write an equation to show the ionization of water; write the ion-product expression for water. Use the ion product to calculate $[H^+]$ and $[OH^-]$ in solution.**

We usually think of water as a nonelectrolyte that contains only molecules of water. However, if we were to measure very carefully, we would find that a few water molecules (1 in 10 million) have formed ions. When water ionizes, a proton is transferred from one water molecule to another to produce a hydronium ion (H_3O^+) and a hydroxide ion (OH^-).

Transfer of Hydrogen

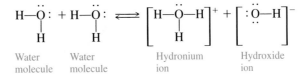

$$H{-}\ddot{O}{:}\ (H) + H{-}\ddot{O}{:}\ (H) \rightleftharpoons \left[H{-}\ddot{O}{-}H\ (H)\right]^+ + \left[{:}\ddot{O}{-}H\right]^-$$

Water molecule | Water molecule | Hydronium ion | Hydroxide ion

The equation for the reaction of water may be written for two water molecules,

$$H_2O + H_2O \rightleftharpoons H_3O^+(aq) + OH^-(aq)$$

or it may be simplified, as follows:

$$H_2O \rightleftharpoons H^+(aq) + OH^-(aq)$$

Ion Product for Water, K_w

In pure water at 25°C, the molar concentration of H^+ and OH^- are each $1.0 \times 10^{-7}\ M$. Brackets are placed around the symbols to indicate their molar concentrations, $[H^+]$ and $[OH^-]$, read as ''the molar concentration of hydrogen ions'' and ''the molar concentration of hydroxide ions,'' respectively.

Molar Concentrations of H^+ and OH^- in Pure Water at 25°C

$$[H^+] = 1.0 \times 10^{-7}\ M$$

$$[OH^-] = 1.0 \times 10^{-7}\ M$$

The concentrations of the ions in water at 25°C are useful in a form called the **ion-product constant, K_w.**

$$K_w = [H^+][OH^-]$$

Substituting the concentration values into the K_w expression, we obtain a K_w value of 1.0×10^{-14}:

$$K_w = [H^+][OH^-] = (1.0 \times 10^{-7}) \times (1.0 \times 10^{-7})$$
$$= 1.0 \times 10^{-14}$$

Table 9.2 Examples of $[H^+]$ and $[OH^-]$ in Acidic, Neutral, and Basic Solutions at 25°C

Type of solution	Acid/Base	$[H^+]$	$[OH^-]$	Ion product (K_w)
Acidic	$[H^+] > [OH^-]$	1.0×10^{-2}	1.0×10^{-12}	1.0×10^{-14}
		1.0×10^{-5}	1.0×10^{-9}	1.0×10^{-14}
Neutral	$[H^+] = [OH^-]$	1.0×10^{-7}	1.0×10^{-7}	1.0×10^{-14}
Basic	$[H^+] < [OH^-]$	1.0×10^{-8}	1.0×10^{-6}	1.0×10^{-14}
		1.0×10^{-11}	1.0×10^{-3}	1.0×10^{-14}

The values for $[H^+]$ and $[OH^-]$ produced by the ionization of water are equal. Therefore, pure water has no acidic or basic properties and is said to be neutral. However, $[H^+]$ increases when an acid is added to water, and $[OH^-]$ decreases in order to maintain the $[H^+][OH^-]$ ion-product value of 1.0×10^{-14}. In any acidic solution, $[H^+]$ is always greater than $[OH^-]$, and the ion product, $[H^+][OH^-]$, is always equal to 1.0×10^{-14}.

We can also apply the ion product to solutions of bases. The addition of a base to water increases $[OH^-]$ and decreases $[H^+]$ to keep the ion product at 1.0×10^{-14}. In basic solutions, $[OH^-]$ is always greater than $[H^+]$, and the K_w ion product remains constant. Table 9.2 gives some examples of $[H^+]$ and $[OH^-]$ in acidic, neutral, and basic solutions.

Using the K_w value of 1.0×10^{-14}, we can calculate the concentration of the ions of an acidic or basic solution.

$$K_w = [H^+][OH^-] = 1.0 \times 10^{-14}$$

Solving for either $[H^+]$ or $[OH^-]$ gives

$$[H^+] = \frac{K_w}{[OH^-]} = \frac{1.0 \times 10^{-14}}{[OH^-]}$$

$$[OH^-] = \frac{K_w}{[H^+]} = \frac{1.0 \times 10^{-14}}{[H^+]}$$

Sample Problem 9.8
Calculate $[H^+]$ in Solution

A basic solution has been prepared with $[OH^-]$ of 1.0×10^{-4} *M*. What is the value for $[H^+]$?

Solution:
We can solve for $[H^+]$ by using the ion product of water,

$$K_w = [H^+][OH^-] = 1.0 \times 10^{-14}$$

and rearranging for $[H^+]$:

$$[H^+] = \frac{1.0 \times 10^{-14}}{[OH^-]}$$

The value for $[OH^-]$ of 1.0×10^{-4} M can now be substituted in the equation to solve for $[H^+]$.

$$[H^+] = \frac{1.0 \times 10^{-14}}{1.0 \times 10^{-4}}$$

$$= 1.0 \times 10^{-10}\ M$$

Study Check:
What is the value for $[H^+]$ of orange juice if its $[OH^-]$ is 2.5×10^{-12} M?

Answer: 4.0×10^{-3} M

Sample Problem 9.9
Calculating $[OH^-]$ in Solution

What is the value for $[OH^-]$ of vinegar if its $[H^+]$ is 2.0×10^{-3} M?

Solution:
Rearranging the ion-product expression for $[OH^-]$ gives

$$[OH^-] = \frac{1.0 \times 10^{-14}}{[H^+]}$$

Substituting the given value of $[H^+]$ completes the calculation.

$$[OH^-] = \frac{1.0 \times 10^{-14}}{2.0 \times 10^{-3}}$$

$$= 5.0 \times 10^{-12}\ M$$

Study Check:
A sample of HCl is prepared with 0.010 mole HCl in 0.5 L of solution. What is $[OH^-]$ of the sample? (*Hint:* Calculate $[H^+]$ first.)

Answer:

$$[H^+] = 2 \times 10^{-2}\ M$$

$$[OH^-] = 5 \times 10^{-13}\ M$$

9.5 The pH of Solutions

Learning Goal **Use the pH of a solution to classify the solution as acidic, neutral, or basic; calculate the pH, $[H^+]$, and $[OH^-]$ of the solution.**

It is usually more convenient to represent the acidity, $[H^+]$, of a solution in terms of **pH.** The pH scale, as shown in Figure 9.12, typically has values that range from 0 to 14. The pH values below 7 correspond to acidic solutions, whereas higher pH values, above 7, correspond to basic solutions. A pH of exactly 7, the midpoint of the pH scale, indicates a solution that is neutral. Table 9.3 lists the pH values for some common substances.

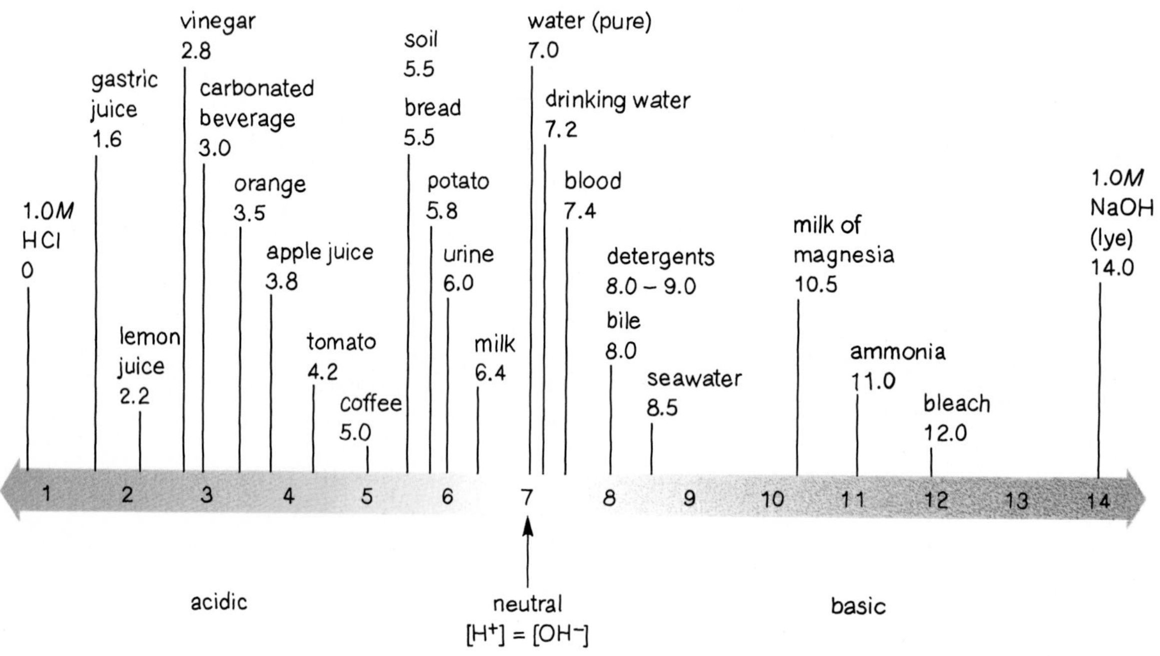

Figure 9.12
Values on the pH scale range from 0 to 14. Values below 7 are acidic, a value of 7 is neutral, and values above 7 are basic.

Table 9.3 pH Values of Some Common Substances

Solution	pH	Solution	pH
1.0 *M* HCl	0	Water (pure)	7.0
Gastric juice	1.6	Blood	7.4
Lemon juice	2.2	Bile	8.0
Vinegar	2.8	Detergents	8.0–9.0
Carbonated beverages	3.0	Milk of magnesia	10.5
Coffee	5.0	Ammonia	11.0
Urine	6.0	Bleach	12.0
Rainwater	6.2	1.0 *M* NaOH	14.0

Sample Problem 9.10
Using pH to Identify Acidic and Basic Solutions

Using the pH given for the following substances, identify them as acidic, basic, or neutral solutions:

a. tomato juice, pH 4.2
b. saliva, pH 6.6
c. antacid, pH 8.5

Solution:

a. The tomato juice is acidic because a pH of 4.2 is in the acidic range on the pH scale.
b. The saliva is acidic because a pH value of 6.6 is below pH 7.
c. The antacid is basic. A pH of 8.5 is in the basic range of the pH scale.

Study Check:
An NaCl solution has a pH of 7.0. Is it acidic, basic, or neutral?

Answer: neutral

The pH Scale

If we compare $[H^+]$, the molar concentration of hydrogen ions, and pH, we find that pH becomes greater as $[H^+]$ gets smaller. In acidic solutions, $[H^+]$ is high and pH is low. In basic solutions, $[H^+]$ is low and pH has a high value. The concentration of the hydrogen ion, $[H^+]$, determines the pH of a solution, which is expressed as

$$\text{pH} = \log \frac{1}{[H^+]} \quad \text{or} \quad -\log [H^+]$$

When $[H^+]$ has a coefficient of 1, the pH is simply the number in the exponent without the negative sign. Table 9.4 gives a comparison of $[H^+]$, $[OH^-]$, and pH values.

HEALTH NOTE

Stomach Acid, HCl

When a person sees, smells, thinks about, and/or tastes food, the gastric glands in the stomach begin to secrete an HCl solution that is strongly acidic. In a single day, a person may secrete as much as 2000 mL of gastric juice.

The HCl in the gastric juice is used to activate a digestive enzyme called pepsin that breaks down proteins in food entering the stomach. The secretion of HCl continues until the stomach has a pH of about 2, which is the optimum pH for activating the digestive enzymes without ulcerating the stomach lining. Normally, large quantities of a viscous mucus are secreted within the stomach to protect it from acid and enzyme damage. ■

Table 9.4 A Comparison of $[H^+]$, $[OH^-]$, and Corresponding pH Values at 25°C

pH	$[H^+]$	$[OH^-]$	
0	10^0	10^{-14}	Acidic
1	10^{-1}	10^{-13}	
2	10^{-2}	10^{-12}	
3	10^{-3}	10^{-11}	
4	10^{-4}	10^{-10}	
5	10^{-5}	10^{-9}	
6	10^{-6}	10^{-8}	
7	10^{-7}	10^{-7}	Neutral
8	10^{-8}	10^{-6}	Basic
9	10^{-9}	10^{-5}	
10	10^{-10}	10^{-4}	
11	10^{-11}	10^{-3}	
12	10^{-12}	10^{-2}	
13	10^{-13}	10^{-1}	
14	10^{-14}	10^0	

Calculating pH

With many scientific calculators, pH is calculated using the (exp) or (EE), the (log), and the change sign, (+/−), keys. For example, for a solution with $[H^+]$ equal to 5.0×10^{-4} *M*,

5.0 (EE) 4 (+/−) (log) (+/−) = 3.30

Enter coefficient — Enter exponent — pH

The number of digits in the decimal portion of the pH is equal to the number of significant figures in the coefficient of $[H^+]$. In this case, there are two digits after the decimal point (.30) because there are two significant figures in the coefficient (5.0).

In the laboratory, the pH of a solution can be determined by using a pH meter or an indicator that shows specific color changes at certain pH values, as shown in Figure 9.13.

Figure 9.13
The dye in red cabbage juice acts as a pH indicator by turning different colors that correspond to different pH values.

Sample Problem 9.11
Calculating the pH of a Solution

Determine the pH of the following solutions:

a. $[H^+] = 1 \times 10^{-5}\ M$
b. $[H^+] = 8 \times 10^{-10}\ M$
c. $[OH^-] = 1 \times 10^{-2}\ M$.

Solution:

a. We can calculate the pH of the solution using the exponent in the value for $[H^+]$.

$$[H^+] = 1 \times 10^{-5} \qquad pH = 5.0$$

b. Entering the following in the calculator gives

8 (EE) 10 (+/−) (log) (+/−) = 9.1 pH

c. Since $[OH^-]$ is given, we first have to calculate the corresponding $[H^+]$. We use the ion product of water:

$$[H^+][OH^-] = 1.0 \times 10^{-14}$$

$$[H^+] = \frac{1.0 \times 10^{-14}}{[OH^-]}$$

$$= \frac{1.0 \times 10^{-14}}{1 \times 10^{-2}}$$

$$= 1 \times 10^{-12}$$

$$pH = 12.0$$

Study Check:
What is the pH of a bleach whose $[OH^-]$ is $5 \times 10^{-4}\ M$? (*Hint:* Calculate $[H^+]$ first.)

Answer:

$$[H^+] = 2 \times 10^{-11}$$

$$pH = 10.7$$

Sometimes we need to determine the $[H^+]$ from the pH. For whole number pH values, we can write $[H^+]$ using the pH as the exponent.

$$[H^+] = 1 \times 10^{-pH}$$

Sample Problem 9.12
Calculating $[H^+]$ from pH

Determine $[H^+]$ for solutions having the following pH values:

a. pH 3.0
b. pH 11.0

Solution:
For pH values that are whole numbers, $[H^+]$ can be written using the pH as the exponent of the hydrogen ion concentration.

$$[H^+] = 1 \times 10^{-pH}$$

a. $[H^+] = 1 \times 10^{-3}\ M$
b. $[H^+] = 1 \times 10^{-11}\ M$

Study Check:
What is $[OH^-]$ of beer if the pH is 5.0?

Answer: $1 \times 10^{-9}\ M$

ENVIRONMENTAL NOTE

Acid Rain

Rain typically has a pH of 6.2. It is slightly acidic because the carbon dioxide in the air combines with water to form carbonic acid. However, in many parts of the world, rain has become considerably more acidic, with pH values as low as 3 being reported. One cause of acid rain is the sulfur dioxide (SO_2) gas produced when coal that contains sulfur is burned.

In the air, the SO_2 gas reacts with oxygen to produce SO_3, which then combines with water to form sulfuric acid, H_2SO_4, a strong acid.

$$S + O_2 \longrightarrow SO_2$$

$$2SO_2 + O_2 \longrightarrow 2SO_3$$

$$SO_3 + H_2O \longrightarrow H_2SO_4$$

In parts of the United States, acid rain has made lakes so acidic, they are no longer able to support fish and plant life. Limestone ($CaCO_3$) is sometimes added to these lakes to neutralize the acid. In Eastern Europe, acid rain has brought about an environmental disaster. Nearly 40% of the forests in Poland have been severely damaged, and some parts of the land are so acidic that crops will not grow. Throughout Europe, monuments made of marble (a form of $CaCO_3$) are deteriorating as acid rain dissolves the marble.

$$2H^+ + CaCO_3 \longrightarrow Ca^{2+} + H_2O + CO_2$$

Efforts to slow or stop the damaging effects of acid rain include the reduction of sulfur emissions. This will require installation of expensive equipment in coal-burning plants to absorb more of the SO_2 gases before they are emitted. In some outdated plants, this may be impossible, and they will need to be closed. It is a difficult problem for engineers and scientists, but one that must be solved. ■

9.6 Ionic Reactions: Acid–Base Neutralization

Learning Goal **Write a balanced equation for the neutralization reaction of an acid and a base.**

We have seen that acidic solutions contain H^+ ions, and basic solutions contain OH^- ions. If we mix an acid with a base, the H^+ ions and OH^- ions combine in an acid–base reaction called **neutralization** to produce water.

$$H^+(aq) + OH^-(aq) \longrightarrow H_2O$$

If the concentration of H^+ is equal to the concentration of OH^-, water will be produced forming a neutralized solution having no acidic or basic properties. Since an acid and a base also contain metal and nonmetal ions, neutralization produces a salt in addition to the water. Recall that a salt is an ionic compound such as KCl, Li_2SO_4, Na_3PO_4.

Consider the neutralization of HCl and NaOH. In solution, HCl exists as $H^+(aq)$ and $Cl^-(aq)$, and NaOH exists as $Na^+(aq)$ and $OH^-(aq)$. The H^+ and OH^- combine to form water, leaving the ions, $Na^+(aq)$ and $Cl^-(aq)$, in the solution.

Acid	+	Base	⟶	Salt	+ Water
H^+		Na^+		Na^+	
	+		⟶		+ H_2O
Cl^-		OH^-		Cl^-	

or

$$HCl(aq) + NaOH(aq) \longrightarrow NaCl(aq) + H_2O$$

Balancing an Acid–Base Neutralization Equation

In another neutralization reaction, sulfuric acid, H_2SO_4, is reacted with sodium hydroxide, NaOH. Since the products of neutralization of an acid and a base are a salt and water, we can write the unbalanced equation as follows:

$$\underset{\text{Acid}}{H_2SO_4(aq)} + \underset{\text{Base}}{NaOH(aq)} \longrightarrow \underset{\text{Salt}}{\text{Salt}} + \underset{\text{Water}}{H_2O}$$

Since the acid has two H^+ ions, two OH^- ions from the base are needed to neutralize the acid. We therefore place a coefficient of 2 in front of NaOH. The neutralization of two H^+ ions and two OH^- ions will result in the formation of two molecules of H_2O.

$$H_2SO_4(aq) + 2NaOH(aq) \longrightarrow \text{Salt} + 2H_2O$$

The ions available for the salt are two Na^+ ions from the base, and the SO_4^{2-} ion from the acid. The formula of the salt, Na_2SO_4, is placed in the equation to give the balanced equation for the neutralization.

$$\underset{\text{Acid}}{H_2SO_4(aq)} + \underset{\text{Base}}{2NaOH(aq)} \longrightarrow \underset{\text{Salt}}{Na_2SO_4(aq)} + \underset{\text{Water}}{2H_2O}$$

Here are some more examples of neutralization reactions between acids and bases:

Acid–Base Neutralization Reactions

$$HNO_3(aq) + KOH(aq) \longrightarrow KNO_3(aq) + H_2O$$

$$2HCl(aq) + Ca(OH)_2(aq) \longrightarrow CaCl_2(aq) + 2H_2O$$

$$H_3PO_4(aq) + 3NaOH(aq) \longrightarrow Na_3PO_4(aq) + 3H_2O$$

Sample Problem 9.13
Writing a Neutralization Equation

Write and balance an equation for the neutralization of HCl and $Ba(OH)_2$.

Solution:
When the reactants, HCl and $Ba(OH)_2$, undergo neutralization, the products are a salt and water.

$$HCl + Ba(OH)_2 \longrightarrow \text{Salt} + H_2O$$

A 2 is placed in front of the HCl to provide two H^+ ions to neutralize the two OH^- ions from the $Ba(OH)_2$. The product of the neutralization is two molecules of H_2O.

$$2HCl + Ba(OH)_2 \longrightarrow \text{Salt} + 2H_2O$$

The ions that remain in solution are the Ba^{2+} ion from the base and two Cl^- ions from the acid. The formula of the salt is $BaCl_2$.

$$2HCl(aq) + Ba(OH)_2(aq) \longrightarrow BaCl_2(aq) + 2H_2O$$

Study Check:
Write an equation for the neutralization of H_3PO_4 with $Mg(OH)_2$.

Answer:

$$2H_3PO_4(aq) + 3Mg(OH)_2(aq) \longrightarrow Mg_3(PO_4)_2(s) + 6H_2O$$

HEALTH NOTE

Antacids Neutralize Excess Stomach Acid

Antacids are substances used to neutralize the effects of excess stomach acid (HCl), and to treat ulcers. Some antacids are mixtures of aluminum hydroxide and magnesium hydroxide. These hydroxides are not very soluble in water, so the levels of available OH^- are not damaging to the intestinal tract. However, aluminum hydroxide has the side effects of producing constipation and binding phosphate in the intestinal tract, which may cause weakness and anorexia. Magnesium hydroxide has a laxative effect. These side effects are less likely when a combination of the antacids is used.

$$Al(OH)_3 + 3HCl \longrightarrow AlCl_3 + 3H_2O$$

$$Mg(OH)_2 + 2HCl \longrightarrow MgCl_2 + 2H_2O$$

Some antacids use calcium carbonate to neutralize excess stomach acid. About 10% of the calcium is absorbed into the bloodstream, where it elevates the levels of serum calcium. Calcium carbonate is not recommended for patients who have peptic ulcers or a tendency to form kidney stones.

$$CaCO_3 + 2HCl \longrightarrow CaCl_2 + H_2O + CO_2(g)$$

Still other antacids contain sodium bicarbonate. This type of antacid has a tendency to increase blood pH and elevate sodium levels in the body fluids. It also is not recommended in the treatment of peptic ulcers.

$$NaHCO_3 + HCl \longrightarrow NaCl + CO_2(g) + H_2O$$

The neutralizing substances in some liquid antacid preparations are given in Table 9.5. ■

Table 9.5 Basic Compounds in Some Antacids

Antacid	Base(s)
Amphojel	$Al(OH)_3$
Milk of Magnesia	$Mg(OH)_2$
Mylanta-II	$Mg(OH)_2$, $Al(OH)_3$
Maalox	$Mg(OH)_2$, $Al(OH)_3$
Di Gel	$Mg(OH)_2$, $Al(OH)_3$
Gelusil M	$Mg(OH)_2$, $Al(OH)_3$
Riopan	$Mg(OH)_2$, $Al(OH)_3$
Bisodol	$CaCO_3$, $Mg(OH)_2$
Titralac	$CaCO_3$
Pepto-Bismol	$CaCO_3$
Tums	$CaCO_3$
Alka-Seltzer	$NaHCO_3$, $KHCO_3$

9.7 Acid–Base Titration

Learning Goal **Calculate the concentration of an acidic solution from the volume and the molarity of the basic solution used to neutralize the acid.**

Suppose we need to find the molarity of an HCl solution. We can do this by a laboratory procedure called a **titration.** In a typical titration, a measured amount of base is added to a certain volume of acid to determine the concentration of the acid. To illustrate, we first place a measured sample, such as 10.0 mL HCl solution, in a flask and add several drops of an **indicator,** usually phenolphthalein, which changes color when pH changes. In an acidic solution, this indicator is colorless. Then we fill a buret with an NaOH solution of a known molarity, such as 1.00 *M*, and carefully add the NaOH to the acidic solution in the flask, as shown in Figure 9.14. Since titration involves neutralization, we need to write a neutralization equation for HCl and NaOH:

$$\mathrm{HCl}(aq) + \mathrm{NaOH}(aq) \longrightarrow \mathrm{NaCl}(aq) + \mathrm{H_2O}$$

To neutralize the H^+ in the acid sample, we must add a volume of base that contains an equal number of moles of OH^-. This occurs at what is called the end point, when all the H^+ is neutralized and one more drop of NaOH turns the indicator in the solution pink. We observe the level of the NaOH in the buret to determine that a volume of 20.0 mL of NaOH was used to reach the neutralization point. From the volume of the NaOH used in the titration, and its molarity, we first calculate the number of moles of NaOH (and therefore the number of moles of HCl) and then the concentration of the HCl.

Moles of NaOH Used

$$20.0\ \cancel{\mathrm{mL}} \times \frac{1\ \cancel{\mathrm{L}}}{1000\ \cancel{\mathrm{mL}}} \times \underbrace{\frac{1.00\ \mathrm{mole}}{1\ \cancel{\mathrm{L}}}}_{\text{Molarity of NaOH}} = 0.0200\ \mathrm{mole\ NaOH}$$

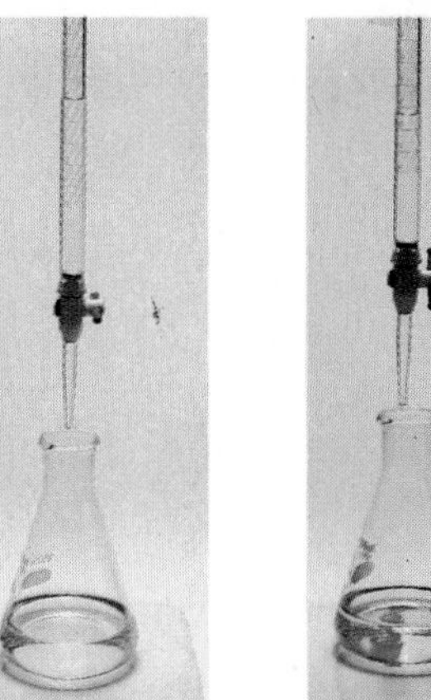

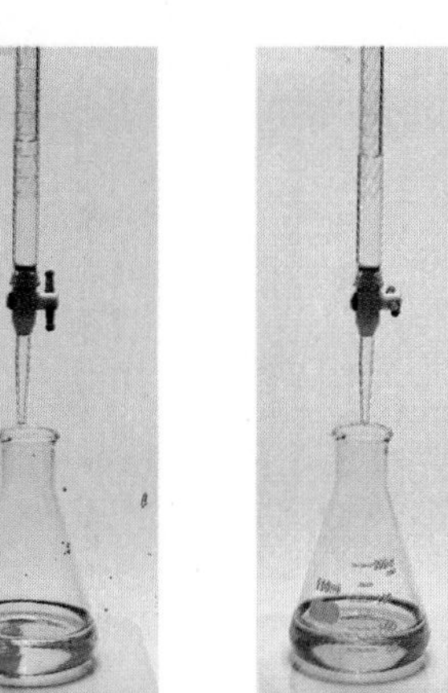

Figure 9.14
The titration of an acid. A known volume of an acid is placed in a flask with an indicator and titrated with a measured volume of NaOH to the neutralization point.

From the equation for neutralization that we have written, we know that 1 mole of NaOH reacts with 1 mole of HCl. Expressing the relationship between HCl and NaOH as conversion factors gives

$$\frac{1 \text{ mole NaOH}}{1 \text{ mole HCl}} \quad \text{and} \quad \frac{1 \text{ mole HCl}}{1 \text{ mole NaOH}}$$

Moles of HCl

$$0.0200 \text{ \sout{mole NaOH}} \times \frac{1 \text{ mole HCl}}{1 \text{ \sout{mole NaOH}}} = 0.0200 \text{ mole HCl}$$

Molarity of HCl

$$M = \frac{\text{Moles}}{\text{Liter}}$$

$$M = \frac{0.0200 \text{ mole HCl}}{0.0100 \text{ L}} = 2.00\ M \text{ HCl}$$

10.0 mL HCl
sample expressed as liters

Thus, the HCl solution used in the titration has a molarity of 2.00 M.

Sample Problem 9.14
Calculating Molarity Using Titration

A 5.00-mL sample of H_2SO_4 is placed in a flask for titration. A total of 35.0 mL of 0.20 M NaOH is needed to reach the neutralization end point. What is the molarity of the acid? The equation for the acid–base neutralization is

$$H_2SO_4(aq) + 2NaOH(aq) \longrightarrow Na_2SO_4(aq) + 2H_2O$$

Solution:
Placing the information in a table gives us

	H_2SO_4	*NaOH*
Volume:	5.00 mL	35.0 mL
Molarity:	?	0.20 M

The number of moles of NaOH is calculated from the volume and the molarity of the NaOH.

$$\text{Moles of NaOH} = 35.0 \text{ \sout{mL}} \times \frac{1 \text{ \sout{L}}}{1000 \text{ \sout{mL}}} \times \frac{0.20 \text{ mole}}{1 \text{ \sout{L} NaOH}}$$
$$= 0.0070 \text{ mole NaOH}$$

From the neutralization equation, we see that 2 moles of NaOH are needed to titrate 1 mole of H_2SO_4.

$$0.0070 \text{ mole NaOH} \times \frac{1 \text{ mole } H_2SO_4}{2 \text{ mole NaOH}} = 0.0035 \text{ mole } H_2SO_4$$

Factor derived from coefficients

Finally, the molarity (concentration) of the acid is calculated from the sample volume (V) and the number of moles of H_2SO_4.

$$V = 5.00 \text{ mL} \times \frac{1 \text{ L}}{1000 \text{ mL}} = 0.00500 \text{ L}$$

$$M = \frac{\text{Moles}}{\text{Liter}}$$

$$= \frac{0.0035 \text{ mole}}{0.00500 \text{ L}} = 0.70\ M\ H_2SO_4$$

Study Check:
What is the volume of 0.100 M NaOH needed to neutralize 5.00 mL of 0.200 M H_3PO_4? (*Hint:* Write the balanced equation for neutralization.)

Answer: 30.0 mL

9.8 Buffers

Learning Goal **Identify the components of a buffer and the role of each component in maintaining the pH of a solution.**

When an acid or a base is added to water, the pH changes drastically. However, when a small amount of an acid or a base is added to a buffered solution, there is very little change in pH. Blood is a solution that is buffered at a pH of 7.35–7.45. If the pH of the blood goes above or below this narrow range, it causes such drastic changes in respiration and metabolism that death can occur. Even though we are constantly producing acids and bases from foods and biological processes, the buffers in the blood absorb additional acid or base, which leaves the pH unchanged. (See Figure 9.15.)

A **buffer** usually contains two substances, a weak acid or a weak base and its salt. For example, a buffer that is important in the blood consists of carbonic acid and its salt, sodium bicarbonate, $NaHCO_3$.

The Weak Acid Part

$$H_2CO_3(aq) \rightleftharpoons H^+(aq) + HCO_3^-(aq)$$

Carbonic acid — Bicarbonate ion

Figure 9.15
Adding an acid or a base to water changes the pH drastically; a buffer resists pH change when an acid or base is added.

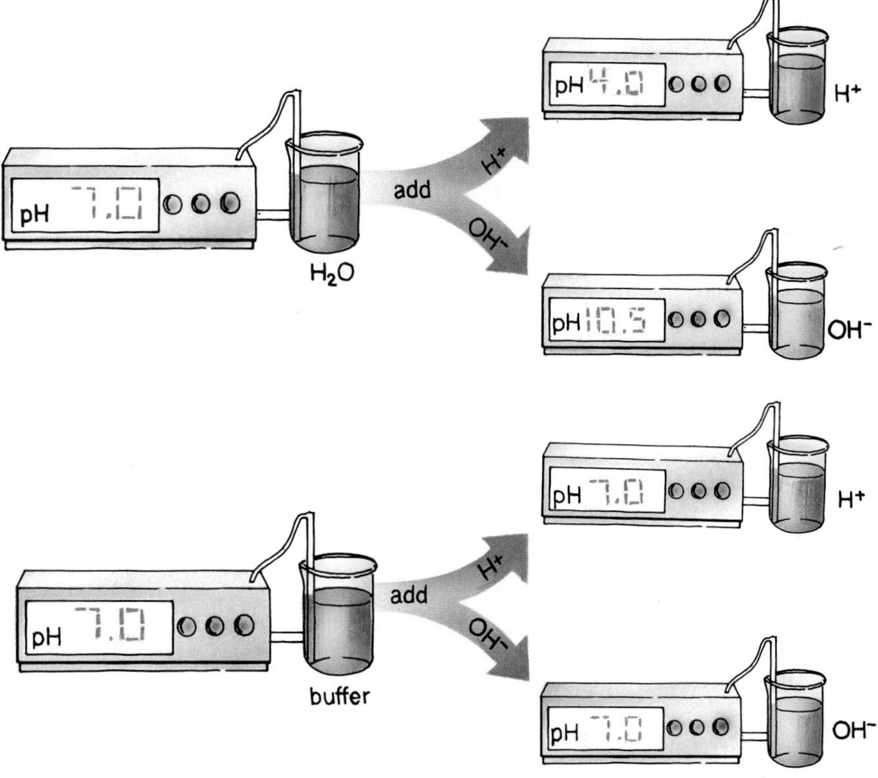

The Salt Part

$$NaHCO_3(s) \longrightarrow Na^+(aq) + HCO_3^-(aq)$$

Sodium bicarbonate (salt of the weak acid) — More bicarbonate ion

To explain how a buffer works, we need to look at the way each part of the buffer helps resist pH change. If some OH^- enters the blood, it is neutralized by the H^+ from the carbonic acid:

Weak Acid Neutralizes OH^-

$$H_2CO_3(aq) + OH^-(aq) \longrightarrow H_2O + HCO_3^-(aq)$$

Carbonic acid in buffer — Hydroxide ion from a base — Bicarbonate ion

In certain diseases, the kidneys fail to remove excess hydrogen ions from the blood. The blood becomes more acidic and its pH falls below 7.35. Then the bicarbonate part of the buffer reacts with the excess H^+ in the blood, as shown in Figure 9.16.

Anion of the Salt Combines With Extra H^+

$$HCO_3^-(aq) + H^+(aq) \longrightarrow H_2CO_3(aq)$$

Bicarbonate ion in buffer — Hydrogen ion — Carbonic acid

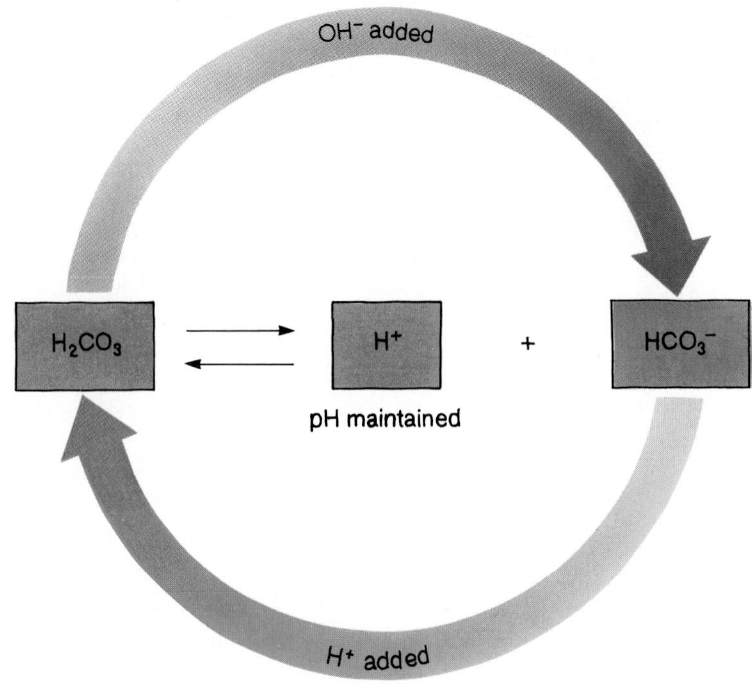

Figure 9.16
The carbonic acid–bicarbonate buffer system.

In summary, when OH^- ions are added to a buffer solution, they are neutralized by the weak acid of the buffer. On the other hand, if additional H^+ ions enter the buffer solution, they combine with the negative ion of the weak acid. Both halves of a buffer are necessary to maintain the pH of the solution.

Sample Problem 9.15
Identifying Buffer Solutions

Indicate whether each of the following substances would make a buffer solution:

a. HCl and NaCl
b. H_3PO_4
c. HF and NaF

Solution:

a. No. This is a solution of a strong acid and its salt. A buffer must contain a weak acid and its salt.
b. No. A weak acid is not sufficient for a buffer.
c. Yes. This mixture consists of the necessary components for a buffer: a weak acid, HF, and its salt, NaF.

HEALTH NOTE

P_{CO_2} and the Carbonate Buffer System in the Blood

In our cells, carbon dioxide (CO_2) is continually produced as an end product of cellular metabolism. Some CO_2 is carried to the lungs for elimination, and the rest dissolves in the body fluids, forming carbonic acid (H_2CO_3) and bicarbonate ion.

$$CO_2(g) + H_2O \longrightarrow H_2CO_3(aq)$$

Carbon dioxide — Carbonic acid

Buffer system

$$H_2CO_3(aq) \longrightarrow H^+(aq) + HCO_3^-(aq)$$

Carbonic acid — Bicarbonate ion

The partial pressure of CO_2 gas, P_{CO_2}, regulates the concentration of carbonic acid, which in turn effects the pH of the blood. Table 9.6 lists the normal values for the components of the blood buffer in arterial blood.

When the amount of CO_2 in the blood rises, the acidity of the blood increases and the pH is lowered. This can happen in the condition called emphysema, in which CO_2 does not diffuse properly through the lung membrane; it can also happen when the activity of the respiratory center in the medulla of the brain is affected by an accident or by depressive drugs. Poor respiration or below-normal gas exchange called hypoventilation also causes an increase in CO_2 in the blood. A higher level of CO_2 produces more carbonic acid in the blood, which produces more H^+, lowering the pH. This condition is called **acidosis.** When the origin of acidosis is respiratory, the condition is called respiratory acidosis. Acidosis caused by conditions that are not respiratory is called metabolic acidosis.

A lower level of CO_2 leads to a higher blood pH, a condition called **alkalosis.** Excitement, trauma, or a high temperature may cause a person to breathe rapidly, a condition called hyperventilation. Since CO_2 is expelled, its partial pressure in the blood falls below normal. The blood buffer brings the CO_2 up to its proper level by converting some carbonic acid to CO_2 and H_2O. This decreases the available H_2CO_3, which is compensated for by H^+ combining with HCO_3^-. The H^+ drops and blood pH rises, causing respiratory alkalosis.

Although the amount of CO_2 in the blood depends on the respiratory processes, the kidneys also control the amount of H^+ that is excreted in the urine. The pH of urine varies considerably and can range from 4.5 to 8.2. In severe conditions, even these mechanisms fail, and critical changes in blood pH occur. Table 9.7 lists some of the conditions that can lead to change in the blood pH, and it includes some possible treatments. ■

Table 9.6
Sample Values for Blood Buffer in Arterial Blood

P_{CO_2}	40 mm Hg
$H_2CO_3(aq)$	1.2 meq/L plasma[a]
HCO_3^-	24 meq/L plasma[a]
pH	7.4

[a]A milliequivalent (meq) is equal to $\frac{1}{1000}$ of an equivalent (equiv), which is the amount of an acid that supplies 1 mole of H^+, or the amount of a base that supplies 1 mole of OH^-.

Table 9.7 Acidosis and Alkalosis: Symptoms, Causes, and Treatments

Respiratory Acidosis: **CO_2 ↑ pH ↓**	
Symptoms:	Failure to ventilate, suppression of breathing, disorientation, weakness, coma
Causes:	Lung disease blocking gas diffusion (e.g., emphysema, pneumonia, bronchitis, and asthma); depression of respiratory center by drugs, cardiopulmonary arrest, stroke, poliomyelitis, or nervous system disorders
Treatment:	Correction of disorder, infusion of bicarbonate
Respiratory Alkalosis: **CO_2 ↓ pH ↑**	
Symptoms:	Increased rate and depth of breathing, numbness, light-headedness, tetany
Causes:	Hyperventilation due to anxiety, hysteria, fever, exercise; reaction to drugs such as salicylate, quinine, and antihistamines; conditions causing hypoxia (e.g., pneumonia, pulmonary edema, and heart disease)
Treatment:	Elimination of anxiety-producing state, rebreathing into a paper bag
Metabolic Acidosis: **H^+ ↑ pH ↓**	
Symptoms:	Increases ventilation, fatigue, confusion
Causes:	Renal disease, including hepatitis and cirrhosis; increased acid production in diabetes mellitus, hyperthyroidism, alcoholism, and starvation; loss of alkali in diarrhea; acid retention in renal failure
Treatment:	Sodium bicarbonate given orally, dialysis for renal failure, insulin treatment for diabetic ketosis
Metabolic Alkalosis: **H^+ ↓ pH ↑**	
Symptoms:	Depressed breathing, apathy, confusion
Causes:	Vomiting, diseases of the adrenal glands, ingestion of excess alkali
Treatment:	Infusion of saline solution, treatment of underlying diseases

9.9 Equivalents and Normality

Learning Goal **Use equivalent weight and normality to determine the number of equivalents of an acid, base, or salt.**

In our discussion of acid–base reactions, we wrote an equation showing that 1 mole of H^+ from an acid neutralizes 1 mole of OH^- from a base and produces 1 mole of water.

$$H^+(aq) + OH^-(aq) \longrightarrow H_2O$$

For acids and bases, we can also state the amount of H^+ or OH^- as **equivalents** (equiv). One equivalent of an acid is the amount of that acid that supplies 1 mole of H^+. One equivalent of a base is the amount of that base that provides 1 mole of OH^-. Using these definitions, we can determine the number of equivalents supplied by some strong acids and bases. (See Table 9.8.)

Table 9.8 Determining Equivalents for Strong Acids and Bases

Moles of Acid	*Moles of H^+ Supplied*	*Equivalents*
1 mole HCl	1 mole H^+	1 equiv HCl
1 mole H_2SO_4	2 moles H^+	2 equiv H_2SO_4
Moles of Base	***Moles of OH^- Supplied***	***Equivalents***
1 mole NaOH	1 mole OH^-	1 equiv NaOH
1 mole $Ca(OH)_2$	2 moles OH^-	2 equiv $Ca(OH)_2$

Equivalent Weight

When working with the mass of an acid or a base in calculations, it is useful to know the **equivalent weight,** which is the amount in grams of the substance that provides 1 equiv of that acid or base. We obtain the equivalent weight by dividing the molar mass of the acid or base by the number of equivalents of H^+ or OH^- it supplies.

$$\text{Equivalent weight} = \frac{\text{Molar mass (g)}}{\text{Number of equivalents}}$$

Sample Problem 9.16
Calculating Equivalent Weight

Calculate the equivalent weight of sulfuric acid, H_2SO_4, when it is completely neutralized.

Solution:
The molar mass of sulfuric acid, H_2SO_4, is 98.0 g. Since the acid supplies $2H^+$, we can determine its equivalent weight as follows:

$$\frac{98.0 \text{ g}}{2 \text{ equiv}} = 49.0 \text{ g/equiv}$$

Therefore, the equivalent weight of sulfuric acid is 49.0 g, since this is the mass of sulfuric acid that will furnish 1 equiv or 1 mole H^+.

Study Check:
What is the equivalent weight of phosphoric acid, H_3PO_4, if it can supply 3 H^+?

Answer: 32.7 g

Normality of Acids and Bases

Normality (*N*) is a way of expressing the concentration of acids and bases in terms of the equivalents of solute in 1 L of solution.

$$\text{Normality } (N) = \frac{\text{Equivalents of solute}}{\text{1 liter of solution}} = \frac{\text{equiv}}{\text{L}}$$

Sample Problem 9.17
Calculating the Normality of a Solution

What is the normality of a basic solution that contains 45 g $Ca(OH)_2$ in 1.0 L of solution?

Solution:
First, we need to determine the number of equivalents of $Ca(OH)_2$ in the solution. The equivalent weight is found by dividing the molar mass of $Ca(OH)_2$, 74.1 g, by 2, since 1 mole supplies 2 OH^-.

$$\frac{74.1 \text{ g } Ca(OH)_2}{2 \text{ equiv}} = 37.1 \text{ g/equiv}$$

The number of equivalents in the 45 g of $Ca(OH)_2$ is

$$45 \text{ g } Ca(OH)_2 \times \frac{1 \text{ equiv}}{37.1 \text{ g}} = 1.2 \text{ equiv}$$

The normality of the solution is calculated as

$$N = \frac{\text{equiv}}{\text{L}} = \frac{1.2 \text{ equiv}}{1.0 \text{ L}} = 1.2\ N$$

Study Check:
What is the normality of an HCl solution if it contains 74 g HCl in 0.250 L of solution?

Answer: 8.0 N

Titrations Using Normality

As with other concentrations, we can express the normality of a solution as a conversion factor. For example, a 3.0 N H_2SO_4 can be written as the following conversion factors:

$$\frac{3.0 \text{ equiv } H_2SO_4}{1 \text{ L solution}} \quad \text{and} \quad \frac{1 \text{ L solution}}{3.0 \text{ equiv } H_2SO_4}$$

In a titration using normality, we can convert between equivalents of acid and equivalents of base, since 1 equiv of acid supplies 1 mole H^+ and 1 equiv of base supplies 1 mole of OH^-.

1 equiv acid = 1 equiv base

Sample Problem 9.18
Using Normality in Titrations

How many liters of 0.25 *N* NaOH are needed to react with 2.0 L of 0.10 *N* H_2SO_4?

Solution:
From the normality of the acid, we calculate the equivalent of the acid as follows:

$$\underset{\text{Volume}}{2.0\ \cancel{L}\ H_2SO_4} \times \underset{\text{Normality}}{\frac{0.10\ \text{equiv}\ H_2SO_4}{1\ \cancel{L}}} = 0.20\ \text{equiv}\ H_2SO_4$$

Since the equivalents of base are equal to the equivalents of acid,

$$0.20\ \cancel{\text{equiv}\ H_2SO_4} \times \frac{1\ \text{equiv NaOH}}{1\ \cancel{\text{equiv}\ H_2SO_4}} = 0.20\ \text{equiv NaOH}$$

The liters of base needed are:

$$0.20\ \cancel{\text{equiv}}\ \text{NaOH} \times \underset{\text{Normality}}{\frac{1\ \text{L NaOH}}{0.25\ \cancel{\text{equiv}}}} = 0.80\ \text{L NaOH}$$

Study Check:
How many milliliters of 6.0 *N* HCl can be neutralized by 450 mL of 2.0 *N* KOH?

Answer: 150 mL

Normalities of Solutions of Electrolytes

The levels of the electrolytes, such as Na^+, Cl^-, K^+, and Ca^{2+}, in body fluids are usually given in terms of equivalents. For ions, an equivalent is defined as the amount of that ion that provides 1 mole of positive or negative charge.

For example, 1 mole of Na^+ ion provides 1 mole of positive charge, so we can say that 1 mole Na^+ is 1 equiv of ionic charge. Similarly, 1 mole of sulfide, S^{2-}, has 2 moles of negative charge. Therefore, 1 mole of S^{2-} has 2 equiv of ionic charge. Table 9.9 lists the equivalents and equivalent weights for some important ions in the body.

Table 9.9 Equivalents of Electrolytes

Ion	Moles of charge	Equivalents	Equivalent weight
Na^+	1 mole charge (+)	1 equiv Na^+	23.0 g
Cl^-	1 mole charge (−)	1 equiv Cl^-	35.5 g
Ca^{2+}	2 mole charge (2+)	2 equiv Ca^{2+}	40.1 g/2 = 20.1 g
Fe^{3+}	3 mole charge (3+)	3 equiv Fe^{3+}	55.8 g/3 = 18.6 g

Sample Problem 9.19
Calculating Equivalents for Salts

How many equivalents are in 75.0 g of Zn^{2+}?

Solution:
Since Zn^{2+} has a 2+ charge, it has 2 equiv/mole. Using the molar mass of zinc, we express the equivalent weight of Zn^{2+} as

$$\frac{65.4 \text{ g } Zn^{2+}}{2 \text{ equiv}} = \frac{32.7 \text{ g } Zn^{2+}}{1 \text{ equiv}}$$

The number of equivalents in 75.0 g Zn^{2+} is

$$75.0 \not{g}\ Zn^{2+} \times \underbrace{\frac{1 \text{ equiv } Zn^{2+}}{32.7 \not{g}}}_{\text{Equivalent weight}} = 2.29 \text{ equiv } Zn^{2+}$$

Study Check:
How many grams of Cl^- are present in 2.0 equiv Cl^-?

Answer: 71 g

HEALTH NOTE

Electrolytes in Body Fluids

The concentrations of electrolytes present in body fluids and in intravenous fluids given to a patient are often expressed in milliequivalents per liter (meq/L) of solution.

1 equiv = 1000 meq

Table 9.10 gives the concentrations of some typical electrolytes in blood plasma. There is a charge balance, since the total number of positive electrolytes is equal to the total number of negative electrolytes. ■

Table 9.10 Some Typical Concentrations of Electrolytes in Blood Plasma

Electrolyte	Concentration (meq/L)	Electrolyte	Concentration (meq/L)
Cations (Positive Ions)		*Anions (Negative Ions)*	
Na^+	138	Cl^-	110
K^+	5	HCO_3^-	30
Mg^{2+}	3	HPO_4^{2-}	4
Ca^{2+}	4	Proteins	6
Total cations	150	Total anions	150

Sample Problem 9.20
Electrolyte Concentration

A typical concentration for Na^+ in the blood is 138 meq/L. How many grams of sodium ion are in 1.00 L of blood?

Solution:
Using the volume and the concentration (in milliequivalents per liter) we can find the number of milliequivalents in 1.00 L of body fluid (1 equiv = 1000 meq).

$$1.00\,\cancel{L} \times \frac{138\,\cancel{meq}}{1\,\cancel{L}} \times \frac{1\text{ equiv}}{1000\,\cancel{meq}} = 0.138\text{ equiv }Na^+$$

The mass of Na^+ is calculated by using the equivalent weight of Na^+.

$$0.138\,\cancel{equiv}\,Na^+ \times \frac{23.0\text{ g }Na^+}{1\,\cancel{equiv}} = 3.17\text{ g }Na^+$$

HEALTH NOTE

Electrolytes in Intravenous Solutions

The use of a specific intravenous solution depends on the nutritional, electrolyte, and fluid needs of the individual patient. Examples of various types of solutions are given in Table 9.11. ■

Table 9.11 Electrolyte Concentrations in Intravenous Replacement

Solution	Electrolytes (meq/L)	Use
Sodium chloride (0.9%)	Na^+ 154, Cl^- 154	Replacement of fluid loss
Potassium chloride with 5% dextrose	K^+ 40, Cl^- 40	Treatment of malnutrition (low potassium levels)
Ringer's solution	Na^+ 147, K^+ 4, Ca^{2+} 4, Cl^- 155	Replacement of fluids and electrolytes lost through dehydration
Maintenance solution with 5% dextrose	Na^+ 40, K^+ 35, Cl^- 40, lactate$^-$ 20, $HPO_4{}^{2-}$ 15	Maintenance of fluid and electrolyte levels
Replacement solution (extracellular)	Na^+ 140, K^+ 10, Ca^{2+} 5, Mg^{2+} 3, Cl^- 103, acetate$^-$ 47, citrate^{3-} 8	Replacement of electrolytes in extracellular fluids

Glossary of Key Terms

acid A substance that produces H^+ in solution.

acidosis A physiological condition in which the blood pH is lower than the normal 7.4.

alkalosis A physiological condition in which the blood pH is higher than the normal 7.4.

base A substance that provides OH^- in an aqueous solution.

buffer A mixture of a weak acid or a weak base and its salt; a buffer resists changes in the pH of a solution.

electrolyte A substance that produces ions when dissolved in water.

equivalent The amount of acid or base that provides 1 mole of H^+ or OH^-; the amount of a salt that supplies 1 mole of positive (+) or negative (−) charge.

equivalent weight The mass in grams of 1 equiv, calculated by dividing the molar mass by the number of equivalents in a mole.

hydronium ion The H_3O^+ ion which is a proton (H^+) bonded to a H_2O molecule.

indicator A substance that is added to a sample during titration and changes color when the pH of the sample changes.

insoluble salt A salt that contains no soluble ions and therefore is not soluble in water. In a mixture, its ions will form a precipitate.

ion-product constant, K_w The ion product for ions in water; $K_w = [H^+][OH^-] = 1.0 \times 10^{-14}$.

neutralization A reaction between an acid and a base to form a salt and water.

nonelectrolyte A substance that dissolves in water as molecules; its solution will not conduct an electrical current.

normality (N) The concentration expressed as equiv/L.

pH A measure of $[H^+]$ in a solution.

precipitation The formation of an insoluble salt when two solutions are mixed.

solubility rules A description of ion combinations that are soluble in water.

soluble salt A salt that contains at least one ion that makes it dissolve in water.

strong acid An acid that is completely ionized in water.

strong base A base that is completely ionized in water.

strong electrolyte An ionic compound or strong acid that ionizes completely when it dissolves in water. Its solution is a conductor of electricity.

titration The addition of a measured amount of a solution such as a base to a certain volume of another solution such as an acid to determine the concentration of the acid.

weak acid An acid that ionizes only slightly in solution.

weak base A base that ionizes only slightly in solution.

weak electrolyte A substance that produces only a few ions along with many molecules when it dissolves in water. Its solution is a weak conductor of electricity.

Problems

Electrolytes *(Goal 9.1)*

9.1 Write a balanced equation for the dissociation of the following salts in water:

a. KCl d. $LiNO_3$
b. Na_2SO_4 e. K_3PO_4
c. $CaCl_2$ f. $Ba(NO_3)_2$

9.2 Indicate whether aqueous solutions of the following will contain ions only, molecules only, or molecules and some ions:

a. KCl, a strong electrolyte d. sodium bromide, a salt
b. glucose, a nonelectrolyte e. urea, a nonelectrolyte
c. acetic acid (vinegar), a weak electrolyte

9.3 Indicate the type of electrolyte represented in the following equations:

a. $KI(s) \xrightarrow{H_2O} K^+(aq) + I^-(aq)$
b. $NH_3(g) + H_2O \rightleftharpoons NH_4^+(aq) + OH^-(aq)$
c. $CH_3OH(l) \xrightarrow{H_2O} CH_3OH(aq)$
d. $C_6H_{12}O_6(s) \xrightarrow{H_2O} C_6H_{12}O_6(aq)$
e. $HF(g) \overset{H_2O}{\rightleftharpoons} H^+(aq) + F^-(aq)$

Ionic Reactions: Formation of a Solid *(Goal 9.2)*

9.4 Write an equation for the neutralization reactions that would produce the following salts:
a. KI b. $MgCl_2$ c. K_2SO_4 d. $LiNO_3$ *e. $Ca_3(PO_4)_2$

9.5 Predict whether the following solutes would be soluble in water:
a. LiCl
b. $PbCl_2$
c. $BaCO_3$
d. K_2O
e. $Fe(NO_3)_3$
f. PbS
g. NaI
h. Na_2S
i. Ag_2O
j. $CaSO_4$
k. $(NH_4)_2CO_3$

9.6 State whether a precipitate would be expected to form if the following solutions are mixed. If a precipitate will form, write a balanced equation for the reaction.
a. $NaCl(aq)$ and $Pb(NO_3)_2(aq)$
b. $K_2S(aq)$ and $NaI(aq)$
c. $Na_2S(aq)$ and $CuCl_2(aq)$
d. $BaCl_2(aq)$ and $Na_2SO_4(aq)$
e. $CaCl_2(aq)$ and $NaNO_3(aq)$
*f. $Na_3PO_4(aq)$ and $MgCl_2(aq)$

Acids and Bases *(Goal 9.3)*

9.7 Identify the following as strong or weak acids:
a. HCl
b. H_2SO_4
c. H_2S
d. HClO

9.8 Write a balanced equation for the ionization of the following acids in water:
a. HCl
b. HNO_3
c. HNO_2, nitrous acid
d. H_2CO_3, carbonic acid

9.9 Write an equation for the dissociation of the following bases in water:
a. LiOH
b. $Mg(OH)_2$
c. KOH

9.10 Write an equation for the reaction of the weak base, ammonia (NH_3), in water.

9.11 Indicate whether each of the following statements is characteristic of an acid or a base:

a. has a sour taste
b. turns red litmus blue
c. reacts with acids
d. has a bitter taste
e. produces hydrogen ions in aqueous solution.

Ionization of Water *(Goal 9.4)*

9.12 a. State the values for $[H^+]$ and $[OH^-]$ in pure water at 25°C.
b. Write the ion-product expression for water and give its value at 25°C.

9.13 Find $[H^+]$ for solutions having the following $[OH^-]$ values:

a. $[OH^-] = 1 \times 10^{-11}\ M$
b. $[OH^-] = 3.0 \times 10^{-5}\ M$
c. $[OH^-] = 1 \times 10^{-1}\ M$
d. $[OH^-] = 8.5 \times 10^{-12}\ M$

***9.14** What is the value for $[H^+]$ of a solution that contains 0.200 g NaOH in 250 mL solution?

9.15 Calculate $[OH^-]$ of a solution when its $[H^+]$ has the following values:

a. $[H^+] = 1 \times 10^{-11}\ M$
b. $[H^+] = 1 \times 10^{-2}\ M$
c. $[H^+] = 2.0 \times 10^{-4}\ M$
d. $[H^+] = 6.0 \times 10^{-8}\ M$

***9.16** What is the value for $[OH^-]$ of a solution that contains 1.26 g HNO_3 in 500.0 mL solution?

The pH of Solutions *(Goal 9.5)*

9.17 State whether the following solutions are acidic, neutral, or basic.

a. blood, pH 7.4
b. coffee, pH 5.5
c. pancreatic juice, pH 8.2
d. vinegar, pH 2.8
e. soda, pH 3.2
f. milk, pH 7.0
g. pH-balanced shampoo, pH 6.0
h. hot tub water, pH 7.8
i. drain cleaner, pH 11.2
j. laundry detergent, pH 9.5

9.18 Arrange the pH values of the following groups in order from the most acidic to the least acidic (most basic):

a. 4.5, 13.0, 0.4, 6.8
b. 1.6, 11.7, 7.1, 2.3, 8.5
c. 14.0, 9.8, 3.3, 4.4, 2.9
d. 8.8, 9.7, 11.4, 13.4, 7.4

9.19 Are the following solutions acidic, neutral, or basic?

a. $[H^+] = 1.0 \times 10^{-4}\ M$
b. $[H^+] = 1.0 \times 10^{-7}\ M$
c. $[OH^-] = 2.0 \times 10^{-3}\ M$
d. $[OH^-] = 1.0 \times 10^{-10}\ M$
e. $[H^+] = 5.0 \times 10^{-2}\ M$
f. $[H^+] = 0.01\ M$
g. $[OH^-] = 1.0 \times 10^{-4}\ M$
h. $[OH^-] = 1.0 \times 10^{-7}\ M$
*i. $[H^+] = 0.0001\ M$
*j. $[OH^-] = 0.001\ M$

9.20 Calculate the pH of each of the solutions in Problem 9.19.

9.21 Complete the following table:

$[H^+]$	$[OH^-]$	pH	Acidic, Basic, or Neutral?
______	1×10^{-6}	______	______
______	______	2	______
1×10^{-5}	______	______	______
______	______	______	Neutral
______	______	10	______

Ionic Reactions: Acid–Base Neutralization *(Goal 9.6)*

9.22 Balance the following acid–base reactions:
a. $HCl + KOH \longrightarrow KCl + H_2O$
b. $HNO_3 + Ca(OH)_2 \longrightarrow Ca(NO_3)_2 + H_2O$
c. $H_3PO_4 + LiOH \longrightarrow Li_3PO_4 + H_2O$
d. $H_2SO_4 + Al(OH)_3 \longrightarrow Al_2(SO_4)_3 + H_2O$

9.23 Write the balanced equation for the neutralization reactions of the following:
a. $NaOH + H_2SO_4 \longrightarrow$
b. $KOH + HNO_3 \longrightarrow$
c. $H_3PO_4 + NaOH \longrightarrow$
d. $H_2CO_3 + Ca(OH)_2 \longrightarrow$

Acid–Base Titration *(Goal 9.7)*

9.24 Calculate the molarity of the acid in each of the following titrations:
a. A 5.0-mL sample of HCl that is titrated with 22.0 mL of 0.50 *M* NaOH.

$$HCl(aq) + NaOH(aq) \longrightarrow NaCl(aq) + H_2O$$

*b. A 25.0-mL sample of H_2SO_4 that is titrated with 38.0 mL of 1.0 *M* NaOH.

$$H_2SO_4(aq) + 2NaOH(aq) \longrightarrow Na_2SO_4(aq) + 2H_2O$$

*c. A 10.0-mL sample of H_3PO_4 that is titrated with 16.0 mL of 1.0 *M* NaOH.

$$H_3PO_4(aq) + 3NaOH(aq) \longrightarrow Na_3PO_4(aq) + 3H_2O$$

Buffers *(Goal 9.8)*

9.25 Which of the following represent a buffer system? Why?
a. HCl and NaCl
b. H_2CO_3
c. NH_3 and NH_4Cl
d. KCl and NaCl
e. H_2CO_3 and $NaHCO_3$

***9.26** Consider the buffer system of acetic acid, $HC_2H_3O_2$, and its salt, $NaC_2H_3O_2$:

$$\underset{\text{Acetic acid}}{HC_2H_3O_2(aq)} \rightleftharpoons H^+(aq) + \underset{\text{Acetate ion}}{C_2H_3O_2^-(aq)}$$

a. What is the purpose of the buffer system?
b. What is the function of the salt, $NaC_2H_3O_2$?
c. Which substance reacts with H^+ added to the buffer?
d. Which substance reacts with OH^- added to the buffer?

Equivalents and Normality *(Goal 9.9)*

9.27 Calculate the equivalent weight of each of the following:
a. HCl
b. $Mg(OH)_2$
c. H_3PO_4
d. $CaCl_2$

9.28 What is the normality of each of the following solutions?
a. 40.0 g H_2SO_4 in 250 mL of solution
b. 25.0 g KOH in 100.0 mL of solution
c. 10.0 g K_2CO_3 in 200.0 mL of solution

***9.29** A 10.0-mL sample of acid is titrated with 35.0 mL of 1.50 *N* NaOH. What is the normality of the acid?

***9.30** How many milliliters of 0.20 *N* NaOH are needed to neutralize 20.0 mL of 0.50 *N* H_3PO_4?

9.31 A physiological saline solution contains 154 meq/L Na^+ and 154 meq/L Cl^-. Calculate the number of grams of Na^+ and of Cl^- in 1 L of the solution.

9.32 A sample of Ringer's solution contains the following concentrations (meq/L) of cations: Na^+ 147, K^+ 4, and Ca^{2+} 4. If Cl^- is the only anion in the solution, what is its concentration in milliequivalents per liter?

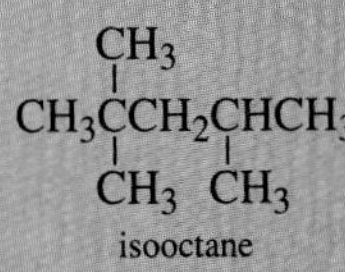

10

Alkanes: An Introduction to Organic Chemistry

Crude oil contains a variety of organic compounds which are separated in the refining process to obtain gasoline and other valuable petroleum products.

Learning Goals

10.1 Identify the properties of a compound as typical of organic or inorganic compounds.

10.2 Write a complete structural formula and a condensed structural formula for an alkane.

10.3 Write the names and the structural formulas for the first 10 straight-chain alkanes.

10.4 Write the IUPAC name of a branched-chain alkane.

10.5 Given the molecular formula of an alkane, write the condensed structural formulas of its isomers.

10.6 Write the name and the structural formula of a cycloalkane.

10.7 Write the name and the structural formula of a haloalkane.

10.8 Write chemical equations for the combustion and the halogenation of alkanes.

SCOPE

The element carbon has a special role in chemistry because it has the capability of forming so many different compounds. Gasoline, coal, medicines, and perfumes are made of organic compounds. The fabrics in your clothes may be naturally occurring organic compounds such as cotton and silk, or they may be synthetic organic compounds such as polyester and nylon. The food we eat is composed of many different organic compounds that supply us with fuel for energy and with the carbon atoms needed for building and repairing the cells of our bodies. In this chapter, we will begin the study of the behavior of carbon compounds, their reactions, and their effects upon our environment.

The word *organic* is an old name given to compounds that scientists believed required a ''vital force'' that could only be produced by living systems. By the early 1800s, however, chemists had shown that urea, a product of protein metabolism in animals, could be prepared from the inorganic compound ammonium cyanate. Since then, several million organic compounds have been prepared in laboratories.

$$\underset{\text{Ammonium Cyanate (inorganic)}}{NH_4CNO} \xrightarrow{\text{heat}} \underset{\text{Urea (organic)}}{H_2N{-}\overset{\overset{\displaystyle O}{\|}}{C}{-}NH_2}$$

10.1 Properties of Organic Compounds

Learning Goal **Identify the properties of a compound as typical of organic or inorganic compounds.**

Most of the chemistry we have studied so far has involved inorganic compounds that were ionic or had polar covalent bonds. In general, these compounds have high melting and boiling points, do not burn in oxygen, and are soluble in water.

In contrast, **organic compounds,** composed mostly of carbon atoms, have covalent bonds. They have low melting and boiling points, burn vigorously, and are soluble in nonpolar solvents. There are a great number of organic compounds because atoms of carbon have a unique ability to bond to other carbon atoms in many different ways. See Table 10.1 for examples of some properties of organic and inorganic compounds.

Table 10.1 Properties of Typical Organic and Inorganic Compounds

Property	Inorganic compound: sodium chloride, NaCl	Organic compound: pentane, C_5H_{12}
Bonding	Ionic	Covalent
Melting point	801°C	−129°C
Boiling point	1413°C	36°C
Soluble in	a polar solvent	a nonpolar solvent
Electrolyte	yes	no
Flammable	no	yes

Sample Problem 10.1
Identifying Properties of Organic Compounds

State whether each of the following properties is typical of an organic or an inorganic compound:

a. soluble in water
b. low boiling point
c. burns in air

Solution:

a. Inorganic; most inorganic compounds are soluble in water.
b. Organic; most organic compounds have low boiling points.
c. Organic; organic compounds are often very flammable.

Study Check:
Would a compound that dissociates into ions in water be an organic or inorganic compound?

Answer: inorganic

10.2 Structural Formulas

Learning Goal **Write a complete structural formula and a condensed structural formula for an alkane.**

The **hydrocarbons** are organic compounds that contain only the elements carbon and hydrogen. Since carbon has four valence electrons, we can write the electron-dot structure of a carbon atom as follows:

$$\cdot\dot{\underset{\cdot}{C}}\cdot$$

To acquire a noble gas structure of eight valence electrons, a carbon atom forms four covalent bonds. In the simplest hydrocarbon, methane (CH_4), there is one carbon atom bonded to four hydrogen atoms. In the electron-dot structure, each shared pair of electrons represents a single bond. In the drawing of the complete **structural formula** for methane, each C—H bond is shown as a single line.

```
    H              H
    ..             |
H : C : H      H — C — H        Methane, CH4
    ..             |
    H              H
Electron-dot   Complete structural
formula        formula
```

Although the structure of methane is written as two-dimensional, methane actually has a three-dimensional shape (shown in Figure 10.1), in which the carbon atom is bonded equally to four hydrogen atoms. The most stable arrangement of the molecule is achieved when the four electron pairs are arranged around the carbon atom as far apart as possible. In methane, the four hydrogen atoms form the corners of a tetrahedron. However, it is more convenient to use the flat formula to represent the molecule.

Figure 10.1
The tetrahedral shape of methane, CH_4.

```
    H
    |
H — C — H
    |
    H
```

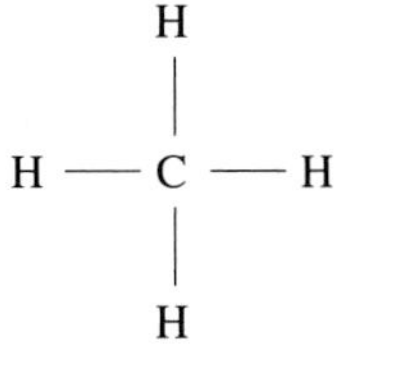

methane

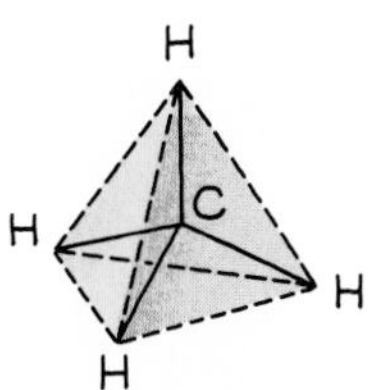

Figure 10.2
A model of ethane, C_2H_6.

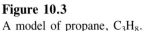

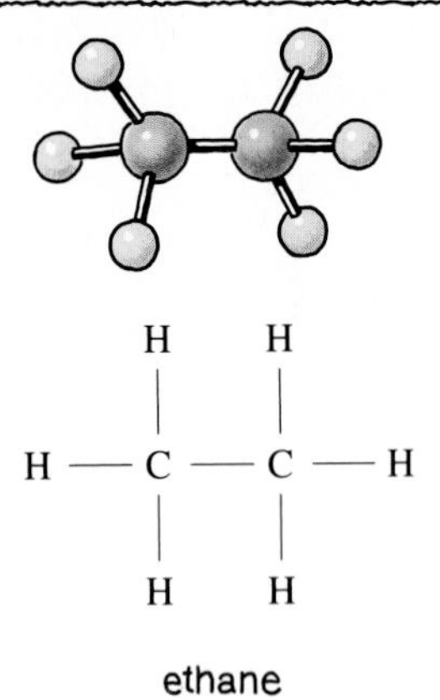

Complete Structural Formulas of Alkanes

The **alkanes** are a group of hydrocarbons that have only single bonds between the carbon atoms. This makes them **saturated hydrocarbons** because each carbon atom is bonded to its maximum of four other atoms. In the structural formula for ethane, C_2H_6, two carbon atoms form a single C—C covalent bond using one valence electron from each carbon atom. That leaves each carbon atom with three valence electrons to form bonds with three hydrogen atoms. Each carbon atom has a total of four bonds, completing its octet.

```
  ·  ·            H  H
 ·C—C·          H—C—C—H     Ethane, C2H6
  ·  ·            H  H
  ↑  ↑
3H· 3H·
```

A molecular model of ethane is illustrated in Figure 10.2.

To write the complete structural formula of propane, C_3H_8, an alkane with three carbon atoms, we must use a total of eight hydrogen atoms to give each carbon atom four bonds:

```
  ·  ·  ·            H  H  H
 ·C—C—C·          H—C—C—C—H     Propane, C3H8
  ·  ·  ·            H  H  H
  ↑   ↑   ↑
3H· 2H· 3H·
```

A molecular model of propane is illustrated in Figure 10.3.

Figure 10.3
A model of propane, C_3H_8.

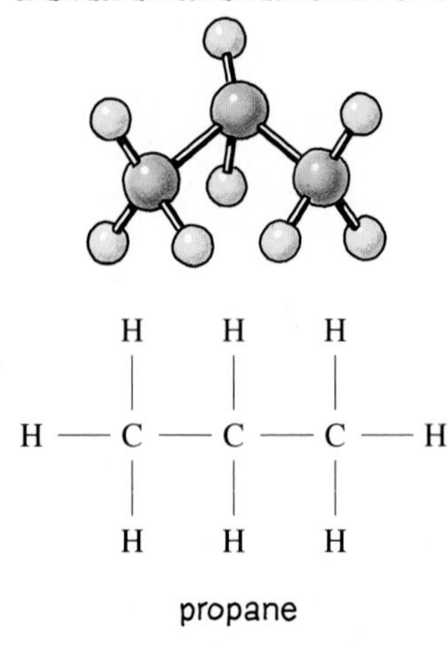

Sample Problem 10.2
Writing Complete Structural Formulas for Alkanes

Complete the following structural formula for pentane, C_5H_{12}, by adding hydrogen atoms.

C—C—C—C—C

Solution:
Hydrogen atoms are added to give each of the carbon atoms a total of four bonds.

```
  H  H  H  H  H
  |  |  |  |  |
H—C—C—C—C—C—H
  |  |  |  |  |
  H  H  H  H  H
```

Condensed Structural Formulas

Although complete structural formulas are helpful it is not always necessary to draw all of the individual C—H and C—C bonds. In a simplified formula called a **condensed structural formula,** each carbon atom and its hydrogen atoms are written as a group. A subscript follows the hydrogen to indicate the number of hydrogen atoms bonded to that carbon atom. Therefore, the C—H single bonds are understood, but not written. A line may be drawn to show each carbon–carbon bond, but it can be omitted. Both complete and condensed structural formulas show the arrangement of the carbon atoms in a molecule. Molecular formulas give only the total number of each kind of atom. Here, for example, are the different types of formulas for propane:

$$\begin{array}{ccccccc} & & & 2H & & & \\ & & & \downarrow & & & \\ & H & & H & & H & \\ & | & & | & & | & \\ H— & C & — & C & — & C & —H \\ & | & & | & & | & \\ & H & & H & & H & \\ & \uparrow & & & & \uparrow & \\ & 3H & & & & 3H & \end{array} \qquad \begin{array}{c} CH_3—CH_2—CH_3 \\ \text{or} \\ CH_3CH_2CH_3 \end{array} \qquad C_3H_8$$

Complete structural formula: shows arrangement of all single bonds

Condensed structural formula: shows arrangement, but no single bonds, or only C—C bonds

Molecular formula: shows no order of atoms, only the number of atoms

Table 10.2 shows the three types of formulas for methane, ethane, and propane.

Table 10.2 Comparison of Types of Formulas

Name	Complete structural formula	Condensed structural formula	Molecular formula
Methane	$\begin{array}{ccc} & H & \\ & \vert & \\ H— & C & —H \\ & \vert & \\ & H & \end{array}$	CH_4	CH_4
Ethane	$\begin{array}{ccccc} & H & & H & \\ & \vert & & \vert & \\ H— & C & — & C & —H \\ & \vert & & \vert & \\ & H & & H & \end{array}$	$CH_3—CH_3$ or CH_3CH_3	C_2H_6
Propane	$\begin{array}{ccccccc} & H & & H & & H & \\ & \vert & & \vert & & \vert & \\ H— & C & — & C & — & C & —H \\ & \vert & & \vert & & \vert & \\ & H & & H & & H & \end{array}$	$CH_3—CH_2—CH_3$ or $CH_3CH_2CH_3$	C_3H_8

10.3 IUPAC Names for Alkanes

Learning Goal **Write the names and the structural formulas for the first 10 straight-chain alkanes.**

At one time, compounds were named in a random fashion. However, as more and more organic compounds were discovered, a systematic way of naming them became necessary. Official rules for naming organic compounds were worked out by the International Union of Pure and Applied Chemistry (IUPAC). However, since there are many common names still in use, they will be included in this discussion, frequently in parentheses after the IUPAC name.

In the IUPAC system, a prefix specifies a particular number of carbon atoms in a carbon chain:

Prefix	*Number of Carbon Atoms*	*Prefix*	*Number of Carbon Atoms*
Meth	1	Hex	6
Eth	2	Hept	7
Prop	3	Oct	8
But	4	Non	9
Pent	5	Dec	10

When the IUPAC rules are used to name organic compounds, the family of the compound is shown by the ending of the name. For example, the names of compounds in the alkane family have an *ane* ending. By attaching the IUPAC prefixes for the number of carbon atoms to the *ane* ending, we can write the names for the first 10 straight-chain alkanes, as shown in Table 10.3. A **straight-chain alkane** is a compound having carbon atoms bonded to carbon atoms in a continuous carbon chain. The alkane names are also used to form the IUPAC names for other organic compounds.

Table 10.3 Ten Straight-Chain Alkanes

Name	Condensed structural formula	Molecular formula
Methane	CH_4	CH_4
Ethane	CH_3CH_3	C_2H_6
Propane	$CH_3CH_2CH_3$	C_3H_8
Butane	$CH_3CH_2CH_2CH_3$	C_4H_{10}
Pentane	$CH_3CH_2CH_2CH_2CH_3$	C_5H_{12}
Hexane	$CH_3CH_2CH_2CH_2CH_2CH_3$	C_6H_{14}
Heptane	$CH_3CH_2CH_2CH_2CH_2CH_2CH_3$	C_7H_{16}
Octane	$CH_3CH_2CH_2CH_2CH_2CH_2CH_2CH_3$	C_8H_{18}
Nonane	$CH_3CH_2CH_2CH_2CH_2CH_2CH_2CH_2CH_3$	C_9H_{20}
Decane	$CH_3CH_2CH_2CH_2CH_2CH_2CH_2CH_2CH_2CH_3$	$C_{10}H_{22}$

Sample Problem 10.3
IUPAC Names for Straight-Chain Alkanes

Write the IUPAC names for the following alkanes:

a. $CH_3CH_2CH_3$
b. $CH_3—CH_2—CH_2—CH_2—CH_2—CH_3$

Solution:

a. propane
b. hexane

Study Check:
Write the condensed structural formula and the name of a straight-chain alkane having five carbon atoms.

Answer: $CH_3—CH_2—CH_2—CH_2—CH_3$ or $CH_3CH_2CH_2CH_2CH_3$; pentane

Three-Dimensional Shapes of Straight-Chain Alkanes

The structural formulas we have written for the alkanes are useful in showing the order of the carbon atoms, but they are two-dimensional and do not represent the actual three-dimensional structure of the carbon chains. Earlier, we saw that methane has a three-dimensional shape called a tetrahedron. When a compound contains several carbon atoms that have single bonds, each carbon atom still has a tetrahedral shape. To form a chain of carbon atoms, each carbon atom bonds at an angle to the next carbon atom, which results in a zigzag pattern of carbon atoms rather than in a straight line. This zigzag pattern of the carbon chain of pentane is shown in Figure 10.4. Although it is convenient to write the structural formulas in a straight line, we must remember that the carbon atoms are always at angles to each other.

Since the carbon atoms in alkanes are connected by single bonds, they can move or rotate freely, which enables the chain to turn in many directions. This free rotation makes it possible for pentane, for example, to have the different patterns illustrated in Figure 10.5.

Figure 10.4
A model of pentane, C_5H_{12}.

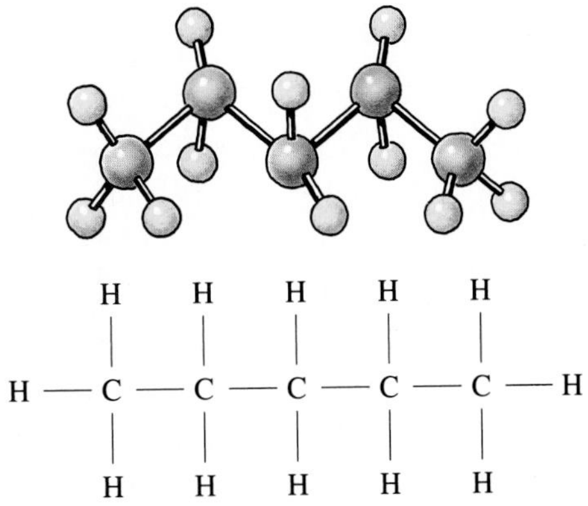

```
      H   H   H   H   H
      |   |   |   |   |
  H — C — C — C — C — C — H
      |   |   |   |   |
      H   H   H   H   H
```

pentane

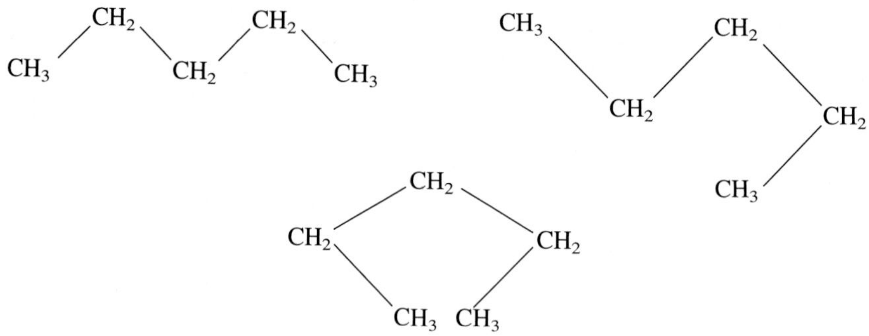

Figure 10.5
Some possible patterns of the carbon chain of pentane due to free rotation of the carbon–carbon single bonds.

CH_3 CH_2 CH_2 CH_2 CH_3

CH_3 CH_2 CH_2 CH_2 CH_3

CH_2 CH_2 CH_2 CH_3 CH_3

ENVIRONMENTAL NOTE

Hydrocarbons in Crude Oil

The first four alkanes—methane, ethane, propane, and butane—are gases at room temperature and pressure and act as anesthetics if inhaled. They are widely used as heating fuels. Alkanes having 5 to 17 carbon atoms are liquids at room temperature; alkanes having 18 or more carbon atoms are solids. Mineral oil is a mixture of liquid hydrocarbons and is used as a cathartic and a lubricant. Motor oil is a mixture of high-molecular-weight liquid hydrocarbons and is used to lubricate the internal parts of engines. Petrolatum, or Vaseline, is a colloidal system in which liquid hydrocarbons having high boiling points are encapsulated in solid hydrocarbons. It is used in ointments and cosmetics, and as a lubricant and a solvent. Figure 10.6 shows some of these hydrocarbons.

All alkanes are nonpolar and insoluble in water. Applying a hydrocarbon mixture such as Vaseline softens the skin because the water in the skin is retained by nonpolar compounds. Most alkanes are less dense than water and float on water. In oil spills, the hydrocarbons remain on the surface, where they cover a large area, rather than sinking to the bottom.

Figure 10.6
Mixtures of hydrocarbons are found in products like these.

Crude oil contains a wide variety of alkanes whose boiling points increase with molar mass. Various products can be obtained through distillation, which separates the oil into fractions with a stepwise increase in temperature. Some products obtained from this fractionation of crude oil are listed in Table 10.4. ■

Table 10.4 Hydrocarbons in a Barrel of Crude Oil

Product	Number of carbon atoms	Temperature range for distillation (°C)
Natural gas	1–4	below 30
Gasoline	5–12	30–200
Kerosene, heating oil	12–15	200–300
Diesel fuel, lubricating oil	15–25	300–400
Asphalt, paraffin wax	over 25	nonvolatile residue

10.4 IUPAC Rules for Naming Branched-Chain Alkanes

Learning Goal **Write the IUPAC name of a branched-chain alkane.**

The alkanes described so far, such as ethane, propane, and butane, have been made of carbon atoms in a straight or continuous chain. However, as the number of carbon atoms in a molecule increases, the carbon chain may get longer, or there may be **branches** or side groups containing one or more carbon atoms attached to the chain. Then the compound is called a **branched-chain alkane.** For example, we can write two different condensed formulas for compounds whose molecular formula is C_4H_{10}. One has a straight chain and the other has a branched chain:

C_4H_{10}

Complete structural formula *Condensed structural formula*

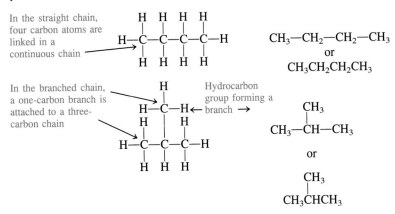

In the straight chain, four carbon atoms are linked in a continuous chain

```
    H  H  H  H
    |  |  |  |
 H—C—C—C—C—H
    |  |  |  |
    H  H  H  H
```

$CH_3—CH_2—CH_2—CH_3$

or

$CH_3CH_2CH_2CH_3$

In the branched chain, a one-carbon branch is attached to a three-carbon chain

Hydrocarbon group forming a branch →

```
       H
       |
    H—C—H
    H  |  H
    |  |  |
 H—C—C—C—H
    |  |  |
    H  H  H
```

```
      CH3
      |
CH3—CH—CH3
```

or

```
  CH3
  |
CH3CHCH3
```

Although they have the same molecular formula, they are different compounds because they have different structures.

In writing the condensed formula of a branched-chain compound, a branch or side group usually appears above or below the carbon chain. Table 10.5 compares complete structural formulas of branched-chain alkanes with their corresponding condensed structural formulas.

Table 10.5 Structural Formulas of Branched-Chain Alkanes

```
Name                 Complete structural      Condensed structural
                     formula                  formula

                           H
                           |
                         H—C—H
                     H     |   H  H                 CH3                     CH3
                     |     |   |  |                 |                       |
2-Methylbutane     H—C—————C———C——C—H        CH3—CH—CH2—CH3    or    CH3CHCH2CH3
                     |     |   |  |
                     H     H   H  H

                         H
                         |
                       H—C—H
                     H   |   H  H  H             CH3                       CH3
                     |   |   |  |  |             |                         |
2,4-Dimethylpentane H—C——C———C——C——C—H      CH3—CH—CH2—CH—CH3   or   CH3CHCH2CHCH3
                     |   |   |  |  |                     |                     |
                     H   H   H  |  H                     CH3                   CH3
                                |
                              H—C—H
                                |
                                H
```

Sample Problem 10.4
Writing a Condensed Structural Formula for a Branched-Chain Compound

Write a condensed structural formula for the compound that has the following complete structural formula:

```
        H           H
        |           |
      H—C—H       H—C—H
    H   |   H  H    |   H
    |   |   |  |    |   |
  H—C———C———C——C————C———C—H
    |   |   |  |    |   |
    H   H   H  H    H   H
```

Solution:
In the condensed structural formula, each carbon atom and its surrounding hydrogen atoms are indicated as a CH group that has a subscript to represent the number of attached hydrogen atoms.

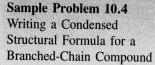

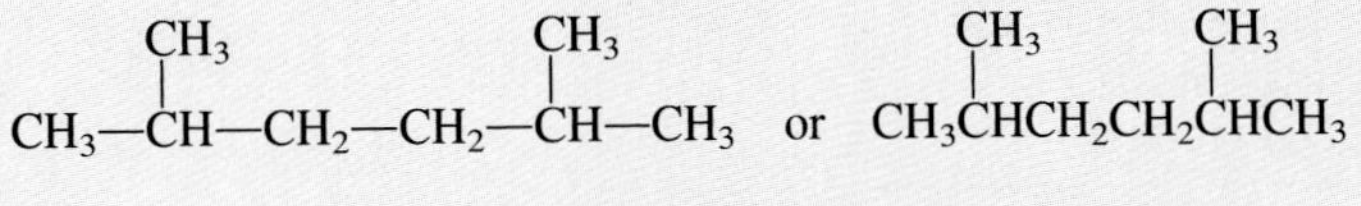

$$\mathrm{CH_3{-}\underset{}{\overset{CH_3}{\overset{|}{C}}}H{-}CH_2{-}CH_2{-}\overset{CH_3}{\overset{|}{C}}H{-}CH_3} \quad \text{or} \quad \mathrm{CH_3\overset{CH_3}{\overset{|}{C}}HCH_2CH_2\overset{CH_3}{\overset{|}{C}}HCH_3}$$

Study Check:
Write a condensed structural formula for the following compound:

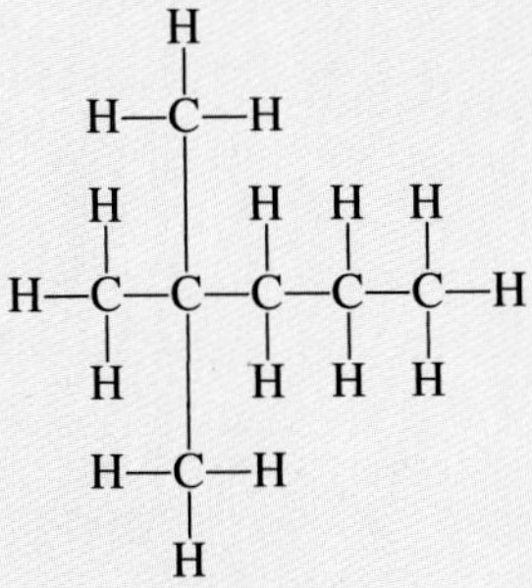

```
           H
           |
        H—C—H
           |
      H    |    H  H  H
      |    |    |  |  |
   H—C—C—C—C—C—H
      |    |    |  |  |
      H    |    H  H  H
           |
        H—C—H
           |
           H
```

Answer

$$\mathrm{CH_3{-}\underset{CH_3}{\underset{|}{\overset{CH_3}{\overset{|}{C}}}}{-}CH_2{-}CH_2{-}CH_3} \quad \text{or} \quad \mathrm{CH_3\underset{CH_3}{\underset{|}{\overset{CH_3}{\overset{|}{C}}}}CH_2CH_2CH_3}$$

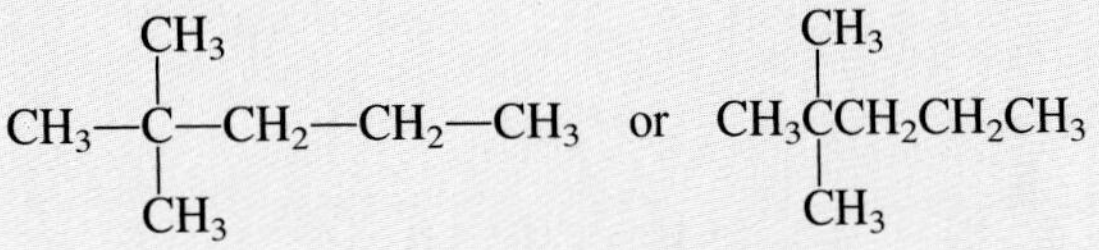

Alkyl Groups

If a hydrogen atom is removed from an alkane, the carbon and hydrogen portion that remains is called an **alkyl group.** Alkyl groups do not exist on their own; they must be attached to something such as a carbon chain. Alkyl groups are named by replacing the *ane* ending present in *alkane* by *yl*. Two of the most common alkyl groups are the methyl group and the ethyl group:

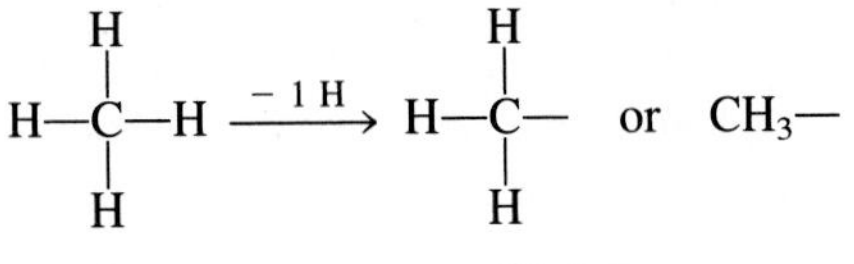

$$\mathrm{H{-}\underset{H}{\underset{|}{\overset{H}{\overset{|}{C}}}}{-}H} \xrightarrow{-1\,\mathrm{H}} \mathrm{H{-}\underset{H}{\underset{|}{\overset{H}{\overset{|}{C}}}}{-}} \quad \text{or} \quad \mathrm{CH_3{-}}$$

Methane — Methyl group

$$\mathrm{H{-}\underset{H}{\underset{|}{\overset{H}{\overset{|}{C}}}}{-}\underset{H}{\underset{|}{\overset{H}{\overset{|}{C}}}}{-}H} \xrightarrow{-1\,\mathrm{H}} \mathrm{H{-}\underset{H}{\underset{|}{\overset{H}{\overset{|}{C}}}}{-}\underset{H}{\underset{|}{\overset{H}{\overset{|}{C}}}}{-}} \quad \text{or} \quad \mathrm{CH_3CH_2{-}}$$

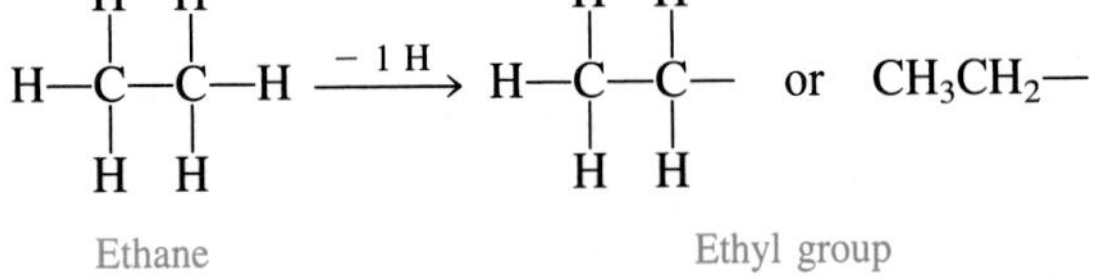

Ethane — Ethyl group

Two alkyl groups are possible from propane, $CH_3CH_2CH_3$. The propyl group attaches to a carbon chain by an end carbon atom, and the isopropyl group attaches at the center atom. (See Figure 10.7.)

Figure 10.7
Models of the alkyl groups derived from methane, ethane, and propane.

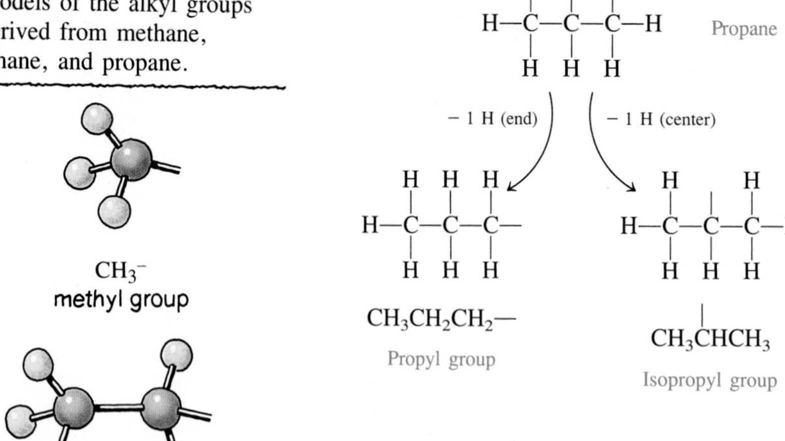

CH_3-
methyl group

CH_3CH_2-
ethyl group

$CH_3CH_2CH_2-$
propyl group

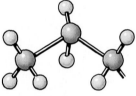

CH_3CHCH_3
isopropyl group

IUPAC Rules for Naming Branched Alkanes

Suppose you wanted to name the following branched-chain alkane:

$$\begin{array}{l} \quad\;\; CH_3 \\ \quad\;\;\, | \\ CH_3-CH-CH_2-CH_2-CH_3 \end{array}$$

We can name this compound using the following IUPAC rules:

Step 1 Find the **parent chain,** which is the longest continuous chain of carbon atoms. Name the parent chain by its alkane name. Since our example has a five-carbon chain, the parent name is *pentane*.

$$\begin{array}{l} \quad\;\; CH_3 \\ \quad\;\;\, | \\ \boxed{CH_3-CH-CH_2-CH_2-CH_3} \end{array}$$

Pentane (from parent chain)

Step 2 Identify each branch, and name hydrocarbon branches as alkyl groups. In this example, a CH_3— is a branch, and it is a methyl group. The name of each branch is placed in front of the name of the parent chain.

$$\begin{array}{l} \quad\;\; \boxed{CH_3} \\ \quad\;\;\, | \\ CH_3-CH-CH_2-CH_2-CH_3 \end{array}$$

Methyl group

Methylpentane

Step 3 Show the location of the alkyl group by assigning numbers to the parent chain, starting from the end nearest a branch. In this example, the parent chain is numbered from left to right, which gives the smallest possible number to the methyl side group. The methyl group is attached to carbon 2 in the chain. If we had numbered incorrectly, from right to left, the carbon attached to the branch would have had a higher number, 4. A hyphen separates the number from the name of the side group; thus, the name of this compound is *2-methylpentane*.

$$\begin{array}{ccccccccc} & & CH_3 & & & & & & \\ & & | & & & & & & \\ CH_3 & - & CH & - & CH_2 & - & CH_2 & - & CH_3 \\ 1 & & 2 & & 3 & & 4 & & 5 \\ 5 & & 4 & & 3 & & 2 & & 1 \end{array}$$

Methyl group

2-Methylpentane

Correct numbering of parent chain

Incorrect numbering gives an incorrect name: 4-methylpentane

It is important to state the location of the alkyl group on the carbon chain here, because there is another methylpentane, which has a methyl group on the third carbon atom:

$$\begin{array}{ccccccccc} & & & & CH_3 & & & & \\ & & & & | & & & & \\ CH_3 & - & CH_2 & - & CH & - & CH_2 & - & CH_3 \\ 1 & & 2 & & 3 & & 4 & & 5 \end{array}$$

3-Methyl group

3-Methylpentane

Step 4 If two or more branches are attached to the parent chain, write the name and location of each in front of the parent name. Remember to number the carbon chain to give the lowest possible numbers to the side groups. When the branches are different, list them in alphabetical order and use hyphens to separate the numbers from the names.

$$\begin{array}{ccccccccccc} & & & & & & \boxed{CH_2CH_3} & & & & \\ & & & & & & | & & & & \\ CH_3 & - & CH_2 & - & CH_2 & - & CH & - & CH & - & CH_3 \\ & & & & & & & & | & & \\ & & & & & & & & \boxed{CH_3} & & \\ 6 & & 5 & & 4 & & 3 & & 2 & & 1 \end{array}$$

Ethyl group

3-Ethyl-2-methylhexane (*Not* 4-ethyl-5-methylhexane)

Methyl group

Correct numbering of this carbon chain goes from right to left to give the lowest numbers to side groups

When branches are identical, a prefix is used to indicate the number of times that branch appears on the carbon chain: *di* (2), *tri* (3), *tetra* (4), and so on. Each of the branches must have a separate number to show its location on the parent chain, even if the branches are attached to the same carbon atom. The numbers for identical branches are separated by commas. The names of the branches, but not the prefixes, determine alphabetical order.

$$\begin{array}{ccccccccc} & & CH_3 & & & & & & \\ & & | & & & & & & \\ CH_3 & - & C & - & CH_2 & - & CH_2 & - & CH_3 \\ & & | & & & & & & \\ & & CH_3 & & & & & & \\ 1 & & 2 & & 3 & & 4 & & 5 \end{array}$$

2,2-Dimethylpentane

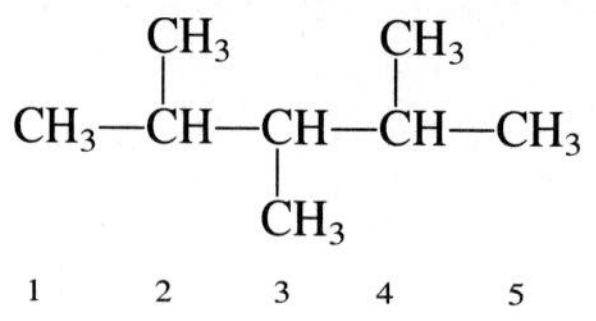

$$\begin{array}{ccccccccc}
 & & CH_3 & & & & CH_3 & & \\
 & & | & & & & | & & \\
CH_3 & - & CH & - & CH & - & CH & - & CH_3 \\
 & & & & | & & & & \\
 & & & & CH_3 & & & & \\
1 & & 2 & & 3 & & 4 & & 5
\end{array}$$

2,3,4-Trimethylpentane

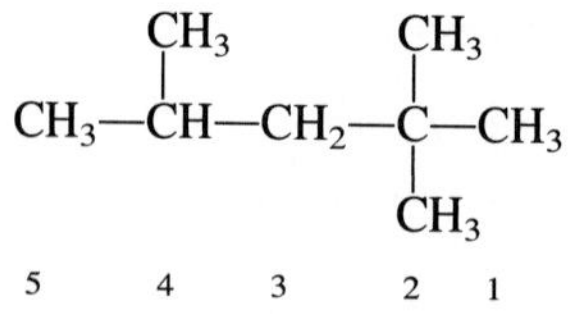

$$\begin{array}{ccccccccc}
 & & CH_3 & & & & CH_3 & & \\
 & & | & & & & | & & \\
CH_3 & - & CH & - & CH_2 & - & C & - & CH_3 \\
 & & & & & & | & & \\
 & & & & & & CH_3 & & \\
5 & & 4 & & 3 & & 2 & & 1
\end{array}$$

2,2,4-Trimethylpentane

Once you start numbering a carbon chain, be sure to continue in that same direction to the other end of the parent chain:

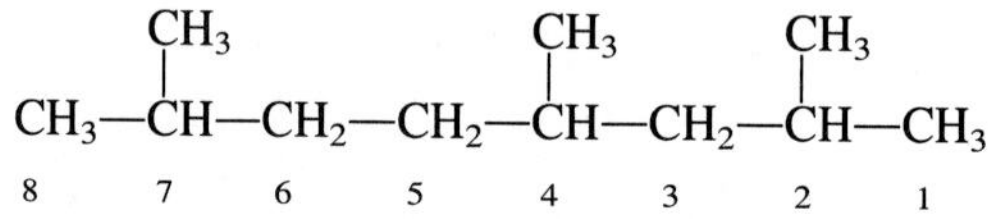

$$\begin{array}{ccccccccccccccc}
 & & CH_3 & & & & & & CH_3 & & & & CH_3 & & \\
 & & | & & & & & & | & & & & | & & \\
CH_3 & - & CH & - & CH_2 & - & CH_2 & - & CH & - & CH_2 & - & CH & - & CH_3 \\
8 & & 7 & & 6 & & 5 & & 4 & & 3 & & 2 & & 1
\end{array}$$

2,4,7-Trimethyloctane

Be sure that you always find the longest continuous carbon chain. Watch out: The longest chain need not be the most obvious horizontal one. Look at the following structure:

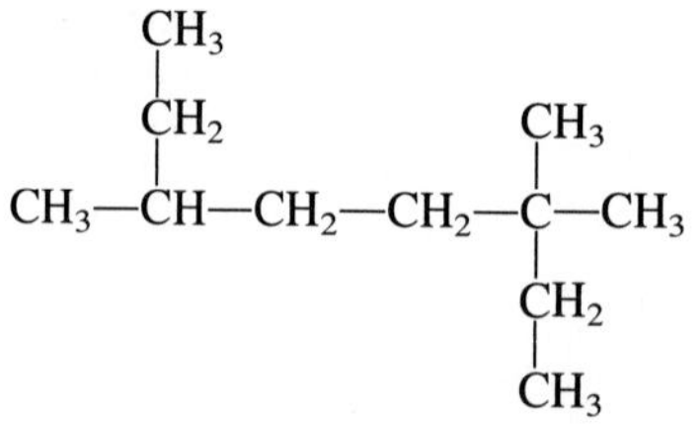

$$\begin{array}{ccccccccccc}
 & & CH_3 & & & & & & & & \\
 & & | & & & & & & & & \\
 & & CH_2 & & & & & & CH_3 & & \\
 & & | & & & & & & | & & \\
CH_3 & - & CH & - & CH_2 & - & CH_2 & - & C & - & CH_3 \\
 & & & & & & & & | & & \\
 & & & & & & & & CH_2 & & \\
 & & & & & & & & | & & \\
 & & & & & & & & CH_3 & &
\end{array}$$

How many carbon atoms are in the parent chain? If you said six, look again. You should be able to find a continuous chain of eight carbon atoms by going around some corners! Numbering the chain will help you find the longest chain. Note that the numbering begins at the end that gives the lowest numbers to the branches.

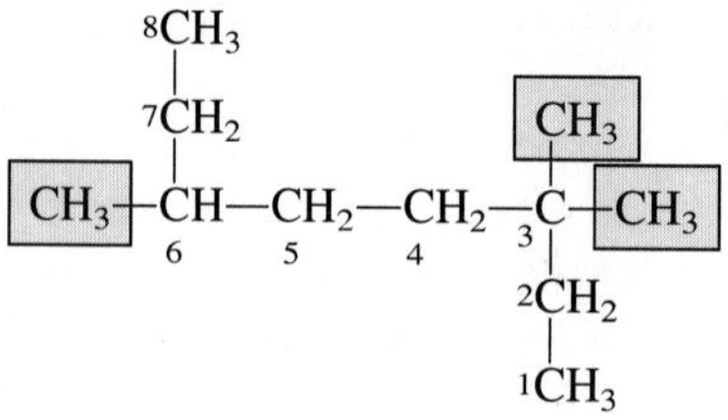

$$\begin{array}{ccccccccccc}
 & & {}^{8}CH_3 & & & & & & & & \\
 & & | & & & & & & & & \\
 & & {}^{7}CH_2 & & & & & & \boxed{CH_3} & & \\
 & & | & & & & & & | & & \\
\boxed{CH_3} & - & CH & - & CH_2 & - & CH_2 & - & C & - & \boxed{CH_3} \\
 & & 6 & & 5 & & 4 & & 3\ | & & \\
 & & & & & & & & {}^{2}CH_2 & & \\
 & & & & & & & & | & & \\
 & & & & & & & & {}^{1}CH_3 & &
\end{array}$$

Branches

3,3,6-Trimethyloctane

Parent chain

Here is a summary of the IUPAC rules for naming branched alkanes:

IUPAC Rules for Naming Branched Alkanes

Step 1 Find the longest continuous chain of carbon atoms and name it as the parent alkane.

Step 2 Determine the names of the alkyl branches.

Step 3 Number the parent chain from the end nearest the branch and indicate the location of each side group on the parent chain.

Step 4 If there are two or more branches, number and name each in alphabetical order by alkyl names. For identical branches, use prefixes of *di, tri, tetra,* and so on to indicate multiples of the same type of branch.

The following may be helpful:

Front Portion	Parent Name	
Side groups are numbered and named alphabetically.	Prefix states the number of carbon atoms in the longest chain.	Ending gives family; an *ane* ending denotes an alkane.

Sample Problem 10.5
Writing IUPAC Names For Branched-Chain Alkanes

Write the IUPAC name for each of the following alkanes:

a. $CH_3—CH_2—CH(CH_2CH_3)—CH_2—CH_3$

b. $CH_3—C(CH_3)(CH_2CH_3)—CH_3$ (vertical chain: CH_3, CH_2, C, CH_3)

c. $CH_3—CH(CH_3)—CH(CH_3)—CH(CH_3)—CH_2—CH_3$

Solution:

a. The parent chain of five carbon atoms is named *pentane*. An ethyl group, $CH_3CH_2—$, is attached at carbon 3 of the parent chain. The name is *3-ethylpentane*.

b. The parent name is *butane,* which is the vertical row of four carbon atoms. Using a prefix of *di* to indicate the two identical methyl branches attached to carbon 2 gives the name *2,2-dimethylbutane*.

c. The parent chain of hexane has three methyl groups attached to carbon atoms 2, 3, and 4; the name is therefore *2,3,4-trimethylhexane* (*not 3,4,5-trimethylhexane*).

Study Check:
Write the IUPAC name for the following branched alkane:

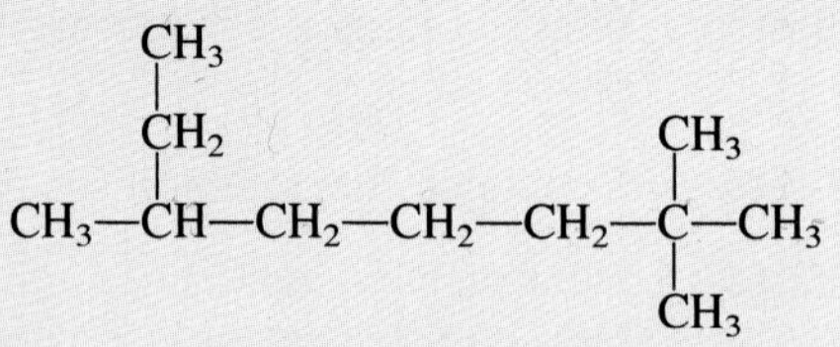

```
        CH3
        |
        CH2                      CH3
        |                        |
CH3—CH—CH2—CH2—CH2—C—CH3
                                 |
                                 CH3
```

Answer: 2,2,6-trimethyloctane

Drawing Structural Formula From the Name

Suppose you are given the name *2,3-dimethylhexane* and asked to write the complete structural formula and the condensed structural formula for that compound. You could proceed as follows:

Step 1 Write the carbon atoms in the parent chain. The *hexane* in the name of the compound tells us that there are six carbon atoms in the parent chain. Write a carbon skeleton of six carbon atoms as the parent chain.

```
—C—C—C—C—C—C—        Hexane
```

Step 2 Number the chain and attach any side groups. The numbers *2,3* in the name indicate that one methyl group, CH_3—, is attached to carbon 2, and the other methyl is attached to carbon 3.

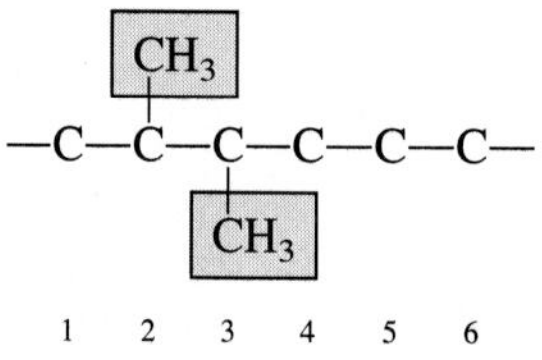

```
     CH3
     |
—C—C—C—C—C—C—        Hexane
        |
        CH3

 1  2  3  4  5  6
```

Step 3 Add hydrogen atoms and write a condensed structural formula.

```
      CH3
   H  |  H  H  H  H                 CH3
   |  |  |  |  |  |                 |
H—C—C—C—C—C—C—H  ⟶  CH3—CH—CH—CH2—CH2—CH3
   |  |  |  |  |  |                      |
   H  H  |  H  H  H                      CH3
         CH3
```

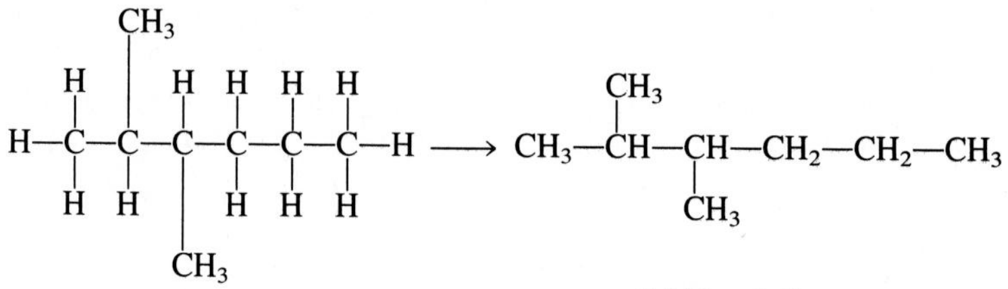

2,3-Dimethylhexane

Sample Problem 10.6
Writing a Condensed Structural Formula From the Name of the Alkane

Write the condensed structural formula for 2,2,4,4-tetramethylpentane.

Solution:

Since the parent name is *pentane,* we first draw a chain of five carbon atoms. Then the four methyl groups, CH_3—, are attached to the parent chain. Two CH_3— groups are attached to carbon 2, and two more are attached to carbon 4.

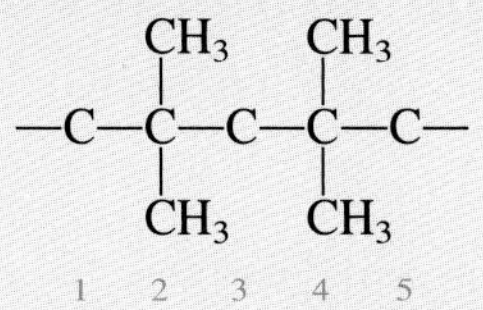

```
       CH3   CH3
        |     |
  —C—C—C—C—C—
        |     |
       CH3   CH3
   1  2  3  4  5
```

The condensed structural formula is written by adding hydrogen atoms to give each carbon atom a total of four bonds.

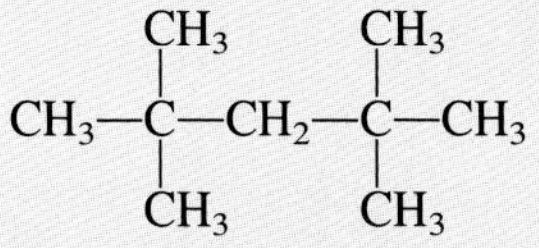

```
       CH3       CH3
        |         |
  CH3—C—CH2—C—CH3
        |         |
       CH3       CH3
```

2,2,4,4-Tetramethylpentane

Study Check:

Write the condensed structural formula of 3-ethylhexane.

Answer

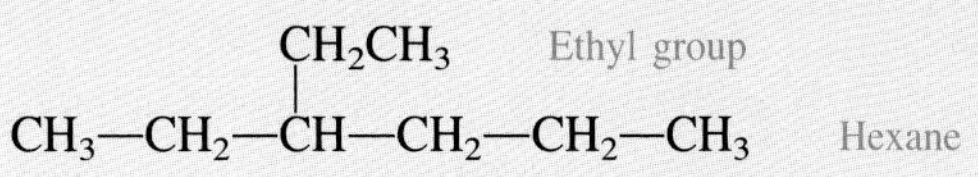

```
              CH2CH3      Ethyl group
               |
  CH3—CH2—CH—CH2—CH2—CH3      Hexane
```

10.5 Structural Isomers

Learning Goal **Given the molecular formula of an alkane, write the condensed structural formulas of its isomers.**

Structural isomers consist of two or more compounds that have the same molecular formula but different arrangements of atoms. For example, as we have seen there are two different compounds with the molecular formula C_4H_{10}. Each has 4 carbon atoms and 10 hydrogen atoms, but one isomer is a straight-chain alkane, and the other isomer is a branched alkane. Although the compounds are composed of the same atoms, their different structures give them different physical and chemical characteristics such as boiling point, solubility, and reactivity. (See Figure 10.8.) A comparison of some of the physical properties of these two isomers is shown in Table 10.6.

Figure 10.8
Models of the structural isomers of C_4H_{10}.

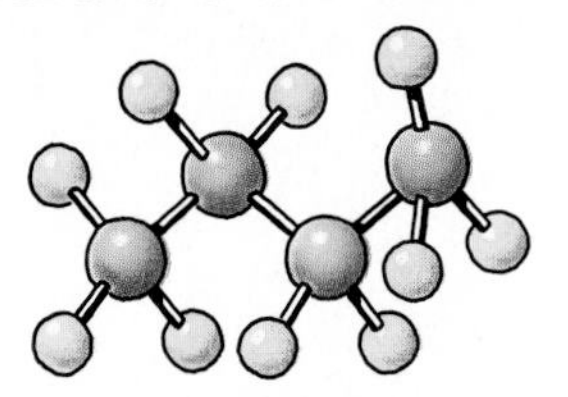

$CH_3CH_2CH_2CH_3$
butane

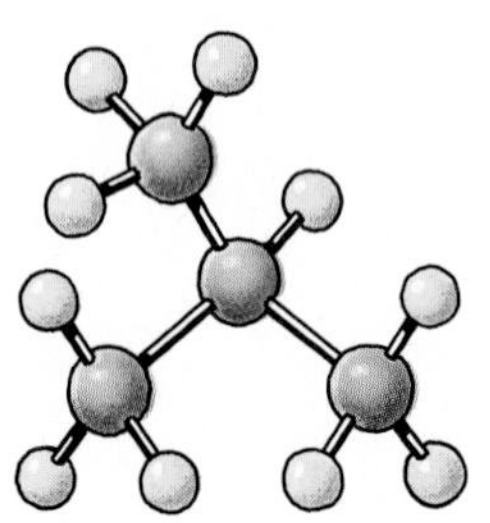

$$\begin{array}{c} CH_3 \\ | \\ CH_3CHCH_3 \end{array}$$
methylpropane

Table 10.6 Some Properties of the Structural Isomers of C_4H_{10}

Property	$CH_3—CH_2—CH_2—CH_3$ **Butane**	$CH_3—\overset{\overset{CH_3}{\vert}}{C}H—CH_3$ **2-Methylpropane**
Molar mass	58.1 g/mole	58.1 g/mole
Melting point	−138°C	−145°C
Boiling point	0°C	−10°C
Density	0.62 g/mL	0.60 g/mL

Drawing Structural Isomers

The molecular formula C_5H_{12} has three structural isomers. One of the isomers is pentane, a straight chain of five carbon atoms:

$$CH_3—CH_2—CH_2—CH_2—CH_3 \quad \text{Pentane}$$

The other isomers of C_5H_{12} are branched alkanes, each having a total of five carbon atoms. One structure can be drawn by attaching a $CH_3—$ branch to a chain of four carbon atoms. Be careful not to attach the methyl group to the ends of the chain, because it would repeat the straight-chain isomer. Placing the $CH_3—$ on carbon 2 gives a different isomer:

$$CH_3—\overset{\overset{CH_3}{|}}{C}H—CH_2—CH_3 \quad \text{2-Methylbutane}$$

If we move the $CH_3—$ to the next carbon in the chain, it gives us the same branched compound, not another isomer.

$$\underset{1}{CH_3}—\underset{2}{\overset{\overset{\boxed{CH_3}}{|}}{C}H}—\underset{3}{CH_2}—\underset{4}{CH_3} \quad \text{is the same compound as} \quad \underset{4}{CH_3}—\underset{3}{CH_2}—\underset{2}{\overset{\overset{\boxed{CH_3}}{|}}{C}H}—\underset{1}{CH_3}$$

2-Methylbutane 2-Methylbutane

The third isomer is drawn by using a parent chain of three carbon atoms and attaching two methyl branches to the center carbon. Any other placement would repeat one of the previous structures.

$$CH_3—\overset{\overset{CH_3}{|}}{\underset{\underset{CH_3}{|}}{C}}—CH_3 \quad \text{2,2-Dimethylpropane}$$

We can check that these compounds are structural isomers by making sure that they have the same molecular formula but different names. This assures us that they are structural isomers and not identical compounds. The three isomers for C_5H_{12} are illustrated in Figure 10.9.

Figure 10.9
Models of the structural isomers of C_5H_{12}.

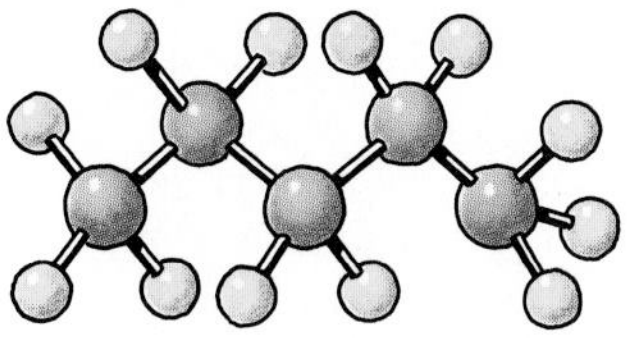

$CH_3CH_2CH_2CH_2CH_3$
pentane

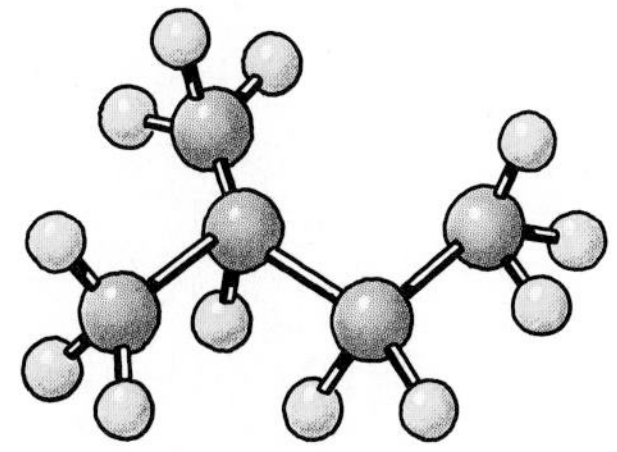

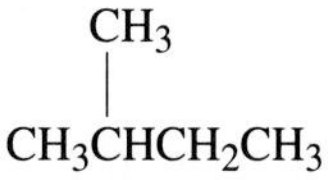

$$\begin{array}{l} \quad\;\, CH_3 \\ \quad\;\;\, | \\ CH_3CHCH_2CH_3 \end{array}$$
methylbutane

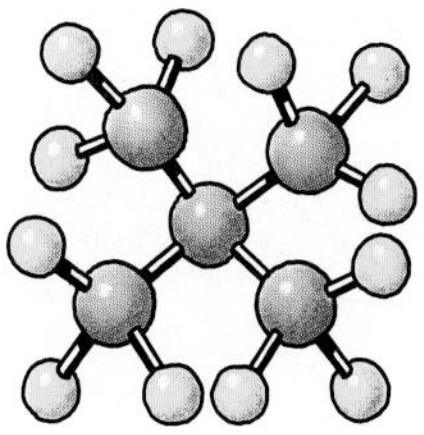

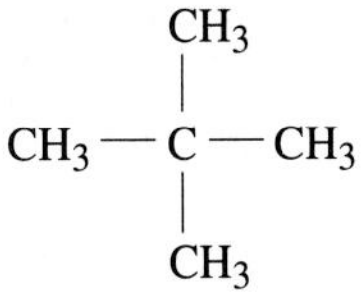

$$\begin{array}{ccccc} & & CH_3 & & \\ & & | & & \\ CH_3 & - & C & - & CH_3 \\ & & | & & \\ & & CH_3 & & \end{array}$$
dimethylpropane

Sample Problem 10.7
Drawing Structural Isomers

There are five isomers that have the molecular formula C_6H_{14}. Draw their condensed structural formulas and give their names.

Solution:
One isomer is the straight-chain alkane with six carbon atoms.

$$CH_3{-}CH_2{-}CH_2{-}CH_2{-}CH_2{-}CH_3 \qquad \text{Hexane}$$

Two isomers can be drawn by attaching a methyl group, CH_3—, to a parent chain of five carbon atoms. In one isomer the CH_3— group is attached to carbon 2, and to carbon 3 in the other.

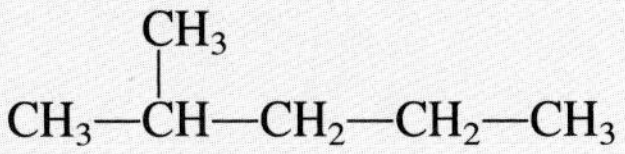

$$\begin{array}{l} \qquad\; CH_3 \\ \qquad\;\; | \\ CH_3{-}CH{-}CH_2{-}CH_2{-}CH_3 \end{array} \qquad \begin{array}{l} \qquad\qquad\quad\; CH_3 \\ \qquad\qquad\quad\;\; | \\ CH_3{-}CH_2{-}CH{-}CH_2{-}CH_3 \end{array}$$

2-Methylpentane 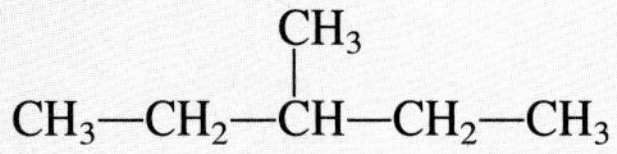 3-Methylpentane

Two more isomers can be drawn by using a parent chain of four carbon atoms and attaching two CH_3— groups to carbon atoms within the chain.

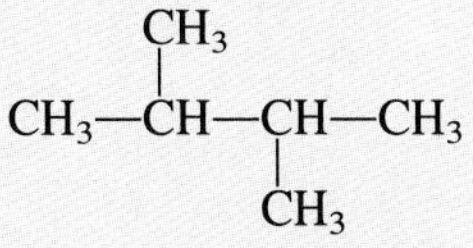

$$\begin{array}{l} \qquad\; CH_3 \\ \qquad\;\; | \\ CH_3{-}CH{-}CH{-}CH_3 \\ \qquad\qquad\;\; | \\ \qquad\qquad\; CH_3 \end{array} \qquad \begin{array}{l} \qquad\; CH_3 \\ \qquad\;\; | \\ CH_3{-}C{-}CH_2{-}CH_3 \\ \qquad\;\; | \\ \qquad\; CH_3 \end{array}$$

2,3-Dimethylbutane 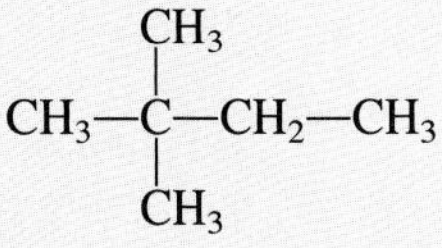 2,2-Dimethylbutane

Any other combination of atoms will give a structure that repeats one of these structural isomers.

Study Check:
Are the following pairs of compounds isomers or identical structures?

a. 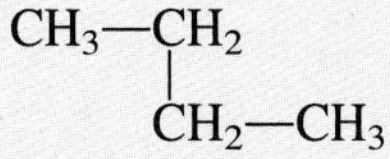
$$\begin{array}{l} CH_3{-}CH_2 \\ \qquad\;\; | \\ \qquad\; CH_2{-}CH_3 \end{array} \quad \text{and} \quad \begin{array}{l} CH_3 \;\; CH_3 \\ \;\; | \qquad | \\ CH_2{-}CH_2 \end{array}$$

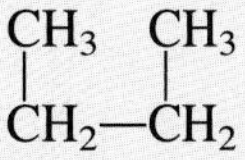

b.
$$\begin{matrix} CH_3 \\ | \\ CH_2-CH_2-CH_3 \end{matrix} \quad \text{and} \quad \begin{matrix} CH_3 \\ | \\ CH-CH_3 \\ | \\ CH_3 \end{matrix}$$

Answer:

a. identical structures (both butane)

b. isomers (butane and 2-methylpropane)

10.6 Cycloalkanes

Learning Goal **Write the name and the structural formula of a cycloalkane.**

If two hydrogen atoms are removed from the ends of an alkane, those carbon atoms can bond together to form a ring structure called a cyclic alkane or a **cycloalkane.** The simplest cycloalkane, cyclopropane, contains three carbon atoms. (See Figure 10.10.)

The structures of cycloalkanes are often simplified to give a geometric shape. No hydrogen atoms are shown, although they are present in the actual compound. For example, the simplified structural formula of cyclopropane is a triangle, as Figure 10.10 illustrates. Each corner of the triangle represents a carbon atom that is attached to two other carbon atoms and two hydrogen atoms.

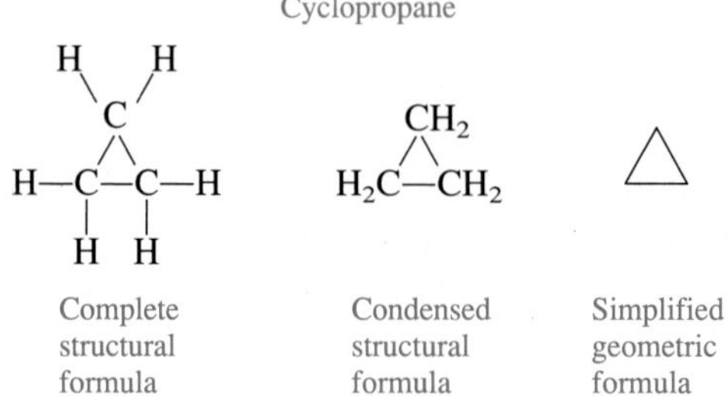

The names of cycloalkanes are derived from the names of straight-chain alkanes by placing the prefix *cyclo* in front of the name of the alkane that has the same number of carbon atoms. Some examples of cycloalkanes are given in Table 10.7.

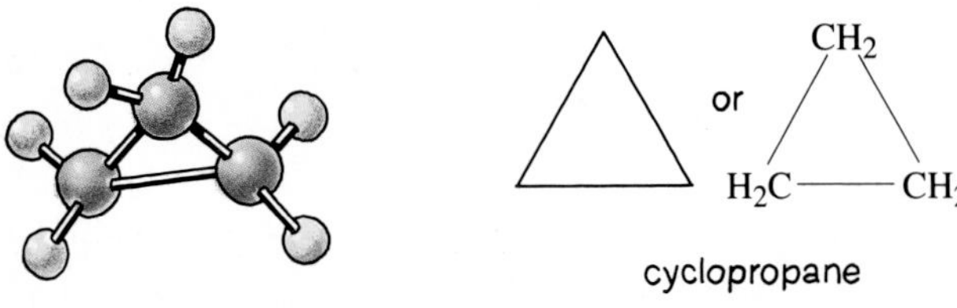

Figure 10.10
A model of the cyclic structure of cyclopropane.

Table 10.7 Cycloalkanes

Cycloalkane	Structural formula	Geometric formula
Cyclopropane	CH_2 / $H_2C—CH_2$ (three-membered ring)	△
Cyclobutane	$H_2C—CH_2$ / $H_2C—CH_2$ (four-membered ring)	□
Cyclopentane	CH_2 / H_2C CH_2 / $H_2C—CH_2$ (five-membered ring)	⬠
Cyclohexane	CH_2 / H_2C CH_2 / H_2C CH_2 / CH_2 (six-membered ring)	⬡

Naming Cycloalkanes That Have Side Groups

Cycloalkanes may contain side groups just as their alkane counterparts do. A hydrocarbon side group on a cycloalkane is named by its alkyl name. When there is only one branch, the ring is not numbered, because all the carbon atoms in the cyclic structure are equal. For example, if a methyl group is attached to any of the carbon atoms of cyclopropane, the identical structural formula of methylcyclopropane results:

Methylcyclopropane

CH_3 = CH_3 = CH_3

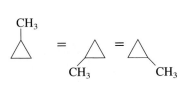

However, if two or more substituents are attached to the cyclic alkane, the ring is numbered in the direction that gives the lowest possible numbers to the side groups. Carbon 1 is assigned to the alkyl group that comes first alphabetically. Here are some examples:

CH_3 CH_3 CH_2CH_3 CH_3 CH_3 CH_3

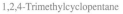

Methylcyclobutane 1-Ethyl-1-methylcyclohexane 1,2,4-Trimethylcyclopentane

Sample Problem 10.8
Naming Cycloalkanes

Give the IUPAC name for the following cycloalkanes:

a. 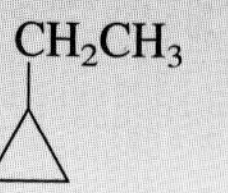 CH_2CH_3 (on a cyclopropane ring)

b. 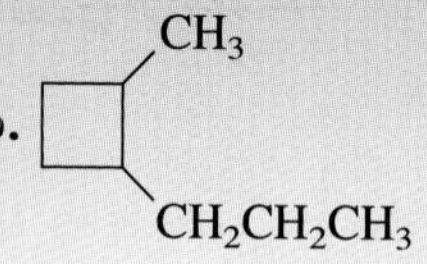 CH_3 and $CH_2CH_2CH_3$ (on a cyclobutane ring)

Solution:

a. The triangle indicates a cyclic alkane that has three carbon atoms, cyclopropane. The side group, —CH_2CH_3, is an ethyl group. The name of this cyclic compound is *ethylcyclopropane*.

b. The square figure represents four carbon atoms in a ring, cyclobutane. The ring is numbered beginning with the methyl group because it comes first alphabetically. Numbering continues around the ring to give the lowest possible numbers for the positions of the side groups, and the name is therefore *1-methyl-2-propylcyclobutane*.

Study Check:
What is the IUPAC name for the following cyclic alkane?

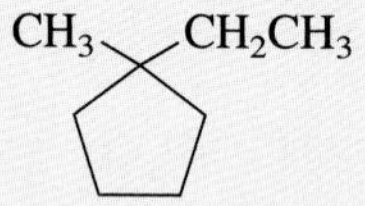

CH_3 CH_2CH_3 (on a cyclopentane ring)

Answer 1-Ethyl-1-methylcyclopentane

Sample Problem 10.9
Writing Structures of Cycloalkanes

Write the structural formula for 1,4-dimethylcyclohexane.

Solution:
The parent name, *cyclohexane,* tells us that the parent chain is a cyclic alkane that has six carbon atoms in a ring. The rest of the name tells us that one methyl group is attached to carbon 1, and another is attached to carbon 4.

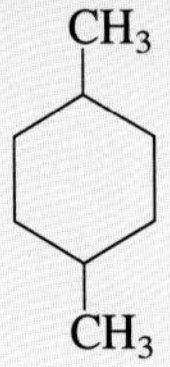

CH_3

CH_3 (on a cyclohexane ring, positions 1 and 4)

10.7 Haloalkanes

Learning Goal **Write the name and the structural formula of a haloalkane.**

When a hydrogen atom of an alkane is replaced with a halogen atom (F, Cl, Br, or I), the resulting compound is called a **haloalkane.** In the IUPAC naming system, the following prefixes are used for the attached halogen atoms:

F fluoro
Cl chloro
Br bromo
I iodo

We name a haloalkane in much the same way as we name branched alkanes. The carbon chain is numbered to give the first halogen atom the lowest number. Then the numbers and names of the halogen atoms and any alkyl groups are listed alphabetically in front of the parent name.

There are some common names that persist in which a simple halogen-containing compound is named as an alkyl halide. In the common name, the alkane portion is named as an alkyl group. Table 10.8 gives both IUPAC and common names (if any) for haloalkanes. (Throughout the discussion of organic chemistry, common names are frequently placed after IUPAC names in parentheses.)

Naming Haloalkanes

Characteristic: halogen atom—F, Cl, Br, I
IUPAC name: haloalkane
Common name: alkyl halide

Table 10.8 Names of Some Haloalkanes

Formula	IUPAC name	Common name
CH_3Cl	Chloromethane	Methyl chloride
$CH_3CH_2CH_2Br$	1-Bromopropane	Propyl bromide
$CH_3CH(F)CH_3$	2-Fluoropropane	Isopropyl fluoride
$CH_3CH(Br)CH_2CH_2Cl$	3-Bromo-1-chlorobutane[a]	
$CH_3CH(CH_3)CH_2C(Cl)_2CH_3$	2,2-Dichloro-4-methylpentane	

[a]Note alphabetical order of side groups.

Sample Problem 10.10
Naming Haloalkanes

Freon 11 and Freon 12 are compounds that have been widely used as refrigerants, coolants in home and car air conditioners, and propellants for aerosol products. They are also known as CFCs or chlorofluorohydrocarbons, a group of compounds known to destroy the ozone in the upper atmosphere. By 2010, many countries including the United States will stop all production of CFCs. What are the IUPAC names for Freon 11 and Freon 12?

```
     Cl            Cl
     |             |
Cl—C—F       F—C—F
     |             |
     Cl            Cl
 Freon 11      Freon 12
```

Solution:
Freon 11, trichlorofluoromethane; Freon 12, dichlorodifluoromethane

Study Check:
Halothane is the commercial name of a haloalkane that is used as an anesthetic. Give the IUPAC name for halothane.

```
    F  Cl
    |  |
F—C—C—H
    |  |
    F  Br
 Halothane
```

Answer: 2-bromo-2-chloro-1,1,1-trifluoroethane

ENVIRONMENTAL NOTE

Pesticides: Chlorinated Cyclic Compounds

Many pesticides are chlorinated cyclic hydrocarbons with commercial names such as aldrin, chlordane, DDT, dieldrin, and lindane. When they are absorbed by the human body, they act as poisons. Symptoms of pesticide poisoning in humans include dizziness, weakness, kidney damage, and nervous system excitability. Respiratory failure, coma, and death may follow.

In the 1940s, DDT was effective in controlling the insects that spread malaria, thypus, and sleeping sickness and was hailed as a miracle. But by the 1950s, many insects had developed resistance to the pesticide, making new pesticides necessary.

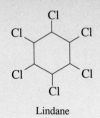

(Structure: cyclohexane ring with one Cl on each carbon)

Lindane

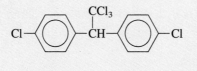

$$Cl-C_6H_4-CH(CCl_3)-C_6H_4-Cl$$

DDT
(dichlorodiphenyltrichloroethane)

Since DDT is a nonpolar compound, it is soluble in nonpolar substances. In lakes and ponds where spraying of DDT occurred, small amounts of the pesticide were ingested and stored in the nonpolar lipids (fats) of microorganisms. Since these microorganisms serve as food for larger organisms and small fish, DDT became more concentrated in the body fat of larger animals as they fed upon these smaller creatures. Predatory birds and animals high in the food chain accumulated the greatest amounts of DDT. Eventually, elevated DDT levels affected their reproduction and contributed to the decline and near extinction of some species of hawks, eagles, falcons, and brown pelicans. Many of the chlorinated hydrocarbons, including DDT, have now been banned from use in widespread spraying.

DDT is also stored in human adipose tissue (body fat) and has been found in human breast milk. ■

10.8 Reactions of Alkanes

Learning Goal **Write chemical equations for the combustion and the halogenation of alkanes.**

The single covalent bonds of alkanes are difficult to break, a feature that makes alkanes unreactive with most chemical substances. However, alkanes do react with oxygen and the halogens.

Combustion

At high temperatures, alkanes burn in oxygen to produce carbon dioxide, water, and heat energy. The methane gas we use to cook our foods and heat our homes, the gasoline that powers our cars, and the fossil fuels used to produce electricity in power plants are all alkanes. Methane makes up 97% of natural gas, the fuel used in gas appliances and in many laboratories. Liquefied propane gas (LPG) is used as a fuel for camping equipment and in homes without direct gas lines.

The reaction of hydrocarbons with oxygen is called **combustion** and the products are CO_2 and H_2O. A great amount of heat energy is also released. We can write the following equations for the reactions of methane and propane with oxygen:

Combustion

$$\text{hydrocarbon} + O_2 \longrightarrow CO_2 + H_2O$$
$$CH_4(g) + 2O_2(g) \longrightarrow CO_2(g) + 2H_2O(g)$$
$$C_3H_8(g) + 5O_2(g) \longrightarrow 3CO_2(g) + 4H_2O(g)$$

Sample Problem 10.11
Writing an Equation for Combustion

Gasoline is a mixture of liquid alkanes, including pentane, C_5H_{12}. Write and balance the equation for the combustion of pentane in the engine of a car.

Solution:
The reactants of this combustion reaction are pentane and oxygen. The products of complete combustion are carbon dioxide and water.

$$C_5H_{12} + 8O_2 \longrightarrow 5CO_2 + 6H_2O$$

Study Check:
Write a balanced equation for the complete combustion of ethane, C_2H_6.

Answer:

$$2C_2H_6 + 7O_2 \longrightarrow 4CO_2 + 6H_2O$$

HEALTH NOTE

Combustion of Glucose in the Body

In the cells of the body, glucose acts as a fuel, reacting with the oxygen we take in and producing heat and energy. Although the metabolic combustion of glucose occurs in many small steps, its end products are carbon dioxide, water, and energy—the same as in the combustion reactions we just looked at.

$$\underset{\text{Glucose (from foods)}}{C_6H_{12}O_6} + \underset{\text{Oxygen (from air)}}{6O_2} \longrightarrow \underset{\text{Carbon dioxide (exhaled)}}{6CO_2} + \underset{\text{Water}}{6H_2O}$$

You may already know that it is dangerous to use gas appliances in a closed room or to run a car in a closed garage where ventilation is not adequate. A limited supply of oxygen causes an incomplete combustion of the fuel, resulting in the production of carbon monoxide and water:

$$\underset{\text{Methane}}{2CH_4} + \underset{\text{Oxygen (limited)}}{3O_2} \longrightarrow \underset{\text{Carbon monoxide}}{2CO} + 4H_2O$$

Carbon monoxide is dangerous when it is inhaled because it passes into the bloodstream where it attaches to hemoglobin. Normally, the hemoglobin carries oxygen from the lungs to the tissues of the body and brings carbon dioxide (CO_2) back to the lungs, where it is released. However, carbon monoxide (CO) binds so tightly to hemoglobin that it is not easily released. The sites where oxygen would bind to hemoglobin are not available. The hemoglobin is unable to transport oxygen, causing oxygen starvation and death if the victim is not treated immediately. ■

Halogenation of Alkanes

If an alkane and a halogen are mixed together in the dark, no reaction occurs. However, if the mixture is heated or brought into the light, a hydrogen atom can be replaced with an atom of fluorine, chlorine, and bromine. A reaction in which one atom replaces another is called a substitution reaction; replacement by a halogen is called a **halogenation** reaction.

Halogenation of Alkanes

$$\text{Alkane} + \text{Halogen} \xrightarrow[\text{or heat}]{\text{Light}} \text{Haloalkane} + \text{Hydrogen halide}$$

$$\begin{array}{c} \text{H} \\ | \\ \text{H—C—H} \\ | \\ \text{H} \end{array} + \text{Cl—Cl} \xrightarrow[\text{or heat}]{\text{Light}} \begin{array}{c} \text{H} \\ | \\ \text{H—C—Cl} \\ | \\ \text{H} \end{array} + \text{H—Cl}$$

$$CH_4 + Cl_2 \xrightarrow[\text{or heat}]{\text{Light}} CH_3Cl + HCl$$

Methane, Chlorine, Chloromethane (methyl chloride), Hydrogen chloride

Halogenation does not need to stop with the substitution of one hydrogen atom. The other hydrogen atoms may also be replaced with halogen atoms to give a mixture of halogenated compounds:

$$CH_3Cl + Cl_2 \xrightarrow[\text{or heat}]{\text{Light}} CH_2Cl_2 + HCl$$

Dichloromethane

$$CH_2Cl_2 + Cl_2 \xrightarrow[\text{or heat}]{\text{Light}} CHCl_3 + HCl$$

Trichloromethane (chloroform)

$$CHCl_3 + Cl_2 \xrightarrow[\text{or heat}]{\text{Light}} CCl_4 + HCl$$

Tetrachloromethane (carbon tetrachloride)

Sample Problem 10.12
Writing Equations for the Halogenation of an Alkane

Write the structural formula of the product formed when one hydrogen atom is replaced by a halogen atom in the following reactions:

a. $CH_3CH_3 + Br_2 \xrightarrow{\text{Light}}$

b. $CH_3CH_2CH_3 + Cl_2 \xrightarrow{\text{Heat}}$

c. $CH_4 + Br_2 \xrightarrow{\text{Dark}}$

Solution:

a. $CH_3CH_3 + Br_2 \xrightarrow{\text{Light}} CH_3CH_2Br + HBr$

b. Propane reacts with chlorine when heated to produce two isomers. In one isomer, a hydrogen atom on an end carbon is replaced by a chlorine atom. In the other isomer, the chlorine atom substitutes for a hydrogen atom on the center carbon. The two isomers are produced in roughly equal amounts.

$$2CH_3CH_2CH_3 + 2Cl_2 \xrightarrow{\text{Light}} \underset{\text{1-Chloropropane}}{CH_3CH_2CH_2Cl} + \underset{\text{2-Chloropropane}}{CH_3\overset{\overset{\displaystyle Cl}{|}}{C}HCH_3} + 2HCl$$

c. No reaction occurs in the dark.

Table 10.9 gives an overview of naming alkanes, and Table 10.10 shows the primary reactions of alkanes.

Table 10.9 Summary of Naming Alkanes

Type	Example	Characteristic	Structure
Alkane	Propane	All single bonds	$CH_3—CH_2—CH_3$
Cycloalkane	Cyclopropane	Cyclic structure with single bonds	△
Haloalkane	2-Bromopropane (Isopropyl bromide)	Hydrogen atom replaced by a halogen atom	$CH_3—\overset{\overset{\displaystyle Br}{|}}{C}H—CH_3$

Table 10.10 Summary of Reactions of Alkanes

Combustion

$$\text{Alkane} + O_2 \longrightarrow CO_2 + H_2O$$

$$CH_4 + 2O_2 \longrightarrow CO_2 + 2H_2O$$

Halogenation (Substitution)

$$\text{Alkane} + \text{Halogen} \xrightarrow[\text{or heat}]{\text{Light}} \text{Haloalkane} + \text{Hydrogen halide}$$

$$CH_4 + Cl_2 \xrightarrow[\text{or heat}]{\text{Light}} CH_3Cl + HCl$$

Glossary of Key Terms

alkane A hydrocarbon that has only single carbon-to-carbon bonds.

alkyl group A side group derived from an alkane by removing one hydrogen atom. The ending of the alkane name is changed to *yl* to name it as an alkyl branch.

branch A side group of carbon or a halogen attached to a parent carbon chain.

branched-chain alkane A hydrocarbon containing at least one side group attached to a parent chain.

combustion A chemical reaction in which an alkane reacts with oxygen to produce CO_2, H_2O, and heat.

complete structural formula A type of formula in which each bond in the compound is shown as a line, C—H or C—C, to give the arrangement of all the atoms in the compound.

condensed structural formula A type of formula that illustrates the arrangement of the carbon atoms in an organic compound by grouping each carbon atom and its hydrogen atoms.

cycloalkane An alkane that exists as a ring or cyclic structure.

haloalkane A derivative of an alkane in which one or more hydrogen atoms have been replaced by a halogen atom.

halogenation A substitution reaction that occurs in the presence of light or heat in which a halogen atom replaces a hydrogen atom in an alkane.

hydrocarbon An organic compound consisting of only carbon and hydrogen atoms.

organic compound Compounds that contain carbon in covalent bonds. Most are nonpolar with low melting and boiling points, insoluble in water, soluble in nonpolar solvents, and flammable.

parent chain The longest continuous chain of carbon atoms in an organic compound.

saturated hydrocarbon An alkane having only single bonds.

straight-chain alkane An alkane consisting of carbon atoms that follow each other in a continuous chain.

structural isomer An organic compound that has the same molecular formula as another organic compound but a different arrangement of atoms.

substitution A reaction in which a hydrogen atom in an organic compound is replaced by another type of atom, usually a halogen.

Problems

Properties of Organic Compounds *(Goal 10.1)*

10.1 Indicate whether each of the following describes an inorganic or an organic compound:

a. a compound that melts at 800°C and conducts electricity when it dissolves in water
b. a liquid that dissolves in cyclohexane
c. a compound that vaporizes at 35°C and burns in oxygen
d. vitamin C, ascorbic acid, $C_6H_8O_6$
e. aluminum hydroxide, $Al(OH)_3$, an antacid
f. aspirin, $C_9H_8O_4$, an analgesic

Structural Formulas *(Goal 10.2)*

10.2 Write the complete structural formulas of the following alkanes by adding the correct number of hydrogen atoms:

a. C—C—C—C—C

b. 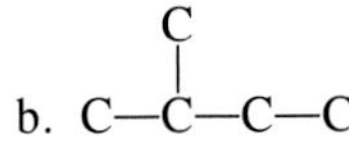

```
        C                         C  C
        |                         |  |
c.  C—C—C             d.  C—C—C—C—C
        |                         |
        C                         C
```

10.3 Write the condensed structural formula for each of the following:

```
                                            H
                                            |
                                          H—C—H
                                            |
     H  H  H  H  H                  H  H    |   H  H
     |  |  |  |  |                  |  |    |   |  |
a. H—C——C——C——C——C—H          c. H—C——C——C——C——C—H
     |  |  |  |  |                  |  |    |   |  |
     H  H  H  H  H                  H  |    H   H  H
                                       |
                                     H—C—H
                                       |
                                       H

     H  H  H                        H  H  H  H  H  H
     |  |  |                        |  |  |  |  |  |
b. H—C——C——C—H                d. H—C——C——C——C——C——C—H
     |  |  |                        |  |  |  |  |  |
     H  H  H                        H  |  H  H  H  H
                                       |
                                     H—C—H
                                       |
                                       H
```

10.4 Write the complete structural formula for each of the following:

a. $CH_3—CH_3$

b. $CH_3—CH_2—CH_2—CH_2—CH_3$

```
              CH3
              |
c.  CH3—CH—CH—CH3
                   |
                   CH3

              CH3
              |
d.  CH3—CH—CH2—CH—CH2—CH2—CH3
                         |
                         CH2—CH3
```

IUPAC Names for Alkanes *(Goal 10.3)*

10.5 Write the IUPAC name for the following alkanes:

a. $CH_3—CH_2—CH_2—CH_3$

b. $CH_3—CH_3$

```
     CH3
     |
c.  CH2—CH2—CH2—CH2
                  |
                  CH2—CH3
```

d. CH_4

10.6 a. What is meant by a straight-chain compound?

b. Do the carbon atoms in a straight-chain alkane actually have a linear structure?

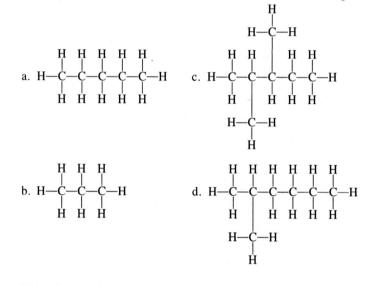

10.7 Write the condensed formula of each of the following:

a. propane
b. heptane
c. pentane
d. octane

IUPAC Rules for Naming Branched-Chain Alkanes *(Goal 10.4)*

10.8 a. How does a straight-chain alkane differ from a branched-chain alkane?
b. Identify the following as straight- or branched-chain alkanes:

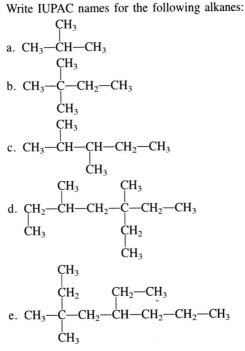

(1) 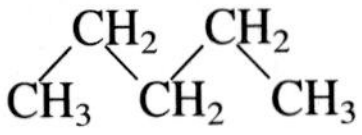 CH_2 and CH_2 (upper row); CH_3, CH_2 and CH_3 (lower row), joined in a zigzag chain

(2) $CH_3—CH_2—CH—CH_2—CH_3$, with $CH_2—CH_3$ on the CH

(3) $CH_2—CH_2—CH_2—CH_2$, with CH_3 above the first CH_2 and $CH_2—CH_3$ below the last CH_2

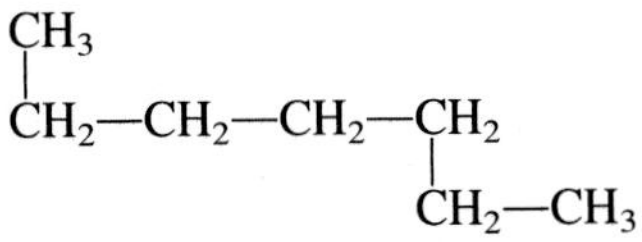

(4) 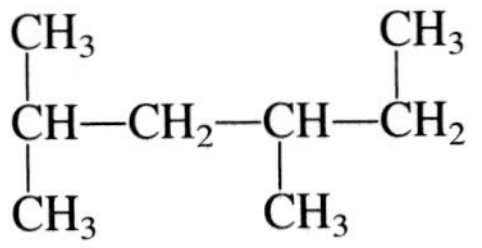 $CH—CH_2—CH—CH_2$, with CH_3 above and CH_3 below the first CH, CH_3 below the second CH, and CH_3 above the last CH_2

(5) 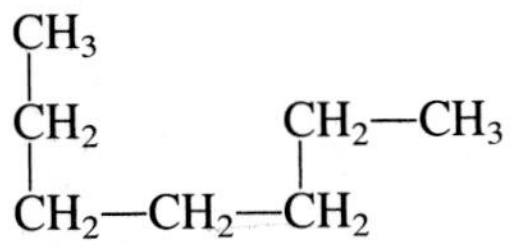 $CH_2—CH_2—CH_2$, with CH_2 and then CH_3 stacked above the first CH_2, and $CH_2—CH_3$ above the last CH_2

10.9 Write IUPAC names for the following alkanes:

a. 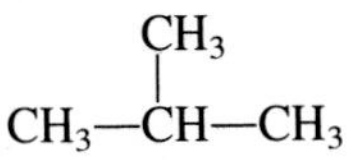 $CH_3—CH—CH_3$, with CH_3 on the CH

b. 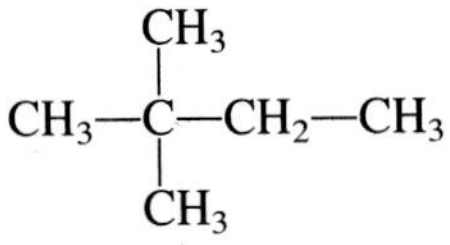 $CH_3—C—CH_2—CH_3$, with CH_3 above and CH_3 below the C

c. $CH_3—CH—CH—CH_2—CH_3$, with CH_3 above the first CH and CH_3 below the second CH

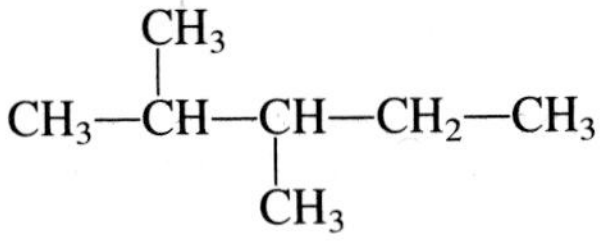

d. 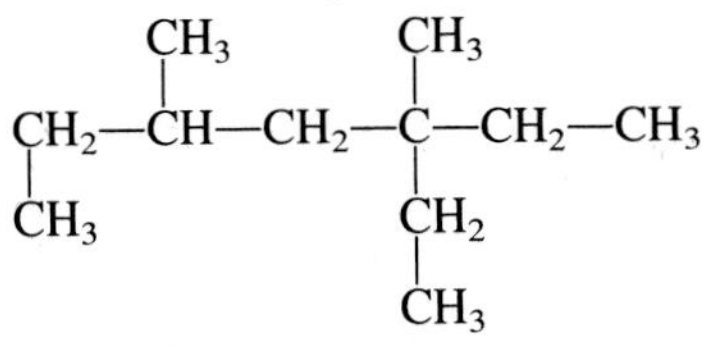 $CH_2—CH—CH_2—C—CH_2—CH_3$, with CH_3 below the first CH_2, CH_3 above the CH, CH_3 above the C, and CH_2 and then CH_3 below the C

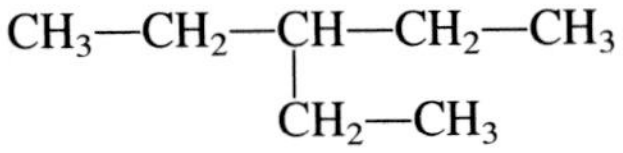

e. 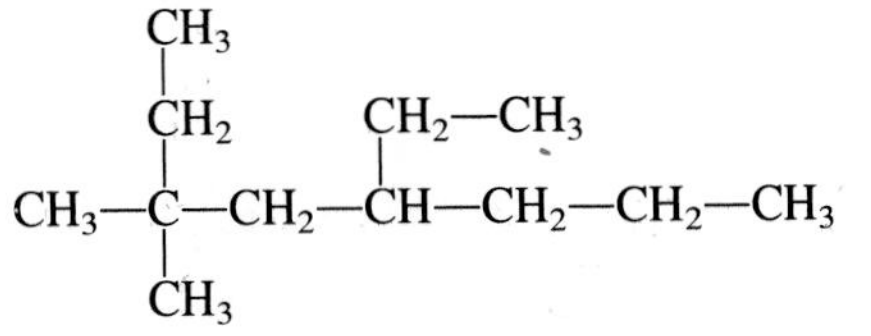 $CH_3—C—CH_2—CH—CH_2—CH_2—CH_3$, with CH_2 and then CH_3 stacked above the C, CH_3 below the C, and $CH_2—CH_3$ above the CH

```
   CH3
   |
   CH—CH2—CH3
   |
   CH2
   |
f. CH2—CH—CH2—CH2—CH—CH3
       |              |
       CH2            CH3
       |
       CH3
```

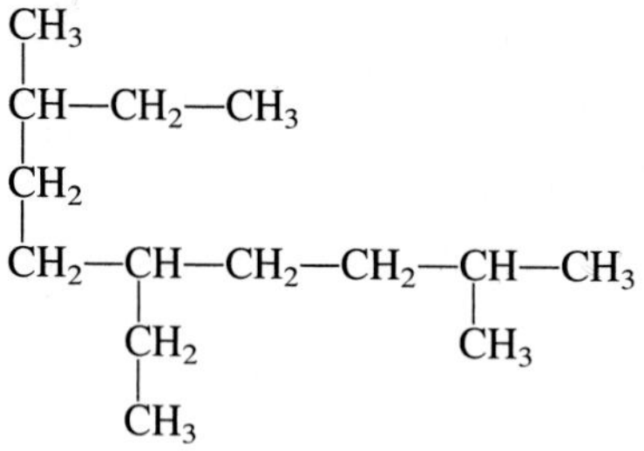

10.10 Write the condensed structural formulas for the following alkanes:

a. 2-methylpentane
b. 3,3-dimethylhexane
c. 2,3-dimethylbutane
d. 3-ethyl-2-methylhexane
e. 2-methyl-4-isopropylheptane
f. 2,2,4,4-tetramethylpentane

Structural Isomers *(Goal 10.5)*

10.11 Write all of the condensed structural formulas and IUPAC names for the three isomers of C_5H_{12}.

10.12 Write the condensed structural formulas and names of all the isomers of heptane that have a five-carbon parent chain.

10.13 Indicate whether the following structural formulas are isomers or identical:

```
                                     CH3
                                     |
a. CH3—CH2—CH2—CH3             CH3—CH—CH3

   CH3           CH3                 CH3
   |             |                   |
b. CH2—CH2—CH2—CH2             CH3—C—CH2—CH3
                                     |
                                     CH3

                         CH3
                         |
c. CH3—CH2—CH3           CH2—CH3

       CH3                         CH3
       |                           |
d. CH3—CH—CH2              CH3—CH2—CH—CH3
           |
           CH3
```

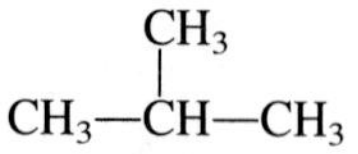

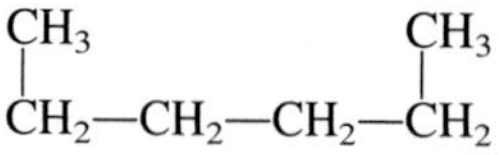

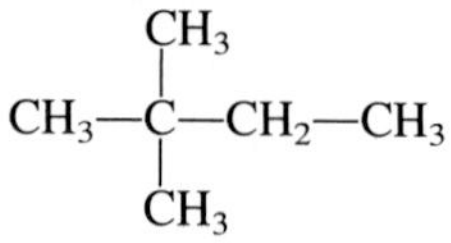

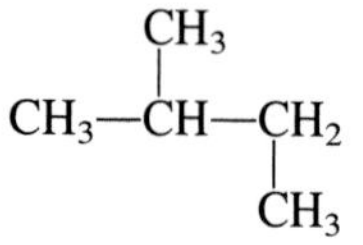

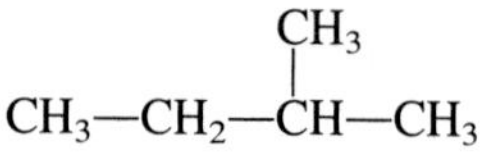

Cycloalkanes *(Goal 10.6)*

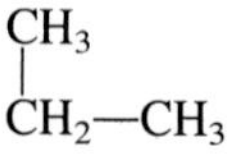

10.14 Give the IUPAC name for the following cycloalkanes:

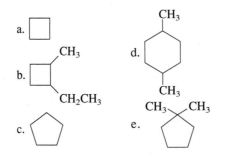

a.

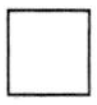

b. 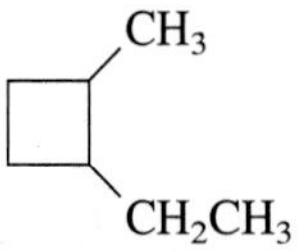 CH_3, CH_2CH_3

c.

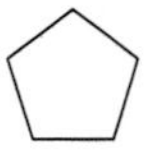

d. 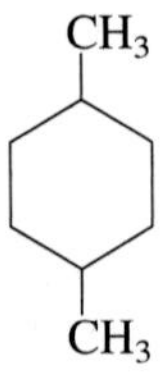 CH_3, CH_3

e. 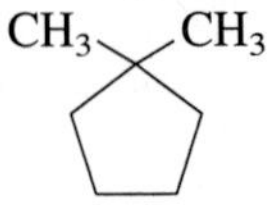 CH_3, CH_3

10.15 Draw the structural formula for each of the following:
a. cyclopentane
b. methylcyclohexane
c. ethylcyclobutane
d. 1,2-dimethylcyclobutane
e. 1,3,5-trimethylcyclohexane

10.16 Draw and name all the isomers of dimethylcyclopentane.

10.17 Draw and name all the cyclic isomers of C_5H_{10}.

Haloalkanes *(Goal 10.7)*

10.18 Write the IUPAC names for the following haloalkanes:

a. $CH_3—CH_2—CH_2—Br$

b. 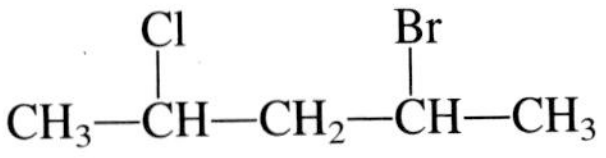 $CH_3—CH(Cl)—CH_2—CH(Br)—CH_3$

c. 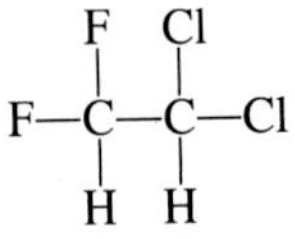 $F—C(F)(H)—C(Cl)(H)—Cl$

d. 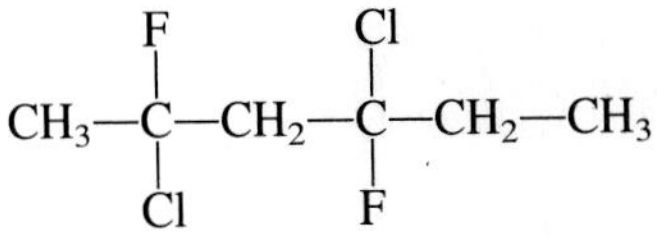 $CH_3—C(F)(Cl)—CH_2—C(Cl)(F)—CH_2—CH_3$

10.19 Chloromethane (methyl chloride) and chloroethane (ethyl chloride) are used as topical anesthetics. When sprayed on the skin, they produce local anesthesia by lowering the temperature in the treated area below 0°C to freeze the nerve endings. This makes local surgery near the surface of the skin painless. What are the structures of chloromethane and chloroethane?

10.20 Write the condensed structural formula for the following:
a. 2-chloropropane
b. 1-bromo-2-chloropropane
c. 1-chloro-1-fluorocyclopropane
d. 2,4-dichloro-3-methylhexane
e. 3,3-dibromo-2-methylpentane
f. 1,1,3,5-tetrachlorocyclohexane

Reactions of Alkanes *(Goal 10.8)*

10.21 Write balanced equations for the complete combustion of the following:
a. methane, CH_4
b. propane, C_3H_8
c. butane, C_4H_{10}
d. cyclobutane, C_4H_8
e. octane, C_8H_{18}

10.22 For the following reactions, write the structural formulas of the products that have one halogen atom:

a. $CH_3CH_3 + Br_2 \xrightarrow{\text{Light}}$

b. $CH_3CH_2CH_3 + Cl_2 \xrightarrow{\text{Light}}$

c. $CH_4 + Cl_2 \xrightarrow{\text{Dark}}$

d. $CH_4 + Br_2 \xrightarrow{\text{Heat}}$

10.23 For the following cycloalkane reactions, draw the structural formulas of the products (if any) having one halogen atom:

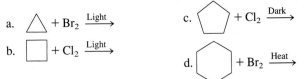

a. △ $+ Br_2 \xrightarrow{\text{Light}}$

b. 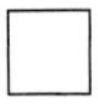 □ $+ Cl_2 \xrightarrow{\text{Light}}$

c. 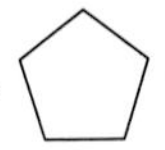 ⬠ $+ Cl_2 \xrightarrow{\text{Dark}}$

d. 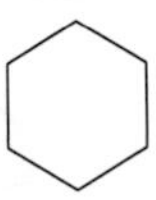 ⬡ $+ Br_2 \xrightarrow{\text{Heat}}$

$-(CH_2-\underset{\displaystyle Cl}{\underset{|}{C}H})_n-$

polyvinyl chloride (PVC)

11

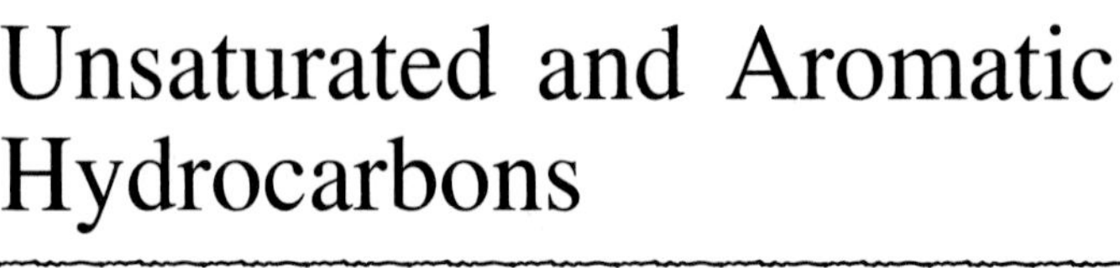

Unsaturated and Aromatic Hydrocarbons

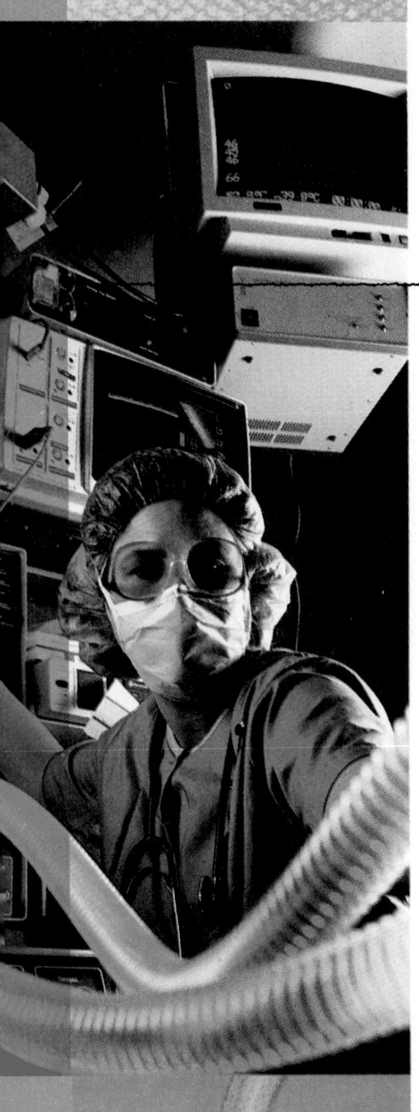

Plastic containers and IV tubing are frequently made of polyethylene which, being relatively inert, will not react with chemicals that are stored in them.

Learning Goals

11.1 Write the name and the structural formula of an alkene.

11.2 Write the structural formulas of an alkene that has cis–trans geometric isomers.

11.3 Write the name and the structural formula of an alkyne.

11.4 Write equations for the addition reactions of alkenes and alkynes.

11.5 Write the name and the structural formula for an aromatic compound.

11.6 Write the products of the substitution reactions of benzene.

SCOPE

In the preceding chapter, we looked at the names, structures, and reactions of saturated hydrocarbons, organic compounds that have only single bonds. In this chapter, we will investigate hydrocarbons that are unsaturated, such as alkenes, which have double bonds, and alkynes, which have triple bonds. We will also look at aromatic compounds, a family of hydrocarbons that contain benzene.

We saw that the single bonds in the alkanes were not very reactive. In contrast, unsaturated hydrocarbons are highly reactive because double and triple bonds are broken more easily than single bonds are. As a result, different atoms or groups of atoms can be added to alkenes and alkynes to prepare several other kinds of organic compounds.

The aromatic compounds of benzene are especially stable because electrons are shared in a six-carbon ring. Benzene and its related compounds behave much like saturated hydrocarbons undergoing substitution reactions rather than like unsaturated compounds undergoing addition reactions.

11.1 Alkenes

Learning Goal **Write the name and the structural formula of an alkene.**

Alkenes are a family of hydrocarbons that contain a double bond. As these new families of organic compounds are introduced, you will find that there is a characteristic group that sets each family apart from the others. This **functional group** is a special feature in the molecule that determines the name and chemical reactivity of the compounds in that family. In the simplest alkene, ethene (or ethylene), C_2H_4, a double bond forms between two carbon atoms when they share four valence electrons. (See Figure 11.1.) An alkene is also called an **unsaturated hydrocarbon.** For example, ethene contains fewer hydrogen atoms than its corresponding saturated alkane, ethane, C_2H_6. The double bond of an alkene molecule has a flat geometry which causes the carbon and hydrogen atoms in the double bond to lie in the same plane.

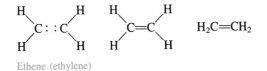

Figure 11.1
A model of the unsaturated hydrocarbon ethene (ethylene), C_2H_4.

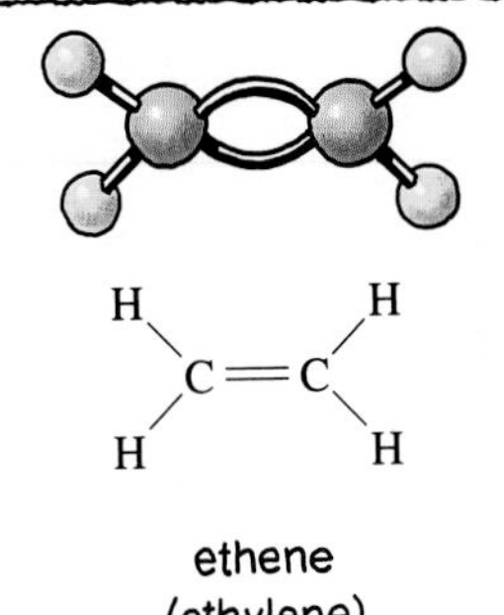

$$H_2C{:}{:}CH_2 \quad (\text{H}_2\text{C}=\text{CH}_2) \quad H_2C=CH_2$$

Ethene (ethylene)

Naming Alkenes

Alkenes are named by using prefixes to represent the number of carbon atoms and the ending *ene* to indicate the double bond. When there is a common name, the ending *ylene* is used.

Naming Alkenes

Functional group: $\rangle C=C \langle$

IUPAC name: Alkene
Common name: Alkylene

Table 11.1 lists some examples of alkenes and their names.

Table 11.1 Naming Some Alkenes

Structural formula	IUPAC name	Common name
$H_2C{=}CH_2$	Ethene	Ethylene
$CH_3{-}CH{=}CH_2$	Propene	Propylene
$CH_3{-}C(CH_3){=}CH_2$	2-Methylpropene	Isobutylene

For higher alkenes, the following steps may be used to give the IUPAC names:

Step 1 Number the carbon chain to give the lowest possible number to the double bond. For example, two butenes can be written for C_4H_8. One has a double bond beginning at carbon 1, and the other has a double bond beginning at carbon 2.

$$\underset{1}{CH_2}{=}\underset{2}{CH}{-}\underset{3}{CH_2}{-}\underset{4}{CH_3} \qquad \underset{1}{CH_3}{-}\underset{2}{CH}{=}\underset{3}{CH}{-}\underset{4}{CH_3}$$

1-Butene   2-Butene

Be sure that you use the longest carbon chain containing the double bond for the parent name.

$$CH_3{-}CH_2{-}\underset{2}{C}(\overset{1}{=}CH_2){-}\underset{3}{CH_2}{-}\underset{4}{CH_3}$$

2-Ethyl-1-butene

The parent chain containing the double bond has four carbon atoms, not five

If the alkene has a ring structure, place the prefix *cyclo* in front of the alkene name.

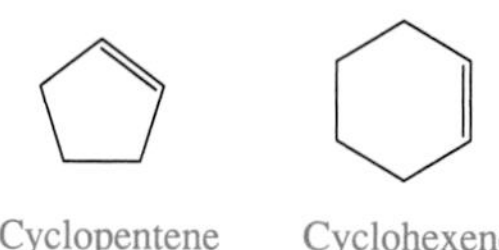

Cyclopentene   Cyclohexene

Step 2 Place the names and numbers of the side groups in front of the alkene name. When numbering side groups, be sure that the parent chain is numbered from the end that is closest to the double bond.

$$\begin{array}{c} CH_3 \\ | \\ CH_3-CH=CH-CH-CH_3 \\ 1 \quad\quad 2 \quad\quad 3 \quad\quad 4 \quad\quad 5 \end{array}$$

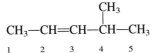

4-Methyl-2-pentene (*not* 2-methyl-3-pentene)

When the chain is numbered from the end nearest the double bond, the double bond begins at carbon 2

When naming a cycloalkene that has a side group, the double bond is numbered as carbon 1 and 2 in the direction around the ring that gives the lowest numbers to the first branch.

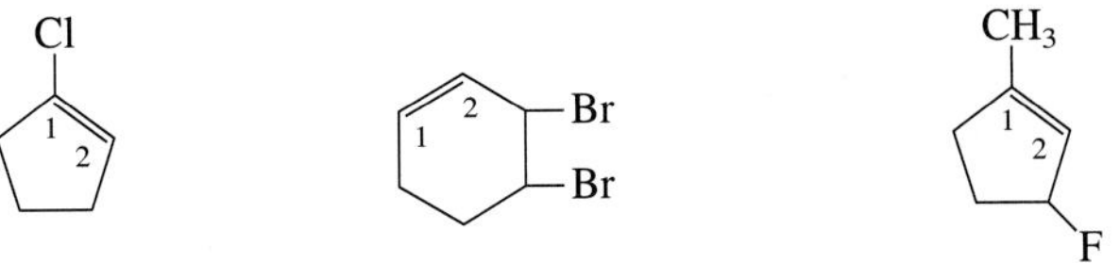

1-Chlorocyclopentene 3,4-Dibromocyclohexene 3-Fluoro-1-methylcyclopentene

Sample Problem 11.1
Naming Alkenes

Write the IUPAC name of each of the following alkenes:

a. $CH_3-CH_2-CH=CH-CH_3$

b. $\begin{array}{c} CH_3 \\ | \\ CH_3-CH=C-CH_3 \end{array}$

c. (cyclobutene ring with CH_2CH_3 substituent)

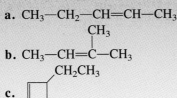

Solution:

a. The parent chain is pentene, an alkene that has five carbon atoms. It is numbered from right to left to make the first carbon atom of the double bond carbon 2. The name of this compound is *2-pentene*.

b. This compound has a four-carbon alkene chain. Numbering from the right side places the double bond and the side group at the second carbon atom. The name of this compound is *2-methyl-2-butene*.

c. This cycloalkene compound has four carbon atoms in the ring, and the parent chain is cyclobutene. The ethyl group is located on the third carbon atom in the ring because numbering begins at the double bond and goes in whatever direction gives the lowest number to the side group. The name of this compound is *3-ethylcyclobutene*.

Study Check:
Give the IUPAC name for the following:

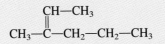

$$\begin{array}{c} CH-CH_3 \\ \| \\ CH_3-C-CH_2-CH_2-CH_3 \end{array}$$

Answer: 3-methyl-2-hexene

Sample Problem 11.2
Naming Cycloalkenes

Write all the structural formulas for chlorocyclopentene and name each.

Solution:

The name indicates a cyclic parent chain with five carbon atoms and a double bond. A single chlorine atom can be placed in three different positions on the cyclopentene ring to give three different isomers:

1-Chlorocyclopentene 3-Chlorocyclopentene 4-Chlorocyclopentene

Study Check:

What is the IUPAC name for the following compound?

Answer: 3,4-dimethylcyclohexene

ENVIRONMENTAL NOTE

Fragrant Alkenes

The odors you associate with cloves and peppermint, or perfumes from roses and lavender, are due to volatile oils that are synthesized by plants.

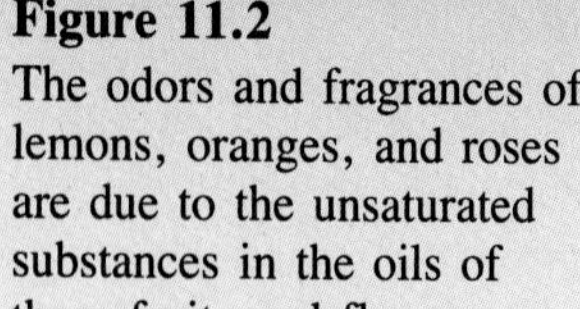

Figure 11.2
The odors and fragrances of lemons, oranges, and roses are due to the unsaturated substances in the oils of these fruits and flowers.

Unsaturated compounds in these oils, formed from units of isoprene, a branched five-carbon alkene, are responsible for the odors.

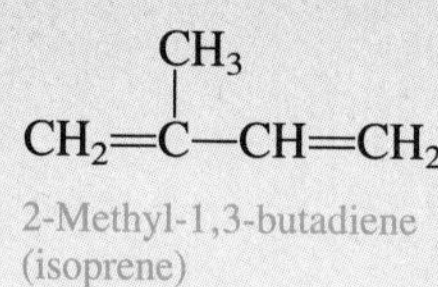

2-Methyl-1,3-butadiene (isoprene)

The structures that follow show the repeating isoprene unit. (See Figure 11.2.)

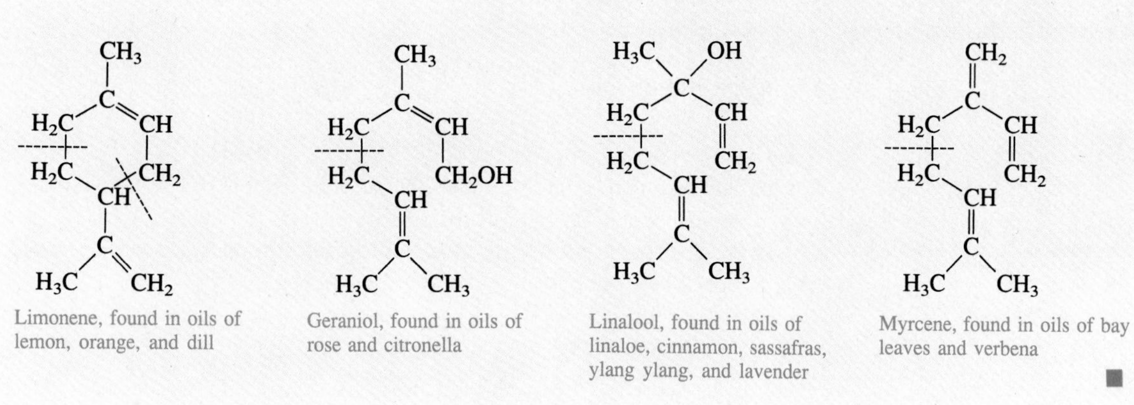

Limonene, found in oils of lemon, orange, and dill
Geraniol, found in oils of rose and citronella
Linalool, found in oils of linaloe, cinnamon, sassafras, ylang ylang, and lavender
Myrcene, found in oils of bay leaves and verbena

11.2 Geometric Isomers

Learning Goal **Write the structures of an alkene that has cis–trans geometric isomers.**

Because the carbon atoms in a double bond are fixed and cannot rotate freely, the groups attached to the double bond remain fixed on one side of the double bond or the other. This limitation leads to cis–trans isomers which have the same order of atoms, but a different arrangement on the double bond. For example, there are cis–trans isomers of 1,2-dichloroethene as shown in Figure 11.3. The cis isomer has the chlorine atoms on the same side of the double bond. In the trans isomer, the chlorine atoms appear on opposite sides of the double bond.

Figure 11.3
Models of the cis and trans isomers of 1,2-dichloroethene.

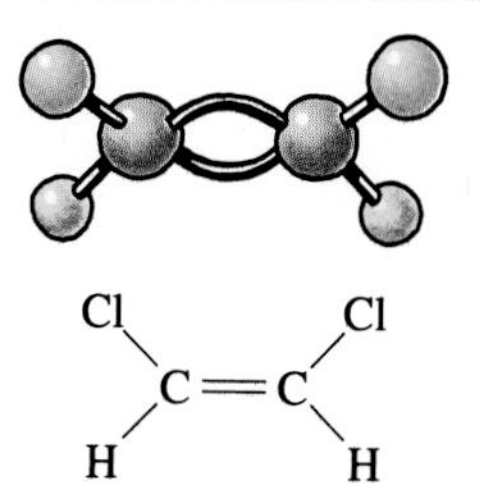

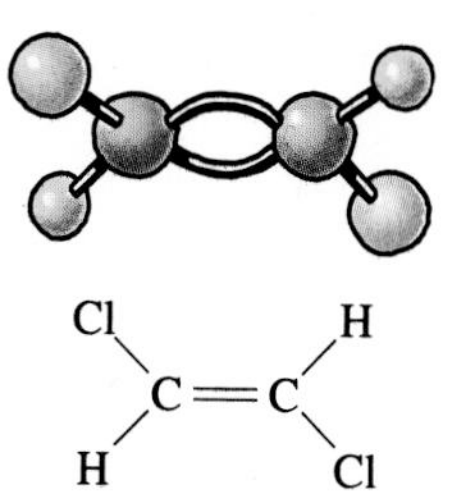

Cis isomer (chlorine atoms on the same side of double bond)

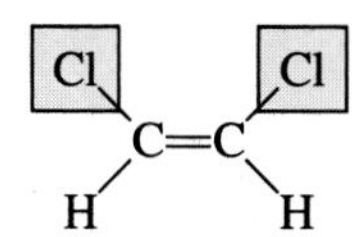

cis-1,2-Dichloroethene

Trans isomer (chlorine atoms on opposite sides of double bond)

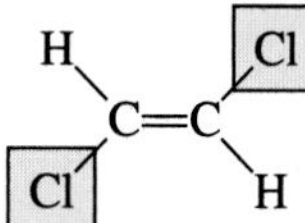

trans-1,2-Dichloroethene

Cis–trans isomers occur when there are different atoms or groups of atoms attached to each carbon atom in the double bond. No cis–trans isomers occur if either carbon atom in the double bond is attached to identical groups or atoms. For example, 1,1-dichloropropene has no cis–trans isomers.

Cl CH_3
C=C
Cl H

1,1-Dichloropropene

Sample Problem 11.3
Cis–Trans Isomers of Alkenes

Indicate whether each of the following can exist as cis–trans isomers. If so, write and name each isomer.

a. $CH_3CH{=}CBr_2$
b. $BrCH{=}CHBr$

Solution:

a. No; one carbon atom of the double bond is attached to two bromine atoms, which are identical atoms.
b. Yes; different atoms are attached to each carbon atom in the double bond.

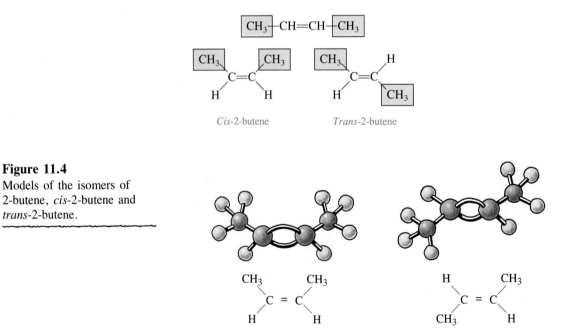

```
Br       Br          Br       H
  \     /              \     /
   C=C                  C=C
  /     \              /     \
H         H          H        Br
```

Cis-1,2-Dibromoethene   *Trans*-1,2-Dibromoethene

Alkenes with four or more carbon atoms can also exist as cis–trans isomers. When the alkyl groups are attached on the same side of the double bond, the compound is a cis isomer. In the trans isomer, the alkyl groups appear on opposite sides of the double bond. The simplest alkene that shows cis–trans isomerism is 2-butene. (See Figure 11.4.)

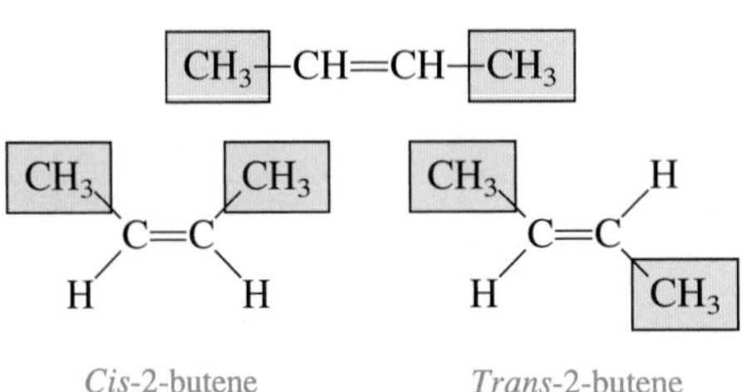

Cis-2-butene   *Trans*-2-butene

Figure 11.4
Models of the isomers of 2-butene, *cis*-2-butene and *trans*-2-butene.

```
CH3       CH3          H         CH3
   \     /               \     /
    C = C                 C = C
   /     \               /     \
H          H          CH3        H
```

cis – 2 – butene   *trans*– 2 – butene

ENVIRONMENTAL NOTE

Pheromones From Insects

Insects emit very small quantities of alkenes called pheromones, which are used to communicate with other members of a species. There are pheromones that warn of danger, call for defense, mark a trail, or are sex attractants. The bioactivity of many of these pheromones often depends on the cis or trans configuration of the double bond in the molecule. Pheromone research is developing ways to use these molecules as an alternative to pesticides. For example, very small amounts of the sex attractant of a certain insect could be used to trap and remove those insects from an agricultural area. ■

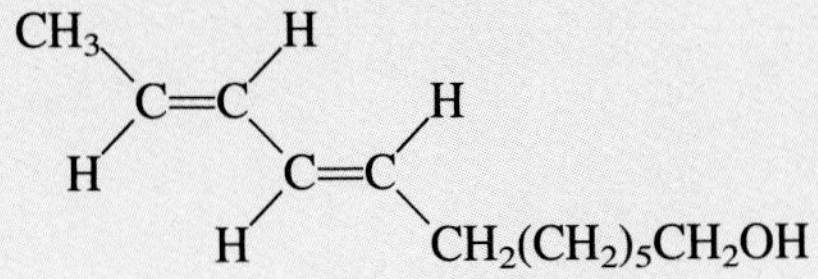

CH_3(H)C=C(H)–(H)C=C(H)$CH_2(CH_2)_5CH_2OH$

Sex attractant for the codling moth

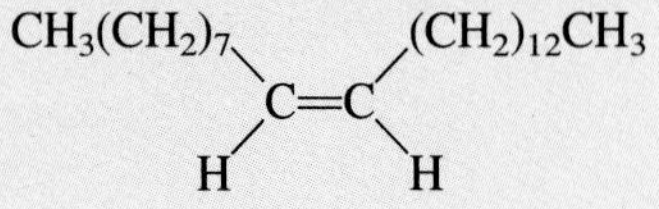

$CH_3(CH_2)_7$(H)C=C(H)$(CH_2)_{12}CH_3$

Sex attractant for female housefly

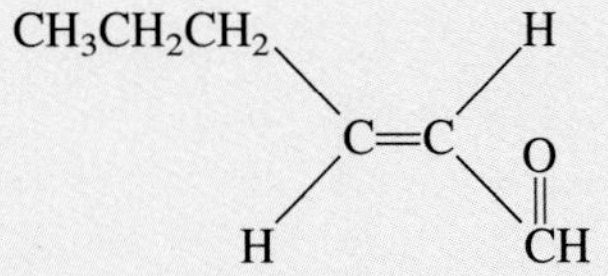

$CH_3CH_2CH_2$(H)C=C(H)CH=O

Alarm signal for ants

$CH_3(CH_2)_{12}$(H)C=C(H)NO_2

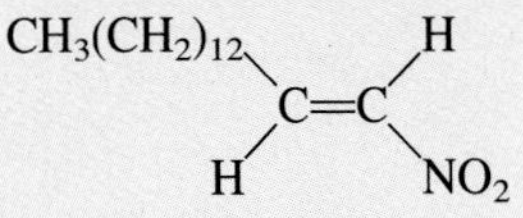

Defense signal for termite

Some more examples of cis–trans isomerism in alkenes follow.

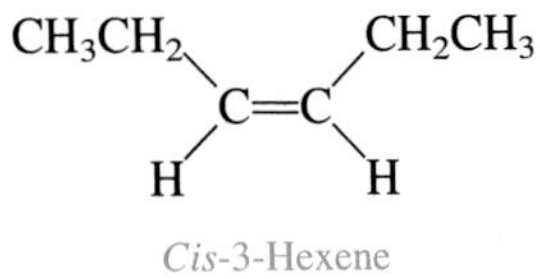

CH_3CH_2(H)C=C(H)CH_2CH_3

Cis-3-Hexene

CH_3CH_2(H)C=C(CH_2CH_3)H

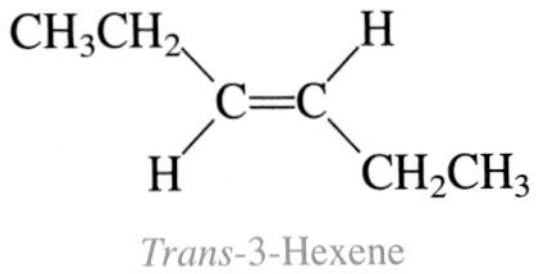

Trans-3-Hexene

Sample Problem 11.4
Writing and Naming Cis–Trans Isomers

Write the structural formulas for the cis–trans isomers of 2-pentene.

Solution:
The formula of 2-pentene is

$$CH_3—CH=CH—CH_2CH_3$$

We can write the cis isomer by placing the alkyl groups on the same side of the double bond, and the trans isomer by placing them on opposite sides of the double bond.

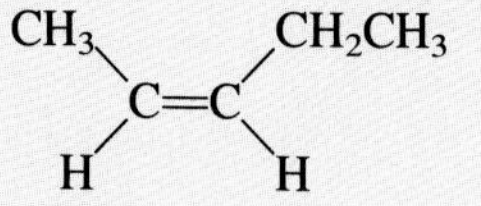

$$\begin{array}{c} CH_3 \quad\;\; CH_2CH_3 \\ C{=}C \\ H \qquad\quad H \end{array} \qquad \begin{array}{c} CH_3 \quad\;\; H \\ C{=}C \\ H \qquad\quad CH_2CH_3 \end{array}$$

Cis-2-Pentene 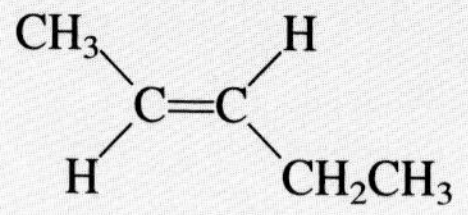 *Trans*-2-Pentene

Study Check:
Name the following isomer:

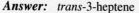

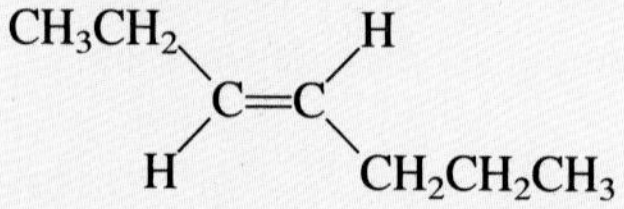

$$\begin{array}{c} CH_3CH_2 \quad\;\; H \\ C{=}C \\ H \qquad\quad CH_2CH_2CH_3 \end{array}$$

Answer: *trans*-3-heptene

HEALTH NOTE

Night Vision

The retinas of our eyes help us see in dim light. Within the retina, there is a compound called rhodopsin that is sensitive to light. Rhodopsin is composed of *cis*-retinal, an unsaturated compound, attached to a protein. When light reaches rhodopsin, the cis isomer of retinal is converted to the trans isomer and no longer fits the protein. The separation of the trans isomer from the protein produces an electrical impulse that the brain perceives as light.

We are able to see objects in dim light because an enzyme converts the trans isomer of the retinal back to the cis isomer, which re-forms the rhodopsin.

If there is a deficiency of rhodopsin in the retina, night blindness may occur. One common cause is a lack of vitamin A in the diet, because retinal is formed from vitamin A. Without a sufficient quantity of retinal, there is not enough rhodopsin formed to enable us to see adequately in dim light. ■

Cis–trans isomers of retinal

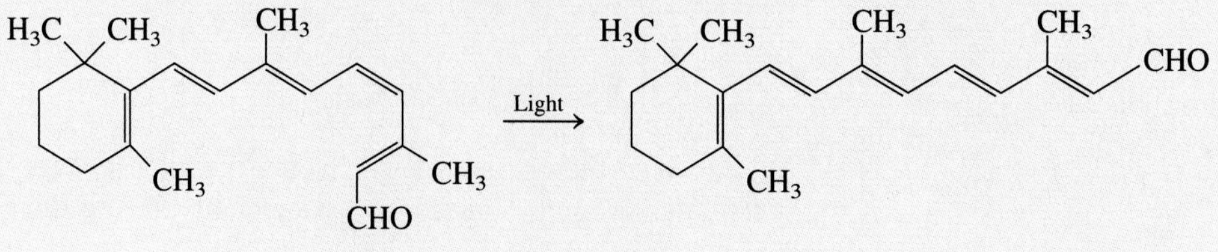

H_3C CH_3 CH_3 CH_3 CH_3 CHO — Light → — H_3C CH_3 CH_3 CH_3 CHO CH_3

Cis isomer Trans isomer

11.3 Alkynes

Learning Goal **Write the name and the structural formula of an alkyne.**

Figure 11.5
A model of acetylene (IUPAC name *ethyne*).

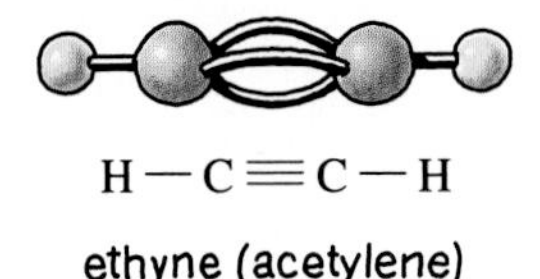

The **alkynes** are another family of unsaturated hydrocarbons. Six electrons are shared between two carbon atoms to form a triple bond. The simplest alkyne is named ethyne in the IUPAC system, but it is commonly known as acetylene. (See Figure 11.5.)

Triple bond

H:C⋮⋮C:H   H—C≡C—H

Ethyne (acetylene)

In the IUPAC system, the names of the alkynes end in *yne*. In longer chains, the alkyne group must be numbered to indicate its location in the molecule. There are no cis–trans isomers for alkynes. Common names for alkynes are related to acetylene, the simplest alkyne.

Naming Alkynes

Functional group: —C≡C—
IUPAC name: Alkyne
Common name: Alkylacetylene

Some examples of alkynes and their names are given in Table 11.2.

Table 11.2 Naming Some Alkynes

Structural formula	IUPAC name	Common name
HC≡CH	Ethyne	Acetylene
$HC{\equiv}C{-}CH_3$	Propyne	Methylacetylene
$CH_3{-}C{\equiv}C{-}CH_3$	2-Butyne	Dimethylacetylene
$CH_3{-}C{\equiv}C{-}CH(CH_3){-}CH_3$ (carbons numbered 1 2 3 4 5)	4-Methyl-2-Pentyne (Side group; Position of triple bond)	

Sample Problem 11.5
Naming Alkynes

Give the IUPAC name for the following alkyne:

$HC{\equiv}C{-}CH(CH_3){-}CH_3$

Solution:
Numbering the carbon chain from the end nearest the triple bond places the side group on carbon 3, so the IUPAC name is *3-methyl-1-butyne*.

$$\begin{array}{c} \quad\quad\quad CH_3 \\ \quad\quad\quad | \\ HC{\equiv}C{-}CH{-}CH_3 \\ 1\ \ 2\ \ 3\ \ \ 4 \end{array}$$ 3-Methyl-1-butyne

Study Check:
Name the following alkyne:

$$\begin{array}{c} CH_3 \\ | \\ CH_3{-}C{-}CH_2{-}C{\equiv}C{-}CH_3 \\ | \\ CH_3 \end{array}$$

Answer: 5,5-dimethyl-2-hexyne

11.4 Chemical Reactions of Unsaturated Hydrocarbons

Learning Goal **Write equations for the addition reactions of alkenes and alkynes.**

The double bonds of alkenes are highly reactive and easily add atoms of hydrogen, a halogen, or a halogen halide. In the **addition** reactions of alkenes, one bond in the double bond is broken in order to form new bonds.

$$>C{=}C< + X-Y \rightarrow -\overset{X}{\underset{|}{\overset{|}{C}}}-\overset{Y}{\underset{|}{\overset{|}{C}}}-$$

Hydrogenation

In the **hydrogenation** reaction, a double or triple bond adds hydrogen to give a saturated hydrocarbon. A **catalyst** such as platinum (Pt) is used to speed up the reaction. It is shown over the arrow in the equation because a catalyst does not appear in the products of the reaction. The general equation for hydrogenation of an alkene can be written as follows:

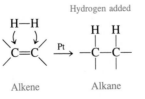

Hydrogen added

$$\underset{\text{Alkene}}{\overset{H-H}{>C{=}C<}} \xrightarrow{Pt} \underset{\text{Alkane}}{-\overset{H}{\underset{|}{\overset{|}{C}}}-\overset{H}{\underset{|}{\overset{|}{C}}}-}$$

Some examples of the hydrogenation of alkenes follow:

$$\underset{\text{Ethene}}{\mathrm{H_2C{=}CH_2}} + \underset{\text{Hydrogen}}{\boxed{\mathrm{H_2}}} \xrightarrow{\text{Pt}} \underset{\text{Ethane}}{\mathrm{H{-}\underset{H}{\overset{\boxed{H}}{C}}{-}\underset{H}{\overset{\boxed{H}}{C}}{-}H}}$$

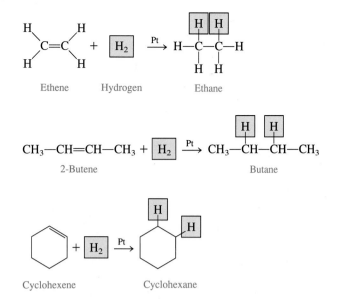

$$\underset{\text{2-Butene}}{\mathrm{CH_3{-}CH{=}CH{-}CH_3}} + \boxed{\mathrm{H_2}} \xrightarrow{\text{Pt}} \underset{\text{Butane}}{\mathrm{CH_3{-}\overset{\boxed{H}}{C}H{-}\overset{\boxed{H}}{C}H{-}CH_3}}$$

$$\underset{\text{Cyclohexene}}{\mathrm{C_6H_{10}}} + \boxed{\mathrm{H_2}} \xrightarrow{\text{Pt}} \underset{\text{Cyclohexane}}{\mathrm{C_6H_{12}}}$$

When an alkyne undergoes hydrogenation, the triple bond reacts with two molecules of hydrogen (H_2). For example, when ethyne (acetylene) adds two molecules of hydrogen, ethane is formed.

$$\underset{\text{Ethyne}}{\mathrm{H{-}C{\equiv}C{-}H}} + \boxed{\mathrm{2H_2}} \xrightarrow{\text{Pt}} \underset{\text{Ethane}}{\mathrm{H{-}\underset{\boxed{H}}{\overset{\boxed{H}}{C}}{-}\underset{\boxed{H}}{\overset{\boxed{H}}{C}}{-}H}}$$

Sample Problem 11.6
Writing Equations for Hydrogenation

Write the equations for the hydrogenation of propene and propyne.

Solution:
When propene reacts with hydrogen in the presence of a platinum catalyst, it adds one molecule of hydrogen to the double bond to give an alkane.

$$\underset{\text{Propene}}{\mathrm{CH_2{=}CH{-}CH_3}} + \underset{\text{Hydrogen}}{\mathrm{H_2}} \xrightarrow{\text{Pt}} \underset{\text{Propane}}{\mathrm{CH_3{-}CH_2{-}CH_3}}$$

When propyne reacts with hydrogen, two molecules of hydrogen can add to the triple bond.

$$\underset{\text{Propene}}{\mathrm{HC{\equiv}C{-}CH_3}} + \underset{\text{Hydrogen}}{\mathrm{2H_2}} \xrightarrow{\text{Pt}} \underset{\text{Propane}}{\mathrm{CH_3{-}CH_2{-}CH_3}}$$

Study Check:
Write an equation for the hydrogenation of cyclopentene using a platinum catalyst.

Answer:

$$\text{Cyclopentene} + H_2 \xrightarrow{Pt} \text{Cyclopentane}$$

Cyclopentene Cyclopentane

HEALTH NOTE

Unsaturated Fats and Hydrogenation

The fats in our diets and adipose tissue contain saturated and unsaturated fatty acids. The most common fatty acids in nature have a carbon chain of 16 or 18 carbon atoms with a carboxylic acid group (—COOH) on the end. Stearic acid is a typical saturated fatty acid found in foods of animal origin, such as whole milk, cheese, meat, and butter.

Long carbon chain

$$\begin{array}{l} \quad CH_2 \; CH_2 \; CH_2 \; CH_2 \; CH_2 \; CH_2 \; CH_2 \; CH_2 \; COOH \\ CH_3 \; CH_2 \; CH_2 \; CH_2 \; CH_2 \; CH_2 \; CH_2 \; CH_2 \; CH_2 \end{array}$$

Carboxylic acid group

Stearic acid, a saturated fatty acid

Oleic acid is a typical unsaturated fatty acid and has a cis double bond at carbon 9:

$$\begin{array}{l} \qquad\qquad\qquad\qquad H \quad\;\; H \\ CH_3 \; CH_2 \; CH_2 \; CH_2 \; C{=}C \quad CH_2 \; CH_2 \; CH_2 \; COOH \\ \quad CH_2 \; CH_2 \; CH_2 \; CH_2 \qquad CH_2 \; CH_2 \; CH_2 \; CH_2 \end{array}$$

Oleic acid, an unsaturated fatty acid

Unsaturated fats are often found in vegetable oil such as corn oil, safflower oil, and sunflower oil.

The process of hydrogenation is used commercially to convert vegetable oils, which contain unsaturated fats and have low melting points, to saturated oils. When hydrogen is added to liquid oils such as corn oil or safflower oil, the product is a more solid fat, such as soft margarine. Additional hydrogenation gives solid fats such as cube margarine and shortening, which are used in cooking. (See Figure 11.6.) ■

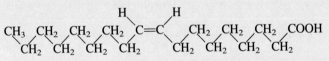

Figure 11.6
Some products prepared by the hydrogenation of unsaturated fatty acids in vegetable oils.

$$\begin{array}{c} H \qquad\quad H \\ C{=}C \\ CH_3(CH_2)_7 \qquad\qquad (CH_2)_7COOH \end{array} + H_2 \xrightarrow{Pt} CH_3(CH_2)_{16}COOH$$

Oleic acid Stearic acid

Halogenation

Alkenes typically react with chlorine or bromine adding atoms of the halogen to the double bond. Such a reaction is called **halogenation.** The halogenation of alkenes occurs readily, without the use of any catalysts. See Figure 11.7.

Halogenation of Alkenes

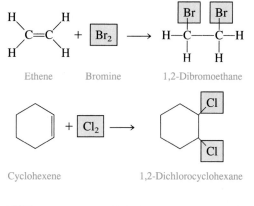

$$H_2C{=}CH_2 + \boxed{Br_2} \longrightarrow H{-}\underset{H}{\overset{\boxed{Br}}{C}}{-}\underset{H}{\overset{\boxed{Br}}{C}}{-}H$$

Ethene Bromine 1,2-Dibromoethane

Cyclohexene + $\boxed{Cl_2}$ ⟶ 1,2-Dichlorocyclohexane (ring with $\boxed{Cl}$ on each of two adjacent carbons)

Cyclohexene 1,2-Dichlorocyclohexane

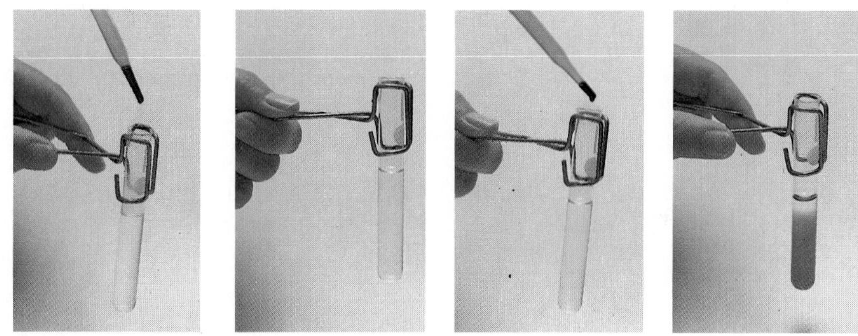

Figure 11.7
When bromine is added to the alkene in the first test tube, the red color is lost immediately as the bromine atoms add to the double bond. In the second test tube, the red color remains because bromine does not react, or reacts slowly, with the alkane.

Sample Problem 11.7
Writing the Products of Halogenation

Write the structural formula of the product in the following reaction:

$$CH_3{-}\overset{CH_3}{\overset{|}{C}}{=}CH_2 + Br_2 \longrightarrow$$

Solution:
The addition of bromine to an alkene places a bromine atom on each of the carbon atoms of the double bond.

$$CH_3{-}\overset{CH_3}{\overset{|}{C}}{=}CH_2 + Br_2 \longrightarrow CH_3{-}\underset{Br}{\overset{CH_3}{C}}{-}\underset{Br}{CH_2}$$

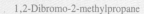

Methylpropene 1,2-Dibromo-2-methylpropane

Study Check:
What is the name of the product formed when chlorine adds to 2-butene?

Answer: 2,3-dichlorobutane

Hydrohalogenation

The hydrogen halides F, HCl, HBr, and HI add to unsaturated sites by a reaction called **hydrohalogenation.** The hydrogen atom bonds to one carbon of the double bond of an alkene, and the halogen atom adds to the other carbon. In the following example, hydrogen chloride adds to ethene.

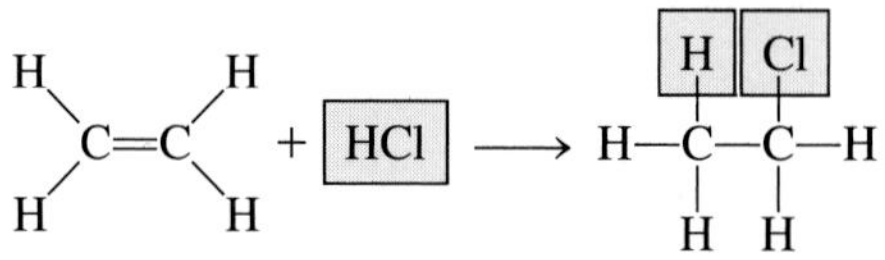

$$H_2C{=}CH_2 + \boxed{HCl} \longrightarrow H{-}\underset{H}{\overset{\boxed{H}}{C}}{-}\underset{H}{\overset{\boxed{Cl}}{C}}{-}H$$

Ethene (ethylene) — Chloroethane (ethyl chloride)

Some more examples of hydrohalogenation follow:

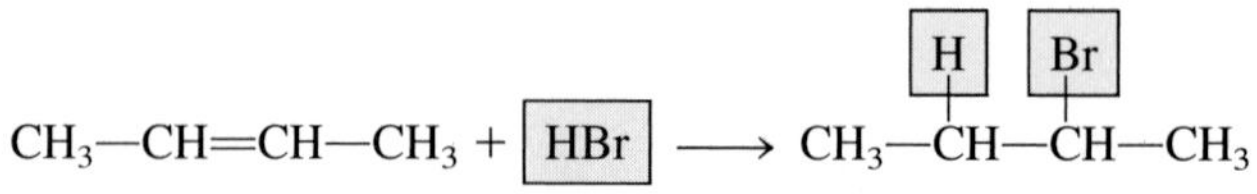

$$CH_3{-}CH{=}CH{-}CH_3 + \boxed{HBr} \longrightarrow CH_3{-}\overset{\boxed{H}}{CH}{-}\overset{\boxed{Br}}{CH}{-}CH_3$$

2-Butene — 2-Bromobutane

Cyclopentene + $\boxed{HI}$ $\longrightarrow$ Iodocyclopentane

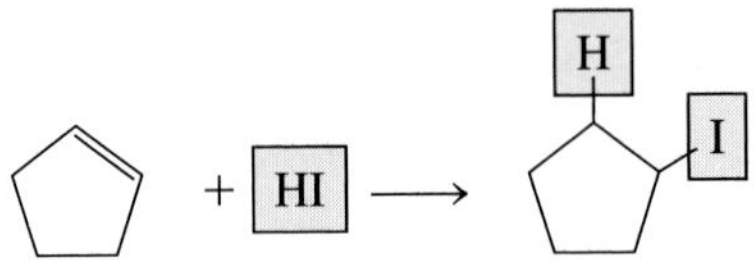

Cyclopentene — Iodocyclopentane

The atoms in a hydrogen halide are not identical. This affects the way that a hydrogen halide adds to a double bond that is not symmetrical. A **symmetrical alkene** has the same number of hydrogen atoms attached to the double bond. An **unsymmetrical alkene** has different numbers of hydrogen atoms attached to the double bond.

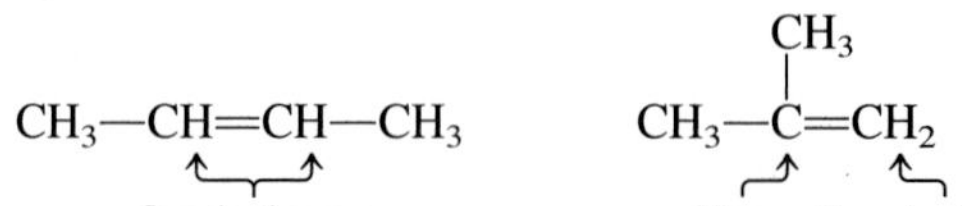

$$CH_3{-}CH{=}CH{-}CH_3$$

One hydrogen

Symmetrical alkene
(same number of hydrogen atoms attached to the carbons in the double bond)

$$CH_3{-}\overset{CH_3}{C}{=}CH_2$$

None Two hydrogens

Unsymmetrical alkene
(different numbers of hydrogen atoms attached to the carbons in the double bond)

For unsymmetrical alkenes, **Markovnikov's rule** is used: When adding a hydrogen halide to an unsymmetrical double bond, the hydrogen atom bonds to the carbon atom that already has the most hydrogen atoms attached. We can follow the addition of HCl to the unsymmetrical double bonds in the following alkenes.

Carbon in the double bond having the most hydrogen atoms

$$CH_3{-}CH{=}CH_2 + \boxed{HCl} \longrightarrow CH_3{-}\underset{|}{\overset{\boxed{Cl}}{C}}H{-}\overset{\boxed{H}}{C}H_2$$

Propene → 2-Chloropropane

Greatest number of hydrogen atoms in the double bond

$$CH_3{-}CH{=}\overset{CH_3}{\overset{|}{C}}{-}CH_3 + \boxed{HCl} \longrightarrow CH_3{-}\underset{\boxed{H}}{CH}{-}\underset{\boxed{Cl}}{\overset{CH_3}{\overset{|}{C}}}{-}CH_3$$

2-Methyl-2-butene → 2-Chloro-2-methylbutane

Sample Problem 11.8
Addition Using Markovnikov's Rule

Write the structural formula for the product of the following reaction:

$$CH_3{-}CH_2{-}CH{=}CH_2 + HBr \longrightarrow$$

Solution:
Since the alkene is unsymmetrical, we follow Markovnikov's rule and add the hydrogen atom to the carbon atom in the double bond that already is attached to the most hydrogen atoms. The bromine atom bonds with the other carbon atom in the double bond.

$$CH_3{-}CH_2{-}CH{=}CH_2 + HBr \longrightarrow CH_3{-}CH_2{-}\overset{Br}{\overset{|}{C}}H{-}\overset{H}{\overset{|}{C}}H_2$$

Carbon in the double bond having the most hydrogen atoms (arrow to CH_2)

1-Butene → 2-Bromobutane

Study Check:
Draw the structural formula and the name of the product obtained when HBr is added to 1-methylcyclopentene.

Answer: H_3C, Br, —H on cyclopentane ring — 1-Bromo-1-methylcyclopentane

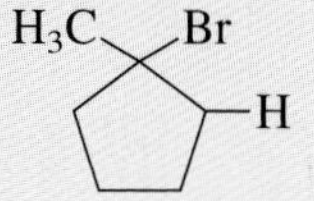

Sample Problem 11.9
Writing a Synthesis Using an Alkene

What alkene would you need to prepare the following compound:

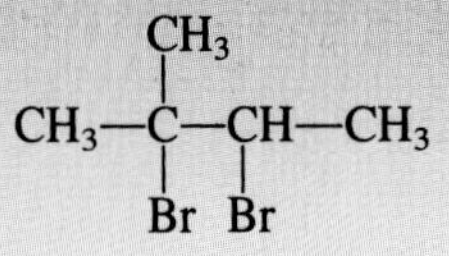

$$CH_3-\overset{\overset{\Large CH_3}{|}}{\underset{\underset{\Large Br}{|}}{C}}-\underset{\underset{\Large Br}{|}}{CH}-CH_3$$

The initial alkene must be 2-methyl-2-butene. The product is prepared by adding bromine.

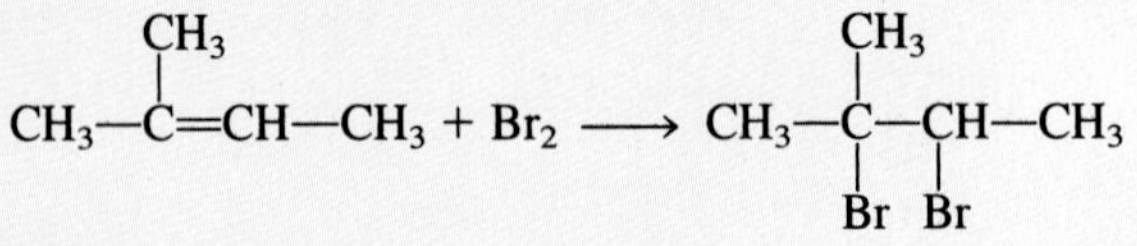

$$CH_3-\overset{\overset{\Large CH_3}{|}}{C}=CH-CH_3 + Br_2 \longrightarrow CH_3-\overset{\overset{\Large CH_3}{|}}{\underset{\underset{\Large Br}{|}}{C}}-\underset{\underset{\Large Br}{|}}{CH}-CH_3$$

Study Check:
Describe the alkene and reagent you would use to prepare 1-chloro-1-methylcyclohexane.

Answer: 1-methylcyclohexene and HCl

Hydration

Water will add to a double bond in the presence of an acid catalyst. This reaction, called **hydration,** is used to prepare alcohols, which have an —OH functional group. For example, water adds to ethene to give the alcohol ethanol.

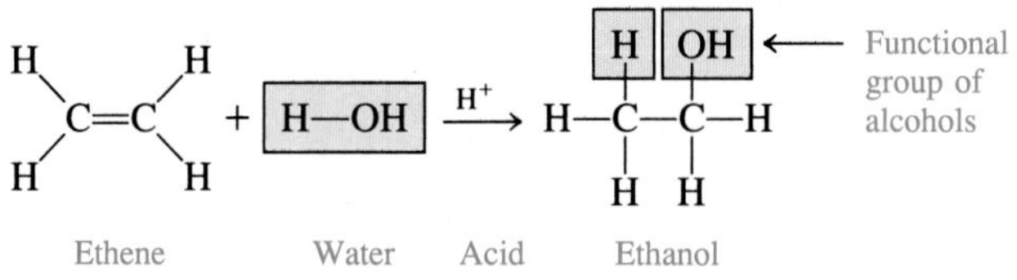

$$\underset{\text{Ethene}}{H_2C=CH_2} + \underset{\text{Water}}{\boxed{H-OH}} \xrightarrow[\text{Acid}]{H^+} \underset{\text{Ethanol}}{H-\overset{\overset{\boxed{H}}{|}}{\underset{\underset{H}{|}}{C}}-\overset{\overset{\boxed{OH}}{|}}{\underset{\underset{H}{|}}{C}}-H}$$

← Functional group of alcohols

The addition of water (H—OH) to an unsymmetrical double bond follows Markovnikov's rule:

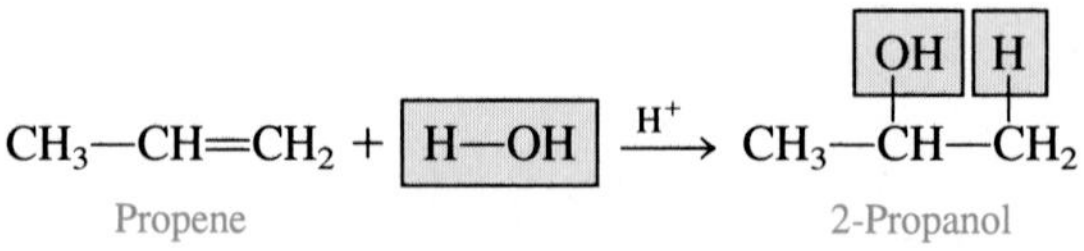

$$\underset{\text{Propene}}{CH_3-CH=CH_2} + \boxed{H-OH} \xrightarrow{H^+} \underset{\text{2-Propanol}}{CH_3-\overset{\overset{\boxed{OH}}{|}}{CH}-\overset{\overset{\boxed{H}}{|}}{CH_2}}$$

Sample Problem 11.10
Writing Products of Hydration

Write structural formulas for the products that will form in the following hydration reactions:

a. $CH_3-CH_2-CH=CH_2 + HOH \xrightarrow{H^+}$

b.  [cyclobutene] $+ HOH \xrightarrow{H^+}$

Solution:

a. Water adds as H— and —OH to the double bond. The alkene is unsymmetrical, so we must use Markovnikov's rule to add the H— to the CH_2 in the double bond, and the —OH to the CH.

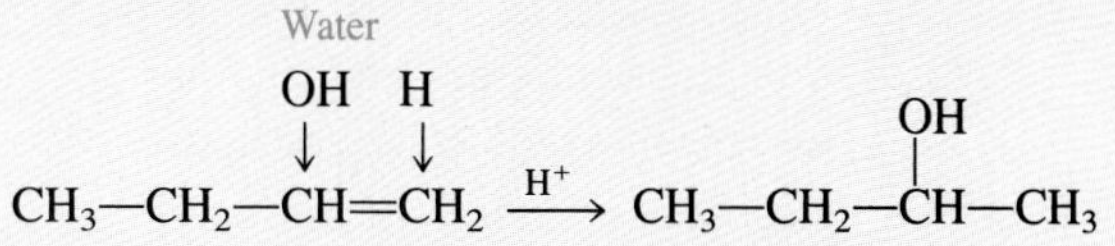

Water (OH, H adding to CH and CH_2 respectively)

$$CH_3{-}CH_2{-}CH{=}CH_2 \xrightarrow{H^+} CH_3{-}CH_2{-}\underset{}{\overset{OH}{\overset{|}{C}}}H{-}CH_3$$

b. In cyclobutene, the H— bonds to one side of the double bond, and the —OH bonds to the other side.

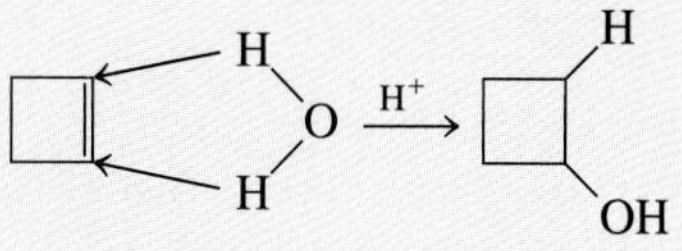

$$\text{cyclobutene} + H{-}O{-}H \xrightarrow{H^+} \text{cyclobutanol (H and OH on adjacent carbons)}$$

Study Check:
Draw the structural formula for the alcohol obtained by the hydration of 2-methyl-2-butene.

Answer:

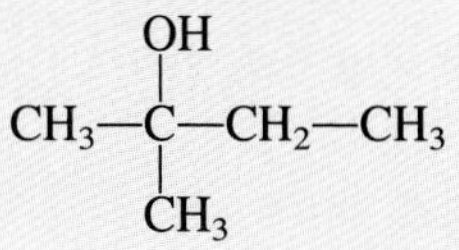

$$CH_3{-}\overset{OH}{\overset{|}{\underset{CH_3}{\underset{|}{C}}}}{-}CH_2{-}CH_3$$

ENVIRONMENTAL NOTE

Polymers From Alkenes

Small alkene molecules can react with each other (polymerize) at high pressure (1000 atm) and temperature (100°C) to produce a variety of polymers that make up the materials we use every day. **Polymers** are long-chain molecules made up of repeating units of smaller molecules called **monomers.** For example, the polymer known as polyethene (or polyethylene), used in the manufacturing of plastic bottles, film, and plastic dinnerware, is made from molecules of ethylene.

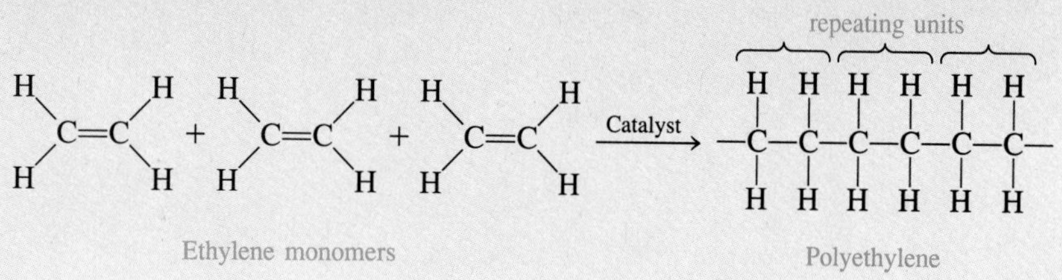

$$H_2C{=}CH_2 + H_2C{=}CH_2 + H_2C{=}CH_2 \xrightarrow{\text{Catalyst}} {-}CH_2{-}CH_2{-}CH_2{-}CH_2{-}CH_2{-}CH_2{-}$$

Ethylene monomers — Polyethylene (repeating units)

Another polymer, known as polyvinyl chloride (PVC), is used as "vinyl" plastic to make such things as plastic pipes and tubing, garbage bags, and garden hoses. PVC is made from molecules of vinyl chloride, a compound that has been identified as a carcinogen by OSHA (the Occupational Safety and Health Administration) and is known to cause liver cancer. People who work with vinyl chloride must be sure to follow strict safety rules that limit their exposure to the carcinogenic compound.

$$\underset{\substack{\text{Chloroethene}\\\text{(vinyl chloride)}}}{CH_2{=}\overset{\overset{Cl}{|}}{C}H + CH_2{=}\overset{\overset{Cl}{|}}{C}H} \longrightarrow \underset{\substack{\text{Polychloroethene}\\\text{(polyvinyl chloride, PVC)}}}{\overbrace{-CH_2-\overset{\overset{Cl}{|}}{C}H}-\overbrace{CH_2-\overset{\overset{Cl}{|}}{C}H-}}$$

Other alkenes have been used to make polymers used as food wrapping, such as Saran, and Teflon, a nonstick coating for pans and cooking utensils. (See Figure 11.8.) Table 11.3 lists some alkenes used in the formation of polymers.

All of these polymers tend to be durable, elastic, and resistant to chemical attack. Although such characteristics make them useful, the polymers in plastic products have single carbon bonds, which makes them unreactive. As a result, plastic polymers do not decompose easily (they are nonbiodegradable) and have become a contributor to pollution. Efforts are being made to make them more degradable. It becomes increasingly important to recycle plastic material, rather than add to our growing landfills. ■

Figure 11.8
Products we use that are made of polymers include Teflon coatings on pots and pans; Saran Wrap for covering food; plastic cups for coffee; garden hoses and raincoats; polypropylene for ski and hiking clothes; and vinyl plastic for bottles.

Table 11.3 Alkenes and Their Polymers

Monomer	Polymer	Uses for polymer
$H_2C{=}CH_2$ Ethene (ethylene)	$-CH_2-CH_2-CH_2-CH_2-$ Polyethylene	Plastic bottles, film, insulation material
$CH_3CH{=}CH_2$ Propene (propylene)	$-CH_2-CH(CH_3)-CH_2-CH(CH_3)-$ Polypropylene	Ski and hiking clothing, carpets, artificial joints, carpeting
$H_2C{=}CH(Cl)$ Chloroethene (vinyl chloride)	$-CH_2-CH(Cl)-CH_2-CH(Cl)-$ Polyvinyl chloride (PVC)	Plastic pipes and tubing, garden hoses, garbage bags
$H_2C{=}CH(Cl)$ + $H_2C{=}C(Cl)-Cl$ Chloroethene (vinyl chloride) 1,1-Dichloroethene	$-CH_2-CH(Cl)-CH_2-C(Cl)_2-CH_2-CH(Cl)-CH_2-C(Cl)_2-$ Saran	Plastic film and wrap
$F-C(F){=}C(F)-F$ Tetrafluoroethene	$-CF_2-CF_2-CF_2-CF_2-$ Teflon	Nonstick coatings
$H_2C{=}CH(C_6H_5)$ Phenylethene (styrene)	$-CH_2-CH(C_6H_5)-CH_2-CH(C_6H_5)-$ Polystyrene	Plastic coffee cups and cartons, insulation

11.5 Aromatic Compounds

Learning Goal **Write the name and the structural formula for an aromatic compound.**

In 1825, Michael Faraday discovered **benzene** in whale oil. When benzene was found to have the formula C_6H_6, scientists were puzzled. They tried unsuccessfully to draw a straight-line structure for this compound. Then, in 1858, August Kekulé suggested that the six carbon atoms formed a cyclic structure having three sets of double bonds. This was also confusing, as double bonds are very reactive, and yet experiments showed that benzene was rather unreactive. It did not undergo the addition reactions typical of alkenes. When it did react, it behaved more like an alkane. In 1865, Kekulé proposed that benzene had two structural formulas and that it alternated between them.

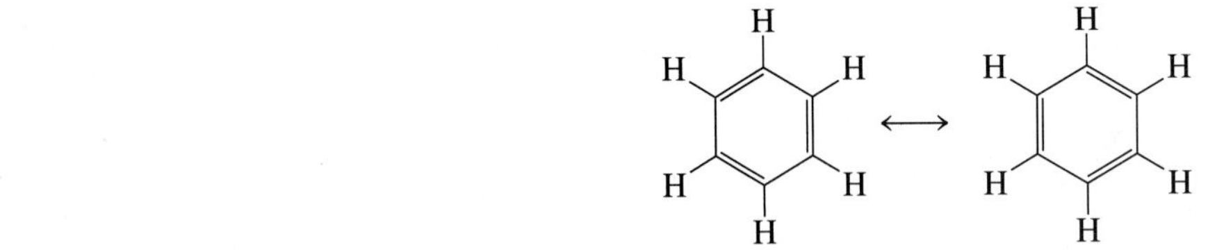

Today, we draw a structural formula for benzene that represents a combination of the Kekulé structures. Scientists reasoned that the six electrons in benzene are shared equally by all of the carbon atoms, which would make the six carbon ring very stable and much less reactive than unsaturated hydrocarbons having isolated double bonds. In this text, we will show benzene as a hexagon with a circle inside to indicate that the electrons are shared equally by the six carbons in the ring.

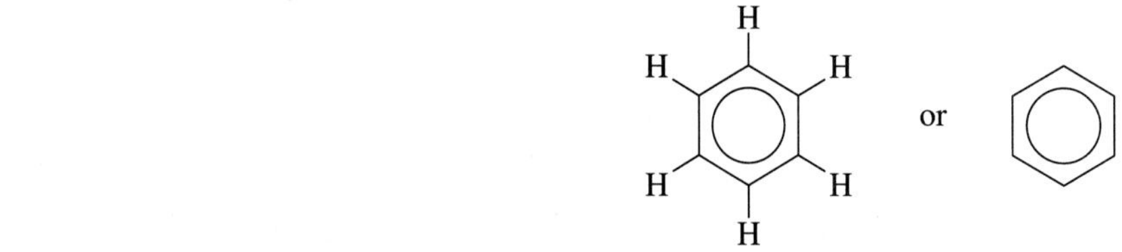

or

benzene

Naming Aromatic Compounds

At one time, compounds having fragrant odors were classified as part of the aromatic family. Today, an **aromatic compound** is defined as one having a benzene ring and chemical properties like benzene. For example, all of the following pain relievers are part of the aromatic family because they contain benzene.

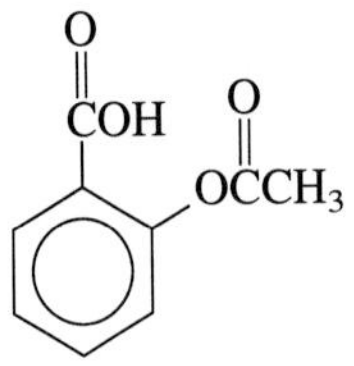

Aspirin (acetylsalicylic acid)

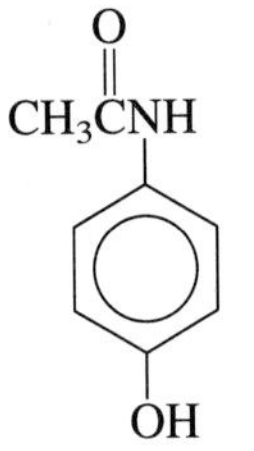

Acetaminophen (p-hydroxyacetanilide)

Ibuprofen [2-(4-isobutylphenyl)-propionic acid]

The IUPAC rule for naming an aromatic compound is to place the name of a side group on the benzene ring in front of the family name *benzene*.

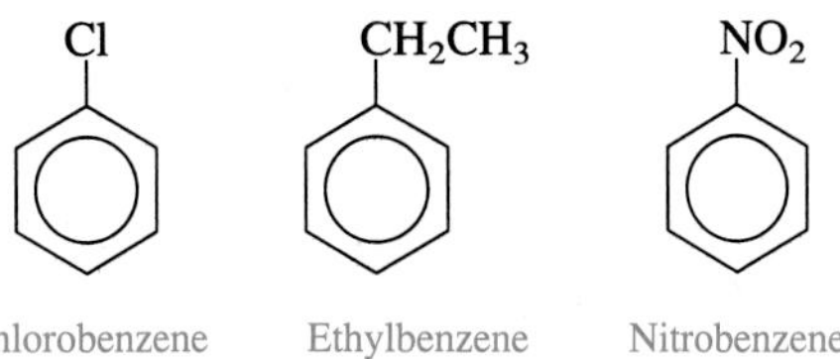

Chlorobenzene Ethylbenzene Nitrobenzene

There are many benzene derivatives that have kept their common names as the IUPAC name:

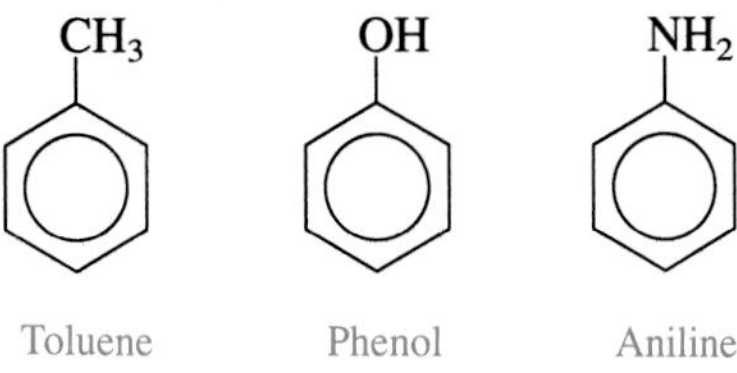

Toluene Phenol Aniline

Sample Problem 11.11
Naming Aromatic Compounds

Name the following organic compound:

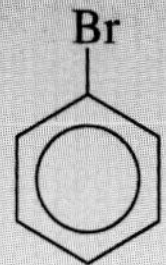

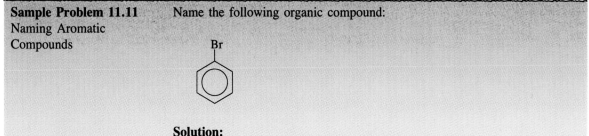

Solution:
The hexagon structure with the circle is the symbol for the compound benzene. The bromine is named as a bromo side group. No number is needed when one side group is attached to benzene. The name is therefore *bromobenzene*.

Study Check:
Write the structural formula for propylbenzene.

Answer:

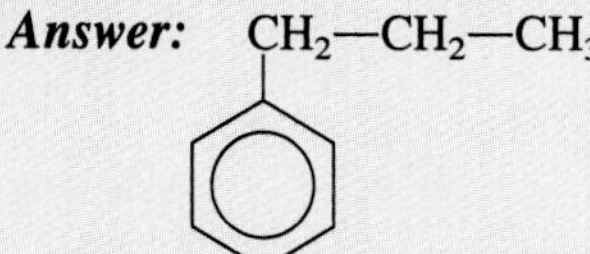

When there are two side groups on benzene, the groups may be numbered to show their placement on the ring. However, a common naming system is often used to indicate their location. The prefix *ortho* (usually abbreviated and placed before the name with a hyphen: *o*-) describes adjacent side groups. When a carbon atom separates the side groups, the prefix *meta* (*m*-) is used. The prefix *para* (*p*-) indicates that there are two carbon atoms between the side groups.

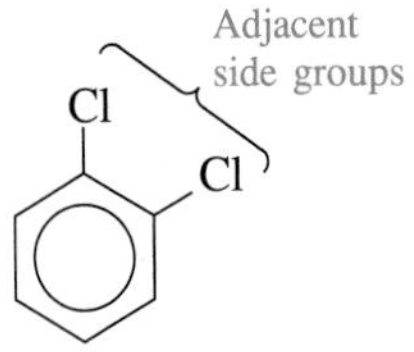

Adjacent side groups

Cl Cl

1,2-Dichlorobenzene
(*o*-dichlorobenzene)

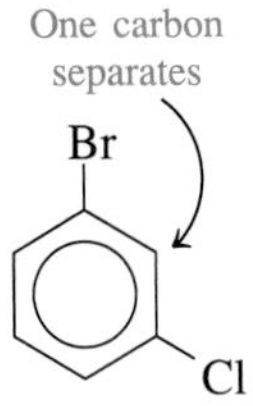

One carbon separates

Br Cl

1-Bromo-3-chlorobenzene
(*m*-bromochlorobenzene)

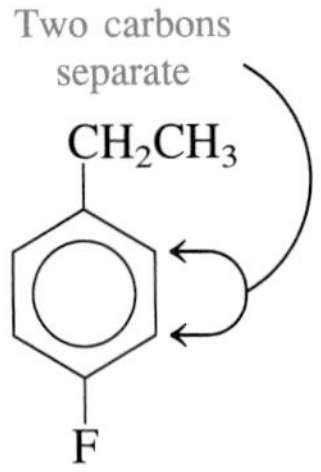

Two carbons separate

CH_2CH_3 F

1-Ethyl-4-fluorobenzene
(*p*-ethylfluorobenzene)

If there are three side groups, the benzene ring is numbered.

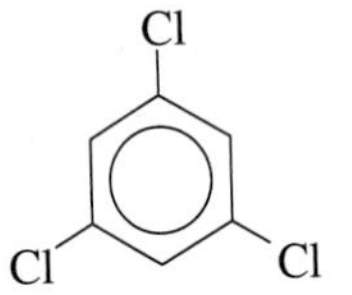

Cl Cl Cl

1,3,5-Trichlorobenzene

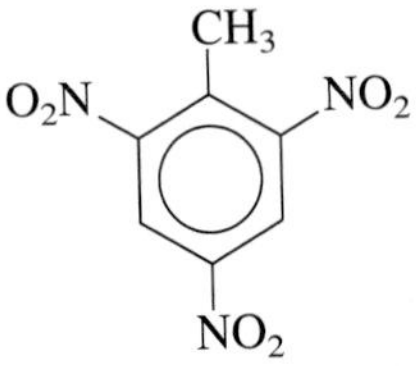

CH_3 O_2N NO_2 NO_2

2,4,6-Trinitrotoluene (TNT)

The name *toluene* is used when a methyl side group is attached to benzene. *Xylene* is used when there are two methyl groups attached to benzene.

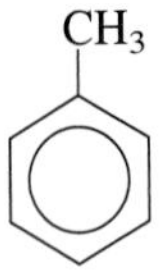

CH_3

Toluene

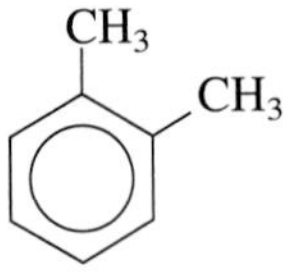

CH_3 CH_3

o-Xylene

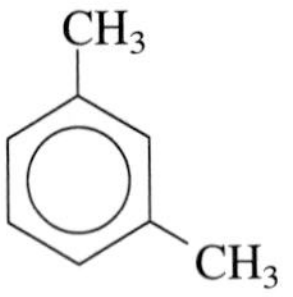

CH_3 CH_3

m-Xylene

Sample Problem 11.12
Naming Aromatic Compounds Having Two Side Groups

Give the IUPAC name and the common name for the following compounds:

a. Br Br

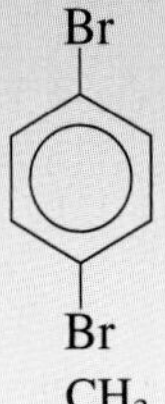

b. CH_3 Cl

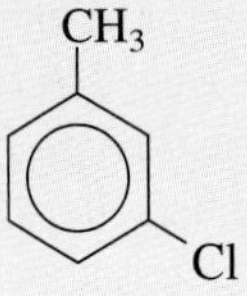

Solution:

a. In the IUPAC system, this benzene ring with two bromine atoms is numbered *1,4-*; the prefix *para* (*p*-) is used for the common name.

IUPAC name: 1,4-Dibromobenzene
Common name: *p*-Dibromobenzene

b. In the IUPAC system, the side groups on the benzene ring are numbered and listed alphabetically; the carbon atom attached to the chlorine atom is carbon 1. When the common name, *toluene,* is used, the carbon atom attached to the methyl side group is designated as carbon 1.

IUPAC name: 3-Chlorotoluene
Common name: *m*-Chlorotoluene

Study Check:

Give the name of the following compound.

CH_3

CH_3

Answer: *m*-xylene

HEALTH NOTE

Benzene Rings in Nature and Cancer

Two or more benzene rings can join together to form larger aromatic systems in which neighboring rings share two or more carbon atoms. Naphthalene has been used in mothballs. Anthracene is used in the manufacture of dyes.

When a compound contains phenanthrene, it is often a carcinogen, a substance known to cause cancer. Several have been identified in soot, coal tar, dyes, burnt meat, automobile exhaust, and cigarette smoke:

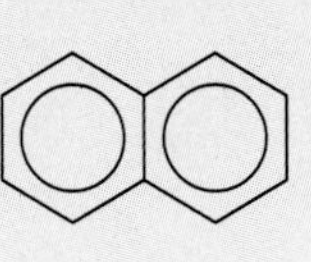
Naphthalene

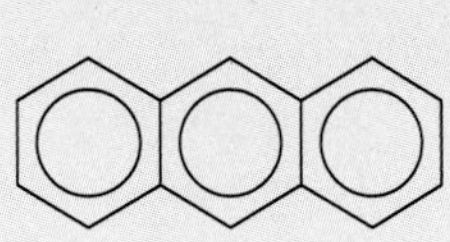
Anthracene

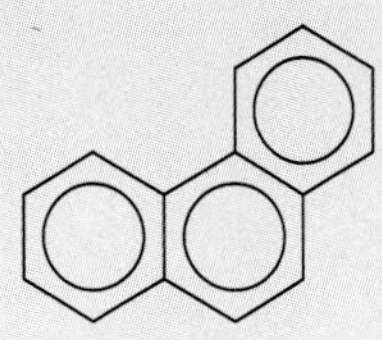
Phenanthrene

Carcinogenic Compounds of Benzene

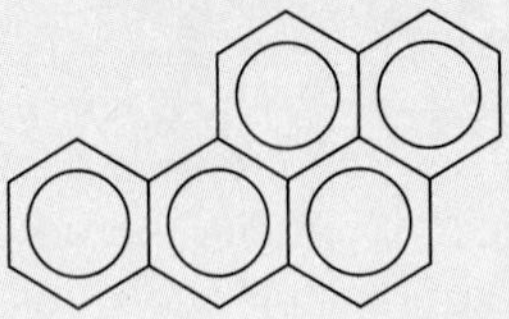

Benzpyrene

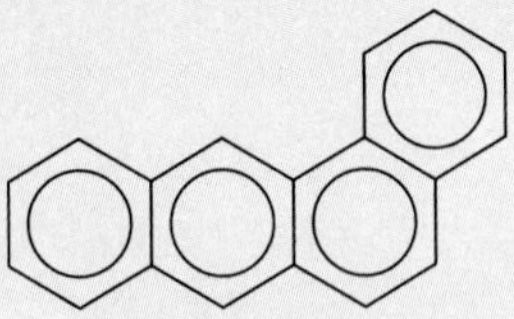

Benzanthracene

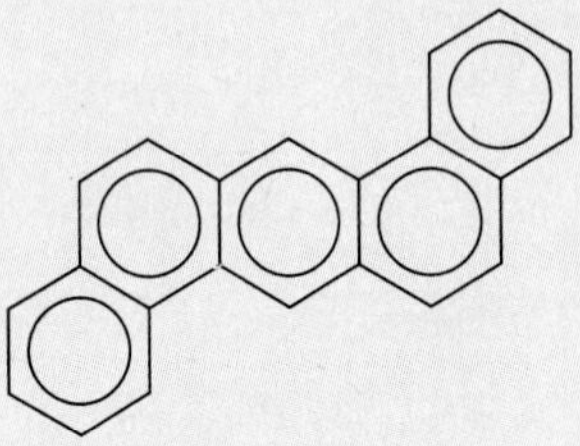

CH_3

CH_3

1,2,5,6-Dibenzanthracene

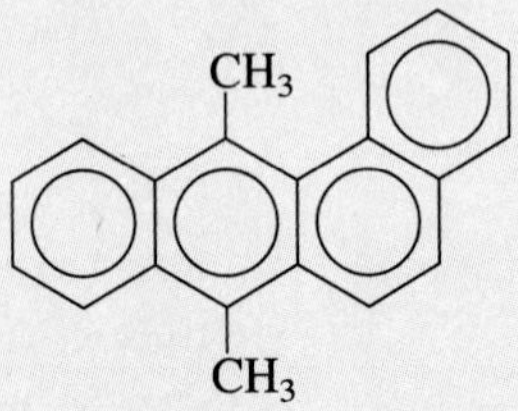

7,12-Dimethylbenzanthracene

Some scientists have suggested that a chemical carcinogen interacts with molecules in a cell in a way that alters the cell's growth. Some think that the DNA in the cells acts as a target for carcinogens that causes a mutation of the genes. Increased exposure to carcinogens increases the chance of DNA alterations in the cells. Some mutations lead to the rapid growth of cells that is typical of cancer. ■

11.6 Chemical Reactions of Benzene

Learning Goal **Write the products of the substitution reactions of benzene.**

We saw earlier that benzene is very stable and does not react easily. When benzene does react, it undergoes substitution like an alkane by replacing side groups for hydrogen atoms.

Halogenation

Halogenation of benzene occurs in the presence of an iron catalyst. Recall that each carbon in the benzene ring contains a hydrogen atom.

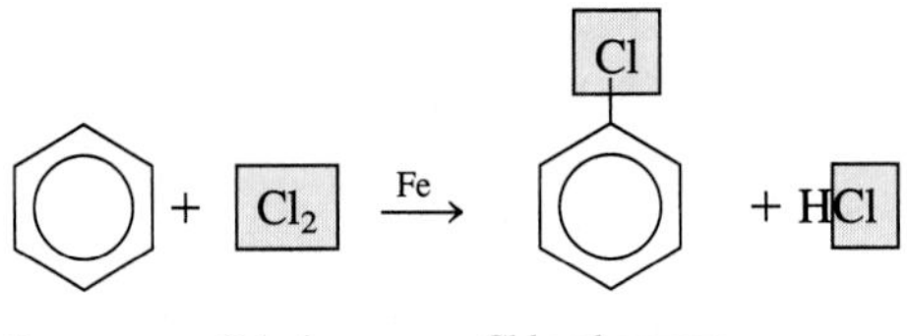

$$C_6H_6 + Cl_2 \xrightarrow{Fe} C_6H_5Cl + HCl$$

Benzene Chlorine Chlorobenzene

Nitration

Benzene reacts with nitric acid in the presence of sulfuric acid. A hydrogen atom in the aromatic ring is replaced by a nitro (—NO_2) group. Such a process is called **nitration.**

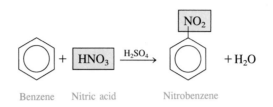

$$C_6H_6 + HNO_3 \xrightarrow{H_2SO_4} C_6H_5NO_2 + H_2O$$

Benzene Nitric acid Nitrobenzene

Alkylation

An alkyl group such as CH_3— can be placed on a benzene ring by **alkylation,** reacting an alkyl halide with benzene in the presence of a catalyst, usually aluminum chloride ($AlCl_3$).

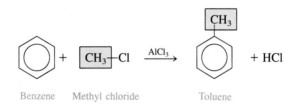

$$C_6H_6 + CH_3\text{—}Cl \xrightarrow{AlCl_3} C_6H_5CH_3 + HCl$$

Benzene Methyl chloride Toluene

Sulfonation

Benzene is not reactive with sulfuric acid. However, if sulfur trioxide (SO_3) is added to the acid, it can replace a hydrogen atom on the benzene ring. This is called **sulfonation.**

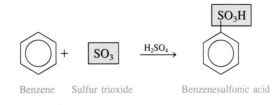

$$C_6H_6 + SO_3 \xrightarrow{H_2SO_4} C_6H_5SO_3H$$

Benzene Sulfur trioxide Benzenesulfonic acid

HEALTH NOTE

Sulfa Drugs

In 1932, sulfanilamide, one of the first antibacterial drugs, was produced by the sulfonation of benzene. It and other sulfa drugs were used to treat bacterial infection before penicillin and other antibiotics were available.

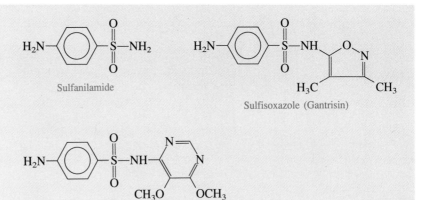

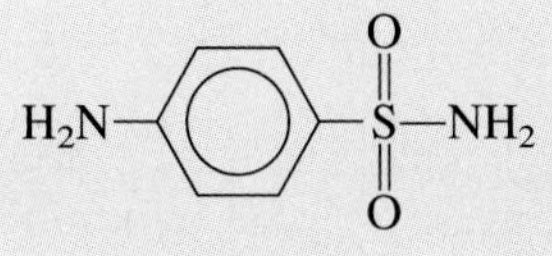

Sulfanilamide

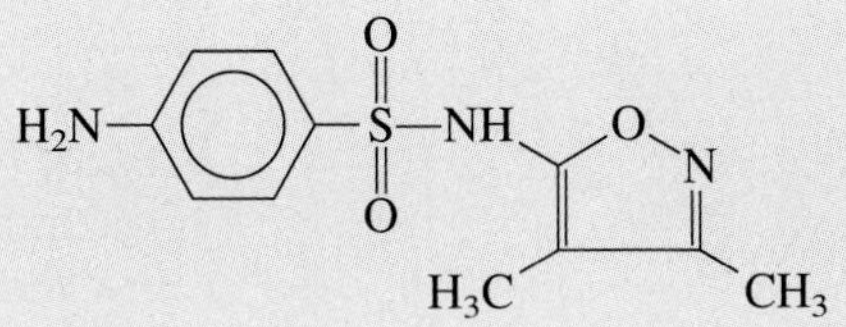

Sulfisoxazole (Gantrisin)

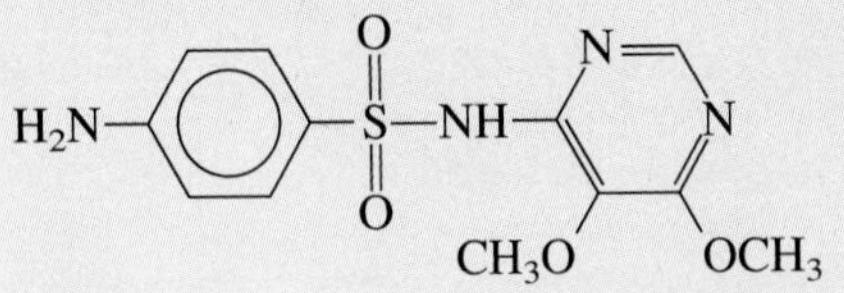

Sulfadoxine

Today, sulfonamides such as sulfisoxazole (Gantrisin) are used to treat urinary tract infections.

Sulfadoxine is another sulfa drug that is used in the treatment of malaria when the malaria is resistant to quinine treatment. ■

Sample Problem 11.13
Chemical Reactions of Benzene

Write the products of the following reactions with benzene:

a. bromination with Br_2

b. alkylation with CH_3CH_2Cl

Solution:

a. In the presence of an iron catalyst, a bromine atom will replace a hydrogen atom on the benzene ring, giving the following product:

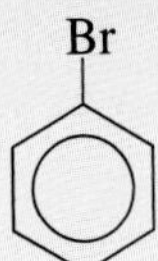

b. In the presence of $AlCl_3$, the ethyl group is placed on the benzene ring, giving the following product:

CH_2CH_3

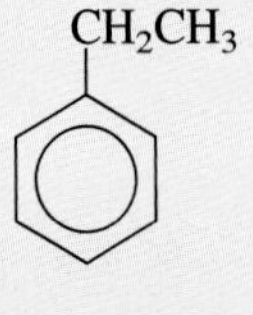

Study Check:

The alkylation of benzene with methyl chloride can give products that have two methyl groups. Write equations for the sequence of reactions that form *p*-xylene.

Answer:

$$C_6H_6 + 2CH_3{-}Cl \xrightarrow{AlCl_3} C_6H_5{-}CH_3 \xrightarrow{AlCl_3} CH_3{-}C_6H_4{-}CH_3 + 2HCl$$

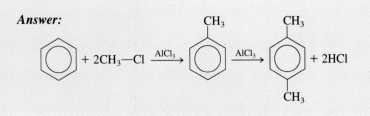

Table 11.4 provides a summary of naming unsaturated hydrocarbons, and Table 11.5 gives a summary of their reactions.

Table 11.4 Summary of Naming Unsaturated and Aromatic Hydrocarbons

Type	Example	Structure
Alkene	Propene (propylene)	$CH_3{-}CH{=}CH_2$
Alkyne	Propyne	$CH_3{-}C{\equiv}CH$
Aromatic	Benzene	(benzene ring)

Table 11.5 Summary of Reactions of Unsaturated and Aromatic Compounds

Hydrogenation

Alkene (or alkyne) + Hydrogen $\xrightarrow{Pt}$ Alkane

$$H_2C{=}CH_2 + H_2 \xrightarrow{Pt} H_3C{-}CH_3$$

$$H{-}C{\equiv}C{-}H + 2H_2 \xrightarrow{Pt} H_3C{-}CH_3$$

Halogenation

Alkene + Halogen $\longrightarrow$ Haloalkane

$$H_2C{=}CH_2 + Br_2 \longrightarrow H_2C(Br){-}CH_2(Br)$$

Hydrohalogenation

Alkene + Hydrogen Halide $\longrightarrow$ Haloalkane

$$H_2C{=}CH_2 + HCl \longrightarrow H_2C(H){-}CH_2(Cl)$$

Hydration

Alkene + HOH $\xrightarrow{H^+}$ Alcohol

$$H_2C{=}CH_2 + HOH \xrightarrow{H^+} H_2C(H){-}CH_2(OH)$$

Halogenation

Benzene + X_2 $\xrightarrow{Fe}$ Halobenzene + HX

$$C_6H_6 + Cl_2 \xrightarrow{Fe} C_6H_5Cl + HCl$$

Nitration

Benzene + HNO_3 $\xrightarrow{H_2SO_4}$ Nitrobenzene + H_2O

$$C_6H_6 + HNO_3 \xrightarrow{H_2SO_4} C_6H_5NO_2 + H_2O$$

Alkylation

Benzene + Alkyl chloride $\xrightarrow{AlCl_3}$ Alkylbenzene + HCl

$$C_6H_6 + CH_3{-}Cl \xrightarrow{AlCl_3} C_6H_5CH_3 + HCl$$

Sulfonation

Benzene + SO_3 $\xrightarrow{H_2SO_4}$ Benzenesulfonic acid

$$C_6H_6 + SO_3 \xrightarrow{H_2SO_4} C_6H_5SO_3H$$

Glossary of Key Terms

addition A reaction in which atoms or groups of atoms bond to a double or triple bond. Addition reactions include the addition of hydrogen (hydrogenation), hydrogen halides, and water (hydration).

alkene An unsaturated hydrocarbon containing a carbon–carbon double bond.

alkylation The substitution of a hydrogen in benzene by an alkyl group by reacting benzene with an alkyl halide, using a catalyst such as aluminum chloride.

alkyne An unsaturated hydrocarbon containing a carbon–carbon triple bond.

aromatic compound A compound that contains at least one benzene ring and has benzene-like properties.

benzene A six-carbon ring with six hydrogen atoms (C_6H_6) and three double bonds. Because the electrons are shared equally by all six carbon atoms, the benzene ring is particularly stable and undergoes only substitution reactions.

catalyst A substance that speeds up a reaction.

cis–trans isomers Structures of alkenes with the same order of atoms but different arrangements on the double bond.

functional group An atom or group of atoms that has a strong influence on the overall chemical behavior of an organic compound.

halogenation The reaction of a halogen (Cl_2 or Br_2) with saturated or unsaturated compounds to form halogen-containing compounds. Double bonds undergo halogenation by addition; alkanes and benzene undergo halogenation by substitution.

hydration An addition reaction in which the components of water, H— and —OH, bond to the carbon–carbon double bond to form an alcohol.

hydrogenation The addition of hydrogen (H_2) to the unsaturated sites of alkenes or alkynes to yield more-saturated hydrocarbons.

hydrohalogenation The addition of a hydrogen halide (HCl, HBr, or HI) to a double or triple bond.

Markovnikov's rule When an unsymmetrical compound (H—X) adds to a double bond, the hydrogen goes to the carbon with the most hydrogens, and X goes to the carbon with the fewest hydrogens.

monomer A small molecule such as an alkene that is the unit that bonds repeatedly to form a polymer.

nitration The substitution of a hydrogen in benzene by the $—NO_2$ group in the presence of sulfuric acid to give a nitrobenzene.

polymer A large molecule formed by the combining of many small molecules called monomers.

sulfonation The substitution of a hydrogen in benzene by the $—SO_3H$ group to give benzenesulfonic acid.

symmetrical alkene An alkene that has the same number of hydrogen atoms attached to either side of the double bond.

unsaturated hydrocarbon A hydrocarbon with one or more double or triple bonds.

unsymmetrical alkene An alkene that has different numbers of hydrogen atoms attached to both sides of the double bond.

Problems

Alkenes *(Goal 11.1)*

11.1 Compare the structural formulas and reactivity of the compounds cyclohexane and cyclohexene.

11.2 Name each of the following alkenes:

a. $CH_2{=}CH_2$ b. $CH_3CH{=}CHCH_2CH_2Br$

c. 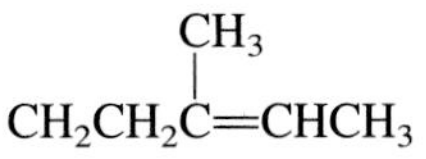 $CH_3CH_2\overset{CH_3}{\overset{|}{C}}{=}CHCH_3$

e. 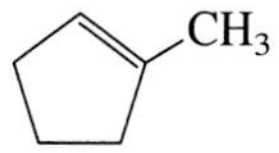 1-methylcyclopentene ring structure with CH_3

d. 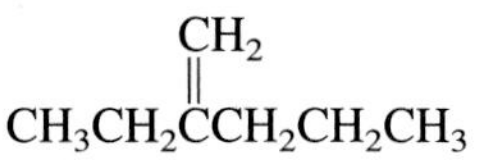 $CH_3CH_2\overset{CH_2}{\overset{\|}{C}}CH_2CH_2CH_3$

f. 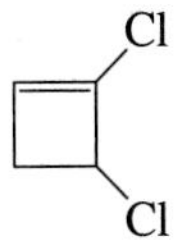 cyclobutene ring with two Cl substituents

11.3 Write the structural formula for each of the following alkenes:

a. propene
b. 2-methylpropene
c. cyclopentene
d. 3-chlorocyclopentene
e. 2,3-dimethylcyclohexene
f. 1,2-dichloro-3-methyl-2-butene

Geometric Isomers *(Goal 11.2)*

11.4 Write the structural formulas for the cis–trans isomers of 2,3-dichloro-2-butene.

11.5 Give the name for each of the following:

a. 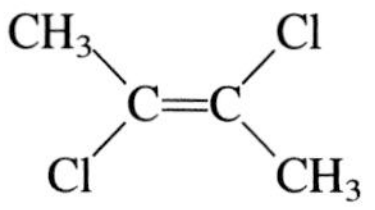 $(CH_3)(Cl)C{=}C(Cl)(CH_3)$ — CH_3 and Cl on the left carbon (CH₃ top, Cl bottom); Cl and CH_3 on the right carbon (Cl top, CH_3 bottom)

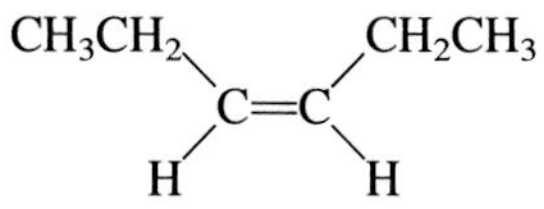

b. $(CH_3CH_2)(H)C{=}C(CH_2CH_3)(H)$ — CH_3CH_2 top left, H bottom left; CH_2CH_3 top right, H bottom right

c. $(CH_3CH_2)(Br)C{=}C(Br)(CH_2CH_2CH_3)$ — CH_3CH_2 top left, Br bottom left; Br top right, $CH_2CH_2CH_3$ bottom right

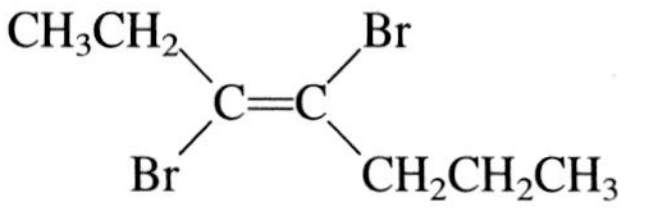

11.6 Compounds having the molecular formula C_4H_8 can be cycloalkanes or alkenes. Draw all six isomers for this formula, including the cis–trans isomers.

11.7 Write the cis–trans isomers of the following:

a. 1,2-dibromoethene
b. 2,3-dichloro-2-pentene
c. 1,2-dibromo-1-butene
d. 2-butene
e. 3-hextene

Alkynes *(Goal 11.3)*

11.8 Name the following alkynes:

a. $HC{\equiv}CH$
b. $CH_3{-}C{\equiv}CH$
c. 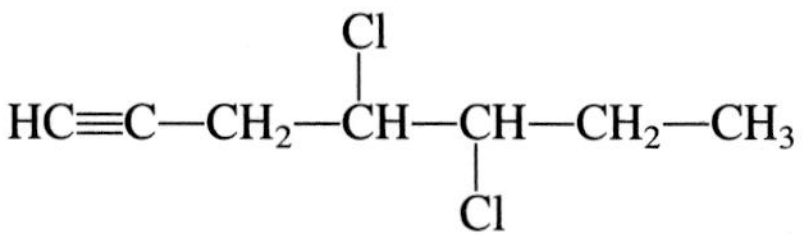 $HC{\equiv}C{-}CH_2{-}\overset{Cl}{\overset{|}{C}H}{-}\underset{Cl}{\underset{|}{C}H}{-}CH_2{-}CH_3$

d. 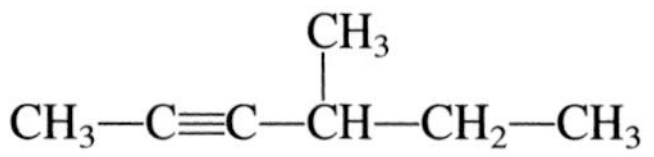
$$\begin{array}{l} \quad\quad\quad\quad\quad\quad\quad CH_3 \\ \quad\quad\quad\quad\quad\quad\quad | \\ CH_3{-}C{\equiv}C{-}CH{-}CH_2{-}CH_3 \end{array}$$

e. 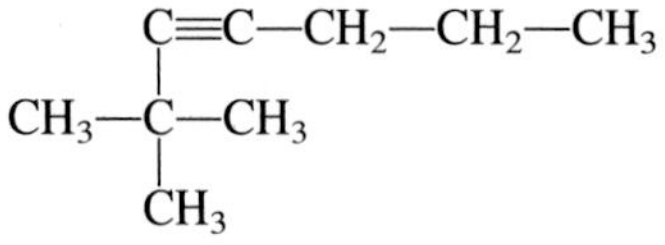
$$\begin{array}{l} \quad\quad\quad C{\equiv}C{-}CH_2{-}CH_2{-}CH_3 \\ \quad\quad\quad | \\ CH_3{-}C{-}CH_3 \\ \quad\quad\quad | \\ \quad\quad\quad CH_3 \end{array}$$

11.9 Draw a structural formula for each of the following alkynes:

a. 2-butyne
b. acetylene
c. 4,4-dichloro-2-pentyne
d. ethylacetylene
e. 3-methyl-1-butyne

Chemical Reactions of Unsaturated Hydrocarbons *(Goal 11.4)*

11.10 Alkenes and alkynes react by addition, but alkanes react by substitution. Explain the differences in these types of reactions.

11.11 Write the structural formula of the product of each of the following reactions of alkenes:

a. $CH_3{-}CH{=}CH{-}CH_3 + H_2 \xrightarrow{Pt}$

b. 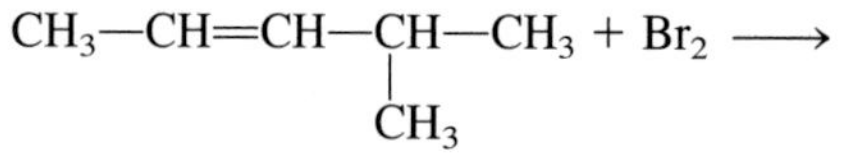
$$\begin{array}{l} CH_3{-}CH{=}CH{-}CH{-}CH_3 + Br_2 \longrightarrow \\ \quad\quad\quad\quad\quad\quad\quad\quad | \\ \quad\quad\quad\quad\quad\quad\quad\quad CH_3 \end{array}$$

c. 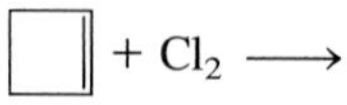 [cyclobutene] $+ Cl_2 \longrightarrow$

d. 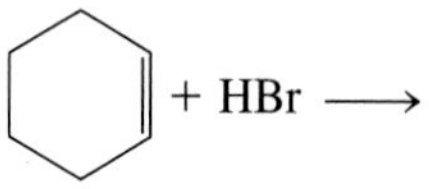 [cyclohexene] $+ HBr \longrightarrow$

e. 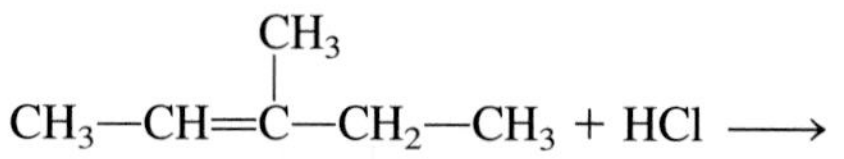
$$\begin{array}{l} \quad\quad\quad\quad\quad\quad\quad CH_3 \\ \quad\quad\quad\quad\quad\quad\quad | \\ CH_3{-}CH{=}C{-}CH_2{-}CH_3 + HCl \longrightarrow \end{array}$$

f. $CH_3{-}CH{=}CH_2 + H_2O \xrightarrow{H^+}$

11.12 Write the structural formula for the product of 1-butyne with H_2 and a platinum catalyst.

11.13 Write an equation using condensed formulas and any catalysts for the following reactions:

a. hydrogenation of 2-methylpropene
b. addition of chlorine to cyclopentene
c. addition of bromine to 2-pentene
d. addition of hydrogen bromide to 2-methyl-2-butene
e. hydration of 1-methylcyclobutene

11.14 Using an appropriate alkene, show how you would produce the following:

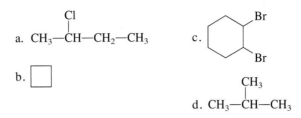

a. $CH_3—CH(Cl)—CH_2—CH_3$

b.

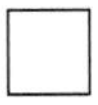

c.

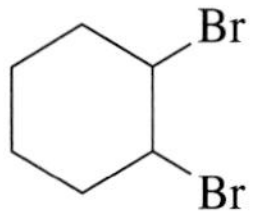

d. $CH_3—CH(CH_3)—CH_3$

Aromatic Compounds *(Goal 11.5)*

11.15 How does benzene differ from cyclohexane and cyclohexene?

11.16 Write the IUPAC name and the common name (if any) for each of the following aromatic compounds:

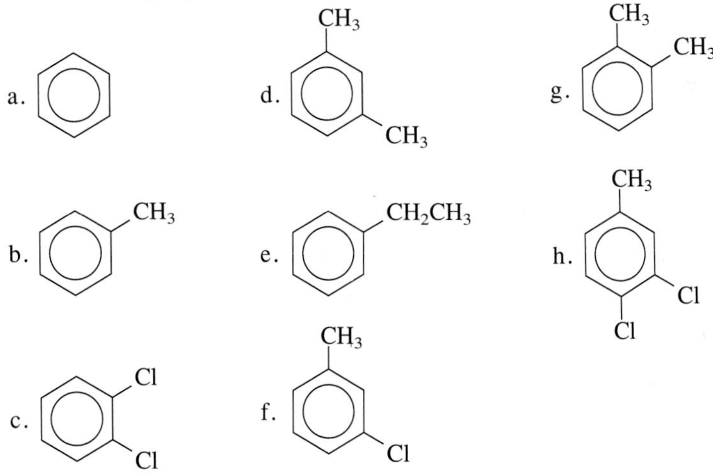

11.17 Write the structural formula of each of the following:

a. benzene
b. *m*-xylene
c. *o*-bromotoluene
d. *m*-dibromobenzene
e. *p*-dichlorobenzene
f. 1,3,5,-tribromobenzene
g. ethylbenzene
h. *p*-ethyltoluene

Chemical Reactions of Benzene *(Goal 11.6)*

11.18 Write an equation for the reaction of benzene with each of the following:

a. Cl_2
b. HNO_3
c. SO_3
d. CH_3CH_2Cl

Review Problems

11.19 Identify each of the following as alkane, haloalkane, alkene, alkyne, or aromatic compound; write the name for each:

a. $CH_3—CH_2—CH{=}CH_2$

b. 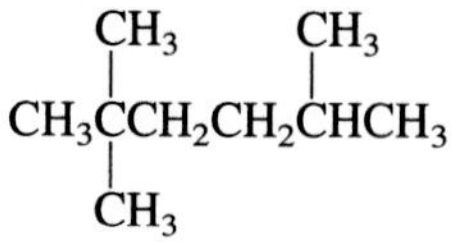 $CH_3C(CH_3)_2CH_2CH_2CH(CH_3)CH_3$ (drawn with CH3 groups above and below the second carbon and above the fifth carbon)

c. 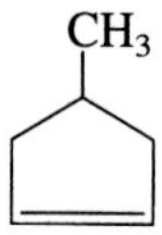 [methylcyclopentene: cyclopentene ring with CH3 substituent]

d. 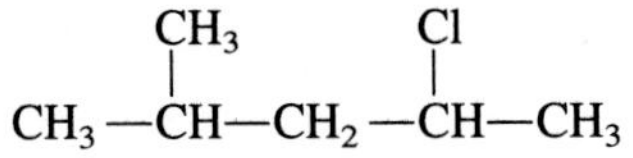 $CH_3—CH(CH_3)—CH_2—CH(Cl)—CH_3$

e. 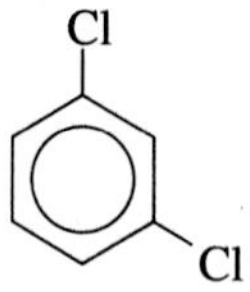 [benzene ring with Cl substituents at 1,3 positions]

f. $CH_3—C{\equiv}C—CH_2—CH_3$

11.20 Write the structural formula of each of the following:

a. 1,2-dimethylcyclopentane
b. *p*-dichlorobenzene
c. propylbromide
d. *o*-chlorotoluene
e. 2,5-dichloro-2-hexene
f. 3-methyl-1-butene
g. *cis*-1,2-dichloro-1-propene
h. 2,4,6-trichlorotoluene
i. 2-butyne

11.21 Write the equation for each of the following reactions:

a. hydrogenation (Pt catalyst) of 1-methylcyclohexene
b. addition of HCl to 2-methyl-2-butene
c. addition of Cl_2 to 2-hexene
d. addition of H_2O to 1-methylcyclopentene
e. nitration of benzene

11.22 Write the structural formula of each of the following and give the reagents you would use to synthesize each:

a. 2,3-dibromopentane
b. toluene
c. chlorocyclohexane
d. 1,2-dichlorobutane
e. cyclobutane

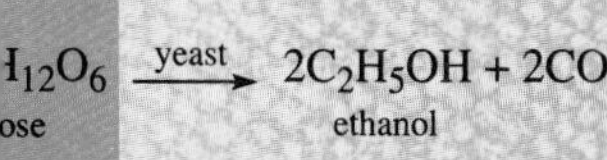

12

Oxygen and Sulfur in Organic Compounds

The process of fermentation which converts glucose to ethanol is one of the earliest known chemical reactions.

Learning Goals

12.1 Give the IUPAC name and the common name (if any) for an alcohol.

12.2 Describe the effect of hydrogen bonding on the boiling points and solubilities of alcohols.

12.3 Write the name and the structural formula of a phenol.

12.4 Write the structural formula of the alkene or the ether produced by the dehydration of an alcohol.

12.5 Write the name and the structural formula of an ether.

12.6 Identify the functional group of a thiol or a sulfide.

12.7 Write the name and the structural formula of an aldehyde or a ketone.

12.8 Given the name or the structural formula of an alcohol, write the aldehyde or the ketone produced by oxidation.

12.9 Write equations for the chemical reactions of aldehydes and ketones; describe some tests for aldehydes.

SCOPE

In this chapter, we will look at some organic compounds that contain oxygen and sulfur. The functional group in alcohols, phenols, and ethers contains a carbon–oxygen single bond; aldehydes and ketones contain a carbon–oxygen double bond. Thiols and sulfides contain sulfur atoms.

The functional group in alcohols is called the hydroxyl group (—OH). Since early times, grains, vegetables, and fruits have undergone fermentation, which produces the ethyl alcohol in alcoholic beverages. Hydroxyl groups are also important in carbohydrates, and in steroids such as cholesterol and estradiol. When the hydroxyl group and a hydrogen atom are removed, an alcohol is dehydrated to form an alkene or an ether. Ethyl ether, a compound that has been used for general anesthesia for over a hundred years, is the best known of the ether family. When an alcohol is oxidized, an aldehyde or a ketone is produced. The simplest aldehyde, formaldehyde, is used for preserving tissue specimens. Several naturally occurring aldehydes such as cinnamon and vanilla have fragrant odors and are used in flavorings.

One of the most commonly used ketones is acetone, which is found in solvents such as fingernail polish remover. Acetone and other ketone bodies are found in the body fluids when large amounts of fats are metabolized. This situation occurs when cells are unable to utilize available carbohydrates, as in diabetes mellitus, high-protein diets, or starvation.

12.1 Alcohols

Learning Goal **Give the IUPAC name and the common name (if any) for an alcohol.**

The **alcohols** are a family of organic compounds that contain a hydroxyl (—OH) group. We can think of an alcohol as a derivative of water in which an alkyl group has replaced one of the hydrogen atoms of water, as can be seen from the structures of the alcohols methanol and ethanol:

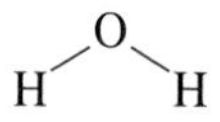

Water

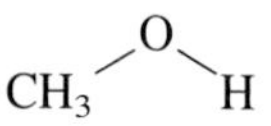

Methanol

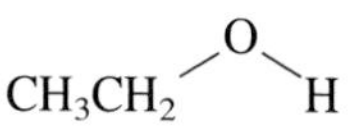

Ethanol

Hydroxyl groups are found in several important compounds in the body, such as cholesterol, glucose, and glycerol.

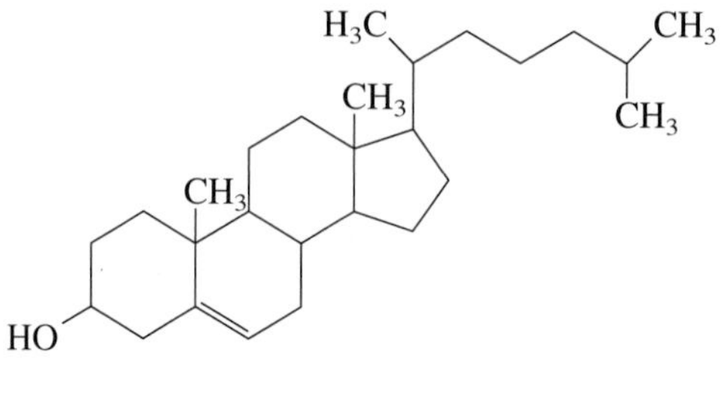

Cholesterol

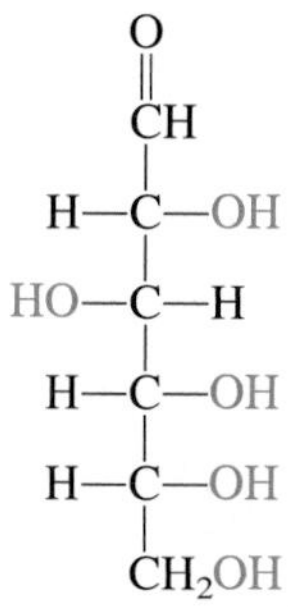

Glucose

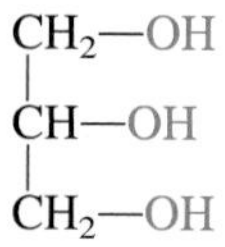

Glycerol

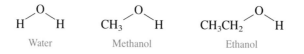

Figure 12.1
Models of methanol and ethanol.

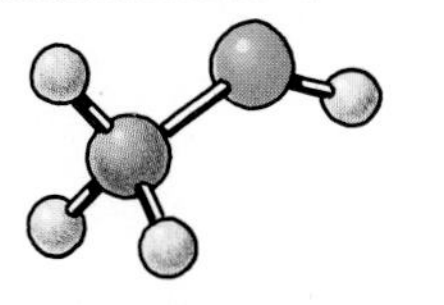

CH_3OH
methanol

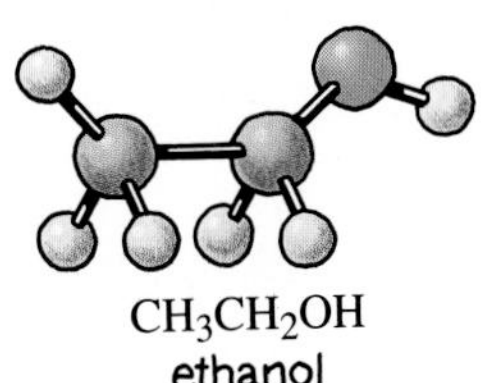

CH_3CH_2OH
ethanol

Naming Alcohols

The IUPAC name of an alcohol is obtained by replacing the *e* of the alkane name of the longest carbon chain attached to the —OH group by *ol*. For example, methanol is the alcohol derived from methane, and ethanol is the alcohol derived from ethane. (See Figure 12.1.) The common names of the simplest alcohols consist of the name of the alkyl group (R) followed by the family name *alcohol*, as in *methyl alcohol* and *ethyl alcohol*.

Naming Alcohols R—OH

Functional group: Hydroxyl, —OH

IUPAC name: Alkanol

Common name: Alkyl alcohol

Some examples of alcohols follow.

```
     H                  H   H
     |                  |   |
 H—C—O—H           H—C—C—O—H
     |                  |   |
     H                  H   H
```

CH_3OH
Methanol
(methyl alcohol)

CH_3CH_2OH
Ethanol
(ethyl alcohol)

In other straight-chain alcohols the chain is numbered to give the —OH group the lowest possible number. The names and numbers for side groups are placed in front of the alcohol name.

```
                              OH                  CH3
                              |                   |
CH3CH2CH2—OH               CH3CHCH3          CH3CHCH2CH2OH
3   2   1                  1   2   3         4   3   2   1
```

$CH_3CH_2CH_2—OH$ (3 2 1)
1-Propanol
(propyl alcohol)

$CH_3CH(OH)CH_3$ (1 2 3)
2-Propanol
(isopropyl alcohol)

$CH_3CH(CH_3)CH_2CH_2OH$ (4 3 2 1)
3-Methyl-1-butanol

A cyclic alcohol is named as a cycloalkanol. The carbon ring is numbered so that the carbon atom bonded to the —OH group is carbon 1. Any other side groups are shown by numbering from carbon 1.

Cyclohexanol
(cyclohexyl alcohol)

2-Methylcyclopentanol

HEALTH NOTE

Alcohols in Health and Medicine

Methanol (CH_3OH), or methyl alcohol, is found in many solvents used to remove paint and in duplicator fluids. It is extremely poisonous. When even small amounts of methanol are ingested, headaches, fatigue, blindness, coma, or death may occur.

Ethanol (CH_3CH_2OH), or ethyl alcohol, one of the earliest known chemicals, is often used in solvents for perfumes, varnishes, and some medicines, such as tincture of iodine. An antiseptic solution that is 70% by volume ethanol and 30% water is used to prepare an area of skin for injection and to sterilize equipment. It destroys bacteria by coagulating protein.

The term *alcohol* commonly refers to the ethyl alcohol produced when yeast is added to sugars present in foods such as potatoes, grapes, rice, barley, and rye.

$$\underset{\text{Glucose}}{C_6H_{12}O_6} \xrightarrow{\text{yeast}} \underset{\text{Ethanol}}{2C_2H_5OH} + 2CO_2$$

When ingested in small amounts, ethanol may produce a feeling of euphoria in the body, although it is a depressant. When larger amounts of ethanol are ingested, mental and physical coordination are impaired. If the ethanol blood concentration exceeds 0.4%, death may occur. A dependency on alcohol may lead to hallucinations, liver disease, gastritis, and psychological disturbances.

Isopropyl alcohol (rubbing alcohol) is used as an astringent because it evaporates rapidly and cools the skin, reducing the size of blood vessels near the surface and decreasing pore size.

$$CH_3\overset{\overset{\displaystyle OH}{|}}{C}HCH_3$$

2-Propanol
(isopropyl alcohol)

Ethylene glycol (1,2-ethanediol) is used as an antifreeze in cooling and heating systems. If ingested, it is extremely poisonous, causing kidney damage, coma, and death.

$$HO—CH_2—CH_2—OH$$

1,2-Ethanediol
(ethylene glycol)

Glycerol (glycerine) is a natural component of fats and oils. It is used in skin lotions, cosmetics, and some soaps as a skin softener.

$$\begin{array}{l} CH_2OH \\ | \\ CHOH \\ | \\ CH_2OH \end{array}$$

1,2,3-Propanetriol
(glycerol)

Menthol is a cyclic alcohol used in throat lozenges and sprays. It causes the mucous membranes to increase their secretions and soothes the respiratory tract.

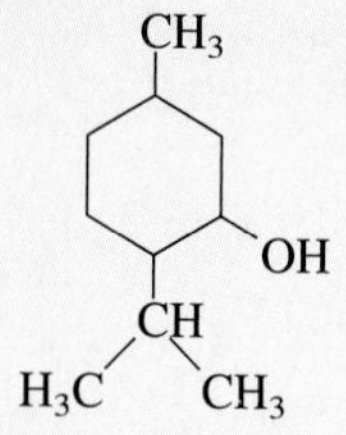

CH₃ — cyclohexane ring — OH; CH(CH₃)(H₃C)

Menthol

Sample Problem 12.1
Naming Alcohols

Write the IUPAC name for each of the following alcohols:

a. 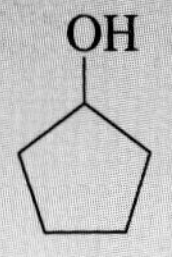 (cyclopentane ring with —OH)

b. 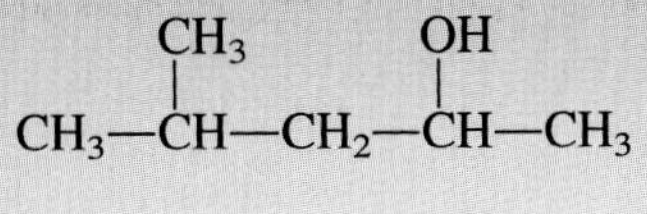
```
        CH3         OH
        |           |
  CH3—CH—CH2—CH—CH3
```

Solution:

a. The —OH group is attached to a cyclopentane ring. Changing the *e* to *ol* gives the IUPAC name of *cyclopentanol.*

b. The parent chain is pentane; the alcohol is named *pentanol.* The carbon chain is numbered to give the position of the —OH group on carbon 2 and the methyl group on carbon 4. The compound is named *4-methyl-2-pentanol.*

Study Check:
Give both IUPAC and common names of the following compound.

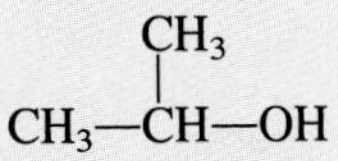

```
      CH3
      |
CH3—CH—OH
```

Answer: 2-propanol (IUPAC name); isopropyl alcohol (common name)

Sample Problem 12.2
Writing Structural Formulas of Alcohols

The alcohol 3,5,5-trimethyl-1-hexanol is used as a plasticizer, a substance added to plastics to keep them pliable. Write its structure.

Solution:
The parent chain of six carbon atoms has an —OH group on carbon 1. Methyl groups are bonded to carbon 3 and to carbon 5.

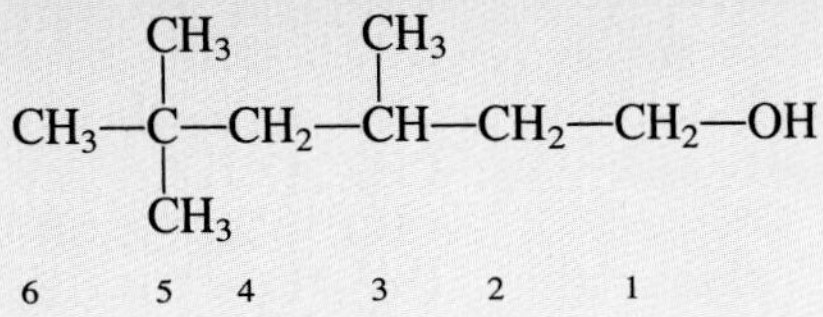

```
       CH3        CH3
       |          |
CH3—C—CH2—CH—CH2—CH2—OH
       |
       CH3
 6     5   4      3     2     1
```

12.2 Physical Properties of Alcohols

Learning Goal **Describe the effect of hydrogen bonding on the boiling points and solubilities of alcohols.**

The electronegative oxygen atom in an alcohol makes the O—H bond very polar. The hydrogen atom is strongly attracted to oxygen atoms in other alcohol molecules. This attraction called hydrogen bonding as illustrated in Figure 12.2 gives alcohols much higher boiling than most other classes of organic compounds. See Table 12.1.

Figure 12.2
Hydrogen bonding occurs between polar hydroxyl groups of methanol molecules.

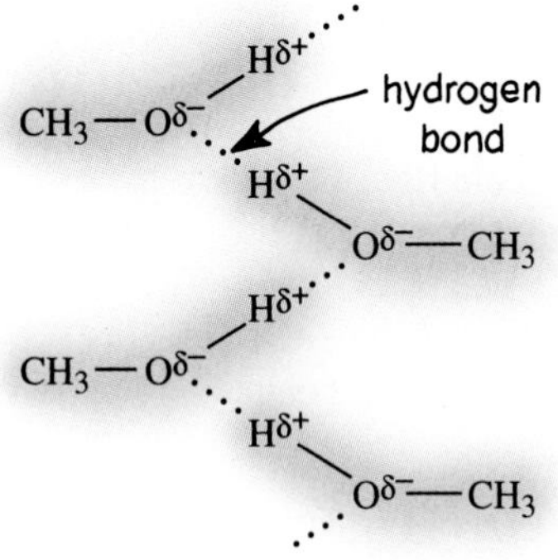

Solubility of Alcohols in Water

Compared to the alkanes, the alcohols with one to four carbon atoms are soluble in water because alcohols can also form hydrogen bonds with water. See Figure 12.3. As the number of carbon atoms increases, the polar —OH group in the alcohol has less effect on solubility and the alcohol acts more like an alkane. Alcohols having five or more carbon atoms are not very soluble in water. The solubilities of some common alcohols are listed in Table 12.2.

Table 12.1 Boiling Points of Some Typical Alcohols and Their Corresponding Alkanes

Name of alkane	Formula	Boiling point (°C)	Name	Formula	Boiling point (°C)
Methane	CH_4	−161	Methanol	CH_3OH	64
Ethane	CH_3CH_3	−88	Ethanol	CH_3CH_2OH	78
Propane	$CH_3CH_2CH_3$	−42	Propanol	$CH_3CH_2CH_2OH$	97
Butane	$CH_3CH_2CH_2CH_3$	0	1-Butanol	$CH_3CH_2CH_2CH_2OH$	117

Figure 12.3
Hydrogen bonding of the polar hydroxyl groups in methanol with water.

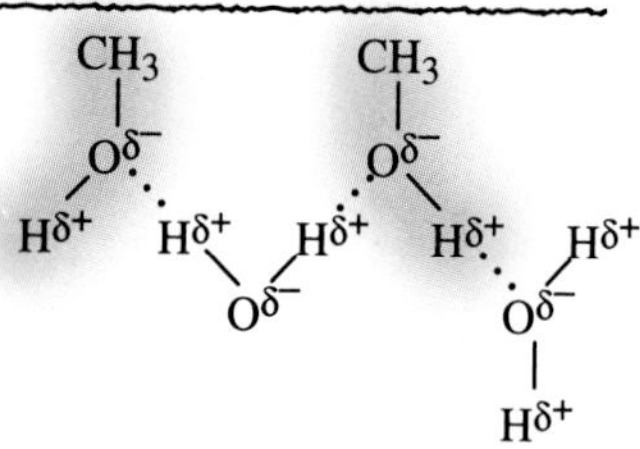

Table 12.2 Solubilities of Some Alcohols in Water

Name	Formula	Solubility (g/100 g H_2O)
Methanol	CH_3OH	any amount
Ethanol	CH_3CH_2OH	any amount
1-Propanol	$CH_3CH_2CH_2OH$	any amount
1-Butanol	$CH_3CH_2CH_2CH_2OH$	8.3
1-Pentanol	$CH_3CH_2CH_2CH_2CH_2OH$	2.7
1-Hexanol	$CH_3CH_2CH_2CH_2CH_2CH_2OH$	0.6

12.3 Phenols

Learning Goal **Write the name and the structural formula of a phenol.**

In phenol, a hydroxyl group is bonded to a benzene ring. The behavior of phenols differs considerably from that of alkyl alcohols. In water, phenol dissociates to give an acidic solution. In fact, an early name for phenol was *carbolic acid*.

HEALTH NOTE

Derivatives of Phenol Are Found in Antiseptics, Essential Oils, and Antioxidants

Derivatives of phenol are used as antiseptics in throat lozenges and mouthwashes. In household disinfectant sprays such as Lysol, the active ingredient is *o*-phenylphenol.

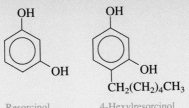

Resorcinol (an antiseptic) | 4-Hexylresorcinol (an antiseptic) | *o*-Phenylphenol (a disinfectant)

Figure 12.4
Derivatives of phenol are found in the essential oils of vanilla, cloves, nutmeg, thyme, and mint.

Several of the essential plant oils that produce the odor or flavor associated with that plant contain derivatives of phenol. Eugenol is found in cloves, isoeugenol in nutmeg, vanillin in the vanilla bean, and thymol in thyme and mint. Thymol has a pleasant, minty taste and is also used in mouthwashes and by dentists to disinfect a cavity in preparation for filling compound. (See Figure 12.4.)

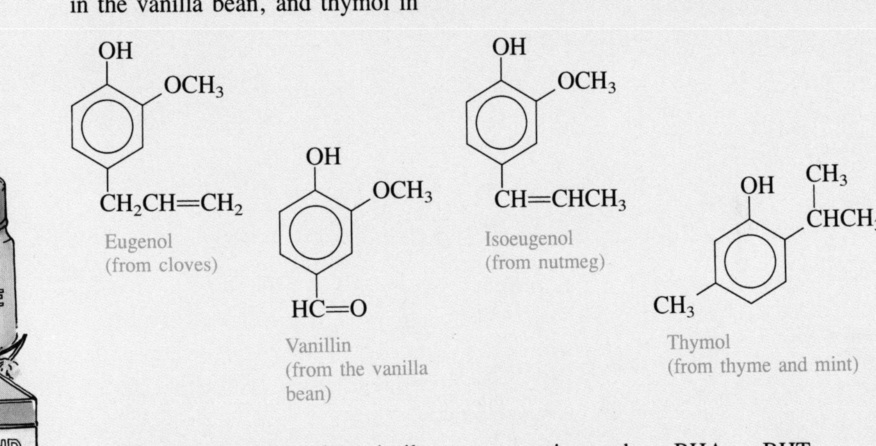

Eugenol (from cloves) | Vanillin (from the vanilla bean) | Isoeugenol (from nutmeg) | Thymol (from thyme and mint)

Foods such as cereals and oils spoil when they react with the oxygen in the air. One of the ways to increase their shelf life is to add a preservative such as BHA or BHT. When these preservatives are present, they react with oxygen to slow food spoilage.

BHA (butylated hydroxyanisole) | BHT (butylated hydroxytoluene)

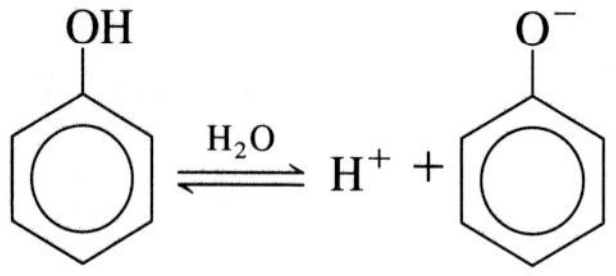

OH $\xrightleftharpoons{H_2O}$ H^+ + O^-

A concentrated solution of phenol is very corrosive and highly irritating to the skin, causing severe burns; ingestion can be fatal. Solutions containing phenol should be used with caution. Dilute solutions of phenol were previously used in hospitals as antiseptics, but they have generally been replaced.

When there are side groups on phenol, the carbon atom attached to the —OH group is designated as carbon 1.

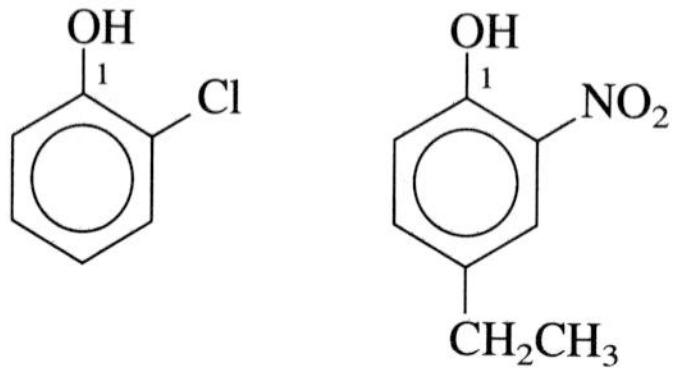

OH 1 Cl — *o*-Chlorophenol

OH 1 NO_2 CH_2CH_3 — 4-Ethyl-2-nitrophenol

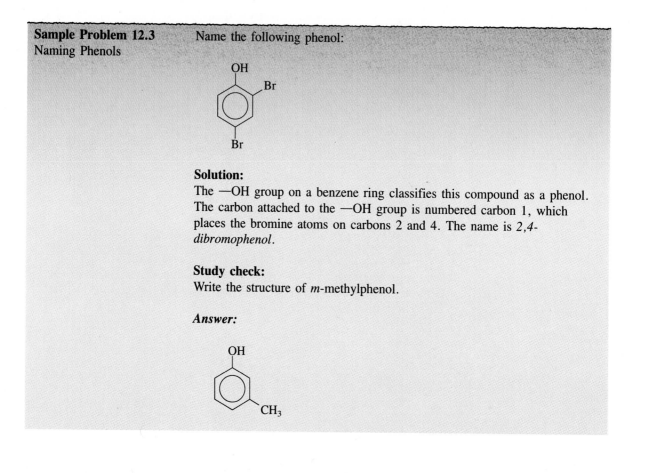

Sample Problem 12.3
Naming Phenols

Name the following phenol:

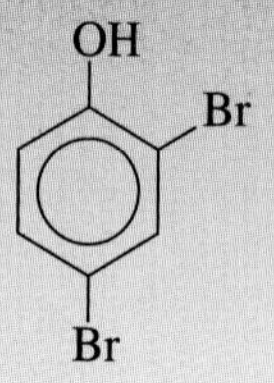

OH Br Br

Solution:
The —OH group on a benzene ring classifies this compound as a phenol. The carbon attached to the —OH group is numbered carbon 1, which places the bromine atoms on carbons 2 and 4. The name is *2,4-dibromophenol*.

Study check:
Write the structure of *m*-methylphenol.

Answer:

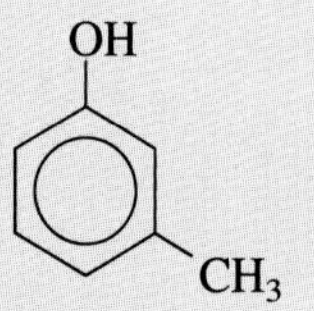

OH CH_3

12.4 Dehydration of Alcohols

Learning Goal **Write the structural formula of the alkene or the ether produced by the dehydration of an alcohol.**

An alcohol can undergo **dehydration,** a reaction in which a molecule of water is removed to form an alkene. This reaction requires heat and an acid catalyst such as H_2SO_4. The water molecule is produced by removing an —H and an —OH from adjacent carbon atoms. A double bond forms between those same carbon atoms to produce an alkene.

Dehydration of an Alcohol

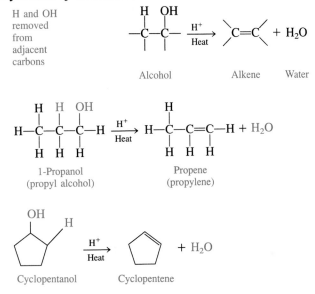

H and OH removed from adjacent carbons

$$\underset{\text{Alcohol}}{-\overset{\overset{\text{H}}{|}}{\underset{|}{\text{C}}}-\overset{\overset{\text{OH}}{|}}{\underset{|}{\text{C}}}-} \xrightarrow[\text{Heat}]{H^+} \underset{\text{Alkene}}{>C=C<} + \underset{\text{Water}}{H_2O}$$

$$\underset{\substack{\text{1-Propanol}\\ \text{(propyl alcohol)}}}{H-\overset{\overset{H}{|}}{\underset{\underset{H}{|}}{C}}-\overset{\overset{H}{|}}{\underset{\underset{H}{|}}{C}}-\overset{\overset{OH}{|}}{\underset{\underset{H}{|}}{C}}-H} \xrightarrow[\text{Heat}]{H^+} \underset{\substack{\text{Propene}\\ \text{(propylene)}}}{H-\overset{\overset{H}{|}}{\underset{\underset{H}{|}}{C}}-\underset{\underset{H}{|}}{C}=\underset{\underset{H}{|}}{C}-H} + H_2O$$

$$\text{Cyclopentanol} \xrightarrow[\text{Heat}]{H^+} \text{Cyclopentene} + H_2O$$

Cyclopentanol   Cyclopentene

When the hydrogen atom of water can be removed from carbon atoms on either side of the alcohol carbon, the major product is the one that comes from removing the hydrogen from the carbon atom that has the lowest number of hydrogen atoms.

Adjacent carbon with fewer H atoms

$$\underset{\text{2-Butanol}}{CH_3-CH_2-\overset{\overset{OH}{|}}{CH}-CH_3} \xrightarrow[\text{Heat}]{H^+} \underset{\substack{\text{2-Butene}\\ \text{(major product:}\\ \text{90\%)}}}{CH_2-CH=CH-CH_3} + \underset{\substack{\text{1-Butene}\\ \text{(minor product:}\\ \text{10\%)}}}{CH_3-CH_2-CH=CH_2}$$

Sample Problem 12.4
Writing Dehydration Equations

Write the structural formula of the dehydration product of the following alcohol:

$$\mathrm{CH_3{-}CH_2{-}\overset{\overset{\Large OH}{|}}{C}H{-}CH_2{-}CH_3 \xrightarrow[\text{Heat}]{H^+}}$$

Solution:
Dehydration of an alcohol occurs in the presence of an acid (H^+) and heat. The H and OH parts of water are taken from adjacent carbon atoms of the alcohol to form the double bond.

$$\underset{\text{3-Pentanol}}{\mathrm{CH_3{-}CH_2{-}\overset{\overset{\Large OH}{|}}{C}H{-}CH_2{-}CH_3}} \xrightarrow[\text{Heat}]{H^+} \underset{\text{2-Pentene}}{\mathrm{CH_3{-}CH_2{-}CH{=}CH{-}CH_3}} + \mathrm{H_2O}$$

Study Check:
What is the name of the dehydration product of cyclopentanol?

Answer: cyclopentene

12.5 Ethers

Learning Goal **Write the name and the structural formula of an ether.**

Figure 12.5
Model of dimethyl ether.

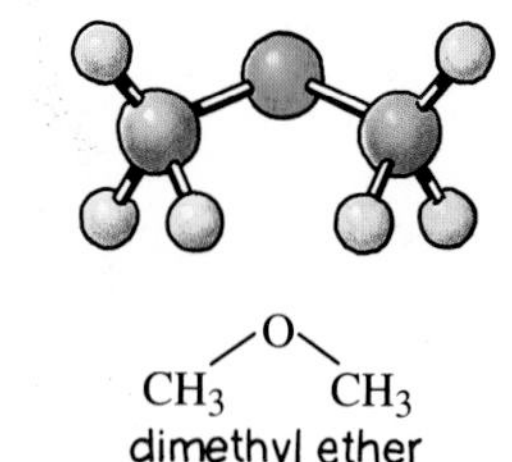

$$\mathrm{CH_3{-}O{-}CH_3}$$
dimethyl ether

Ethers are a family of organic compounds that can be thought of as derivatives of water. In an alcohol, one hydrogen atom of a water molecule is replaced by an alkyl (R) group; in an ether, both hydrogen atoms in water are replaced. (See Figure 12.5.)

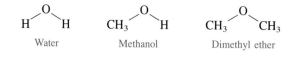

$$\underset{\text{Water}}{\mathrm{H{-}O{-}H}} \qquad \underset{\text{Methanol}}{\mathrm{CH_3{-}O{-}H}} \qquad \underset{\text{Dimethyl ether}}{\mathrm{CH_3{-}O{-}CH_3}}$$

Naming Ethers

In the IUPAC system, the smaller alkyl group is named as an alkoxy group that is attached to the parent alkane chain. However, most ethers use their common names, in which the alkyl groups attached to the oxygen atom are listed in alphabetical order followed by the word *ether*. If the groups are identical, the prefix *di* may be used, but it is optional.

Naming Ethers R—O—R

Functional group: —O—

IUPAC name: Alkoxy alkane

Common name: Alkyl alkyl ether
or dialkyl ether

$CH_3—O—CH_3$	$CH_3CH_2—O—CH_3$	$CH_3CH_2—O—CH_2CH_2CH_3$
Methoxy methane (dimethyl ether; methyl ether)	Methoxy ethane (ethyl methyl ether)	1-Ethoxy propane (ethyl propyl ether)

$$\begin{array}{c} \quad OCH_3 \\ \quad | \\ CH_3CHCH_2CH_3 \end{array}$$

2-Methoxy butane

When there is a benzene ring in an ether, it is named as a phenyl group. The simplest aromatic ether, methyl phenyl ether, is commonly called *anisole*.

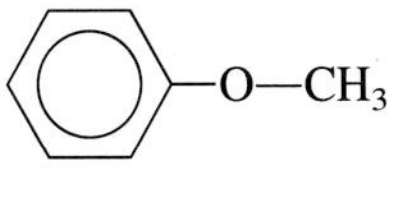

$C_6H_5—O—CH_3$ — Methyl phenyl ether (anisole)

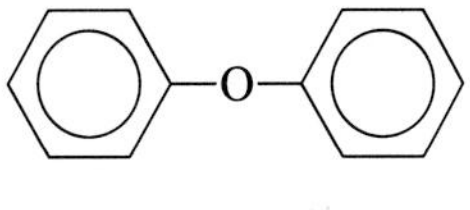

$C_6H_5—O—C_6H_5$ — Diphenyl ether (phenyl ether)

Sample Problem 12.5
Naming Ethers

Name the following ether:

$CH_3—O—CH_2CH_3$

Solution:
In this ether, a methyl ($CH_3—$) and an ethyl ($CH_3CH_2—$) group are attached to the oxygen atom. Naming the alkyl groups alphabetically gives the name of *ethyl methyl ether*.

In the IUPAC system, the methyl group attached to the oxygen atom is named *methoxy*. Since it is attached to a two-carbon chain, the IUPAC name is *methoxy ethane*.

Study Check:
What are two common names for the following ether?

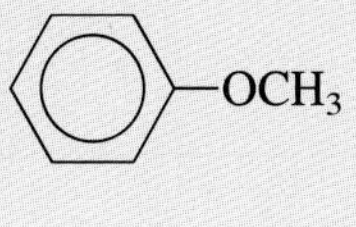

$C_6H_5—OCH_3$

Answer: Methyl phenyl ether; anisole

Preparation of an Ether

An ether can be prepared by heating an alcohol in the presence of an acid catalyst. Two alcohol molecules join, splitting out a water molecule. (The OH group of one alcohol molecule and an H from the other alcohol form the water molecule.) The remaining portions of the alcohol molecules join, forming the ether.

$$\underset{\text{Alcohol}}{R{-}O{-}H} + \underset{\text{Alcohol}}{H{-}O{-}R} \xrightarrow[\text{Heat}]{H^+} \underset{\text{Ether}}{R{-}O{-}R} + \underset{\text{Water}}{H{-}OH}$$

$$CH_3{-}OH + HO{-}CH_3 \xrightarrow[\text{Heat}]{H^+} CH_3{-}O{-}CH_3 + H_2O$$

This may also be written as

$$2CH_3OH \xrightarrow[\text{Heat}]{H^+} CH_3{-}O{-}CH_3 + H_2O$$

Methanol (methyl alcohol) — Methoxy methane (methyl ether)

Sample Problem 12.6
Ether Formation

Write the structural formula of the ether produced by the following dehydration:

$$2CH_3CH_2OH \xrightarrow[\text{Heat}]{H^+}$$

Ethanol

Solution:
The structural formula of the ether product is written by attaching the alkyl groups of the reacting alcohols to an oxygen atom. The removal of an H and an OH produces a water molecule.

Ether link ↓

$$CH_3CH_2OH + HOCH_2CH_3 \xrightarrow[\text{Heat}]{H^+} CH_3CH_2{-}O{-}CH_2CH_3 + H_2O$$

Ethanol (ethyl alcohol) — Ethoxy ethane (ethyl ether)

Boiling Points of Ethers

Ethers have lower boiling points than alcohols of the same molecular weight because there are no hydrogen atoms bonded to the oxygen atom to form hydrogen bonds with other ether molecules. (See Table 12.3.)

HEALTH NOTE

Ethers as Anesthetics

A general anesthetic causes a loss of all sensation by inhibiting the ability of nerve cells to send signals of pain to the brain. (See Figure 12.6.) Diethyl ether, or "ether," was used as an anesthetic for over a hundred years. It has anesthetic properties over a wide concentration range, and minimal side effects. However, it is extremely volatile at room temperature, and there is always a danger of combustion. For this reason, the use of diethyl ether as an anesthetic was discontinued. Other ethers that have been used as anesthetics are Enflurane, Vinethene, and methoxyflurane. They are less flammable than diethyl ether, but because of side effects many have been replaced by halothane.

Figure 12.6
Anesthetics used during surgery produce a loss of all sensation.

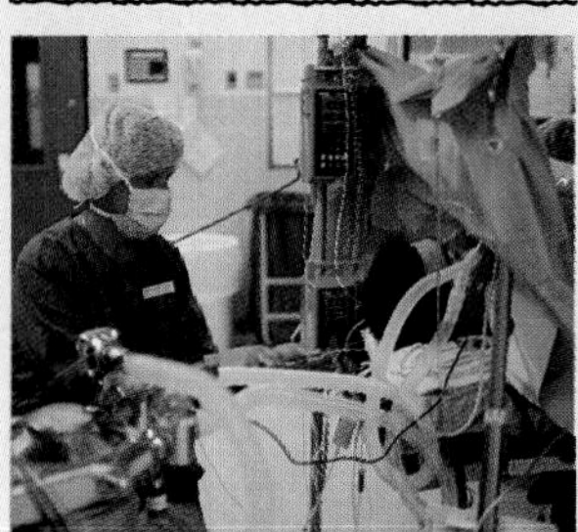

$CH_2{=}CH{-}O{-}CH{=}CH_2$

Vinethene

$F{-}CHF{-}O{-}CF_2{-}CHF{-}Cl$ (structural formula: F above and H below the first C; F above and F below the second C; F above and H below the third C)

Enflurane

$Cl{-}CHBr{-}CF_2{-}F$ (structural formula: Br above and H below the first C; F above and F below the second C)

Halothane

$Cl{-}CHCl{-}CF_2{-}O{-}CH_2{-}H$ (structural formula: Cl above and H below the first C; F above and F below the second C; H above and H below the third C)

Penthrane

Table 12.3 Boiling Points for Some Alcohols and Ethers Having the Same Molecular Formulas

Alcohol	Boiling point (°C)	Ether	Boiling point (°C)
CH_3CH_2OH	78	CH_3OCH_3	−23
$CH_3CH_2CH_2OH$	82	$CH_3CH_2OCH_3$	11
$CH_3CH_2CH_2CH_2OH$	117	$CH_3CH_2OCH_2CH_3$	35
$CH_3CH_2CH_2CH_2CH_2OH$	138	$CH_3CH_2CH_2OCH_2CH_3$	88

Solubility of Ethers

Ethers are also less soluble in water than alcohols are, but more soluble than alkanes. The shorter-chain ethers such as methyl ether and ethyl ether are somewhat soluble in water because the polar oxygen in an ether can form a hydrogen bond with water. However, ethers that contain larger alkane portions are not soluble in water.

$$H^{\delta+}{-}O^{\delta-}{-}H^{\delta+}\cdots O^{\delta-}(R){-}R$$

Hydrogen bond

Water Ether

Uses of Ethers

Ethers—diethyl ether in particular—are used as solvents because they are less reactive than many other organic materials and do not react with the substances they dissolve. Also, their low boiling points allow them to be easily separated from the solutes. However, they are extremely flammable and the utmost care must be used when working with them.

12.6 Thiols and Sulfides

Learning Goal **Identify the functional group of a thiol or a sulfide.**

The **thiols** (also called mercaptans) are similar to alcohols but have —SH functional groups rather than —OH groups. They can be thought of as derivatives of hydrogen sulfide.

H—S—H	CH_3—S—H
Hydrogen sulfide	Methanethiol (methylmercaptan)

In the IUPAC system, thiols are named by adding the suffix *thiol* to the parent name. The location of the —SH group is shown by numbering the parent chain. The common name is obtained by writing the alkyl group name followed by the word mercaptan.

Figure 12.7
The characteristic odors of skunks and garlic are due to thiols and sulfides.

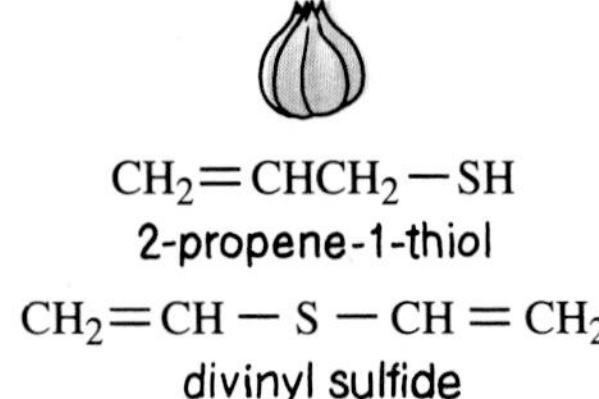

Naming Thiols R—SH

Functional group: —SH

IUPAC name: Alkanethiol

Common name: Alkyl mercaptan

CH_3—SH	CH_3CH_2—SH	$CH_3CH(SH)CH_2CH(CH_3)CH_3$
Methanethiol (methyl mercaptan)	Ethanethiol (ethyl mercaptan)	4-Methyl-2-pentanethiol

Many short-chain thiols have strong, disagreeable odors. (See Figure 12.7.) The smell of natural gas is not that of the methane, which is odorless, but of a small amount of ethanethiol that has been added to help detect a gas leak. The odor that a skunk emits as a protective device is due to two thiols. The characteristic odor of onions is due to 1-propanethiol, which is also a lachrymator, a substance that will make your eyes tear.

$CH_3CH_2CH_2SH$	$CH_3CH(CH_3)CH_2CH_2SH$	H(H_3C)C=C(H)CH_2SH
1-Propanethiol (in onions)	3-Methyl-1-butanethiol (in skunk odor)	*trans*-2-Butene-1-thiol (in skunk odor)

Sulfides

Sulfides are analogous to ethers because they contain a —S— group in place of an oxygen atom. The naming of a sulfide is similar to ether naming. The names of the alkyl groups attached to the sulfide link are listed followed by the word *sulfide*.

$CH_3—S—CH_3$ Dimethyl sulfide

$CH_3—S—CH_2CH_3$ Ethyl methyl sulfide

The distinctive aroma of garlic is due to a thiol and a sulfide.

Odor of garlic

$CH_2{=}CH—CH_2—SH$ 2-Propene-1-thiol

$CH_2{=}CH—S—CH{=}CH_2$ Divinyl sulfide

Figure 12.8
In a perm, the disulfide bonds in hair protein are reduced so they break apart, and then oxidized so they form disulfide bonds again.

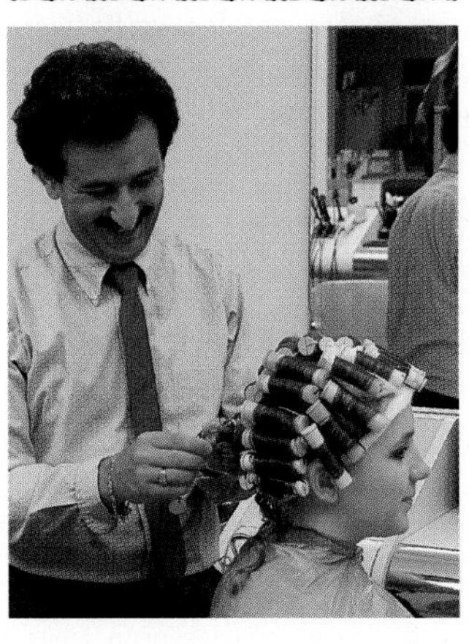

Disulfides

When two sulfur atoms link together to give a —S—S— group, the compound is called a disulfide.

$CH_3—S—S—CH_3$ Dimethyl disulfide

Disulfides are formed by the reaction of two thiols. Such a reaction occurs in hair when a permanent wave is given. Much of the protein in the hair is cross-linked by disulfide bonds. When a permanent is given, a reducing substance is put on the hair to break apart the disulfide bonds. Then, the hair is wrapped around curlers and an oxidizing substance is used that causes the disulfide bonds to form between protein strands that take the shape of the curlers. (See Figure 12.8.)

Oxidation of Thiols

$$CH_3—S—H + H—S—CH_3 \xrightarrow{\text{Oxidation}} CH_3—S—S—CH_3 + H_2O$$

Sample Problem 12.7
Identifying Thiols and Sulfides

Identify each of the following sulfur compounds as a thiol, a sulfide, or a disulfide:

a. $CH_3CH_2—S—CH_3$ **b.** CH_3SH **c.** $CH_3—S—S—CH_3$

Solution:

a. A sulfur atom makes this a sulfide.
b. The —SH group makes this a thiol.
c. Two sulfur atoms makes this a disulfide.

12.7 Aldehydes and Ketones

Learning Goal **Write the name and the structural formula of an aldehyde or a ketone.**

Aldehydes and ketones contain a functional group called the carbonyl group, one of the most important functional groups in organic chemistry and biochemistry. In **aldehydes,** the carbonyl group is attached to a carbon atom and a hydrogen atom. (In formaldehyde, it is bonded to two hydrogen atoms.) In **ketones,** the carbonyl group is attached to two carbon atoms. (See Figure 12.9.)

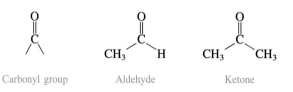

Figure 12.9
Models of aldehydes and ketones: formaldehyde, acetaldehyde, and acetone.

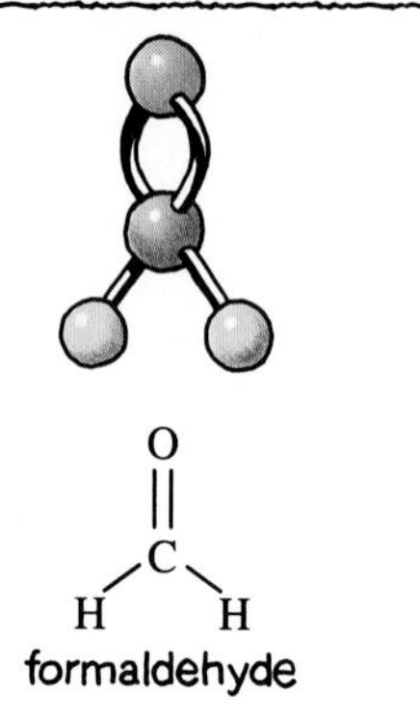

O
‖
CH_3CH
acetaldehyde

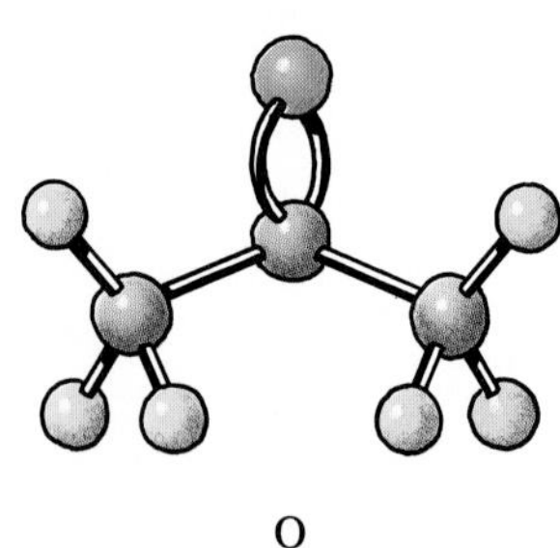

O
‖
CH_3CCH_3
acetone

O ‖ C (with two bonds)	O ‖ CH_3—C—H	O ‖ CH_3—C—CH_3
Carbonyl group	Aldehyde	Ketone

Naming Aldehydes

In the IUPAC system, an aldehyde is named by replacing the *e* in the parent alkane name with *al*. For small aldehydes, common names are generally used. The common names of the first four aldehydes use the prefixes *form, acet, propion,* and *butyr* followed by the suffix, *aldehyde*.

Naming Aldehydes ***R—CH(=O)***

Functional group: —CH(=O)

IUPAC name: Alkanal

Common name: (prefix) Aldehyde

H—C(=O)—H	CH_3—CH(=O)	CH_3—CH_2—CH(=O)
Methanal (formaldehyde)	Ethanal (acetaldehyde)	Propanal (propionaldehyde)

The carbonyl group in an aldehyde is always the first carbon in the chain; no number is needed. Side groups are located by numbering the carbon chain from the carbonyl group. In the common name, the position of a side group is indicated by the Greek letters α (alpha), β (beta), γ (gamma), and δ (delta). The alpha carbon is the carbon next to the carbonyl group.

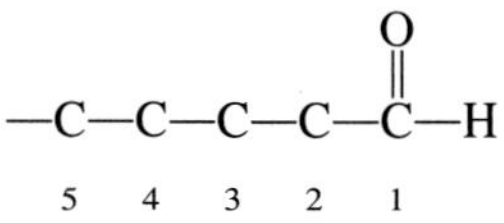

$$-\overset{}{C}-C-C-C-\overset{O}{\overset{\|}{C}}-H$$
5 4 3 2 1

Numbering the aldehyde in the IUPAC system

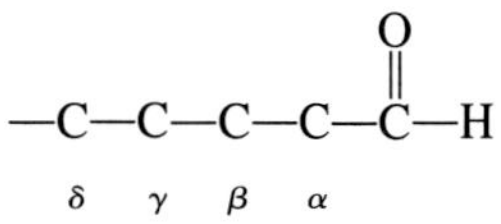

$$-C-C-C-C-\overset{O}{\overset{\|}{C}}-H$$
δ γ β α

The common names of aldehydes use Greek letters

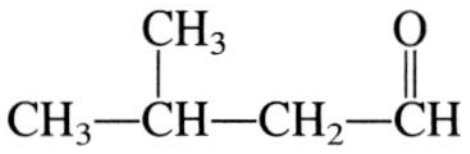

$$CH_3-\overset{CH_3}{\overset{|}{C}H}-CH_2-\overset{O}{\overset{\|}{C}H}$$

3-Methylbutanal
(β-methylbutyraldehyde)

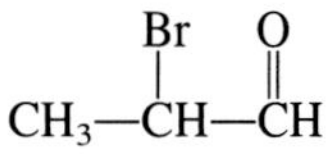

$$CH_3-\overset{Br}{\overset{|}{C}H}-\overset{O}{\overset{\|}{C}H}$$

2-Bromopropanal
(α-bromopropionaldehyde)

The aldehyde of benzene is usually called benzaldehyde.

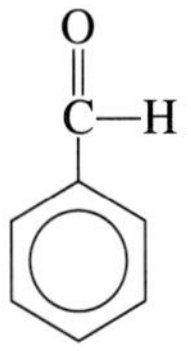

$$C_6H_5-\overset{O}{\overset{\|}{C}}-H$$

Benzaldehyde

Sample Problem 12.8
Naming Aldehydes

Give the IUPAC name and the common name for the following aldehyde:

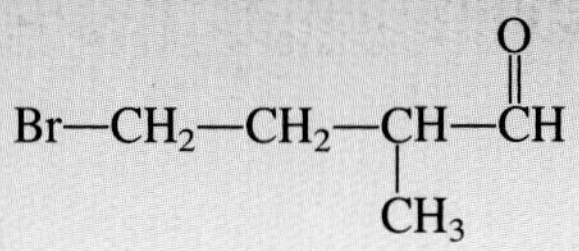

$$Br-CH_2-CH_2-\underset{CH_3}{\underset{|}{C}H}-\overset{O}{\overset{\|}{C}H}$$

Solution:
The IUPAC name based on the parent chain of butane, numbered from the carbonyl group, is *4-bromo-2-methylbutanal.* Since the parent chain of this aldehyde consists of four carbon atoms, its common name is *butyraldehyde*. Using Greek letters to locate the side groups gives the common name *γ-bromo-α-methylbutyraldehyde*.

Study Check:
The simplest sugar is called glyceraldehyde. From its structure, give its IUPAC and common name. (As a side group, an —OH is called hydroxy.)

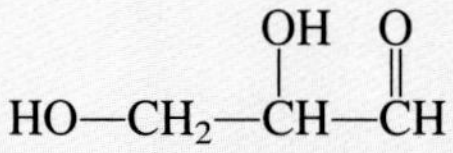

$$HO-CH_2-\overset{OH}{\overset{|}{C}H}-\overset{O}{\overset{\|}{C}H}$$

Answer: 2,3-dihydroxypropanal (IUPAC name); α,β-dihydroxypropionaldehyde (common name)

Naming Ketones

In the IUPAC system, a ketone is named by using the parent name of the longest carbon chain containing the carbonyl group and changing the *e* of the alkane name to *one*. If a ketone has five or more carbon atoms, the carbonyl group is numbered. In a cyclic ketone, the carbonyl group is carbon 1. In the common names of ketones, the alkyl groups bonded to the carbonyl group are named followed by the word *ketone*.

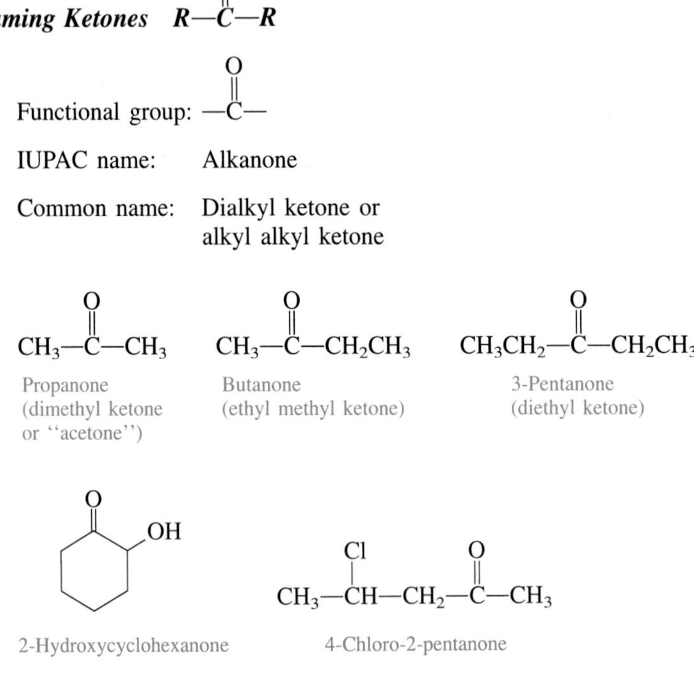

Many ketones are good solvents for organic compounds and find use in products such as paint strippers as shown in Figure 12.10.

Figure 12.10
Acetone is the solvent in fingernail polish remover, ethyl methyl ketone is used in paint removers, and Formalin is used to preserve tissues.

Sample Problem 12.9
Naming Ketones

Write the IUPAC and the common name for the following ketone:

$$CH_3CH_2CH_2-\overset{\overset{\displaystyle O}{\|}}{C}-CH_2CH_3$$

Solution:
For the IUPAC name, we use the entire carbon chain of six carbon atoms that includes the ketone group, giving the name *hexanone*. Since the carbonyl group is on the third carbon atom in the chain, the ketone is named as *3-hexanone*. The common name for this ketone, determined by alphabetically listing the alkyl groups attached to the carbonyl group, is *ethyl propyl ketone*.

Study Check:
What is the IUPAC name of the following ketone?

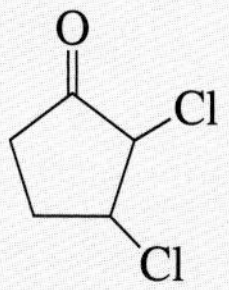

O
Cl
Cl

Answer: 2,3-dichlorocyclopentanone

Sample Problem 12.10
Writing the Structural Formula of a Ketone

Write the structural formula of 4-methyl-2-pentanone.

Solution:
The parent name indicates a chain of five carbons with a carbonyl group on the second carbon. A methyl group is attached to the fourth carbon.

$$\underset{1}{CH_3}-\underset{2}{\overset{\overset{\displaystyle O}{\|}}{C}}-\underset{3}{CH_2}-\underset{4}{\overset{\overset{\displaystyle CH_3}{|}}{CH}}-\underset{5}{CH_3}$$

Ketone group on carbon 2

Methyl group on carbon 4

A five-carbon parent chain is numbered from the end nearest the carbonyl group

Physical Properties of Aldehydes and Ketones

The carbonyl groups of aldehydes and ketones give them sufficient polarity to make their boiling points higher than those of nonpolar organic compounds, such as alkanes and ethers. However, no hydrogen bonding occurs in aldehydes or ketones, so their boiling points are lower than corresponding alcohols. In Table 12.4, the boiling points of some organic substances of comparable molar mass are shown.

Table 12.4 Boiling Points of Selected Organic Substances of Similar Mass

Substance	Molar mass	Formula	Boiling point (°C)
Pentane	72	$CH_3CH_2CH_2CH_2CH_3$	36
Diethyl ether	74	$CH_3CH_2—O—CH_2CH_3$	35
Butyraldehyde	72	$CH_3CH_2CH_2\overset{\overset{\displaystyle O}{\|}}{C}H$	76
2-Butanone	72	$CH_3\overset{\overset{\displaystyle O}{\|}}{C}CH_2CH_3$	80
1-Butanol	74	$CH_3CH_2CH_2CH_2OH$	118

Aldehydes and ketones having one to four carbons are soluble in water because the polar oxygen can hydrogen bond with water, pulling the small alkane chains into solution.

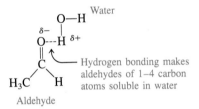

Water
O—H
δ−
O---H δ+
C
H_3C H
Hydrogen bonding makes aldehydes of 1–4 carbon atoms soluble in water
Aldehyde

HEALTH NOTE

Aldehydes and Ketones in Health

A 40 percent aqueous solution of formaldehyde called formalin is used as a germicide and as a preservative for tissues. Formaldehyde is also used in the manufacturing of paper, insulation materials, and cosmetics, including some shampoos. There is concern that formaldehyde is a potential carcinogen and should be eliminated from household substances.

Propanone (also called acetone or dimethyl ketone), is an excellent solvent for many organic materials. It is found in paint removers and fingernail polish remover. In the body, acetone may be produced when large amounts of fats are metabolized for energy. This happens in uncontrolled diabetes, fasting, and high-protein diets. The odor of acetone can be detected on the person's breath.

The monosaccharides or sugars are compounds having an aldehyde or ketone group and several hydroxyl groups.

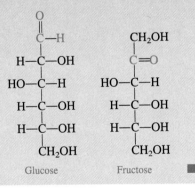

Glucose:
O
‖
C—H
H—C—OH
HO—C—H
H—C—OH
H—C—OH
CH_2OH

Fructose:
CH_2OH
C=O
HO—C—H
H—C—OH
H—C—OH
CH_2OH

Glucose Fructose

12.8 Preparation of Aldehydes and Ketones

Learning Goal **Given the name or the structural formula of an alcohol, write the aldehyde or the ketone produced by oxidation.**

Aldehydes and ketones can be prepared from alcohols. However, before we can discuss these reactions, it is necessary to classify alcohols.

Classification of Alcohols

Alcohols are classified as primary (1°), secondary (2°), or tertiary (3°), according to the number of alkyl groups bonded to the carbon atom attached to the —OH group. As we have seen, the letter R in the structural formula of an alcohol represents an alkyl group.

A **primary** (1°) **alcohol** has one alkyl group attached to the carbon bonded to the —OH group. Methanol is included in the primary alcohol category. In a **secondary** (2°) **alcohol,** there are two alkyl groups attached to the carbon bonded to —OH. When there are three alkyl groups, an alcohol is classified as a **tertiary** (3°) **alcohol.** The prefix *tert* (abbreviated *t*) is used in the common name to indicate a tertiary alcohol. Table 12.5 illustrates the classification of alcohols and gives examples of each.

Table 12.5 Classification of Alcohols

Primary (1°) Alcohol (One Alkyl Group)	Secondary (2°) Alcohol (Two Alkyl Groups)	Tertiary (3°) Alcohol (Three-Alkyl Groups)
R—C(H)(H)—OH (alkyl group: R; alcohol carbon: C)	R—C(H)(R)—OH	R—C(R)(R)—OH
Examples		
$CH_3—CH_2—OH$	$CH_3—CH(OH)—CH_3$	$CH_3—C(OH)(CH_3)—CH_3$
Ethanol (ethyl alcohol)	2-Propanol (isopropyl alcohol)	2-Methyl-2-propanol (*tert*-butyl alcohol)

Sample Problem 12.11
Classifying Alcohols

Classify the following as primary (1°), secondary (2°), or tertiary (3°) alcohols:

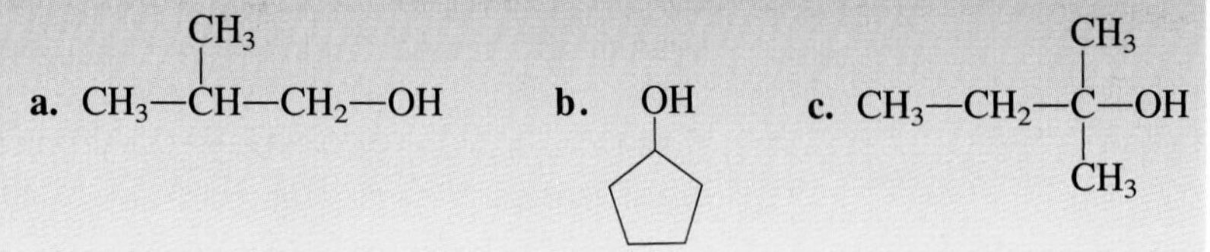

a. $CH_3-\underset{}{\overset{CH_3}{\overset{|}{C}}}H-CH_2-OH$ b. OH on cyclopentane ring c. $CH_3-CH_2-\underset{CH_3}{\underset{|}{\overset{CH_3}{\overset{|}{C}}}}-OH$

Solution:

a. The carbon atom bonded to the —OH group is attached to one alkyl group. This is a primary (1°) alcohol.
b. The carbon bonded to the —OH group is bonded to two carbon atoms in the cyclic structure. This is a secondary (2°) alcohol.
c. The carbon bonded to the —OH group is attached to three alkyl groups. This is a tertiary (3°) alcohol.

Study Check:
Classify the following alcohol:

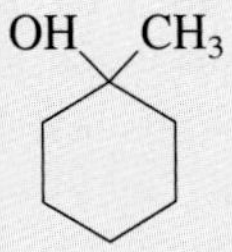

(cyclohexane ring bearing OH and CH_3 on the same carbon)

Answer: This is a tertiary (3°) alcohol.

Oxidation of Alcohols

The **oxidation** of alcohols occurs when two hydrogen atoms are removed by an oxidizing agent such as potassium dichromate, $K_2Cr_2O_7$. A hydrogen is removed from the —OH group, and one from its carbon producing a product with a carbonyl group. The hydrogens and an oxygen provided by the oxidizing agent give the product, water.

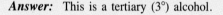

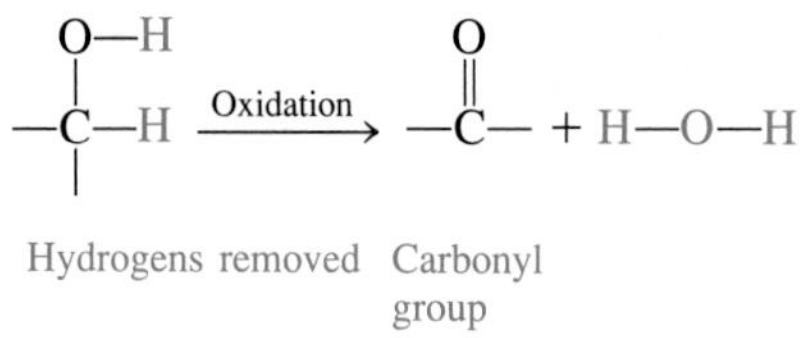

$$-\underset{|}{\overset{O-H}{\overset{|}{C}}}-H \xrightarrow{\text{Oxidation}} -\overset{O}{\overset{\|}{C}}- + H-O-H$$

Hydrogens removed Carbonyl group

Preparation of Aldehydes

When a primary alcohol is oxidized, the product is an aldehyde.

$$\text{Primary alcohol} \xrightarrow{\text{Oxidation}} \text{Aldehyde} + \text{H—O—H}$$

$$CH_3—\overset{\overset{\displaystyle O—H}{|}}{\underset{\underset{\displaystyle H}{|}}{C}}—H \xrightarrow{\text{Oxidation}} CH_3—\overset{\overset{\displaystyle O}{\|}}{C}—H + H_2O$$

Ethanol (ethyl alcohol) — Ethanal (acetaldehyde)

$$CH_3CH_2CH_2CH_2OH \xrightarrow{\text{Oxidation}} CH_3CH_2CH_2\overset{\overset{\displaystyle O}{\|}}{C}H + H_2O$$

1-Butanol (butyl alcohol) — Butanal (butyraldehyde)

Oxidation of Secondary Alcohols

When a secondary alcohol is oxidized, two hydrogen atoms are removed to produce a ketone.

$$\text{Secondary alcohol} \xrightarrow{\text{Oxidation}} \text{Ketone} + \text{H—O—H}$$

$$CH_3—\overset{\overset{\displaystyle O—H}{|}}{\underset{\underset{\displaystyle H}{|}}{C}}—CH_3 \xrightarrow{\text{Oxidation}} CH_3—\overset{\overset{\displaystyle O}{\|}}{C}—CH_3 + H_2O$$

2-Propanol (isopropyl alcohol) — Propanone (dimethyl ketone)

$$C_6H_{11}—OH \xrightarrow{\text{Oxidation}} C_6H_{10}{=}O + H_2O$$

Cyclohexanol — Cyclohexanone

Tertiary alcohols do not oxidize because there is no hydrogen atom to remove from the carbon bonded to the —OH group. Therefore, tertiary alcohols cannot produce carbonyl groups.

$$\text{Tertiary alcohol} \xrightarrow{\text{Oxidation}} \text{no reaction}$$

$$CH_3—\overset{\overset{\displaystyle CH_3}{|}}{\underset{\underset{\displaystyle CH_3}{|}}{C}}—OH \xrightarrow{\text{Oxidation}} \text{no oxidation occurs}$$

No hydrogen atom to remove

Sample Problem 12.12
Writing Oxidation Products

Draw structural formulas of the products that form when the following alcohol undergoes oxidation:

$$CH_3{-}CH_2{-}\underset{}{\overset{\displaystyle OH}{\overset{|}{C}}}H{-}CH_3 \xrightarrow{\text{Oxidation}}$$

Solution:
Since the reactant is a secondary alcohol, oxidation produces a ketone and water.

$$\underset{\text{2-Butanol}}{CH_3{-}CH_2{-}\overset{\displaystyle OH}{\overset{|}{C}}H{-}CH_3} \xrightarrow{\text{Oxidation}} \underset{\substack{\text{2-Butanone}\\ \text{(ethyl methyl ketone)}}}{CH_3{-}CH_2{-}\overset{\displaystyle O}{\overset{\|}{C}}{-}CH_3} + H_2O$$

Study Check:
Using equations, show the steps that convert propene to propanone.

Answer:

Step 1 Hydration using Markovnikov's rule

$$CH_3{-}CH{=}CH_2 + H_2O \xrightarrow{H^+} CH_3{-}\overset{\displaystyle OH}{\overset{|}{C}}H{-}CH_3$$

Step 2 Oxidation of the secondary alcohol

$$CH_3{-}\overset{\displaystyle OH}{\overset{|}{C}}H{-}CH_3 \xrightarrow{\text{Oxidation}} CH_3{-}\overset{\displaystyle O}{\overset{\|}{C}}{-}CH_3 + H_2O$$

12.9 Reactions of Aldehydes and Ketones

Learning Goal **Write equations for the chemical reactions of aldehydes and ketones; describe some tests for aldehydes.**

Oxidation of Aldehydes

An aldehyde can be easily oxidized to a carboxylic acid. However, ketones are resistant to further oxidation. The family of carboxylic acids is discussed in Chapter 13.

$$\text{Aldehyde} \xrightarrow{\text{Oxidation}} \text{Carboxylic acid}$$

$$CH_3-\overset{\overset{\displaystyle O}{\|}}{C}-H \xrightarrow{\text{Oxidation}} CH_3-\overset{\overset{\displaystyle O}{\|}}{C}-OH$$

Ethanal (acetaldehyde) — Ethanoic acid (acetic acid)

$$CH_3-\overset{\overset{\displaystyle O}{\|}}{C}-CH_3 \xrightarrow{\text{Oxidation}} \text{no further oxidation}$$

Ketone

Tests for Aldehydes

Since aldehydes oxidize easily, some mild oxidizing agents can be used to identify aldehydes. The **Tollens' test** uses Tollens' reagent, an alkaline solution of Ag^+. As the aldehyde is oxidized, the silver ions are reduced to metallic silver, which forms a mirror on the inside of the test tube indicating the presence of the aldehyde.

$$CH_3-\overset{\overset{\displaystyle O}{\|}}{C}-H + Ag^+ \longrightarrow CH_3-\overset{\overset{\displaystyle O}{\|}}{C}-OH + Ag$$

Acetaldehyde — Silver ion — Acetic acid — Forms a silver mirror

In the **Benedict's** and **Fehling's tests,** alkaline solutions of copper(II) ions are added to aldehydes. As the aldehyde is oxidized, the Cu^{2+} ions are reduced to Cu^+, which gives a brick-red precipitate of copper(I) oxide, Cu_2O. Some carbohydrates such as glucose contain aldehyde groups that give positive tests with Benedict's and Fehling's reagents. Such carbohydrates that reduce copper(II) ion are called reducing sugars.

Oxidation of aldehyde group in sugar

$$\begin{array}{c} O \\ \| \\ C-H \\ | \\ H-C-OH \\ | \\ HO-C-H \\ | \\ H-C-OH \\ | \\ H-C-OH \\ | \\ CH_2OH \end{array} + 2Cu^{2+} \xrightarrow{OH^-} \begin{array}{c} O \\ \| \\ C-OH \\ | \\ H-C-OH \\ | \\ HO-C-H \\ | \\ H-C-OH \\ | \\ H-C-OH \\ | \\ CH_2OH \end{array} + Cu_2O(s)$$

Glucose (a reducing sugar) — Blue — Gluconic acid — Brick-red

HEALTH NOTE

Oxidation of Alcohol in the Body

In the liver, enzymes oxidize ethanol to acetaldehyde. This in turn is oxidized to acetic acid, which is finally converted to carbon dioxide and water. Although ethanol can eventually be detoxified by the liver, the formation of its aldehyde and acidic intermediates causes considerable damage within the cells of the liver.

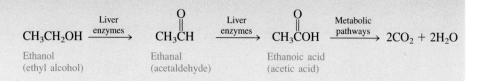

$$\underset{\text{Ethanol (ethyl alcohol)}}{CH_3CH_2OH} \xrightarrow[\text{enzymes}]{\text{Liver}} \underset{\text{Ethanal (acetaldehyde)}}{CH_3\overset{O}{\overset{\|}{C}}H} \xrightarrow[\text{enzymes}]{\text{Liver}} \underset{\text{Ethanoic acid (acetic acid)}}{CH_3\overset{O}{\overset{\|}{C}}OH} \xrightarrow[\text{pathways}]{\text{Metabolic}} 2CO_2 + 2H_2O$$

The toxicity of methanol is caused by its oxidation to formaldehyde, a substance that affects the retina of the eye. Additional oxidation produces formic acid, which is not readily eliminated from the body. Increased levels of formic acid can lower blood pH so severely that a coma is induced.

$$\underset{\text{Methanol}}{CH_3OH} \xrightarrow[\text{enzymes}]{\text{Liver}} \underset{\text{Methanal (formaldehyde)}}{H\overset{O}{\overset{\|}{C}}H} \xrightarrow[\text{enzymes}]{\text{Liver}} \underset{\text{Methanoic acid (formic acid)}}{H\overset{O}{\overset{\|}{C}}OH}$$

Methanol poisoning may be treated in the hospital by giving ethanol to the patient. The enzymes in the liver pick up ethanol molecules to oxidize instead of methanol molecules. This allows the methanol to be eliminated from the body without forming its dangerous oxidation product, formaldehyde. ■

Reduction (Hydrogenation) of Aldehydes and Ketones

In hydrogenation, hydrogen adds to the carbonyl groups of aldehydes and ketones in the presence of a metal catalyst such as platinum or nickel. The addition of hydrogen causes the **reduction** of aldehydes to primary alcohols and ketones to secondary alcohols.

Hydrogenation of Aldehydes Produces Primary (1°) Alcohols:

$$\underset{\text{Propanal (propionaldehyde)}}{CH_3{-}CH_2{-}\overset{O}{\overset{\|}{C}}H} + H_2 \xrightarrow{Pt} \underset{\text{1-Propanol (propyl alcohol)}}{CH_3{-}CH_2{-}\overset{OH}{\overset{|}{C}}H_2}$$

Hydrogenation of Ketones Produces Secondary (2°) Alcohols:

$$CH_3{-}\overset{\overset{\displaystyle O}{\|}}{C}{-}CH_3 + H_2 \xrightarrow{Pt} CH_3{-}\underset{\underset{\displaystyle H}{|}}{\overset{\overset{\displaystyle OH}{|}}{C}}{-}CH_3$$

Propanone (dimethyl ketone) — 2-Propanol (isopropyl alcohol)

Sample Problem 12.13
Hydrogenation of Carbonyl Groups

Write the product for the reduction of cyclopentanone.

Solution:
The reacting molecule is a cycloketone that has five carbon atoms. The carbonyl group will add hydrogen to form the corresponding alcohol.

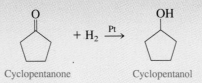

Cyclopentanone $+ H_2 \xrightarrow{Pt}$ Cyclopentanol

Study Check:
What is the name of the product obtained from the hydrogenation of propionaldehyde?

Answer: 1-propanol (IUPAC name); propyl alcohol (common name)

Table 12.6 provides a summary of naming the oxygen and sulfur compounds covered in this chapter, and Table 12.7 gives a summary of their reactions.

Table 12.6 Summary of Naming Organic Oxygen and Sulfur Compounds

Class	IUPAC name	Common name	Structure
Alcohol	Ethanol	Ethyl alcohol	$CH_3CH_2{-}OH$
Phenol	Benzenol	Phenol	$C_6H_5{-}OH$ (benzene ring with OH)
Ether	Methoxy methane	Dimethyl ether	$CH_3{-}O{-}CH_3$
Thiol	Methanethiol	Methyl mercaptan	$CH_3{-}SH$
Sulfide	Dimethyl sulfide		$CH_3{-}S{-}CH_3$
Aldehyde	Propanal	Propionaldehyde	$CH_3CH_2{-}\overset{\overset{\displaystyle O}{\|}}{C}H$
Ketone	Propanone	Dimethyl ketone (acetone)	$CH_3{-}\overset{\overset{\displaystyle O}{\|}}{C}{-}CH_3$

Table 12.7 Summary of Reactions for Organic Oxygen and Sulfur Compounds

Alcohols

Dehydration

$$\text{Alcohol} \xrightarrow{H^+} \text{Alkene} + H_2O$$

$$\overset{\overset{\displaystyle H}{|}}{C}H_2\overset{\overset{\displaystyle OH}{|}}{C}H_2 \xrightarrow{H^+} CH_2{=}CH_2 + H_2O$$

Ether formation

$$\text{Two alcohols} \xrightarrow{H^+} \text{Ether} + H_2O$$

$$CH_3{-}OH + HO{-}CH_3 \xrightarrow{H^+} CH_3{-}O{-}CH_3 + H_2O$$

Oxidation

Alcohol (1°):

$$CH_3{-}CH_2OH \xrightarrow{\text{Oxidation}} CH_3{-}\overset{\overset{\displaystyle O}{\|}}{C}H$$

Alcohol (2°):

$$CH_3{-}\overset{\overset{\displaystyle OH}{|}}{C}H{-}CH_3 \xrightarrow{\text{Oxidation}} CH_3{-}\overset{\overset{\displaystyle O}{\|}}{C}{-}CH_3$$

Aldehydes

Oxidation

$$\text{Aldehyde} \xrightarrow{\text{Oxidation}} \text{Carboxylic acid}$$

$$CH_3{-}\overset{\overset{\displaystyle O}{\|}}{C}H \xrightarrow{\text{Oxidation}} CH_3{-}\overset{\overset{\displaystyle O}{\|}}{C}OH$$

Hydrogenation (Reduction)

$$\text{Aldehyde} + H_2 \xrightarrow{Pt} \text{Alcohol (1°)}$$

$$CH_3{-}\overset{\overset{\displaystyle O}{\|}}{C}H + H_2 \xrightarrow{Pt} CH_3{-}CH_2{-}OH$$

Ketones

Hydrogenation (Reduction)

$$\text{Ketone} + H_2 \xrightarrow{Pt} \text{Alcohol (2°)}$$

$$CH_3{-}\overset{\overset{\displaystyle O}{\|}}{C}{-}CH_3 + H_2 \xrightarrow{Pt} CH_3{-}\overset{\overset{\displaystyle OH}{|}}{C}H{-}CH_3$$

Glossary of Key Terms

alcohol An organic compound, R—OH, that contains the hydroxyl (—OH) group.

aldehyde An organic compound, $R-\overset{\overset{O}{\|}}{C}H$, that has the carbonyl functional group on the end carbon in the chain.

Benedict's Test A test that indicates the presence of an aldehyde by the reduction of copper(II) ion to give a red compound of copper(I) oxide, Cu_2O.

dehydration A reaction in which water is removed from an alcohol in the presence of an acid to form alkenes or ethers.

ether An organic compound, R—O—R, in which the functional group, an oxygen atom, is bonded to two hydrocarbon groups.

Fehling's Test A test that uses an alkaline solution of copper(II) ions to indicate the presence of aldehydes.

ketone An organic compound, $R-\overset{\overset{O}{\|}}{C}-R$, that contains the carbonyl functional group on an inside carbon of the chain.

oxidation The loss of two hydrogen atoms, as in the oxidation of alcohols to aldehydes or ketones, or the addition of an oxygen atom, as in the oxidation of aldehydes to carboxylic acids.

primary alcohol An alcohol that has one alkyl group (or none) attached to the alcohol carbon atom.

reduction The addition of hydrogen to a carbonyl bond. Aldehydes reduce to yield primary alcohols; ketones yield secondary alcohols.

secondary alcohol An alcohol that has two alkyl groups attached to the alcohol carbon atom.

sulfide An analog of an ether in which an —S— replaces the —O— of the ether.

tertiary alcohol An alcohol that has three alkyl groups attached to the alcohol carbon atom.

thiol An analog of an alcohol that contains an —SH group in place of the —OH group of the alcohol.

Tollens' Test A test that indicates the presence of an aldehyde by the formation of a silver mirror.

Problems

Alcohols *(Goal 12.1)*

12.1 Write the IUPAC name for each of the following alcohols:

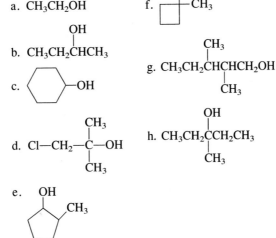

a. CH_3CH_2OH

b. $CH_3CH_2\underset{}{\overset{\overset{OH}{|}}{C}}HCH_3$

c. (cyclohexane ring)—OH

d. $Cl-CH_2-\underset{\underset{CH_3}{|}}{\overset{\overset{CH_3}{|}}{C}}-OH$

e. (cyclopentane ring with OH and adjacent CH_3)

f. (cyclobutane ring with OH and CH_3 on the same carbon)

g. $CH_3CH_2\overset{\overset{CH_3}{|}}{C}H\underset{\underset{CH_3}{|}}{C}HCH_2OH$

h. $CH_3CH_2\underset{\underset{CH_3}{|}}{\overset{\overset{OH}{|}}{C}}CH_2CH_3$

12.2 Write the condensed structural formula of each of the following alcohols:
a. 1-propanol
b. ethyl alcohol
c. 2,4-dimethyl-2-pentanol
d. 2-bromo-3-methyl-1-butanol
e. 3-ethylcyclopentanol
f. 2,4-dichlorocyclohexanol
g. isopropyl alcohol

Physical Properties of Alcohols *(Goal 12.2)*

12.3 Predict the higher boiling points in the following pairs of compounds:
a. ethane or ethanol
b. propane or propanol
c. ethanol or propanol

12.4 Give an explanation for the following:
a. Ethanol is soluble in water, but ethane is not.
b. Propanol is more soluble in water than is butanol.
c. Hexanol is not very soluble in water.

Phenols *(Goal 12.3)*

12.5 Identify the phenols in the following compounds:

a. $CH_3C(=O)-NH-C_6H_4-OH$

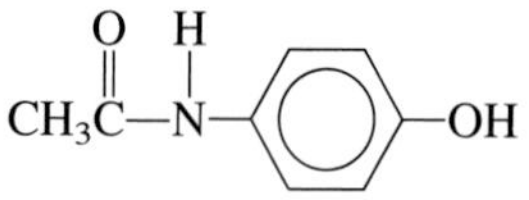

Tylenol

b. OH, O_2N, NO_2, NO_2 on benzene ring

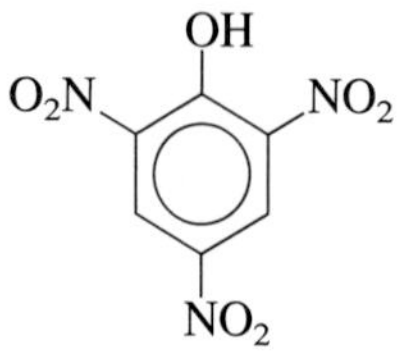

Picric acid

c. $NHCCH_3$ (with O), OC_2H_5 on benzene ring

Phenacetin

d. OH, OH, $CH_2CH_2CH_2CH_2CH_2CH_3$ on benzene ring

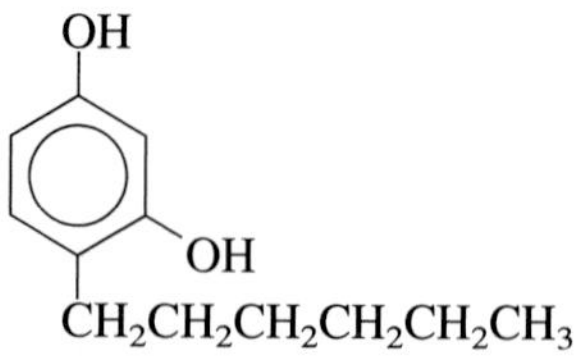

Hexylresorcinol

e. NH_2, $CH_2CHCOOH$, OH on benzene ring

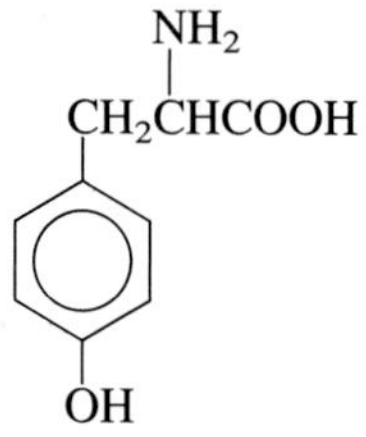

Tyrosine

12.6 Name the following phenols:

a. OH, NO_2, NO_2 on benzene ring

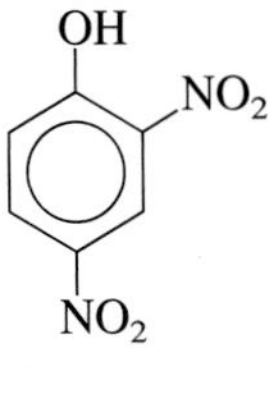

b. OH, Cl on benzene ring

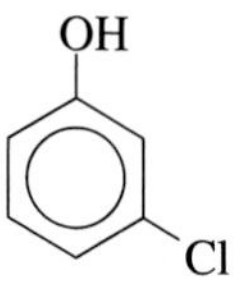

c. OH, CH_2CH_3 on benzene ring

12.7 Write a structural formula for each of the following:
a. 3-bromo-4-methylphenol
b. *m*-ethylphenol
c. *p*-phenylphenol
d. 2,4,6-trichlorophenol
e. 3,4-dichloro-2-methylphenol

Dehydration of Alcohols *(Goal 12.4)*

12.8 Complete the following dehydration reactions:

a. $CH_3\overset{\displaystyle OH}{\overset{|}{C}}HCH_2CH_3 \xrightarrow[\text{Heat}]{H^+}$

b. ______________ $\xrightarrow[\text{Heat}]{H^+} CH_3\underset{\displaystyle CH_3}{\underset{|}{C}}HCH{=}CH_2$

c. ______________ $\xrightarrow[\text{Heat}]{H^+}$ (cyclobutene)

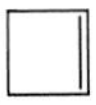

d. 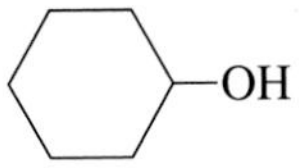 (cyclohexane ring)—OH $\xrightarrow[\text{Heat}]{H^+}$

e. $CH_3\overset{\displaystyle CH_3}{\overset{|}{C}}HCH_2CH_2OH \xrightarrow[\text{Heat}]{H^+}$

12.9 Write equations for the following reactions:
a. hydration of cyclopentene
b. hydration of 2-methyl-2-butene
c. dehydration of cyclobutanol
d. dehydration of 4-methyl-2-pentanol

12.10 Draw the structural formula of the alcohol needed to prepare each of the following compounds:
a. $CH_3—O—CH_3$
b. $CH_2{=}CH—CH_2—CH_3$

Ethers *(Goal 12.5)*

12.11 Write the names of each of the following ethers:

a. $CH_3—O—CH_2CH_3$

b. 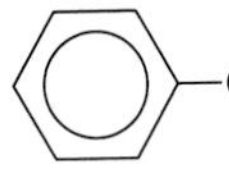 (benzene ring)—$O—CH_3$

c. $CH_3CH_2CH_2—O—CH_3$

d. 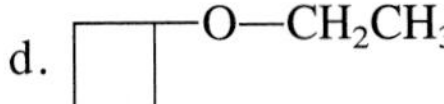

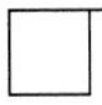

12.12 Write the structural formula of the following ethers:
a. methyl propyl ether
b. ethyl cyclopropyl ether
c. diphenyl ether
d. dimethyl ether
e. diethyl ether

12.13 Write the structural formulas of ether products of the following dehydration reactions:

a. $2CH_3OH \xrightarrow[\text{Heat}]{H^+}$ b. 2 (cyclohexyl)$-OH \xrightarrow[\text{Heat}]{H^+}$

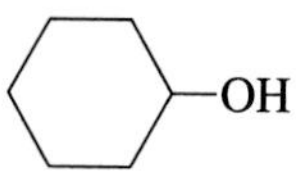

12.14 Based on structure, arrange the following in order of increasing boiling points: 1-propanol, ethyl methyl ether, and butane.

Thiols and Sulfides *(Goal 12.6)*

12.15 Identify the following as thiol, sulfide, or disulfide:

a. $CH_3{-}S{-}CH_3$

b. 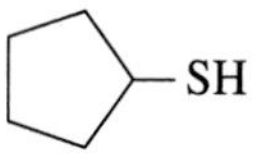 (cyclopentyl)$-SH$

c. $CH_3CH_2\overset{\overset{\large SH}{|}}{C}HCH_3$

d. $CH_3CH_2{-}S{-}S{-}CH_2CH_3$

12.16 Write the structural formulas of the following sulfur-containing compounds:
a. 1-propanethiol
b. diethyl disulfide
c. 3-methyl-1-butanethiol
d. dimethyl sulfide

Aldehydes and Ketones *(Goal 12.7)*

12.17 Write the IUPAC and common names (if any) for the following:

a. $CH_3\overset{\overset{\large O}{\|}}{C}H$

b. $H{-}\overset{\overset{\large O}{\|}}{C}{-}H$

c. $CH_3{-}\overset{\overset{\large Cl}{|}}{C}H{-}CH_2{-}CH_2{-}\overset{\overset{\large O}{\|}}{C}H$

d. $CH_3CH_2CH_2CH_2\overset{\overset{\large O}{\|}}{C}H$

e. $CH_3\overset{\overset{\large CH_3}{|}}{C}HCH_2\overset{\overset{\large O}{\|}}{C}H$

12.18 Write the structural formula of each of the following aldehydes:

a. propionaldehyde
b. formaldehyde
c. ethanal
d. β-chlorobutyraldehyde
e. 3-methylpentanal
f. benzaldehyde

12.19 Write the names of the following ketones:

a. $CH_3CH_2\overset{\overset{\large O}{\|}}{C}CH_3$

b. $CH_3CH_2CH_2\overset{\overset{\large O}{\|}}{C}CH_2CH_3$

c.  cyclopentanone ring (C=O on cyclopentane)

d. $CH_3\overset{\overset{\large Br}{|}}{C}HCH_2\overset{\overset{\large O}{\|}}{C}CH_3$

e. 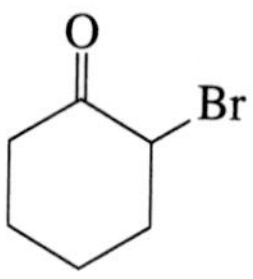 cyclohexanone ring with Br on the carbon adjacent to C=O

12.20 Write the structural formula of each of the following ketones:

a. diethyl ketone
b. acetone
c. 3-ethylcyclohexanone
d. methyl isopropyl ketone

12.21 Identify each of the following compounds as an alcohol, an ether, an aldehyde, or a ketone. Give a name for each.

a. $CH_3CH_2\overset{\overset{\displaystyle O}{\|}}{C}H$

b. $CH_3\overset{\overset{\displaystyle OH}{|}}{C}HCH_3$

c. $CH_3CH_2CH_2OH$

d. $CH_3CH_2CH_2OCH_3$

e. $CH_3\overset{\overset{\displaystyle O}{\|}}{C}CH_3$

f. 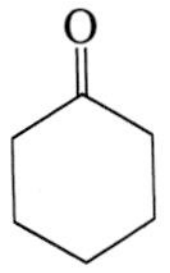 (cyclohexanone ring structure)

Preparation of Aldehydes and Ketones *(Goal 12.8)*

12.22 Classify the alcohols in Problem 12.1 as primary (1°), secondary (2°), or tertiary (3°).

12.23 Write the oxidation products of the following:

a. $CH_3CH_2CH_2CH_2OH \xrightarrow{\text{Oxidation}}$

b. $CH_3CH_2\overset{\overset{\displaystyle OH}{|}}{C}HCH_2CH_3 \xrightarrow{\text{Oxidation}}$

c. $CH_3\overset{\overset{\displaystyle CH_3}{|}}{C}HCH_2OH \xrightarrow{\text{Oxidation}}$

d. 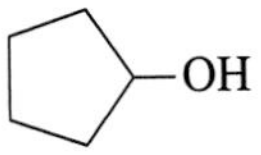 (cyclopentane ring)—OH $\xrightarrow{\text{Oxidation}}$

e. $CH_3CH_2\underset{\underset{\displaystyle CH_3}{|}}{\overset{\overset{\displaystyle CH_3}{|}}{C}}OH \xrightarrow{\text{Oxidation}}$

12.24 Write an equation for each of the following reactions:

a. oxidation of cyclopentanol
b. dehydration of cyclopentanol
c. oxidation of 1-propanol
d. dehydration of 1-propanol

12.25 Draw the structural formula of the alcohol that will give each of the following oxidation products:

a. formaldehyde
b. cyclopentanone
c. 2-butanone
d. benzaldehyde
e. 3-methylcyclohexanone

Reactions of Aldehydes and Ketones *(Goal 12.9)*

12.26 Indicate the compound in the following pairs that will form a brick-red precipitate with Benedict's reagent:

a. pentane or pentanal
b. propanone or propionaldehyde
c. formaldehyde or methane

12.27 Give the structural formula for the hydrogenation product of each of the following:

a. $CH_3\overset{\overset{O}{\|}}{C}CH_3$

b. $CH_3CH_2CH_2CH_2\overset{\overset{O}{\|}}{C}H$

c. $CH_3\overset{\overset{O}{\|}}{C}CH_2\overset{\overset{CH_3}{|}}{C}HCH_3$

d. 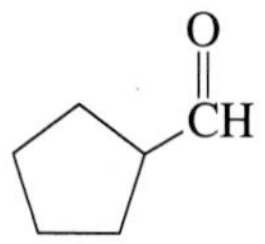 (cyclopentane ring)–$\overset{\overset{O}{\|}}{C}H$

e. 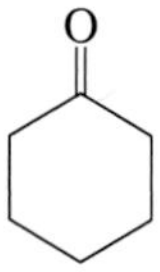 (cyclohexanone: cyclohexane ring with =O)

Review Problems

12.28 Identify each of the following as an alkane, alkene, alkyne, aromatic, haloalkane, alcohol, phenol, ether, thiol, sulfide, aldehyde, or ketone:

a. $CH_3\underset{\underset{CH_3}{|}}{\overset{\overset{CH_3}{|}}{C}}CH_2CH_2\overset{\overset{CH_3}{|}}{C}HCH_3$

b. $CH_3\overset{\overset{Cl}{|}}{C}HCH_2\overset{\overset{Br}{|}}{C}HCH_2CH_3$

c. $CH_3\overset{\overset{CH_3}{|}}{C}HCH_2\overset{\overset{OH}{|}}{C}HCH_3$

d. $CH_3CH{=}CH\overset{\overset{Br}{|}}{C}HCH_2CH_3$

e. 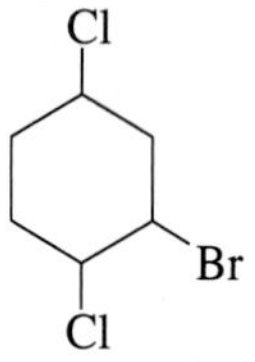 (cyclohexane ring with Cl, Br, and Cl substituents)

f. $CH_3CH_2\overset{\overset{SH}{|}}{C}HCH_3$

g. $CH_3C{\equiv}CCH_3$

h. 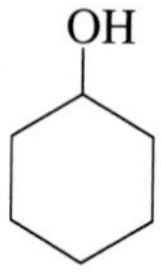 (cyclohexane ring with OH)

i. $CH_3OCH_2CH_3$

j. $CH_3{-}S{-}CH_3$

k. 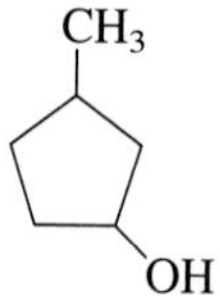 (cyclopentane ring with CH_3 and OH)

l. $CH_3\underset{\underset{CH_3}{|}}{\overset{\overset{CH_3}{|}}{C}}OCH_3$

m. 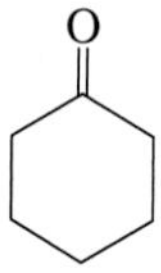 (cyclohexane ring with =O)

12.29 Draw the structural formula of each of the following:

a. 2-methylcyclopentanone
b. *p*-dichlorobenzene
c. 2-chloropropanal
d. *m*-bromophenol
e. diethyl ether
f. methanethiol
g. *β*-chlorobutyraldehyde
h. 3-methyl-1-butanol

12.30 Sometimes, several steps are need to prepare a compound. Using a combination of the reactions we have studied, indicate how you might prepare the following from the starting substance given. (For example, 2-propanol could be prepared from 1-propanol by first dehydrating the alcohol to give propene and then hydrating it again to give 2-propanol according to Markovnikov's rule.)

a. Prepare 2-methyl-2-propanol from 2-methyl-1-propanol.
b. Prepare butanone from 1-butene.
c. Prepare cyclopentene from cyclopentanone.

12.31 Identify the functional groups in the following polyfunctional molecules:

a. estradiol

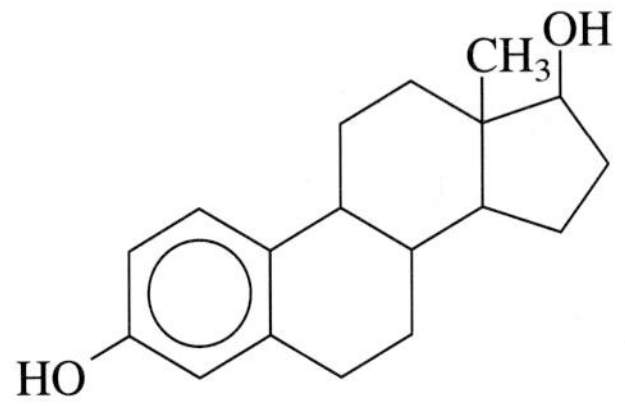

CH_3 OH

HO

b. tetrahydrocortisone

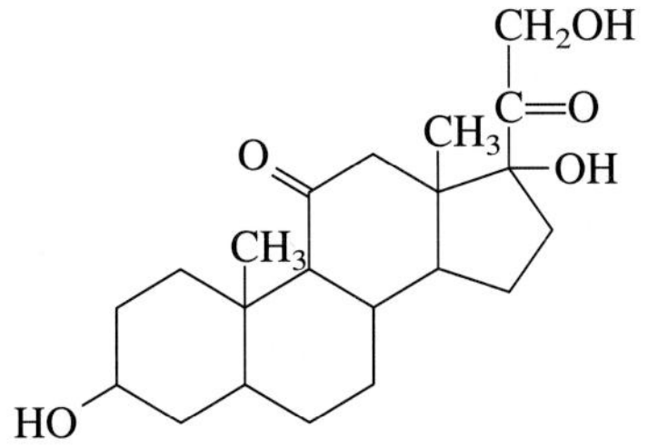

CH_2OH
C=O
CH_3 —OH
O
CH_3
HO

c. testosterone

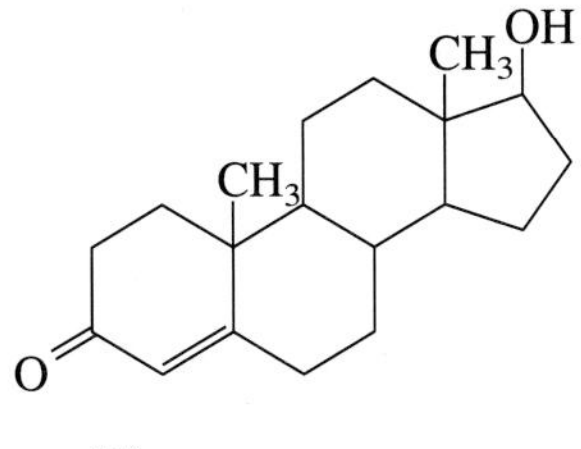

CH_3 OH
CH_3
O

d. vanillin

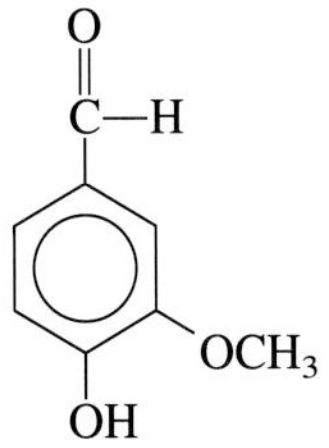

O
‖
C—H

OCH_3
OH

Caffeine is one of the nitrogen-containing compounds called alkaloids which have physiological effects on the body.

13

Carboxylic Acids, Esters, and Nitrogen-Containing Compounds

Learning Goals

13.1 Write the name and structural formula of a carboxylic acid.

13.2 Discuss the solubility, melting and boiling points, ionization in water, and neutralization of carboxylic acids.

13.3 Write the structural formula and name of the ester produced by the reaction of an alcohol and a carboxylic acid.

13.4 Write the structural formula of the products of ester hydrolysis and saponification.

13.5 Write the name and structural formula of a simple amine.

13.6 Identify heterocyclic amines and describe some of their physiological activity.

13.7 Write equations for the ionization and neutralization of amines.

13.8 Write the name and structural formula of the amide produced from a carboxylic acid and an amine.

SCOPE

When you think of an acid, you think of a substance that has a sour taste, produces hydronium ions in water, and turns blue litmus red. These are the properties of carboxylic acids, which are weak acids because they ionize only slightly in aqueous solutions. The vinegar you use in your salad is a carboxylic acid (acetic acid); the tartness of citrus fruits is due to another carboxylic acid, citric acid.

When a carboxylic acid combines with an alcohol, an ester is produced. Aspirin is an ester. Some esters with pleasant aromas and flavors are found in fruits and flowers. Fats and oils are esters of the alcohol glycerol and fatty acids.

Amines, amino acids, and amides contain nitrogen. Many, like adrenaline and the amphetamines, are physiologically active. Alkaloids, nitrogen-containing compounds such as curare, belladonna, and digitalis, are produced naturally in plants and also have physiological activity. Some amides, such as phenobarbital, are used as sedatives and in anticonvulsant medications.

13.1 Carboxylic Acids

Learning Goal **Write the name and structural formula of a carboxylic acid.**

Carboxylic acids contain a functional group called a **carboxyl group,** which consists of a hydroxyl group attached to a carbonyl group. Sometimes, it is convenient to use the abbreviation —COOH for the carboxyl group.

$$-\overset{\overset{\displaystyle O}{\|}}{C}-OH \quad \text{or} \quad -COOH \qquad R-\overset{\overset{\displaystyle O}{\|}}{C}-O-H$$

Carboxyl group — Carboxylic acid

Naming Carboxylic Acids

A carboxylic acid is named in the IUPAC system by replacing the *-e* of the alkane name by *-oic acid*.

Naming Carboxylic Acids $R-\overset{\overset{\displaystyle O}{\|}}{C}OH$

Functional group: $-\overset{\overset{\displaystyle O}{\|}}{C}OH$

IUPAC name: Alkanoic acid

Common names: Formic acid, acetic acid, propionic acid, butyric acid

However, common names derived from their natural source are used extensively for most carboxylic acids. Formic acid is released under the skin from a bee sting or other insect bites. Acetic acid is the oxidation product of the ethanol in wines and apple cider. The resulting solution of acetic acid is known as vinegar. (See Table 13.1.) Some models of carboxylic acids are shown in Figure 13.1.

Figure 13.1
Models of formic acid, acetic acid, and propionic acid.

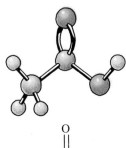

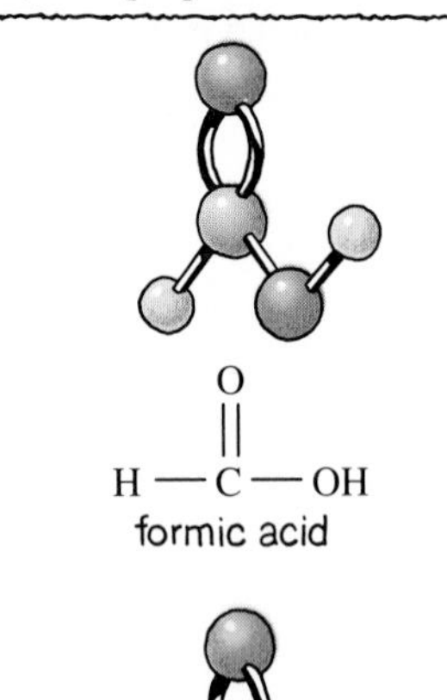

$H{-}\overset{\overset{O}{\|}}{C}{-}OH$
formic acid

$CH_3{-}\overset{\overset{O}{\|}}{C}{-}OH$
acetic acid

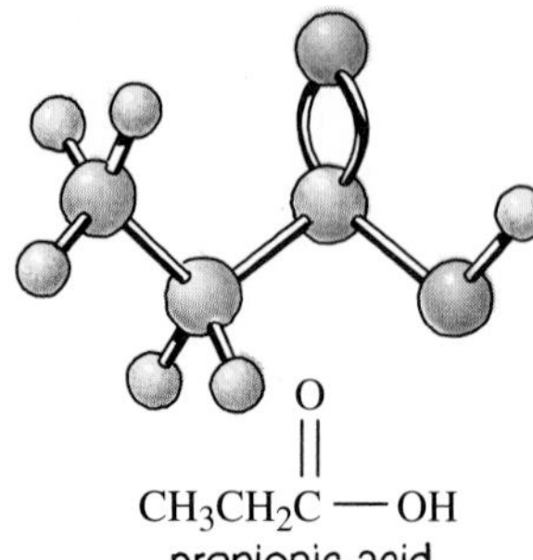

$CH_3CH_2\overset{\overset{O}{\|}}{C}{-}OH$
propionic acid

Table 13.1 Names and Natural Sources of Some Carboxylic Acids

Structure	IUPAC name	Common name	Natural source
$H\overset{\overset{O}{\|}}{C}OH$	Methanoic acid	Formic acid	Ant stings (Lat. *formica,* ''ant'')
$CH_3\overset{\overset{O}{\|}}{C}OH$	Ethanoic acid	Acetic acid	Vinegar (Lat. *acetum,* ''sour'')
$CH_3CH_2\overset{\overset{O}{\|}}{C}OH$	Propanoic acid	Propionic acid	Dairy products (Greek. *pro,* ''first''; *pion,* ''fat'')
$CH_3CH_2CH_2\overset{\overset{O}{\|}}{C}OH$	Butanoic acid	Butyric acid	Rancid butter (Lat. *butyrum,* ''butter'')

When using the IUPAC name, the carboxyl carbon is carbon 1. In the common names of acids, the Greek letters α (alpha), β (beta), and γ (gamma) are used to show the location of side groups on the carbon atoms next to the carboxyl group. Name ketone groups as *keto,* hydroxyl groups as *hydroxy.*

$$-\underset{\uparrow}{C}-\underset{\uparrow}{C}-\underset{\uparrow}{C}-\underset{\uparrow}{\overset{\overset{O}{\|}}{C}}-OH$$

IUPAC	4	3	2	1
Common	γ	β	α	

$$CH_3{-}\overset{\overset{O}{\|}}{C}{-}CH_2{-}\overset{\overset{O}{\|}}{C}{-}OH$$

3-Ketobutanoic acid
(β-ketobutyric acid)

$$CH_3{-}\overset{\overset{OH}{|}}{CH}{-}\overset{\overset{O}{\|}}{C}{-}OH$$

2-Hydroxypropanoic acid
(α-hydroxypropionic acid, lactic acid)

The carboxylic acid of benzene is named benzoic acid.

$$C_6H_5{-}\overset{\overset{O}{\|}}{C}OH$$

Benzoic acid

Sample Problem 13.1
Naming Carboxylic Acids

Write a name for each of the following carboxylic acids:

a. 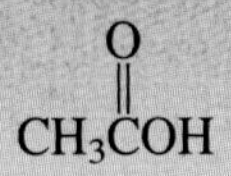 $CH_3\overset{\overset{\displaystyle O}{\|}}{C}OH$

b. 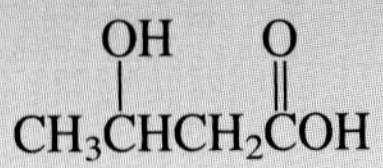 $CH_3\overset{\overset{\displaystyle OH}{|}}{C}HCH_2\overset{\overset{\displaystyle O}{\|}}{C}OH$

Solution:

a. This is a carboxylic acid having two carbon atoms. In the IUPAC system, the *e* in ethane is replaced by *oic acid,* ethanoic acid. Its common name is *acetic acid.*

b. This carboxylic acid has a hydroxyl group on the third carbon. In the common name, the Greek letter β specifies the second carbon atom from the carbonyl group.

IUPAC name: 3-Hydroxybutanoic acid
Common name: β-Hydroxybutyric acid

Study Check:
Name the following:

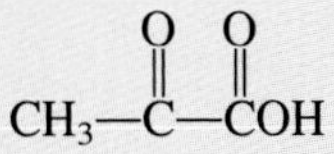

$CH_3—\overset{\overset{\displaystyle O}{\|}}{C}—\overset{\overset{\displaystyle O}{\|}}{C}OH$

Answer: 2-ketopropanoic acid; α-ketopropionic acid

Preparation of Carboxylic Acids

In the preceding chapter, we stated that aldehydes can be easily oxidized to give carboxylic acids. In the oxidation, an oxidizing agent such as $K_2Cr_2O_7$ provides the oxygen for the carboxyl group.

$$\text{aldehyde} \xrightarrow{\text{Oxidation}} \text{carboxylic acid}$$

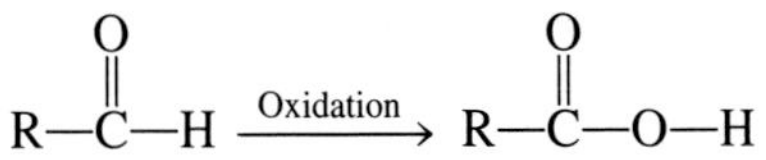

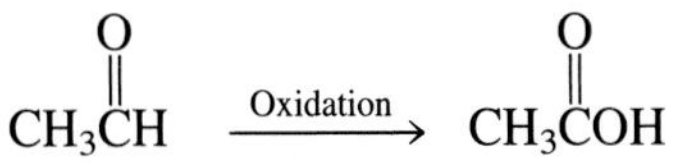

$$R—\overset{\overset{\displaystyle O}{\|}}{C}—H \xrightarrow{\text{Oxidation}} R—\overset{\overset{\displaystyle O}{\|}}{C}—O—H$$

$$CH_3\overset{\overset{\displaystyle O}{\|}}{C}H \xrightarrow{\text{Oxidation}} CH_3\overset{\overset{\displaystyle O}{\|}}{C}OH$$

Ethanal (acetaldehyde) → Ethanoic acid (acetic acid)

Sample Problem 13.2
Preparation of Carboxylic Acids

Draw the structures of the carboxylic acids formed from the following oxidation reactions:

a. $\overset{O}{\overset{\|}{HCH}} \xrightarrow{\text{Oxidation}}$

b. $CH_3CH_2CH_2OH \xrightarrow{\text{Oxidation}}$

Solution:

a. The oxidation of formaldehyde will produce formic acid. The product can be written by replacing the hydrogen atom of the aldehyde by a hydroxyl group (—OH).

$$\underset{\substack{\text{Methanal}\\\text{(formaldehyde)}}}{\overset{O}{\overset{\|}{HC}}\text{—H}} \xrightarrow{\text{Oxidation}} \underset{\substack{\text{Methanoic acid}\\\text{(formic acid)}}}{\overset{O}{\overset{\|}{HC}}\text{—OH}}$$

b. A primary alcohol will oxidize to an aldehyde, which can oxidize further to a carboxylic acid.

$$\underset{\substack{\text{1-propanol}\\\text{(propyl alcohol)}}}{CH_3CH_2CH_2OH} \xrightarrow{\text{Oxidation}} \underset{\substack{\text{Propanal}\\\text{(proprionaldehyde)}}}{CH_3CH_2\overset{O}{\overset{\|}{C}}H} \xrightarrow{\text{Oxidation}} \underset{\substack{\text{Propanoic acid}\\\text{(propionic acid)}}}{CH_3CH_2\overset{O}{\overset{\|}{C}}OH}$$

Study Check:
Write the structural formula of the carboxylic acid produced by the oxidation of both the hydroxyl groups in 1,3-propandiol, HO—$CH_2CH_2CH_2$—OH

Answer: $HO\overset{O}{\overset{\|}{C}}CH_2\overset{O}{\overset{\|}{C}}OH$

The fatty acids in fats and oils are long-chain carboxylic acids that have an even number of carbon atoms. Fatty acids can be saturated or unsaturated. Some typical fatty acids are described in Table 13.2.

Arachidonic acid is a long-chain carboxylic acid with 20 carbon atoms. It is important in the synthesis of biological chemicals such as prostaglandins, which are found in body tissues and appear to be part of the body's immune system. In addition, they regulate the secretion of gastric acid, cause contraction of smooth muscle, and affect blood pressure.

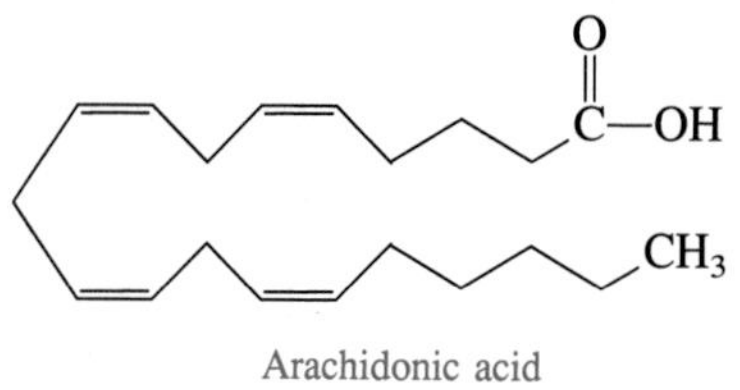

O
‖
C—OH
CH₃

Arachidonic acid

Table 13.2 Some Fatty Acids and Their Sources

Structure	Common name	Source
Saturated fatty acids		
$CH_3(CH_2)_{10}COOH$	Lauric acid	Coconut oil, palm kernel oil
$CH_3(CH_2)_{12}COOH$	Myristic acid	Nutmeg
$CH_3(CH_2)_{14}COOH$	Palmitic acid	Palm oil
$CH_3(CH_2)_{16}COOH$	Stearic acid	Beef and mutton tallow, cocoa butter
Unsaturated fatty acids		
$CH_3(CH_2)_7CH{=}CH(CH_2)_7COOH$	Oleic acid	Olive, peanut oil
$CH_3(CH_2)_4CH{=}CHCH_2CH{=}CH(CH_2)_7COOH$	Linoleic acid	Soybean, safflower, sunflower seed oil
$CH_3CH_2CH{=}CHCH_2CH{=}CHCH_2CH{=}CH(CH_2)_7COOH$	Linolenic acid	Herring, linseed oil

HEALTH NOTE

Some Naturally Occurring Carboxylic Acids

Many of the carboxylic acids that appear in metabolic pathways may contain more than one carboxyl group. The names of acids with two carboxyl groups are obtained by adding the suffix *dioic acid* to the alkane name. However, many dicarboxylic acids use their common names.

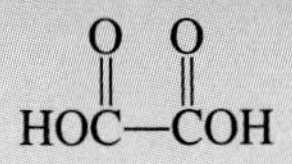

$$HO\overset{O}{\overset{\|}{C}}-\overset{O}{\overset{\|}{C}}OH$$

Ethanedioic acid
(oxalic acid)

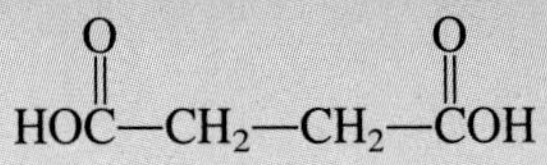

$$HO\overset{O}{\overset{\|}{C}}-CH_2-CH_2-\overset{O}{\overset{\|}{C}}OH$$

Butanedioic acid
(succinic acid)

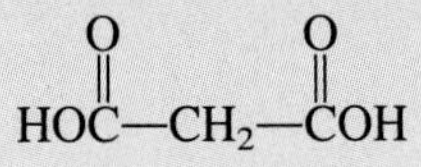

$$HO\overset{O}{\overset{\|}{C}}-CH_2-\overset{O}{\overset{\|}{C}}OH$$

Propanedioic acid
(malonic acid)

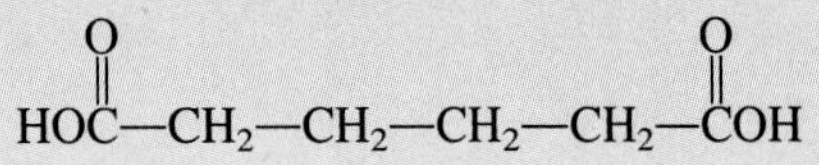

$$HO\overset{O}{\overset{\|}{C}}-CH_2-CH_2-CH_2-CH_2-\overset{O}{\overset{\|}{C}}OH$$

Hexanedioic acid
(adipic acid)

Citric acid, found in many fruits, contributes to their tart taste.

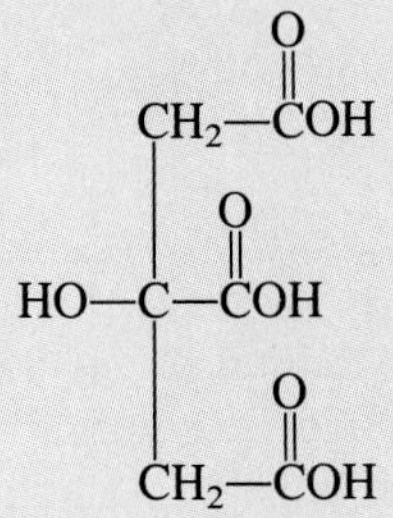

$$\begin{array}{l} \quad CH_2-\overset{O}{\overset{\|}{C}}OH \\ \quad | \\ HO-C-\overset{O}{\overset{\|}{C}}OH \\ \quad | \\ \quad CH_2-\overset{O}{\overset{\|}{C}}OH \end{array}$$

Citric acid ■

13.2 Some Properties of Carboxylic Acids

Learning Goal **Discuss the solubility, melting and boiling points, ionization in water, and neutralization of carboxylic acids.**

The carboxyl group contains polar O—H and polar C=O bonds. As a result, several hydrogen atoms form between carboxylic acid molecules to give them some of the highest melting and boiling points found in families of organic compounds. (See Table 13.3.)

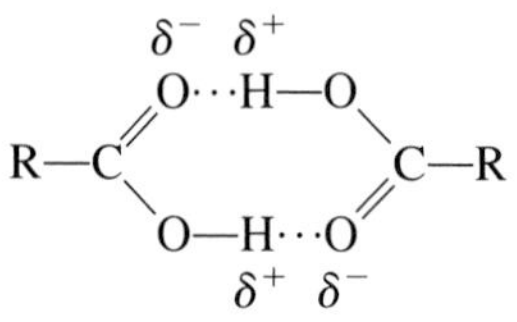

$$\begin{array}{ccccc} & \delta^- & \delta^+ & & \\ & O\cdots & H—O & & \\ R—C\!\!\!\!\diagup\!\!\!\!\diagup & & & \diagdown & \\ \diagdown & & & & C—R \\ & O—H\cdots & O & \diagup\!\!\!\!\diagup & \\ & \delta^+ & \delta^- & & \end{array}$$

Hydrogen bonding between two carboxylic acids

Table 13.3 Melting and Boiling Points of Some Selected Organic Compounds

Structural formula	Family	Melting point (°C)	Boiling point (°C)
$CH_3CH_2CH_3$	Alkane	−188	−42
CH_3CH_2Cl	Haloalkane	−136	12
$CH_3CH_2CH_2OH$	Alcohol	−127	97
$CH_3CH_2\overset{\overset{O}{\Vert}}{C}H$	Aldehyde	−81	49
$CH_3\overset{\overset{O}{\Vert}}{C}OH$	Carboxylic acid	17	118

Solubility in Water

Carboxylic acids with one to four carbon atoms are very soluble in water because the carboxyl group forms hydrogen bonds with water molecules; acids with five or more carbon atoms are mostly insoluble. (See Table 13.4.)

Table 13.4 Solubilities of Some Carboxylic Acids

Carboxylic acid	Structural formula	Solubility in water (g/100 g)
Formic acid	$HCOOH$	any amount
Acetic acid	CH_3COOH	any amount
Propionic acid	CH_3CH_2COOH	any amount
Butyric acid	$CH_3CH_2CH_2COOH$	any amount
Pentanoic acid	$CH_3CH_2CH_2CH_2COOH$	5
Hexanoic acid	$CH_3CH_2CH_2CH_2CH_2COOH$	1

Ionization of Carboxylic Acids

Like the inorganic acids, carboxylic acids taste sour, ionize to give H^+ ions, turn pH paper red, and neutralize bases. Since carboxylic acids are weak acids, only a few of the acid molecules ionize in aqueous solution. The negative ion that forms is named by replacing *ic acid* with *ate,* exactly what we did in the inorganic section on acids. Recall that the negative ion formed in the ionization of nitric acid (HNO_3) is nitrate (NO_3^-). When sulfuric acid (H_2SO_4) ionizes in water, it forms hydrogen ions and the sulfate ion (SO_4^{2-}).

Ionization of a Carboxylic Acid in Water

$$\underset{\text{Carboxylic acid}}{R\text{—}\overset{\overset{\displaystyle O}{\|}}{C}\text{—}OH} \underset{}{\overset{H_2O}{\rightleftharpoons}} \underset{\text{Carboxylate ion}}{R\text{—}\overset{\overset{\displaystyle O}{\|}}{C}O^-} + H^+$$

$$\underset{\text{Acetic acid}}{CH_3\overset{\overset{\displaystyle O}{\|}}{C}OH} \overset{H_2O}{\rightleftharpoons} \underset{\text{Acetate ion}}{CH_3\overset{\overset{\displaystyle O}{\|}}{C}O^-} + H^+$$

Sample Problem 13.3
Ionization of Carboxylic Acids in Water

Write the equation for the ionization of propionic acid in water.

Solution:
The ionization of propionic acid produces a hydrogen ion and a carboxylate ion.

$$\underset{\substack{\text{Propanoic acid}\\ \text{(propionic acid)}}}{CH_3CH_2\overset{\overset{\displaystyle O}{\|}}{C}OH} \overset{H_2O}{\rightleftharpoons} \underset{\substack{\text{Propanoate ion}\\ \text{(propionate ion)}}}{CH_3CH_2\overset{\overset{\displaystyle O}{\|}}{C}O^-} + H^+$$

Neutralization of a Carboxylic Acid

A carboxylic acid is neutralized by a base to give a salt and water. The salt is named by giving the metal ion first followed by the carboxylate ion name.

Neutralization

carboxylic acid + base ⟶ carboxylate salt + H_2O

$$CH_3\overset{\overset{\displaystyle O}{\|}}{C}OH + NaOH \longrightarrow CH_3\overset{\overset{\displaystyle O}{\|}}{C}O^- \; Na^+ + H_2O$$

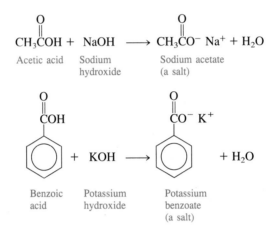

Acetic acid | Sodium hydroxide | Sodium acetate (a salt)

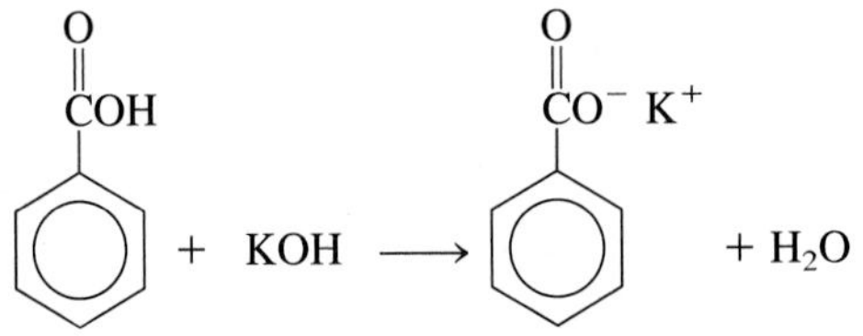

$$C_6H_5\overset{\overset{\displaystyle O}{\|}}{C}OH + KOH \longrightarrow C_6H_5\overset{\overset{\displaystyle O}{\|}}{C}O^- \; K^+ + H_2O$$

Benzoic acid | Potassium hydroxide | Potassium benzoate (a salt)

Sample Problem 13.4
Neutralization of a Carboxylic Acid

Write the equation for the neutralization of propionic acid with sodium hydroxide.

Solution:
The neutralization of an acid with a base produces the salt of the acid and water.

$$CH_3CH_2\overset{\overset{\displaystyle O}{\|}}{C}OH + NaOH \longrightarrow CH_3CH_2\overset{\overset{\displaystyle O}{\|}}{C}O^- \; Na^+ + H_2O$$

Propionic acid | Sodium hydroxide | Sodium propionate (salt)

Study Check:
What carboxylic acid will give potassium butyrate when it is neutralized by KOH?

Answer:

$$CH_3CH_2CH_2\overset{\overset{\displaystyle O}{\|}}{C}OH$$

Butyric acid

13.3 Esters

Learning Goal **Write the structural formula and name of the ester produced by the reaction of an alcohol and a carboxylic acid.**

Figure 13.2
A model of methyl acetate, an ester.

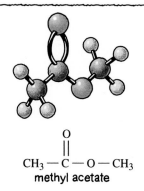

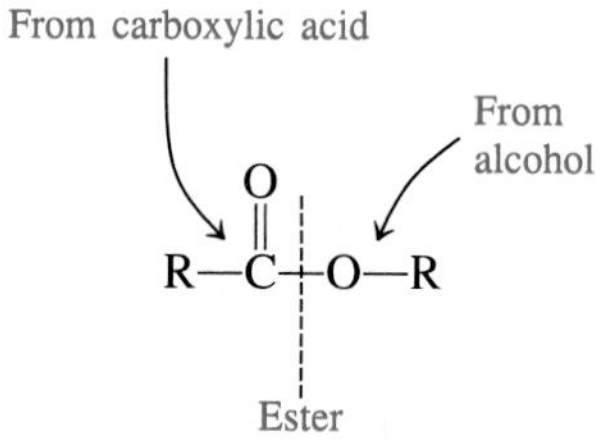

$$CH_3-\overset{\overset{O}{\|}}{C}-O-CH_3$$

methyl acetate

When a carboxylic acid reacts with an alcohol, an ester is produced. A model of the ester methyl acetate is illustrated in Figure 13.2.

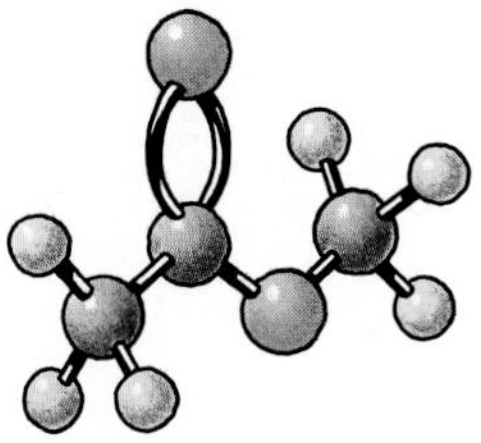

From carboxylic acid — From alcohol

$$R-\overset{\overset{O}{\|}}{C}-O-R$$

Ester

$$CH_3-\overset{\overset{O}{\|}}{C}-O-CH_3$$

Methyl acetate

Naming Esters

An ester is named as two words. The first word is the alkyl group attached to the oxygen of the ester group. The second word is derived from the name of the carboxylic acid used to form the ester. The ending *ic acid* is replaced with the suffix *ate*. The common name which is used more often uses the common name of the carboxylic acid.

Naming Esters $\boldsymbol{R-\overset{\overset{O}{\|}}{C}-OR}$

Functional group: $-\overset{\overset{O}{\|}}{C}O-$
IUPAC name: Alkyl carboxylate
Common name: Alkyl carboxylate

$CH_3CH_2\overset{\overset{O}{\|}}{C}OCH_3$	$CH_3CH_2CH_2\overset{\overset{O}{\|}}{C}OCH_2CH_3$	$C_6H_5\overset{\overset{O}{\|}}{C}OCH_2CH_3$
	Name of alkyl group	
Methyl	Ethyl	Ethyl
	Name of carboxylic acid	
IUPAC: Propanoic acid	Butanoic acid	Benzoic acid
Common: Propionic acid	Butyric acid	Benzoic acid
	Name of ester	
IUPAC: Methyl propanoate	Ethyl butanoate	Ethyl benzoate
Common: Methyl propionate	Ethyl butyrate	Ethyl benzoate

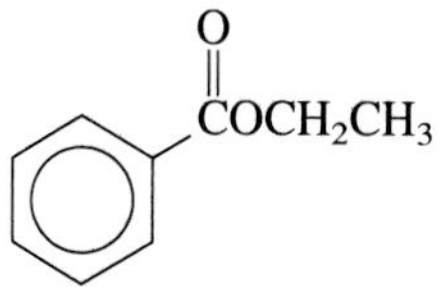

Sample Problem 13.5
Naming Esters

Write the names of the following esters:

a. $CH_3CH_2\overset{O}{\overset{\|}{C}}{-}OCH_2CH_3$ **b.** 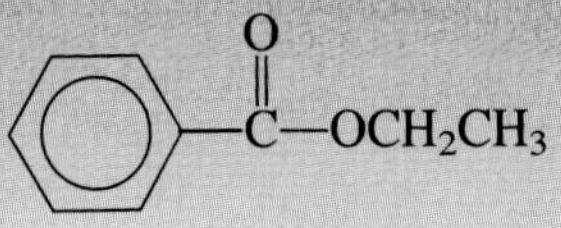 $C_6H_5{-}\overset{O}{\overset{\|}{C}}{-}OCH_2CH_3$

Solution:

a. The alkyl portion is ethyl and the carboxylic acid is propanoic (propionic) acid. Changing the *ic acid* ending to *ate* gives the IUPAC name of *ethyl propanoate,* and the common name of *ethyl propionate*.

b. The alkyl portion is ethyl and the carboxylic acid portion is from benzoic acid. To name the ester, change the ending of the acid to *ate* to give *ethyl benzoate*.

Study Check:
Draw the structural formula of methyl butyrate.

Answer:

$$CH_3CH_2CH_2\overset{O}{\overset{\|}{C}}OCH_3$$

Preparation of Esters

In the **esterification** reaction, a carboxylic acid reacts with an alcohol in the presence of an acid catalyst to give an ester. For example, acetic acid reacts with methyl alcohol in the presence of concentrated sulfuric acid to give methyl acetate and water.

Esterification

$$\underset{\text{Carboxylic acid}}{R{-}\overset{O}{\overset{\|}{C}}{-}O{-}H} + \underset{\text{Alcohol}}{H{-}OR} \xrightarrow{H^+} R{-}\overset{O}{\overset{\|}{C}}{-}OR + \underset{\text{Water}}{H_2O}$$

$$\underset{\text{Acetic acid}}{CH_3{-}\overset{O}{\overset{\|}{C}}{-}OH} + \underset{\text{Methyl alcohol}}{CH_3OH} \xrightarrow{H^+} \underset{\text{Methyl acetate}}{CH_3{-}\overset{O}{\overset{\|}{C}}{-}OCH_3} + H_2O$$

$$\underset{\text{Formic acid}}{H{-}\overset{O}{\overset{\|}{C}}{-}OH} + \underset{\text{Propyl alcohol}}{CH_3CH_2CH_2OH} \xrightarrow{H^+} \underset{\text{Propyl formate}}{H{-}\overset{O}{\overset{\|}{C}}{-}OCH_2CH_2CH_3}$$

HEALTH NOTE

Esters in Health and Medicine

Some esters are in common use, such as aspirin, oil of wintergreen, vitamin C, oils, and waxes.

Aspirin (acetylsalicylic acid), is an analgesic (pain reliever), antipyretic (fever reducer), and anti-inflammatory agent. It can be formed from salicylic acid as follows:

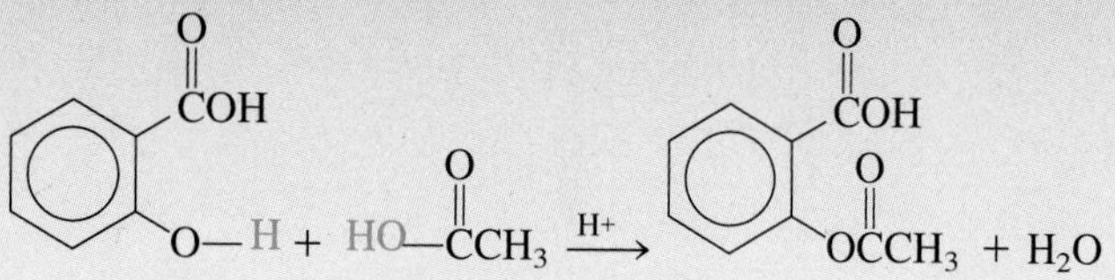

$$\text{C}_6\text{H}_4(\text{COOH})\text{O—H} + \text{HO—}\overset{\text{O}}{\overset{\|}{\text{C}}}\text{CH}_3 \xrightarrow{\text{H}^+} \text{C}_6\text{H}_4(\text{COOH})\text{O}\overset{\text{O}}{\overset{\|}{\text{C}}}\text{CH}_3 + \text{H}_2\text{O}$$

Salicylic acid — Acetic acid — Acetylsalicylic acid, aspirin

Oil of wintergreen, or methyl salicylate, has a spearmint odor and flavor. It is used in mint flavorings and in skin ointments where it acts as a counterirritant, producing heat to soothe sore muscles.

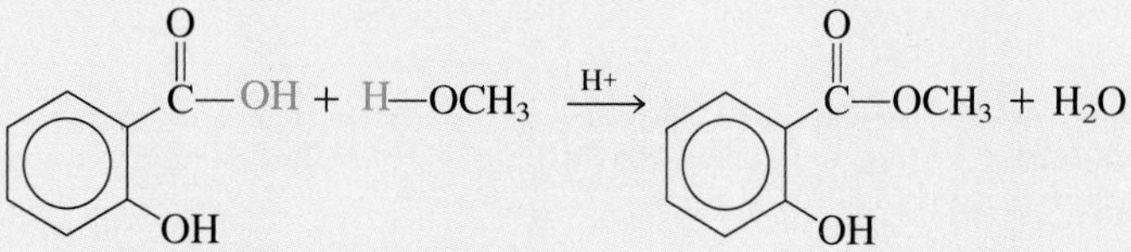

$$\text{C}_6\text{H}_4(\text{OH})\overset{\text{O}}{\overset{\|}{\text{C}}}\text{—OH} + \text{H—OCH}_3 \xrightarrow{\text{H}^+} \text{C}_6\text{H}_4(\text{OH})\overset{\text{O}}{\overset{\|}{\text{C}}}\text{—OCH}_3 + \text{H}_2\text{O}$$

Salicylic acid — Methyl alcohol — Methyl salicylate (oil of wintergreen)

Vitamin C, or ascorbic acid, is a water-soluble vitamin found in fresh fruits and vegetables. Since both the acid and alcohol group are on the same molecule, an ester bond forms within the molecule to give a cyclic structure. Vitamin C is used as a preservative in foods and to prevent the browning of fruit.

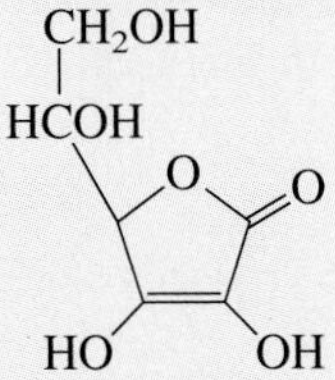

CH_2OH | HCOH attached to a five-membered lactone ring (ring O, C=O, HO–C=C–OH)

Vitamin C (ascorbic acid)

Waxes produced in animals and plants include beeswax, used to make candles and polishes, and carnauba wax used in car and floor waxes. They are esters of long-chain carboxylic acids and alcohols.

$$\text{CH}_3(\text{CH}_2)_{14}\overset{\text{O}}{\overset{\|}{\text{C}}}\text{—O(CH}_2)_{29}\text{CH}_3$$

Beeswax

$$\text{CH}_3(\text{CH}_2)_{24}\overset{\text{O}}{\overset{\|}{\text{C}}}\text{O(CH}_2)_{29}\text{CH}_3$$

Carnauba wax

Fats and oils are esters of glycerol, a trihydroxy alcohol, and fatty acids. In fats from animal sources, most of the fatty acids are saturated; in oils from vegetables, more of the fatty acids contain double bonds.

Ester bonds

$$\begin{array}{l} H_2C\text{—}OH + H\text{—}O\text{—}\overset{O}{\overset{\|}{C}}(CH_2)_{16}CH_3 \\ \quad| \\ H\text{—}C\text{—}OH + H\text{—}O\text{—}\overset{O}{\overset{\|}{C}}(CH_2)_{16}CH_3 \\ \quad| \\ H_2C\text{—}OH + H\text{—}O\text{—}\overset{O}{\overset{\|}{C}}(CH_2)_{16}CH_3 \end{array} \longrightarrow \begin{array}{l} H_2C\text{—}O\overset{O}{\overset{\|}{C}}(CH_2)_{16}CH_3 \\ \quad| \\ H\text{—}C\text{—}O\overset{O}{\overset{\|}{C}}(CH_2)_{16}CH_3 \\ \quad| \\ H_2C\text{—}O\overset{O}{\overset{\|}{C}}(CH_2)_{16}CH_3 \end{array} + 3H_2O$$

Glycerol (trihydroxy alcohol) Stearic acid (a fatty acid) Tristearin (a saturated fat)

■

Sample Problem 13.6
Writing Esterification Equations

Write the structural formulas of the products of the following esterification:

$$C_6H_5\text{—}\overset{O}{\overset{\|}{C}}OH + CH_3OH \xrightarrow{H+}$$

Benzoic acid Methyl alcohol

Solution:
In the presence of an acid, a carboxylic acid and an alcohol form an ester. The structural formula of the ester can be written by replacing the hydrogen in the carboxylic acid with the alkyl (CH_3—) portion of the alcohol.

$$C_6H_5\text{—}\overset{O}{\overset{\|}{C}}\text{—}OH + CH_3OH \xrightarrow{H+} C_6H_5\text{—}\overset{O}{\overset{\|}{C}}\text{—}OCH_3 + H_2O$$

Benzoic acid Methyl alcohol Methyl benzoate

Study Check:
What alcohol reacts with acetic acid to form propyl acetate,

$$CH_3\overset{O}{\overset{\|}{C}}\text{—}OCH_2CH_2CH_3?$$

Answer: propyl alcohol, $CH_3CH_2CH_2OH$

ENVIRONMENTAL NOTE

Odors and Flavors of Esters

Esters produce the pleasant odors and flavors associated with many fruits and flowers. Several of these esters are listed in Table 13.5. See Figure 13.3. ■

Figure 13.3
Esters are responsible for part of the odor and flavor of oranges, bananas, pears, pineapples, and strawberries.

Table 13.5 Esters in Fruits and Flowers

Flavor/odor	Structural formula	Common name
Rum	$HC(=O)OCH_2CH_3$	Ethyl formate
Raspberry	$HC(=O)OCH_2CH(CH_3)CH_3$	Isobutyl formate
Pear	$CH_3C(=O)OCH_2CH_2CH_3$	Propyl acetate
Banana	$CH_3C(=O)OCH_2CH_2CH(CH_3)CH_3$	Isoamyl acetate
Orange	$CH_3C(=O)O(CH_2)_7CH_3$	Octyl acetate
Pineapple	$CH_3CH_2CH_2C(=O)OCH_2CH_3$	Ethyl butyrate
Apricot	$CH_3CH_2CH_2C(=O)O(CH_2)_4CH_3$	Pentyl butyrate

13.4 Hydrolysis and Saponification of Esters

Learning Goal **Write the structural formulas of the products of ester hydrolysis and saponification.**

The term *lysis* refers to a splitting or breakdown of a substance by water. In **hydrolysis,** an ester reacts with water in the presence of a strong acid catalyst. The ester is split to give the carboxylic acid and alcohol. In the body, enzymes (biological catalysts) direct the process of digestion, a series of hydrolysis reactions in which large, complex foodstuffs are split or broken down into smaller molecules.

Acid Hydrolysis of an Ester

Ester bond broken

$$R{-}\overset{\overset{O}{\|}}{C}{\downarrow}{-}O{-}R + H{-}OH \xrightarrow{H^+} R{-}\overset{\overset{O}{\|}}{C}{-}OH + H{-}OR$$

Ester Water Carboxylic acid Alcohol

$$CH_3\overset{\overset{O}{\|}}{C}{-}OCH_3 + H{-}OH \xrightarrow{H^+} CH_3\overset{\overset{O}{\|}}{C}{-}OH + H{-}OCH_3$$

Methyl acetate Water Acetic acid Methyl alcohol

Sample Problem 13.7
Ester Hydrolysis

Aspirin that is stored may undergo hydrolysis in the presence of water and heat. What are the hydrolysis products of aspirin? Why does a bottle of old aspirin smell like vinegar?

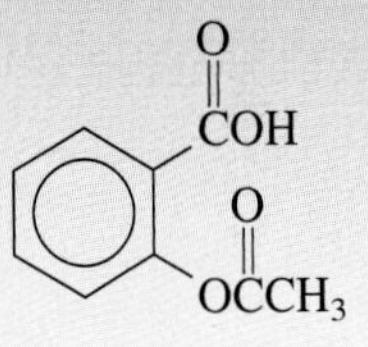

$$C_6H_4(\overset{\overset{O}{\|}}{C}OH)(O\overset{\overset{O}{\|}}{C}CH_3)$$

Aspirin

Solution:
To write the hydrolysis products, separate the compound at the ester bond. Complete the formula of the carboxylic acid by adding —OH from water to the carbonyl group and the —H to complete the alcohol.

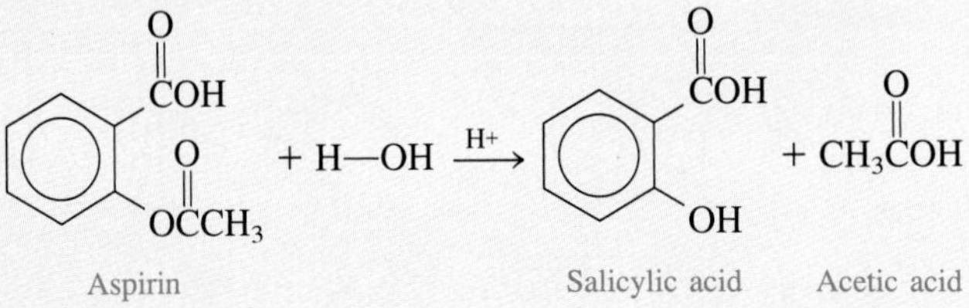

$$C_6H_4(\overset{\overset{O}{\|}}{C}OH)(O\overset{\overset{O}{\|}}{C}CH_3) + H{-}OH \xrightarrow{H+} C_6H_4(\overset{\overset{O}{\|}}{C}OH)(OH) + CH_3\overset{\overset{O}{\|}}{C}OH$$

Aspirin Salicylic acid Acetic acid

Acetic acid gives the vinegar odor to a sample of aspirin that has hydrolyzed.

Study Check:
What products would you expect in the hydrolysis of ethyl propionate in acid?

Answer:

$$CH_3CH_2\overset{\overset{O}{\|}}{C}OH + CH_3CH_2OH$$

Propionic acid Ethyl alcohol

Saponification

Saponification is a reaction that occurs when an ester is heated with a strong base such as NaOH or KOH. The products of the base hydrolysis are the salt of the acid and the alcohol.

Saponification of an Ester

$$\underset{\text{Ester}}{R{-}\overset{O}{\overset{\|}{C}}{-}OR} + \underset{\text{Sodium hydroxide}}{NaOH} \xrightarrow{\text{Heat}} \underset{\text{Salt}}{R{-}\overset{O}{\overset{\|}{C}}O^-Na^+} + \underset{\text{Alcohol}}{R{-}OH}$$

$$\underset{\text{Methyl propionate}}{CH_3CH_2\overset{O}{\overset{\|}{C}}OCH_3} + NaOH \xrightarrow{\text{Heat}} \underset{\text{Sodium propionate}}{CH_3CH_2\overset{O}{\overset{\|}{C}}O^-Na^+} + \underset{\text{Methyl alcohol}}{CH_3OH}$$

Sample Problem 13.8
Writing Saponification Equations

Write the structural formulas for the products of the following saponification reaction:

$$CH_3\overset{O}{\overset{\|}{C}}{-}OCH_2CH_2CH_3 + NaOH \xrightarrow{\text{Heat}}$$

Solution:
An ester reacting with a base (NaOH) will give the salt of the acid and an alcohol. In the salt, the sodium ion from the base replaces the hydrogen atom in the carboxylic acid; the other product is propyl alcohol.

$$\underset{\text{Propyl acetate}}{CH_3\overset{O}{\overset{\|}{C}}{-}OCH_2CH_2CH_3} + \underset{\text{Sodium hydroxide}}{NaOH} \xrightarrow{\text{Heat}} \underset{\text{Sodium acetate}}{CH_3\overset{O}{\overset{\|}{C}}O^-\ Na^+} + \underset{\text{Propyl alcohol}}{CH_3CH_2CH_2OH}$$

Study Check:
What are the hydrolysis products when ethyl butyrate is heated with KOH?

Answer:

$$\underset{\text{Potassium butyrate}}{CH_3CH_2CH_2\overset{O}{\overset{\|}{C}}{-}O^-\ K^+} + \underset{\text{Ethyl alcohol}}{HOCH_2CH_3}$$

HEALTH NOTE

Fats and Soaps

Fats are esters, and in the body they are hydrolyzed to fatty acids and glycerol during the digestion process. Enzymes from the pancreas, called lipases, are mixed with the fats from foodstuffs in the small intestine to hydrolyze the fats.

$$
\begin{array}{lcl}
CH_2-O-\overset{\overset{O}{\|}}{C}-C_{17}C_{35} & & CH_2-OH \\
\;| & & \;| \\
CH-O-\overset{\overset{O}{\|}}{C}-C_{17}C_{35} + 3H_2O & \xrightarrow{\text{Lipases}} & CH-OH + 3C_{17}H_{35}-\overset{\overset{O}{\|}}{C}-OH \\
\;| & & \;| \\
CH_2-O-\overset{\overset{O}{\|}}{C}-C_{17}H_{35} & & CH_2-OH
\end{array}
$$

Tristearin (a fat) Glycerol Stearic acid (a fatty acid)

Soaps are the salts of long-chain fatty acids produced when a fat is heated with a strong base such as NaOH or KOH.

$$
\begin{array}{lcl}
CH_2-O-\overset{\overset{O}{\|}}{C}-C_{17}H_{35} & & CH_2-OH \\
\;| & & \;| \\
CH-O-\overset{\overset{O}{\|}}{C}-C_{17}H_{35} + 3NaOH & \xrightarrow{\text{Heat}} & CH-OH + 3Na^{+}\;{}^{-}O-\overset{\overset{O}{\|}}{C}-C_{17}H_{35} \\
\;| & & \;| \\
CH_2-O-\overset{\overset{O}{\|}}{C}-C_{17}H_{35} & & CH_2-OH
\end{array}
$$

Ionic Nonpolar

A soap

Tristearin Glycerol Sodium stearate

Within the soap molecule, there is an ionic section that is hydrophilic, or ''water-loving,'' and a long, nonpolar carbon chain that is hydrophobic, or ''water-fearing.'' When soap molecules are mixed with a fat or oil, the nonpolar ends are attracted to the nonpolar fat or oil particles. The fat or oil is pulled into the water by the solubility of the ionic ends of the soap molecules in water. (See Figure 13.4.) ■

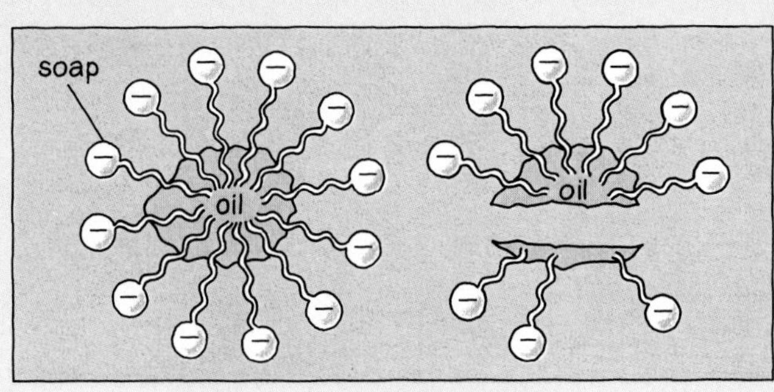

Figure 13.4
Fats and grease are attracted to the nonpolar ends of soap molecules, and then they are removed from clothes or dishes by the polar ends of the soap molecules dissolving in water.

13.5 Amines

Learning Goal **Write the name and structural formula of a simple amine.**

Amines are derivatives of ammonia in which one, two, or three hydrogen atoms are replaced by alkyl groups. In a primary (1°) amine, the nitrogen atom is attached to one alkyl group; in a secondary (2°) amine, the nitrogen atom is attached to two carbon groups; and in a tertiary (3°) amine, all three hydrogen atoms in ammonia have been replaced.

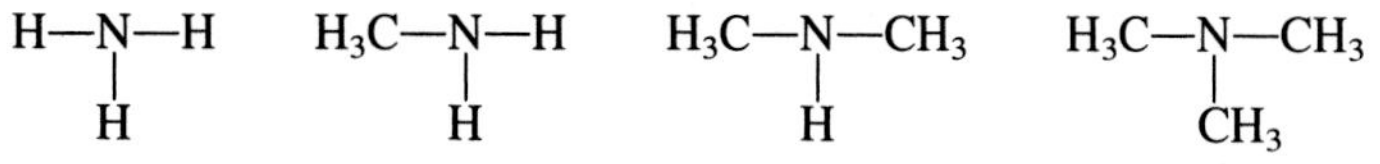

H—N(—H)—H	H_3C—N(—H)—H	H_3C—N(—H)—CH_3	H_3C—N(CH_3)—CH_3
Ammonia	Primary amine	Secondary amine	Tertiary amine

Sample Problem 13.9
Classifying Amines

Classify the following amines as primary, secondary, or tertiary:

a. 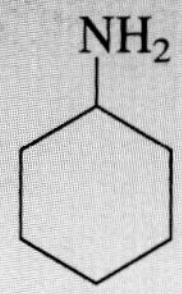 NH_2 attached to a cyclohexane ring **b.** 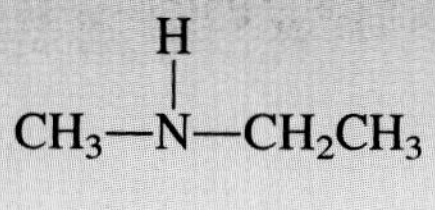 CH_3—N(H)—CH_2CH_3

Solution:

a. This is a primary amine because one of the hydrogen atoms of NH_3 has been replaced.

b. When two alkyl groups are attached to the nitrogen atom, the compound is a secondary amine.

Polarity of Amines

The N—H bond of primary and secondary amines is polar and forms hydrogen bonds with other amines. They have boiling points higher than those of the alkanes but not as high as alcohols and carboxylic acids because the N—H bond is not as polar as the O—H bond. Tertiary amines cannot form hydrogen bonds.

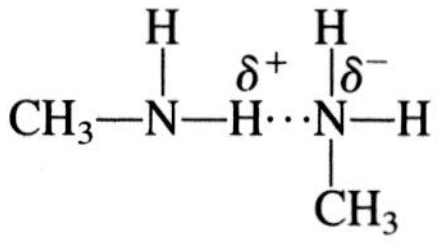

$$CH_3\text{—}\underset{}{\overset{H}{\overset{|}{N}}}\text{—}\overset{\delta^+}{H}\cdots\overset{H}{\overset{|\delta^-}{\underset{\underset{CH_3}{|}}{N}}}\text{—}H$$

Hydrogen bonding of a primary amine

Amines also form hydrogen bonds with water, but only the smaller amines with one to four carbon atoms are soluble. Amines that have more than five carbon atoms have limited solubility.

Figure 13.5
Models of amines: methylamine and dimethylamine.

Hydrogen bonding makes small amines soluble in water

$$CH_3—\overset{\displaystyle H}{\overset{|}{N}}—H^{\delta+}\cdots\overset{\displaystyle H}{\overset{|}{O}}{}^{\delta-}—H$$

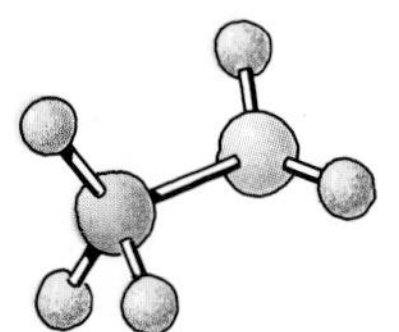

$CH_3—NH_2$
methylamine

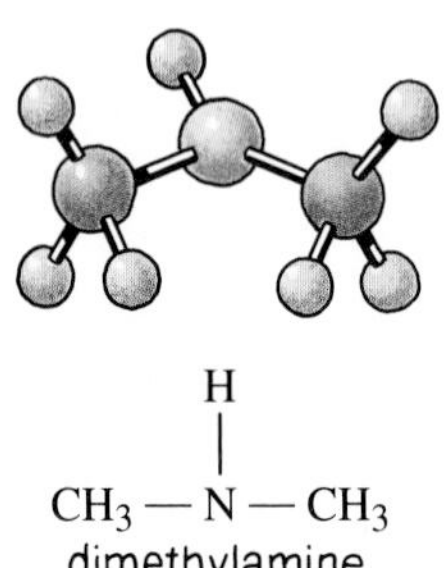

$$CH_3—\overset{\displaystyle H}{\overset{|}{N}}—CH_3$$
dimethylamine

Naming Amines

Amines are generally named by their common names which consist of the names of each alkyl group attached to the nitrogen atom followed by the suffix *amine*. See Figure 13.5. Methylamine and ethylamine have odors similar to ammonia; higher alkyl amines have fishy odors.

Naming Amines R—NH₂

Functional group $—\overset{|}{N}—$

Common name: Alkyl amine

$$\underset{\text{Methylamine}}{CH_3—\overset{\displaystyle H}{\overset{|}{N}}—H} \qquad \underset{\text{Ethylmethylamine}}{CH_3CH_2—\overset{\displaystyle H}{\overset{|}{N}}—CH_3} \qquad \underset{\text{Dimethylpropylamine}}{CH_3CH_2CH_2—\overset{\displaystyle CH_3}{\overset{|}{N}}—CH_3}$$

The amine of benzene is named aniline which is used in the names of its derivatives. When alkyl groups are attached to the nitrogen atom in aniline, each is indicated by using N— before the name of each group.

Aniline N-methylaniline N,N-dimethylaniline

Sample Problem 13.10
Naming Amines

Write a common name for each of the following amines:

a. $CH_3CH_2—NH_2$

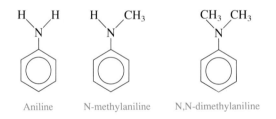

b. $CH_3—\overset{\displaystyle CH_3}{\overset{|}{N}}—CH_3$

c. CH_3, CH_2CH_3 on N attached to benzene ring

Solution:

a. This amine has one ethyl group attached to the nitrogen atom; its name is *ethylamine*.

b. This amine has three methyl groups attached to the nitrogen atom; its name is *trimethylamine*.

c. This amine is a derivative of aniline and has a methyl and an ethyl group attached to the nitrogen atom; its name is *N-ethyl-N-methylaniline*.

Study Check:
Draw the structure of ethylpropylamine.

Answer:

$$CH_3CH_2-\overset{\displaystyle H}{\overset{|}{N}}-CH_2CH_2CH_3$$

HEALTH NOTE

Amines in Health and Medicine

In response to allergic reactions or to injured cells, the body produces increased levels of histamine that cause blood vessels to dilate and increase the permeability of the cells affected. Redness and swelling occurs in the area. Administering an antihistamine such as diphenylhydramine helps block the effects of histamines.

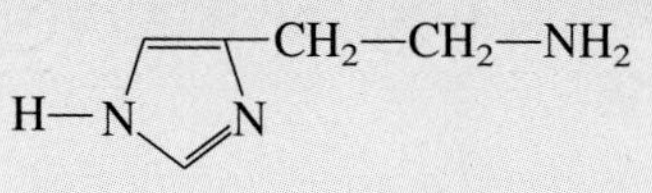

H—N(imidazole ring)N with —$CH_2-CH_2-NH_2$

Histamine

$(C_6H_5)_2CH-O-CH_2-CH_2-N(CH_3)-CH_3$

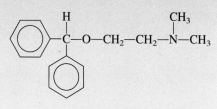

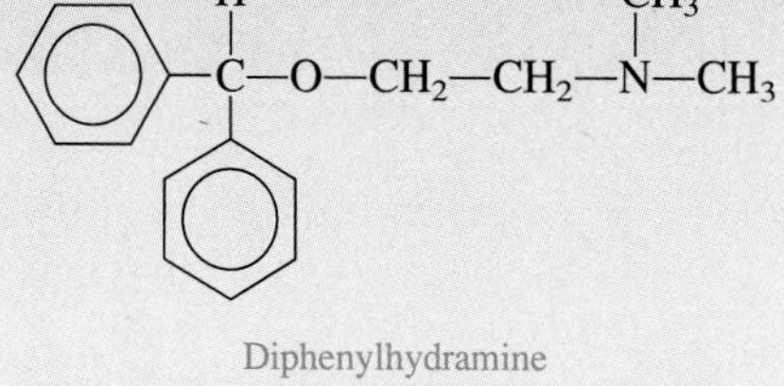

Diphenylhydramine

Epinephrine (adrenaline) and norepinephrine (noradrenaline) are used in remedies for colds, hay fever, and asthma. They contract the capillaries in the mucous membranes of the respiratory passage. In crisis situations, epinephrine, released by the adrenal medulla, raises the blood glucose level in preparation for "fight or flight," and moves the blood to the muscles.

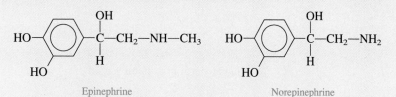

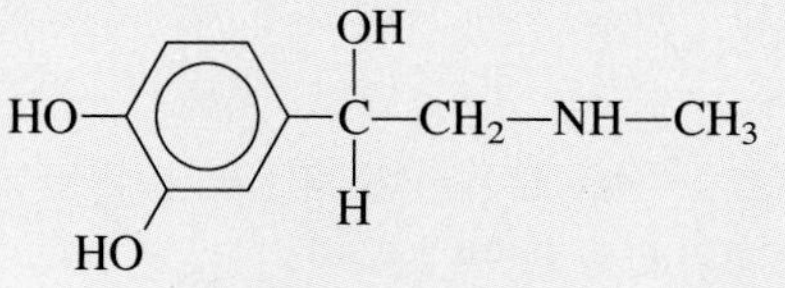

$(HO)_2C_6H_3-CH(OH)-CH_2-NH-CH_3$

Epinephrine

$(HO)_2C_6H_3-CH(OH)-CH_2-NH_2$

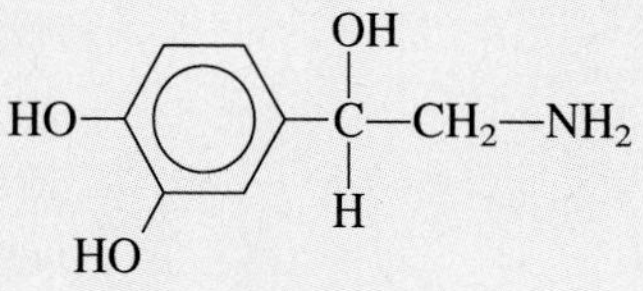

Norepinephrine

Benzedrine (amphetamine), and Neo-Synephrine (phenylephrine) are medications used to reduce respiratory congestion from colds, hay fever, and asthma. Sometimes, benzedrine is taken internally to combat the desire to sleep, but it has side effects and can be habit-forming.

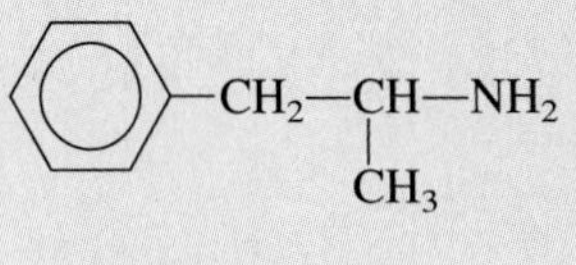

$C_6H_5-CH_2-CH(CH_3)-NH_2$

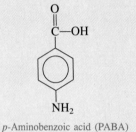

Benzedrine, amphetamine

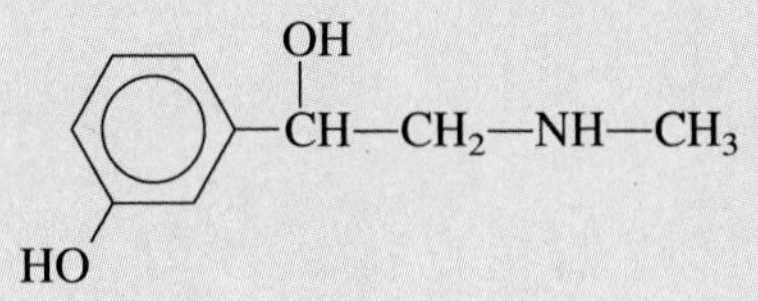

$HO-C_6H_4-CH(OH)-CH_2-NH-CH_3$

Neo-Synephrine, phenylephrine

p-Aminobenzoic acid, or PABA, used in suntan preparations protects the skin by absorbing ultraviolet radiation. ■

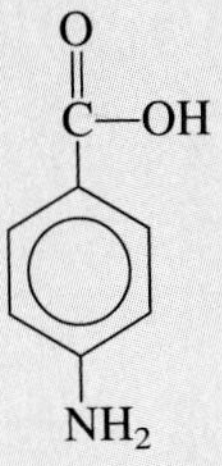

$HOOC-C_6H_4-NH_2$

p-Aminobenzoic acid (PABA)

13.6 Heterocyclic Amines and Physiological Activity

Learning Goal **Identify heterocyclic amines and describe some of their physiological activity.**

When a nitrogen atom is part of a ring structure, the compound is a heterocyclic amine. The simplest five-member rings include pyrrolidine, pyrrole, and imidazole, which has two nitrogen atoms. Both pyrrole and imidazole are unsaturated.

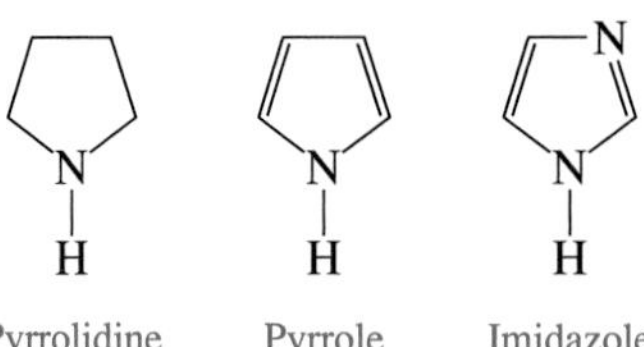

Pyrrolidine Pyrrole Imidazole

The pyrrole ring is part of the porphyrin system found in chlorophyll and the heme portion of hemoglobin.

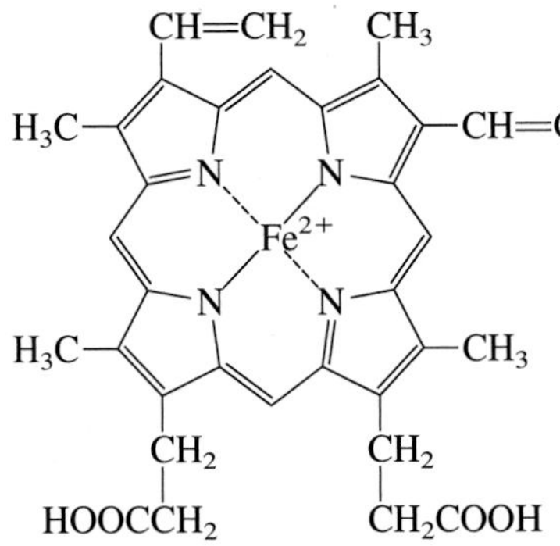

Heme

Pyridine is a heterocyclic amine that has a benzene-like structure with one nitrogen atom in the ring. Purine and pyrimidine have ring structures that contain two or more nitrogen atoms. They are important building blocks of the nucleic acids DNA and RNA.

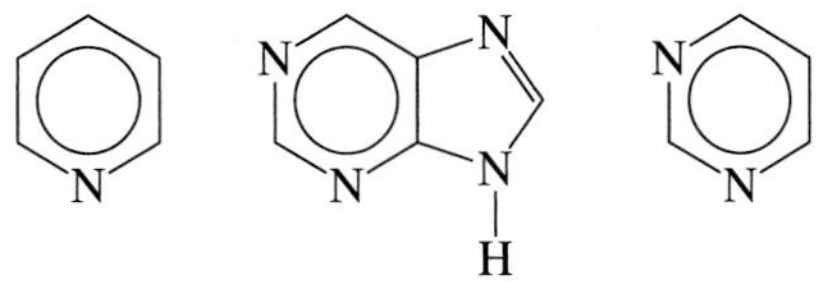

Pyridine Purine Pyrimidine

Sample Problem 13.11
Identifying Heterocyclic Amines

Which of the following is a heterocyclic amine?

a. Benzedrine

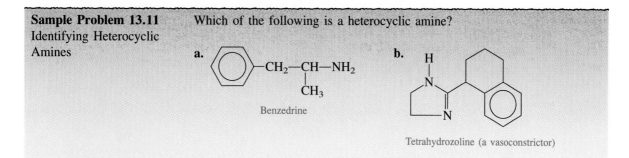

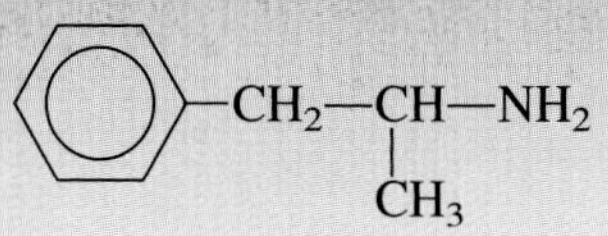

b. Tetrahydrozoline (a vasoconstrictor)

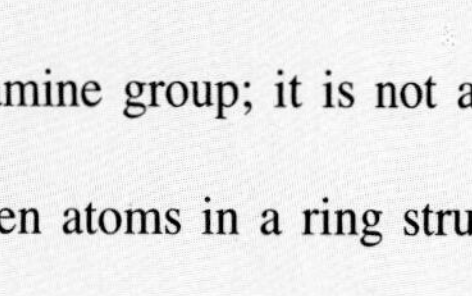

Solution:

a. Benzedrine contains a primary amine group; it is not a heterocyclic amine.

b. Tetrahydrozoline contains nitrogen atoms in a ring structure; it is a heterocyclic amine.

Physiological Activity of Amines

Alkaloids are physiologically active nitrogen-containing compounds produced by plants. The term *alkaloid* refers to the basic (alkali-like) characteristics of these amines. Certain alkaloids find use in anesthetics and as hallucinogens, although many have habituating effects. (See Figure 13.6.)

Figure 13.6
Alkaloids, nitrogen-containing compounds having physiological activity, may be obtained from such plants as belladonna (atropine). Other alkaloids may be obtained from tobacco (nicotine), coca (cocaine), cactus (mescaline), and poppies (morphine).

Quinine obtained from the bark of the cinchona tree has been used in the treatment of malaria since the 1600s:

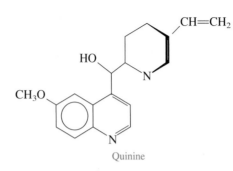

[Structure: HO, CH_3O, N, N, $CH{=}CH_2$]

Quinine

Nicotine found in the leaves of the tobacco plant and coniine from hemlock are extremely toxic alkaloids.

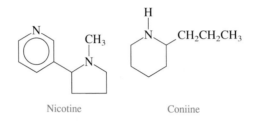

[Structures: N, CH_3, N; H, N, $CH_2CH_2CH_3$]

Nicotine Coniine

Atropine from belladonna and cocaine from the coca plant are used in low concentrations as anesthetics for eye and sinus procedures. However, higher doses produce euphoria followed by depression and a desire for additional quantities of the drug. Chemists have altered the structures of substances such as atropine and cocaine in an effort to develop synthetic alkaloids such as procaine, which is used for local anesthesia. The synthetic products retain the anesthetic qualities of the natural alkaloid without the addictive side effects.

[Structures: CH_3, N, O–C(=O)–CH(CH_2OH)–phenyl; CH_3, N, $\overset{O}{\parallel}COCH_3$, O–C(=O)–phenyl]

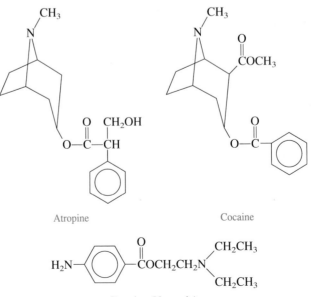

Atropine Cocaine

$$H_2N-C_6H_4-\overset{O}{\overset{\parallel}{C}}OCH_2CH_2N(CH_2CH_3)_2$$

Procaine (Novocain)

For many centuries, morphine, an alkaloid found in poppies, has been used as a painkiller. However, it also has strong hallucinogenic and addictive side effects. A synthetic alkaloid, meperidine, or Demerol, was developed as a painkiller with a chemical structure similar to morphine but with reduced side effects. One area of research in pharmacology is the synthesis of a morphine-like compound that can be used safely as a painkiller with no side effects. Codeine and heroin have structures very similar to morphine with similar side effects.

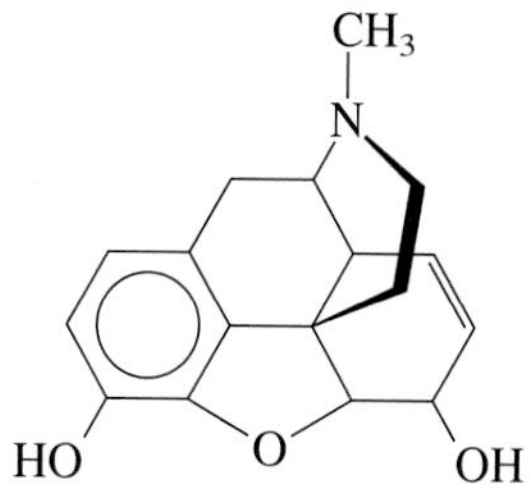

Morphine
(opium)

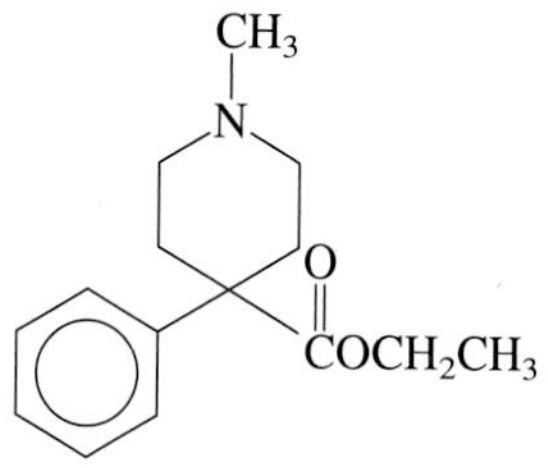

Demerol
(meperidine)

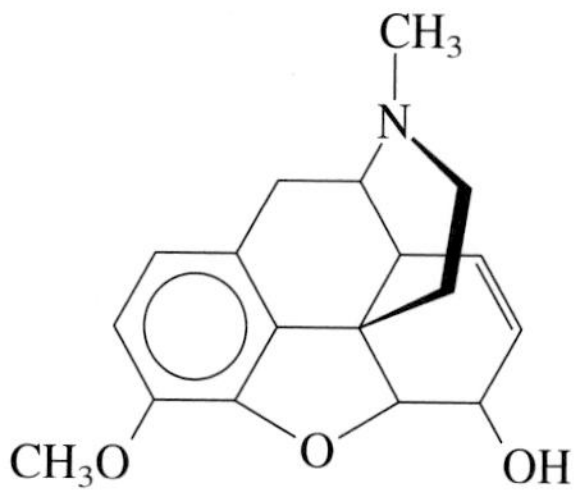

Codeine

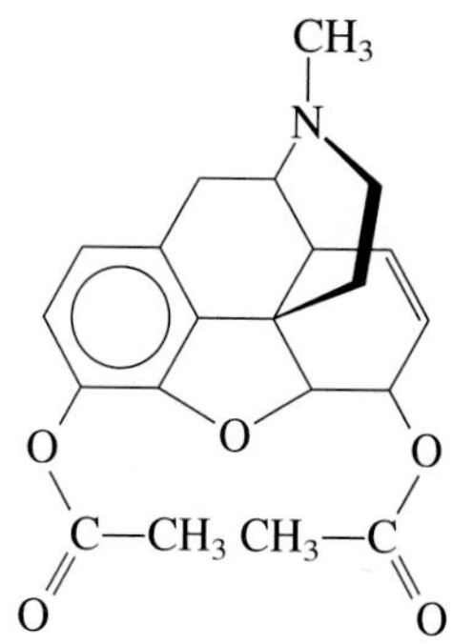

Heroin

Alkaloids are prevalent among the compounds known as hallucinogens. Examples include mescaline, from the peyote cactus, and LSD (lysergic acid diethylamide), prepared from lysergic acid that is produced by a fungus that grows on rye.

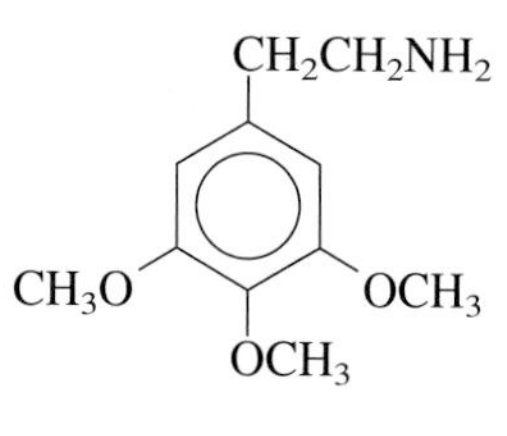

Mescaline

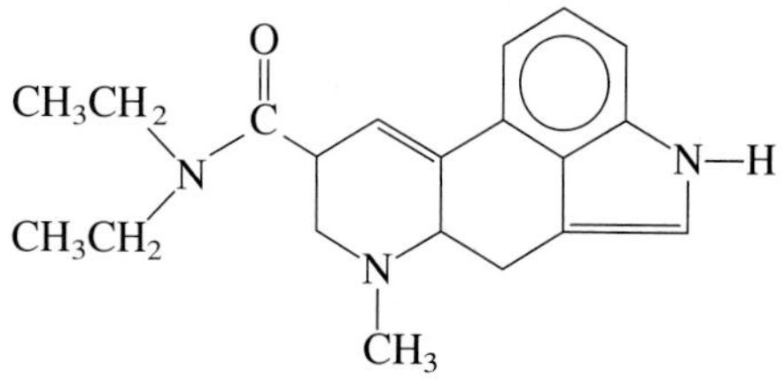

Lysergic acid diethylamide
(LSD)

Other heterocyclic amines acts as tranquilizers by reducing the transmission of nerve impulses to the brain. Low levels of serotonin in the brain appear to be associated with depressive states. Reserpine has been used as a sedative for psychotic patients, and chlorpromazine has been effective as a medication for schizophrenia.

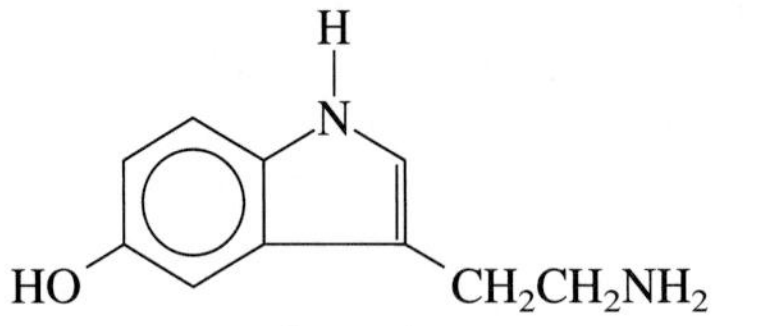

Serotonin

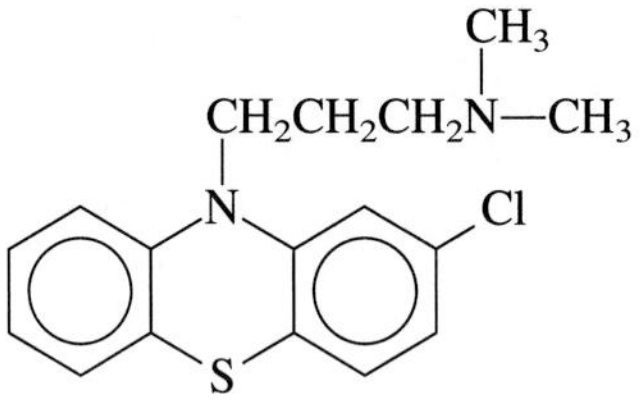

Chlorpromazine

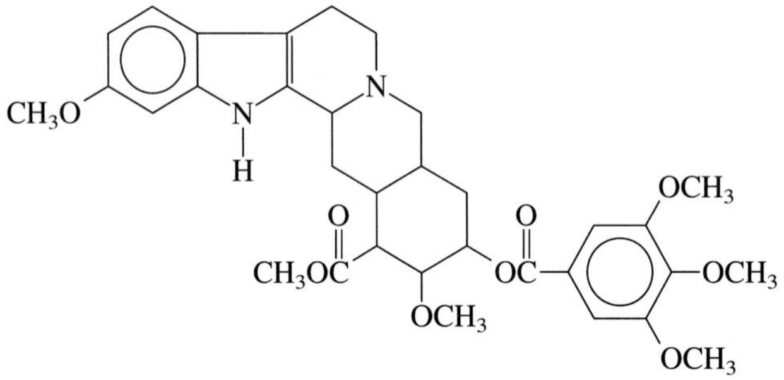

Reserpine

Sample Problem 13.12
Physiologically Active Amines

Caffeine, a compound found in coffee beans and tea, is a stimulant of the central nervous system. Why is it a heterocyclic amine? Why is it an alkaloid?

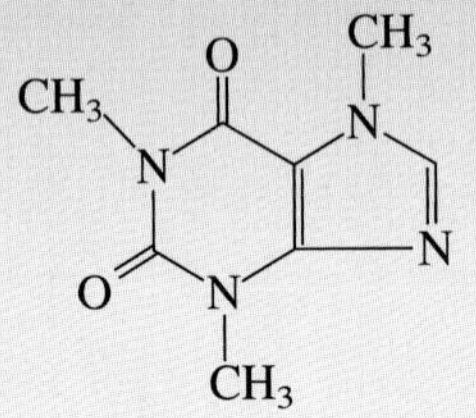

Caffeine

Solution:
Caffeine is a heterocyclic amine because its structure contains nitrogen atoms in a ring. It is an alkaloid because it is produced by coffee trees and has physiological activity.

13.7 Reactions of Amines

Learning Goal **Write equations for the ionization and neutralization of amines.**

Ammonia acts as a weak base because its unshared electron pair can accept a proton from water.

$$\mathrm{H{-}\ddot{N}(H){-}H + H{-}OH \rightleftharpoons H{-}\overset{H}{\underset{H}{N^{+}}}{-}H + OH^-}$$

Ammonia — Ammonium hydroxide

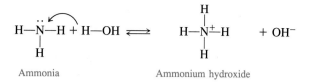

In a similar way, an amine also acts as a weak base. The ionization produces an ammonium-type ion and a hydroxide ion.

Ionization of Amines in Water

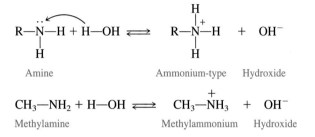

$$\mathrm{R{-}\ddot{N}(H){-}H + H{-}OH \rightleftharpoons R{-}\overset{H}{\underset{H}{N^{+}}}{-}H + OH^-}$$

Amine — Ammonium-type — Hydroxide

$$\mathrm{CH_3{-}NH_2 + H{-}OH \rightleftharpoons CH_3{-}\overset{+}{N}H_3 + OH^-}$$

Methylamine — Methylammonium — Hydroxide

Sample Problem 13.13
Ionization of Amines

Write an equation that shows propylamine acting as a weak base in water.

Solution:
The amine acts as a weak base by accepting a proton from water to give a propylammonium ion and a hydroxide ion.

$$\mathrm{CH_3CH_2CH_2{-}NH_2 + H{-}OH \rightleftharpoons CH_3CH_2CH_2{-}\overset{+}{N}H_3 + OH^-}$$

Propylamine — Propylammonium — Hydroxide

Study Check:
Write an equation for the reaction of ethylmethylamine with water.

$$\mathrm{CH_3CH_2{-}\underset{H}{N}{-}CH_3 + H{-}OH \rightleftharpoons CH_3CH_2{-}\overset{H}{\underset{H}{N^{+}}}{-}CH_3 + OH^-}$$

Neutralization of Amines

When you squeeze lemon juice on fish, you counteract the fishy odor of the basic amines in the fish by neutralizing them with an acid. When an amine reacts with an acid, a salt is formed. The proton (H^+) from the acid bonds with the unshared pair of electrons on the nitrogen atom. The only product is the salt; no water is formed.

Neutralization of Amines

$$\text{R—}\overset{\text{H}}{\underset{\text{H}}{\text{N}}}\text{:} + \text{H—Cl} \longrightarrow \text{R—}\overset{\text{H}}{\underset{\text{H}}{\overset{+}{\text{N}}}}\text{—H Cl}^-$$

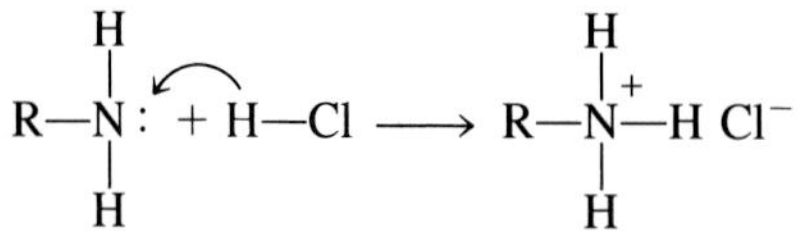

Amine Acid Salt

$$CH_3CH_2\text{—}NH_2 + HCl \longrightarrow CH_3CH_2\text{—}\overset{+}{N}H_3\ Cl^-$$

Ethylamine Hydrochloric acid Ethylammonium chloride

The salt is named by replacing the *amine* suffix name by *ammonium* followed by the name of the negative ion.

Sample Problem 13.14
Neutralization of Amines

Write an equation for the reaction of dimethylamine and HBr.

Solution:

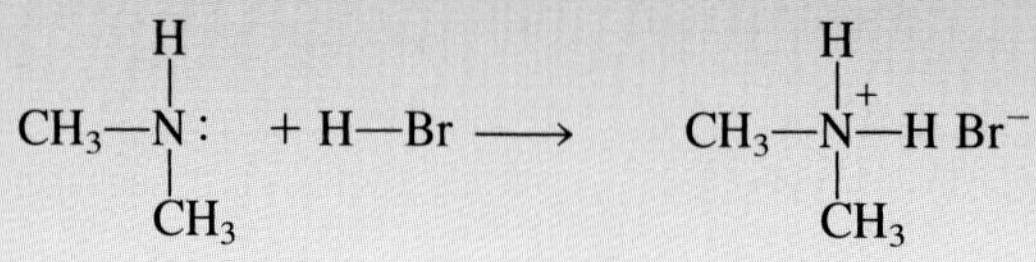

$$CH_3\text{—}\overset{H}{\underset{CH_3}{N}}\text{:} + H\text{—}Br \longrightarrow CH_3\text{—}\overset{H}{\underset{CH_3}{\overset{+}{N}}}\text{—}H\ Br^-$$

Dimethylamine Dimethylammonium bromide

Study Check:
What is the salt formed from the reaction of trimethylamine and HCl?

Answer:

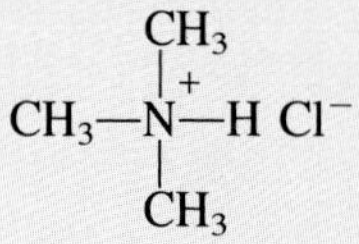

$$CH_3\text{—}\overset{CH_3}{\underset{CH_3}{\overset{+}{N}}}\text{—}H\ Cl^-$$

Trimethylammonium chloride

13.8 Amides

Learning Goal **Write the name and structural formula of the amide produced from a carboxylic acid and an amine.**

Amides are derivatives of carboxylic acids in which the hydroxyl (—OH) portion of the acid is replaced by an amino (—NH_2) group.

$$R\text{—}\overset{O}{\overset{\|}{C}}\text{—}OH \qquad R\text{—}\overset{O}{\overset{\|}{C}}\text{—}NH_2$$

Carboxylic acid Amide

Figure 13.7
Model of ethanamide (acetamide), an amide.

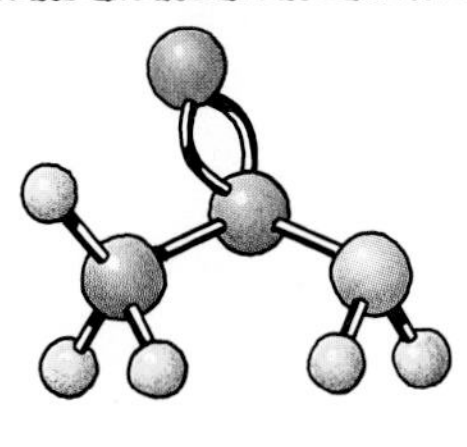

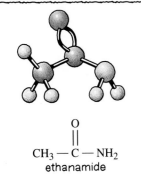

$$CH_3-\overset{\overset{\displaystyle O}{\|}}{C}-NH_2$$

ethanamide

Naming Amides

The amides are named as derivatives of carboxylic acids by changing the *ic* or *oic acid* ending to *amide*. Common names are used for simple amides. (See Figure 13.7.)

Naming Amides: $R-\overset{\overset{\displaystyle O}{\|}}{C}-NH_2$

Functional group:	$-\overset{\overset{\displaystyle O}{\|}}{C}-NH_2$
IUPAC name:	Alkanamide
Common name:	Prefixamide

$$H\overset{\overset{\displaystyle O}{\|}}{C}-NH_2 \qquad CH_3\overset{\overset{\displaystyle O}{\|}}{C}-NH_2 \qquad CH_3CH_2\overset{\overset{\displaystyle O}{\|}}{C}-NH_2 \qquad CH_3CH_2CH_2\overset{\overset{\displaystyle O}{\|}}{C}-NH_2$$

Methanamide (formamide) — Ethanamide (acetamide) — Propanamide (propionamide) — Butanamide (butyramide)

When there are alkyl groups attached to the nitrogen atom, their names are preceded by the letter *N*.

$$CH_3\overset{\overset{\displaystyle O}{\|}}{C}-\overset{\overset{\displaystyle H}{|}}{N}-CH_3 \qquad CH_3CH_2\overset{\overset{\displaystyle O}{\|}}{C}-\underset{\underset{\displaystyle CH_3}{|}}{N}-CH_3$$

N-methylethanamide (*N*-methylacetamide) — *N*,*N*-dimethylpropanamide (*N*,*N*-dimethylpropionamide)

Sample Problem 13.15
Writing the Structural Formula of an Amide

Write the structural formula of *N*-methylbenzamide.

Solution:
Benzamide is the amide of benzoic acid. There is a methyl (CH_3-) group attached to the nitrogen atom of the amide group.

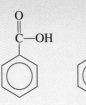

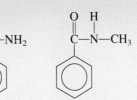

$$C_6H_5-\overset{\overset{\displaystyle O}{\|}}{C}-OH \qquad C_6H_5-\overset{\overset{\displaystyle O}{\|}}{C}-NH_2 \qquad C_6H_5-\overset{\overset{\displaystyle O}{\|}}{C}-\overset{\overset{\displaystyle H}{|}}{N}-CH_3$$

Benzoic acid — Benzamide — *N*-methylbenzamide

Study Check:
Name the following amide:

$$CH_3CH_2CH_2\overset{\overset{\displaystyle O}{\|}}{C}{-}NHCH_2CH_3$$

Answer: *N*-ethylbutanamide; *N*-ethylbutyramide

Formation of Amides

An amide can be prepared by heating a carboxylic acid with ammonia or an amine. A molecule of water is eliminated, and the fragments of the two molecules combine to form the amide, much like ester formation. This reaction, called **amidation,** results in the formation of an amide bond.

Amidation

$$R{-}\overset{\overset{\displaystyle O}{\|}}{C}{-}OH + H{-}\overset{\overset{\displaystyle H}{|}}{N}{-}H \xrightarrow{\text{Heat}} R{-}\overset{\overset{\displaystyle O}{\|}}{C}{-}\overset{\overset{\displaystyle H}{|}}{N}{-}H + H_2O$$

Carboxylic acid Ammonia Amide

$$R{-}\overset{\overset{\displaystyle O}{\|}}{C}{-}OH + H{-}\overset{\overset{\displaystyle H}{|}}{N}{-}R \xrightarrow{\text{Heat}} R{-}\overset{\overset{\displaystyle O}{\|}}{C}{-}\overset{\overset{\displaystyle H}{|}}{N}{-}R + H_2O$$

Carboxylic acid Amine *N*-alkylamine

$$CH_3{-}\overset{\overset{\displaystyle O}{\|}}{C}{-}OH + H{-}NH_2 \xrightarrow{\text{Heat}} CH_3{-}\overset{\overset{\displaystyle O}{\|}}{C}{-}\overset{\overset{\displaystyle H}{|}}{N}{-}H + H_2O$$

Acetic acid Ammonia Acetamide

$$CH_3{-}\overset{\overset{\displaystyle O}{\|}}{C}{-}OH + H{-}\overset{\overset{\displaystyle H}{|}}{N}{-}CH_3 \xrightarrow{\text{Heat}} CH_3{-}\overset{\overset{\displaystyle O}{\|}}{C}{-}\overset{\overset{\displaystyle H}{|}}{N}{-}CH_3 + H_2O$$

Acetic acid Methylamine *N*-methylacetamide

Peptide Bonds in Proteins

Amino acids contain both a carboxylic acid group and an alpha (α) amino group in the same molecule. About 20 amino acids are used in the structures of proteins. In our cells, the amino group from one amino acid forms an amide bond with the carboxylic acid group of another amino acid. Long chains of amino acids are linked together through these amide bonds to make the proteins and enzymes of the body. In proteins these amide linkages are called peptide bonds.

Amino Acids

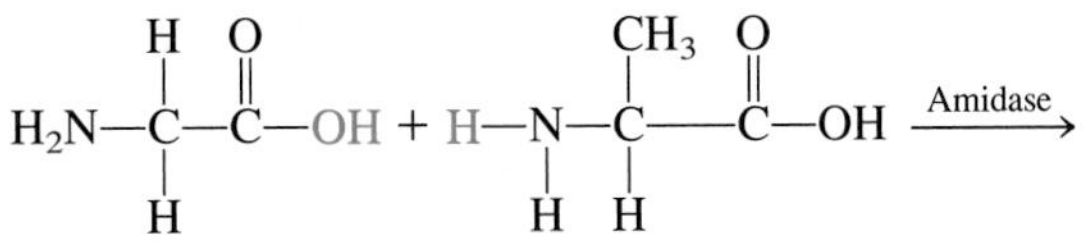

$$H_2N-\underset{\underset{H}{|}}{\overset{\overset{H}{|}}{C}}-\overset{\overset{O}{\|}}{C}-OH + H-\underset{\underset{H}{|}}{N}-\underset{\underset{H}{|}}{\overset{\overset{CH_3}{|}}{C}}-\overset{\overset{O}{\|}}{C}-OH \xrightarrow{\text{Amidase}}$$

Glycine Alanine

Amide bond "Peptide bond"

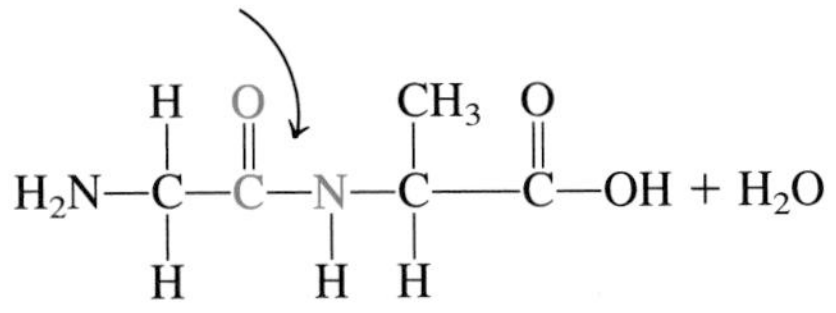

$$H_2N-\underset{\underset{H}{|}}{\overset{\overset{H}{|}}{C}}-\overset{\overset{O}{\|}}{C}-\underset{\underset{H}{|}}{N}-\underset{\underset{H}{|}}{\overset{\overset{CH_3}{|}}{C}}-\overset{\overset{O}{\|}}{C}-OH + H_2O$$

Glycylalanine, a dipeptide

Sample Problem 13.16
Amidation

Write the structural formula of the amide produced by heating benzoic acid and methylamine.

Solution:
The formula of the amide product can be written by attaching the carbonyl group from the acid to the nitrogen atom of the amine. The —OH is removed from the acid and an —H from the amine to form water.

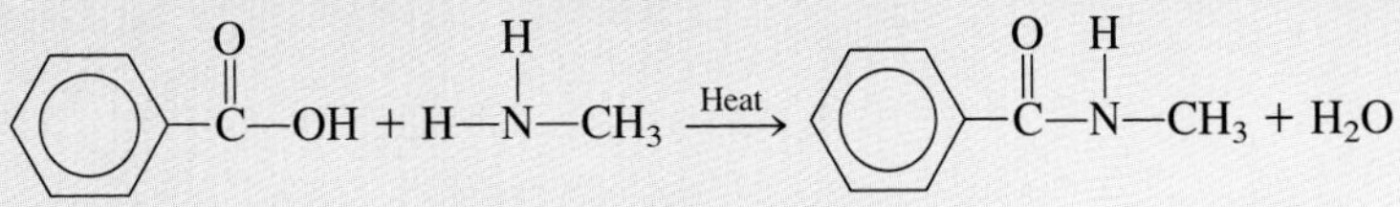

$$C_6H_5-\overset{\overset{O}{\|}}{C}-OH + H-\overset{\overset{H}{|}}{N}-CH_3 \xrightarrow{\text{Heat}} C_6H_5-\overset{\overset{O}{\|}}{C}-\overset{\overset{H}{|}}{N}-CH_3 + H_2O$$

Benzoic acid Methylamine *N*-methylbenzamide

Study Check:
What acid and amine are needed to prepare *N*-propylbenzamide?

Answer: benzoic acid and propylamine

Hydrolysis of Amides

Amides are neutral substances and not very reactive in aqueous solutions. They can undergo hydrolysis with water, but the reaction requires considerable heating time. Hydrolysis of an amide yields a carboxylic acid and ammonia or an amine.

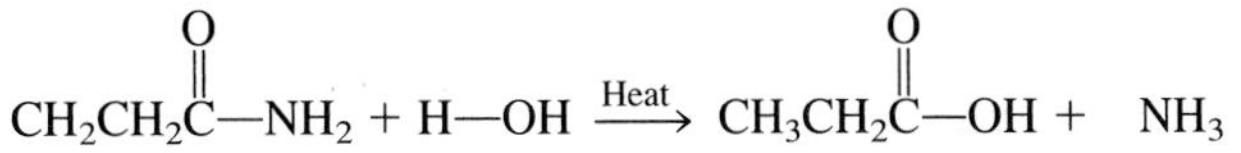

$$CH_2CH_2\overset{\overset{O}{\|}}{C}-NH_2 + H-OH \xrightarrow{\text{Heat}} CH_3CH_2\overset{\overset{O}{\|}}{C}-OH + NH_3$$

Propionamide Propionic acid Ammonia

HEALTH NOTE

Amides in Health and Medicine

The simplest natural amide is urea, an end product of protein metabolism in the body. The kidneys remove urea from the blood and provide for its excretion in urine. If the kidneys malfunction, urea is not removed and builds to toxic level—a condition called uremia.

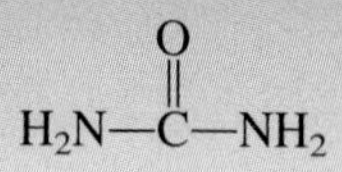

$$H_2N—\overset{\overset{\displaystyle O}{\|}}{C}—NH_2$$

Urea

Synthetic amides have found use as sweeteners and drugs. Saccharin is a very powerful sweetener and is used as a sugar substitute. Aspirin substitutes including phenacetin and acetaminophen (*p*-hydroxyacetanilide) act as analgesics and antipyretics but may have less anti-inflammatory effect.

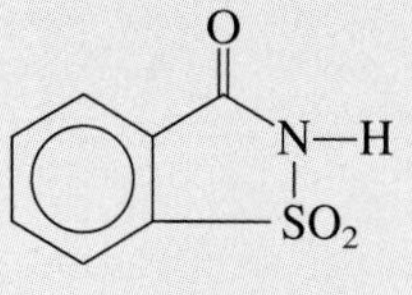

Saccharin 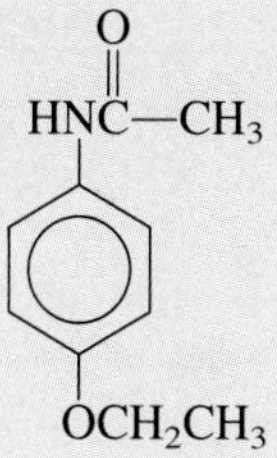   Phenacetin 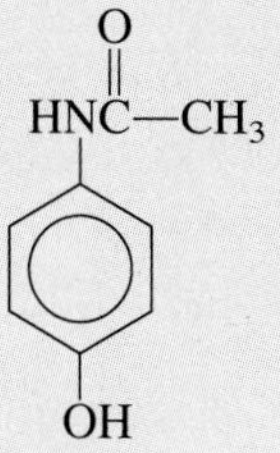   Acetaminophen

Many barbiturates are cyclic amides that act as sedatives in small dosages or sleep inducers in larger dosages. They are often habit-forming. Barbiturate drugs include phenobarbital (Luminal), pentobarbital (Nembutal) and secobarbital (Seconal).

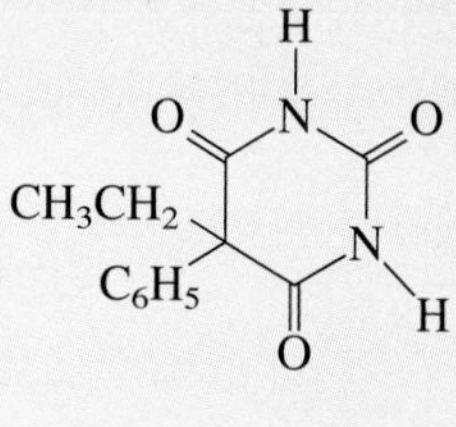

Luminal (phenobarbital) 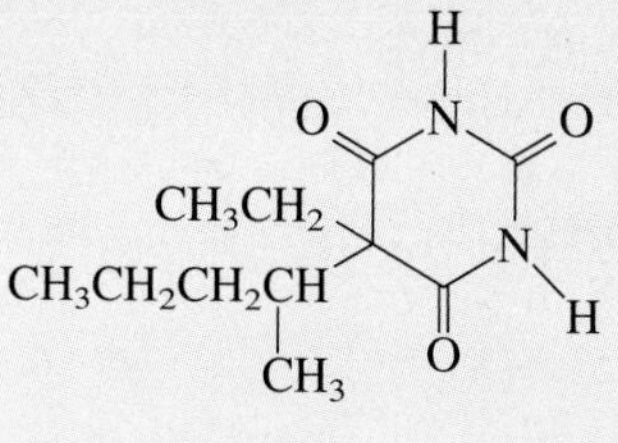   Nembutal (pentobarbital) 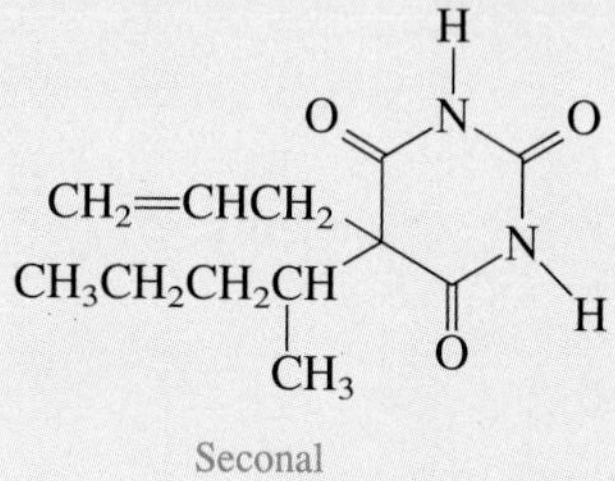   Seconal (secobarbital)

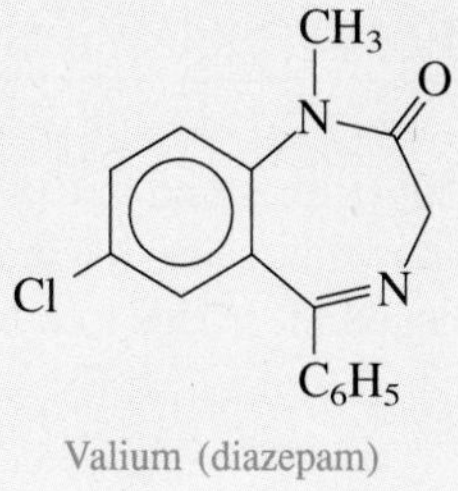

Valium (diazepam)

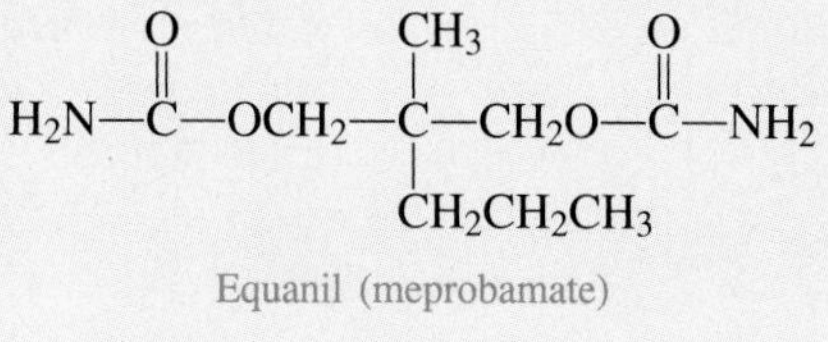

$$H_2N—\overset{\overset{\displaystyle O}{\|}}{C}—OCH_2—\underset{\underset{\displaystyle CH_2CH_2CH_3}{|}}{\overset{\overset{\displaystyle CH_3}{|}}{C}}—CH_2O—\overset{\overset{\displaystyle O}{\|}}{C}—NH_2$$

Equanil (meprobamate)

Sample Problem 13.17
Hydrolysis of an Amide

Write the products of the hydrolysis of the following amide:

$$CH_3\overset{\overset{O}{\|}}{C}-\overset{\overset{H}{|}}{N}CH_3$$

Solution:
In hydrolysis, the products are the carboxylic acid and the amine.

$$CH_3\overset{\overset{O}{\|}}{C}-\overset{\overset{H}{|}}{N}CH_3 + H_2O \xrightarrow{\text{Heat}} CH_3\overset{\overset{O}{\|}}{C}OH + CH_3NH_2$$

Summary of Naming

Class	Structure	IUPAC name	Common name
Carboxylic acid	$CH_3\overset{\overset{O}{\|}}{C}-OH$	Ethanoic acid	Acetic acid
Salt	$CH_3\overset{\overset{O}{\|}}{C}-O^- K^+$	Potassium ethanoate	Potassium acetate
Ester	$CH_3-\overset{\overset{O}{\|}}{C}O-CH_3$	Methyl ethanoate	Methyl acetate
Amine	$CH_3CH_2-NH_2$		Ethylamine
Salt	$CH_3CH_2-\overset{+}{N}H_3\ Cl^-$		Ethylammonium chloride
Amide	$CH_3\overset{\overset{O}{\|}}{C}-NH_2$	Ethanamide	Acetamide

Summary of Reactions

Aldehydes

Oxidation

aldehyde $\xrightarrow{\text{Oxidation}}$ carboxylic acid

$$CH_3-\overset{\overset{O}{\|}}{C}H \xrightarrow{\text{Oxidation}} CH_3-\overset{\overset{O}{\|}}{C}OH$$

Carboxylic Acids

Ionization

carboxylic acid $\xrightarrow{H_2O}$ carboxylate ion + H^+

$$CH_3\overset{\overset{O}{\|}}{C}OH \xrightarrow{H_2O} CH_3\overset{\overset{O}{\|}}{C}O^- + H^+$$

Carboxylic Acids (continued)

Neutralization

carboxylic acid + base $\longrightarrow$ carboxylate salt + H_2O

$$CH_3\overset{\overset{\displaystyle O}{\|}}{C}OH + NaOH \longrightarrow CH_3\overset{\overset{\displaystyle O}{\|}}{C}O^- \ Na^+ + H_2O$$

Esterification

carboxylic acid + alcohol $\xrightarrow{H^+}$ ester + H_2O

$$CH_3\overset{\overset{\displaystyle O}{\|}}{C}OH + CH_3OH \xrightarrow{H^+} CH_3\overset{\overset{\displaystyle O}{\|}}{C}OCH_3 + H_2O$$

Amidation

carboxylic acid + ammonia $\xrightarrow{Heat}$ amide + H_2O

$$CH_3\overset{\overset{\displaystyle O}{\|}}{C}OH + NH_3 \xrightarrow{Heat} CH_3\overset{\overset{\displaystyle O}{\|}}{C}{-}NH_2 + H_2O$$

carboxylic acid + amine $\xrightarrow{Heat}$ amide + H_2O

$$CH_3\overset{\overset{\displaystyle O}{\|}}{C}OH + NH_2{-}CH_3 \xrightarrow{Heat} CH_3\overset{\overset{\displaystyle O}{\|}}{C}{-}NHCH_3 + H_2O$$

Esters

Hydrolysis

ester + H_2O $\xrightarrow{H^+}$ carboxylic acid + alcohol

$$CH_3\overset{\overset{\displaystyle O}{\|}}{C}OCH_3 + H_2O \xrightarrow{H^+} CH_3\overset{\overset{\displaystyle O}{\|}}{C}OH + CH_3OH$$

Saponification

ester + base $\longrightarrow$ salt + alcohol

$$CH_3\overset{\overset{\displaystyle O}{\|}}{C}OCH_3 + NaOH \longrightarrow CH_3\overset{\overset{\displaystyle O}{\|}}{C}O^- \ Na^+ + CH_3OH$$

Amines

Ionization

amine + H_2O $\rightleftarrows$ ammonium ion and OH^-

$CH_3{-}NH_2 + H_2O \longrightarrow CH_3{-}NH_3{}^+OH^-$

Neutralization

amine + acid $\longrightarrow$ ammonium salt

$CH_3{-}NH_2 + HCl \longrightarrow CH_3{-}NH_3{}^+Cl^-$

Amidation

carboxylic acid + amine $\xrightarrow{Heat}$ amide + H_2O

$$CH_3\overset{\overset{\displaystyle O}{\|}}{C}OH + NH_3 \xrightarrow{Heat} CH_3\overset{\overset{\displaystyle O}{\|}}{C}NH_2 + H_2O$$

Amides

Hydrolysis

amide + H—OH $\xrightarrow{\text{Heat}}$ carboxylic acid + ammonia (or amine)

$$CH_3\overset{\overset{O}{\|}}{C}NH_2 + H_2O \xrightarrow{\text{Heat}} CH_3\overset{\overset{O}{\|}}{C}OH + NH_3$$

Glossary of Key Terms

alkaloid A plant extract that contains heterocyclic amines that have physiological effects.

amidation Formation of an amide from a carboxylic acid and ammonia or an amine.

amide An organic compound containing the carbonyl group attached to an amino group or a substituted nitrogen atom:

$$R-\overset{\overset{O}{\|}}{C}-NH_2 \qquad R-\overset{\overset{O}{\|}}{C}-\overset{\overset{H}{|}}{N}-R$$

amine An organic compound in which the hydrogen atoms of ammonia, NH_3, are replaced with one, two, or three hydrocarbon groups.

carboxyl group A functional group found in carboxylic acids.

$$-\overset{\overset{O}{\|}}{C}-OH$$

O Carbonyl, Hydroxyl = Carboxyl group

carboxylic acid A family of organic compounds containing the carboxyl group, $R-\overset{\overset{O}{\|}}{C}OH$ or R—COOH.

ester An organic compound containing a $-\overset{\overset{O}{\|}}{C}-O-$ between two alkyl groups.

$$R-\overset{\overset{O}{\|}}{C}-O-R$$

esterification The formation of an ester from a carboxylic acid and an alcohol with the elimination of a molecule of water in the presence of an acid catalyst.

heterocyclic amine A cyclic structure that contains one or more nitrogen atoms.

hydrolysis A splitting apart of a larger molecule into two smaller molecules by the addition of water.

saponification The hydrolysis of an ester with a strong base to produce the salt of the acid and the alcohol. With fats, saponification with a strong base produces soaps that are the salts of long-chain fatty acids and glycerol.

Problems

Carboxylic Acids *(Goal 13.1)*

13.1 Write the IUPAC name and a common name (if any) for the following carboxylic acids:

a. $CH_3\overset{\overset{O}{\|}}{C}OH$ b. $CH_3CH_2CH_2\overset{\overset{O}{\|}}{C}OH$

c. $HCOH$ (with C=O)

d. 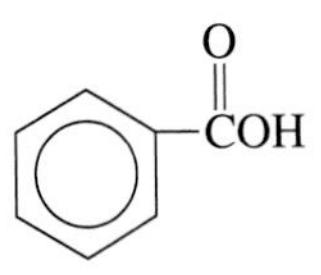 C_6H_5—COH (benzene ring attached to C(=O)OH)

e. CH_3CH_2COH (with C=O)

f. $Cl—CH_2CH_2COH$ (with C=O)

g. $CH_3—CH_2—CH(CH_3)—COH$ (with C=O)

h. $CH_3—CH(OH)—COH$ (with C=O)

13.2 Write the structural formulas of the expected oxidation products of the following:

a. CH_3CH (C=O) $\xrightarrow{\text{Oxidation}}$

b. 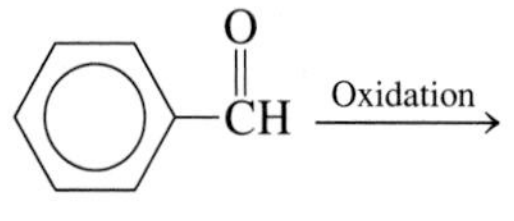 C_6H_5—CH (benzene ring attached to C=O) $\xrightarrow{\text{Oxidation}}$

c. $CH_3CH(CH_3)CH_2CH$ (C=O) $\xrightarrow{\text{Oxidation}}$

*d. $CH_3CH_2CH_2OH \xrightarrow{\text{Oxidation}}$

e. HCH (C=O) $\xrightarrow{\text{Oxidation}}$

f. 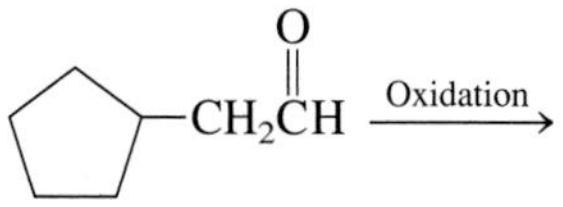 cyclopentyl—CH_2CH (C=O) $\xrightarrow{\text{Oxidation}}$

13.3 Draw structures for the following carboxylic acids:

a. β-methylbutyric acid
b. 3-chloropentanoic acid
c. 4-bromobutanoic acid
d. α-hydroxypropionic acid
e. 3-bromobenzoic acid
f. *p*-aminobenzoic acid

Some Properties of Carboxylic Acids *(Goal 13.2)*

13.4 a. Describe the hydrogen bonding between two acetic acid molecules.
b. Why does the hydrogen bonding in carboxylic acids give high melting and boiling points?

13.5 In the following pairs, which compound would have the higher boiling point? Explain.

a. $CH_3CH_2CH_2Cl$ and CH_3CH_2COH (C=O)

b. $CH_3CH_2CH_2OH$ and CH_3COH (C=O)

13.6 In the following pairs, which compound would be more soluble in water?

a. $CH_3CH_2—O—CH_3$ and CH_3COH (C=O)

b. CH_3CH_2COH (C=O) and $CH_3CH_2CH_2CH_2OH$

13.7 Write an equation for the ionization of the following carboxylic acids in water.

a. $CH_3\overset{O}{\overset{\|}{C}}OH$

b. 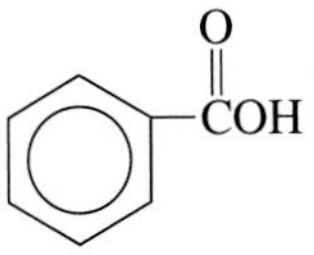 $C_6H_5-\overset{O}{\overset{\|}{C}}OH$ (benzene ring)

c. $CH_3CH_2\overset{O}{\overset{\|}{C}}OH$

13.8 Write an equation for the reaction of the carboxylic acids in Problem 13.7 with potassium hydroxide.

Esters *(Goal 13.3)*

13.9 Circle the ester bond(s) in the following compounds:

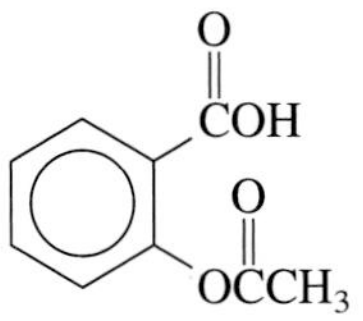

Benzene ring bearing $\overset{O}{\overset{\|}{C}}OH$ and $O\overset{O}{\overset{\|}{C}}CH_3$

Acetylsalicylic acid (aspirin)

$CH_3(CH_2)_{14}-\overset{O}{\overset{\|}{C}}O-(CH_2)_{29}CH_3$

Beeswax

$CH_3\overset{O}{\overset{\|}{C}}OCH_2CH_2CH_2CH_2CH_3$

Pentylacetate (in bananas)

13.10 Write the structural formulas of the esters formed by reacting the following carboxylic acids with methyl alcohol:

a. benzoic acid
b. 2-methylpropanoic acid
c. α-chloroacetic acid

13.11 Draw the structural formulas of the esters formed by the reactions of the following carboxylic acids and alcohols:

a. $CH_3CH_2\overset{O}{\overset{\|}{C}}OH + HO\underset{}{\overset{CH_3}{\overset{|}{C}}}HCH_3 \xrightarrow{H^+}$

b. 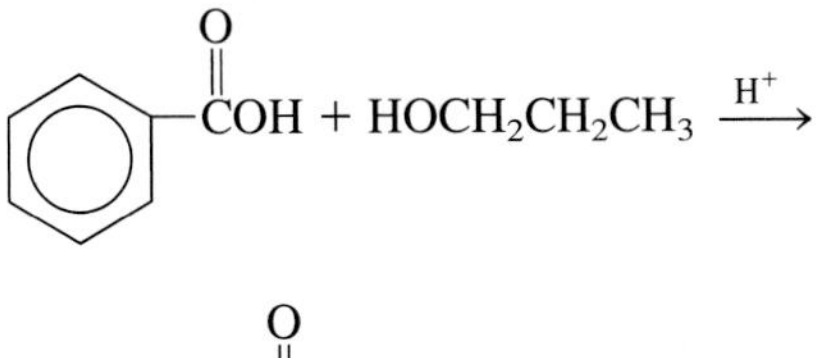 $C_6H_5-\overset{O}{\overset{\|}{C}}OH + HOCH_2CH_2CH_3 \xrightarrow{H^+}$ (benzene ring)

c. $CH_3CH_2CH_2\overset{O}{\overset{\|}{C}}OH + HO-C_6H_5 \xrightarrow{H^+}$ (benzene ring)

*d. $2CH_3\overset{O}{\overset{\|}{C}}OH + HO{-}CH_2CH_2CH_2{-}OH \xrightarrow{H^+}$

*e. 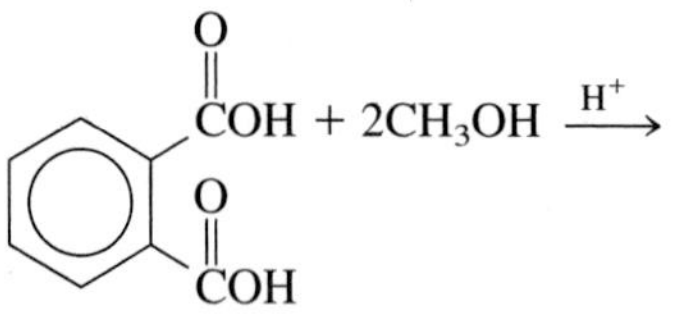 $C_6H_4(\overset{O}{\overset{\|}{C}}OH)_2 + 2CH_3OH \xrightarrow{H^+}$

13.12 The structural formula of the compound salicylic acid is shown below.
a. What two reactive functional groups are present?

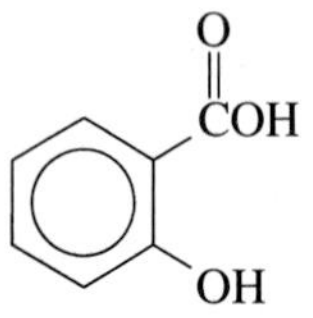

$C_6H_4(\overset{O}{\overset{\|}{C}}OH)(OH)$

b. Write the structure of aspirin, the ester product that forms when salicylic acid reacts with acetic acid.
c. Write the structure of methyl salicylate, oil of wintergreen, formed when salicylic acid forms an ester with methyl alcohol.

13.13 Write the IUPAC name and a common name (if any) of each of the following esters:

a. $CH_3\overset{O}{\overset{\|}{C}}OCH_3$

b. $CH_3\overset{O}{\overset{\|}{C}}OCH_2CH_3$

c. $H\overset{O}{\overset{\|}{C}}OCH_2CH_2CH_3$

d. 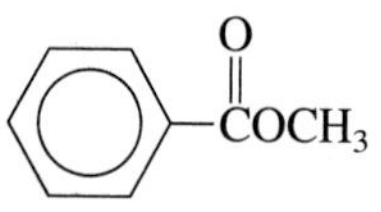 $C_6H_5{-}\overset{O}{\overset{\|}{C}}OCH_3$

e. $CH_3CH_2CH_2CH_2\overset{O}{\overset{\|}{C}}OCH_2CH_3$

f. 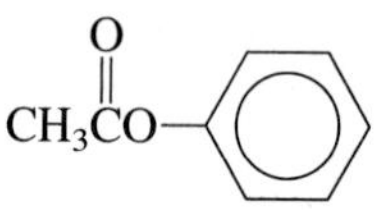 $CH_3\overset{O}{\overset{\|}{C}}O{-}C_6H_5$

g. 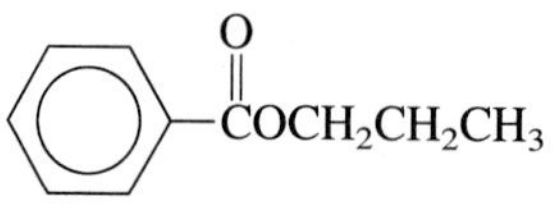 $C_6H_5{-}\overset{O}{\overset{\|}{C}}OCH_2CH_2CH_3$

13.14 Many esters are the source of flavors and odors of fruits. Write the structure for the following esters used as flavoring agents:
a. ethyl butyrate, pineapple
b. ethyl formate, rum
*c. pentyl butyrate, apricot
*d. octyl acetate, orange

Hydrolysis and Saponification of Esters *(Goal 13.4)*

13.15 Draw the structural formulas of the products from the acid-catalyzed hydrolysis of the following compounds:

a. $HC(=O)—O—CH_2CH_2CH_3 + H_2O \xrightarrow{H^+}$

b. $CH_3CH_2—C(=O)—O—C_6H_5 + H_2O \xrightarrow{H^+}$

c. $CH_3C(=O)—O—CH_2CH_2CH_2CH_3 + H_2O \xrightarrow{H^+}$

d. $C_6H_5—C(=O)—OCH_2CH_3 + H_2O \xrightarrow{H^+}$

13.16 Draw the structural formulas of the products from saponification of the following compounds:

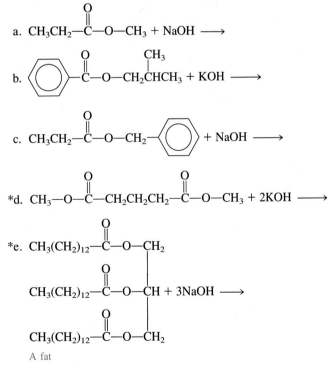

a. $CH_3CH_2—C(=O)—O—CH_3 + NaOH \longrightarrow$

b. $C_6H_5—C(=O)—O—CH_2CH(CH_3)CH_3 + KOH \longrightarrow$

c. $CH_3CH_2—C(=O)—O—CH_2—C_6H_5 + NaOH \longrightarrow$

*d. $CH_3—O—C(=O)—CH_2CH_2CH_2—C(=O)—O—CH_3 + 2KOH \longrightarrow$

*e.
$$\begin{array}{l} CH_3(CH_2)_{12}—C(=O)—O—CH_2 \\ \qquad\qquad\qquad\qquad\qquad\quad | \\ CH_3(CH_2)_{12}—C(=O)—O—CH + 3NaOH \longrightarrow \\ \qquad\qquad\qquad\qquad\qquad\quad | \\ CH_3(CH_2)_{12}—C(=O)—O—CH_2 \end{array}$$

A fat

Amines *(Goal 13.5)*

13.17 There are three amine isomers with the molecular formula C_3H_9N. Name each, and classify each as a primary, secondary, or tertiary amine.

13.18 Classify the following amines as primary, secondary, or tertiary amines:

a. 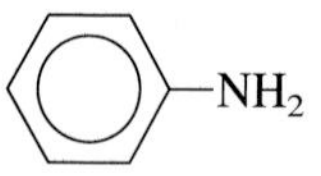 C_6H_5—NH_2

b. CH_3CH_2—N(H)—CH_2CH_3

c. $CH_3CH(CH_3)CH_2$—NH_2

d. 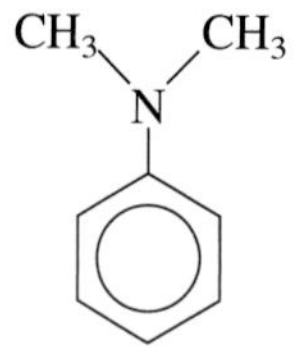 CH_3 CH_3 N C_6H_5

e. 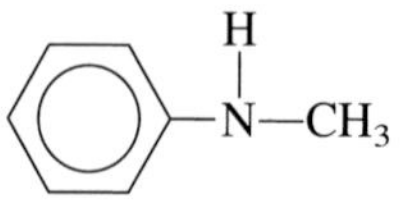 C_6H_5—N(H)—CH_3

13.19 Write the common name for each of the following amines:

a. CH_3CH_2—NH_2

b. $CH_3CH_2CH_2CH_2$—NH_2

c. 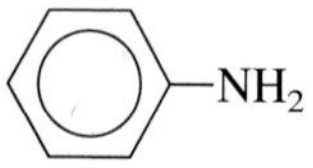 C_6H_5—NH_2

d. CH_3CH_2—N(H)—CH_2CH_3

e. CH_3CH_2—N(CH_3)—$CH_2CH_2CH_3$

f. 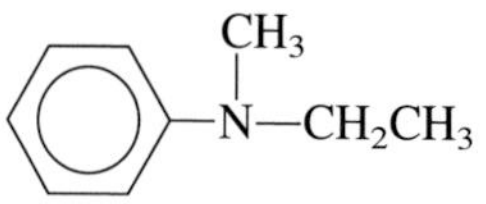 C_6H_5—N(CH_3)—CH_2CH_3

g. 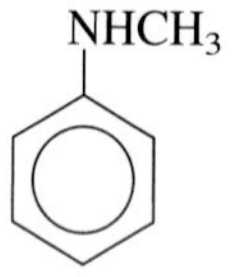 $NHCH_3$ C_6H_5

13.20 Draw the structures of the following amines:

a. ethylamine
b. *N*-methylaniline
c. *p*-chloroaniline
d. dimethylamine
e. methylisopropylamine
f. cyclopentylamine

Heterocyclic Amines and Physiological Activity *(Goal 13.6)*

13.21 What is a heterocyclic amine?

13.22 The following physiologically active compounds are heterocyclic amines. Name the heterocyclic ring in each.

a. Methioprim (tumor antagonist)

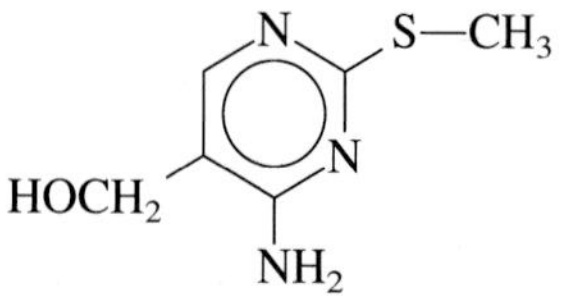

N S—CH_3 N $HOCH_2$ NH_2

b. Pyrrolnitrin (antifungicide)

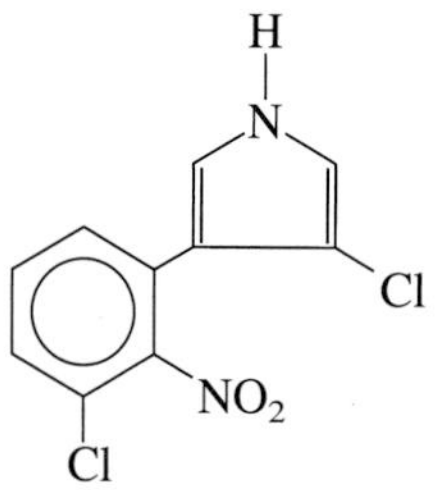

H N Cl NO_2 Cl

c. Pyrimethamine (antimalarial)

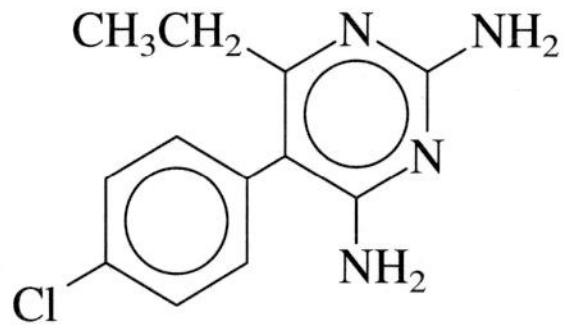

CH_3CH_2 N NH_2 N Cl NH_2

e. Nicotine

N N CH_3

d. Adrenoglomerulotropin (hormone)

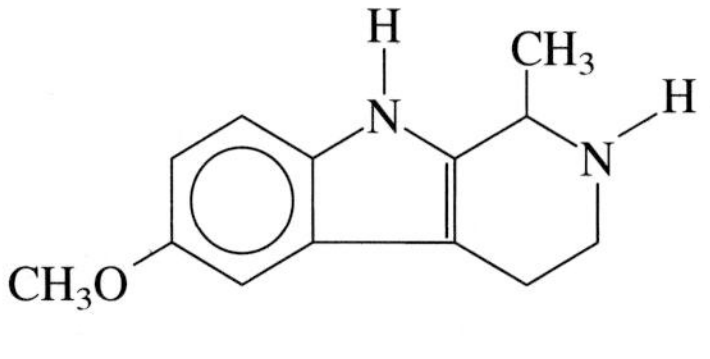

H N CH_3 H N CH_3O

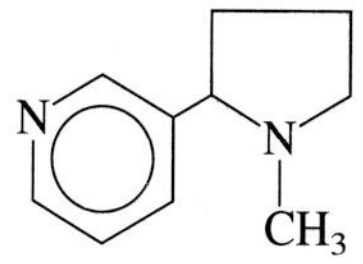

13.23 Iproniazid is used as an antidepressant. Name the families of organic compounds represented in iproniazid.

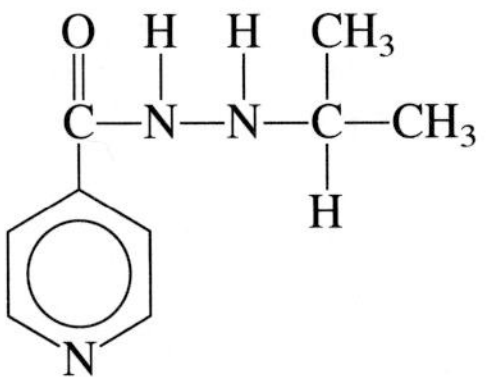

O H H CH_3 C—N—N—C—CH_3 H N

Reactions of Amines *(Goal 13.7)*

13.24 Write an equation for each of the following compounds acting as bases in aqueous solutions.

a. $CH_3—NH_2$

b. $CH_3—N(CH_3)—H$

c. 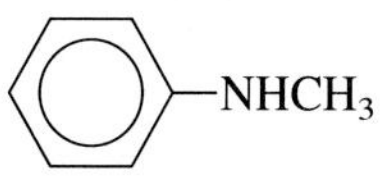 $C_6H_5—NHCH_3$

13.25 Draw the structural formula of the salt obtained from the reactions of the following amines with acids:

a. $CH_3CH_2—NH_2 + HCl \longrightarrow$

b. 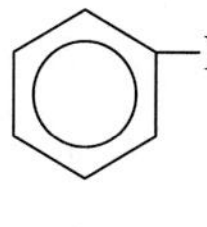 $C_6H_5—NH_2 + HBr \longrightarrow$

c. $CH_3CH_2—NH—CH_2CH_3 + HCl \longrightarrow$

d. $CH_3—N(CH_3)—CH_3 + HNO_3 \longrightarrow$

Amides *(Goal 13.8)*

13.26 Write the names for the following amides:

a. $CH_3C(=O)-NH_2$

b. $CH_3CH_2CH_2C(=O)-NH_2$

c. 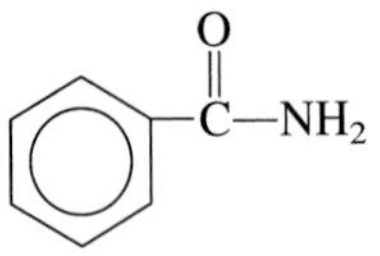 $C_6H_5-C(=O)-NH_2$

d. $CH_3C(=O)-N(CH_3)-CH_3$

e. $CH_3CH_2CH_2C(=O)-NHCH_2CH_3$

13.27 Draw the structural formula of the following amides:

a. propanamide
b. *N*-methylpentamide
c. 3-methylbutyramide
d. *N,N*-dimethylbenzamide

13.28 Identify the numbered functional groups in the following compounds as alcohol, phenol, ether, carboxylic acid, ester, amine, or amide:

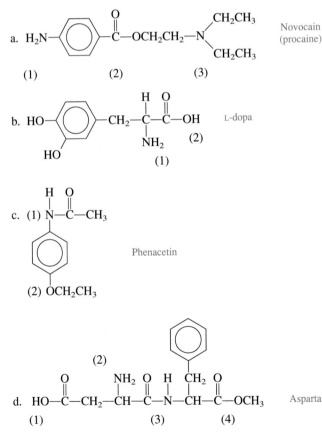

a. 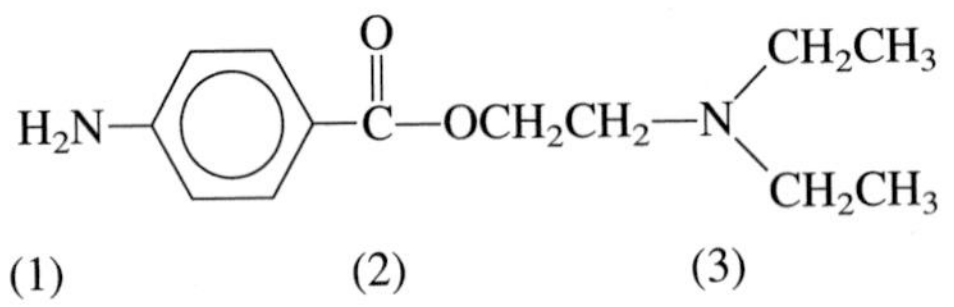 $H_2N-C_6H_4-C(=O)-OCH_2CH_2-N(CH_2CH_3)_2$ — Novocain (procaine)
(1) (2) (3)

b. 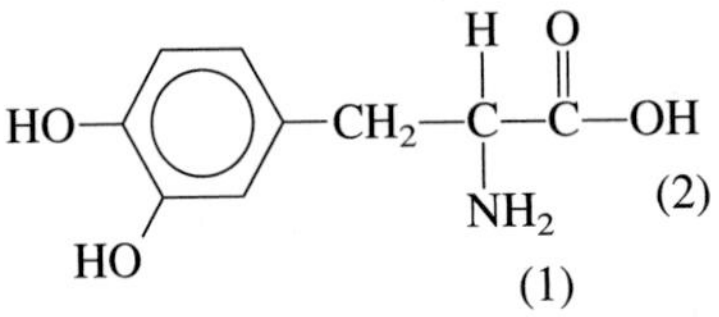 $HO-C_6H_3(HO)-CH_2-C(H)(NH_2)-C(=O)-OH$ — L-dopa
NH_2 (1); $C(=O)-OH$ (2)

c. 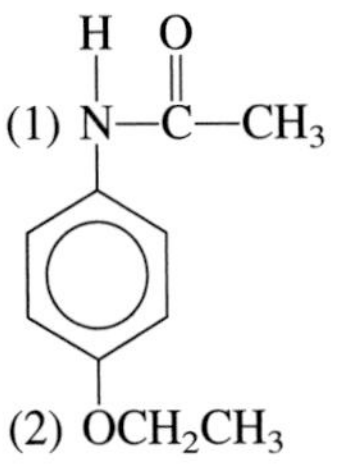 (1) $N(H)-C(=O)-CH_3$ attached to a benzene ring bearing (2) OCH_2CH_3 — Phenacetin

d. 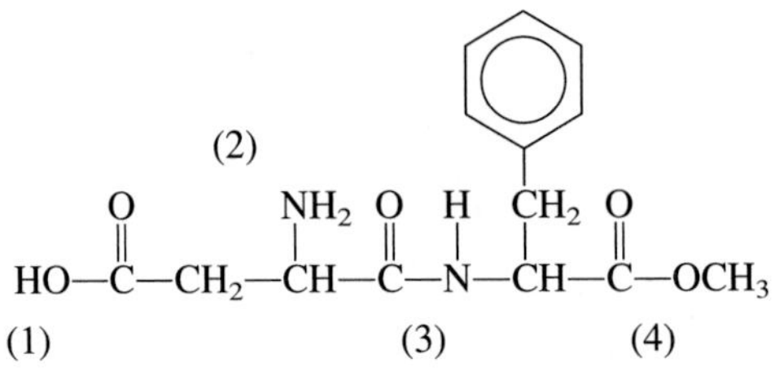 $HO-C(=O)-CH_2-CH(NH_2)-C(=O)-N(H)-CH(CH_2C_6H_5)-C(=O)-OCH_3$ — Aspartame (a sweetener)
(1) $HO-C(=O)-$; (2) NH_2; (3) $-C(=O)-N(H)-$; (4) $-C(=O)-OCH_3$

13.29 Draw the structural formula of the amide formed in the following amidation reactions:

a. $CH_3\overset{O}{\overset{\|}{C}}OH + NH_3 \xrightarrow{\text{Heat}}$

b. $CH_3CH_2\overset{O}{\overset{\|}{C}}OH + NH_3 \xrightarrow{\text{Heat}}$

c. $CH_3\overset{O}{\overset{\|}{C}}OH + NH_2CH_2CH_3 \xrightarrow{\text{Heat}}$

d. $CH_3CH_2\overset{O}{\overset{\|}{C}}OH + \underset{CH_3}{\overset{H}{N}}—CH_3 \xrightarrow{\text{Heat}}$

e. $H_2N—\underset{H}{\overset{H}{C}}—\overset{O}{\overset{\|}{C}}—OH \quad + \quad H_2N—\underset{H}{\overset{H}{C}}—\overset{O}{\overset{\|}{C}}—OH \xrightarrow{\text{Heat}}$

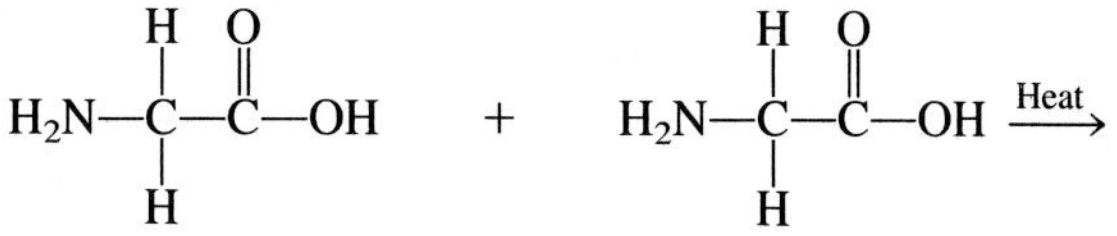

Glycine Amino acids Glycine

13.30 Write the structural formulas of the carboxylic acid and amine needed for the formation of the following amides:

a. $CH_3\overset{O}{\overset{\|}{C}}—NHCH_3$

b. $C_6H_5—\overset{O}{\overset{\|}{C}}—NH_2$ (benzene ring)

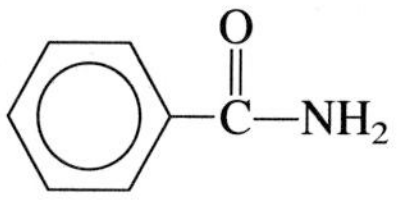

c. $CH_3CH_2\overset{O}{\overset{\|}{C}}—NH—C_6H_5$ (benzene ring)

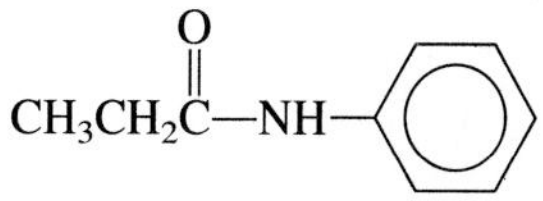

13.31 Write the structural formulas of the hydrolysis products for the following amides:

a. $CH_3\overset{O}{\overset{\|}{C}}—NHCH_3 + H_2O \xrightarrow{\text{Heat}}$

b. $C_6H_5—\overset{O}{\overset{\|}{C}}—NH_2 + H_2O \xrightarrow{\text{Heat}}$ (benzene ring)

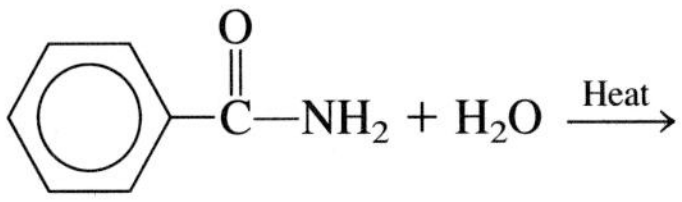

c. $HO—C_6H_4—\overset{H}{N}—\overset{O}{\overset{\|}{C}}CH_3 + H_2O \xrightarrow{\text{Heat}}$ (para-substituted benzene ring)

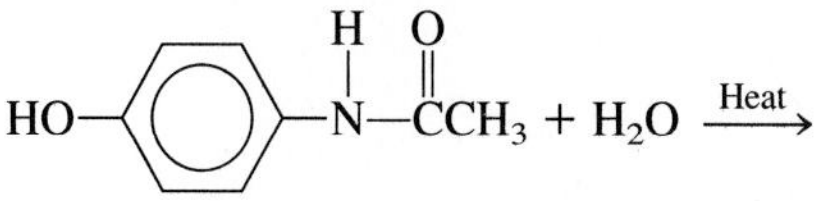

Acetaminophen

d. $CH_3CH_2\overset{O}{\overset{\|}{C}}—NHCH_3 + H_2O \xrightarrow{\text{Heat}}$

13.32 Using a reference book such as the *Merck Index,* look up the structural formulas of the following medicinal drugs and list the functional groups in the compounds. You may need to refer to the cross-index of names in the back of the book.

a. Keflex, an antibiotic
b. Inderal, a β-channel blocker used to treat heart irregularities
c. Ibuprofen, anti-inflammatory agent
d. Aldomet (methyldopa)
e. Percodan, narcotic pain reliever
f. Triamterene, diuretic

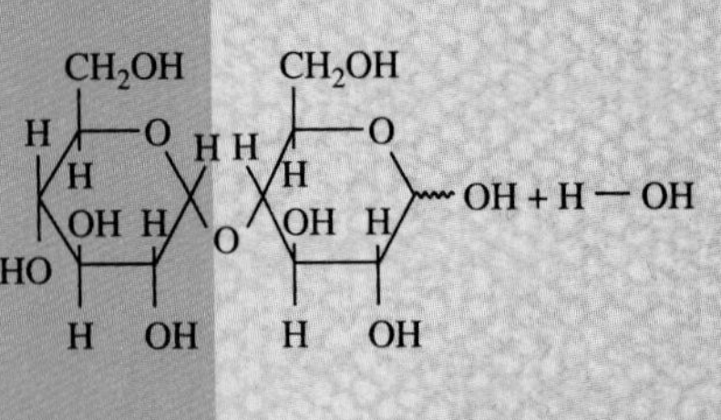

14

Carbohydrates

Learning Goals

14.1 Given the structural formula of a monosaccharide, classify it as an aldo- or ketotriose, tetrose, pentose, or hexose.

14.2 Identify the chiral carbon(s) in an asymmetric molecule.

14.3 Write the Fischer projection for the D or L isomer of a monosaccharide.

14.4 Draw Haworth structures for some of the dietary monosaccharides.

14.5 Describe the oxidation and glycoside formation of monosaccharides.

14.6 For the disaccharides maltose, lactose, and sucrose, identify the monosaccharide units, the type of glycosidic bond, and sources of each.

14.7 Given the name or structure of a polysaccharide, describe its monosaccharide units, glycosidic linkages, and sources.

14.8 Predict the reactions of carbohydrates in the Benedict's, fermentation, and iodine tests.

Complex carbohydrates found in grains such as wheat and corn serve as a major source of energy that maintains a constant level of glucose in the blood.

SCOPE

The foodstuffs in our diets such as bread, pasta, potatoes, rice, and beans contain carbohydrates. Carbohydrates are also called **saccharides,** which comes from the Latin word *saccharum,* ''sugar.'' The processes of digestion break apart the bonds of the polysaccharides in these foods to give monosaccharides, primarily glucose, a carbohydrate that provides most of our cellular energy. Cellulose is a polysaccharide used by plants to build rigid cell walls. Although it is not digestible by humans, it does play an important role in our diets by providing fiber. Our diets may also include sucrose, table sugar, and lactose, a sugar found in milk.

The National Academy of Sciences has recommended that one-half of our calories be obtained from the starches which are complex carbohydrates. Because they are digested more slowly, a constant level of glucose is maintained in the blood over a longer period of time. Foods high in complex carbohydrates include potatoes, corn, rice, and wheat.

14.1 Classification of Carbohydrates

Learning Goal **Given the structural formula of a monosaccharide, classify it as an aldo- or ketotriose, tetrose, pentose, or hexose.**

Carbohydrates are made of carbon, hydrogen, and oxygen. They are produced by green plants and bacteria in a process known as photosynthesis. In green leaves, photosynthesis occurs when energy from sunlight is absorbed and used to convert carbon dioxide (CO_2) and water into carbohydrates such as glucose ($C_6H_{12}O_6$), and oxygen (O_2).

Photosynthesis in Plants

$$6CO_2 + 6H_2O \xrightarrow{\text{sunlight}} \underset{\text{Glucose}}{C_6H_{12}O_6} + 6O_2$$

In body tissues, glucose is oxidized in metabolic reactions known as respiration. Chemical energy is released to do work in the cells along with the products carbon dioxide and water. Photosynthesis and respiration complete the carbon cycle by storing energy from the sun that is made available to our cells as carbohydrates are metabolized. (See Figure 14.1.)

Respiration in Animals

$$\underset{\text{Glucose}}{C_6H_{12}O_6} + 6O_2 \longrightarrow 6CO_2 + 6H_2O + \text{energy}$$

Types of Carbohydrates

The simplest carbohydrates are the **monosaccharides,** typically composed of three to six carbon atoms. **Disaccharides** consist of two monosaccharides, and

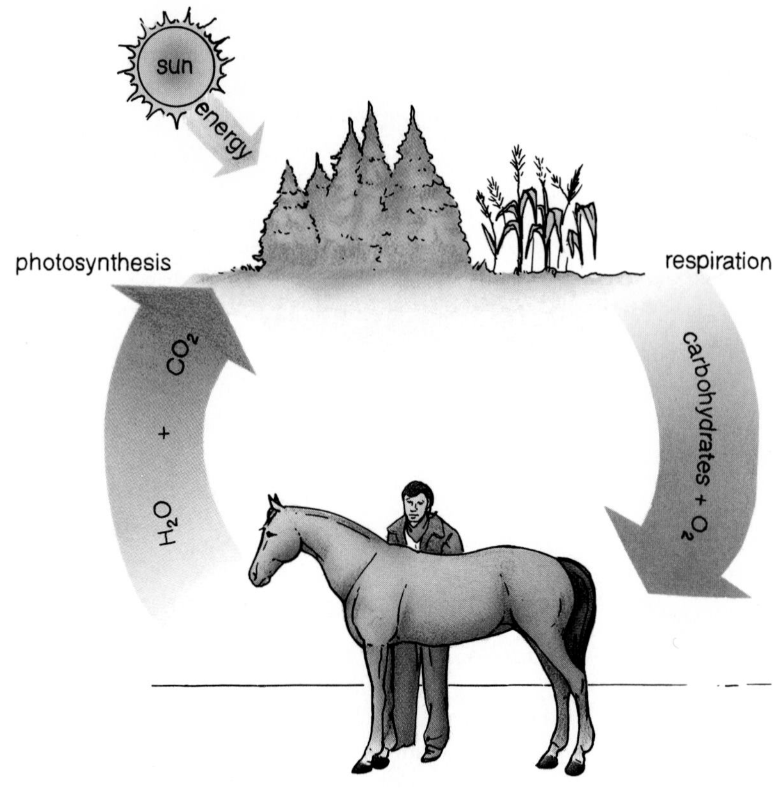

Figure 14.1
The carbon cycle in nature depicts the interdependence of photosynthesis and respiration.

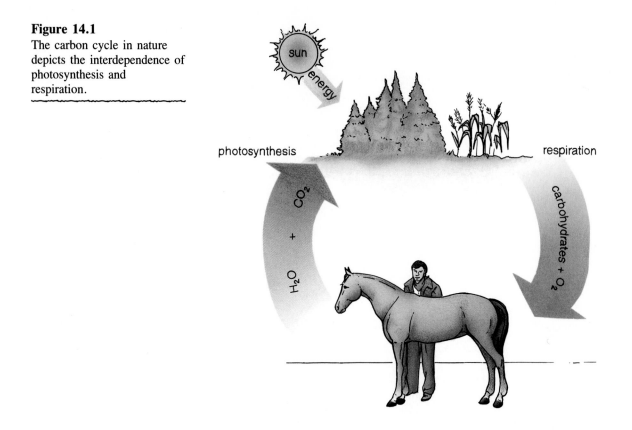

polysaccharides are complex carbohydrates containing many monosaccharides. When disaccharides and polysaccharides are hydrolyzed, they give monosaccharides.

$$\text{Monosaccharides} \xrightarrow{H_2O} \text{No hydrolysis}$$

$$\text{Disaccharides} \xrightarrow{H_2O} \text{Two monosaccharides}$$

$$\text{Polysaccharides} \xrightarrow{H_2O} \text{Many monosaccharides}$$

Classification of Monosaccharides

Monosaccharides have a general formula, $(CH_2O)_n$. Because the amount of hydrogen is twice the oxygen, sugars were once thought to be hydrates of carbon, thus the name *carbohydrate*. Now we describe a carbohydrate as a compound that has an aldehyde or ketone group and many hydroxyl (—OH) groups.

The names of the carbohydrates usually end in *ose*. A monosaccharide with an aldehyde group is classified as an **aldose;** a **ketose** is a monosaccharide that contains a ketone group. The number of carbon atoms in a monosaccharide is indicated by a prefix, so a triose contains three carbons, a tetrose contains four carbons, a pentose has five carbons, and a hexose contains six carbons. The prefixes may be combined to specify both the carbonyl group and the number of carbon atoms. For example, ribose is an aldopentose, a five-carbon aldose, and fructose is a ketohexose, a six-carbon ketose. Most of the naturally occurring monosaccharides are aldoses with five or six carbon atoms.

Glyceraldehyde	Erythrose	Ribose	Glucose	Fructose
CHO	CHO	CHO	CHO	CH_2OH
H—C—OH	H—C—OH	H—C—OH	H—C—OH	C=O
CH_2OH	H—C—OH	H—C—OH	HO—C—H	HO—C—H
	CH_2OH	H—C—OH	H—C—OH	H—C—OH
		CH_2OH	H—C—OH	H—C—OH
			CH_2OH	CH_2OH
$C_3H_6O_3$	$C_4H_8O_4$	$C_5H_{10}O_5$	$C_6H_{12}O_6$	$C_6H_{12}O_6$
(aldotriose)	(aldotetrose)	(aldopentose)	(aldohexose)	(ketohexose)

Sample Problem 14.1
Monosaccharides

Classify the following monosaccharides according to the carbonyl group and number of carbon atoms:

a. Ribulose	b. Glyceraldehyde	c. Mannose
CH_2OH	CHO	CHO
C=O	H—C—OH	HO—C—H
H—C—OH	CH_2OH	HO—C—H
H—C—OH		H—C—OH
CH_2OH		H—C—OH
		CH_2OH

Solution:

a. Ribulose is a ketopentose because it has a ketone group (C=O) and five carbon atoms ($C_5H_{10}O_5$).

b. Glyceraldehyde, a three-carbon monosaccharide, is an aldotriose.

c. Mannose with an aldehyde group and six carbon atoms $C_6H_{12}O_6$ is an aldohexose.

Study Check:

Classify the monosaccharide erythulose.

CH_2OH
C=O
H—C—OH
CH_2OH

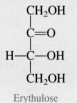

Erythulose

Answer: ketotetrose

14.2 Stereoisomers

Learning Goal **Identify the chiral carbon(s) in an asymmetric molecule.**

Stereoisomers are compounds with the same structural formula, but different three-dimensional arrangements. The cis-trans isomers of alkenes were an example of one type of stereoisomer. Now we will study steroisomers of carbohydrates by first looking at the symmetry of some common objects.

Symmetry

A drinking glass, a ball, a plate, and a tennis racket all have symmetry. A symmetrical object can be divided to give two identical halves. When one half is placed over the other, all of the parts match exactly. We say that the two halves are superimposable.

Chiral Objects

An object that is **chiral** cannot be divided into identical halves. Consider your hands. If you place the palms of your hands together, you can only match the thumbs and the fingers. The left-hand cannot be matched up completely with the right-hand. The hands are **asymmetrical** and nonsuperimposable. See Figure 14.2. Other examples of chiral objects include left and right shoes, golf clubs, and gloves. Objects that are symmetrical are called **achiral** because they have identical halves that are superimposable.

Figure 14.2
Symmetrical objects have two equal halves; asymmetrical objects do not.

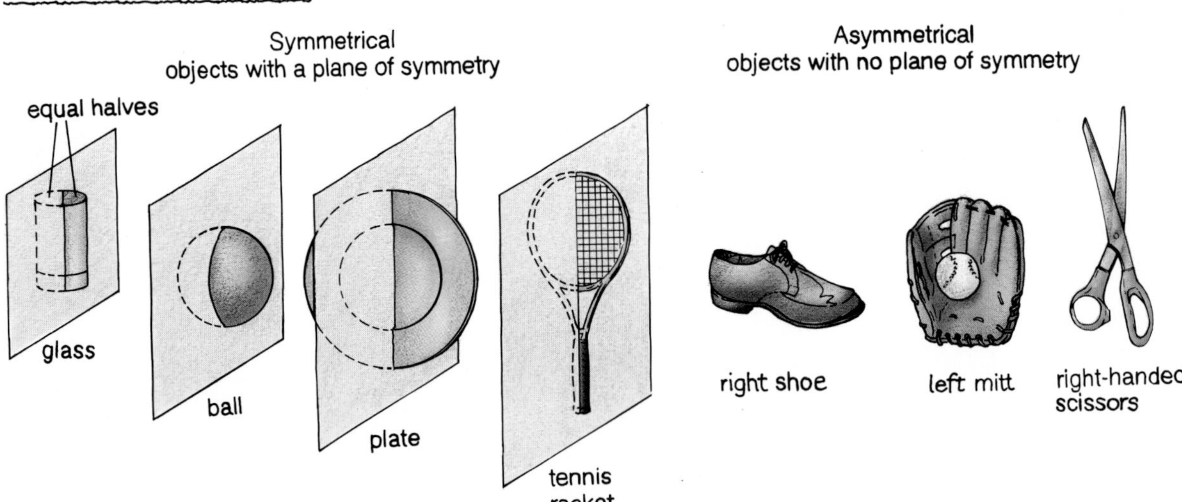

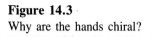

Figure 14.3
Why are the hands chiral?

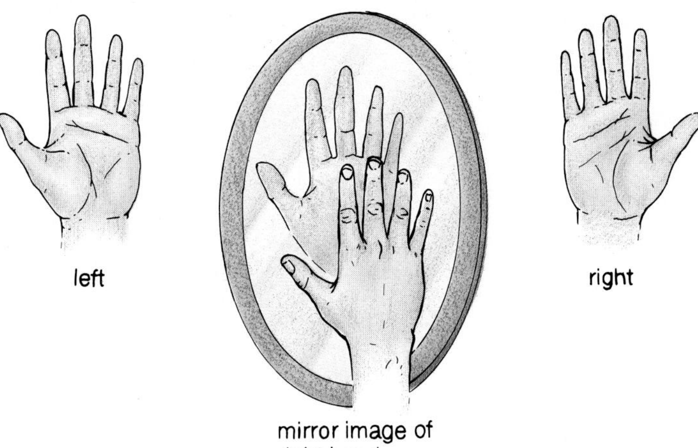

Mirror Images

If you look at one of your hands in a mirror, its image reflects your other hand. We say that a left hand is a mirror image of a right hand. See Figure 14.3. Both chiral and achiral objects have mirror images, but the mirror image of a chiral object is not identical. A left shoe is different than its mirror image, the right shoe; a left-handed golf club is not identical to its mirror image, the right-handed club.

Sample Problem 14.2
Chiral and Achiral Objects

Classify each of the following objects as chiral or achiral:

- **a.** Left ear
- **b.** A hairbrush
- **c.** A plain vase
- **d.** Catchers mitt
- **e.** A plain soupspoon

Solution:

- **a.** Chiral; the left ear is the mirror image of the right ear.
- **b.** Achiral; a hairbrush can be divided into two equal halves.
- **c.** Achiral; a plain vase has two identical halves that are superimposable.
- **d.** Chiral; a left-handed mitt is the mirror image and nonsuperimposable with a right-handed mitt.
- **e.** Achiral; a plain soupspoon has two identical halves.

Chiral Molecules

A molecule is chiral if it has at least one carbon atom called a **chiral carbon** that is attached to four different atoms or groups. The following compound is chiral.

Figure 14.4
The mirror images of a chiral molecule cannot be superimposed.

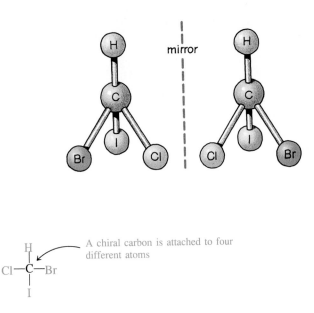

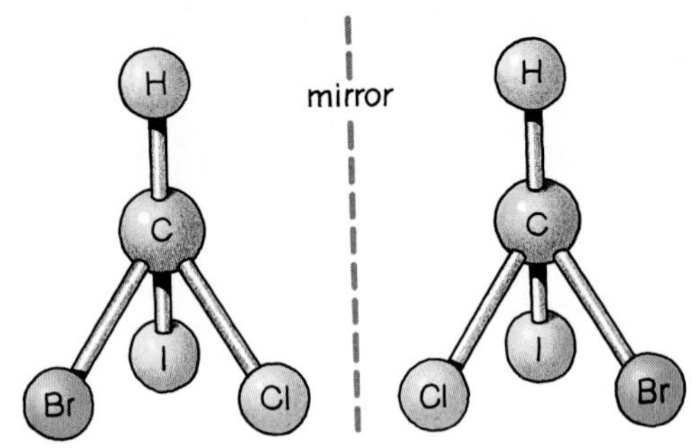


```
     H
     |  ↙ A chiral carbon is attached to four
Cl—C—Br    different atoms
     |
     I

A chiral molecule
```

Like a chiral object, a chiral molecule has a mirror image that is different and nonsuperimposable. In Figure 14.4, we see a three-dimensional model of the chiral molecule and its mirror image shown. If we match up the hydrogen and iodine atoms, the bromine and chlorine atoms appear on opposite sides. The mirror images of chiral compounds are called **stereoisomers.**

If a carbon atom in a molecule is attached to two or more identical atoms, the molecule is **achiral.** Some examples of chiral and achiral molecules follow:

```
        Identical atoms                  Chiral carbons
     H            Cl               H              H
     |            |                |              |
  H—C—H       Cl—C—H          Cl—C—F     CH3—C—OH
     |            |                |              |
     Cl           F                I            CH2CH3

  Achiral      Achiral          Chiral         Chiral
```

Sample Problem 14.3
Chiral Molecules

Indicate whether each of the following compounds is chiral or achiral:

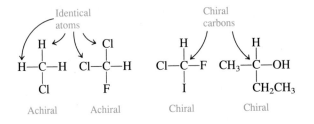

```
       H                  H
       |                  |
a. H—C—Cl      b. HO—C—CH3
       |                  |
       Cl                 NH2
```

Solution:

a. This molecule is achiral because two of the atoms attached to the carbon atom are the same.

b. Because four different groups are attached to the central carbon, this molecule is chiral.

Study Check:
Identify the chiral carbon atom in 2-butanol.

```
     H  OH H  H
     |  |  |  |
  H—C—C—C—C—H
     |  |  |  |
     H  H  H  H

     1  2  3  4
```

Answer: Carbon 2 is the only chiral carbon in the molecule.

```
     H  OH    H  H
     |  |     |  |
  H—C—C—————C—C—H
     |  |     |  |
     H  H     H  H
     1  2     3  4
```

Fischer Projections

Consider the mirror images of 2-butanol in Figure 14.5. If we imagine that we are looking at the carbon atom between the H and OH, it appears that the hydrocarbon groups are behind the page, whereas the H and OH come forward. On paper, the groups in back are connected to the chiral carbon by dashed lines, and those in front by solid wedges. This three-dimensional view can be represented as a **Fischer projection.** A chiral carbon is at the center of a vertical and horizontal line. The groups that project toward the viewer are placed on the ends of the horizontal line, and the groups that are in back are at the top and bottom of the vertical line.

Sample Problem 14.4
Drawing Fischer Projections

Draw the Fischer projection of the mirror image of the following chiral compound:

```
       Br
       |
  HO———+——H
       |
       Cl
```

Solution:
The mirror image is drawn by keeping the same groups at the top and the bottom, and placing the —OH and —H groups on opposite sides of the chiral carbon atom.

```
       Br             Br
       |              |
 (HO)——+——(H)    (H)——+——(OH)
       |              |
       Cl             Cl
                 Mirror image
```

Study Check:
Draw Fischer projections for the mirror images of 1-chloro-1-ethanol.

Answer:

```
       CH3              CH3
        |                |
  HO----+----H     H-----+----OH
        |                |
        Cl               Cl
```

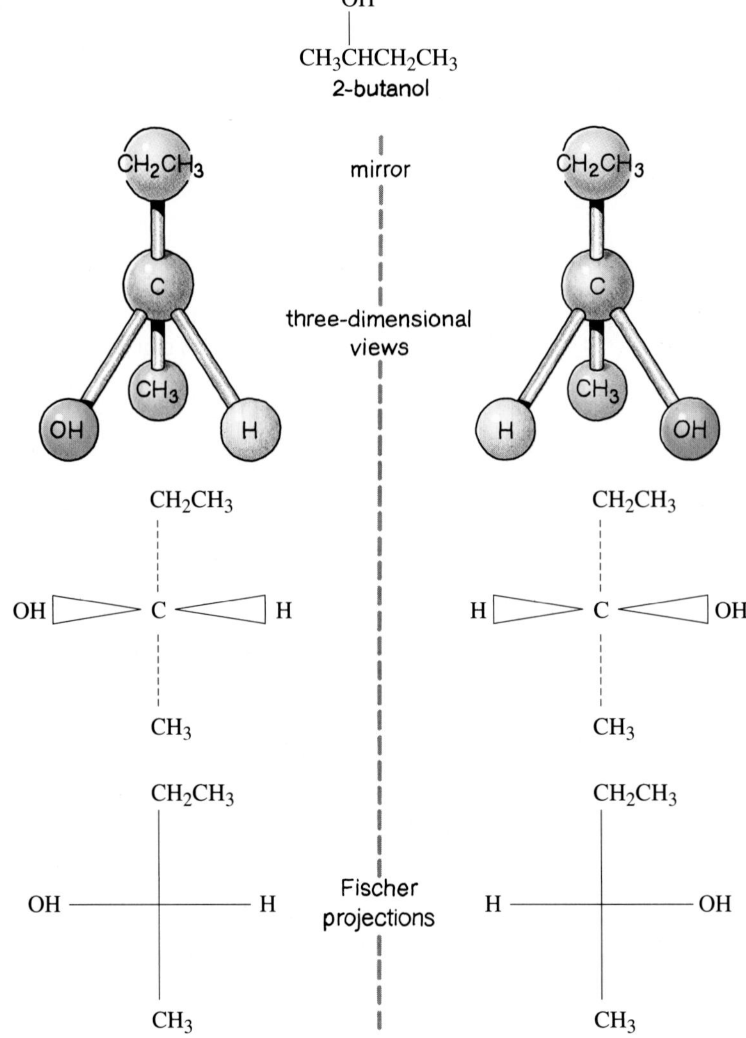

Figure 14.5
Mirror images of 2-butanol in three-dimensional views and the Fischer projection.

Optical Activity

Although stereoisomers contain the same order of atoms, their different spatial arrangements cause them to react differently with plane-polarized light. When ordinary light is polarized, the light waves vibrate in only one plane. As the polarized light interacts with a solution of a chiral compound, the plane of light rotates and emerges at an angle to the original plane. (See Figure 14.6.) Since only chiral compounds can rotate the plane of polarized light, they are said to be **optically active.** An optically active substance that rotates light to the right is dextrorotatory. One that rotates light to the left is levorotatory. Because the mirror images of chiral compounds rotate plane-polarized light in opposite directions, they are called **optical isomers.** Table 14.1 summarizes some of the features of chiral and achiral compounds.

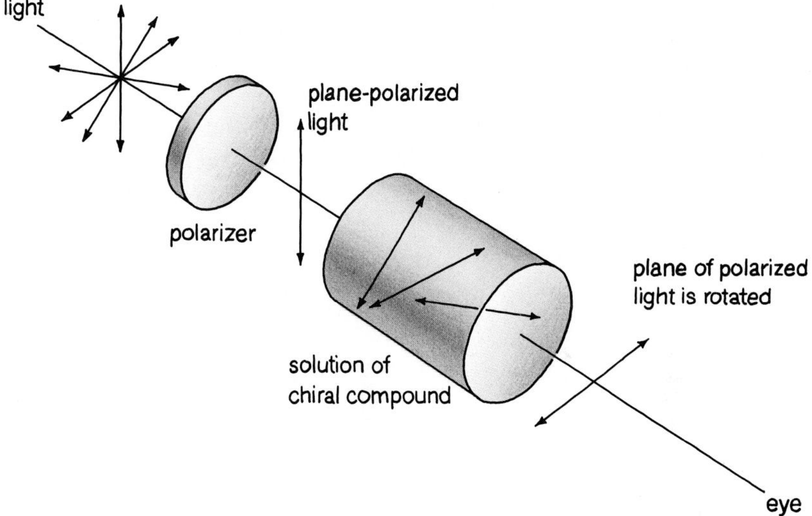

Figure 14.6
Plane-polarized light changes direction as it interacts with an optical isomer (a chiral molecule) and emerges at a different angle.

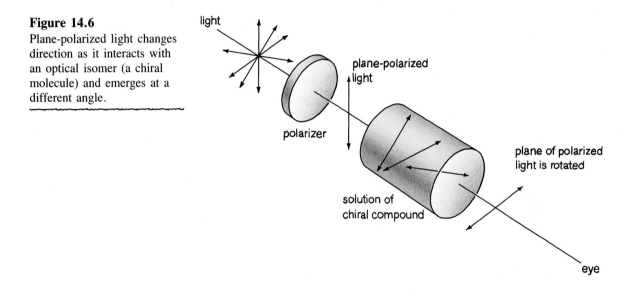

Table 14.1 Features of Chiral and Achiral Compounds

Chiral compounds	Achiral compounds
Asymmetrical	Symmetrical
Cannot be divided into equal halves	Can be divided into equal halves
Contain one or more chiral carbons	Contain only achiral carbons
Nonsuperimposable mirror images	Superimposable mirror images
Optically active	Not optically active

14.3 Chiral Sugars

Learning Goal **Write the Fischer projection for the D or L isomer of a monosaccharide.**

One of the simplest sugars, glyceraldehyde, is a chiral molecule. By convention, the Fischer projection of a monosaccharide is written with the carbonyl group at the top and the —CH_2OH group at the bottom. In the Fischer projections of the mirror images of glyceraldehyde, the letter L- identifies the stereoisomer with the —OH group written on the left of the chiral carbon. In D-glyceraldehyde, the —OH is written on the right. (See Figure 14.7.)

CHO
|
CHOH
|
CH_2OH
Glyceraldehyde

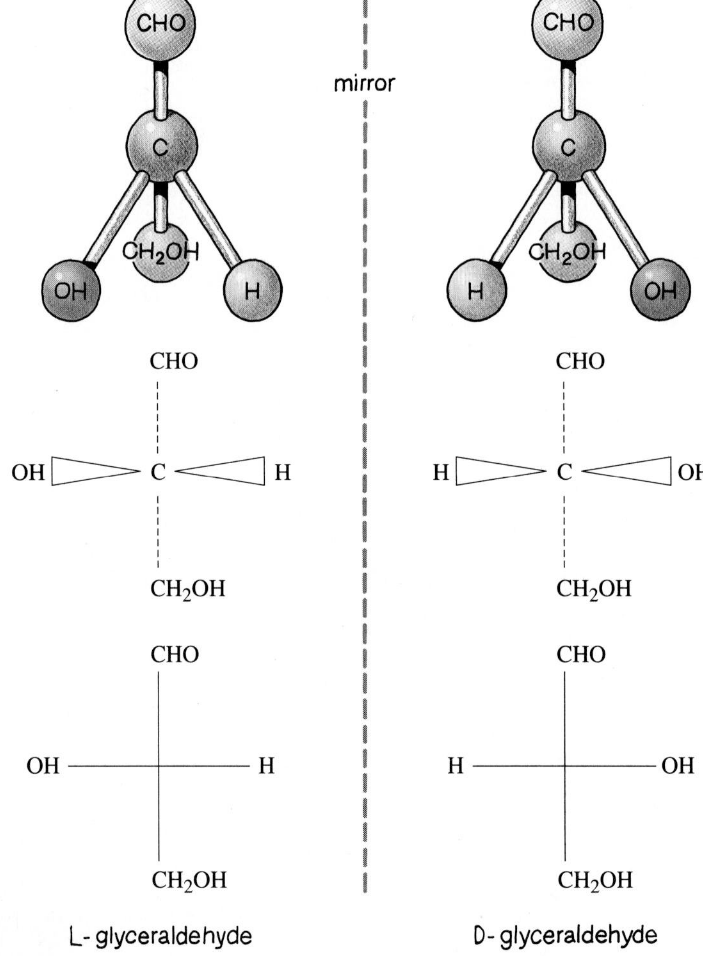

Figure 14.7
The D- and L-isomers of glyceraldehyde.

Fischer Projections of the Mirror Images

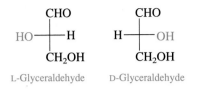

L-Glyceraldehyde	D-Glyceraldehyde
CHO	CHO
HO—C—H	H—C—OH
CH_2OH	CH_2OH

In carbohydrates with longer carbon chains, there will be several chiral carbons. Then it is the chiral carbon at the bottom of the Fischer projection that determines the D- or L- isomer. For example, for glucose, carbon 5 is the bottom chiral carbon, because carbon 6 in $—CH_2OH$ is not chiral. When the hydroxyl(—OH) group on carbon 5 is on the right in the Fischer projection, the isomer is D-glucose. In L-glucose, the OH— on carbon 5 is on the left.

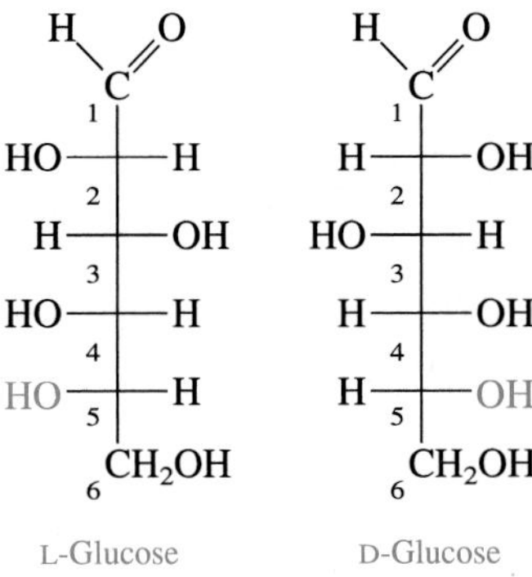

L-Glucose D-Glucose

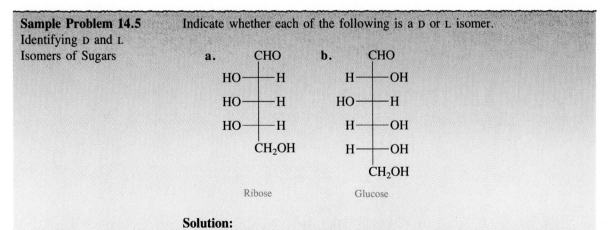

Sample Problem 14.5
Identifying D and L Isomers of Sugars

Indicate whether each of the following is a D or L isomer.

a. Ribose	**b.** Glucose
CHO	CHO
HO—C—H	H—C—OH
HO—C—H	HO—C—H
HO—C—H	H—C—OH
CH_2OH	H—C—OH
	CH_2OH

Solution:

a. Our guide to designating a D or L isomer is the position of the hydroxyl group (—OH) on the bottom chiral carbon. In ribose, the hydroxyl group on carbon 4 makes it L-ribose.

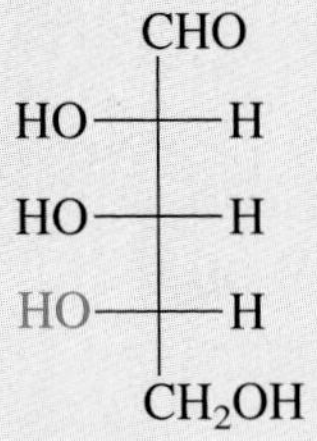

```
        CHO
   HO───┼───H
   HO───┼───H
   HO───┼───H
        CH2OH
```

L-Ribose

b. In glucose, the hydroxyl (—OH) on the right of the bottom chiral carbon makes it D-glucose.

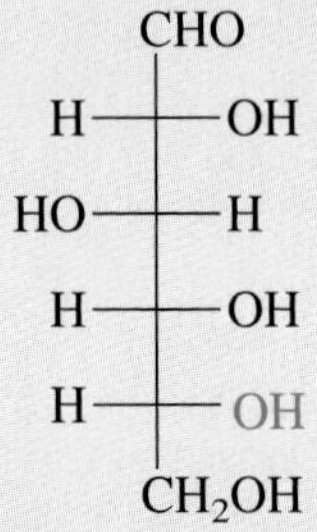

```
        CHO
    H───┼───OH
   HO───┼───H
    H───┼───OH
    H───┼───OH
        CH2OH
```

D-Glucose

Study Check:
Name the following as the D or L isomers:

a.

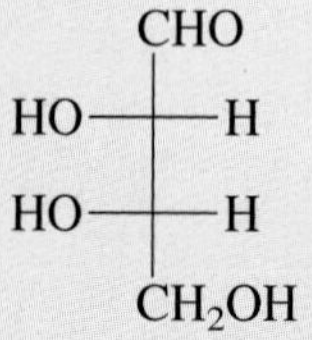

```
        CHO
   HO───┼───H
   HO───┼───H
        CH2OH
```

Erythrose

b.

```
        CH2OH
        C=O
   HO───┼───H
    H───┼───OH
    H───┼───OH
        CH2OH
```

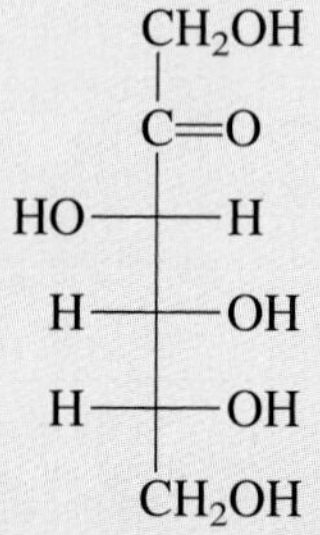

Fructose

Answer: a. L-Erythrose b. D-Fructose

14.4 Monosaccharides and Their Haworth Structures

Learning Goal **Draw Haworth structures for some of the dietary monosaccharides.**

HEALTH NOTE

Optical Isomers in Nature

Most compounds in biological systems consist of only one of the isomer forms. Rarely do both the D and L forms of a biomolecule exist together. It may be that enzymes developed for one stereoisomer were not reactive with the other. Most biologically active monosaccharides are D isomers. The monosaccharide known as blood sugar is D-glucose. Among the amino acids, only the L isomers such as L-alanine and L-lysine are biologically active. The enzymes in the body cannot use the D isomers of amino acids. The L isomer of dopamine is used in the treatment of Parkinson's disease; the D isomer has no biological effect. One form of LSD strongly affects the production of serotonin in the brain and causes hallucinations. The other isomer causes little change. ■

Many of the important monosaccharides are pentoses and hexoses.

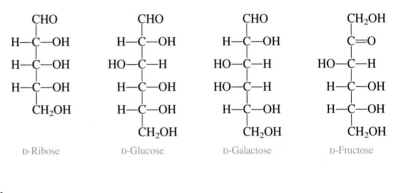

D-Ribose	D-Glucose	D-Galactose	D-Fructose
CHO	CHO	CHO	CH_2OH
H—C—OH	H—C—OH	H—C—OH	C=O
H—C—OH	HO—C—H	HO—C—H	HO—C—H
H—C—OH	H—C—OH	HO—C—H	H—C—OH
CH_2OH	H—C—OH	H—C—OH	H—C—OH
	CH_2OH	CH_2OH	CH_2OH

Glucose

The most abundant hexose in our diet is D-glucose, which has a molecular formula $C_6H_{12}O_6$. Also known as dextrose, grape sugar, and blood sugar, glucose is found in fruits, vegetables, corn syrup, and honey. It is the building block of such complex carbohydrates as starch and cellulose.

In the blood, glucose normally has a concentration of 70–90 mg/dL. However, the amount of glucose in the blood depends on the time that has passed since eating. In the first hour after a meal, the level of glucose rises to about 130 mg/dL, and then it decreases over the next 2–3 hr as it is utilized in the tissues. Some glucose is stored as glycogen in the liver and muscle.

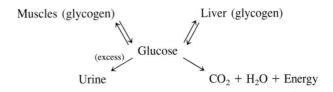

Muscles (glycogen) ⇄ Glucose ⇄ Liver (glycogen)

Glucose → (excess) Urine

Glucose → $CO_2 + H_2O$ + Energy

HEALTH NOTE

Hyperglycemia and Hypoglycemia

To evaluate blood glucose levels, a doctor may order a glucose tolerance test. The patient fasts for 12 hr, and then drinks a solution containing 100 g of glucose. A blood sample is taken immediately, followed by more blood samples each half-hour for 2 hr, and then every hour for a total of 5 hr. If the blood glucose exceeds 130 mg/dL and remains high, hyperglycemia may be indicated. An example of a disease that can cause hyperglycemia is diabetes mellitus, which occurs when the pancreas is unable to produce sufficient quantities of insulin. As a result, glucose cannot enter the cells and accumulates in the blood to levels as high as 350 mg/dL. Symptoms of diabetes in people under the age of 40 include thirst, excessive urination, increased appetite, and weight loss. In older persons, diabetes is sometimes associated with excessive weight gain.

When a person is hypoglycemic, the blood glucose level rises and then decreases rapidly to levels as low as 40 mg/dL. In some cases, hypoglycemia is caused by overproduction of insulin by the pancreas. Low blood glucose can cause dizziness, general weakness, and muscle tremors. Sometimes, a diet is prescribed that consists of several small meals high in protein and low in carbohydrate. Some hypoglycemic patients are finding success with balanced diets that include more complex carbohydrates rather than simple sugars. (See Figure 14.8.) ■

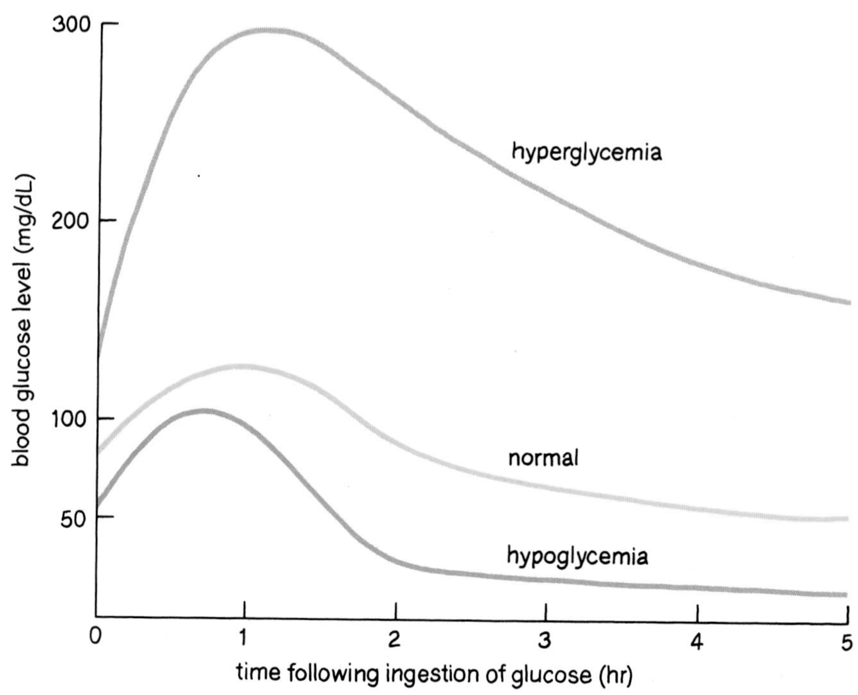

Figure 14.8
Blood glucose levels following ingestion of 100 g glucose, for normal, hyperglycemic, and hypoglycemic conditions.

Hemiacetals

When an aldehyde reacts with an alcohol, a **hemiacetal** is formed. The oxygen of the alcohol adds to the carbonyl group of the aldehyde.

$$\underset{\text{Aldehyde}}{R-\overset{\overset{\displaystyle O}{\|}}{C}-H} + \underset{\text{Alcohol}}{H-OR} \longrightarrow \underset{\text{Hemiacetal}}{R-\overset{\overset{\displaystyle O-H}{|}}{\underset{\underset{\displaystyle H}{|}}{C}}-OR}$$

$$\underset{\text{Ethanal}}{CH_3-\overset{\overset{\displaystyle O}{\|}}{C}-H} + \underset{\text{Methanol}}{H-OCH_3} \longrightarrow \underset{\text{Hemiacetal}}{CH_3-\overset{\overset{\displaystyle O-H \leftarrow \text{New alcohol group}}{|}}{\underset{\underset{\displaystyle H}{|}}{C}}-OCH_3}$$

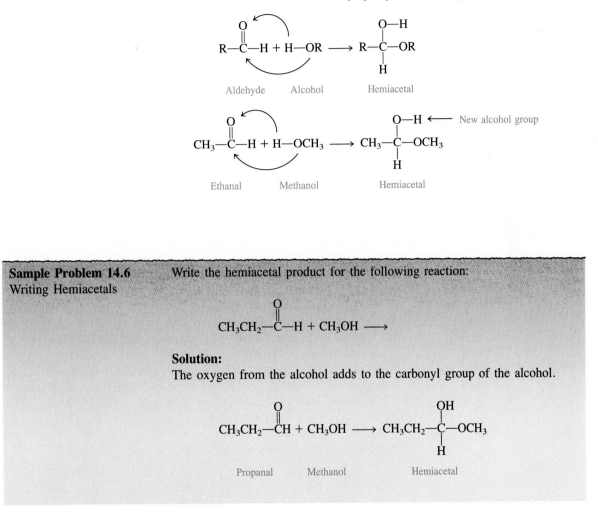

Sample Problem 14.6
Writing Hemiacetals

Write the hemiacetal product for the following reaction:

$$CH_3CH_2-\overset{\overset{\displaystyle O}{\|}}{C}-H + CH_3OH \longrightarrow$$

Solution:
The oxygen from the alcohol adds to the carbonyl group of the alcohol.

$$\underset{\text{Propanal}}{CH_3CH_2-\overset{\overset{\displaystyle O}{\|}}{C}H} + \underset{\text{Methanol}}{CH_3OH} \longrightarrow \underset{\text{Hemiacetal}}{CH_3CH_2-\overset{\overset{\displaystyle OH}{|}}{\underset{\underset{\displaystyle H}{|}}{C}}-OCH_3}$$

Hemiacetal Bonds in Monosaccharides

Because aldopentoses and aldohexoses contain both aldehyde groups and hydroxyl groups, they can form hemiacetal groups within the same molecule. In most aldohexoses, the hydroxyl (—OH) group on carbon 5 reacts with the aldehyde group of the same molecule. The result, a cyclic structure with six atoms, is the most prevalent form of the monosaccharides. The Fischer projection can also be written as a Haworth structure to show the formation of the cyclic hemiacetal structure.

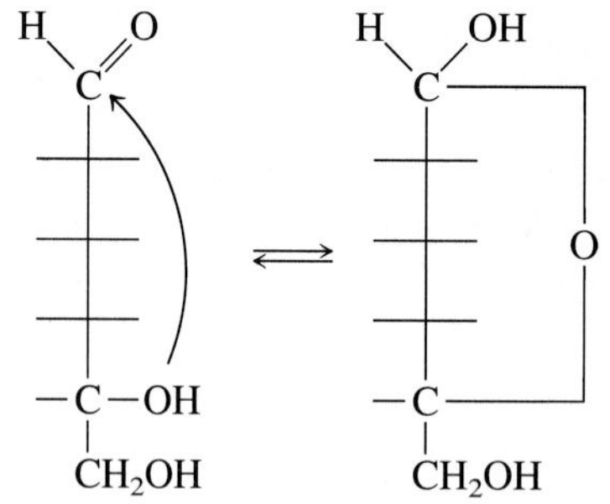

Open-chain of a D-Sugar

Hemiacetal

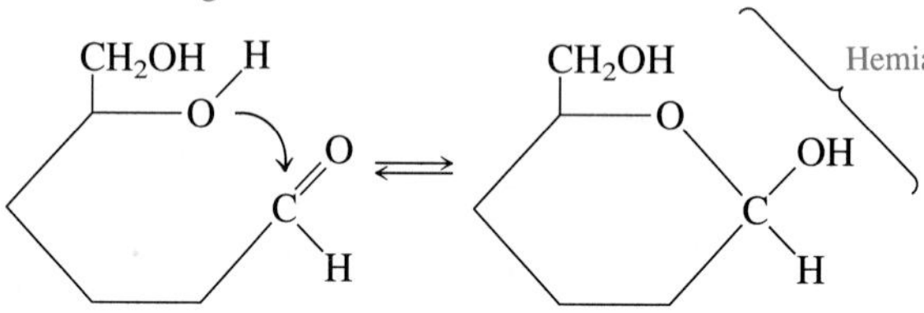

Hemiacetal group

Haworth Structure of Glucose

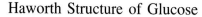

To draw the **Haworth structure** for the cyclic form of D-glucose, we can use the following rules:

1. Number the carbon atoms in the Fischer projection from the top.
2. Turn the Fischer projection on its right side and fold the chain into a hexagon shape.
3. Rotate carbon 5 to its —OH group can react with the aldehyde group. This places the $—CH_2OH$ group in D sugars above carbon 5.
4. Place the —OH groups on carbons 2, 3, and 4. The OH groups on the left of the Fischer projection are written up and the OH groups on the right are written down.
5. Complete the ring structure by writing the hemiacetal bond.

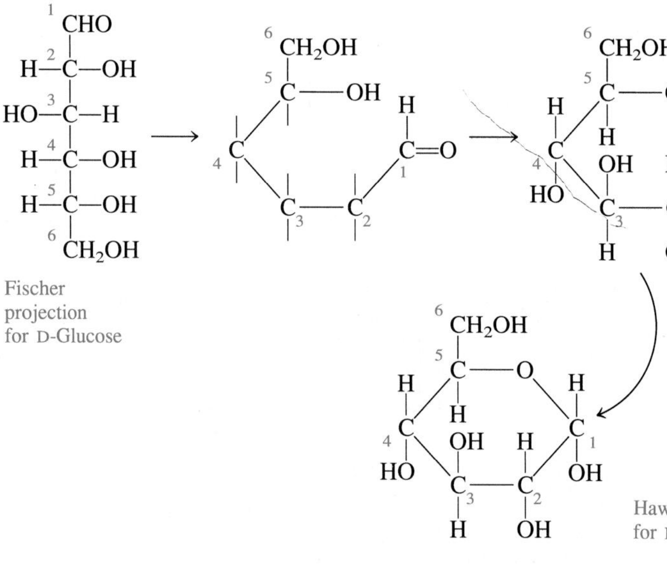

Fischer projection for D-Glucose

Haworth structure for D-Glucose

Anomers

When the hemiacetal forms, a new hydroxyl (—OH) group appears on carbon 1. Carbon 1 is now a chiral carbon that has two possible isomers. The new hydroxyl group can be written up or down. When it is drawn downward, the cyclic isomer is designated as an α-D-sugar; if it is written up, it is the β-D-sugar. Because the α and β isomers differ only at carbon 1, they are called **anomers,** and carbon 1 is called the anomeric carbon.

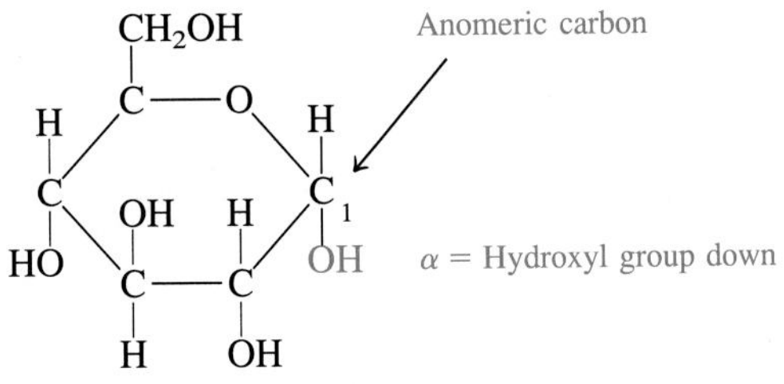

Anomeric carbon

α = Hydroxyl group down

α-D-Glucose

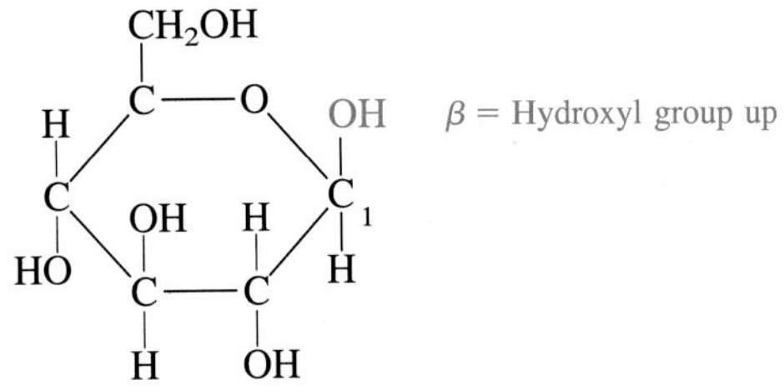

β = Hydroxyl group up

β-D-Glucose

Sample Problem 14.7
Drawing Haworth Structures for Sugars

Draw the β anomer for D-mannose, a carbohydrate found in immunoglobulins:

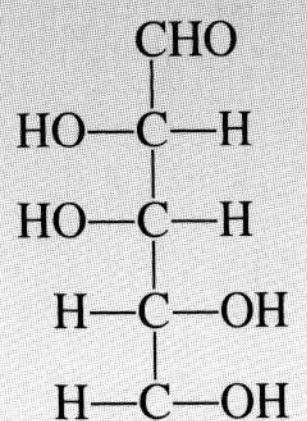

```
      CHO
       |
  HO—C—H
       |
  HO—C—H
       |
   H—C—OH
       |
   H—C—OH
       |
      CH2OH
```

D-Mannose

Solution:
First, number the carbon atoms in the open chain starting at the aldehyde group. Draw and number the carbon and oxygen atoms in a six-atom ring structure. Write the —OH groups on the left of the open chain above the ring, and the —OH groups on the right below. At carbon 1, write the new hydroxyl group upward to give the β-D-mannose anomer.

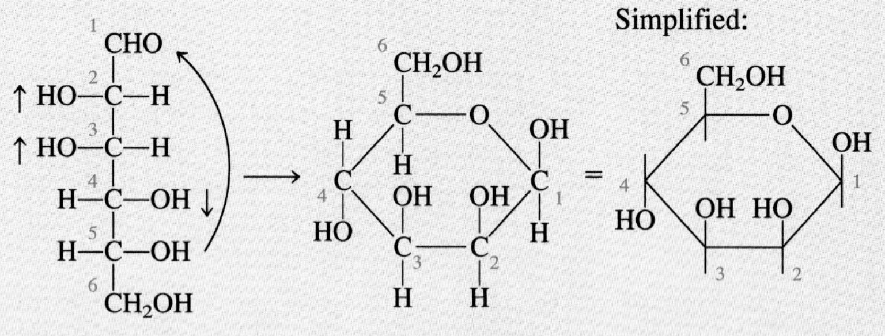

$$\begin{array}{l} {}^{1}CHO \\ \uparrow HO-{}^{2}C-H \\ \uparrow HO-{}^{3}C-H \\ H-{}^{4}C-OH \downarrow \\ H-{}^{5}C-OH \\ {}^{6}CH_2OH \end{array}$$

D-Mannose

β-D-Mannose

Simplified:

Study Check:

Write the cyclic hemiacetal structure of a monosaccharide called α-D-gulose.

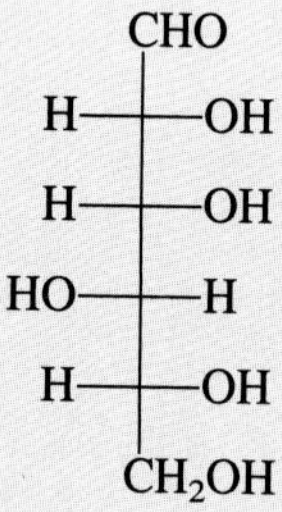

$$\begin{array}{c} CHO \\ H-C-OH \\ H-C-OH \\ HO-C-H \\ H-C-OH \\ CH_2OH \end{array}$$

D-Glucose

Answer:

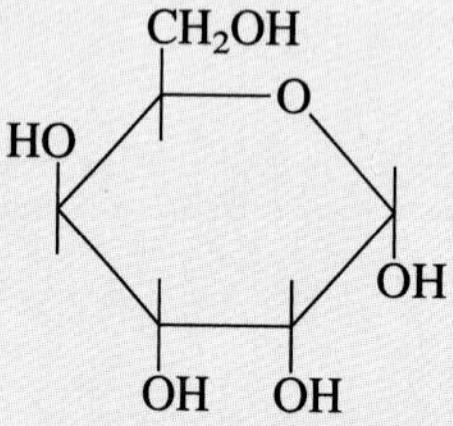

α-D-Glucose

Mutarotation of Anomers

When polarized light is passed through a fresh solution of α-D-glucose, it initially gives an optical rotation of 19° while a fresh solution of β-D-glucose gives a rotation of 112°. As the solutions stand, the optical activity of each changes until both give a reading of 52.7°. The change in optical rotation occurs as the anomers undergo mutarotation. Mutarotation is the process in which the ring structure of α- or β-D-glucose forms an open chain and then closes again as the hemiacetal reforms. As the molecule opens and closes, the hydroxyl (—OH) group shifts from the α to the β position and back again. Although the open chain is an essential part of mutarotation, only a small amount is present at any given time. Eventually a mixture is obtained that is about 36% α-D-glucose and 64% β-D-glucose, which gives the final reading (52.7°) of the optical activity.

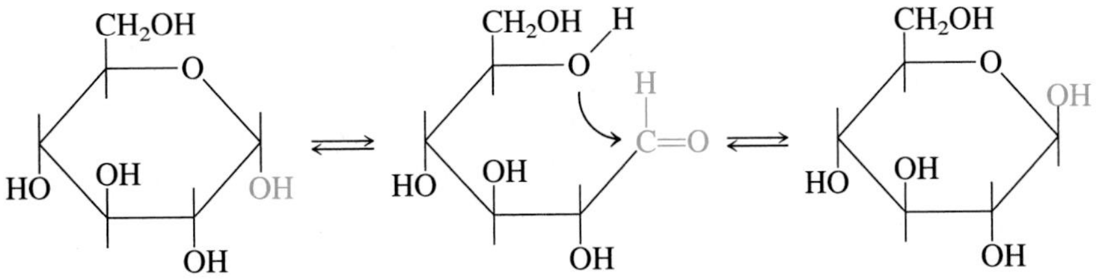

α-D-Glucose cyclic (36%)

D-Glucose open chain (trace)

β-D-Glucose cyclic (64%)

Galactose

Galactose is an aldohexose that does not occur free in nature; it is obtained as a hydrolysis product of the disaccharide lactose, a sugar found in milk and milk products. Galactose is the prevalent monosaccharide in the cellular membranes of the brain and nervous system.

In a condition called **galactosemia,** an infant cannot metabolize galactose. As galactose accumulates in the blood and tissues, the child suffers from cataracts, mental retardation, and cirrhosis. The treatment for galactosemia is the removal of all galactose-containing foods, mainly milk and milk products, from the diet. If this is done immediately after birth, no ill effects occur.

Haworth Structures for α- and β-D-Galactose

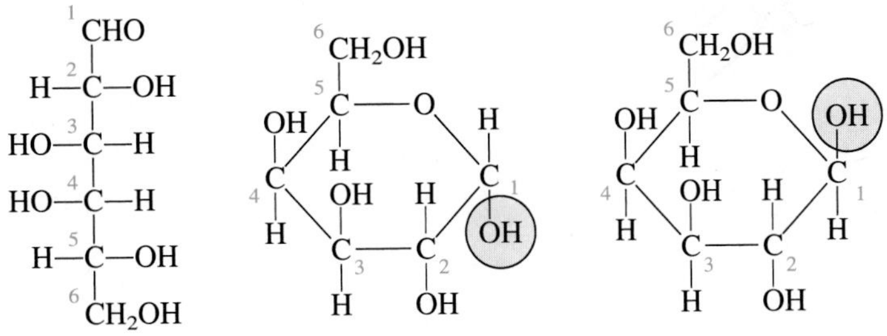

D-Galactose

α-D-Galactose

β-D-Galactose

Simplified Structures:

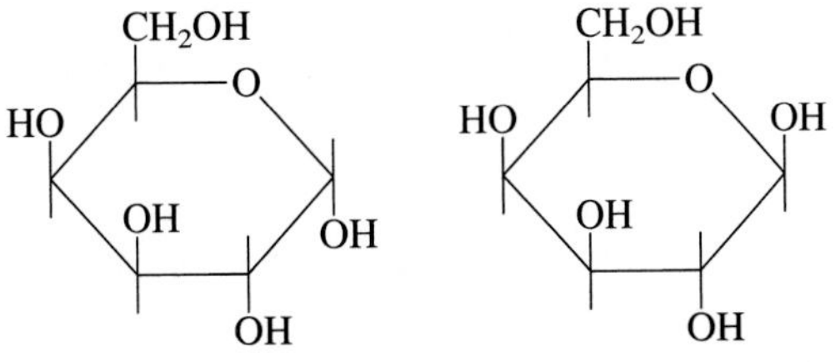

α-D-Galactose

β-D-Galactose

Fructose

Fructose is the sweetest of the carbohydrates, twice as sweet as the most common sweetener, sucrose. This makes fructose popular with dieters since less fructose and therefore fewer calories are needed to provide a pleasant taste. After fructose enters the bloodstream, it is converted to its isomer, glucose, to produce energy in the cells.

Fructose is found in fruit juices and honey; it is also called levulose and fruit sugar. Fructose is also obtained as a hydrolysis product of sucrose, the disaccharide that is also known as table sugar.

In contrast to glucose and galactose, fructose is a ketohexose. The cyclic structure for fructose contains a **hemiketal** group formed when a hydroxyl group in the sugar adds to the ketone group. The general reaction for hemiketal formation is

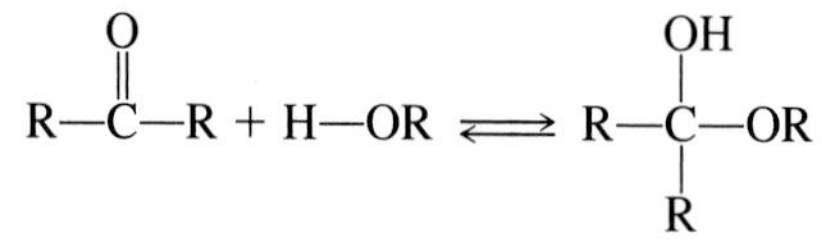

Ketone Alcohol Hemiketal

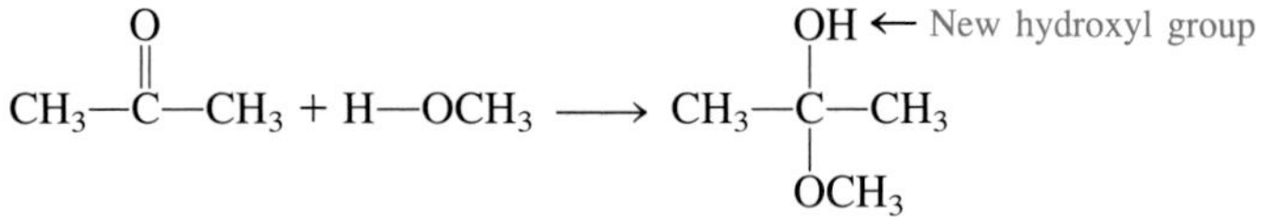

Propanone Methanol Hemiketal

The most common hemiketal for fructose occurs between the hydroxyl (—OH) group on carbon 5 and the keto group on carbon 2.

Haworth Structures for D-Fructose

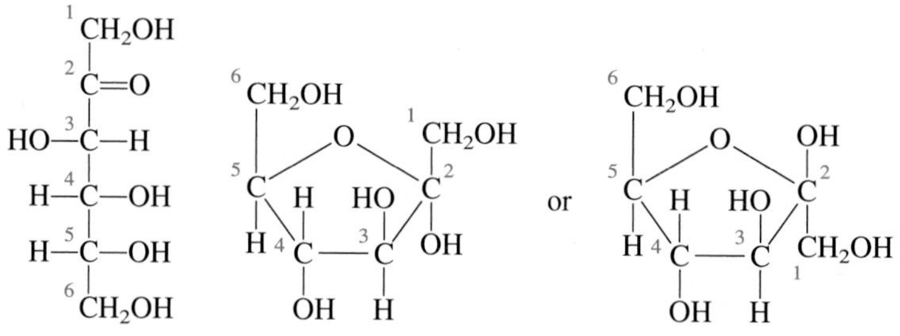

D-Fructose α-D-Fructose β-D-Fructose

Simplified Structures:

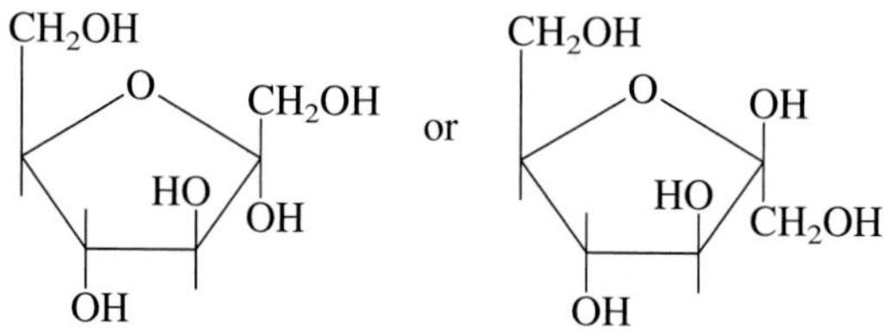

α-D-Fructose β-D-Fructose

14.5 Properties of Monosaccharides

Learning Goal **Describe the oxidation and glycoside formation of monosaccharides.**

Although monosaccharides exist primarily in the cyclic forms, they can be oxidized by Tollens' or Benedict's reagents. It is the aldehyde group that is oxidized as the sugar goes through the open chain during mutarotation. In Chapter 12, we saw that oxidation of aldehydes with Tollens' reagent reduces silver ion (Ag^+) to silver (Ag). With Benedict's (or Fehling's) reagent, cupric ion (Cu^{2+}) is reduced to cuprous ion (Cu^+). A sugar that reacts with Tollens' or Benedict's reagent is called a **reducing sugar.**

$$\begin{array}{c} O \\ \| \\ C{-}H \\ | \\ H{-}C{-}OH \\ | \\ HO{-}C{-}H \\ | \\ H{-}C{-}OH \\ | \\ H{-}C{-}OH \\ | \\ CH_2OH \end{array} + 2Cu^{2+} \xrightarrow{\text{Oxidation}} \overset{\text{Oxidized}}{\begin{array}{c} O \\ \| \\ C{-}OH \\ | \\ H{-}C{-}OH \\ | \\ HO{-}C{-}H \\ | \\ H{-}C{-}OH \\ | \\ H{-}C{-}OH \\ | \\ CH_2OH \end{array}} + \overset{\text{reduced}}{Cu_2O(s)}$$

Open chain of D-Glucose, a reducing sugar — Blue — D-Gluconic acid — Brick-red

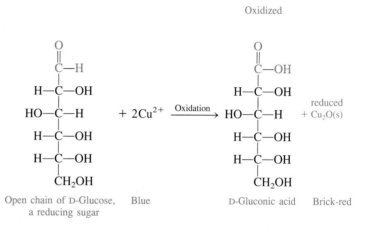

Fructose, a hemiketal, is also a reducing sugar. In the open-chain form, a rearrangement between the hydroxyl group on carbon 1 and the ketone group provides an aldehyde group that can be oxidized.

$$\begin{array}{c} {}^{1}CH_2OH \\ | \\ {}_{2}C{=}O \\ | \end{array} \xrightleftharpoons{\text{rearrangement}} \begin{array}{c} H \quad O \\ \diagdown \;\; /\!\!/ \\ C \\ {}^{1}| \\ H{-}C{-}OH \\ {}^{2}| \end{array}$$

D-Fructose (ketose) — D-Glucose (aldose)

Figure 14.9
A test strip is used to determine the presence and level of glucose in the urine by comparing the color change to a color chart on the container.

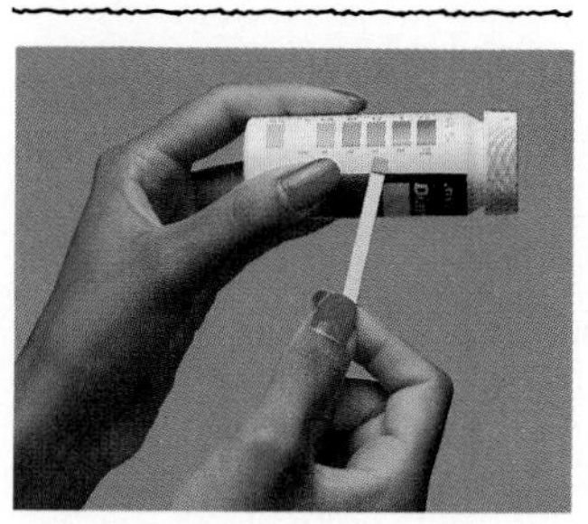

Another clinical test that is more specific for glucose uses the enzyme glucose oxidase on a urine test strip. The oxidase enzyme converts glucose to gluconic acid and hydrogen peroxide, H_2O_2. The peroxide reacts with a dye in the test strip to give different colors. The level of glucose present in the urine is found by matching the color produced to a color chart found on the container. See Figure 14.9.

HEALTH NOTE

Testing for Glucose in Urine

Normally, glucose is filtered by the kidneys and reabsorbed by the bloodstream. However, if the glucose blood level exceeds about 160 mg/dL, the kidney cannot reabsorb it all, and glucose spills over into the urine, a condition known as glucosuria.

The Benedict's test can be used to determine the presence of glucose in urine. The amount of cuprous oxide (Cu_2O) formed is proportional to the amount of reducing sugar present in the urine. If the sugar level is high, the specimen turns yellow or bright orange. Lower levels of reducing sugar turn the solution green.

In the hospital, a similar reaction occurs when a sample of urine is tested for glucose. Table 14.2 lists some colors associated with the concentration of glucose in the urine. ■

Table 14.2 Glucose Test Results

	Glucose present	
Color	**(%)**	**(mg/dL)**
Blue	<0.1	<100
Blue-green	0.25	250
Green	0.50	500
Yellow	1.00	1000
Orange	2.00	2000

Glycosides

A hemiacetal or hemiketal group reacts with an alcohol to form an **acetal.**

$$CH_3-\underset{H}{\overset{OH}{\underset{|}{\overset{|}{C}}}}-OCH_3 + H-OCH_3 \xrightarrow{H^+} CH_3-\underset{H}{\overset{OCH_3}{\underset{|}{\overset{|}{C}}}}-OCH_3 + H_2O$$

Hemiacetal Alcohol Acetal

$$CH_3-\underset{OCH_3}{\overset{OH}{\underset{|}{\overset{|}{C}}}}-CH_3 + H-OCH_3 \xrightarrow{H^+} CH_3-\underset{OCH_3}{\overset{OCH_3}{\underset{|}{\overset{|}{C}}}}-CH_3 + H_2O$$

Hemiacetal Alcohol Acetal

The hemiacetal or hemiketal group of a monosaccharide also reacts with alcohols to form acetals. The acetal products of carbohydrates are called **glycosides,** and the linkage between the hydroxyl group on the sugar and the alcohol is called a **glycosidic bond.**

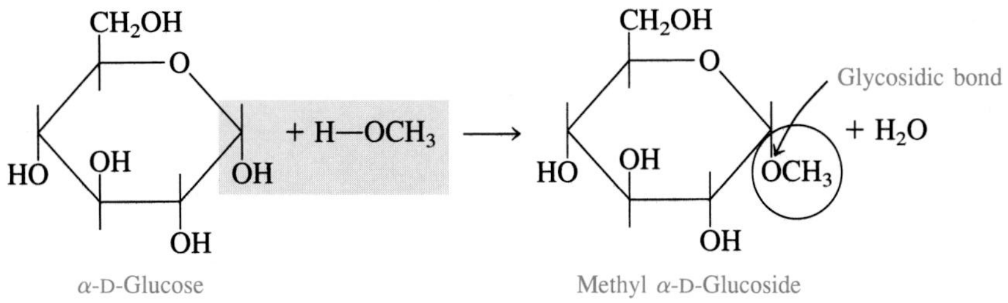

CH$_2$OH, O, HO, OH, OH, OH + H—OCH$_3$ ⟶ CH$_2$OH, O, HO, OH, OCH$_3$, OH + H_2O

Glycosidic bond

α-D-Glucose — Methyl α-D-Glucoside

Sample Problem 14.8
Glycosides

Draw the structure of the glycoside formed when methyl alcohol reacts with the hydroxyl group on carbon 1 of β-D-galactose.

Solution:
In the structure of β-D-galactose, the hydroxyl group on carbon 1 forms an ether bond with methyl alcohol.

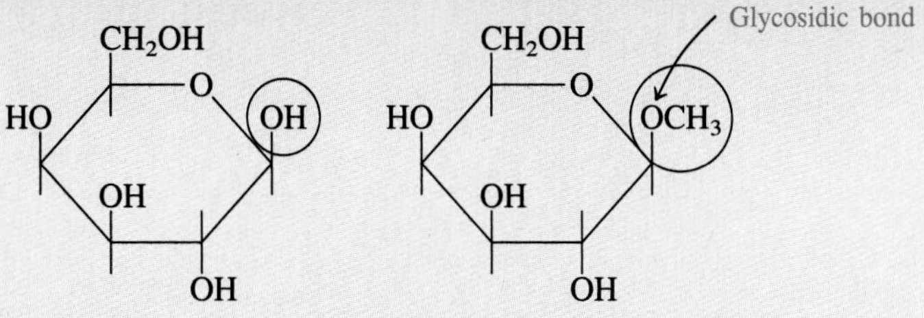

Glycosidic bond

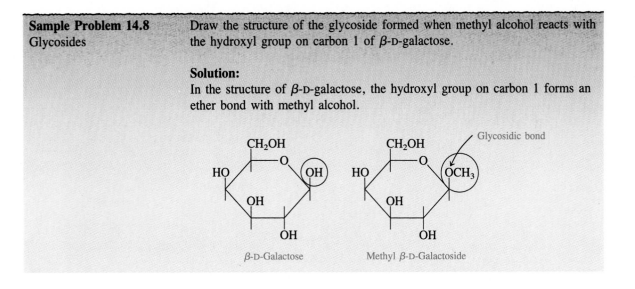

β-D-Galactose — Methyl β-D-Galactoside

14.6 Disaccharides

Learning Goal **For the disaccharides maltose, lactose, and sucrose, identify the monosaccharide units, the type of glycosidic bond, and sources of each.**

A disaccharide is a sugar composed of two monosaccharide units with a molecular formula of $C_{12}H_{22}O_{11}$. When the dietary disaccharides maltose, lactose, and sucrose are hydrolyzed, two monosaccharide are obtained.

Disaccharide	+ H_2O ⟶	*two monosaccharides*
Maltose	+ H_2O ⟶	D-Glucose + D-Glucose
Lactose	+ H_2O ⟶	D-Glucose + D-Galactose
Sucrose	+ H_2O ⟶	D-Glucose + D-Fructose

In a disaccharide, the most typical glycosidic bond forms between the anomeric —OH on carbon 1 and the —OH group on carbon 4 of the second monosaccharide. The resulting glycosidic bond is labeled as an α-1,4-bond or β-1,4-bond depending on the isomer that reacted.

α- and β-1,4-Glycosidic Bonds in a Disaccharide

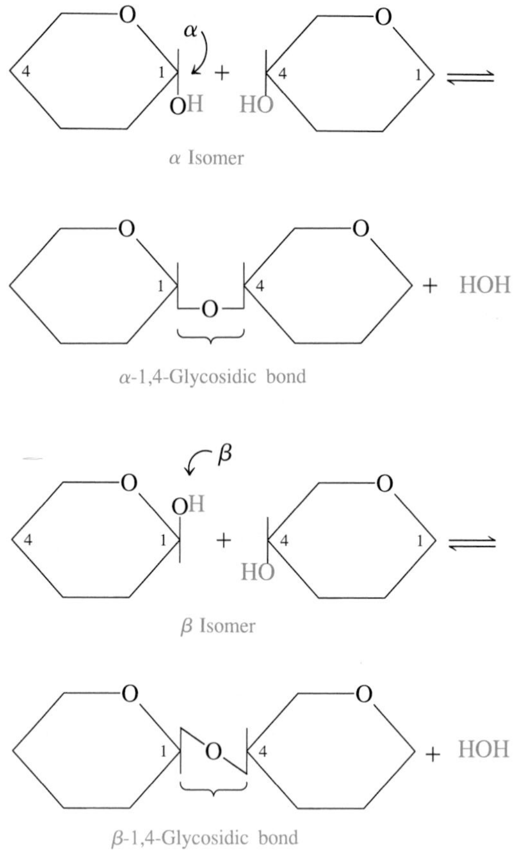

α Isomer

α-1,4-Glycosidic bond

β Isomer

β-1,4-Glycosidic bond

Maltose

Maltose, or malt sugar, is a disaccharide obtained from the hydrolysis of starch. When maltose in barley and other grains is hydrolyzed by yeast enzymes, glucose is produced that can undergo fermentation to give ethanol and carbon dioxide. Maltose is used in cereals, candies, and the brewing of beverages.

The maltose molecule is composed of two D-glucose sugars linked by a glycosidic bond between the α—OH group on carbon 1 of the first glucose and the OH on carbon 4 of the second glucose molecule. The bond is an α-1,4-glycoside bond. The free —OH on the anomeric carbon gives both α-maltose and β-maltose, although β-maltose is more prevalent. Maltose is a reducing sugar because mutarotation occurs at the anomeric carbon to give an aldehyde group available for oxidation.

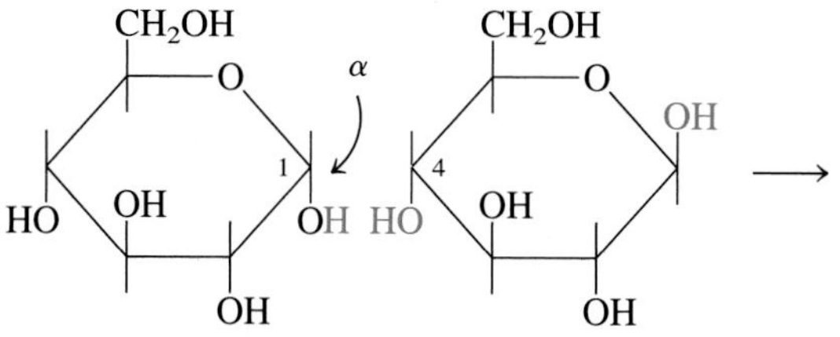

α-Glucose β-Glucose

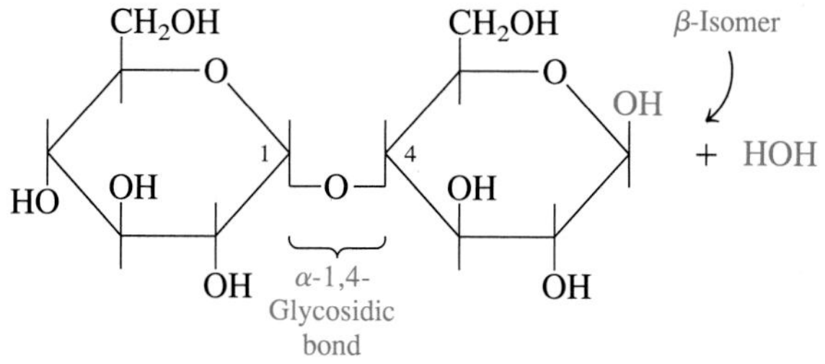

β-Isomer + HOH

α-1,4-Glycosidic bond

β-Maltose

Lactose

Lactose, or milk sugar, is a disaccharide found in milk and milk products. It makes up 6–8% of human milk and about 4–5% of cow's milk and is used in products that attempt to duplicate mother's milk. Some people do not produce sufficient quantities of an enzyme needed to hydrolyze lactose. The sugar remains undigested in the stomach and intestinal tract, causing abdominal cramps and diarrhea.

Lactose consists of a β-1,4-glycosidic bond between β-galactose and D-glucose. The mutarotation of the hemiacetal group on the anomeric carbon of glucose gives the α-lactose and β-lactose isomers. It also provides an aldehyde group for oxidation, which means that lactose is a reducing sugar.

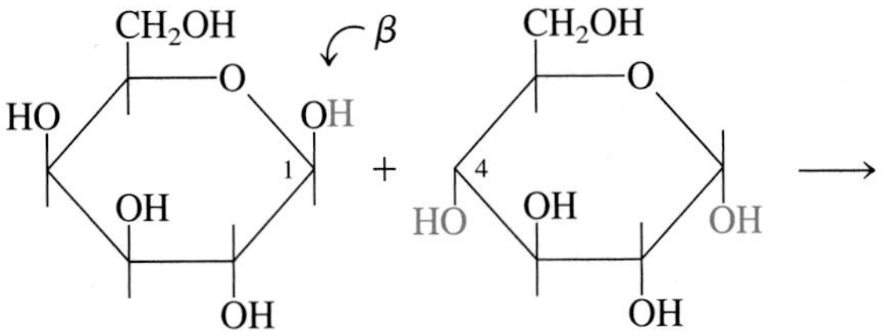

β-Galactose α-Glucose

Figure 14.10
Sucrose, one of the most abundant carbohydrates, is obtained from sugar cane and sugar beets.

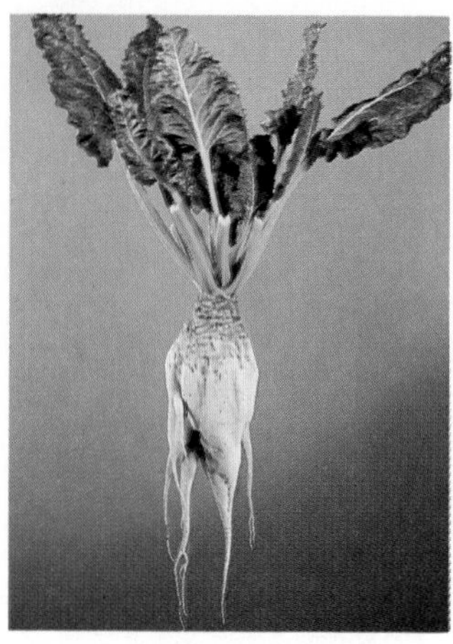

β-1,4-Glycosidic bond

+ HOH

(α Isomer)

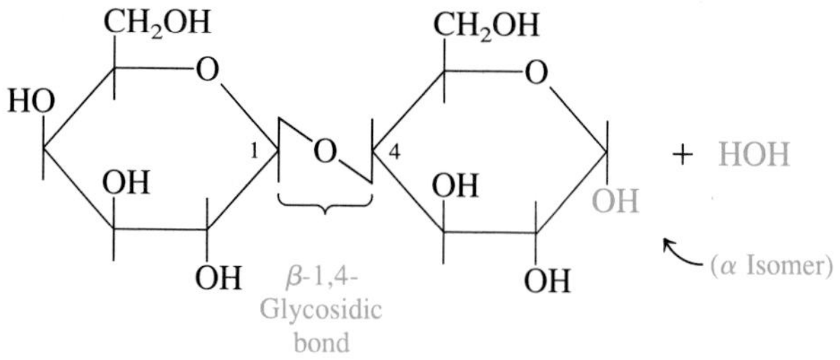

α-Lactose

Sucrose

The disaccharide **sucrose,** table sugar, is a product of sugar cane and sugar beets. See Figure 14.10. It is produced in greater quantities than is any other organic compound. Some estimates indicate that each person in the United States consumes an average of 100 lb of sucrose every year either by itself or in a variety of food products.

Sucrose consists of an α-1,2-glycosidic bond between carbon 1 of D-glucose and carbon 2 of a D-fructose molecule. In sucrose, the glycosidic bond ties up both anomeric carbons; there is no mutarotation. Without a free aldehyde group, it cannot be oxidized. Sucrose is not a reducing sugar.

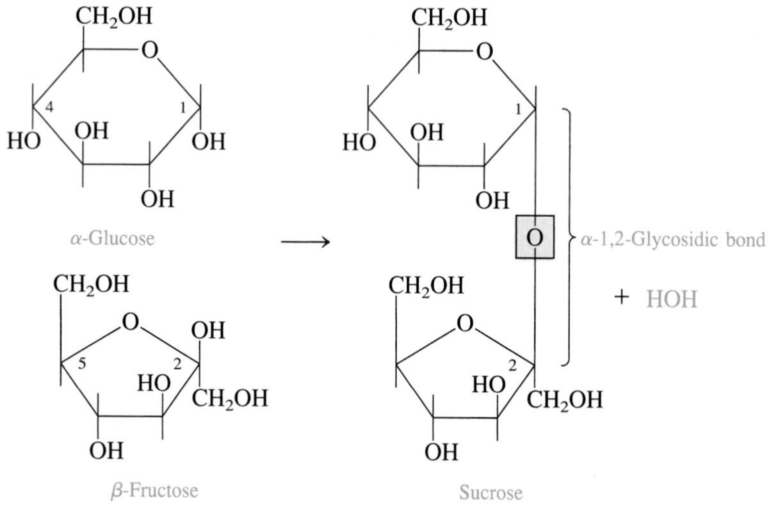

α-Glucose → Sucrose

β-Fructose

α-1,2-Glycosidic bond

+ HOH

Sample Problem 14.9
Glycosidic Bonds in Disaccharides

Specify the type of glycosidic bond, the monosaccharide units, and the name of the following disaccharide.

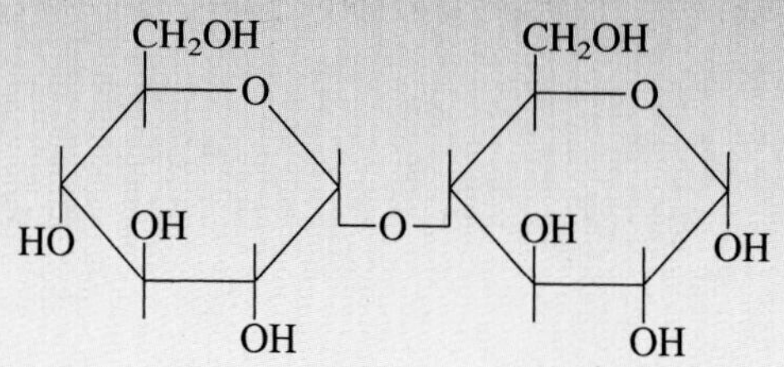

Solution:
Two glucose molecules linked by an α-1,4-glycosidic bond form the dissaccharide maltose. When the free —OH group on the anomeric carbon 1 is written downward, it is the α isomer, or α-maltose.

Study Check:
List the monosaccharides and type of glycosidic bonds in maltose, lactose, and sucrose.

Answer: In maltose, two glucose units are bonded by an α-1,4-glycosidic bond. In lactose, a β-1,4-glycosidic bond links galactose to glucose. In sucrose, an α-1,2-glycosidic bond links glucose and fructose.

14.7 Polysaccharides

Learning Goal **Given the name or structure of a polysaccharide, describe its monosaccharide units, glycosidic linkages, and sources.**

A **polysaccharide** is a long-chain polymer of monosaccharides. Three important polysaccharides—starch, cellulose, and glycogen—are all polymers of D-glucose, which differ in the type of glycosidic bonds and the amount of branching in the molecule.

Starch

Starch, a storage form of glucose in plants, is found in rice, wheat, potatoes, grains, and cereals. There are two kinds of starch, amylose and amylopectin. **Amylose** consists of D-glucose molecules connected by α-1,4-glycosidic bonds. A typical polymer of amylose may contain from 250 to 4000 glucose units.

Amylopectin makes up as much as 80% of starch. It is similar to amylose except that there are branches on the glucose chains. At every 20 or 25 glucose units along the chain, there is a branch of glucose molecules attached by an α-1,6-glycosidic bond between carbon 1 of the branch and carbon 6 in the chain. (See Figure 14.11.)

Glycogen

Glycogen, or animal starch, is the storage form of glucose found in the liver and muscle of animals. It is hydrolyzed in the cells at a rate that maintains the blood level of glucose and provides energy between meals. The structure of glycogen is very similar to that of amylopectin found in plants. The glucose chain contains α-1,4-glycosidic bonds, and branches are attached by α-1,6-glycosidic bonds. However, glycogen contains many more glucose units and is more highly branched with branches occurring about every 10 glucose units.

Figure 14.11
The polysaccharides amylose and amylopectin.

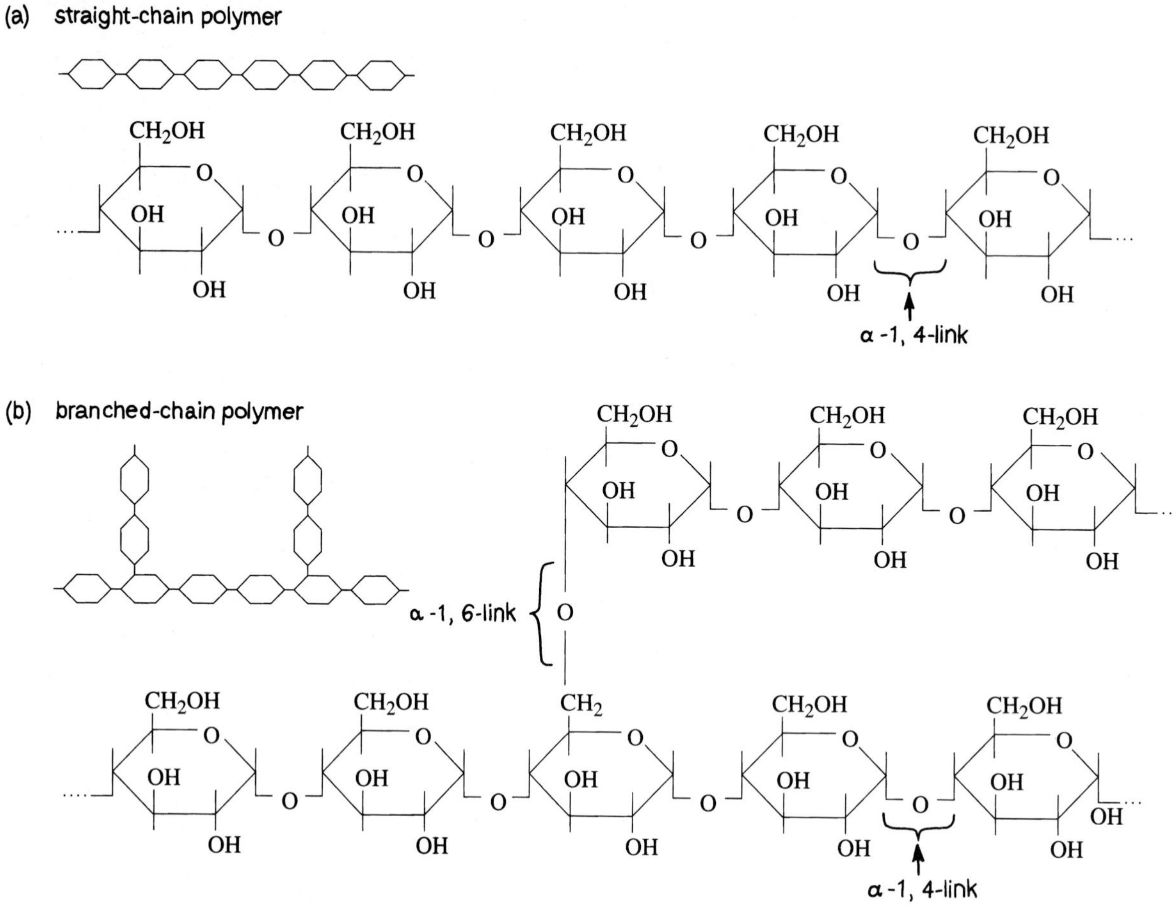

(a) straight-chain polymer

α-1, 4-link

(b) branched-chain polymer

α-1, 6-link

α-1, 4-link

Cellulose

Cellulose is the major structural material of plant cells. Cotton is almost pure cellulose. In cellulose, glucose molecules form a long unbranched chain similar to that of amylose. However, the glucose units in cellulose are linked by β-1,4-glycosidic bonds. The long unbranched glucose chains in cellulose are aligned in parallel rows that are held in place by hydrogen bonds between the hydroxyl groups. This gives a rigid structure to the cell walls in wood and fiber. (See Figure 14.12.)

We obtain glucose from the starches in our diets because enzymes in our saliva and pancreatic juices hydrolyze the α-1,4-glycosidic bonds. However, there are no enzymes in humans that hydrolyze the β-1,4-glycosidic bonds of cellulose; we cannot digest cellulose. Some animals such as goats and cows, and insects like termites are able to obtain glucose from cellulose. Their digestive systems contain bacteria with enzymes that can hydrolyze β-1,4-glycosidic bonds. Table 14.3 summarizes the various carbohydrates in this chapter.

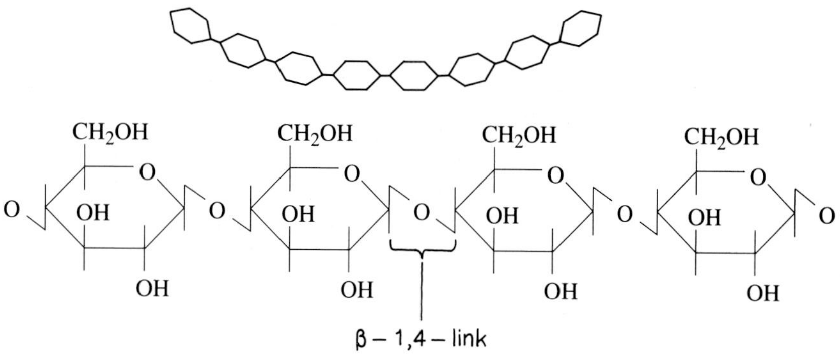

Figure 14.12
Cellulose, composed of β-1, 4-glycosidic bonds, builds cell walls in plants and makes up nearly 100% of cotton.

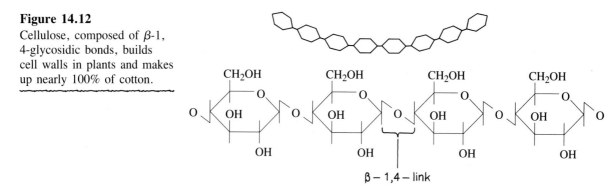

Table 14.3 Summary of Carbohydrates

Carbohydrate family	Compound	Food sources	Monosaccharides
Monosaccharides	Glucose	Fruit juices, honey, vegetables, corn syrup Starch hydrolysis	
	Galactose	Milk and milk products Lactose hydrolysis	
	Fructose	Fruit juices, honey, Sucrose hydrolysis	
Disaccharides	Maltose	Germinating grains Starch hydrolysis	Glucose + glucose
	Lactose	Milk and milk products	Glucose + galactose
	Sucrose	Sugar cane, sugar beets	Glucose + fructose
Polysaccharides	Amylose	Rice, wheat, grains, cereals	Many glucose molecules; straight-chain α-1,4 bonds
	Amylopectin	Rice, wheat, grains, cereals	Many glucose molecules; branched-chain α-1,4 and α-1,6 bonds
	Glycogen	Liver, muscles	Many glucose molecules; branched-chain α-1,4 and α-1,6 bonds
	Cellulose	Plant fiber	Many glucose molecules in a straight chain of β-1,4 bonds; not digestible by humans

Sample Problem 14.10
Polysaccharides

Describe the monosaccharide units, the structure, and the food sources of amylose.

Solution:
Amylose is a straight-chain polymer of glucose molecules connected by α-1,4-glycosidic bonds. Amylose is found in plants such as rice, wheat, and other grains.

Study Check:
How is cellulose different from amylose?

Answer: Amylose and cellulose are both unbranched polymers of D-glucose. However, glucose units in cellulose are linked by β-1,4-glycosidic bonds rather than the α-1,4-glycosidic bonds found in amylose.

HEALTH NOTE

Fiber in the Diet

The term *dietary fiber* includes all plant material that is not digestible by humans, including cellulose. Food sources of dietary fiber include whole grains, bran, fruits, and vegetables.

Fiber in the diet aids in the formation of bulk in the intestinal tract, which increases the absorption of water along the tract. The uptake of water has a laxative effect by producing softer stools. It also increases the rate at which digestive wastes move through the intestinal tract, which lessens the time the intestine comes in contact with any ingested carcinogens. Some forms of diverticulitis (inflammation of the colon) have been relieved by increasing the quantity of fiber in the diet.

The absorptive effects of fiber may also be beneficial in weight maintenance. Fiber increases the bulk in the stomach and intestines without contributing to caloric intake. Fiber may also absorb some of the carbohydrate and cholesterol from the diet, thus decreasing the quantity that diffuses through the intestinal walls. ■

HEALTH NOTE

Modified Sugars Have Important Roles

Some alterations of monosaccharide units give modified sugars that are important in cellular structure, connective tissues, anticoagulants, and antibiotics.

Two modified sugars have structures similar to glucose. In glucuronic acid, carbon 6 has been oxidized to a carboxylic acid. In glucosamine and *N*-acetylglucosamine, a hydroxyl group has been replaced by an amine.

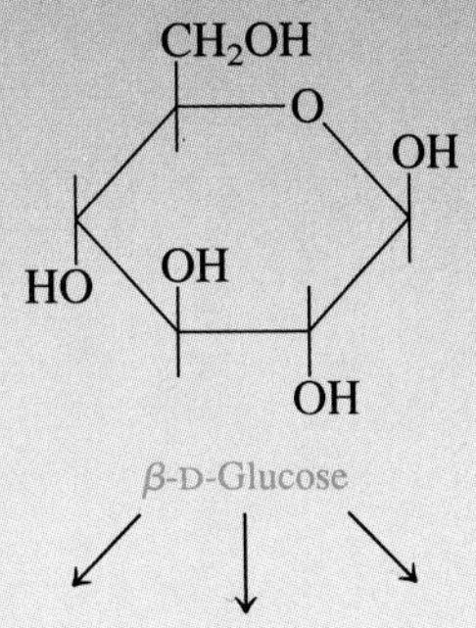

CH_2OH, O, OH, HO, OH, OH

β-D-Glucose

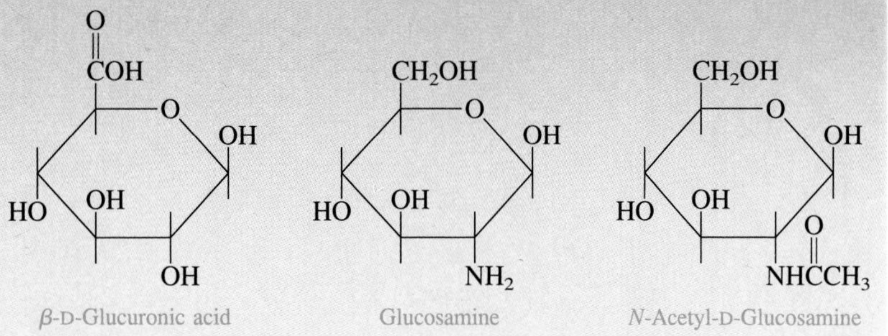

β-D-Glucuronic acid Glucosamine *N*-Acetyl-D-Glucosamine

Chitin, a structural polysaccharide found in insect skeletons and shells of crabs and lobsters, is a polymer of *N*-acetyl-D-glucosamine. (See Figure 14.13.)

Figure 14.13
Chitin, a polymer of *N*-acetyl-D-glucosamine, is a structural material in crab and lobster shells.

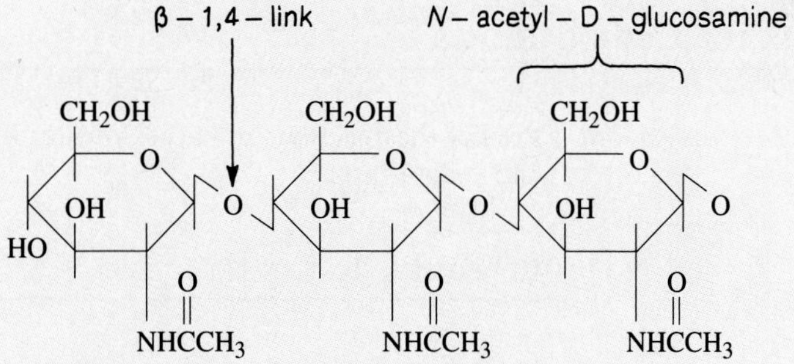

Connective tissues are held together by a viscous matrix of mucopolysaccharides such as hyaluronic acid, a polymer with alternating units of D-glucuronic acid and *N*-acetyl-D-glucosamine.

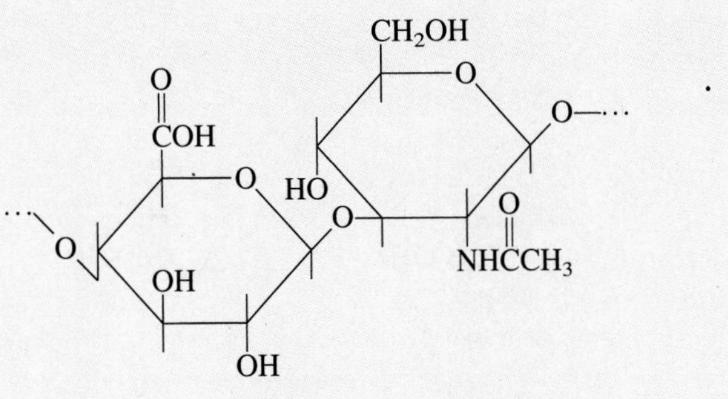

Units in a polymer of hyaluronic acid

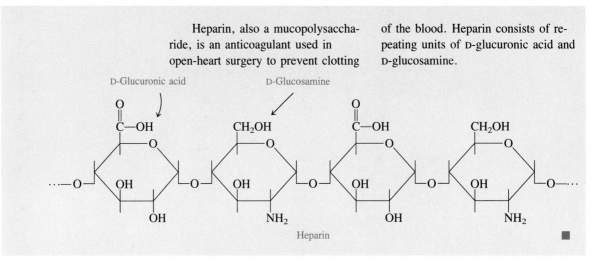

Heparin, also a mucopolysaccharide, is an anticoagulant used in open-heart surgery to prevent clotting of the blood. Heparin consists of repeating units of D-glucuronic acid and D-glucosamine.

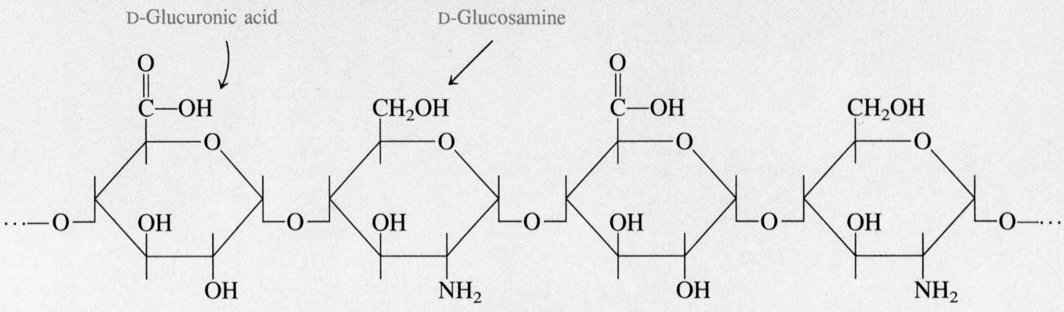

D-Glucuronic acid
D-Glucosamine

Heparin

14.8 Tests for Carbohydrates

Learning Goal **Predict the reactions of carbohydrates in the Benedict's, fermentation, and iodine tests.**

Benedict's Test

In the **Benedict's test,** cupric hydroxide, $Cu(OH)_2$, is added to the sugar solution, and the mixture is heated for 5 min. Any aldehyde group of the open chain is oxidized and cupric ion (Cu^{2+}) is reduced to cuprous (Cu^{+}) ion. The blue solution changes to green or orange as the reddish-orange precipitate of cuprous oxide forms.

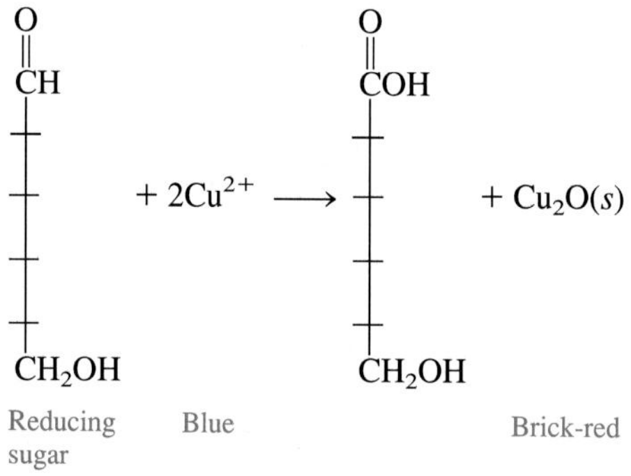

$$\text{CHO}\cdots\text{CH}_2\text{OH} + 2\text{Cu}^{2+} \longrightarrow \text{COOH}\cdots\text{CH}_2\text{OH} + \text{Cu}_2\text{O}(s)$$

Reducing sugar — Blue — Brick-red

Monosaccharides such as glucose, galactose, and fructose and most disaccharides such as maltose and lactose (but not sucrose) are reducing sugars. Sucrose is not a reducing sugar and does not react with Benedict's reagent. Polysaccharides also show no reaction with Benedict's reagent because a polysaccharide does not provide free aldehyde groups.

Fermentation Test

When treated with yeast, the hexoses glucose and fructose (but not galactose) will undergo fermentation.

$$\underset{\text{Glucose or fructose}}{C_6H_{12}O_6} \xrightarrow{\text{yeast}} \underset{\text{Ethanol}}{2C_2H_5OH} + 2CO_2(g)$$

The disaccharides, maltose and sucrose, also undergo fermentation because yeast contains enzymes for their hydrolysis to give glucose and fructose. However, lactose will not ferment because the enzyme lactase required for its hydrolysis is not present in yeast. Fermentation does not occur with polysaccharides. This reaction constitutes a **fermentation test** for glucose, fructose, maltose, and sucrose.

Iodine Test

The **iodine test** causes the polysaccharide amylose in starch to react strongly with iodine to form a characteristic deep blue-black complex. Cellulose, glycogen, and amylopectins produce reddish-purple and brown colors. Such color does not develop with any of the mono- or disaccharides.

Sample Problem 14.11
Tests for Carbohydrates

Indicate the results that fructose, sucrose, and amylose will give in the iodine, fermentation, and Benedict tests.

Solution:
Iodine will not produce a color change with fructose or sucrose, while amylose will give a blue-black color. The fermentation test will produce $CO_2(g)$ with fructose and glucose, but not with amylose. Benedict's reagent will turn brick-red with fructose, but not with amylose or sucrose.

	Tests		
Carbohydrate	**Benedict's**	**Fermentation**	**Iodine**
Fructose	Brick-red color	CO_2 gas	No color change
Sucrose	No color change	CO_2 gas	No color change
Amylose	No color change	No gas bubbles	Blue-black color

Study Check:
What reaction would you expect lactose to give with Benedict's reagent, yeast, and iodine?

Answer: Lactose, a disaccharide, will give a brick-red color change with Benedict's reagent and no reaction with yeast or iodine.

Summary of Reactions

Hemiacetal formation

Aldehyde + Alcohol ⟶ Hemiacetal

$$CH_3{-}\overset{\overset{\displaystyle O}{\|}}{C}H + H{-}OCH_3 \longrightarrow CH_3{-}\underset{\underset{\displaystyle H}{|}}{\overset{\overset{\displaystyle OH}{|}}{C}}{-}OCH_3$$

Hemiketal formation

Ketone + Alcohol ⟶ Hemiketal

$$CH_3{-}\overset{\overset{\displaystyle O}{\|}}{C}{-}CH_3 + H{-}OCH_3 \longrightarrow CH_3{-}\underset{\underset{\displaystyle CH_3}{|}}{\overset{\overset{\displaystyle OH}{|}}{C}}{-}OCH_3$$

Acetal formation

Hemiacetal or Hemiketal + Alcohol ⟶ Acetal + H_2O

$$CH_3{-}\underset{\underset{\displaystyle H\ (CH_3)}{|}}{\overset{\overset{\displaystyle OH}{|}}{C}}{-}OCH_3 + H{-}OCH_3 \longrightarrow CH_3{-}\underset{\underset{\displaystyle H\ (CH_3)}{|}}{\overset{\overset{\displaystyle OCH_3}{|}}{C}}{-}OCH_3 + H_2O$$

Sugar + Alcohol ⟶ Glycoside + H_2O

CH₂OH, O, HO, OH, OH, OH + H—OCH₃ ⟶ CH₂OH, O, HO, OH, OCH₃, OH + H₂O

Glycosidic bond

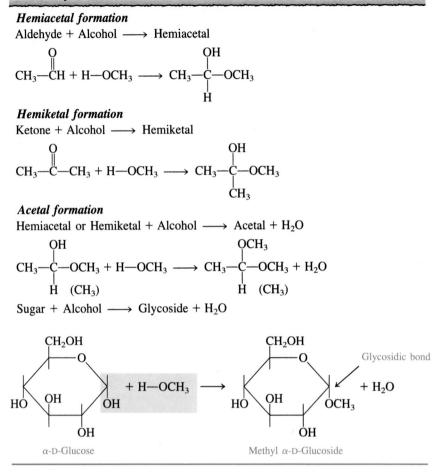

α-D-Glucose   Methyl α-D-Glucoside

Glossary of Key Terms

acetal The product of the addition of an alcohol to a hemiacetal or hemiketal.

achiral An object or molecule that has symmetry; it contains no chiral carbon atoms.

aldose Monosaccharides that contain an aldehyde group.

amylopectin A branch-chain polymer of glucose units produced in plants.

amylose An unbranched polymer of starch produced in plants composed of glucose units in α-1,4-glycosidic bonds.

anomers Two isomers, α and β, of monosaccharides formed when the carbonyl group becomes a chiral carbon because of hemiacetal or hemiketal formation.

asymmetric An object or molecule that cannot be divided to give two identical halves. The mirror images of asymmetric objects or molecules are not superimposable.

Benedict's test A chemical test for monosaccharides and disaccharides (except sucrose) in which blue Cu^{2+} is reduced to brick-red $Cu_2O(s)$.

carbohydrate A polyhydroxy compound that contains an aldehyde or ketone group.

cellulose A polysaccharide composed of glucose units in an unbranched polymer linked by β-1,4-glycosidic bonds that cannot be hydrolyzed by humans.

chiral Objects that have right- and left-hand forms which are not superimposable.

chiral carbon A carbon atom that is attached to four different atoms or groups of atoms; a mirror image can be drawn that is not superimposable.

disaccharide A carbohydrate composed of two monosaccharides.

fermentation test A test for glucose, fructose, maltose, or sucrose, in which the sugar reacts with enzymes in yeast to give ethanol and carbon dioxide gas.

Fischer projection A representation of the chiral carbons in a molecule. Vertical lines are attached to groups that go back, and horizontal lines are attached to the groups that project toward the viewer.

fructose A monosaccharide found in honey and fruit juices; it is combined with glucose in sucrose. Also called levulose and grape sugar.

galactose A monosaccharide that occurs combined with glucose in lactose.

glucose The most prevalent of the monosaccharides in the diet. Found in fruits, vegetables, corn syrup, and honey. Also known as blood sugar and dextrose. Combines in glycosidic bonds to form most of the polysaccharides.

glycogen A polysaccharide formed in the liver and muscles for the storage of glucose as an energy reserve. It is composed of glucose in a highly branched polymer.

glycoside The acetal product of a monosaccharide reacting with an alcohol or another sugar.

glycosidic bond The acetal bond that forms when an alcohol or the hydroxyl group of a monosaccharide adds to a hemiacetal. It is the type of bond that links monosaccharide units in di- or polysaccharides.

Haworth structure The ring structure of a sugar in which —OH groups on the left of the open chain are written above the carbon, and —OH groups on the right of the open chain are written below.

hemiacetal The product of the addition of an alcohol to the carbonyl group of an aldehyde.

hemiketal The product of the addition of an alcohol to the carbonyl group of a ketone.

iodine test A test for amylose that forms a blue-black color after iodine is added to the sample.

ketose A monosaccharide that contains a ketone group.

lactose A disaccharide consisting of glucose and galactose found in milk and milk products.

maltose A disaccharide consisting of two glucose units; it is obtained from the hydrolysis of starch and in germinating grains.

monosaccharide A simple carbohydrate typically composed of three to six carbon atoms.

optical activity The interaction of a chiral molecule with plane polarized light that causes the light to bend. The angle at which the light emerges is rotated to the left or the right.

optical isomers Mirror images of a chiral compound that rotate plane-polarized light in opposite directions.

polysaccharide A polymer of many monosaccharide units, usually glucose. Polysaccharides differ in the types of glycosidic bonds and the amount of branching in the molecule.

reducing sugar A carbohydrate with a free aldehyde group capable of reducing cupric ion to give a positive test with Benedict's reagent.

saccharide A term from the Latin word *saccharum,* meaning ''sugar''; it is used to describe the carbohydrate family.

stereoisomers Compounds with the same structural formulas, but different three-dimensional arrangements of their atoms.

sucrose A disaccharide composed of glucose and fructose; a nonreducing sugar, commonly called table sugar or ''sugar.''

Problems

Classification of Carbohydrates *(Goal 14.1)*

14.1 What is the difference between an aldose and a ketose?

14.2 Classify each of the following monosaccharides:

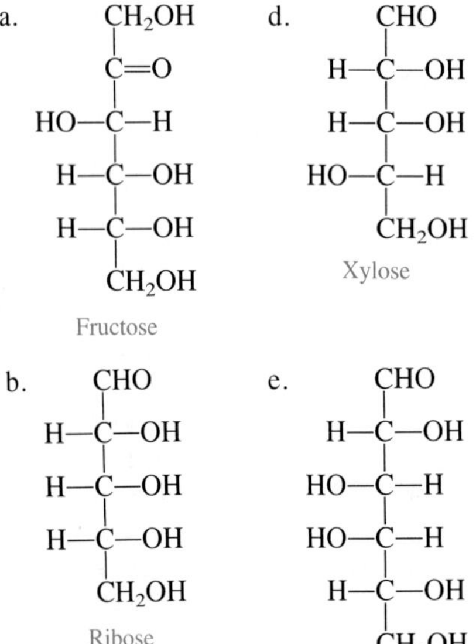

c.
CH_2OH
C=O
CH_2OH
Dihydroxyacetone

14.3 Draw the structure of a monosaccharide that is a ketotetrose.

Stereroisomers *(Goal 14.2)*

14.4 Indicate whether the following objects are chiral (asymmetrical), or achiral (symmetrical). Consider each object to be without markings or printing.

a. A table tennis ball
b. A glove
c. An envelope
d. An unsharpened pencil
e. A tennis shoe
f. A golf club
g. A door
h. A coffee cup

14.5 Indicate which of the following compounds is chiral:

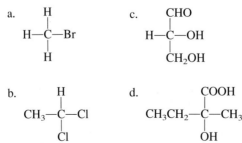

a.
H
H—C—Br
H

b.
H
CH_3—C—Cl
Cl

c.
CHO
H—C—OH
CH_2OH

d.
COOH
CH_3CH_2—C—CH_3
OH

e. CHO / $\mathrm{H{-}C{-}CH_2CH_3}$ / H (Fischer-type: CHO on top, H on left, CH_2CH_3 on right, H below)

f. COOH / $\mathrm{H{-}C{-}OH}$ / $\mathrm{CH_3}$

Lactic acid

g. HO, HO on benzene ring $-\mathrm{C(OH)(H){-}CH_2NHCH_3}$

Epinephrine

14.6 Lactic acid is optically active. What does that mean?

COOH / H—+—OH / $\mathrm{CH_3}$

lactic acid

Chiral Sugars *(Goal 14.3)*

14.7 For each of the following sugars state whether each is the D or L isomer and write the mirror image.

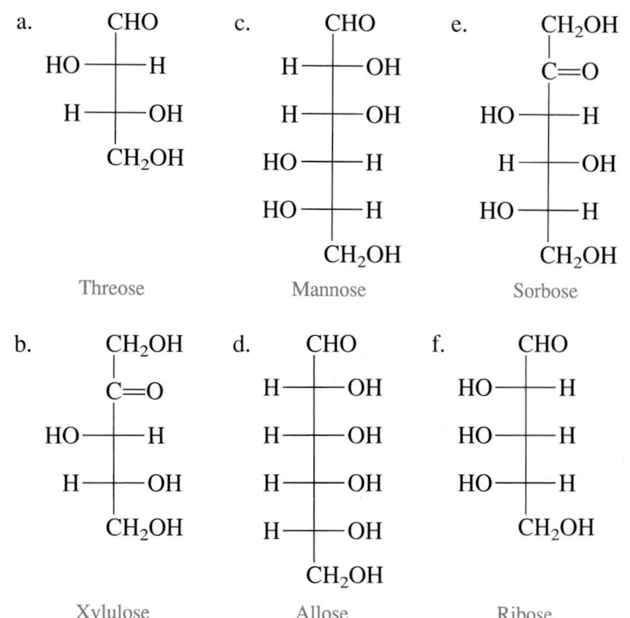

Monosaccharides and Their Haworth Structures *(Goal 14.4)*

14.8 Write the hemiacetal product of the following:

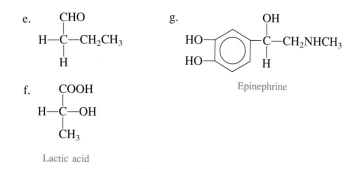

$$\mathrm{CH_3CH_2\overset{\overset{\displaystyle O}{\|}}{C}H + HOCH_3 \longrightarrow}$$

14.9 What carbon atoms are typically involved in the formation of a hemiacetal of an aldohexose?

14.10 Draw the Haworth structures for the α and β anomers of the following aldohexoses:

a. 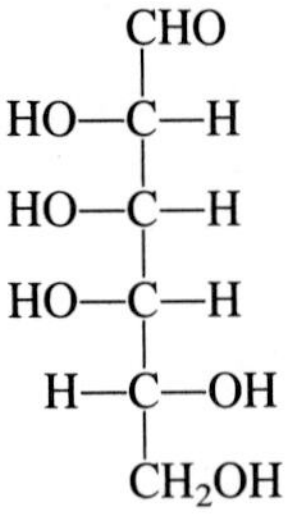
```
      CHO
       |
  HO—C—H
       |
  HO—C—H
       |
  HO—C—H
       |
   H—C—OH
       |
      CH2OH
```
D-Talose

b. 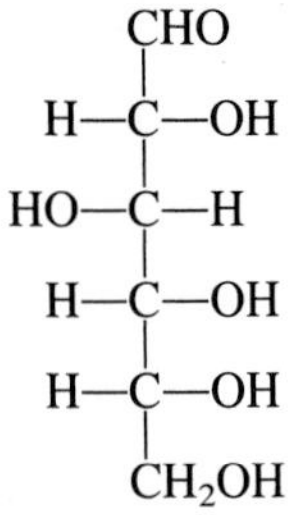
```
      CHO
       |
   H—C—OH
       |
  HO—C—H
       |
   H—C—OH
       |
   H—C—OH
       |
      CH2OH
```
D-Glucose

c. 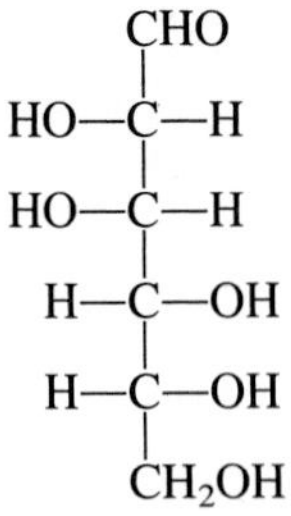
```
      CHO
       |
  HO—C—H
       |
  HO—C—H
       |
   H—C—OH
       |
   H—C—OH
       |
      CH2OH
```
D-Mannose

14.11 Write the hemiketal formed by the following reaction:

$$CH_3\overset{\displaystyle O}{\overset{\|}{C}}CH_3 + HOCH_2CH_3 \longrightarrow$$

14.12 The following are the Haworth structures of some D-sugars. Mark the new chiral carbon and name the sugar including the α or β form.

a. CH_2OH O CH_2OH HO OH OH

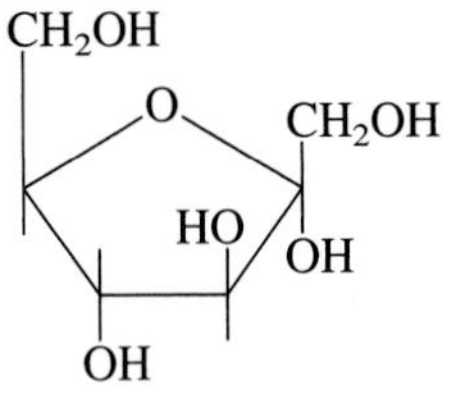

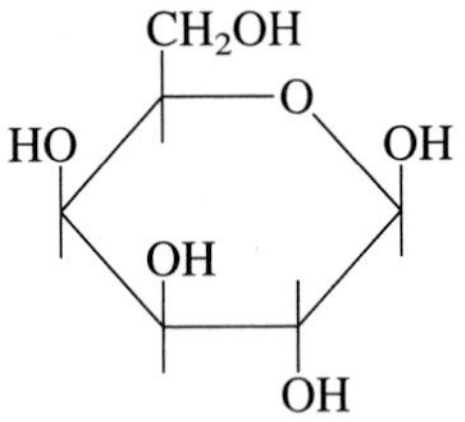

b. CH_2OH O HO OH OH OH

c. CH_2OH O HO OH OH OH

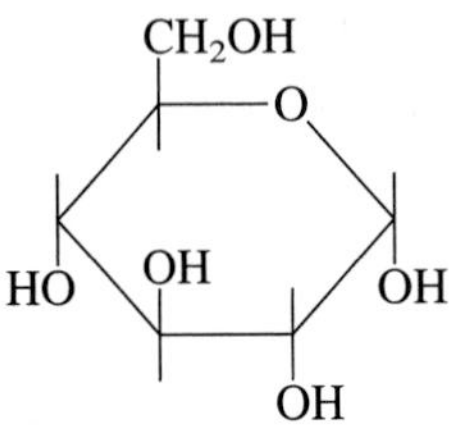

d. CH_2OH O HO OH OH OH

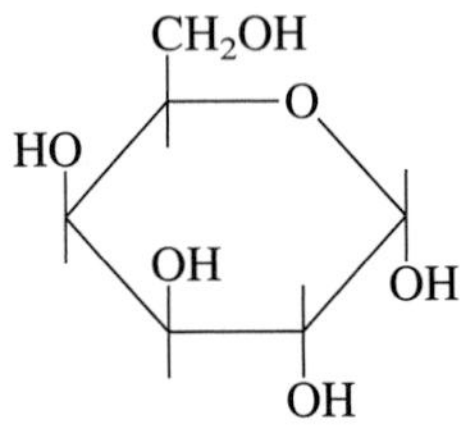

14.13 Draw the Haworth structures of
a. β-D-glucose b. β-D-fructose

14.14 Using Haworth structures, write an equation for the mutarotation of galactose.

14.15 Describe some sources of
a. glucose b. fructose c. galactose

Properties of Monosaccharides *(Goal 14.5)*

14.16 What is a reducing sugar?

14.17 What are the products when D-galactose reacts with Benedict's reagent?

14.18 Why does fructose, a ketohexose, act as a reducing sugar with Benedict's reagent?

***14.19** D-Sorbitol used as a sweetener is found in seaweed and many kinds of berries. When D-sorbitol is oxidized, it forms D-glucose. What is the formula of D-sorbitol?

***14.20** Write the α and β glycoside products for each of the following sugars when they react with CH_3OH.

a. D-glucose
b. D-galactose

***14.21** Raffinose is a trisaccharide found in Australian manna and in cottonseed meal. It is composed of three different sugars and is not a reducing sugar. Identify the sugars in raffinose.

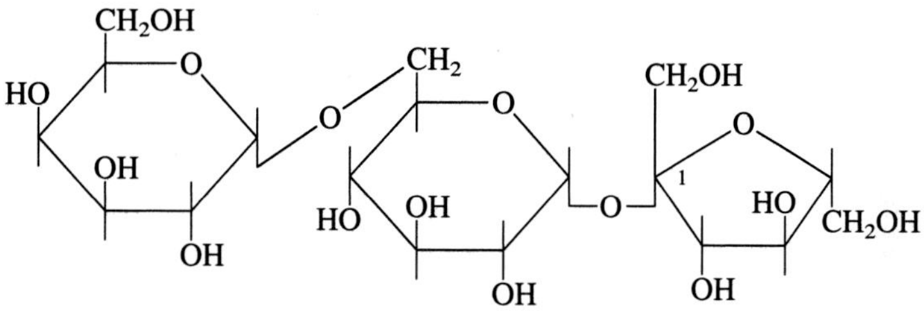

Disaccharides *(Goal 14.6)*

14.22 For each of the following disaccharides state: the monosaccharide units produced by hydrolysis, the type of glycosidic bond, the name of the disaccharide (including α or β form), and whether it is a reducing sugar or not.

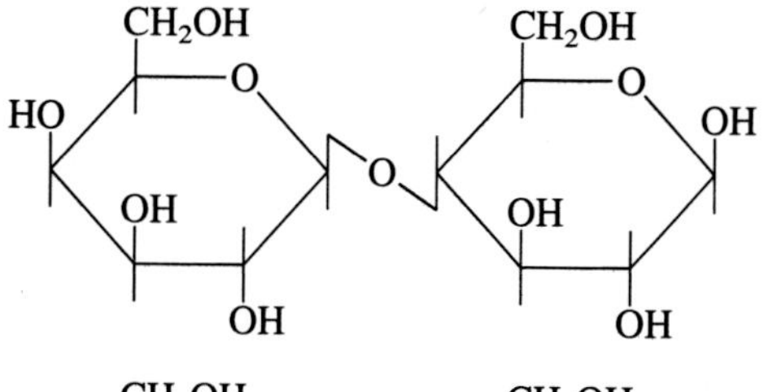

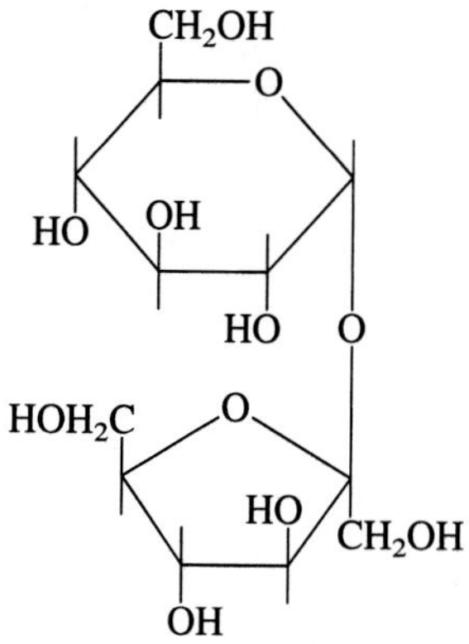

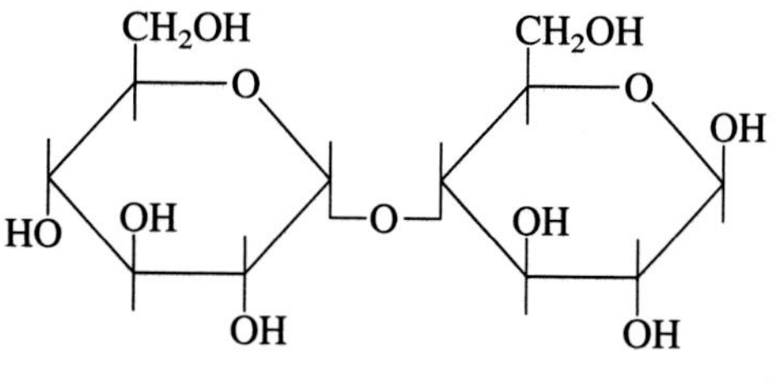

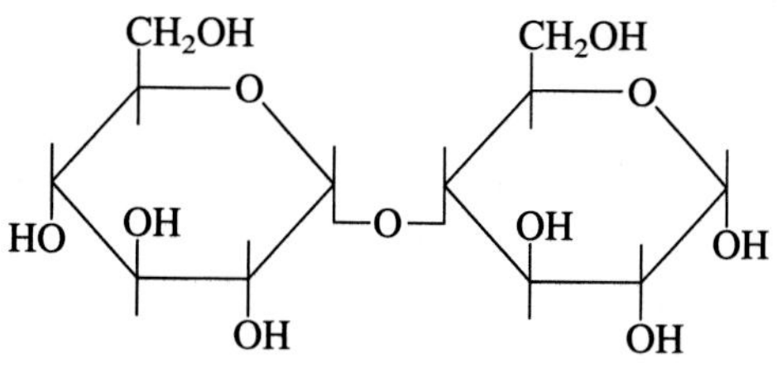

14.23 β-Cellobiose is a disaccharide obtained from the hydrolysis of cellulose. It is quite similar to maltose except it has a β-1,4-glycosidic bond. What is the structure of β-cellobiose?

Polysaccharides *(Goal 14.7)*

14.24 Indicate the structural features of the following polysaccharides: (a) the type of monosaccharide, (b) straight-chain or branched-chain, (c) the type(s) of glycosidic bond(s) in each.

Polysaccharide	*Monosaccharide α- or β-glucose*	*Straight- or branched-chain*	*Glycosidic bond(s) α-1,4, β-1,4, and/or α-1,6*
Amylose	______	______	______
Amylopectin	______	______	______
Glycogen	______	______	______
Cellulose	______	______	______

14.25 Identify the type of polysaccharide represented by each of the following segments:

a.

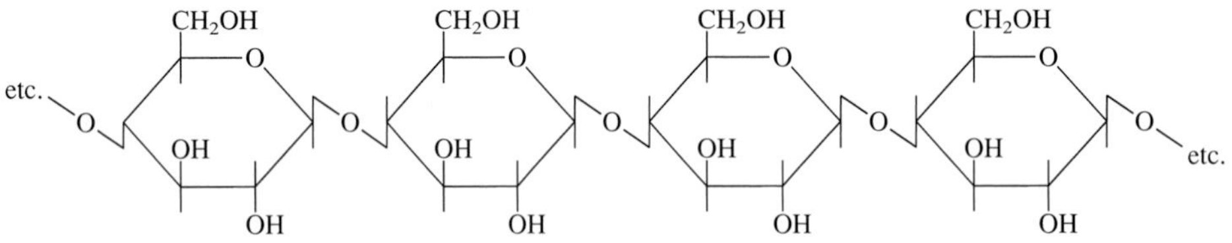

b.

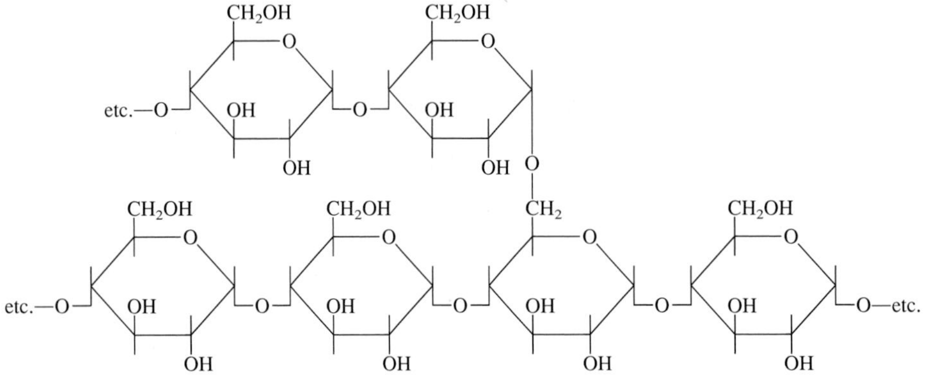

c.

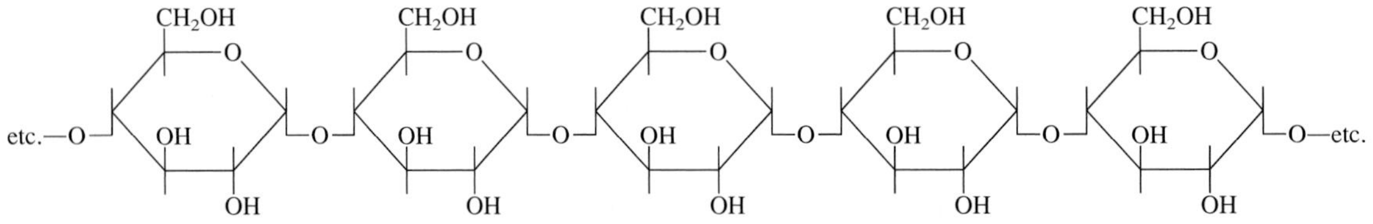

14.26 Why can humans digest starch but not cellulose when both contain D-glucose?

Tests for Carbohydrates *(Goal 14.8)*

14.27 Describe the test reagents, the sugars that react, and the results for the following carbohydrate tests:

a. Benedict's test
b. fermentation test
c. iodine test

14.28 Indicate the results of the following tests for the sugars indicated:

Carbohydrate	*Iodine test*	*Fermentation*	*Benedict's test*
Galactose			
Maltose			
Sucrose			
Starch			
Lactose			

14.29 Give the name of a carbohydrate for each of the following descriptions:

a. a disaccharide that is not a reducing sugar
b. a disaccharide that occurs as a breakdown product of starch
c. a sugar composed of glucose and fructose
d. the storage form of carbohydrate in plants
e. the storage form of carbohydrates in animals
f. a carbohydrate that is used for structure by plants
g. a monosaccharide that combines with glucose to form lactose
h. a monosaccharide that does not ferment
i. a disaccharide composed of two glucose units
j. a carbohydrate that is not digestible by humans

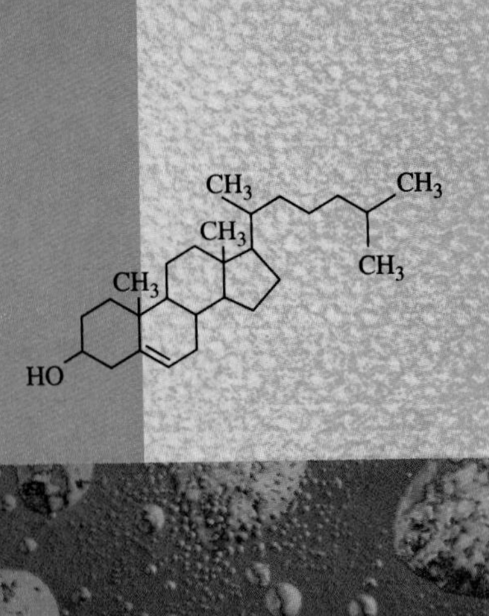

15

Lipids

Learning Goals

15.1 Identify a fatty acid as saturated or unsaturated. Predict whether that fatty acid is liquid or solid at room temperature.

15.2 Write the structural formula of a wax or a triglyceride produced when a fatty acid reacts with a long-chain alcohol or glycerol.

15.3 Given the structure or name of a triglyceride, describe melting points, write the products of hydrogenation, acid or enzyme hydrolysis, oxidation, and saponification.

15.4 Describe the components of phosphoglycerides, sphingolipids, and glycolipids.

15.5 Given the structural formula or name of a compound, determine if it is related to the terpene or steroid family of lipids.

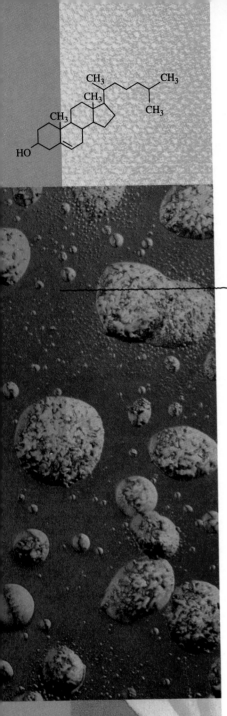

High levels of serum cholesterol have been associated in many studies with the accumulation of lipid deposits within the coronary arteries, restricting the flow of blood to the tissue.

SCOPE

Lipids are a family of biomolecules with seemingly unrelated structures. However, their common feature is that most are not very soluble in water. When we talk of fats, waxes, vegetable oils, steroids, or fat-soluble vitamins, we are discussing lipids. Complex lipids, such as triglycerides (fats) and phospholipids, can be hydrolyzed to simpler substances including fatty acids. Simple lipids, such as steroids and terpenes, do not contain fatty acids and cannot be hydrolyzed. In the body, that ever-present adipose tissue known as body fat is a stored form of energy that also provides protection and insulation for internal organs.

Today there is much concern about the effects of saturated fats and cholesterol in our diets. Many reports indicate a strong association between saturated fats and cholesterol levels and diseases such as diabetes and high blood pressure. Many researchers believe that a high level of cholesterol in the blood is related to arteriosclerosis, a condition in which deposits of lipid materials accumulate in the coronary blood vessels. These plaques restrict the flow of blood to the tissue, causing necrosis (death) of the tissue. In the heart, this could result in a myocardial infarction (heart attack).

Recent studies indicate that saturated fats may play an important role in certain cancers, particularly cancer of the breast and colon, and possibly of the pancreas, ovaries, and prostate. As a result of these findings, nutritionists and the American Institute for Cancer Research recommend diets with fewer calories, less fat and cholesterol, and more fiber and starch.

15.1 Fatty Acids

Learning Goal **Identify a fatty acid as saturated or unsaturated. Predict whether that fatty acid is liquid or solid at room temperature.**

Fatty acids are long-chain carboxylic acids that are insoluble in water. The fatty acids in fats and oils typically have an even number of carbon atoms; 12–18 carbon atoms are the most common in biological systems. **Saturated** fatty acids contain only single bonds, whereas **unsaturated** fatty acids have one or more double bonds. Table 15.1 lists some typical fatty acids in lipids.

Properties of Fatty Acids

The saturated fatty acids have straight-chain structures that allow their molecules to fit close together and form strong attractions. (See Figure 15.1.) As a result, they have high melting points because energy is needed to break the bonds between the molecules and melt the fatty acid. The melting point increases for longer carbon chains. Saturated fats are usually solids at room temperatures.

In contrast, most unsaturated fatty acids contain cis double bonds that cause one or more bends in the carbon chain. As a result, unsaturated fatty acids have irregular shapes, do not fit closely together, and have fewer bonds between the molecules. This makes their melting points so low that they are liquids at room temperature. In general, the melting point of an unsaturated fatty acid is lower as the number of double bonds increases.

Table 15.1 Some Fatty Acids Found in Fats

Number of carbon atoms	Structural formula	Melting point (°C)	Common name	Source
Saturated fatty acids				
12	$CH_3(CH_2)_{10}\overset{O}{\overset{\|}{C}}OH$	44	Lauric acid	Coconut
14	$CH_3(CH_2)_{12}\overset{O}{\overset{\|}{C}}OH$	54	Myristic acid	Nutmeg
16	$CH_3(CH_2)_{14}\overset{O}{\overset{\|}{C}}OH$	63	Palmitic acid	Palm
18	$CH_3(CH_2)_{16}\overset{O}{\overset{\|}{C}}OH$	70	Stearic acid	Animal fat
Unsaturated fatty acids				
18	$CH_3(CH_2)_7CH{=}CH(CH_2)_7\overset{O}{\overset{\|}{C}}OH$	14	Oleic acid	Olive, corn
18	$CH_3(CH_2)_4CH{=}CHCH_2CH{=}CH(CH_2)_7\overset{O}{\overset{\|}{C}}OH$	−5	Linoleic acid	Soybean, safflower, sunflower
18	$CH_3(CH_2)_4CH{=}CHCH_2CH{=}CHCH_2CH{=}CH(CH_2)_4\overset{O}{\overset{\|}{C}}OH$	−11	Linolenic acid	Herring, linseed

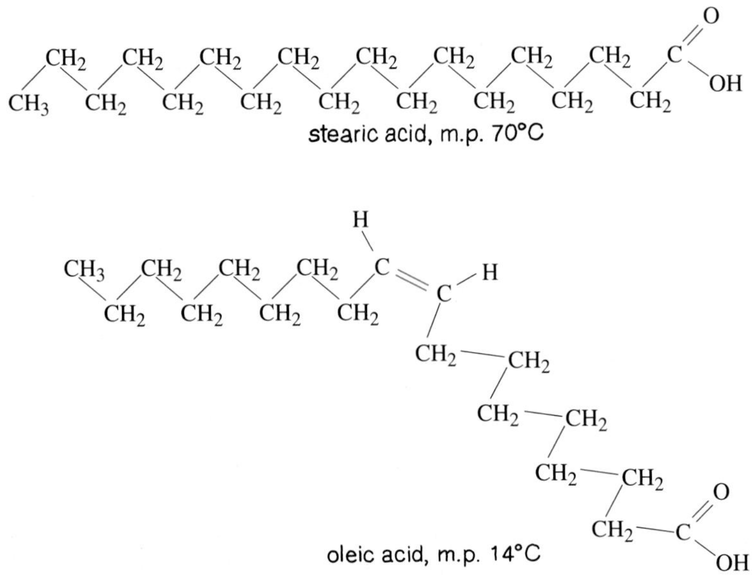

Figure 15.1
Saturated stearic acid has a straight chain, while unsaturated oleic acid has a bend in the carbon chain where the cis double bond occurs.

Essential Fatty Acids

The human body is capable of synthesizing most fatty acids from carbohydrates. However, humans and other mammals cannot synthesize fatty acids that have more than one double bond, such as linoleic acid and linolenic acid. These fatty acids are called essential fatty acids because they must be provided by the diet.

A deficiency of essential fatty acids can cause skin dermatitis in infants. However, the role of fatty acids in adult nutrition is not well understood. Adults do not usually have a deficiency of essential fatty acids.

Sample Problem 15.1
Structures and Properties of Fatty Acids

Consider the structural formula of oleic acid,

$$CH_3(CH_2)_7CH{=}CH(CH_2)_7\overset{\overset{\large O}{\|}}{C}OH$$

a. Why is the substance called an acid?
b. How many carbon atoms are in oleic acid?
c. Is it a saturated or unsaturated fatty acid?
d. Is it most likely to have a high or low melting point?
e. Would it be soluble in water?

Solution:

a. Oleic acid contains a carboxylic acid group.
b. It contains 18 carbon atoms.
c. It is an unsaturated fatty acid.
d. It has a rather low melting point.
e. No, its long hydrocarbon chain makes it insoluble in water.

Study Check:
Palmitoleic acid is fatty acid with the following formula:

$$CH_3(CH_2)_5CH{=}CH(CH_2)_7\overset{\overset{\large O}{\|}}{C}OH$$

a. How many carbon atoms are in palmitoleic acid?
b. Is it a saturated or unsaturated fatty acid?
c. Is it most likely to have a high or low melting point?

Answer: a. 16 b. unsaturated c. low (0.5°C.)

15.2 Waxes, Fats, and Oils

Learning Goal **Write the structural formula of a wax or a triglyceride produced when a fatty acid reacts with a long-chain alcohol or glycerol.**

HEALTH NOTE

Prostaglandins

Prostaglandins are hormone-like substances found in very low amounts in most tissues and fluids of the body. The parent structure, prostanoic acid, which has 20 carbon atoms, is formed from an unsaturated fatty acid called arachidonic acid. Small changes in the structure of prostanoic acid produce the different prostaglandins.

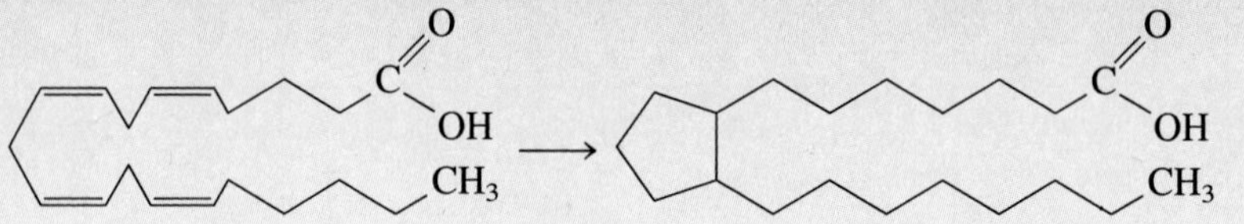

Arachidonic acid → Prostanoic acid

Prostaglandins produced in the tissues enter the blood, where they regulate several physiological activities. Some increase blood pressure as vasoconstrictors, and others lower blood pressure as vasodilators. There are also prostaglandins that stimulate contraction and relaxation of smooth muscle. Still others play a role in reproduction, respiration, fat metabolism, and nerve impulse transmission. Some also cause inflammation, pain, and fever. These effects are inhibited by anti-inflammatory agents such as aspirin and acetaminophen, which act by preventing the formation of the prostaglandins in the tissues.

The prostaglandins PGE_2 and $PGF_{2\alpha}$ are used medically to treat hypertension, to increase uterine contractions, and to relieve bronchial asthma.

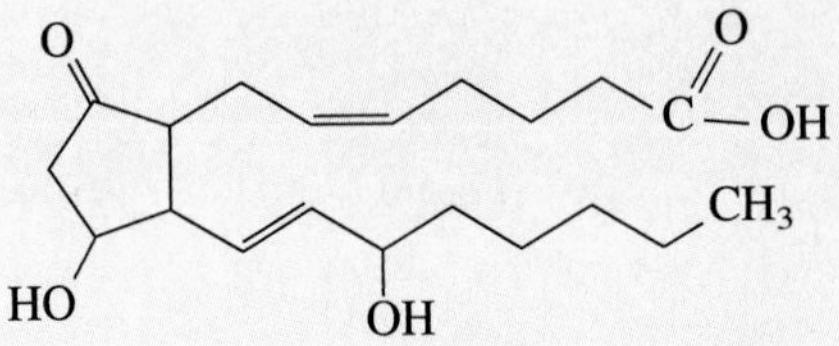

PGE_2

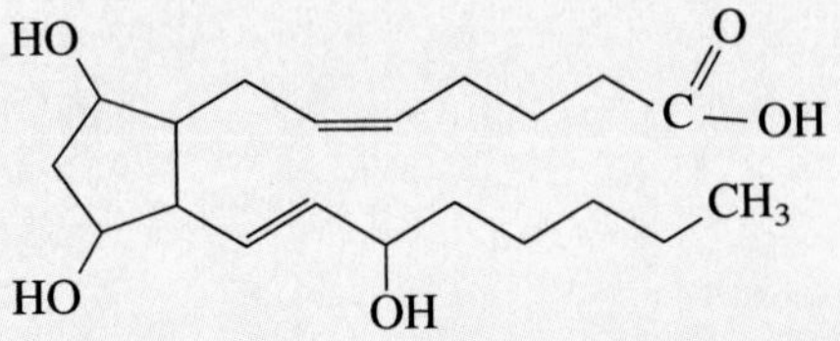

$PGF_{2\alpha}$

Waxes

A **wax** is an ester of a saturated fatty acid and a long-chain alcohol, each containing from 14 to 30 carbon atoms. A typical wax is a solid that melts easily.

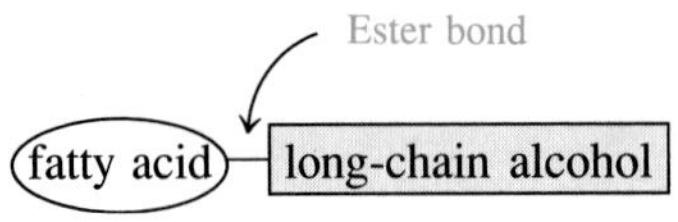

Ester bond

fatty acid — long-chain alcohol

Wax

Figure 15.2
Waxes are found in creams and lotions, furniture, car and shoe polishes, candles, and cosmetics.

Waxes are found in many plants and animals. Wax coatings on fruits and on the leaves and stems of plants help to prevent loss of water and damage from pests. Waxes on the skin, fur, and feathers of animals and birds provide a waterproof coating. Lanolin, a wax obtained from wool, is used in creams and lotions to aid retention of water, which softens the skin. Beeswax obtained from honeycombs, and carnauba wax obtained from palm trees, are used in furniture, car, and floor polishes, and waxes. The oil of the sperm whale contains a wax called spermaceti, which is used in making candles and cosmetics. (See Figure 15.2.) Table 15.2 provides examples of some of these waxes.

Table 15.2 Some Typical Waxes

Type	Structural formula	Source	Uses
Beeswax	$CH_3(CH_2)_{14}{-}\overset{O}{\overset{\|}{C}}{-}O{-}(CH_2)_{29}CH_3$	Honeycomb	Candles, shoe polish, wax paper
Carnauba wax	$CH_3(CH_2)_{24}{-}\overset{O}{\overset{\|}{C}}{-}O{-}(CH_2)_{29}CH_3$	Brazilian palm tree	Waxes for furniture, cars, floors, shoes
Spermaceti wax	$CH_3(CH_2)_{14}{-}\overset{O}{\overset{\|}{C}}{-}O{-}(CH_2)_{15}CH_3$	Sperm whale	Candles, soaps, cosmetics, ointments

Figure 15.3
Vegetable oils such as olive oil, corn oil, and safflower oil contain unsaturated fats. Olives contain 15–30% olive oil, found in both the skin and pulp of the small olive fruit. Corn oil, which makes up one-half of the germ portion of the corn kernel, floats to the surface when kernels are cracked and softened in water.

Fats and Oils

Fats and **oils,** the most prevalent form of lipids, include butterfat, lard, body fat, and vegetable oils such as olive oil and corn oil. See Figure 15.3. Also known as **triglycerides,** fats and oils are esters of glycerol and three fatty acids.

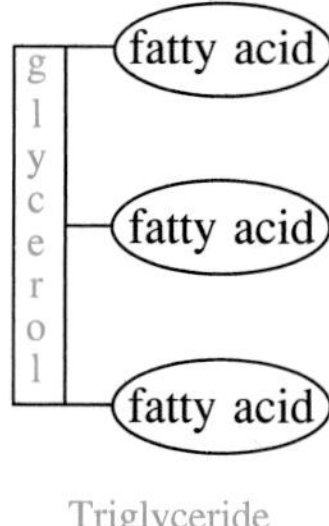

Triglyceride

For example, in tristearin, a simple triglyceride, three stearic acid molecules form ester bonds with glycerol.

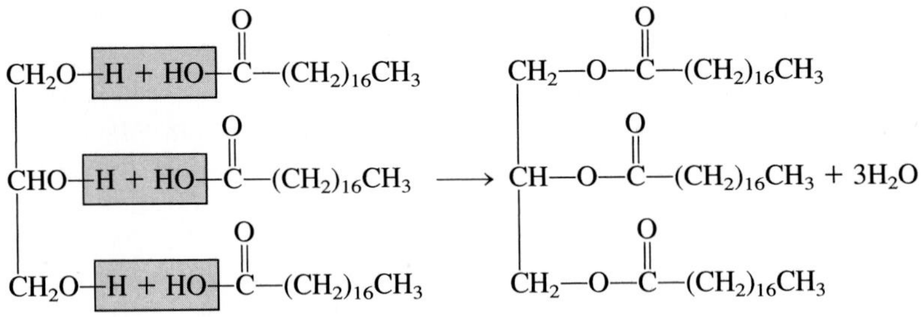

$$\begin{array}{l} CH_2O\text{—}\boxed{H + HO}\text{—}\overset{O}{\overset{\|}{C}}\text{—}(CH_2)_{16}CH_3 \\ \quad | \\ CHO\text{—}\boxed{H + HO}\text{—}\overset{O}{\overset{\|}{C}}\text{—}(CH_2)_{16}CH_3 \\ \quad | \\ CH_2O\text{—}\boxed{H + HO}\text{—}\overset{O}{\overset{\|}{C}}\text{—}(CH_2)_{16}CH_3 \end{array} \longrightarrow \begin{array}{l} CH_2\text{—}O\text{—}\overset{O}{\overset{\|}{C}}\text{—}(CH_2)_{16}CH_3 \\ \quad | \\ CH\text{—}O\text{—}\overset{O}{\overset{\|}{C}}\text{—}(CH_2)_{16}CH_3 \\ \quad | \\ CH_2\text{—}O\text{—}\overset{O}{\overset{\|}{C}}\text{—}(CH_2)_{16}CH_3 \end{array} + 3H_2O$$

Glycerol | 3 Stearic acid molecules | Tristearin, a fat

However, most fats and oils are **mixed triglycerides** that contain two or three different fatty acids. For example, a mixed triglyceride might be made from lauric acid, myristic acid, and palmitic acid. One possible structure for the mixed triglyceride follows:

A Mixed Triglyceride

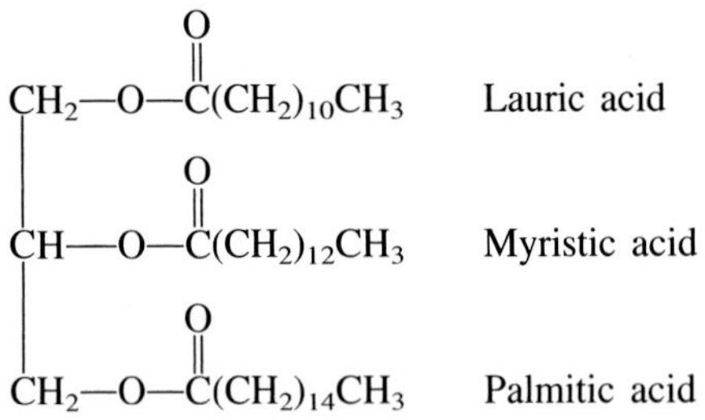

$$\begin{array}{ll} CH_2\text{—}O\text{—}\overset{O}{\overset{\|}{C}}(CH_2)_{10}CH_3 & \text{Lauric acid} \\ \quad | & \\ CH\text{—}O\text{—}\overset{O}{\overset{\|}{C}}(CH_2)_{12}CH_3 & \text{Myristic acid} \\ \quad | & \\ CH_2\text{—}O\text{—}\overset{O}{\overset{\|}{C}}(CH_2)_{14}CH_3 & \text{Palmitic acid} \end{array}$$

A mixed triglyceride

Sample Problem 15.2
Writing Structures for Triglycerides

Draw the structural formula of triolein, a simple triglyceride.

Solution:
Triolein is the triglyceride of glycerol and three oleic acid molecules. Each fatty acid is attached by an ester bond to one of the hydroxyl groups in glycerol.

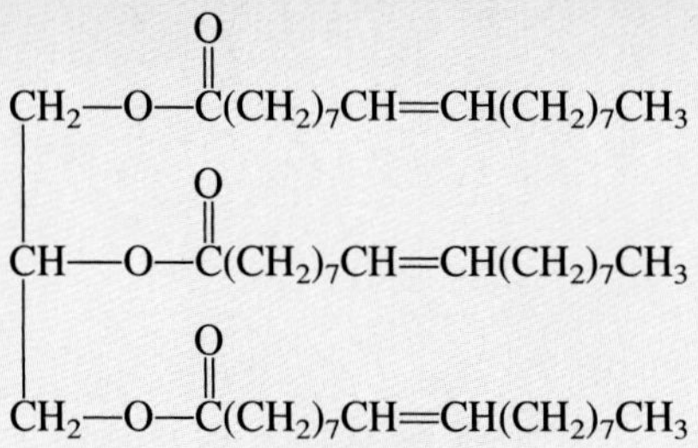

$$\begin{array}{l} CH_2\text{—}O\text{—}\overset{O}{\overset{\|}{C}}(CH_2)_7CH\text{=}CH(CH_2)_7CH_3 \\ \quad | \\ CH\text{—}O\text{—}\overset{O}{\overset{\|}{C}}(CH_2)_7CH\text{=}CH(CH_2)_7CH_3 \\ \quad | \\ CH_2\text{—}O\text{—}\overset{O}{\overset{\|}{C}}(CH_2)_7CH\text{=}CH(CH_2)_7CH_3 \end{array}$$

Triolein

Study Check:
Write the structure of trimyristin.

Answer:

$$
\begin{array}{l}
\quad\quad\quad\quad\quad\;\; O \\
\quad\quad\quad\quad\quad\;\; \| \\
CH_2-O-C(CH_2)_{12}CH_3 \\
| \quad\quad\quad\quad\quad O \\
| \quad\quad\quad\quad\quad \| \\
CH-O-C(CH_2)_{12}CH_3 \\
| \quad\quad\quad\quad\quad O \\
| \quad\quad\quad\quad\quad \| \\
CH_2-O-C(CH_2)_{12}CH_3
\end{array}
$$

HEALTH NOTE

Nondairy Creamers Can Boost Cholesterol

Some persons concerned about their cholesterol intake have switched to nondairy creamers. However, many popular nondairy creamers are made from coconut oil (check the label), which is a highly saturated fat; in fact, it is more saturated than cream. Consumption of coconut oil in the nondairy creamer has the same effect as consuming any other saturated fat; it raises the cholesterol level. Even whole milk raises cholesterol levels less than a nondairy creamer made from coconut oil.

The typical amount of nondairy creamer used in a cup of coffee is about 1 tablespoon, which contains about 2 g of saturated fat. A person who drinks 5 cups of coffee in a day obtains 10 g of saturated fat just from the nondairy creamer. That is half of the daily recommended amount of saturated fats, which is about 20 g. Liquid nondairy creamers made from soybean oil are less saturated, but they still contain the same amount of fat. Suggestions for a low-fat cream substitute include powdered milk, evaporated milk, low-fat or skim milk, and even whole milk. ■

15.3 Properties of Triglycerides

Learning Goal **Given the structure or name of a triglyceride, describe melting points, write the products of hydrogenation, acid or enzyme hydrolysis, oxidation, and saponification.**

Melting Points of Triglycerides

Fats and oils are both composed of triglycerides. The term fat refers to a solid triglyceride, usually from animal sources such as meat, whole milk, butter and margarine, eggs, and cheese. A few triglycerides produced by plants such as palm oil and coconut oil contain mostly saturated fatty acids. However, the molecules are shorter in length, which makes coconut oil a liquid at room temperature. The percentage of fat in some selected foods and the comparison of the saturated and unsaturated fatty acids in that fat are shown in Figure 15.4.

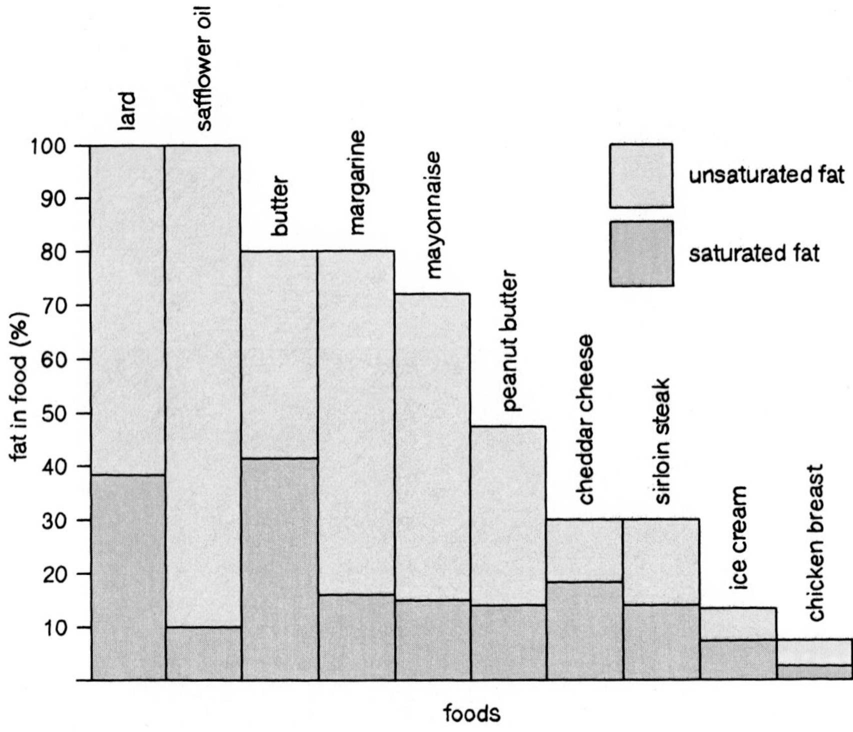

Figure 15.4
Percent total fat and saturated and unsaturated fatty acids in some selected foods.

The term **oil** refers to liquid triglycerides from plant sources. Olive oil and peanut oil are monounsaturated oils because they contain large amounts of oleic acid, which has one double bond. Oils from corn, cottonseed, safflower, sunflower, and sesame seeds are polyunsaturated oils, because they contain large amounts of fatty acid with two or more double bonds. See Table 15.3.

Table 15.3 Typical Saturated & Unsaturated Fatty-Acid Composition of Some Common Oils

		% Fatty acid by weight		
Source	**Mp**	**Saturated**	**Monounsaturated**	**Polyunsaturated**
Animal fats				
Butterfat	32	59	31	10
Lard	31	43	47	10
Human fat	15	35	52	13
Vegetable oils				
Coconut oil	25	91	8	1
Palm oil	35	47	43	10
Olive oil	−6	10	85	5
Corn oil	−20	15	50	34
Soybean oil	−16	13	29	58
Safflower oil	−18	7	19	74

Animal fats usually contain more saturated fatty acids than do vegetable oils. Since saturated fatty acids have higher melting points than unsaturated fatty acids, the melting points of saturated fats are also higher. They are usually solid at room temperature. Since an oil contains more unsaturated fatty acids, it is usually a liquid at room temperature. The melting points of some common fats and oils and their fatty acid composition are listed in Table 15.3.

Hydrogenation

The **hydrogenation** of unsaturated fatty acids in triglycerides converts double bonds to single, saturated bonds by the addition of hydrogen. The hydrogen is bubbled through hot oil in the presence of a nickel catalyst.

$$-CH{=}CH- + \boxed{H_2} \xrightarrow{Ni} -\overset{\boxed{H\ \ H}}{\underset{H\ \ H}{C-C}}-$$

$$\begin{array}{l} CH_2-O-\overset{O}{\overset{\|}{C}}-(CH_2)_7CH{=}CH(CH_2)_7CH_3 \\ | \\ CH-O-\overset{O}{\overset{\|}{C}}-(CH_2)_7CH{=}CH(CH_2)_7CH_3 \\ | \\ CH_2-O-\overset{O}{\overset{\|}{C}}-(CH_2)_7CH{=}CH(CH_2)_7CH_3 \end{array} + 3H_2 \xrightarrow{Ni} \begin{array}{l} CH_2-O-\overset{O}{\overset{\|}{C}}-(CH_2)_{16}CH_3 \\ | \\ CH-O-\overset{O}{\overset{\|}{C}}-(CH_2)_{16}CH_3 \\ | \\ CH_2-O-\overset{O}{\overset{\|}{C}}-(CH_2)_{16}CH_3 \end{array}$$

Triolein (unsaturated) Tristearin (saturated)

In commercial hydrogenation, the addition of hydrogen is stopped before all the double bonds in an oil become saturated. Complete hydrogenation gives a very brittle product, while the partial hydrogenation of a liquid vegetable oil changes it to a soft, semisolid fat. As the fat becomes more saturated, the melting point increases. Control of the degree of hydrogenation gives the various types of partially hydrogenated vegetable oil products on the market today—soft margarines, solid stick margarines, and solid shortenings. (See Figure 15.5.) Although these products now contain more saturated fatty acids than the original oils, they contain no cholesterol, unlike similar products from animal sources such as butter and lard.

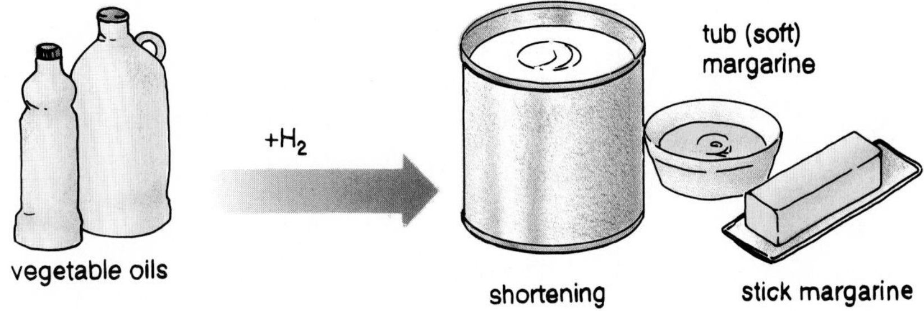

Figure 15.5
Soft margarines, stick margarines, and solid shortenings are produced by the partial hydrogenation of vegetable oils.

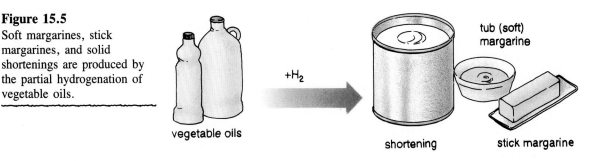

Hydrolysis

Triglycerides are hydrolyzed (split by water) in the presence of strong acids or digestive enzymes called **lipases.** The products of **hydrolysis** of the ester bonds are glycerol and three fatty acids. The polar glycerol is soluble in water, but the fatty acids with their long hydrocarbon chains are not.

Water adds to ester bonds

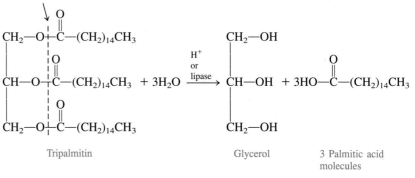

$$\begin{array}{l} CH_2{-}O{-}\overset{O}{\overset{\|}{C}}{-}(CH_2)_{14}CH_3 \\ | \\ CH{-}O{-}\overset{O}{\overset{\|}{C}}{-}(CH_2)_{14}CH_3 \\ | \\ CH_2{-}O{-}\overset{O}{\overset{\|}{C}}{-}(CH_2)_{14}CH_3 \end{array} + 3H_2O \xrightarrow[\text{lipase}]{H^+ \text{ or}} \begin{array}{l} CH_2{-}OH \\ | \\ CH{-}OH \\ | \\ CH_2{-}OH \end{array} + 3HO{-}\overset{O}{\overset{\|}{C}}{-}(CH_2)_{14}CH_3$$

Tripalmitin — Glycerol — 3 Palmitic acid molecules

Sample Problem 15.3
Products of Fat Hydrolysis

Write the equation for the reaction of the enzyme lipase that hydrolyzes trilaurin during the digestion process.

Solution:
The fat trilaurin is an ester of glycerol and three molecules of lauric acid,

$$CH_3(CH_2)_{10}\overset{O}{\overset{\|}{C}}OH.$$

$$\begin{array}{l} CH_2{-}O{-}\overset{O}{\overset{\|}{C}}{-}(CH_2)_{10}CH_3 \\ | \\ CH{-}O{-}\overset{O}{\overset{\|}{C}}{-}(CH_2)_{10}CH_3 \\ | \\ CH_2{-}O{-}\overset{O}{\overset{\|}{C}}{-}(CH_2)_{10}CH_3 \end{array} + 3H_2O \xrightarrow[\text{lipase}]{H^+ \text{ or}} \begin{array}{l} CH_2{-}OH \\ | \\ CH{-}OH \\ | \\ CH_2{-}OH \end{array} + 3HO{-}\overset{O}{\overset{\|}{C}}{-}(CH_2)_{10}CH_3$$

Trilaurin — Glycerol — 3 Lauric acid molecules

Study Check:
What is the name of the product formed when a triglyceride containing oleic acid and linoleic acid is completely hydrogenated?

Answer: tristearin

Saponification

When a fat is heated with a strong base such as sodium hydroxide, **saponification** of the fat gives glycerol and the sodium salts of the fatty acids, which are soaps. When NaOH is used, a solid soap is produced that can be molded into a desired shape; KOH produces a softer, liquid soap. The softness of the soap is also related to the amount of unsaturation of the fatty acids. Oils that are polyunsaturated produce softer soaps. Names like ''coconut'' or ''avocado shampoo'' tell you the source of the oil used in the hydrolysis reaction.

$$\text{Fat or oil} + \text{Strong base} \longrightarrow \text{Glycerol} + \text{Soaps (salts of fatty acids)}$$

$$\begin{array}{l} CH_2-O-\overset{\overset{\displaystyle O}{\|}}{C}(CH_2)_{16}CH_3 \\ | \\ CH-O-\overset{\overset{\displaystyle O}{\|}}{C}(CH_2)_{16}CH_3 \\ | \\ CH_2-O-\overset{\overset{\displaystyle O}{\|}}{C}(CH_2)_{16}CH_3 \end{array} + 3NaOH \longrightarrow \begin{array}{l} CH_2-OH \\ | \\ CH-OH \\ | \\ CH_2-OH \end{array} + 3Na^+ \; {}^-O-\overset{\overset{\displaystyle O}{\|}}{C}-(CH_2)_{16}CH_3$$

Tristearin — Sodium hydroxide — Glycerol — Sodium stearate, a soap

Oxidation

A fat or oil becomes rancid when its double bonds are oxidized by oxygen and microorganisms. The oxidation products are short-chain fatty acids and aldehydes that have very disagreeable odors.

$$-CH{=}CH- \xrightarrow{\text{Oxidation}} -\overset{\overset{\displaystyle O}{\|}}{C}-H + H-\overset{\overset{\displaystyle O}{\|}}{C}- \xrightarrow{\text{Oxidation}} -\overset{\overset{\displaystyle O}{\|}}{C}-OH + HO-\overset{\overset{\displaystyle O}{\|}}{C}-$$

Unsaturated fatty acids — Short-chain aldehydes — Short-chain carboxylic acids

Figure 15.6
After strenuous exercise, increased body temperature and perspiration promote the oxidation of oils that accumulate on the surface of the skin.

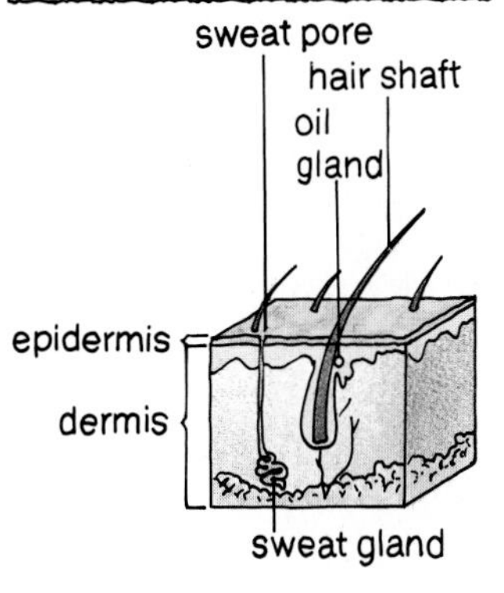

If a vegetable oil has no antioxidant present, such as vitamin E, BHA, or BHT, it will oxidize easily. You can detect an oil that has become rancid by its unpleasant odor. If an oil is covered tightly and stored in a refrigerator, the process of oxidation can be slowed down and the oil will last longer.

Oxidation also occurs in the oils that accumulate on the surface of the skin during heavy exercise. (See Figure 15.6.) The relatively high temperature of the body and the presence of microorganisms on the skin promote rapid oxidation of these oils as they are exposed to oxygen and water. The resulting short-chain aldehydes and acids account for the odors associated with workouts and heavy perspiration.

HEALTH NOTE

Cleaning Action of Soap

Because a soap is the salt of a long-chain fatty acid, the two ends of a soap molecule have very different polarities. The long carbon chain end is nonpolar and hydrophobic (water-fearing). It is soluble in nonpolar substances such as oil or grease, but it is not soluble in water. The carboxylate salt end is polar and hydrophilic (water-loving). It is very soluble in water, but not in oils or grease.

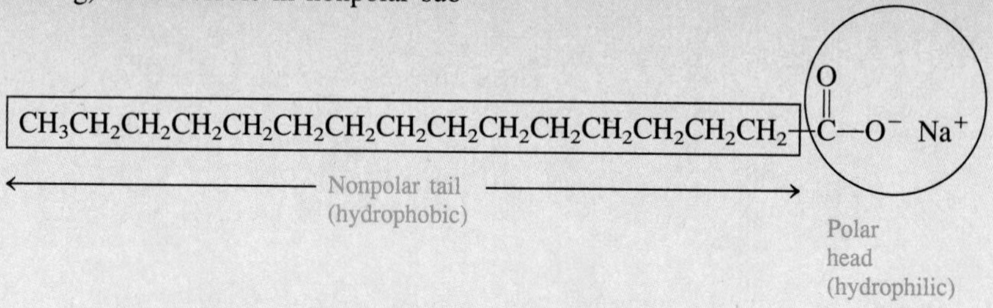

$$\underbrace{CH_3CH_2CH_2CH_2CH_2CH_2CH_2CH_2CH_2CH_2CH_2CH_2CH_2CH_2CH_2CH_2}_{\text{Nonpolar tail (hydrophobic)}}\underbrace{-\overset{\overset{\displaystyle O}{\|}}{C}-O^- \ Na^+}_{\text{Polar head (hydrophilic)}}$$

The dual polarity of a soap (salt of a fatty acid)

When a soap is used to clean grease or oil, the nonpolar carbon chains of the soap molecules dissolve in the nonpolar fats and oils, while the polar carboxylate ends dissolve in water. In this way, soap molecules coat the oil or grease, forming clusters called micelles. Because the outer surfaces of the micelles are hydrophilic, small globules of oil and fat are dispersed in the water layer and can be rinsed away. (See Figure 15.7.)

One of the problems of using soaps is that the polar carboxylate group reacts with ions in hard water such as Ca^{2+}, Mg^{2+}, and Fe^{3+} and forms water-insoluble substances.

$$\underset{\text{Stearate ion}}{2CH_3(CH_2)_{16}\overset{\overset{\displaystyle O}{\|}}{C}-O^-} + \underset{\text{Magnesium ion in water}}{Mg^{2+}} \longrightarrow \underset{\text{Magnesium stearate (insoluble)}}{(CH_3(CH_2)_{16}\overset{\overset{\displaystyle O}{\|}}{C}-O^-)_2\ Mg^{2+}}$$

These insoluble products produce a dull coating or scum on fabrics and dishes. To solve this problem, detergents that do not form insoluble products with hard water were developed. One type of detergent consists of long nonpolar hydrocarbon chains attached to a sulfate group that form soluble calcium and magnesium salts. They are also biodegradable, which means they break down into nonpolluting products in the environment, as do soaps.

$$\underbrace{CH_3CH_2CH_2CH_2CH_2CH_2CH_2CH_2CH_2CH_2CH_2CH_2}_{\text{(hydrophobic)}}\underbrace{-O-\overset{\overset{\displaystyle O}{\|}}{\underset{\underset{\displaystyle O}{\|}}{S}}-O^-\ Na^+}_{\text{(hydrophilic)}}$$

Sodium lauryl sulfate

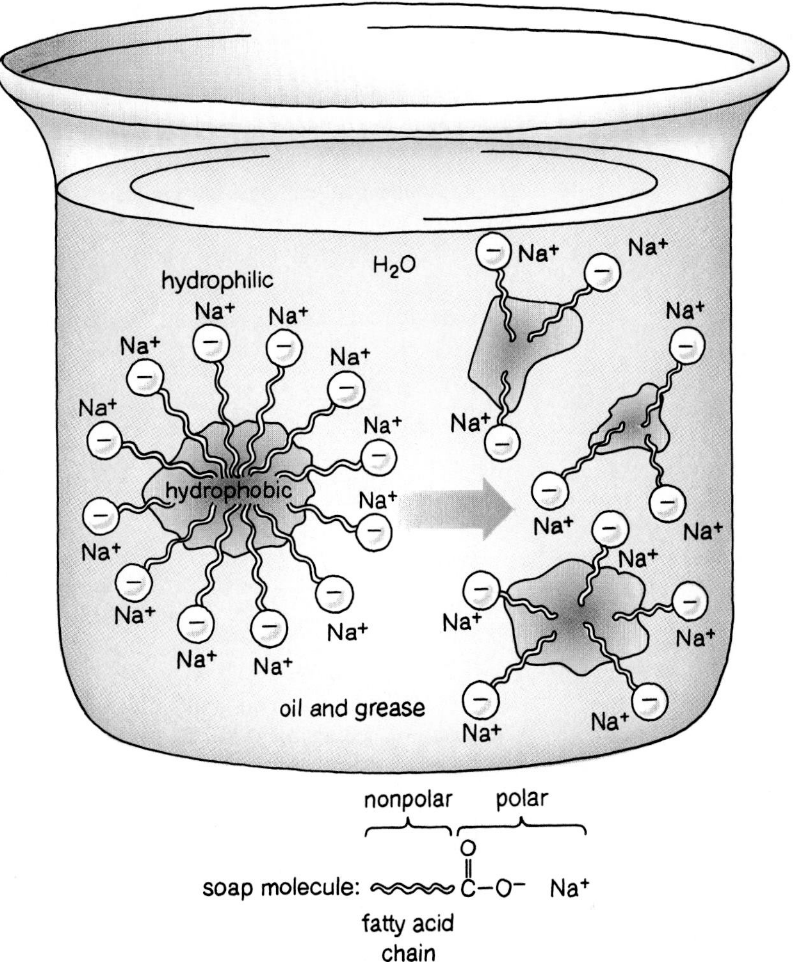

Figure 15.7
The cleaning action of soap. The nonpolar portion of the soap molecule dissolves in the grease and oil accompanying dirt on clothing and dishes. The polar ends of the soap molecules are attracted to water and pull the grease or oil into the aqueous solution.

15.4 Phospholipids

Learning Goal **Describe the components of phosphoglycerides, sphingolipids, and glycolipids.**

The phospholipids are a family of lipids containing glycerol, fatty acids and a phosphate group. Unlike most other lipids, phospholipids contain polar and nonpolar sections. This enables them to play a role in the transport of less soluble lipids through the bloodstream. They are also major components in the structure of cell membranes.

Phosphoglycerides

Phosphoglycerides are the most abundant lipids in cell membranes, where they play an important role in cellular permeability. They make up much of the myelin sheath that protects the nerve cells. In the body fluids, they combine with the less polar triglycerides and cholesterol to make them more soluble as they are transported in the body.

Like the triglycerides, phosphoglycerides contain glycerol and fatty acids. They also contain a phosphate group and an amino alcohol. Different fatty acids, saturated and unsaturated, are found in phosphoglycerides.

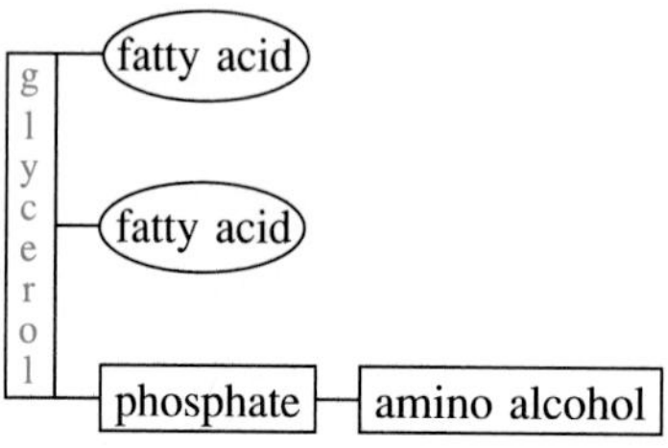

Phosphoglyceride

The most common amino alcohols found in phosphoglycerides are choline, serine, and ethanolamine.

$$HO{-}CH_2{-}CH_2{-}\overset{+}{N}(CH_3)_3$$

Choline

$$HO{-}CH_2{-}CH(\overset{+}{N}H_3){-}COO^-$$

Serine

$$HO{-}CH_2{-}CH_2{-}\overset{+}{N}H_3$$

Ethanolamine

Lecithins and **cephalins** are two types of phosphoglycerides that are particularly abundant in brain and nerve tissues as well as egg yolks, wheat germ, and yeast.

Lecithins contain choline, and cephalins contain ethanolamine, and sometimes serine. In the following structural formulas, palmitic and stearic acids are used. The ionization of the phosphate and amino group provides a polar section that makes phosphoglycerides such as lecithin and cephalin soluble in water.

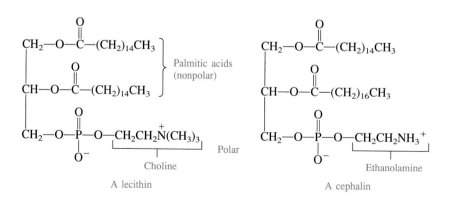

$$\begin{array}{l} CH_2{-}O{-}\overset{O}{\overset{\|}{C}}{-}(CH_2)_{14}CH_3 \\ | \\ CH{-}O{-}\overset{O}{\overset{\|}{C}}{-}(CH_2)_{14}CH_3 \\ | \\ CH_2{-}O{-}\overset{O}{\overset{\|}{P}}(O^-){-}O{-}CH_2CH_2\overset{+}{N}(CH_3)_3 \end{array}$$

Palmitic acids (nonpolar)

Polar

Choline

A lecithin

$$\begin{array}{l} CH_2{-}O{-}\overset{O}{\overset{\|}{C}}{-}(CH_2)_{14}CH_3 \\ | \\ CH{-}O{-}\overset{O}{\overset{\|}{C}}{-}(CH_2)_{16}CH_3 \\ | \\ CH_2{-}O{-}\overset{O}{\overset{\|}{P}}(O^-){-}O{-}CH_2CH_2NH_3^+ \end{array}$$

Ethanolamine

A cephalin

Sample Problem 15.4
Drawing Phospholipid Structures

Draw the structure of cephalin, using stearic acid for the fatty acids and serine for the amino alcohol. Describe each of the components in the phosphoglyceride.

Solution:
In general, phosphoglycerides are composed of a glycerol molecule in which two carbon atoms are attached to fatty acids such as stearic acid. The third carbon atom is attached in an ester bond to phosphate linked to an amino alcohol. In this example, the amino alcohol is serine.

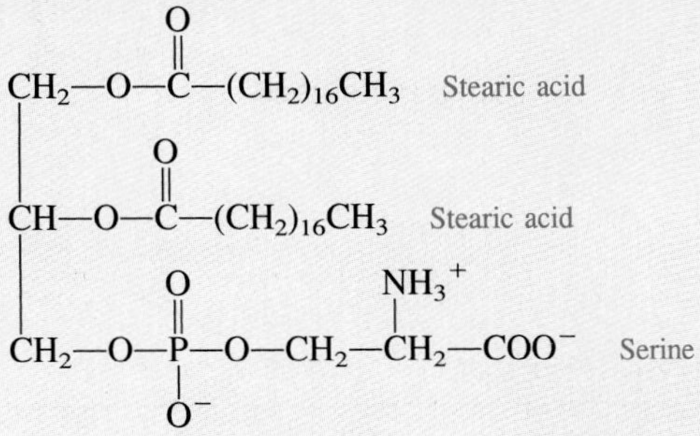

$$
\begin{array}{l}
CH_2{-}O{-}\overset{\overset{\displaystyle O}{\|}}{C}{-}(CH_2)_{16}CH_3 \quad \text{Stearic acid} \\
|\\
CH{-}O{-}\overset{\overset{\displaystyle O}{\|}}{C}{-}(CH_2)_{16}CH_3 \quad \text{Stearic acid} \\
|\\
CH_2{-}O{-}\underset{\underset{\displaystyle O^-}{|}}{\overset{\overset{\displaystyle O}{\|}}{P}}{-}O{-}CH_2{-}\overset{\overset{\displaystyle NH_3^+}{|}}{CH}{-}COO^- \quad \text{Serine}
\end{array}
$$

Cephalin

Study Check:
Phospholipids containing saturated fatty acids can pack together tightly to form rigid membranes, while phospholipids with unsaturated fatty acids form more flexible membranes. Explain.

Answer: Saturated fatty acids have a straight-chain structure that permits the phospholipids to line up close together. However, because unsaturated fatty acids are bent at the double bonds, they would have a nonregular, looser arrangement in a membrane. Such a membrane is more flexible.

One of the functions of phospholipids in cellular membranes is to separate one fluid compartment from another. Every phospholipid has both a polar and nonpolar portion, which means that one end is attracted to water while the other end is repelled. A current model called the **fluid-mosaic model** suggests that a cell membrane consists of a double row of phospholipid molecules called the bilipid layer. To form this bilipid layer, a double row of phospholipids are aligned with their nonpolar, hydrophobic (water-fearing) hydrocarbon chains at the center. The polar, hydrophilic (water-loving) groups that can hydrogen bond with water form the surfaces of the membrane.

This bilipid layer acts as a barrier that separates the contents inside a membrane from the surrounding fluids. It provides selective permeability, allowing nonpolar molecules such as oxygen and carbon dioxide to pass through, but is impermeable to ions and most polar molecules.

Embedded in the bilipid layer are proteins that provide tunnels through which ions and polar molecules can pass. Proteins on the surface of the bilipid layer also act as receptors for hormones, neurotransmitters, antibiotics, and other chemicals that modify cellular activities. (See Figure 15.8.)

Figure 15.8
Fluid mosaic model of a cell membrane. A double row of phospholipids form a barrier with polar ends in contact with fluids and nonpolar ends at the center away from the water. Proteins embedded in the bilipid layer permit passage of polar substances and act as receptors.

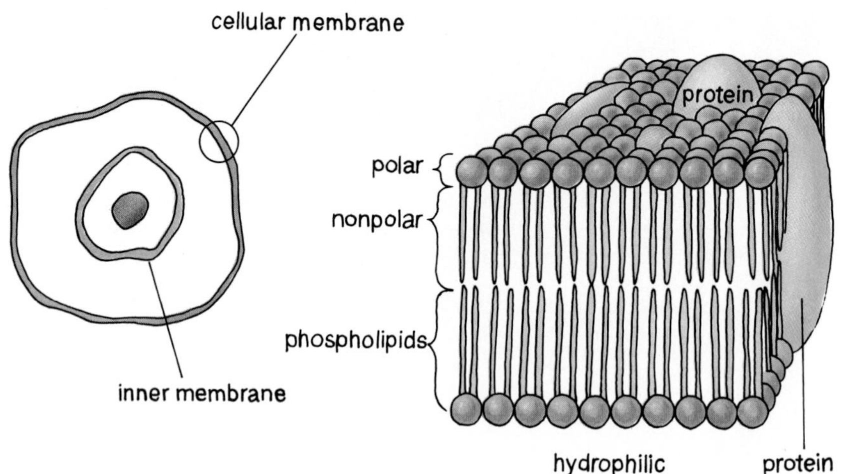

Sphingolipids

The **sphingolipids** are a group of phospholipids that are abundant in brain and nerve tissues. The sphingolipids are esters of the amino alcohol sphingosine instead of glycerol.

$$\begin{array}{l} CH_3(CH_2)_{12}-CH=CH-CH-OH \\ \qquad\qquad\qquad\qquad\qquad\quad | \\ \qquad\qquad\qquad\qquad\qquad\quad CH-NH_2 \\ \qquad\qquad\qquad\qquad\qquad\quad | \\ \qquad\qquad\qquad\qquad\qquad\quad CH_2-OH \end{array}$$

Sphingosine

One of the most abundant sphingolipids is sphingomyelin. It is the white matter of the myelin sheath, a coating surrounding the nerve cells that increases the speed of nerve impulses and insulates and protects the nerve cells. In sphingomyelin, the sphingosine is linked by an amide bond to a fatty acid, and to a phosphate ester of choline, an amino alcohol.

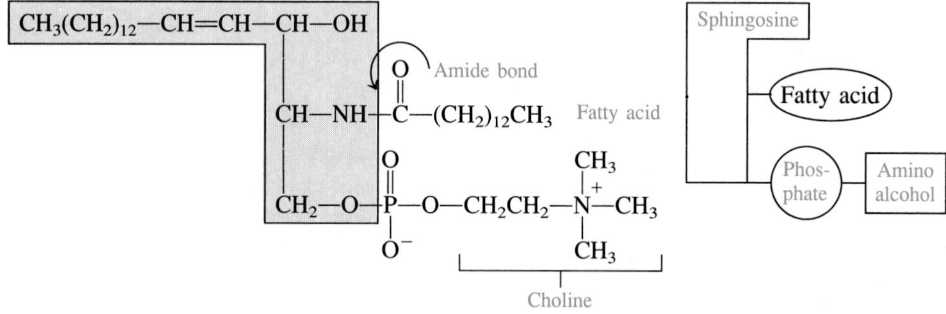

Sphingomyelin, a sphingolipid

Glycolipids

Glycolipids are abundant in the brain and in the myelin sheaths of nerves. Some glycolipids contain glycerol attached to two fatty acids and a monosaccharide, usually galactose.

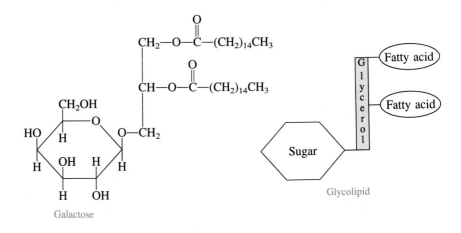

$$\begin{array}{l} CH_2-O-\overset{\displaystyle O}{\overset{\|}{C}}-(CH_2)_{14}CH_3 \\ | \\ CH-O-\overset{\displaystyle O}{\overset{\|}{C}}-(CH_2)_{14}CH_3 \\ | \\ CH_2-O-\text{galactose} \end{array}$$

Galactose

Glycolipid

Glycerol — Fatty acid, Fatty acid; Sugar

Cerebrosides, another type of glycolipid, are similar to the sphingolipids. Cerebrosides contain sphingosine, a fatty acid with 24 carbon atoms, and a monosaccharide, which is usually D-galactose and sometimes D-glucose. Gangliosides are similar to cerebrosides but contain two monosaccharides, such as glucose and galactose. Gangliosides are important in cell membranes as receptors for hormones, viruses, and several drugs.

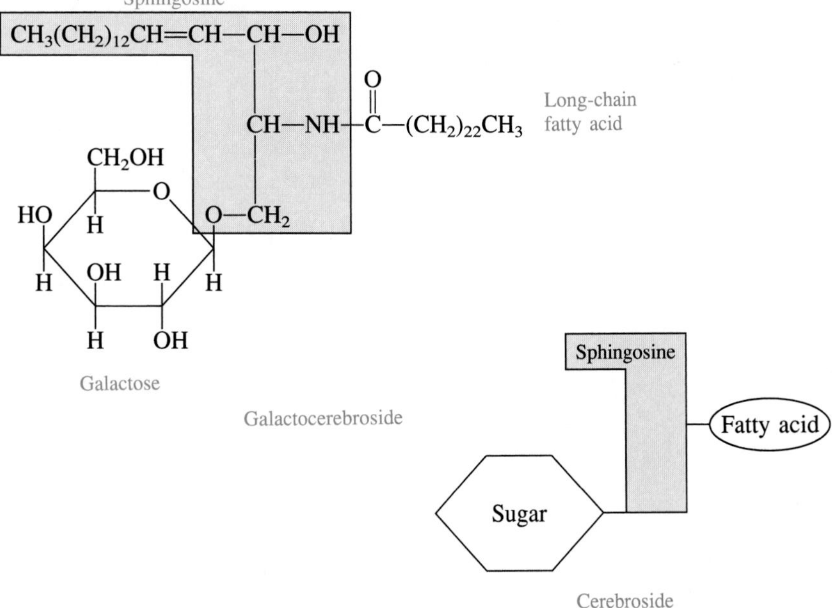

HEALTH NOTE

Lipid Diseases (Lipidoses)

Lipid diseases, or **lipidosis,** involves the excessive accumulation of a phospholipid, sphingolipid, or ganglioside in the brain, liver, and spleen. The lipid accumulates because there is a deficiency or absence of an enzyme needed for their breakdown.

In Gaucher's disease, there is an accumulation of a glycolipid called glucocerebroside. An enzyme known as β-glucosidase, needed to hydrolyze glucose from the glycolipid, is absent. As the glycolipid accumulates, the spleen and liver become enlarged. The cerebroside also accumulates in the bone marrow, causing an enlargement of the bone-marrow cells—an effect that is used as a diagnostic tool.

In Niemann-Pick disease, phosphatidylcholine and sphingomyelin accumulate in the spleen and liver, which become enlarged. In this disease, there is a deficiency of the enzyme called sphingomyelinase, which is needed for the hydrolysis of sphingomyelin. The accumulation of sphingomyelins in the brain leads to mental retardation and death in early infancy.

In Tay-Sachs disease, gangliosides containing glucose and galactose accumulate in the brain. In Tay-Sachs disease, the gangliosides cannot be broken down because the enzyme that does this, called hexosaminidase, is absent. The accumulation of the gangliosides in the brain causes death within the first year of life. ■

15.5 Terpenes and Steroids

Learning Goal **Given the structural formula or name of a compound, determine if it is related to the terpene or steroid family of lipids.**

Terpenes and steroids are two important families of lipids that do not contain fatty acids. They are included in the lipid family because they are not soluble in water.

Terpenes

Terpenes are lipids found in the oils of some plants and flowers that give them their characteristic odors and colors. For example, the smell of a geranium, mint, a lemon or pine tree, and the colors of carrots and tomatoes are due to terpenes.

Terpenes are composed of two or more sections containing five carbon atoms related to isoprene.

$$CH_2{=}\overset{\overset{\displaystyle CH_3}{|}}{C}{-}CH{=}CH_2$$

Isoprene

or (skeletal formula of isoprene, with CH_3 branch)

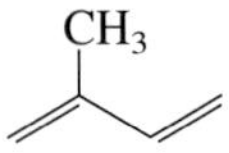

In the following terpenes, two or more isoprene units are present. Dashed lines are used to separate the isoprene units.

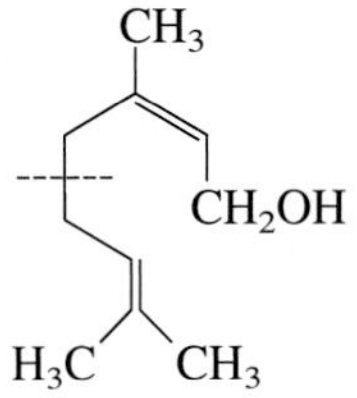

CH_3 CH_2OH H_3C CH_3

Geraniol
(roses)

CH_3 H_2C C CH_3

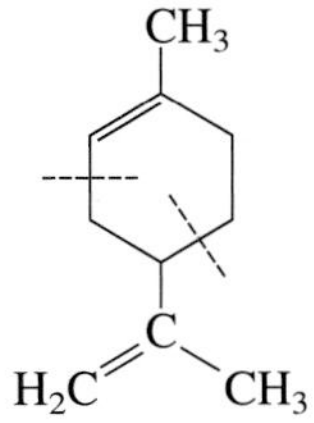

Limonene
(lemon, oranges)

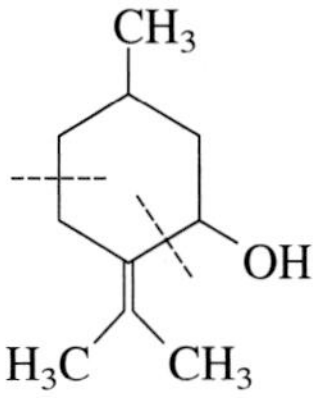

CH_3 OH H_3C CH_3

Menthol
(mint)

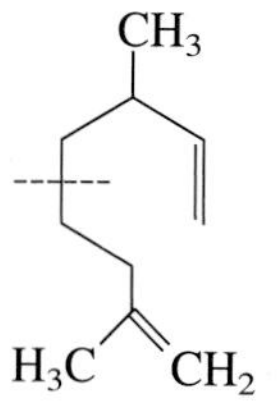

CH_3 H_3C CH_2

Ocimene
(basil)

O CH CH_3 H_3C CH_3

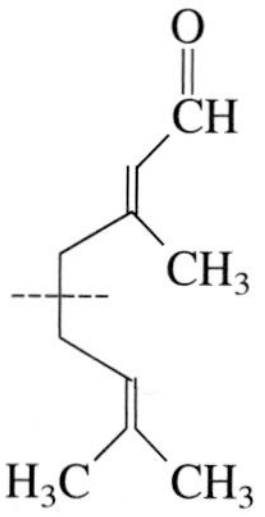

Citral
(lemongrass)

Vitamins A, E, and K

The fat-soluble vitamins such as vitamins A, E, and K are derived from terpenes. Vitamin A is required for the formation of a pigment in the retina necessary for night vision.

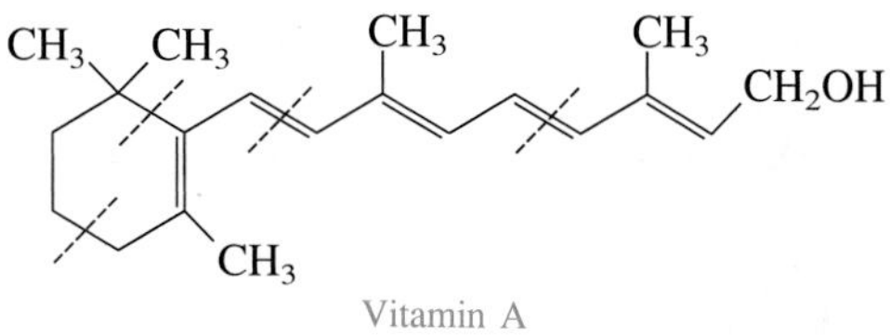

CH_3 CH_3 CH_3 CH_3 CH_2OH CH_3

Vitamin A
(retinol)

Vitamins E and K belong to a subgroup of the terpenes called tocopherols because they have aromatic ring structures attached to a side chain of isoprene units. Vitamin E may prevent the oxidation of unsaturated fatty acids in cellular membranes. Vitamin K is needed in the formation of prothrombin for blood clotting. A further discussion of vitamins is found in Chapter 17.

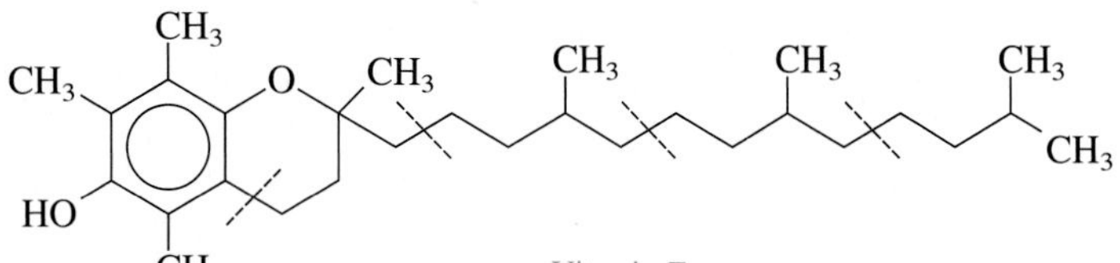

CH_3 CH_3 O CH_3 CH_3 CH_3 CH_3 CH_3 HO CH_3

Vitamin E

Steroids

The word steroid comes from the Latin word *stereos,* meaning "solid." **Steroids** are compounds consisting of a fused-ring structure called the steroid nucleus. By attaching various groups to the steroid nucleus, a variety of steroid compounds are formed. These include cholesterol, one of the most important steroid compounds in the body, male and female sex hormones, steroid hormones, bile acids, and vitamin D.

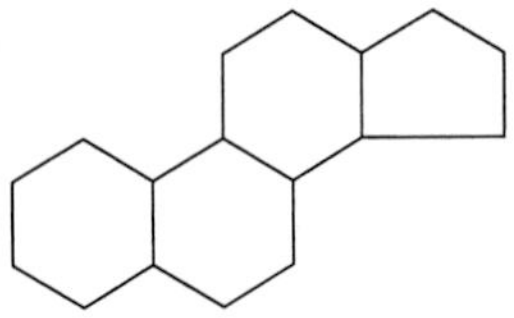

Steroid nucleus

Cholesterol

Cholesterol is called a sterol because it contains an alcohol group (—OH) in addition to some methyl groups, and a branched chain of carbon atoms.

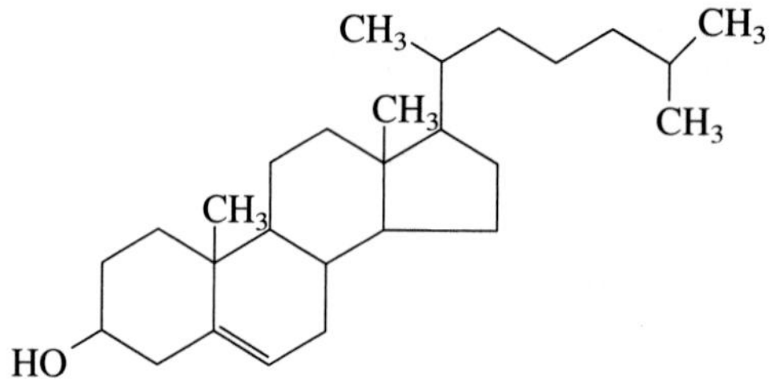

CH_3 CH_3 CH_3 CH_3 CH_3 HO

Cholesterol

Sample Problem 15.5
Structure of Cholesterol

Consider the structure of cholesterol:

CH_3 CH_3 CH_3 CH_3 CH_3 HO

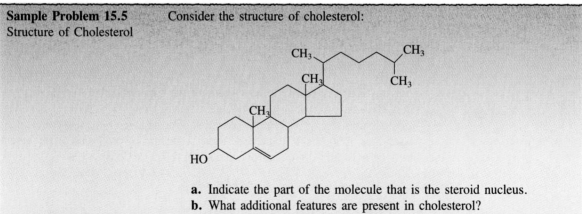

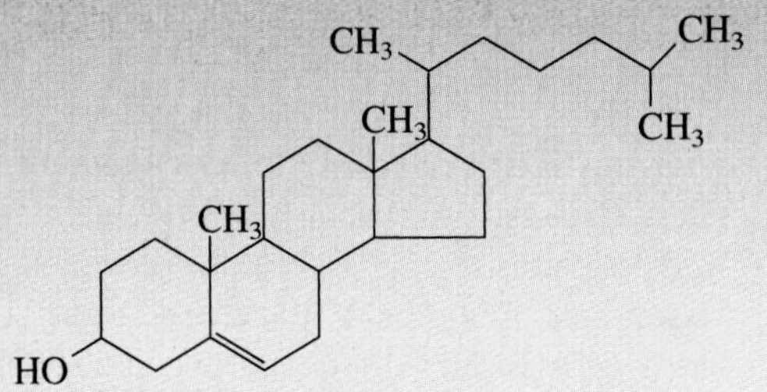

a. Indicate the part of the molecule that is the steroid nucleus.
b. What additional features are present in cholesterol?
c. Indicate the part of the molecule that classifies cholesterol as a sterol.
d. Why is cholesterol in the lipid family?

Solution:

a. The four-ring system represents the steroid nucleus.

b. The structure for cholesterol contains several attached side groups; an alcohol group (—OH), two methyl groups (—CH_3), and a branched carbon side chain along with an unsaturated site in one of the cyclic rings.

c. The alcohol group is responsible for the sterol classification.

d. Cholesterol is not soluble in water; it is classified with the lipid family.

HEALTH NOTE

Cholesterol in the Body

Cholesterol is a component of cellular membranes, myelin sheath, and brain and nerve tissue. It is also found in the liver, bile salts, and skin, where it forms vitamin D. In the adrenal gland, it is used to synthesize steroid hormones. While cholesterol in the body is obtained from eating meats, milk, and eggs, it is also synthesized by the liver from fats, carbohydrates, and proteins. There is no cholesterol in vegetable and plant products.

If a diet is high in cholesterol, the liver produces less. A typical daily American diet includes 400–500 mg of cholesterol, one of the highest in the world. The American Heart Association has recommended that we consume no more than 300 mg of cholesterol a day. The cholesterol content of some typical foods is listed in Table 15.4.

Table 15.4 Cholesterol Content of Some Foods

Food	Serving size	Cholesterol (mg)
Liver (beef)	3 oz	370
Egg	1	250
Lobster	3 oz	175
Fried chicken	$3\frac{1}{2}$ oz	130
Hamburger	3 oz	85
Chicken (no skin)	3 oz	75
Fish (salmon)	3 oz	40
Butter	1 tablespoon	30
Whole milk	1 cup	35
Skim milk	1 cup	5
Margarine	1 tablespoon	0

When cholesterol exceeds its saturation levels in the bile, gallstones may form. Gallstones are composed of almost 100% cholesterol with some calcium salts, fatty acids, and phospholipids. High levels of cholesterol are associated with the accumulation of lipid deposits (plaques) that line and narrow the coronary arteries. Clinically, cholesterol levels are considered elevated if the total plasma cholesterol level exceeds 200–230 mg/dL.

Suggestions for reducing a high serum cholesterol level include decreasing the intake of foods containing cholesterol and changing the type of fat ingested. Some research indicates that saturated fats in the diet may stimulate the production of cholesterol by the liver. A diet that is low in saturated fats as well as cholesterol appears to be helpful in reducing the serum cholesterol level.

Other factors that may also contribute to heart disease are genetic history, lack of exercise, smoking, obesity, diabetes, sex, and age. ■

Steroid Hormones

The sex hormones and the adrenocortical hormones are closely related in structure to cholesterol. The estrogens are produced in the ovaries to regulate the menstrual cycle. They are also responsible for female secondary sex characteristics. The male hormones, such as testosterone, are called androgens. Testosterone is responsible for male secondary sex characteristics.

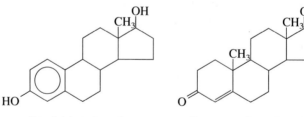

OH
CH_3
HO
Estradiol (an estrogen)

OH
CH_3
CH_3
O
Testosterone (an androgen)

Corticosteroids

Some of the hormones produced by the adrenal gland are corticosteroids. A mineralcorticoid such as aldosterone is responsible for electrolyte and water balance by the kidneys. A glucocorticoid such as cortisol decreases the amount of protein synthesized by muscle and stimulates the conversion of amino acids to glucose. More steroid hormones are discussed in Chapter 17.

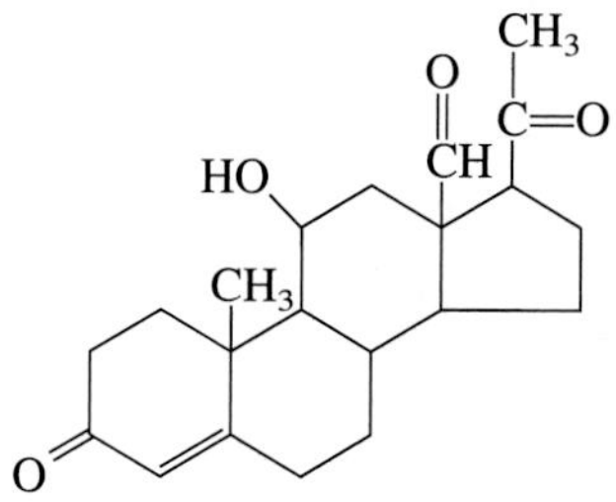

Aldosterone

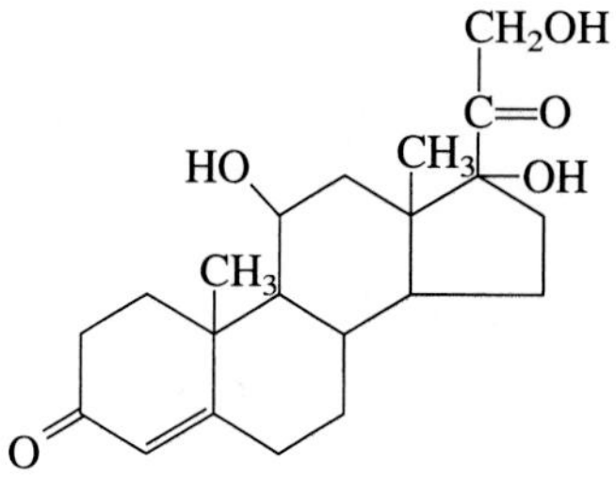

Cortisol

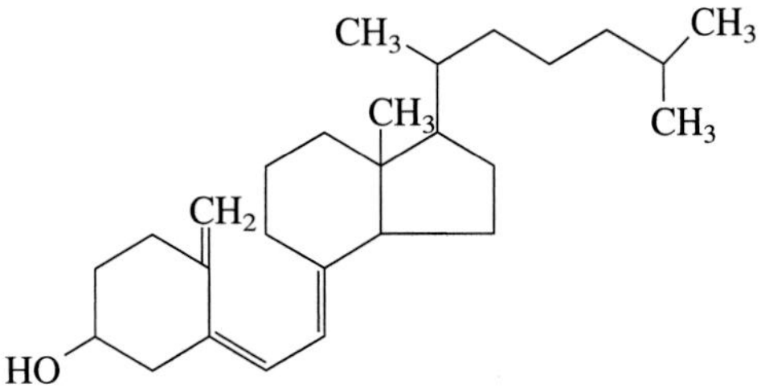

Vitamin D_3
(cholecalciferol)

Vitamin D

Vitamin D is a fat-soluble vitamin that is a member of the sterol family. It is produced in the skin from reactions of sunlight and a cholesterol-type compound. It also occurs naturally in some foods such as sardines, salmon, and egg yolks. Vitamin D increases the absorption of calcium from the intestinal tract and regulates the amount of calcium deposited in the bones and teeth. More about vitamin D is discussed in Chapter 17.

Bile Acids

Bile acids such as cholic acid are synthesized in the liver from cholesterol. Their water-soluble Na^+ and K^+ salts, called bile salts, act like detergents in the small intestine. They keep cholesterol in solution and emulsify fats, forming smaller fat globules that are more accessible to the lipases (fat-digesting enzymes).

a. Vitamin A

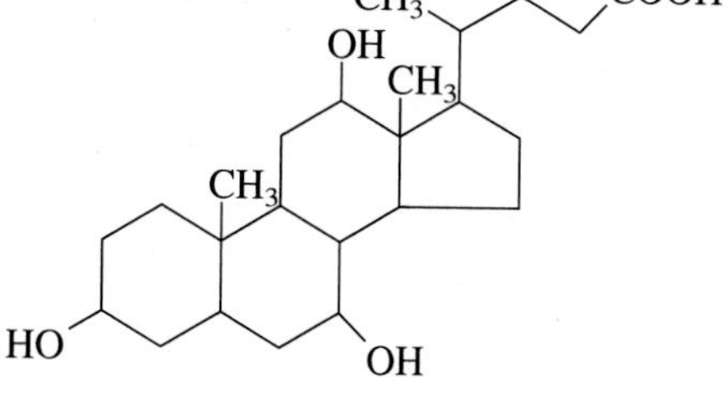

Cholic acid (a bile acid)

Sample Problem 15.6
Identifying Terpenes and Steroids

Classify the following lipids as part of the steroid or terpene family:

a. [steroid structure: HO on aromatic ring, CH_3 groups at ring junctions, OH on five-membered ring]

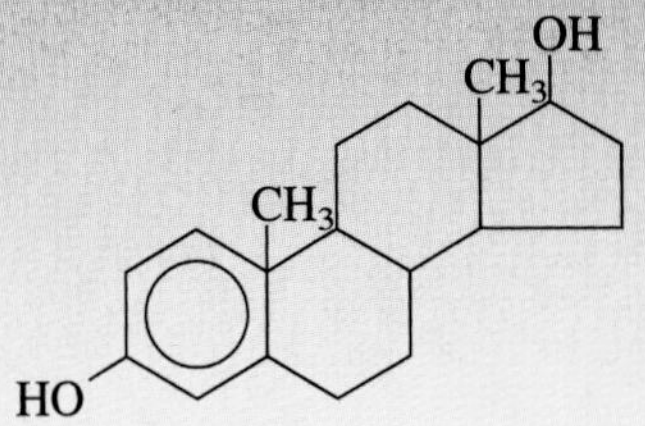

b. $CH_2{=}C(CH_3){-}CH_2{-}CH_2{-}CH_2{-}CH(CH_3){-}CH_2{-}CH_2OH$

Solution:

a. This structure contains a steroid nucleus, which makes it a part of the steroid family.

b. This structure consists of two isoprene units, which makes it a part of the terpene family.

Study Check:

Vitamin A is not similar in structure to vitamin D. Why are they both labeled fat-soluble vitamins?

Answer: Vitamin A is a terpene, while vitamin D is in the steroid family. Both are lipids because they are not soluble in water.

HEALTH NOTE

Lipoproteins: Transporting Lipids in Solution

Because triglycerides and cholesterol are nonpolar, they are insoluble in blood and are not transported easily. To improve their solubility, they attach to proteins and phospholipids, forming a polar complex called a **lipoprotein.** (See Figure 15.9.)

Lipoproteins vary in size and in the type of lipid carried. They can be separated by differences in density. The VLDL (very-low-density lipoproteins) carry triglycerides from the intestine by way of the liver to the capillaries, where the triglycerides are hydrolyzed. The fatty acids produced are used for the synthesis of fats (triglycerides) within the adipose tissues.

The LDL (low-density lipoproteins) are composed primarily of cholesterol, which is carried to cells for the synthesis of cell membranes and steroid hormones. About 20% of serum cholesterol is HDL (high-density lipoproteins), a form that carries cholesterol back to the liver, where it can be mixed with bile and excreted. (See Table 15.5.)

Since high cholesterol levels have been associated with the onset of atherosclerosis and heart disease, the

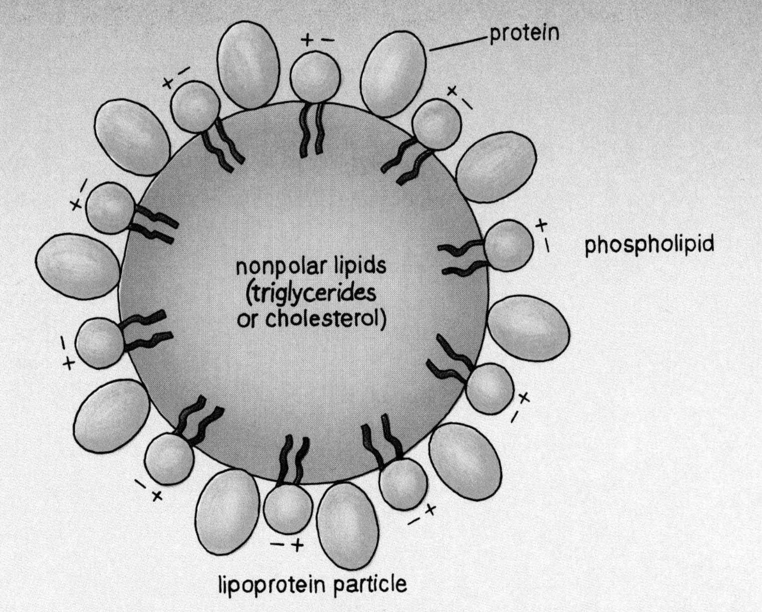

Figure 15.9
A lipoprotein particle; a form in which nonpolar lipids are carried through the bloodstream.

Table 15.5 Some Properties of Plasma Lipoproteins

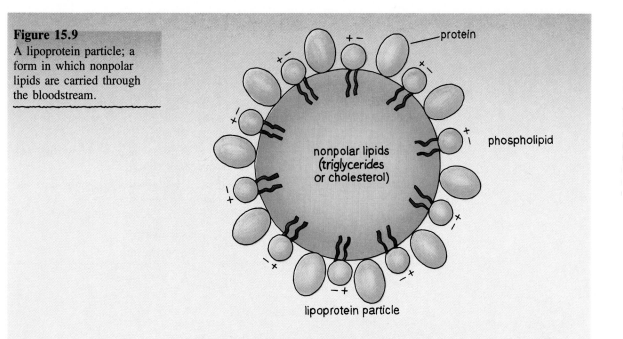

Type	Protein (%)	Phospholipid (%)	Function
VLDL	10	20	Carries dietary triglycerides to capillaries for hydrolysis
LDL	25	20	Carries cholesterol to cells for use in synthesis; allows accumulation
HDL	50	30	Carries cholesterol to liver for removal

Figure 15.10
Human artery clogged with fat.

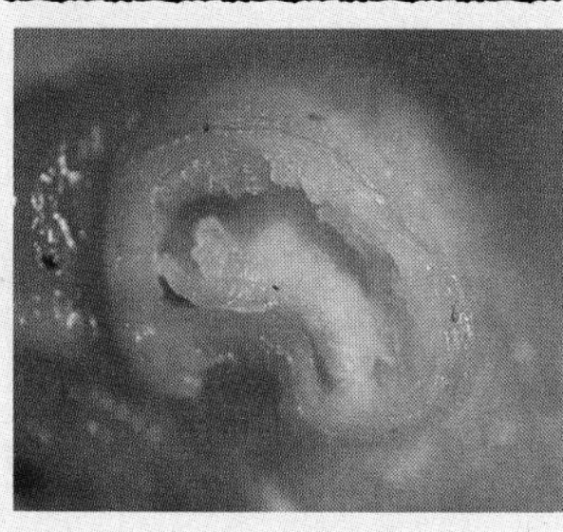

measurement of the LDL and HDL blood serum levels is of much interest to medical researchers.

When the LDL serum levels are increased, there is an increase in the triglyceride and cholesterol serum levels. This increases the risk of myocardial infarctions (heart attacks). Cholesterol can deposit in the blood vessels, causing restriction of blood flow. (See Figure 15.10.) However, high levels of HDL are associated with decreased risk of myocardial infarctions because cholesterol is being removed from the cells and excreted. This lowers the amount of serum cholesterol and lessens the risk of plaque formation within the blood vessels. HDL levels appear to increase in persons who exercise regularly and eat less saturated fats. They are higher in women whose estrogen levels are adequate. ■

Classification of Lipids

Lipids	Composition
Complex	
Waxes	Fatty acids and long-chain alcohols
Triacylglycerols	Fatty acids and glycerol
Phosphoglycerides	Fatty acids, glycerol, phosphate, and an amino alcohol
Sphingolipids	Fatty acids, sphingosine, phosphate, and an amino alcohol
Glycolipids	Fatty acids, glycerol or sphingosine, and one or two monosaccharides
Simple	
Terpenes	Isoprenoid units
Steroids	Fused-ring structure

Key Terms

cephalin A phospholipid found in brain and nerve tissues that incorporates the amino alcohol serine or ethanolamine.

cerebroside A glycolipid consisting of sphingosine, a fatty acid with 24 carbon atoms, and a monosaccharide (usually D-galactose).

cholesterol The most prevalent of the steroid compounds found in cellular membranes; needed for the synthesis of vitamin D, hormones, and bile acids.

fat Another term for solid triglycerides.

fatty acid A long-chain carboxylic acid found in fats.

fluid mosaic model A model of a cell membrane in which phospholipids are arranged as a bilipid layer interspersed with proteins arranged at different depths.

glycolipid A phospholipid that contains a monosaccharide.

hydrogenation The addition of hydrogen to unsaturated fats.

hydrolysis The splitting of bonds by water, usually in the presence of acid or enzymes; in fats, hydrolysis breaks apart the ester bonds between glycerol and fatty acids.

lecithin A phospholipid containing choline.

lipase An enzyme that hydrolyzes fats in the digestive tract.

lipid A family of compounds that is nonpolar in nature and not soluble in water; includes fats, waxes, phospholipids, steroids, and terpenes.

lipidosis A genetic disease in which there is a deficiency of an enzyme for the hydrolysis of a lipid. The lipid accumulates to toxic levels.

lipoproteins A combination of nonpolar lipids and proteins to form a polar complex that can be transported through body fluids.

mixed triglyceride A fat or oil that contains two or more different fatty acids.

oil Another term for liquid triglycerides.

phosphoglyceride A polar lipid containing two fatty acids, a phosphate group, and an amino group such as choline, serine, or ethanolamine.

prostaglandin One of a number of compounds derived from arachidonic acid that regulate several physiological processes.

saponification The reaction of fats with strong bases producing salts of the fatty acids, known as soaps, and glycerol.

saturated Saturated lipids composed of saturated fatty acids that have no double bonds; they have higher melting points than unsaturated lipids, and are usually solid at room temperatures.

sphingolipid A phospholipid in which the alcohol sphingosine has replaced glycerol.

steroid A type of lipid composed of a multicyclic ring system; there are no fatty acids in a steroid.

terpene A type of lipid derived from a five-carbon unsaturated unit called isoprene.

triglyceride A family of lipids composed of three fatty acids bonded through ester bond to glycerol; a trihydroxy alcohol.

unsaturated A lipid or fatty acid that contains double bonds.

wax The ester of a long-chain alcohol and a long-chain saturated fatty acid.

Problems

Fatty Acids *(Goal 15.1)*

15.1 Describe some similarities and differences in the structures of a saturated fatty acid and an unsaturated fatty acid.

15.2 Why does stearic acid, which has 18 carbon atoms, have a melting point (70°C) that is much higher than that of linoleic acid (−5°C)?

15.3 Write the structure of each of the following saturated fatty acids:
a. myristic acid, 14 carbon atoms
b. palmitic acid, 16 carbon atoms
c. stearic acid, 18 carbon atoms

15.4 For the following fatty acids, state whether:
a. they are saturated or unsaturated
b. their source is animal or vegetable
c. they are liquid or solid at room temperature

$$CH_3(CH_2)_4CH{=}CHCH_2CH{=}CH(CH_2)_7\overset{\overset{\displaystyle O}{\|}}{C}OH$$

Linoleic acid

$$CH_3(CH_2)_{14}\overset{\overset{\displaystyle O}{\|}}{C}OH$$

Palmitic acid

Waxes, Fats, and Oils *(Goal 15.2)*

15.5 Draw the structure of each of the following waxes:
a. one component of beeswax, formed from myricyl alcohol, $CH_3(CH_2)_{29}OH$, and palmitic acid
b. jojoba wax, formed from arachidonic acid, a 20-carbon saturated fatty acid, and 1-docosanol, $CH_3(CH_2)_{20}CH_2OH$

15.6 a. Draw the structure of glycerol.
b. A triglyceride forms only stearic acid and glycerol upon hydrolysis. Draw its structure.

15.7 What fatty acids are present in the following triglyceride?

$$
\begin{array}{l}
CH_2-O-\overset{\overset{\large O}{\|}}{C}-(CH_2)_{14}CH_3 \\
| \\
CH-O-\overset{\overset{\large O}{\|}}{C}-(CH_2)_7CH=CHCH_2CH=CH(CH_2)_4CH_3 \\
| \\
CH_2-O-\overset{\overset{\large O}{\|}}{C}-(CH_2)_7CH=CH(CH_2)_7CH_3
\end{array}
$$

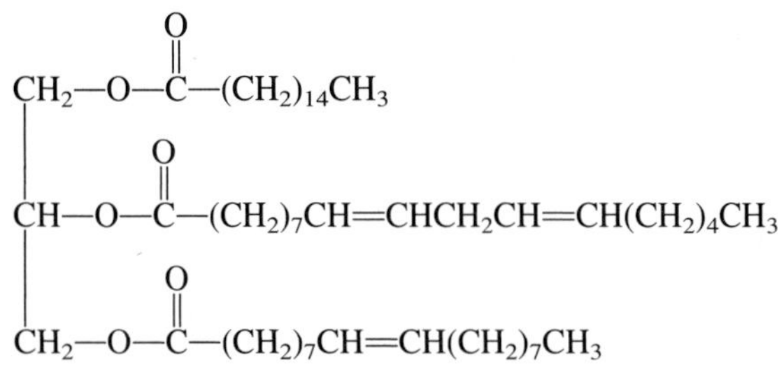

***15.8** A mixed triglyceride contains two palmitic acid molecules to every oleic acid molecule. Write a possible structure for the compound. What other structure is possible?

Properties of Triglycerides *(Goal 15.3)*

15.9 Safflower oil is called a polyunsaturated oil, whereas olive oil is a monounsaturated oil. Explain.

15.10 Why does olive oil have a lower melting point than butterfat?

15.11 Why does coconut oil, a vegetable oil, have a melting point similar to those of animal fats?

15.12 What is the percentage of unsaturated fatty acids in coconut oil, compared to that in olive oil?

15.13 Determine which fat or oil in each pair will have the most unsaturation:

a. triolean or tristearin
b. a triglyceride having two molecules of oleic acid and one molecule of linoleic acid, or a triglyceride having two molecules of linoleic acid and one molecule of oleic acid

15.14 Some typical meals at fast-food restaurants are listed here. Calculate the number of kilocalories from fat and the percentage of total kilocalories due to fat (1 g of fat = 9 kcal). Would you expect the fats to be mostly saturated or unsaturated? Why?

a. a chicken dinner, 830 kcal, 46 g fat
b. cheeseburger (quarter-pounder), 518 kcal, 29 g fat
c. pepperoni pizza (three slices), 560 kcal, 18 g fat
d. beef burrito, 466 kcal, 21 g fat
e. deep-fried fish (three pieces), 477 kcal, 28 g fat

15.15 Because peanut oil floats on the top of peanut butter, many brands of peanut butter are hydrogenated. A solid product then forms that is mixed into the peanut butter and does not separate. What is the product of the complete hydrogenation of the following triglyceride in peanut oil?

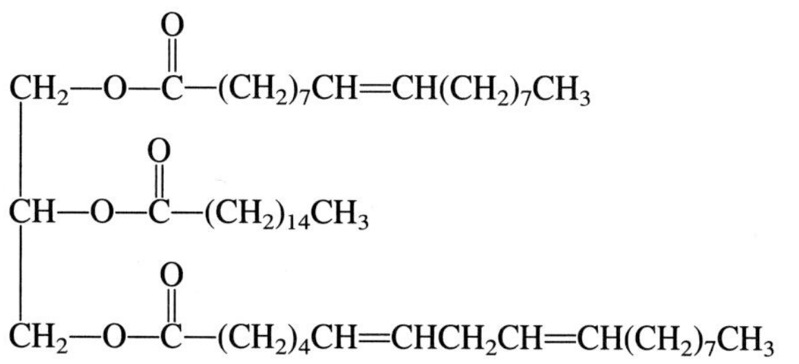

$$
\begin{array}{l}
\quad\quad\quad\;\; O \\
\quad\quad\quad\;\; \| \\
CH_2-O-C-(CH_2)_7CH{=}CH(CH_2)_7CH_3 \\
| \quad\quad\quad\;\; O \\
| \quad\quad\quad\;\; \| \\
CH-O-C-(CH_2)_{14}CH_3 \\
| \quad\quad\quad\;\; O \\
| \quad\quad\quad\;\; \| \\
CH_2-O-C-(CH_2)_4CH{=}CHCH_2CH{=}CH(CH_2)_7CH_3
\end{array}
$$

15.16 A label on a cube of margarine states that it contains partially hydrogenated corn oil.

a. How has the liquid corn oil been changed?
b. Why is the margarine product solid?

15.17 A label on a bottle of 100% sunflower seed oil states that it is lower in saturated fats than all the leading oils. The saturated fats in sunflower oil are as follows: 6% palmitic acid, 2% stearic acid, and 1% archidic acid (a 20-carbon saturated fatty acid).

a. How does the percentage of saturated fats in sunflower seed oil compare to that of safflower and corn oil? (See Table 15.3.)
b. Is the claim valid?

15.18 Write the equation for the acid hydrolysis of each of the following:

a. trimyristin
b. triolein

15.19 Write the equation for the NaOH saponification of each of the following:

a. tripalmitin
b. triolein

15.20 Why should a bottle of vegetable oil that has no preservatives be tightly covered and refrigerated?

15.21 List the general products of the following reactions:

a. complete hydrogenation of an unsaturated triglyceride
b. acid or enzyme hydrolysis of a triglyceride
c. saponification of a triglyceride

Phospholipids *(Goal 15.4)*

15.22 How are phosphoglycerides different from triglycerides?

15.23 Why are phosphoglycerides soluble in water?

15.24 A phosphoglyceride contains two molecules of palmitic acid, and the amino alcohol ethanolamine. Draw the structure of the compound. What is another name for this type of phospholipid?

Palmitic acid $CH_3(CH_2)_{14}\overset{\overset{O}{\|}}{C}-OH$

Ethanolamine $HO-CH_2-CH_2-\overset{+}{N}H_3$

15.25 How does a lecithin differ from a cephalin?

15.26 How do sphingosines differ from phosphoglycerides?

15.27 How are phospholipids similar to soaps?

15.28 How does the polarity of the phospholipids contribute to their function in cell membranes?

15.29 Both a lecithin and a sphingomyelin contain the amino alcohol choline. How do they differ in structure? Are they both phospholipids?

Terpenes and Steroids *(Goal 15.5)*

15.30 a. What is the characteristic structural feature of a steroid?
b. What is the characteristic structural feature of a terpene?

15.31 Identify each of the following as a steroid or a terpene:

a. CH_3 CH_3 CH_3 CH_2OH CH_3 CH_3

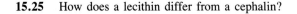

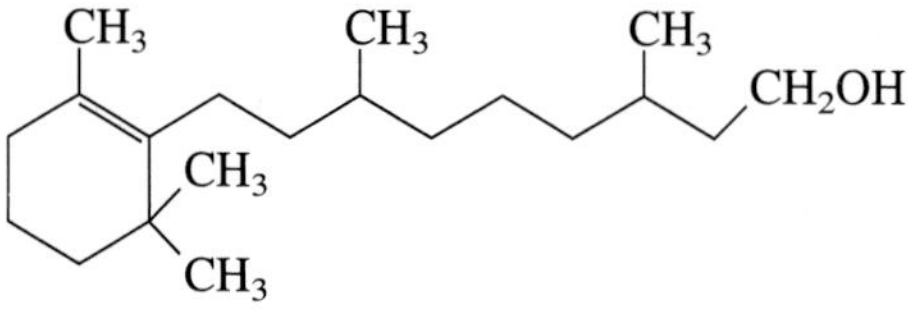

b. Androsterone (a male hormone)

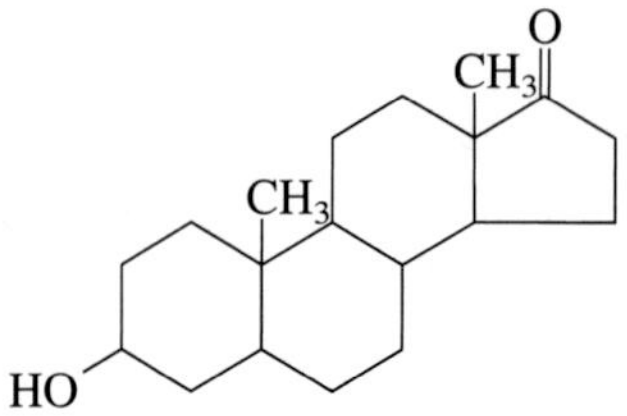

O CH_3 CH_3 HO

c. Vitamin K (important in blood clotting)

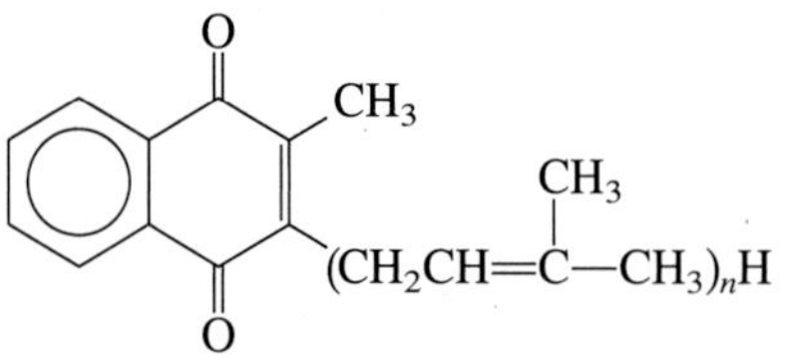

O CH_3 CH_3 $(CH_2CH{=}\overset{|}{C}{-}CH_3)_nH$ where n = 5–10 O

d. Cortisol (a steroid hormone)

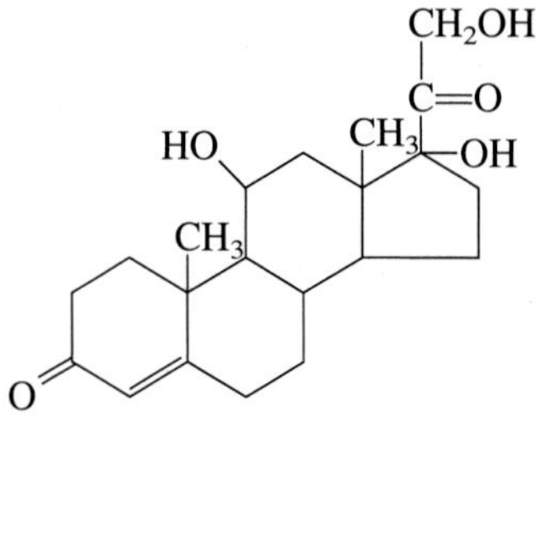

CH_2OH $C{=}O$ HO CH_3 OH CH_3 O

e. Cholic acid

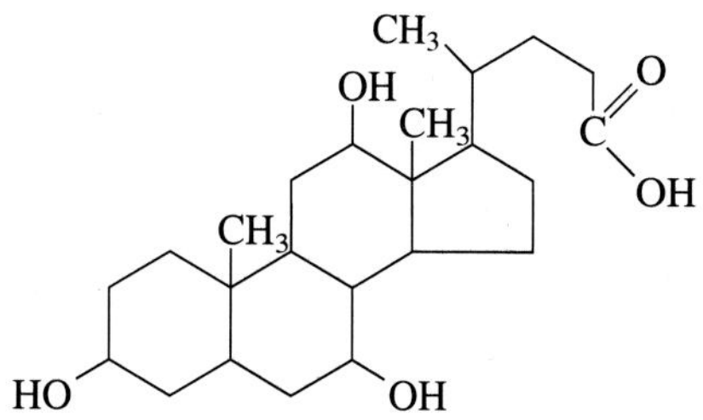

CH_3 OH CH_3 CH_3 O C OH HO OH

15.32 Classify each of the following structures as triglyceride, wax, phospholipid, sphingolipid, terpene, or steroid.

a. $CH_3(CH_2)_{14}\overset{\overset{\displaystyle O}{\|}}{C}—O(CH_2)_{29}CH_3$

b. testosterone

c. 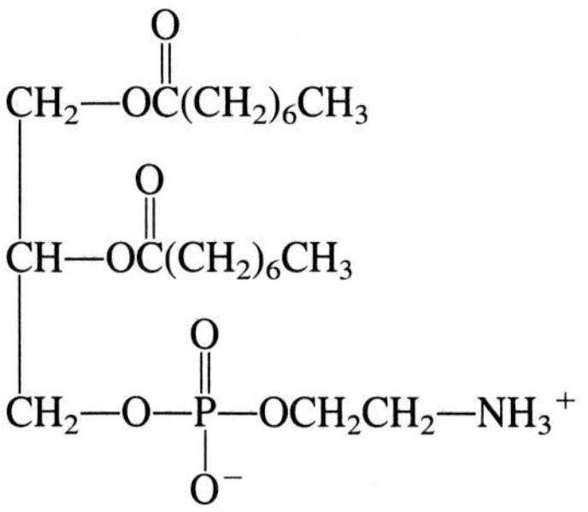
$$\begin{array}{l} CH_2—O\overset{\overset{\displaystyle O}{\|}}{C}(CH_2)_6CH_3 \\ | \\ CH—O\overset{\overset{\displaystyle O}{\|}}{C}(CH_2)_6CH_3 \\ | \\ CH_2—O—\overset{\overset{\displaystyle O}{\|}}{\underset{\underset{\displaystyle O^-}{|}}{P}}—OCH_2CH_2—NH_3^+ \end{array}$$

d. oil of caraway

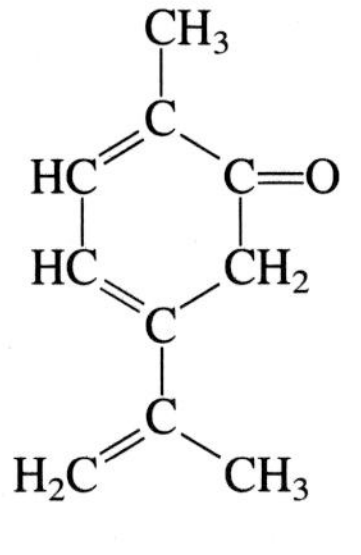

CH_3, C, HC, $C{=}O$, HC, CH_2, C, C, H_2C, CH_3

e. 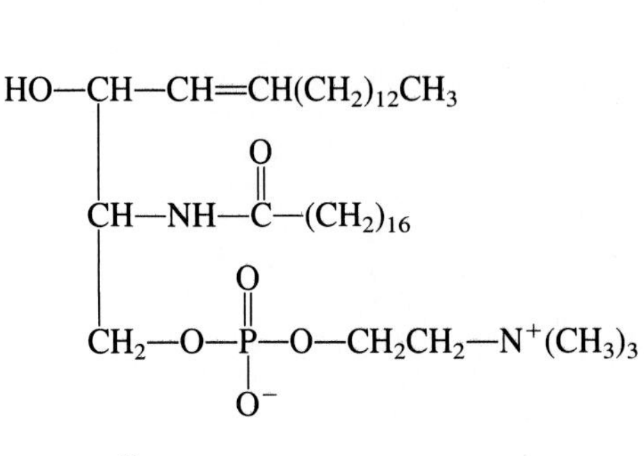
$$\begin{array}{l} HO—CH—CH{=}CH(CH_2)_{12}CH_3 \\ \quad\;\; | \\ \quad\;\; CH—NH—\overset{\overset{\displaystyle O}{\|}}{C}—(CH_2)_{16} \\ \quad\;\; | \\ \quad\;\; CH_2—O—\overset{\overset{\displaystyle O}{\|}}{\underset{\underset{\displaystyle O^-}{|}}{P}}—O—CH_2CH_2—N^+(CH_3)_3 \end{array}$$

f. 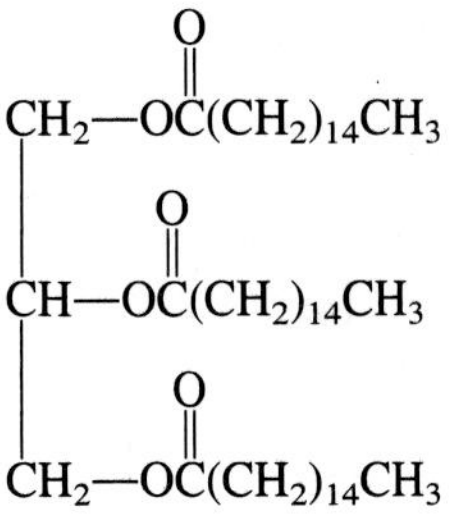
$$\begin{array}{l} CH_2—O\overset{\overset{\displaystyle O}{\|}}{C}(CH_2)_{14}CH_3 \\ | \\ CH—O\overset{\overset{\displaystyle O}{\|}}{C}(CH_2)_{14}CH_3 \\ | \\ CH_2—O\overset{\overset{\displaystyle O}{\|}}{C}(CH_2)_{14}CH_3 \end{array}$$

15.33 Identify the structural features in each of the following lipids as fatty acids, salts of fatty acids, glycerol, phosphate, long-chain alcohol, amino alcohol, sphingosine, monosaccharide, steroid nucleus, and/or isoprene units.

a. beeswax
b. cholesterol
c. lecithin
d. tripalmitin
e. oil of lemon
f. soap
g. glycolipid
h. testosterone
i. vitamin E

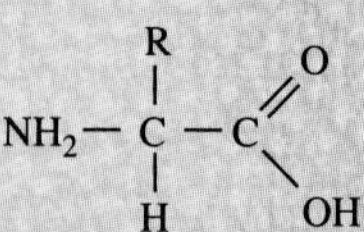

16

Proteins

Athletes need protein to build their muscles, to produce the hemoglobin that carries oxygen in the bloodstream, and to build enzymes and hormones that direct metabolic activities.

Learning Goals

16.1 Write the structural formula of an amino acid.
16.2 Distinguish between complete and incomplete protein.
16.3 Write the ionized form of an amino acid in acid, in base, and at the isoelectric point.
16.4 Write the structural formula for a peptide.
16.5 Determine the primary structure of a polypeptide.
16.6 Describe the secondary structures of proteins: the alpha helix, the pleated sheet, and the triple helix.
16.7 Describe the tertiary and quaternary structures of proteins.
16.8 Classify proteins by structure and function.
16.9 Discuss denaturation, hydrolysis, electrophoresis, and chemical tests for proteins.

SCOPE

The word *protein* is derived from the Greek word *proteios,* meaning "first." Made of amino acids, proteins provide structure in membranes, build cartilage and connective tissue, transport oxygen in blood and muscle, direct biological reactions as enzymes, defend the body against infection, and control metabolic processes as hormones. They can even be a source of energy.

Compared to many of the compounds we have studied, protein molecules can be gigantic. Insulin has a molecular weight of 5700, hemoglobin a molecular weight of about 64,000. Some virus proteins are still larger, having molecular weights of over 40 million. Yet all proteins are polymers made up of about 20 different kinds of amino acids. Each kind of protein is composed of a certain number of amino acids arranged in a certain order. It is this order of amino acids that determines the characteristics of each protein and its biological action.

16.1 Amino Acids

Learning Goal **Write the structural formula of an amino acid.**

Proteins are composed of building blocks called **amino acids.** Every amino acid contains two functional groups, an amino group ($—NH_2$) and a carboxylic acid group (—COOH). The amino group is bonded to the alpha (α) carbon (the carbon adjacent to the carboxylic acid group), which makes them alpha amino acids. Various side chains (R) are attached to the alpha carbon, giving each amino acid unique characteristics.

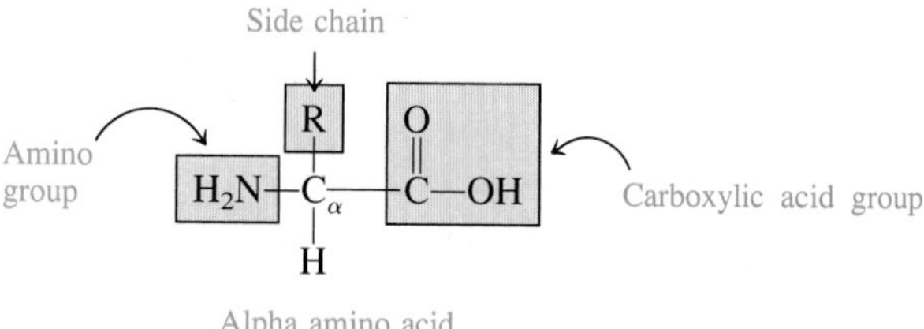

Alpha amino acid

Optical Isomerism

With the exception of glycine, all of the amino acids are chiral and optically active. The D or L configuration of an amino acid is determined by comparing the amino group to the OH in glyceraldehyde. All naturally occurring amino acids have been found to be L amino acids.

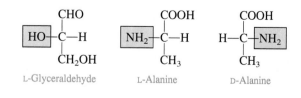

CHO	COOH	COOH
HO—C—H	NH_2—C—H	H—C—NH_2
CH_2OH	CH_3	CH_3
L-Glyceraldehyde	L-Alanine	D-Alanine

Classification of Amino Acids

Amino acids are classified according to their side chains or R groups as nonpolar, polar, acidic, or basic. **Nonpolar** amino acids contain a hydrocarbon side chain that makes them **hydrophobic. Polar** amino acids are **hydrophilic** because their side groups are attracted to water. For example, serine, threonine, and tyrosine are polar because they contain an —OH in their side chains. Acidic and basic amino acids contain carboxylic acid or amino side chains that ionize at certain pH values.

Nonpolar Amino Acids

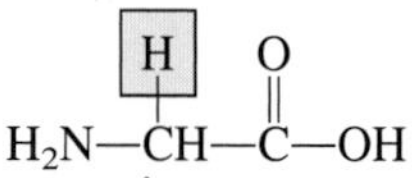

Glycine (Gly)

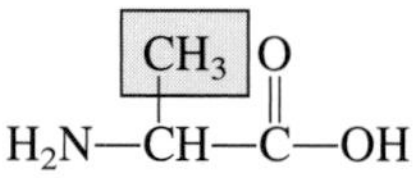

Alanine (Ala)

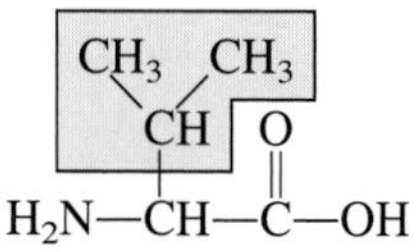

Valine (Val)

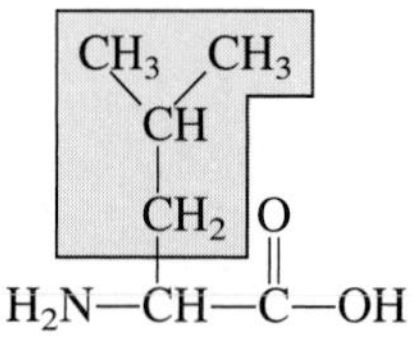

Leucine (Leu)

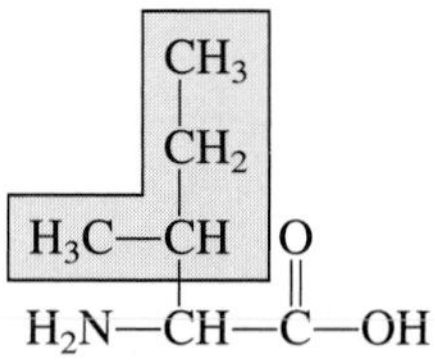

Isoleucine (Ile)

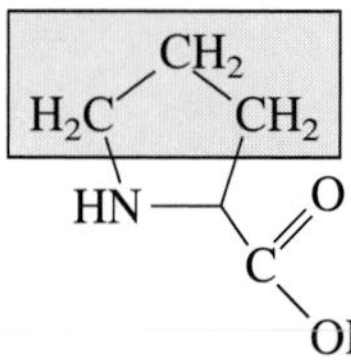

Proline (Pro)

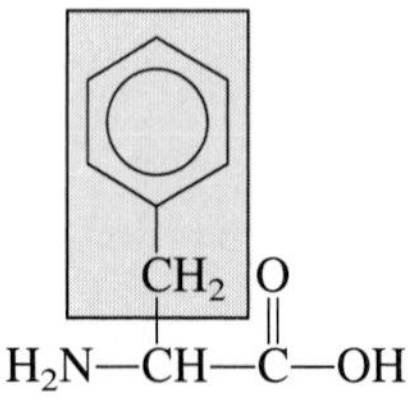

Phenylalanine (Phe)

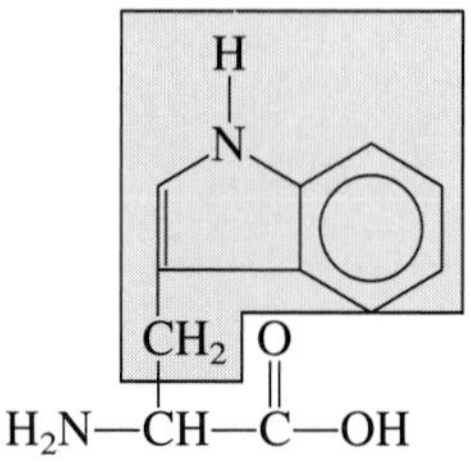

Tryptophan (Trp)

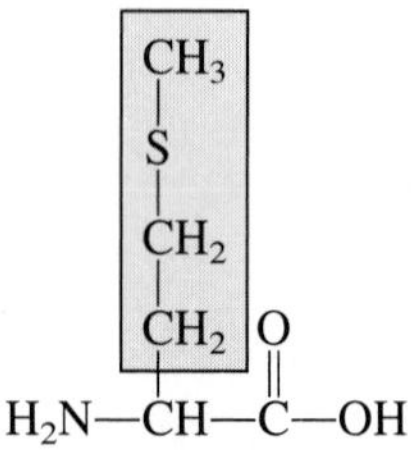

Methionine (Met)

Polar Amino Acids

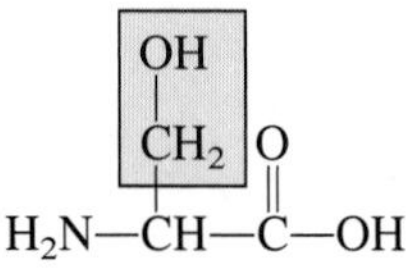

Serine (Ser)

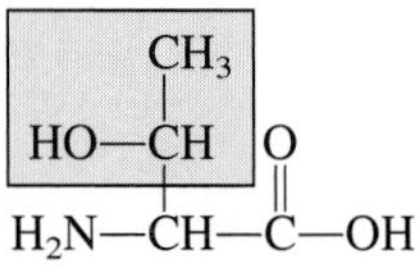

Threonine (Thr)

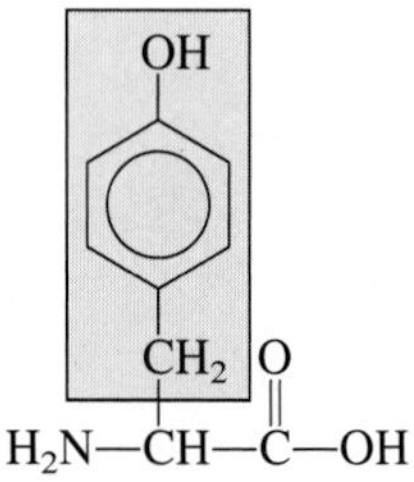

Tyrosine (Tyr)

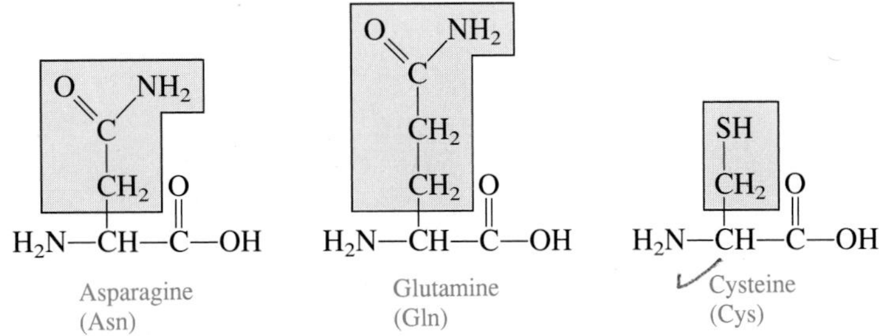

Asparagine (Asn) Glutamine (Gln) Cysteine (Cys)

Acidic Amino Acids

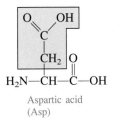

Aspartic acid (Asp) Glutamic acid (Glu)

Basic Amino Acids

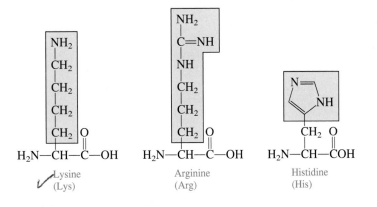

Lysine (Lys) Arginine (Arg) Histidine (His)

Sample Problem 16.1
Structural Formulas of Amino Acids

Write the structural formulas for the following amino acids:
a. alanine (R = $-CH_3$) b. serine (R = $-CH_2OH$)

Solution:
The structural formulas are written by attaching the side chain (R) to the alpha carbon of the general structural formula of an amino acid.

Alpha amino acid

$$\text{a. } NH_2-\underset{\boxed{CH_3}}{CH}-\overset{O}{\overset{\|}{C}}-OH$$

Alanine (Ala)

$$\text{b. } NH_2-\underset{\boxed{\underset{OH}{CH_2}}}{CH}-\overset{O}{\overset{\|}{C}}-OH$$

Serine (Ser)

Study Check:
Why is aspartic acid classified as an acidic amino acid?

Answer: The side chain in aspartic acid contains a carboxylic acid group that can ionize.

16.2 Essential Amino Acids

Learning Goal **Distinguish between complete and incomplete protein.**

Table 16.1
Essential Amino Acids

Arginine	Phenylalanine
Histidine	Threonine
Isoleucine	Tryptophan
Leucine	Valine
Lysine	
Methionine	

Of the 20 amino acids used to build proteins in the body, 10 cannot be synthesized from carbohydrates or lipids in the body. These **essential amino acids** must be obtained from the proteins in the diet. See Table 16.1.

Complementary Proteins

A **complete protein** contains all of the essential amino acids in the proper amounts. An **incomplete** protein is low in one or more of the essential amino acids, usually lysine, tryptophan, or methionine. Except for gelatin, proteins from animal sources are complete, whereas proteins from vegetable sources are incomplete, as Table 16.2 shows.

In a diet that includes animal protein, such as meat, milk, eggs, or cheese, all the essential amino acids are supplied. However, because these foods also contain saturated fats and cholesterol, many people obtain their protein from a diet of grains and vegetables. Such a diet must combine foods that have **complementary proteins** to ensure that all the essential amino acids are provided. For example, rice is deficient in lysine, and beans are deficient in methionine and tryptophan. However, when they are served together, they complement each other and provide all the essential amino acids. See Figure 16.1.

Diseases of Protein Deficiency

When one or more of the essential amino acids is missing in the diet, the synthesis of proteins in the body is severely limited. In a condition known as kwashiorkor, the number of calories in the diet is sufficient, but protein is incomplete. Without all the essential amino acids, a child's growth is halted, the color and

Table 16.2 Examples of Complete and Incomplete Protein Sources

Source	Type of protein	Amino acid deficiency
Animal protein		
Egg	Complete	None
Milk	Complete	None
Meat, fish, poultry	Complete	None
Gelatin	Incomplete	Trp
Vegetable protein		
Grains		
Wheat	Incomplete	Lys
Corn	Incomplete	Lys, trp
Rice (brown, white)	Incomplete	Lys
Oats	Incomplete	Lys
Legumes		
Beans	Incomplete	Met, trp
Peas	Incomplete	Met
Nuts		
Almonds	Incomplete	Lys, trp
Walnuts	Incomplete	Lys, trp

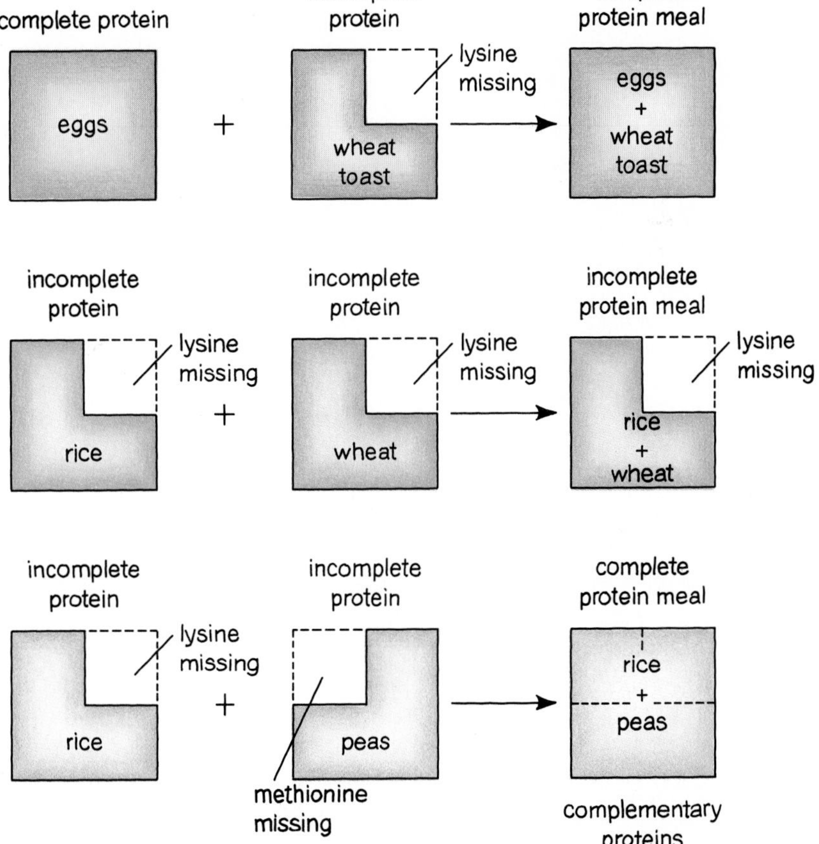

Figure 16.1
All the essential amino acids are provided in a meal that consists of a complete protein of animal origin or complementary proteins of vegetable origin.

texture of hair are altered, a fatty liver develops, and the abdomen shows a characteristic enlargement due to edema. When the diet is also low in calories kwashiorkor is combined with marasmus, causing a wasting of body mass without the edema.

Sample Problem 16.2
Complementary Proteins

Using Table 16.2, indicate whether each of the following combinations of proteins is complete or incomplete.

a. corn and rice
b. corn, rice, and eggs

Solution:

a. Corn is a vegetable protein that is missing lysine and tryptophan. Rice is a vegetable protein that is missing lysine. Since they both lack lysine, the combination is still incomplete.
b. Although corn and rice both lack lysine, eggs are a complete protein and will provide all of the essential amino acids including lysine. The combination is a complete protein.

Study Check:
Will a meal of rice and beans provide a complete protein? Why?

Answer: Yes. Rice and beans are missing different essential amino acids, but their combination provides all the necessary amino acids.

16.3 Properties of Amino Acids

Learning Goal **Write the ionized form of an amino acid in acid, in base, and at the isoelectric point.**

Amino acids are solid at room temperature, have high melting points, and are soluble in water. These properties are quite different from those of amines or carboxylic acids. See Table 16.3.

Table 16.3 Comparison of an Amine, a Carboxylic Acid, and an Amino Acid Having Similar Molar Mass

An amine	A carboxylic acid	An amino acid
$CH_3CH_2CH_2CH_2—NH_2$	$CH_3CH_2—\overset{O}{\overset{\parallel}{C}}—OH$	$H_2N—CH_2—\overset{O}{\overset{\parallel}{C}}—OH$
Butylamine	Propanoic acid	Glycine
Molar mass 73	Molar mass 74	Molar mass 75
Melting point −49°C	Melting point −21°C	Melting point 262°C

The high melting points of amino acids are similar to the melting points of ionized or saltlike compounds. An amino acid ionizes when a proton is transferred from the carboxylic acid group to the amino group. In glycine shown below, the carboxyl group becomes negatively charged, and the amino group acquires a positive charge. This ionized form of the amino acid called a dipolar ion or **zwitterion** has an overall charge of zero.

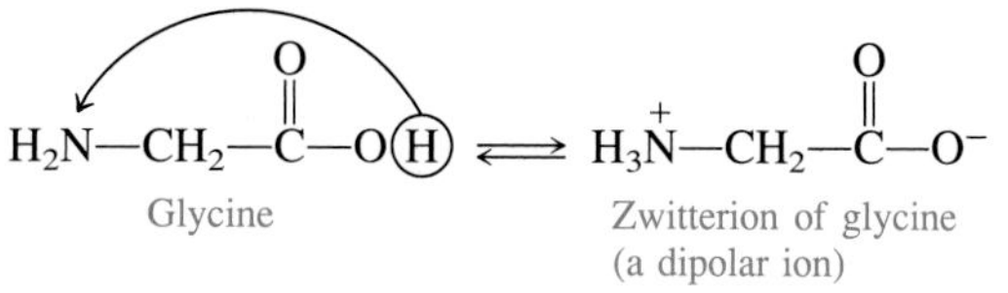

$$H_2N{-}CH_2{-}\overset{O}{\overset{\|}{C}}{-}O(H) \rightleftharpoons H_3\overset{+}{N}{-}CH_2{-}\overset{O}{\overset{\|}{C}}{-}O^-$$

Glycine

Zwitterion of glycine (a dipolar ion)

The zwitterions of some other amino acids follow.

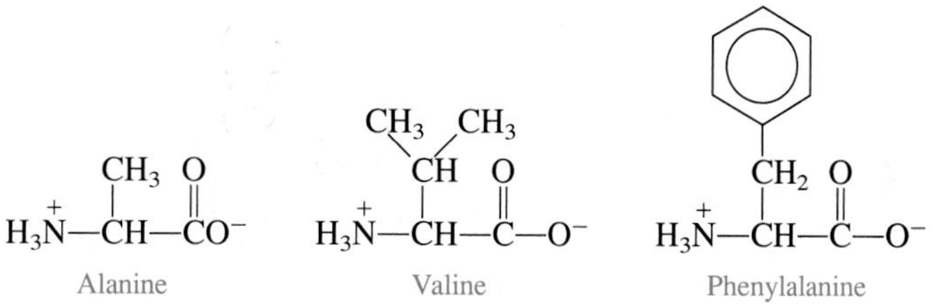

Alanine: $H_3\overset{+}{N}{-}CH(CH_3){-}\overset{O}{\overset{\|}{C}}O^-$

Valine: $H_3\overset{+}{N}{-}CH(CH(CH_3)_2){-}\overset{O}{\overset{\|}{C}}{-}O^-$

Phenylalanine: $H_3\overset{+}{N}{-}CH(CH_2C_6H_5){-}\overset{O}{\overset{\|}{C}}{-}O^-$

Sample Problem 16.3
Writing Zwitterions of Amino Acids

Draw the structural formula for the zwitterion of serine.

Solution:
In the zwitterion, there is a negative charge on the carboxylate group and a positive charge on the ammonium group.

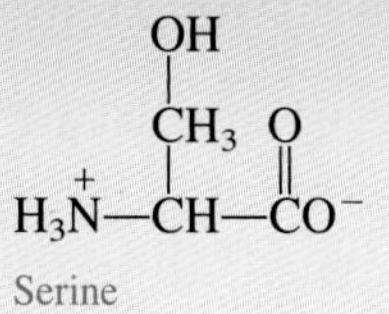

$H_3\overset{+}{N}{-}CH(CH_3{-}OH){-}\overset{O}{\overset{\|}{C}}O^-$

Serine

Study Check:
Write the structural formula of the zwitterion of threonine.

Answer:

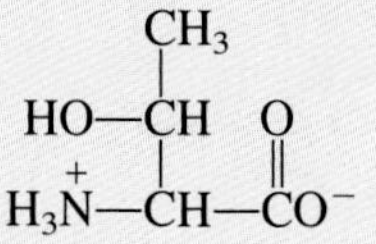

$H_3\overset{+}{N}{-}CH(CH(OH)CH_3){-}\overset{O}{\overset{\|}{C}}O^-$

Isoelectric Point

The zwitterion form of an amino acid exists at a pH value called the **isoelectric point.** At this pH, the molecule is electrically neutral and would not move toward a positive or negative electrode. Nonpolar amino acids have isoelectric points at pH values close to 7. The basic amino acids reach their isoelectric points at much higher pH values, whereas acidic amino acids become electrically neutral at low pH values. Table 16.4 lists isoelectric points for some typical amino acids.

Table 16.4 Isoelectric Points of Some Amino Acids

Amino acids	pH
Aspartic acid (acidic)	2.8
Glutamic acid (acidic)	3.2
Serine (polar)	5.7
Methionine (nonpolar)	5.8
Glycine (nonpolar)	6.0
Valine (nonpolar)	6.0
Alanine (nonpolar)	6.1
Lysine (basic)	9.7
Arginine (basic)	10.8

At pH values below or above their isoelectric points, zwitterions act as acids and bases because they are able to donate or accept protons (H^+). Suppose that some acid is added to a glycine solution at its isoelectric pH value of 6.0. The zwitterion of glycine picks up H^+ and becomes an ion with a positive charge.

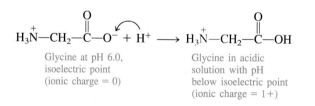

$$H_3\overset{+}{N}—CH_2—\overset{\overset{\large O}{\|}}{C}—O^- + H^+ \longrightarrow H_3\overset{+}{N}—CH_2—\overset{\overset{\large O}{\|}}{C}—OH$$

Glycine at pH 6.0, isoelectric point (ionic charge = 0)

Glycine in acidic solution with pH below isoelectric point (ionic charge = 1+)

However, when base is added to the zwitterion, the OH^- is neutralized by an H^+ from the glycine. In a base, glycine becomes a negative ion.

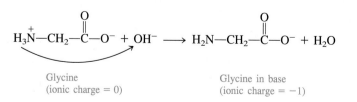

$$H_3\overset{+}{N}—CH_2—\overset{\overset{\large O}{\|}}{C}—O^- + OH^- \longrightarrow H_2N—CH_2—\overset{\overset{\large O}{\|}}{C}—O^- + H_2O$$

Glycine (ionic charge = 0)

Glycine in base (ionic charge = −1)

We can summarize the changes in ionic charge with the addition of acid or base to the solution of an amino acid at its isoelectric point.

$$\text{pH below} \xleftarrow{H^+} \text{isoelectric point} \xrightarrow{OH^-} \text{pH above}$$

$$\text{ion } (+) \longleftarrow \text{zwitterion}(+)(-) \longrightarrow \text{ion } (-)$$

Sample Problem 16.4
Amino Acids in Acid or Base

Write the ionic structure of valine that would be expected in a solution that has a pH of 4.0.

Solution:
In an acidic solution with a pH below 6.0 (isoelectric point of valine), the COO^- of valine picks up a proton.

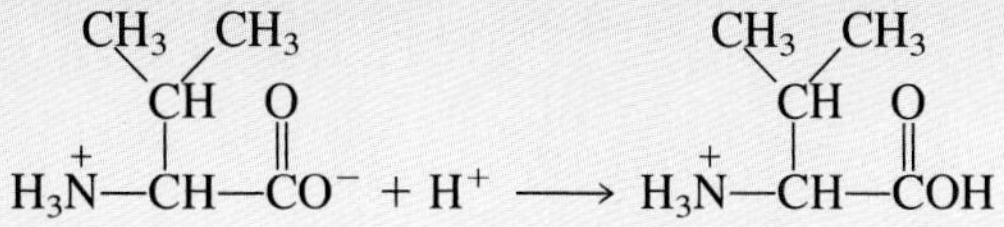

$$H_3\overset{+}{N}-CH(CH(CH_3)_2)-\overset{O}{\overset{\|}{C}}O^- + H^+ \longrightarrow H_3\overset{+}{N}-CH(CH(CH_3)_2)-\overset{O}{\overset{\|}{C}}OH$$

Study Check:
Draw the structure of the ionized form of valine in a solution that has a pH of 10.

Answer:

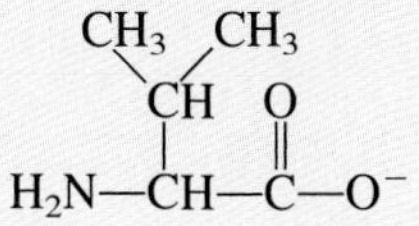

$$H_2N-CH(CH(CH_3)_2)-\overset{O}{\overset{\|}{C}}-O^-$$

16.4 Peptides

Learning Goal **Write the structural formula for a peptide.**

A **peptide** is a chain of amino acids held together by peptide bonds. **Peptide bonds** are amide bonds that occur between the carboxyl group of one amino acid and the amino group of the next with the loss of water.

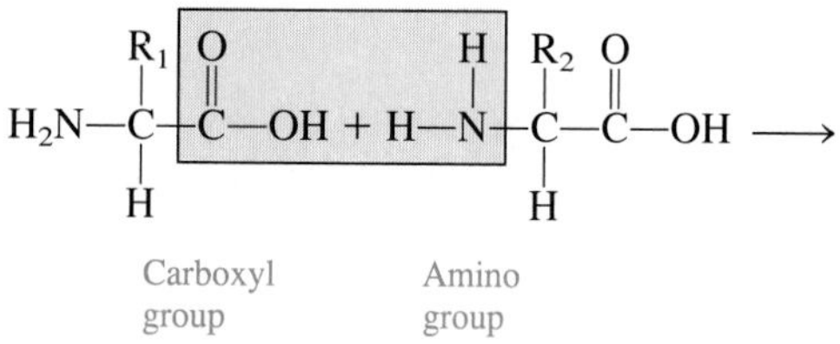

$$H_2N-\underset{H}{\overset{R_1}{C}}-\overset{O}{\overset{\|}{C}}-OH + H-\overset{H}{N}-\underset{H}{\overset{R_2}{C}}-\overset{O}{\overset{\|}{C}}-OH \longrightarrow$$

Carboxyl group

Amino group

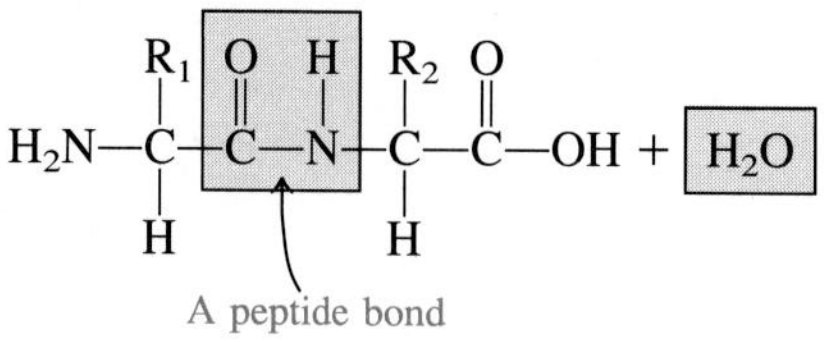

$$H_2N-\underset{H}{\overset{R_1}{C}}-\boxed{\overset{O}{\overset{\|}{C}}-\overset{H}{N}}-\underset{H}{\overset{R_2}{C}}-\overset{O}{\overset{\|}{C}}-OH + \boxed{H_2O}$$

A peptide bond

A peptide bond joining two amino acids produces a dipeptide. The N-terminal end is the amino acid with the free amino group ($-NH_2$); the amino acid with the free carboxyl group is the C-terminal end. Starting from the N-terminal end, the amino acids in a peptide are named with *yl* endings followed by the full name of the amino acid at the C-terminal end.

Formation of a Dipeptide

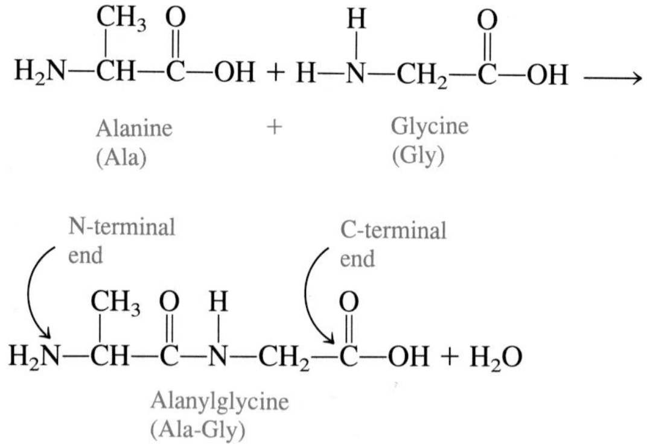

$$H_2N-\overset{CH_3}{CH}-\overset{O}{\overset{\|}{C}}-OH + H-\overset{H}{N}-CH_2-\overset{O}{\overset{\|}{C}}-OH \longrightarrow$$

Alanine (Ala) + Glycine (Gly)

N-terminal end — C-terminal end

$$H_2N-\overset{CH_3}{CH}-\overset{O}{\overset{\|}{C}}-\overset{H}{N}-CH_2-\overset{O}{\overset{\|}{C}}-OH + H_2O$$

Alanylglycine
(Ala-Gly)

Sample Problem 16.5
Writing Structural Formulas of Dipeptides

Write a structural formula for the dipeptide valylserine.

Solution:
The N-terminal amino acid is valine followed by serine. A peptide bond links the carboxyl group of valine and the amino group of serine.

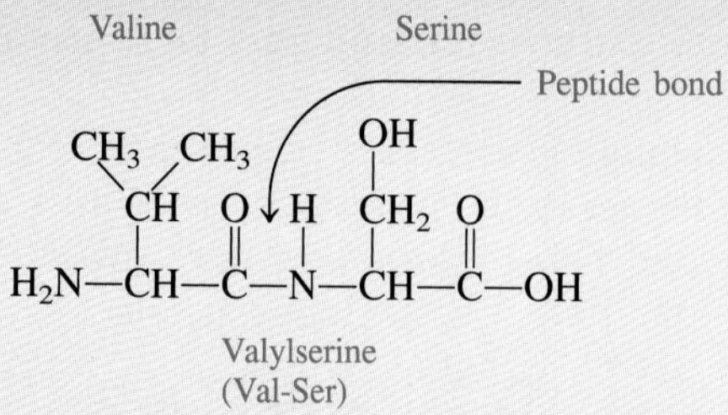

Valine — Serine — Peptide bond

$$H_2N-\underset{}{\overset{CH(CH_3)_2}{CH}}-\overset{O}{\overset{\|}{C}}-\overset{H}{N}-\overset{CH_2OH}{CH}-\overset{O}{\overset{\|}{C}}-OH$$

Valylserine
(Val-Ser)

Study Check:
Aspartame is an artificial sweetener 200 times sweeter than sucrose. It is a methyl ester of a dipeptide. What are the amino acids in aspartame?

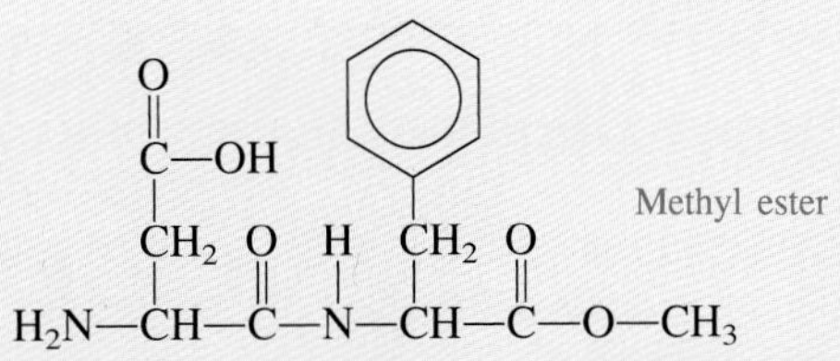

$$H_2N-\underset{\underset{\underset{\overset{|}{\underset{O}{\overset{\|}{C}}-OH}}{|}}{CH_2}}{CH}-\overset{O}{\overset{\|}{C}}-\overset{H}{\overset{|}{N}}-\overset{\overset{C_6H_5}{|}}{\overset{CH_2}{CH}}-\overset{O}{\overset{\|}{C}}-O-CH_3$$

Methyl ester

Aspartame

Answer: aspartic acid and phenylalanine

Amino acids are added to a peptide by forming peptide bonds with the C-terminal amino acid. For example, adding serine to alanylglycine forms a tripeptide, alanylglycylserine (Ala-Gly-Ser).

Formation of a Tripeptide

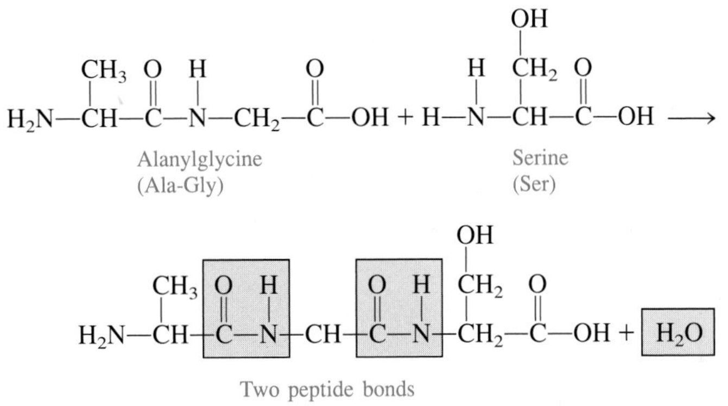

$$H_2N-\overset{CH_3}{\overset{|}{CH}}-\overset{O}{\overset{\|}{C}}-\overset{H}{\overset{|}{N}}-CH_2-\overset{O}{\overset{\|}{C}}-OH + H-\overset{H}{\overset{|}{N}}-\overset{\overset{OH}{|}}{\overset{CH_2}{\overset{|}{CH}}}-\overset{O}{\overset{\|}{C}}-OH \longrightarrow$$

Alanylglycine (Ala-Gly) Serine (Ser)

$$H_2N-\overset{CH_3}{\overset{|}{CH}}-\boxed{\overset{O}{\overset{\|}{C}}-\overset{H}{\overset{|}{N}}}-CH-\boxed{\overset{O}{\overset{\|}{C}}-\overset{H}{\overset{|}{N}}}-\overset{\overset{OH}{|}}{CH_2}-CH_2-\overset{O}{\overset{\|}{C}}-OH + \boxed{H_2O}$$

Two peptide bonds

Alanylglycylserine (Ala-Gly-Ser)

Sample Problem 16.6
Naming a Tripeptide

Identify the amino acids in the following tripeptide and give its name.

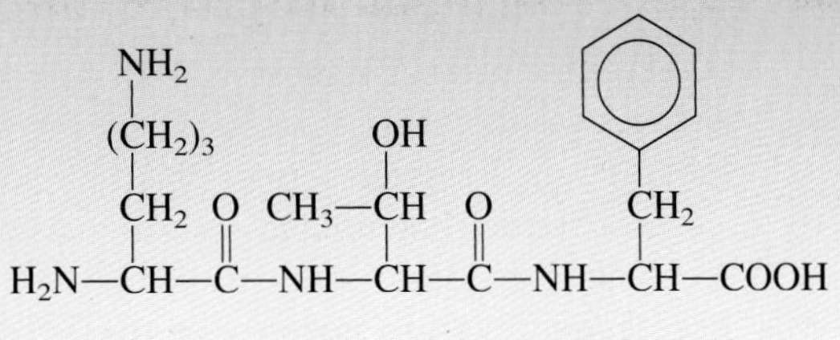

$$H_2N-\overset{\overset{\overset{NH_2}{|}}{\overset{(CH_2)_3}{|}}}{\overset{CH_2}{\overset{|}{CH}}}-\overset{O}{\overset{\|}{C}}-NH-\overset{\overset{OH}{|}}{\overset{CH_3-CH}{\overset{|}{CH}}}-\overset{O}{\overset{\|}{C}}-NH-\overset{\overset{C_6H_5}{|}}{\overset{CH_2}{\overset{|}{CH}}}-COOH$$

Solution:
From the side chains, we can identify the amino acids as lysine, threonine, and phenylalanine. It is named lysylthreonylphenylalanine(Lys-Thr-Phe).

16.5 Primary Protein Structure

Learning Goal **Determine the primary structure of a polypeptide.**

Table 16.5 Tripeptides Possible with Glutamic Acid, Histidine, and Proline

Glu-Pro-His
Glu-His-Pro
Pro-Glu-His
Pro-His-Glu
His-Glu-Pro
His-Pro-Glu

Longer chains of amino acids are called **polypeptides.** When there are more than 50 amino acids in the chain, the polypeptide is called a **protein.** The sequence of amino acids linked by peptide bonds is known as the **primary structure** of the polypeptide or protein. Consider a tripeptide that stimulates the thyroid to release thyroxine, a thyroid hormone. By using acid hydrolysis, all the peptide bonds in the protein can be broken and the resulting mixture analyzed to determine the number and kinds of amino acids. Upon hydrolysis, we find that this tripeptide contains histidine (His), glutamic acid (Glu), and proline (Pro). However, this does not tell us their sequence in the tripeptide. Six tripeptide sequences are possible from these three amino acids, as seen in Table 16.5.

One method used to determine the amino acid sequence is to add Sanger's reagent, 2,4-dinitrofluorobenzene, to a sample of the tripeptide. This reagent reacts with the N-terminal amino acid. Following hydrolysis, the amino acid attached to 2,4-dinitrobenzene can be identified as the first amino acid in our tripeptide.

NO_2

O_2N—⟨benzene ring⟩—F + Glu-His-Pro ⟶

2,4-Dinitrofluorobenzene Tripeptide

NO_2

O_2N—⟨benzene ring⟩—Glu-His-Pro $\xrightarrow{\text{hydrolysis}}$

NO_2

O_2N—⟨benzene ring⟩—Glu + His + Pro

Identify glutamic acid as the N-terminal amino acid

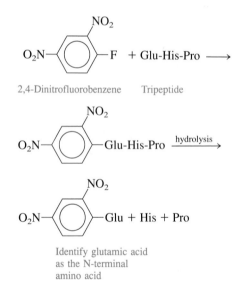

Similar methods remove and identify the amino acids from the N-terminal end, one amino acid at a time. However, such a method is tedious and useful only for small peptides, up to about 20 amino acids.

Another method used to analyze the amino acid sequence is partial hydrolysis of the peptide to give smaller peptide fragments. For example, a partial hydrolysis of our tripeptide might give His-Pro and Glu-His. By overlapping the peptide pieces where amino acids match, we can determine that the primary structure of our tripeptide is Glu-His-Pro.

Overlapping of Peptide Fragments

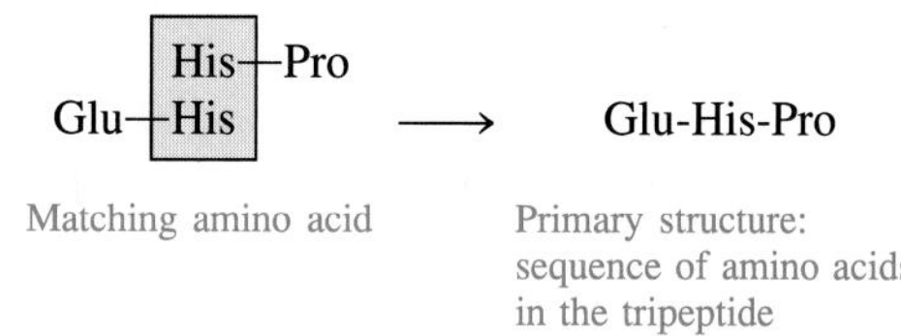

Matching amino acid

Primary structure: sequence of amino acids in the tripeptide

Sample Problem 16.7
Amino Acid Sequence

The complete hydrolysis of a pentapeptide gave a solution containing five different amino acids: Leu, Arg, Phe, Ala, and Ser. Upon reacting the pentapeptide with Sanger's reagent, the hydrolysis products included arginine bonded to 2,4-dinitrobenzene. Upon partial hydrolysis, two peptide fragments were obtained, Ser-Ala-Phe and Arg-Leu-Ser. What is the sequence of amino acids in the pentapeptide?

Solution:
The Arg-2,4-dinitrobenzene product indicates that arginine is the N-terminal amino acid. When the amino acids are matched by overlapping the peptide pieces, the following order is indicated:

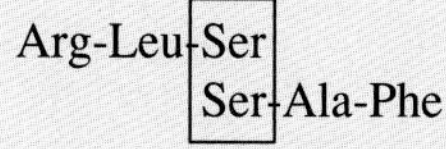

Sequence: Arg-Leu-Ser-Ala-Phe

Study Check:
Complete hydrolysis of a hexapeptide gives Ala(2), Val, Gly, Met, and His. The N-terminal amino acid was alanine. Partial hydrolysis gave the fragments of Val-Ala-Gly, His-Met-Val, and Ala-His-Met. What is the amino acid order in the hexapeptide?

Answer: Ala-His-Met-Val-Ala-Gly

The primary structures for many proteins have been found using similar identification methods. For example, Figure 16.2 shows the primary structure of human insulin, two polypeptide chains held together by disulfide bonds. The sequence of the amino acids is shown by the order of the three-letter abbreviations. The N-terminal amino acid is written on the left and the C-terminal amino acid is on the right.

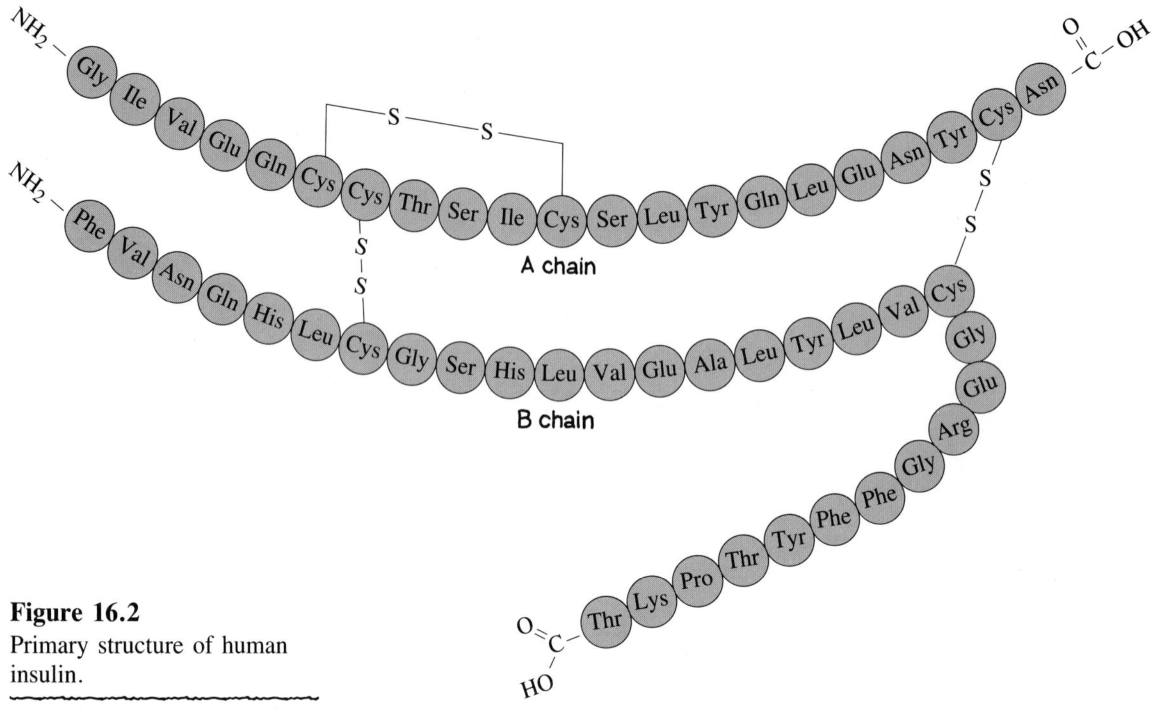

Figure 16.2
Primary structure of human insulin.

HEALTH NOTE

Endorphins, the Body's Natural Opiates

Painkillers known as endorphins and enkephalins are produced naturally in the body. They are polypeptides that bind to the same receptors in the brain as does morphine to give relief from pain. This effect appears to be responsible for the runner's high, the temporary loss of pain when severe injury occurs, and the analgesic effects of acupuncture.

Four groups of endorphins, α, β, γ, and δ, have been identified. Alpha (α)-endorphin contains 16 amino acids, β-endorphin is a polypeptide containing 31 amino acids, γ-endorphin has 17 amino acids, and δ-endorphin has 27 amino acids.

The enkephalins are pentapeptides, the smallest molecules with opiate activity. The enkephalins are found in the thalamus and in parts of the spinal cord that transmit pain impulses. The amino acid sequence of an enkephalin is found in the longer amino acid sequence of the endorphins.

α-Endorphin

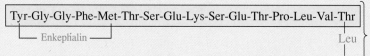

Tyr-Gly-Gly-Phe-Met-Thr-Ser-Glu-Lys-Ser-Glu-Thr-Pro-Leu-Val-Thr

Enkephalin

Leu

Glu-Gly-Lys-Lys-Tyr-Ala-Asn-Lys-Ile-Ile-Ala-Asn-Lys-Phe

β-Endorphin

Substance P, a polypeptide with 11 amino acids, has been found to transmit pain impulses to the brain. Endorphins may act to prevent the release of substance P, which would account for their sedating effects. ■

16.6 Secondary Protein Structures

Learning Goal **Describe the secondary structures of proteins; the alpha helix, the pleated sheet, and the triple helix.**

Figure 16.3
The alpha helix, a secondary protein structure.

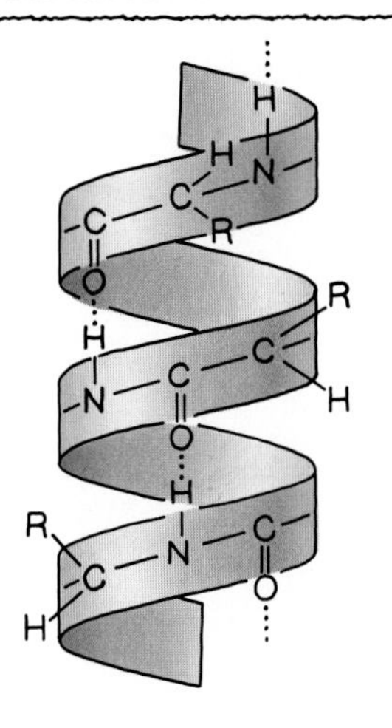

The secondary structures of proteins are brought about by hydrogen bonding within a peptide chain or between several chains. Polypeptides containing amino acids with bulky side chains typically are wound into a helical or corkscrew shape called an **alpha helix** (α-helix). Hydrogen bonds form between a hydrogen in a N—H group and the oxygen of a C=O group in the next turn of the helix. Because many hydrogen bonds occur within the polypeptide, the chain is pulled into a tight coil that looks like a spring or telephone cord. The side chains of the amino acids in the polypeptide extend outward from the alpha helix backbone. See Figure 16.3.

O
‖
—C—N—
|
H
⋮ ← hydrogen bonding between peptide bonds along a polypeptide chain
O
‖
—C—N—
|
H

Figure 16.4
In hair and wool, the secondary structure of the fibrous protein alpha keratin contains three or seven alpha helixes wrapped together.

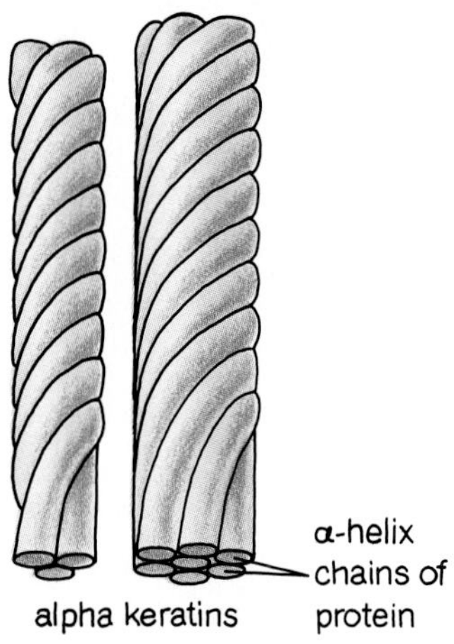

Fibrous proteins called alpha keratins make up the tough fibers of hair, wool, skin, nails, and horns of animals. A fibril consists of three or seven alpha helixes that have been tightly coiled like a rope. See Figure 16.4. When wet, the fibrils in wool stretch as hydrogen bonds are broken and then return to the original shape as the bonds re-form as the wool dries. In hair, the alpha helical chain are held together by disulfide linkages between the side groups of the many cysteines in hair protein.

Pleated Sheet

Another type of secondary structure called the **pleated sheet** is found in fibroin, the fibrous protein of silk. Peptide chains are held together side by side by hydrogen bonds that form between them. Within this sheet of peptide chains, the most prevalent amino acids, glycine, alanine, and serine are bonded at angles that allow their small side chains to extend outward. This results in a series of pleats giving the name pleated sheet. The hydrogen bonds that hold the pleated sheets tightly in place account for the strength and durability of silk. See Figure 16.5.

Figure 16.5
Two polypeptide chains in a pleated sheet.

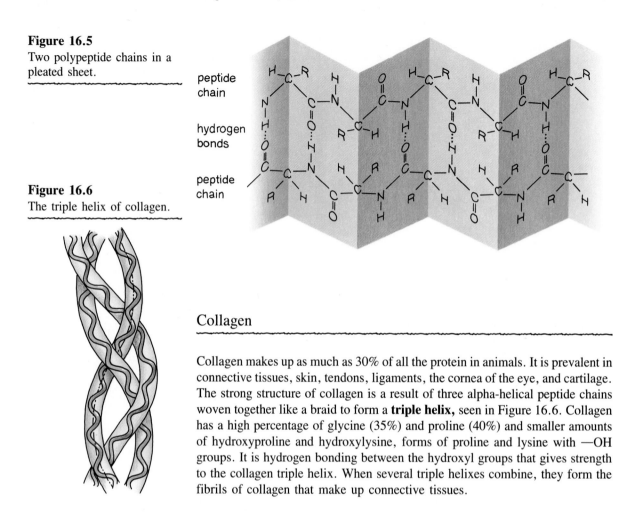

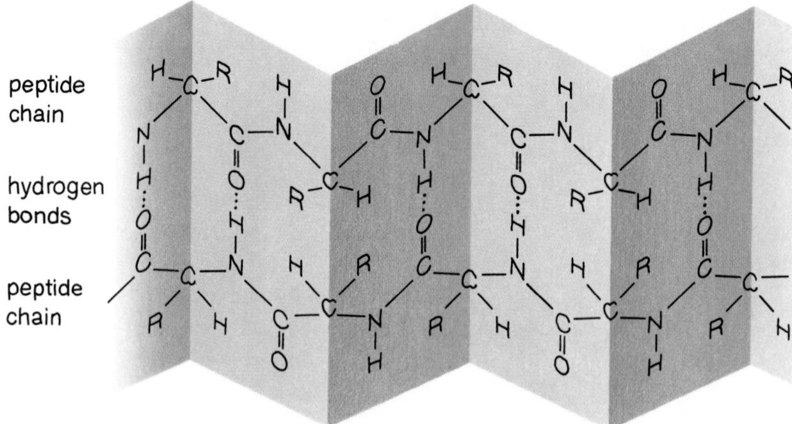

Figure 16.6
The triple helix of collagen.

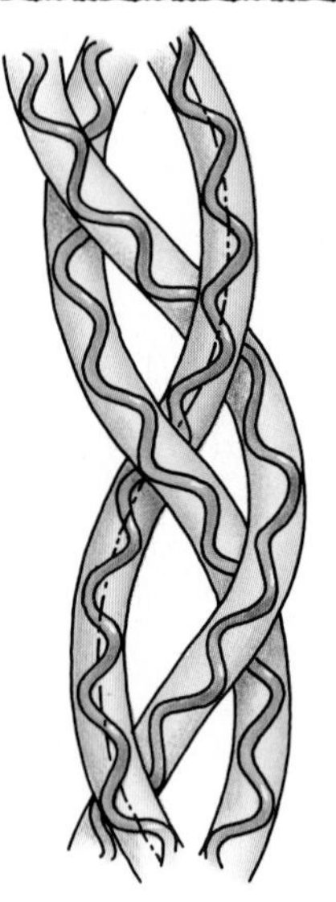

Collagen

Collagen makes up as much as 30% of all the protein in animals. It is prevalent in connective tissues, skin, tendons, ligaments, the cornea of the eye, and cartilage. The strong structure of collagen is a result of three alpha-helical peptide chains woven together like a braid to form a **triple helix,** seen in Figure 16.6. Collagen has a high percentage of glycine (35%) and proline (40%) and smaller amounts of hydroxyproline and hydroxylysine, forms of proline and lysine with —OH groups. It is hydrogen bonding between the hydroxyl groups that gives strength to the collagen triple helix. When several triple helixes combine, they form the fibrils of collagen that make up connective tissues.

Sample Problem 16.8
Identifying Secondary Structures

Indicate the type of secondary structure (alpha helix, pleated sheet, or triple helix) described in each of the following statements:

a. a coiled peptide chain held in place by hydrogen bonding between the peptide bonds within the same peptide chain

b. hydrogen bonding that occurs between peptide chains arranged side by side.

Solution:

a. alpha helix.

b. pleated sheet.

Study Check:
What is the structure of collagen?

Answer: In collagen, three polypeptide chains are wound together to form a triple helix. Hydrogen bonds between the chains maintain the structure of the triple helix.

16.7 Tertiary and Quaternary Protein Structures

Learning Goal **Describe the tertiary and quaternary structures of proteins.**

Unlike the fibrous proteins of hair, wool, silk, or collagen, globular proteins fold up on themselves forming unique three-dimensional structures that are soluble in water. Examples include insulin, hemoglobin, enzymes, and antibodies.

In forming the **tertiary structure,** in which the polypeptide chain folds upon itself, amino acids having polar, hydrophilic side chains are on the outside of the protein, where they can be in contact with the water environment. At the same time, amino acids with nonpolar hydrophobic side chains group together in the center of the protein. See Figure 16.7.

Cross-Linkages

The arrangement of the protein in its tertiary structure is stabilized by cross-links that result from interactions among the various side chains that come into the same area of the folded protein chain. For example, disulfide bonds form between the sulfur atoms of two cysteine side chains.

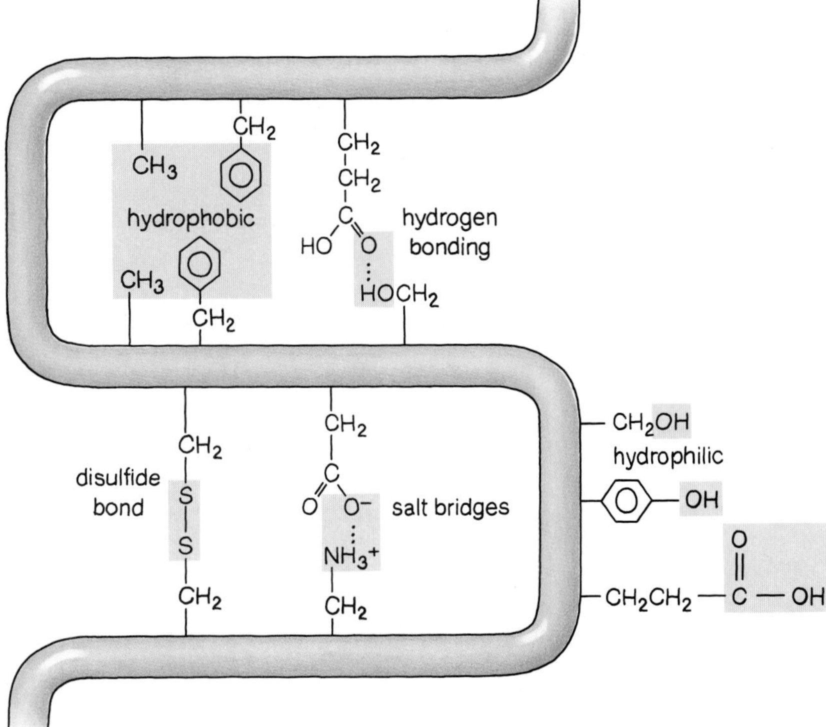

Figure 16.7
Linkages between the side chains of amino acids stabilize the tertiary structure of a globular protein.

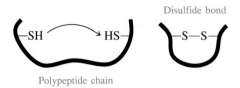

Disulfide bond

$-SH \quad HS-$ $\quad -S-S-$

Polypeptide chain

Ionic bonds, or salt bridges, form between the $-COO^-$ of the side chain of the acidic amino acids, aspartic acid or glutamic acid, and the $-NH_3^+$ of the side chain of the basic amino acids lysine or arginine.

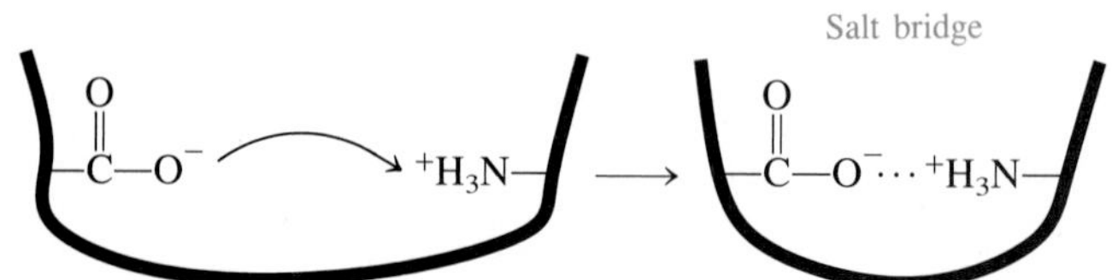

Hydrogen bonds form between a polar side chain such as a carbonyl group ($-C{=}O$) and a hydroxyl ($-OH$) or amino ($-NH_2$) in another amino acid.

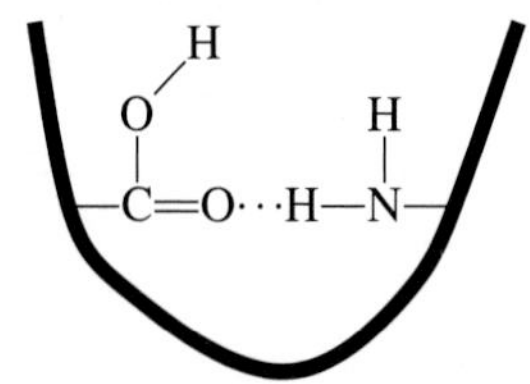

Sample Problem 16.9
Cross-Links in Tertiary Structures

What type of interaction would you expect between the side chains of the following amino acids?

a. cysteine and cysteine
b. glutamic acid and lysine

Solution:

a. Since cysteine contains a —SH side chain, a disulfide bond will form.

$$-SH + HS- \longrightarrow -S-S-$$

Disulfide bond

b. The acidic side group of glutamic acid will form an ionic bond or salt bridge with the basic side group of lysine.

$$-\overset{\overset{\displaystyle O}{\|}}{C}-O^- + H_3\overset{+}{N}- \longrightarrow -\overset{\overset{\displaystyle O}{\|}}{C}-O^-\cdots H_3\overset{+}{N}-$$

From glutamic acid — From lysine — Ionic bond or salt bridge

Study Check:
Where would you expect to find valine and leucine in the tertiary structure of a protein?

Answer: Since both valine and leucine are hydrophobic, we would expect to find them in the center of the protein structure.

Myoglobin

Myoglobin, a polypeptide, stores oxygen in muscle. High concentrations of myoglobin have been found in the muscles of sea mammals such as seals and whales that stay under the water for long periods of time. Myoglobin, as seen in Figure 16.8, contains 153 amino acids in a single polypeptide chain. About 75% of its polypeptide chain exists as an alpha helix. A compact spherical shape is the result of the folding of the polypeptide chain, including its helical portions, into a tertiary structure. In one part of the molecule, there is a heme group, which has an iron(II) ion in its center. It is this heme group and the surrounding nonpolar amino acids that form a pocket for storing oxygen.

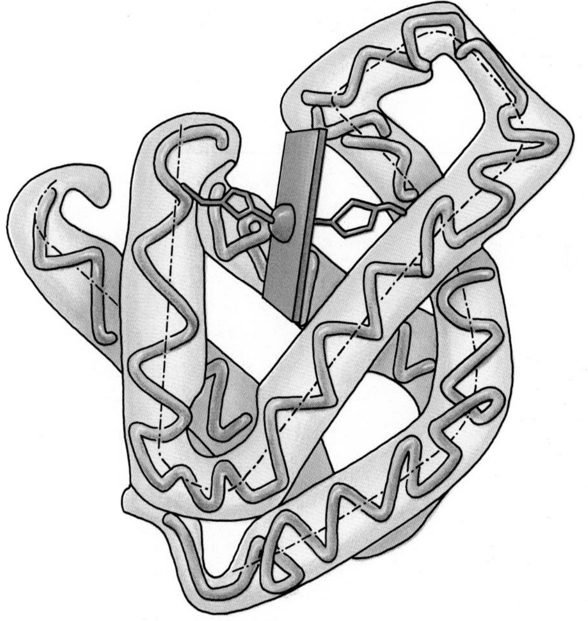

Figure 16.8
Tertiary structure of myoglobin.

Hemoglobin—A Quaternary Structure

Hemoglobin is a globular protein that carries oxygen from the lungs through the bloodstream to body tissues. It consists of four peptide chains or subunits, two α-chains and two β-chains, as shown in Figure 16.9. When a protein requires

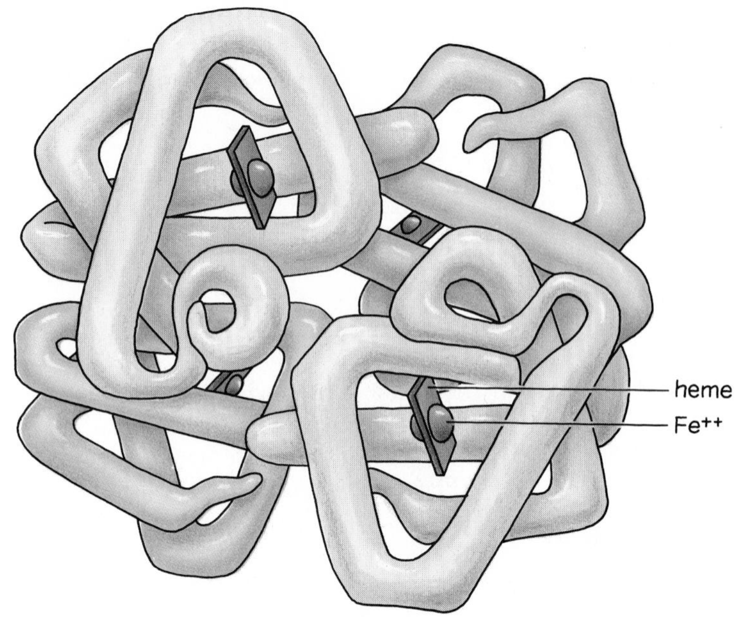

Figure 16.9
The structure of hemoglobin contains four peptide subunits, each containing a heme prosthetic group where oxygen is stored.

two or more peptide subunits for its biological function, the grouping is called a **quaternary structure.** In hemoglobin, all four subunits must be present for the hemoglobin to function as an oxygen carrier. Each subunit contains a heme group, so the complete quaternary structure of hemoglobin can transport up to four molecules of oxygen.

It is interesting to note that hemoglobin and myoglobin have similar biological functions. Hemoglobin carries oxygen in the blood, whereas myoglobin carries oxygen in muscle. Myoglobin is a single polypeptide chain with a molecular weight of 17,000, about one-fourth the molecular weight of hemoglobin (64,000). In addition, the tertiary structure of the single polypeptide myoglobin is almost identical to the tertiary structure of a single subunit of hemoglobin. Myoglobin transports just one molecule of oxygen, just as each subunit of hemoglobin carries one oxygen molecule. The similarity in tertiary structures allows both proteins to bind and release oxygen in a similar manner. Table 16.6 and Figure 16.10 summarize the structural levels of proteins.

Table 16.6 Summary of Structural Levels in Proteins

Structural level	Characteristics
Primary	The sequence of amino acids
Secondary	A coiled or braided shape such as an alpha helix, pleated sheet, or triple helix
Tertiary	A folding of an alpha helix into a compact, three-dimensional shape stabilized by cross-linkages
Quaternary	A combination of 2 or more polypeptide chains that forms a large, biologically active protein

Figure 16.10
Characteristics of the four structural levels of proteins.

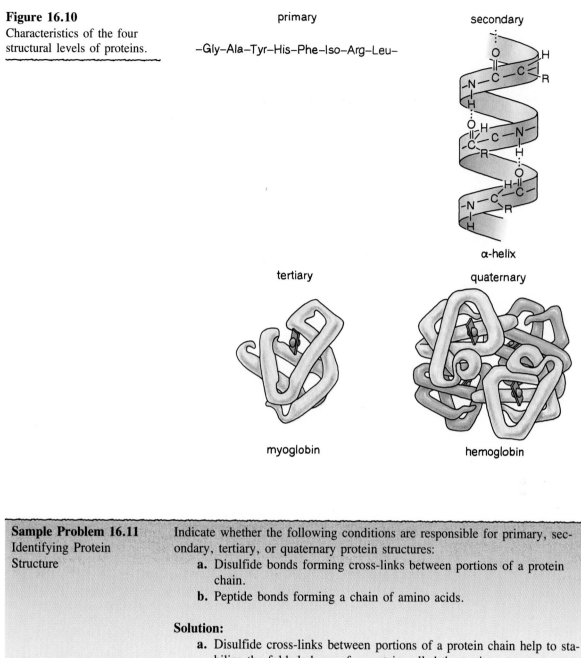

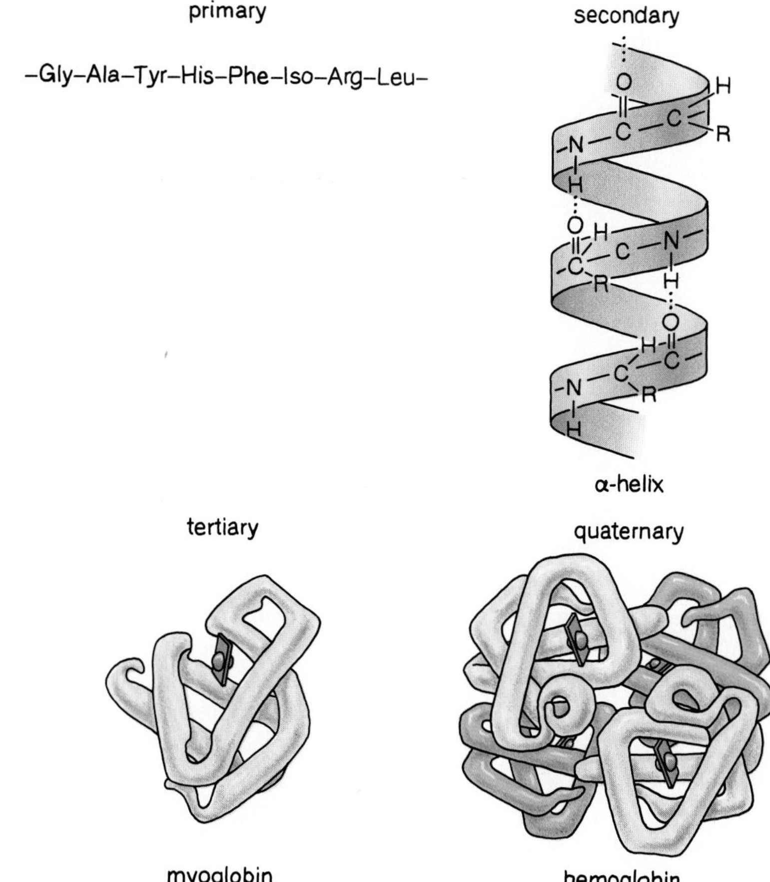

Sample Problem 16.11
Identifying Protein Structure

Indicate whether the following conditions are responsible for primary, secondary, tertiary, or quaternary protein structures:

a. Disulfide bonds forming cross-links between portions of a protein chain.

b. Peptide bonds forming a chain of amino acids.

Solution:

a. Disulfide cross-links between portions of a protein chain help to stabilize the folded shape of a protein called the tertiary structure.

b. The peptide bonds that hold the amino acids in a specific sequence determine the primary structure.

Study Check:

What structural level is represented by the grouping of two subunits in insulin?

Answer: quaternary

HEALTH NOTE

Sickle-Cell Anemia

Sickle-cell anemia is a disease caused by an abnormality in the shape of one of the subunits of the hemoglobin protein. In the beta chain, the sixth amino acid, glutamic acid, which is polar, is replaced by valine, a nonpolar amino acid. The change in polarity of the side chain severely affects the tertiary structure of the hemoglobin chain and greatly alters its ability to function. The affected red blood cells change from a rounded shape to a crescent shape, like a sickle, which interferes with their ability to transport adequate quantities of oxygen. See Figure 16.11.

The sickled cells are removed from circulation more rapidly than normal red blood cells, causing anemia and low oxygen pressure in the tissues. Clumps of sickled cells clog the capillaries, where they cause great pain and organ damage. Critically low oxygen levels may occur in the affected tissues.

Normal beta chain: Val-His-Leu-Thr-Pro-Glu-Glu-Lys-

Sickled beta chain: Val-His-Leu-Thr-Pro-Val-Glu-Lys-

In sickle-cell anemia, both genes for the altered hemoglobin must be inherited. However, there are only a few sickle-cells found in persons who carry one gene for sickle-cell hemoglobin, a condition that is known to provide protection from malaria. ■

Figure 16.11
Comparison of the shape of sickled cells with normal red blood cells. The sickled shape interferes with the ability of hemoglobin to store and carry oxygen.

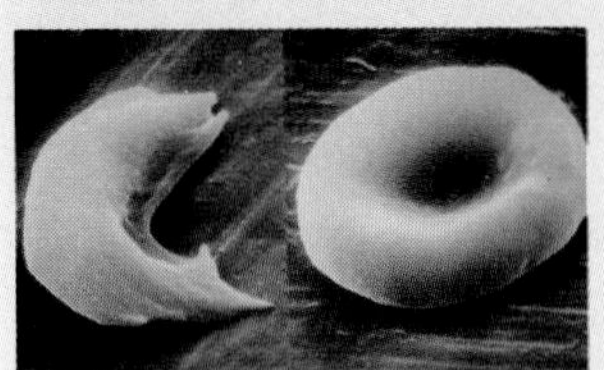

16.8 Classification of Proteins

Learning Goal **Classify proteins by structure and function.**

Proteins have been classified according to structural features and function. **Fibrous proteins,** which are insoluble in water, make up the tough fibers of skin, cartilage, nails, and collagen of connective tissue. In this category, we include the alpha keratins, pleated sheet of silk fibroin, and collagen. **Globular** proteins are water-soluble and so compact that they are nearly spherical in shape. They function in the cells as enzymes, hormones, and antibodies.

Proteins may be classified further by their composition. Simple proteins consist only of a polypeptide chain. Conjugated proteins contain an additional component called a prosthetic group that is not an amino acid or peptide. Some examples are given in Table 16.7.

Classification of Proteins by Function

Proteins perform many functions in the body. Some, called enzymes, regulate biological reactions such as digestion and cellular metabolism. Other proteins

Table 16.7 Classification of Proteins by Composition

Simple proteins Type	Examples
Albuminoids	Keratin in skin, hair, nails Collagen in cartilage
Albumins	Egg albumin Serum albumin
Globulins	Antibodies
Histones	Chromatin in chromosomes

Conjugated proteins Type	Examples	Prosthetic group
Glycoproteins	Interferon	Carbohydrate
Lipoproteins	Lipoproteins	Triglyceride, phospholipid, cholesterol
Metalloproteins	Hemoglobin	Metal ion
Nucleoproteins	Ribosomes	Nucleic acids
Phosphoproteins	Casein	Phosphate

form structural components such as cartilage, hair, and nails. Still other proteins, hemoglobin and myoglobin, carry oxygen in the blood and muscle. Table 16.8 gives examples of proteins that are classified by their functions in biological systems.

Table 16.8 Classification of Proteins by Function

Classification of protein	Examples	Function
Structural	Collagen Keratin	Builds tendons and cartilage Forms hair, skin, wool, and nails
Contractile	Myosin Actin	Muscle contraction Muscle contraction
Transport	Hemoglobin Lipoproteins	Transports oxygen in bloodstream Transport lipids through the body fluids
Hormonal	Insulin Growth hormone Estradiol Testosterone	Increases glucose metabolism Regulates body growth Stimulates female sex characteristics Stimulates male sex characteristics
Catalytic	Sucrase Trypsin Dehydrogenase	Catalyzes hydrolysis of sucrose Catalyzes hydrolysis of protein Converts ethanol to acetaldehyde
Antibodies	Immunoglobulins	Recognizes and destroys bacterial or viral antigens
Storage	Casein Ferritin	Stores protein as milk Stores iron in liver for the production of red blood cells

Sample Problem 16.11
Classifying Proteins by Function

Match the following functions with the protein:

a. casein **b.** lipase **c.** hemoglobin

1. catalyzes metabolic reactions of lipids
2. carries oxygen in the bloodstream
3. stores amino acids in milk

Solution:

a. 3. The protein casein stores amino acids in milk.
b. 1. The protein lipase is an enzyme that catalyzes metabolic reactions of lipids.
c. 2. The transport protein hemoglobin carries oxygen through the bloodstream.

16.9 Properties of Proteins

Learning Goal **Discuss denaturation, hydrolysis, electrophoresis, and chemical tests for proteins.**

The biological function of a protein depends upon its three-dimensional shape. However, changes called **denaturation** occur when the hydrogen bonds and cross-linkages that maintain the protein structure are broken. Factors that can cause denaturation include heat, acid, base, and heavy metal ions. When the protein is denatured, it unfolds like a loose piece of spaghetti. With the loss of its shape, it is no longer biologically active. (See Figure 16.12.)

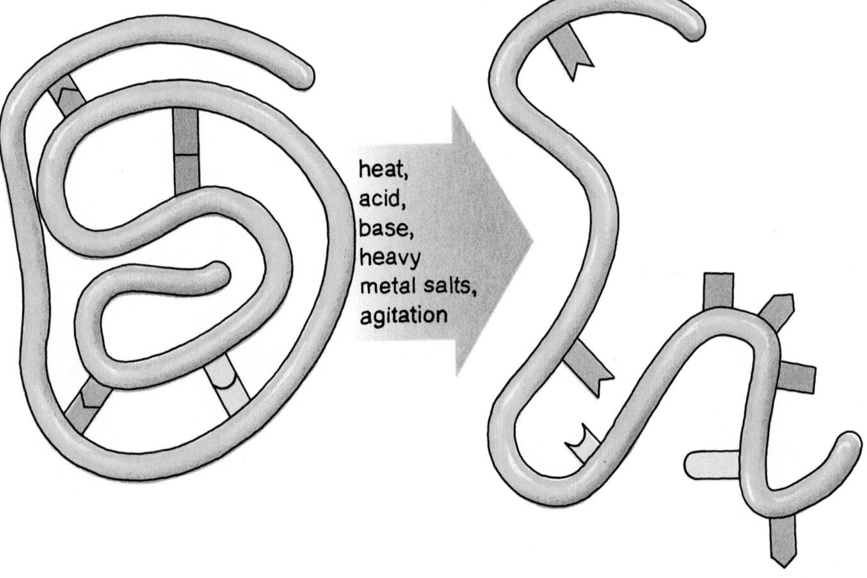

Figure 16.12
Denaturation of a protein. Denaturing agents such as heat, acid, base, and heavy-metal salts cause a disruption of the tertiary structure, destroy its globular shape, and render the protein inactive.

When denaturation occurs under mild conditions, the protein may be restored to its original shape and biological activity by carefully reversing the conditions. However, most changes are so drastic that the denaturation cannot be reversed. The protein becomes solid and precipitates out of solution.

Heat

The conditions at which proteins function most effectively are called optimum conditions. The optimum temperature for most proteins is 37°C (body temperature). Few proteins can remain biologically active above 50°C. Both heat and ultraviolet light denature proteins because they increase kinetic energy which disrupts hydrogen bonds and attractions between nonpolar side groups. Whenever you cook food, you are using heat to denature protein. An egg that is placed in boiling water (100°C) is denatured because heat coagulates the proteins in the egg.

High temperatures are often used to disinfect surgical instruments, gowns, and gloves. When these materials are subjected to the high temperatures of an autoclave, the proteins of any bacteria present will be denatured, making the bacteria inactive.

Acids and Bases

Adding an acid or a base to a protein solution changes the ionic charges of the acidic and basic side groups. As a result, salt bridges are disrupted. For example, when acid is added to milk, the pH is lowered to the isoelectric point of the milk protein casein. The casein becomes insoluble in water, clumps together (curdles), and separates. In the preparation of yogurt and cheese, a bacteria is added to milk. This bacteria produces lactic acid which denatures the milk protein.

Tannic acid is a weak acid used in burn ointments to cause a coagulation of the proteins at the site of the burn. The solid protein that forms provides a protective cover, which prevents further loss of fluid from the burn.

Organic Solvents

Solvents such as ethanol and isopropyl alcohol are used as disinfectants because they cause denaturation by forming their own hydrogen bonds with the protein. A 70% solution of ethanol can pass through the cell walls of bacteria and cause coagulation of proteins within the cell. When an alcohol swab is used to clean wounds or to prepare the skin for an injection, it destroys bacteria, thus sterilizing the skin and preventing infection.

Heavy-Metal Ions

The heavy-metal ions Ag^+, Pb^{2+}, and Hg^{2+} react with the disulfide bonds of protein and the acidic amino acids. A loss of disulfide bonds and salt bridges denatures the protein, and it precipitates out of solution. Sometimes salts of heavy metals are used as antiseptics. For example, in hospitals dilute (1%) solutions of $AgNO_3$ have been placed in the eyes of newborn babies to destroy the bacteria that cause gonorrhea.

If heavy-metals ions are ingested, they can severely disrupt body proteins, especially in the stomach. An antidote for their accidental ingestion is a high-protein food such as milk, eggs, or cheese. Proteins in the food tie up the heavy-metal ions until the stomach can be pumped or vomiting can be induced.

Agitation

The whipping of cream and the beating of egg whites into meringue are examples of mechanical agitation that denatures a protein. The violent whipping action causes the protein to stretch until the cross-linkages break apart. Table 16.9 summarizes the factors that can denature protein and their effects.

Table 16.9 Factors That Denature Protein

Factors	Effect on protein structure
Heat	Disrupts hydrogen bonds and hydrophobic attractions; can cause coagulation
Acids and bases	Disrupt salt bridges
Organic solvents	Disrupt hydrogen bonds
Heavy-metal ions (Ag^+, Pb^{2+}, and Hg^{2+})	React with disulfide bonds and acidic amino acids
Agitation	Stretches a protein until cross-links break forming a solid

Sample Problem 16.12
Effects of Denaturation

Explain what happens to the tertiary structure of a protein when it is placed in an acidic solution.

Solution:
An acid causes denaturation of the protein. When a hydrogen ion adds to an acidic side chain ($—COO^-$), the group ($—COOH$) loses its ability to form a salt bridge. A loss in ionic attractions causes the tertiary structure to unfold and lose its shape.

Hydrolysis

The hydrolysis of a protein occurs in the presence of acid or an enzyme such as pepsin or trypsin. The hydrolysis of a protein produces smaller peptides and eventually individual amino acids.

$$H_2N{-}CH_2{-}\overset{\overset{O}{\|}}{C}{-}\underset{\underset{H}{|}}{N}{-}\overset{\overset{CH_3}{|}}{CH}{-}\overset{\overset{O}{\|}}{C}{-}OH + H_2O \xrightarrow[]{H^+ \text{ or Enzyme}}$$

Glycylalanine (Gly-Ala)

$$H_2N{-}CH_2{-}\overset{\overset{O}{\|}}{C}{-}OH + H_2N{-}\overset{\overset{CH_3}{|}}{CH}{-}\overset{\overset{O}{\|}}{C}{-}OH$$

Glycine (Gly) Alanine (Ala)

Hydrolysis is the reaction that occurs when you digest protein. In the stomach, enzymes split the protein chain into smaller polypeptides. In the small intestine, other enzymes called peptidases hydrolyze the fragments to give amino acids, which can be absorbed into the bloodstream.

Stomach:

$$\text{Proteins} \xrightarrow[\text{pH 1–2}]{} \text{Polypeptides}$$

Small intestine:

$$\text{Polypeptide} \xrightarrow[\text{pH 8–9}]{\text{Peptidases, Trypsin}} \text{Amino acids}$$

Electrophoresis

Table 16.10 Isoelectric Points of Some Proteins

Proteins	pH
Egg albumin	4.6
Casein (milk)	4.6
Gelatin	4.8
Insulin	5.3
Hemoglobin	6.8
Myoglobin	7.0
Chymotrypsin	9.5
Lysozyme	11.0

Proteins are not very soluble at their isoelectric points. In proteins, it is the side chains, particularly of the acidic and basic amino acids, that determine the pH at which they are electrically neutral. Table 16.10 lists isoelectric points for some proteins.

Electrophoresis is used in the hospital laboratory to separate proteins by their different isoelectric points. A protein mixture is placed at the center of a paper strip or gel between two electrodes, as shown in Figure 16.13. A buffer solution of a certain pH is applied to the paper or gel. As an electric current is passed through the mixture, positively charged proteins move toward the negative electrode and negatively charged proteins move toward the positive electrode.

The proteins are separated because they migrate at different rates toward an electrode, depending on their amount of ionic charge. Any protein at its isoelectric point is electrically neutral and does not move during electrophoresis. The paper strip is then sprayed or dipped in reagents that make the separated proteins visible. They are identified by their direction and rate of migration in the electric field.

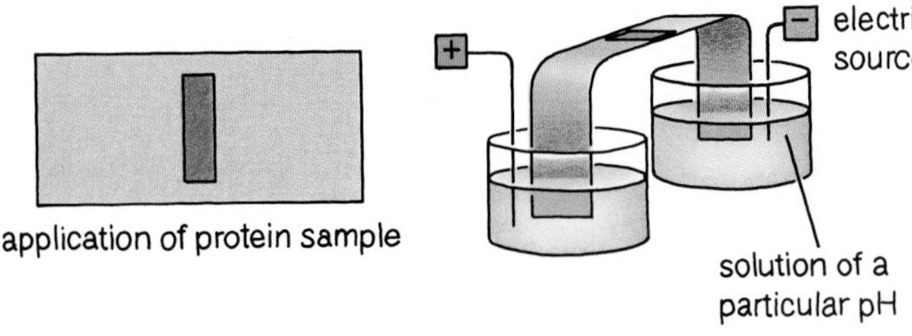

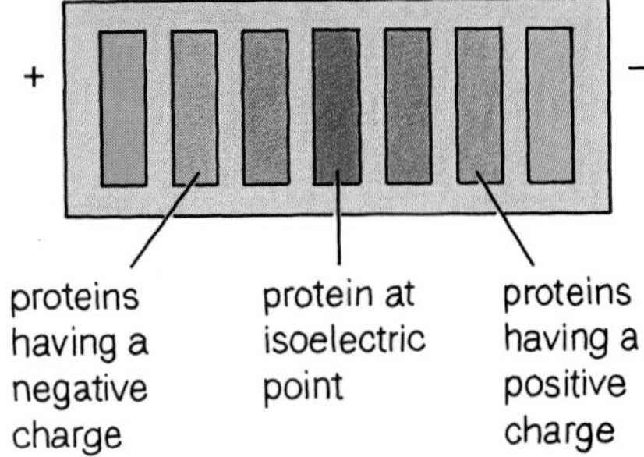

Figure 16.13
A protein having a positive charge moves toward the negative electrode; a protein having a negative charge moves toward the positive electrode; an electrically neutral protein is at its isoelectric point and does not move.

Testing for Amino Acids and Proteins

Amino acids, peptide bonds, and some side groups of amino acids react with certain reagents to give very distinctive colors. In the biuret test, a peptide or protein is treated with biuret reagent ($CuSO_4$ and NaOH). The cupric ion (Cu^{2+}) forms a violet to blue color when it reacts with peptides that have two or more peptide bonds. No reaction occurs when amino acids or dipeptides are treated with biuret reagent.

In the ninhydrin test, amino acids with free amino groups produce a blue to purple color when heated with ninhydrin. If paper chromatography is used to separate the amino acids, the formation of blue spots on the paper after spraying with ninhydrin identifies the individual amino acids. Proline and hydroxyproline, two amino acids that have substituted amino groups, give a yellow color with ninhydrin.

In the xanthoproteic test, a protein containing amino acids with a benzene ring, tryptophan, tyrosine, or phenylalanine, gives a yellow color with nitric acid, HNO_3. If you spill nitric acid on your skin, a yellow spot appears because these amino acids are present in skin proteins.

In the sulfur test, the sulfur-containing amino acids cysteine and cystine react with lead acetate to form a black precipitate of lead sulfide, PbS. No reaction occurs with methionine.

Glossary of Key Terms

alpha helix A secondary level of protein structure consisting of hydrogen bonds within a polypeptide chain that forms a structure that is coiled like a corkscrew.

amino acid The building block of proteins, consisting of an amino group, a carboxylic acid group, and a side chain that is unique.

complementary protein A grouping of proteins that provides all the essential amino acids.

complete protein A protein that provides all of the essential amino acids in the proper amounts.

denaturation The loss of secondary and tertiary protein structure caused by heat, agitation, acids, bases, organic solvents, or heavy metals.

electrophoresis The use of electric current to separate proteins with different isoelectric points.

essential amino acid An amino acid that must be supplied by the diet because the body is unable to synthesize the amounts required each day.

fibrous protein A protein that is insoluble in water; consists of polypeptide chains with alpha helixes, pleated sheets, or triple helixes that make up the fibers of hair, wool, skin, nails, and silk.

globular protein A protein that acquires a compact shape from attractions between side chains of the amino acids in the protein.

hydrophilic amino acid An amino acid having polar, acidic, or basic side chains that are attracted to water; water-loving.

hydrophobic amino acid An amino acid having hydrocarbon side chains that are ''water-fearing'' and not soluble in water.

incomplete protein A protein, usually from vegetables and grains, that is deficient in one or more essential amino acids.

isoelectric point The pH at which an amino acid or protein becomes electrically neutral.

nonpolar amino acid An amino acid that is not soluble in water because it contains a nonpolar side chain.

peptide The combination of two or more amino acids joined by amide bonds, dipeptide, tripeptide, etc.

peptide bond The amide bond that links amino acids in polypeptides and proteins.

pleated sheet A secondary level of protein structure that consists of hydrogen bonds between polypeptide chains, forming a regular zigzag pattern.

polar amino acid An amino acid that is soluble in water because its side chain contains a polar group such as hydroxyl (−OH), thiol, a carbonyl, amino, or carboxyl.

polypeptide A polymer of many amino acids joined by peptide bonds.

primary structure The sequence of the amino acids linked by peptide bonds in a protein.

protein A term used for polypeptides with 50 or more amino acids in the peptide chain.

quaternary structure A protein structure in which two or more protein units combine to form an active protein.

tertiary structure The folding of a protein into a compact structure maintained by the interaction of side groups such as salt bridges and disulfide bonds.

triple helix The protein structure found in collagen consisting of three polypeptide chains woven together like a braid.

zwitterion The dipolar form of an amino acid consisting of two oppositely charged ionic regions.

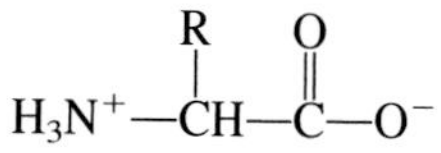

$$H_3N^+\text{—}\overset{\displaystyle R}{\overset{|}{C}H}\text{—}\overset{\displaystyle O}{\overset{\|}{C}}\text{—}O^-$$

Problems

Amino Acids *(Goal 16.1)*

16.1 Describe the features of an amino acid.

16.2 Draw the structural formula of an amino acid that contains the following:

a. sulfur
b. an alcohol side chain
c. a basic side chain
d. an acidic side chain
e. a benzene ring
f. a hydrocarbon side chain

16.3 Give the name and abbreviation for each of the following amino acids:

a. 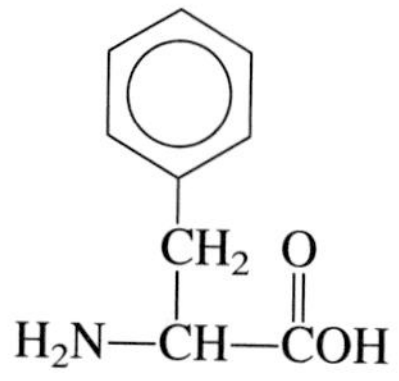 $H_2N\text{—}\overset{\displaystyle C_6H_5\text{—}CH_2}{\overset{|}{C}H}\text{—}\overset{\displaystyle O}{\overset{\|}{C}}OH$ (side chain: benzene ring—CH_2)

b. 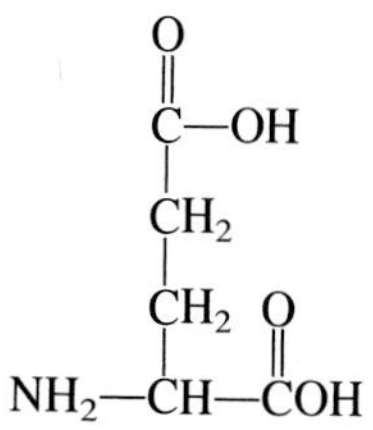 $NH_2\text{—}\overset{\displaystyle HO\text{—}\overset{O}{\overset{\|}{C}}\text{—}CH_2\text{—}CH_2}{\overset{|}{C}H}\text{—}\overset{\displaystyle O}{\overset{\|}{C}}OH$

c. $H_2N—\overset{\overset{CH_3\ \ CH_3}{\diagdown\ \diagup}}{\underset{|}{CH}}$ ...

$$\text{c. } H_2N-\underset{\displaystyle |}{\overset{\displaystyle (CH_3)_2CH}{CH}}-\overset{\displaystyle O}{\overset{\|}{C}}OH \qquad \text{d. } NH_2-\overset{\displaystyle CH_3-CH_2-\overset{|}{C}H-CH_3}{\overset{|}{C}H}-\overset{\displaystyle O}{\overset{\|}{C}}OH$$

16.4 Indicate whether each side group of the amino acids in Problem 16.3 is hydrophobic or hydrophilic.

Essential Amino Acids *(Goal 16.2)*

16.5 What is an essential amino acid?

16.6 Seeds and vegetables are often deficient in one or more essential amino acids. Using the following table, state whether the following combinations would provide complementary proteins:

Source	*Lysine*	*Tryptophan*	*Methionine*
Oatmeal	low	ok	ok
Rice	low	ok	ok
Garbanzo beans	ok	low	ok
Lima beans	ok	low	low
Cornmeal	low	ok	ok

a. rice and garbanzo beans
b. lima beans and cornmeal
c. a salad of garbanzo beans and lima beans
d. rice and lima beans
e. rice and oatmeal
f. cornmeal and lima beans

16.7 What disease can develop when the diet does not include all of the essential amino acids?

Properties of Amino Acids *(Goal 16.3)*

16.8 a. What is a zwitterion?
b. How does an amino acid ionize?
c. What is the isoelectric point of an amino acid?

16.9 Write the zwitterions of the following amino acids:
a. glycine d. tyrosine
b. phenylalanine e. methionine
c. serine f. leucine

***16.10** The isoelectric point of leucine occurs at a pH of 6.0. Indicate the ionic structure of leucine at the following pH values:
a. pH 2.0 d. pH 8.0
b. pH 4.0 e. pH 10.0
c. pH 6.0

***16.11** Aspartic acid has an isoelectric point of 2.8. Indicate its ionic structure at the pH values given in Problem 16.10.

Peptides *(Goal 16.4)*

16.12 Draw the structural formulas of the following dipeptides:
a. Ala-Cys b. Serylisoleucine c. Met-Gln

16.13 Write out two possible structural formulas and names for the dipeptides that could form from the following:
a. glycine and leucine b. alanine and tyrosine

16.14 Draw the structural formulas for the following tripeptides:
a. Gly-Ala-Ala
b. Met-Tyr-Phe
c. Glucylvalylisoleucine

16.15 Write the structural formulas for the following tetrapeptides:
a. Val-Ser-Cys-Gln b. Asp-Val-Gly-Tyr

Primary Protein Structure *(Goal 16.5)*

16.16 Determine the amino acid sequence for the following peptides from the N-terminal amino acid and the fragments obtained from partial hydrolysis:

Amino acids	*N-terminal amino acid*	*Peptide fragments*
a. Glu, Phe, Tyr	Tyr	Phe-Glu + Tyr-Phe
b. Ala, Cys, His, Val	Cys	Cys-Val + Val-Ala + Ala-His
c. Arg, Glu, Gly, Leu, Thr	Thr	Arg-Leu-Glu + Thr-Gly-Arg + Gly-Arg-Leu
*d. Cys, Gly, Lys(2), Met, Val	Lys	Lys-Val-Gly + Lys-Met + Met-Cys-Lys

Secondary Protein Structures *(Goal 16.6)*

16.17 Describe the alpha helix structure of a polypeptide.

***16.18** Silk fibroin forms a pleated sheet. How does it differ from the alpha helix?

16.19 a. Describe the structure of collagen.
b. Compare collagen and alpha keratin.

Tertiary and Quaternary Protein Structures *(Goal 16.7)*

16.20 In myoglobin, about half of the 153 amino acids have nonpolar side chains.
a. Where would you expect those amino acids to be located in the tertiary structure?

b. Where would you expect the polar side chains to be?
c. Why would the globular tertiary structure of myoglobin be more soluble in water than the pleated sheet of silk?

16.21 What type of cross-linkages would you expect from the following side chains in a tertiary structure?
a. two cysteines
b. glutamic acid and lysine
c. serine and aspartic acid
d. two leucines

16.22 A portion of a polypeptide chain contains the following sequence of amino acids

-Leu-Val-Cys-Asp-

a. Which amino acid in the sequence can form a disulfide cross-link?
b. Which amino acid(s) would most likely be found inside the protein structure? Why?
c. Which amino acid(s) would be found on the outside of the protein? Why?
d. How does the primary structure of a protein affect its tertiary structure?

16.23 What are some differences between the following pairs?
a. secondary and tertiary protein structures
b. essential and nonessential amino acids
c. polar and nonpolar amino acids
d. di- and tripeptides
e. a salt bridge and a disulfide bond
f. fibrous and globular proteins
g. alpha helix and pleated sheet
h. alpha helix and triple helix
i. tertiary and quaternary proteins

16.24 State whether the following statements apply to primary, secondary, tertiary, or quaternary protein structure:
a. Side chains interact to form cross-links such as disulfide bonds or salt bridges.
b. Two protein units combine to form an active protein.
c. Peptide bonds hold amino acids together in a polypeptide chain.
d. Several peptide chains are held together by hydrogen bonds between adjacent chains.
e. Hydrogen bonding between carbonyl oxygen atoms and nitrogen atoms of amide groups causes a peptide to coil.
f. Hydrophobic side chains seeking a nonpolar environment move toward the inside portions of the chain.
g. Protein chains of collagen form a triple helix.

Classification of Proteins *(Goal 16.8)*

16.25 Classify the following as fibrous or globular proteins.

a. lipase, an enzyme
b. muscle fibers
c. hemoglobin
d. hair
e. collagen
f. antibodies

16.26 Classify the following as simple or conjugated proteins.
a. globulins
b. hemoglobin
c. lipoprotein
d. collagen
e. keratins in hair

16.27 Match the following function of proteins with the examples listed below:
(1) catalytic
(2) structural
(3) transport
(4) storage
(5) protective
(6) hormonal
(7) contractile
(8) toxin
a. hemoglobin, oxygen carrier in the blood
b. collagen, a major component of tendons and cartilage
c. keratin, a protein found in hair
d. amylase, an enzyme that hydrolyzes starch
e. insulin, a hormone needed for glucose utilization
f. antibodies, proteins that disable foreign proteins
g. casein, milk protein
h. lipases, enzymes that hydrolyze lipids

Properties of Proteins *(Goal 16.9)*

16.28 An egg placed in hot water at 100°C is soft boiled in about 3 min.
a. What changes have occurred in the protein structure of the protein in the egg white?
b. What parts of the protein structure remain unchanged?

16.29 How would the strong acid in the stomach affect the structure of protein eaten during a meal?

16.30 Prior to an injection, the skin is cleansed with an alcohol swab. How does the alcohol affect the bacteria on the skin?

16.31 How do each of the following take advantage of the denaturation reactions of proteins?
a. adding $AgNO_3$ to the eyes of newborn infants
b. placing tannic acid on a burn
c. heating milk to 60°C to make yogurt
d. placing surgical instruments in a 120°C autoclave
e. treating seeds with a $HgCl_2$ solution

16.32 Write the hydrolysis products of the following peptides:

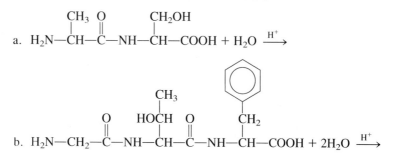

a. $H_2N{-}\underset{\large CH_3}{CH}{-}\overset{O}{\overset{\|}{C}}{-}NH{-}\underset{\large CH_2OH}{CH}{-}COOH + H_2O \xrightarrow{H^+}$

(In a., CH_3 and CH_2OH are printed above the CH groups.)

b. $H_2N{-}CH_2{-}\overset{O}{\overset{\|}{C}}{-}NH{-}\underset{HOCH(CH_3)}{CH}{-}\overset{O}{\overset{\|}{C}}{-}NH{-}\underset{CH_2C_6H_5}{CH}{-}COOH + 2H_2O \xrightarrow{H^+}$

(In b., the side chains HOCH–CH_3 and CH_2–phenyl (benzene ring) are printed above the CH groups.)

16.33 How does denaturation of a protein differ from its hydrolysis?

***16.34** A protein mixture is placed on a gel for electrophoresis. The proteins have the following isoelectric points: albumin, 4.9, hemoglobin, pH 6.8, and lysozyme, 11.0. A buffer of pH 6.8 is placed on the gel.

a. Which protein will migrate toward the positive electrode?
b. Which protein will migrate toward the negative electrode?
c. Which protein will remain where it was originally placed and not migrate at all?

16.35 Describe the chemical test you would use to determine the presence of each of the following:

a. tripeptide
b. tyrosine
c. cysteine
d. protein
e. proline

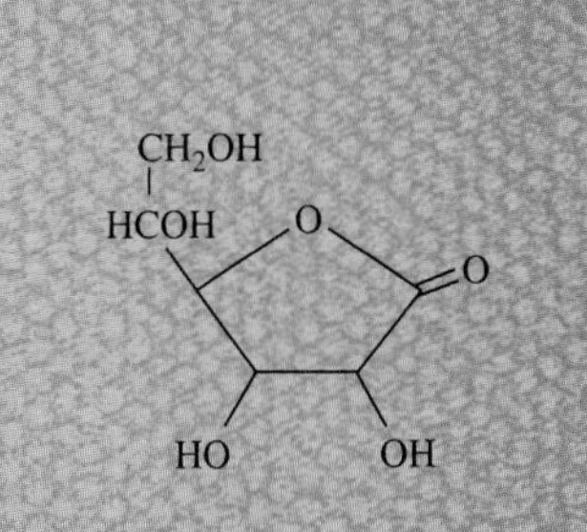

17

Enzymes, Vitamins, and Hormones

Learning Goals

17.1 Describe the role of an enzyme in an enzyme-catalyzed reaction.

17.2 Name and classify an enzyme.

17.3 Describe the effects of substrate and enzyme concentration, temperature, and pH on enzyme activity.

17.4 Describe the effect of an inhibitor upon enzyme activity.

17.5 Describe the enzymes that hydrolyze carbohydrates, lipids, and proteins during digestion.

17.6 Describe a vitamin in terms of water or fat solubility and biological function.

17.7 Describe some steroid and peptide hormones.

*tamin C, an ester found in
esh fruits and vegetables,
: necessary for the catalytic
tivity of certain enzymes.*

SCOPE

Enzymes are proteins that increase the rate of biological reactions within the cells of the body. Enzymes digest our food, contract our muscles, and synthesize hemoglobin. They catalyze reactions that supply all the necessary biomolecules and energy for our survival.

Our diets contain carbohydrates, fats, and proteins, but the cells in our bodies cannot use them until they go through the process of digestion, the chemical breakdown of food catalyzed by digestive enzymes. Polysaccharides are hydrolyzed to monosaccharides, fats are broken down to glycerol and fatty acids, and proteins are hydrolyzed to amino acids. These products are small enough to be absorbed through the cell walls for use in biological processes and growth.

Our diets also provide minerals and vitamins, substances required by many enzymes for biological activity. There was a time when sailors on long sea voyages became ill with scurvy, a condition characterized by skin hemorrhages and extreme tenderness in the joints. It was discovered that scurvy was caused by a deficiency of vitamin C. By eating limes and lemons on long voyages, the sailors did not come down with scurvy.

Hormones are chemical substances that relay messages from an organ of the body to another part of the body. Chemically related to steroids or peptides, hormones released into the bloodstream increase the rate of particular cellular reactions.

17.1 Enzyme Action

Learning Goal **Describe the role of an enzyme in an enzyme-catalyzed reaction.**

In the laboratory, we can break down polysaccharides, fats, or proteins by using a strong acid or base, high temperatures, and long reaction times. However, such drastic conditions are destructive to biological systems. Instead, our cells contain enzymes that carry out the same reactions quickly and under milder conditions. By definition, an **enzyme** is a protein that catalyzes biological reactions. A reactant must first obtain a certain amount of energy before it can take part in a chemical reaction. This is true even if the reaction takes place in our cells.

Consider an analogy. Suppose you are going to take a walk. However, the place you want to go is on the other side of a steep hill. You will need a certain amount of energy to activate yourself so that you can get to the top of the hill. Once there, you will be able to walk easily down the other side to your destination (the *product* of your walk). Now suppose someone tells you there is a tunnel that goes through the hill. By taking this alternate route, you will not need to use as much energy to reach your destination. That makes it possible for you to get there faster.

An enzyme is like this tunnel. It allows a chemical reaction to proceed in a different way with less energy. Enzymes catalyze reactions because they lower the energy needed by the reactant. For example, the hydrolysis of a protein in the diet would occur so slowly without a catalyst that it would not meet the body's requirements for amino acids. However, when a digestive enzyme such as pepsin or trypsin is present, the hydrolysis occurs at a much faster rate.

Figure 17.1
The energy needed to activate the substrates CO_2 and H_2O is lowered by the enzyme carbonic anhydrase.

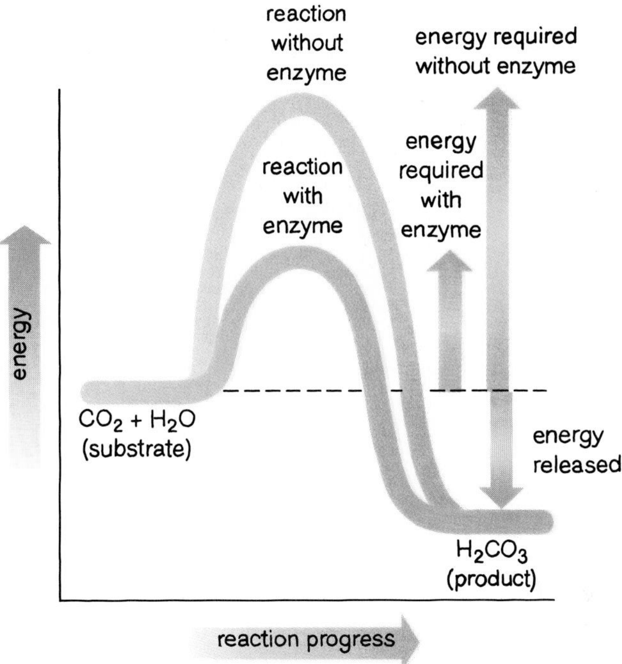

In the blood, an enzyme called carbonic anhydrase converts large amounts of carbon dioxide and water to carbonic acid. It makes the reaction go 10 million times faster than it goes without the catalyst. The energy diagram in Figure 17.1 illustrates the effect of an enzyme catalyst.

Lock-and-Key Model for Enzyme-Catalyzed Reactions

In an enzyme-catalyzed reaction, an enzyme (**E**) must first combine with a reaction, the **substrate** (**S**). Most enzymes catalyze one kind of reaction, in which they break or form a particular bond. When an enzyme reacts with only one substrate, it is said to exhibit **substrate specificity.**

The entire three-dimensional structure of the enzyme is responsible for recognizing and binding to a specific substrate. The combination is called an **enzyme–substrate complex** (**ES**). The substrate fits into a pocket formed by a small group of amino acids called the **active site.** This is called a **lock-and-key model** because the substrate must fit into the active site the way a key fits into a lock. (See Figure 17.2.) With the substrate at the active site, a chemical reaction occurs that involves breaking or forming bonds of the substrate. The products that form are no longer attracted to the active site and leave the enzyme. The enzyme goes on to catalyze the same reaction with other substrate molecules.

$$\underset{\text{Enzyme + Substrate}}{E + S} \rightleftharpoons \underset{\text{Enzyme–Substrate complex}}{ES} \longrightarrow \underset{\text{Enzyme + Product}}{E + P}$$

Figure 17.2
A substrate binds to the active site of an enzyme. Reaction at the active site breaks bonds or forms new bonds to give new products which are released.

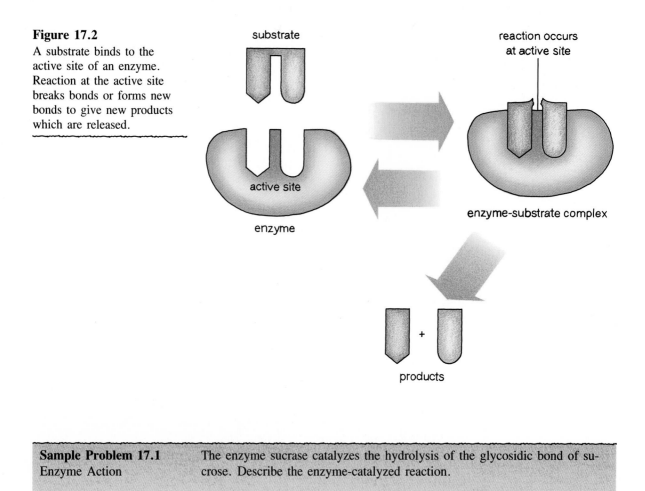

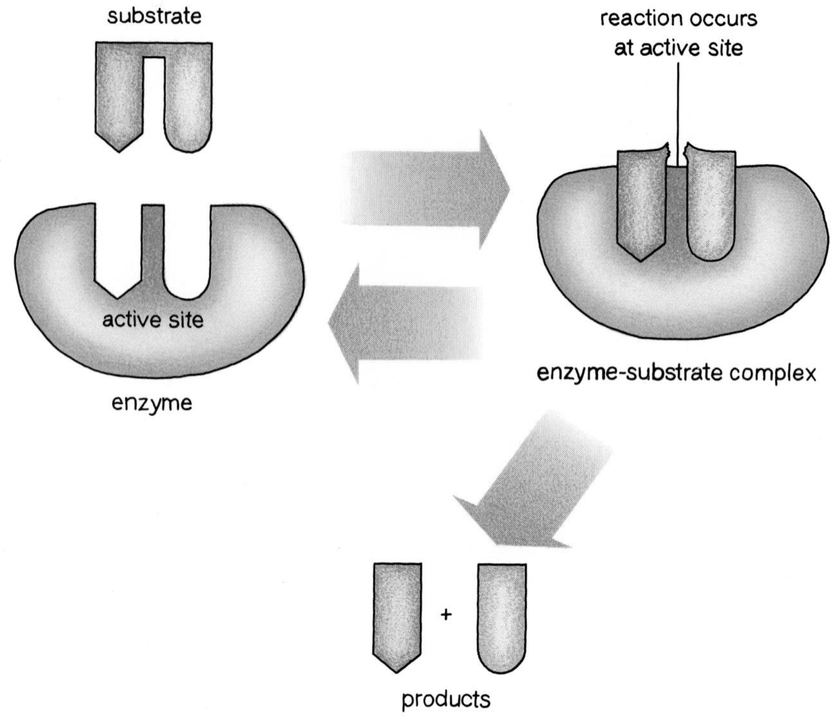

Sample Problem 17.1
Enzyme Action

The enzyme sucrase catalyzes the hydrolysis of the glycosidic bond of sucrose. Describe the enzyme-catalyzed reaction.

Solution:
In the hydrolysis of sucrose, the enzyme and substrate form an enzyme–substrate complex. At the active site, the glycosidic bond of sucrose is broken to give glucose and fructose.

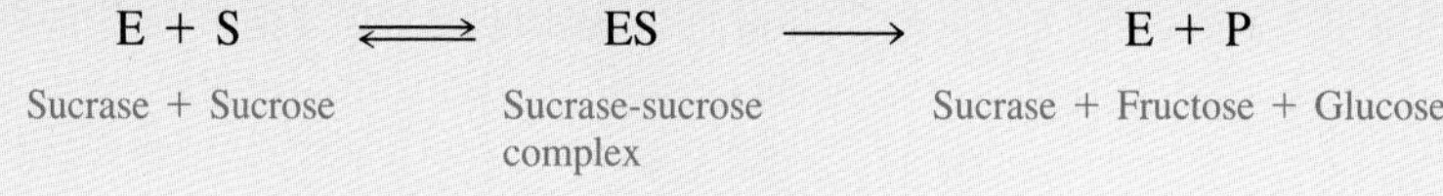

$$E + S \rightleftharpoons ES \longrightarrow E + P$$

Sucrase + Sucrose — Sucrase-sucrose complex — Sucrase + Fructose + Glucose

17.2 Names and Classification of Enzymes

Learning Goal **Name and classify an enzyme.**

The common names of many enzymes are derived from the names of their substrates by changing the end of the name to *ase*. The enzyme that hydrolyzes sucrose is called sucrase, and the enzymes that hydrolyze lipids are called lipases.

Substrate	*Name of Enzyme*
Sucrose	Sucrase
Lipid	Lipase
Maltose	Maltase

Enzymes have also been named for the type of reaction they catalyze. An enzyme that removes CO_2 is called decarboxylase. The prefix *de* means "to remove."

Reaction	*Name*
Removes CO_2	Decarboxylase
Adds H_2O	Hydrase
Splits ester bonds	Esterase, hydrolase

Some of the trivial names in use do not describe a substrate or a reaction. For example, pepsin, renin, and trypsin are all names of enzymes that hydrolyze proteins.

Sample Problem 17.2
Common Names of Enzymes

Predict the common name of an enzyme that would act on the following substrates:

a. lactose
b. urea

Solution:

a. The common name of an enzyme is obtained by replacing the ending of the substrate name with *ase*. The enzyme for lactose would be called lactase.
b. The name of the enzyme for urea would be urease.

Study Check:
What is the name of an enzyme that removes hydrogen from a substrate?

Answer: dehydrogenase

Types of Enzymes

Some enzymes such as pepsin are *simple* proteins that consist only of a polypeptide chain. It is the tertiary protein structure of a simple enzyme that makes it biologically active. Other enzymes are *conjugated* proteins in which a protein portion is inactive without a cofactor. The cofactor is the nonprotein portion of an enzyme, such as a metal ion that is necessary for enzyme activity. Table 17.1 lists some examples of enzymes and their metal ion cofactors. If the cofactor is an organic compound, usually a vitamin, it is called a **coenzyme.**

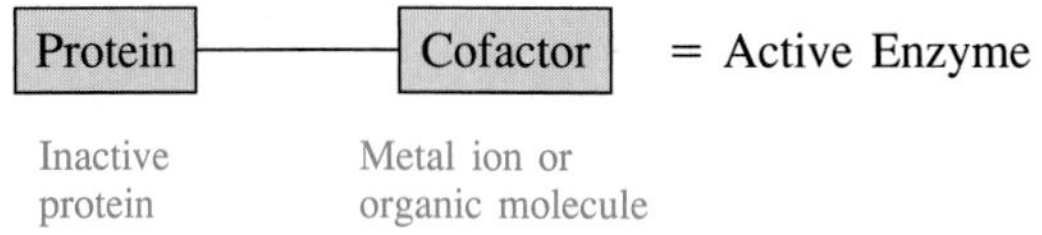

Table 17.1 Metal Ions Required by Some Enzymes

Metal Ion	Enzymes
Ca^{2+}	Thromboplastin
Cu^{2+}	Tyrosinase, cytochrome oxidase
Fe^{2+}, Fe^{3+}	Cytochrome oxidase, catalase, dehydrogenases
Mg^{2+}	Pyruvate kinase
Mn^{2+}	Arginase, pyruvate carboxylase, phosphatases, succinic dehydrogenase, glycosyl transferases, cholinesterase
K^{+}	Pyruvate kinase
Zn^{2+}	Carbonic anhydrase, carboxy peptidase, lactic dehydrogenase, alcohol dehydrogenase

Sample Problem 17.3
Types of Enzymes

Classify the following enzymes as simple or conjugated proteins:

a. This enzyme requires Mg^{2+} for biological activity.
b. This enzyme consists of protein only.

Solution:

a. An enzyme requiring a metal ion Mg^{2+} is a conjugated protein.
b. An enzyme consisting only of protein is a simple protein.

Study Check:

An enzyme requires vitamin B_1 for biological activity. How would it be classified?

Answer: conjugated protein

Isoenzymes in Medical Diagnosis

Different cells in the body produce enzymes for the same type of reactions. Although the proteins are similar, they are not identical. Enzymes that catalyze the same reactions but vary slightly in structure are called **isoenzymes.** For example, there are five isoenzymes for lactate dehydrogenase (LDH), an enzyme that converts lactic acid to pyruvic acid. Each LDH consists of a quaternary structure containing four subunits. In heart muscle, the most prevalent subunit is designated as H. In skeletal muscle, the major subunit is designated as M. Because the LDH isoenzymes in different tissues have different combinations of subunits, they have different isoelectric points and can be separated by electrophoresis.

In healthy tissues, enzymes such as the LDH isoenzymes are contained within cellular membranes. However, if the cells of a particular organ are damaged, the contents including the enzymes spill into the blood. By identifying the isoenzyme that becomes elevated in the blood serum, it is possible to determine which type of tissue has been damaged. For example, liver diseases can be detected by a rise in the serum LDH_5 level. When a myocardial infarction (MI), or heart attack, damages the cells in heart muscle, there is an increase in the serum LDH_1 level.

Isoenzyme	LDH_1	LDH_2	LDH_3	LDH_4	LDH_5
Subunits	H_4	H_3M	H_2M_2	HM_3	M_4
Abundant in	Heart, kidneys	Heart, kidneys, brain, red blood cells	Kidneys, brain	Spleen	Liver, skeletal muscle

Another isoenzyme used diagnostically is creatine kinase (CK), which consists of two types of subunits. One subunit is prevalent in the brain (B) and the other predominates in skeletal muscle (M). In normal patients only the CK_3 is present in small amounts in the blood serum. However, in a patient who has suffered a myocardial infarction (MI), the levels of CK_2 are elevated soon after the heart attack.

Isoenzyme	CK_1	CK_2	CK_3
Subunits	BB	MB	MM
Abundant in	Brain, lung	Heart	Skeletal muscle, red blood cells

Tissue damage is also assessed by transaminases such as glutamate oxaloacetate transaminase (GOT) and glutamate pyruvate transaminase (GPT). Immediately following a myocardial infarction, there is a significant increase in GOT in the blood serum, and the level rises rapidly in the first 6–12 hr. In severe cases, it can reach 20 times normal value. The serum GOT returns to normal levels in 3–5 days. (See Figure 17.3.) In diseases of the liver, both serum GOT and GPT levels are elevated. By assessing the levels of these two enzymes, a doctor can determine whether damage has occurred in the heart or liver. Table 17.2 lists some enzymes used to diagnose diseases of certain organs.

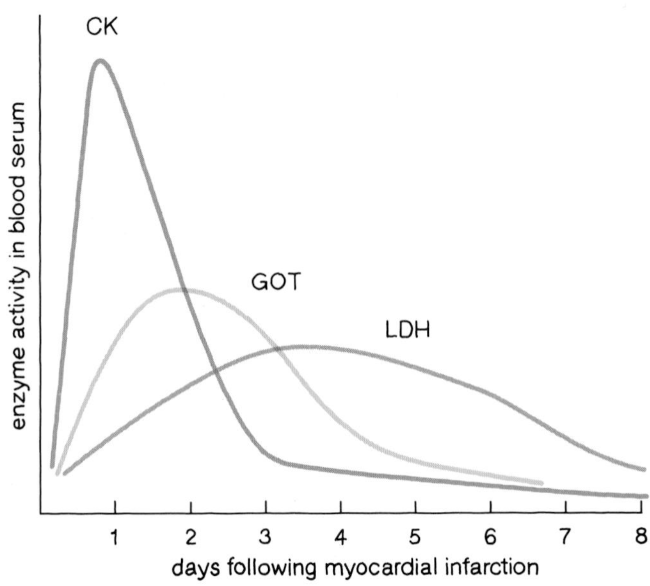

Figure 17.3
Normally, blood serum levels of GOT, CK, and LDH are low. After a heart attack, their levels rise. The serum levels of these enzymes can aid the assessment of the severity of heart damage.

Table 17.2 Serum Enzymes Used in Diagnosis of Tissue Damage

Organ	Condition	Diagnostic enzymes
Heart	Myocardial infarction (MI)	Lactate dehydrogenase (LDH_1) Creatine kinase (CK_2) Glutamic oxaloacetic transaminase (GOT)
Liver	Cirrhosis, carcinoma, hepatitis	Glutamic pyruvic transaminase (GPT) Lactate dehydrogenase (LDH_5) Alkaline phosphatase (ALP) Glutamic oxaloacetic transaminase (GOT)
Bone	Rickets, carcinoma	Alkaline phosphatase (ALP)
Pancreas	Pancreatic diseases	Amylase Cholinesterase Lipase (LPS)
Prostate	Carcinoma	Acid phosphatase (ACP)

Classification of Enzymes

Because enzymes often have several common names, a systematic method of classifying enzymes has been established. (See Table 17.3.) This system organizes enzymes according to the general type of reaction that they catalyze. However, this naming system has not completely replaced the prevalent use of the common names.

Table 17.3 Classification of Enzymes

Class	General reaction	Examples	Reactions catalyzed
Oxidoreductases	Oxidation/reduction	Oxidases Dehydrogenases Reductases	Oxidation Remove hydrogen Add hydrogen

Example:

$$\underset{\text{Ethanol}}{CH_3CH_2OH} \xrightarrow{\text{Alcohol dehydrogenase}} \underset{\text{Acetaldehyde}}{CH_3\overset{\overset{\displaystyle O}{\|}}{C}H} + 2H$$

Class	General reaction	Examples	Reactions catalyzed
Transferases	Move a functional group	Transaminases Kinases	Transfer of amino groups Transfer of phosphate

Examples:

$$\underset{\text{Alanine}}{CH_3-\overset{\boxed{NH_2}}{\overset{|}{C}H}-COOH} + \underset{\alpha\text{-Ketoglutaric acid}}{HOOC-\overset{\overset{\displaystyle O}{\|}}{C}-CH_2CH_2-COOH} \xrightarrow{\text{Transaminase}}$$

$$\underset{\text{Glutamic acid}}{HOOC-\overset{\boxed{NH_2}}{\overset{|}{C}H}-CH_2CH_2-COOH} + \underset{\text{Pyruvic acid}}{CH_3-\overset{\overset{\displaystyle O}{\|}}{C}-COOH}$$

$$\underset{\text{Phosphoenolpyruvate (PEP)}}{\begin{array}{l} O \\ \| \\ C-O^- \quad \boxed{\begin{array}{l} O^- \\ | \\ -P{=}O \\ | \\ O^- \end{array}} \\ | \\ C-O- \\ \| \\ CH_2 \end{array}} + \boxed{ADP} \xrightarrow[\text{Mg}^{2+}]{\text{Pyruvate kinase}} \underset{\text{Pyruvate}}{\begin{array}{l} O \\ \| \\ C-O^- \\ | \\ C{=}O \\ | \\ CH_3 \end{array}} + \boxed{ATP}$$

Class	General reaction	Examples	Reactions catalyzed
Hydrolases	Hydrolysis of bonds	Peptidases Lipases Amylase Phosphatase	Hydrolyze peptide bonds Hydrolyze ester bonds in triglycerides Split 1,4-glycosidic bonds in amylose Hydrolyze phosphate groups

Example:

$$\underset{\text{(lipid)}}{\text{Triglyceride}} + 3H_2O \xrightarrow{\text{Lipase}} 3 \text{ fatty acids} + \text{glycerol}$$

Class	General reaction	Examples	Reactions catalyzed
Lyases	Remove small groups to form double bond; add small groups to double bonds	Decarboxylases Dehydrases Deaminases	Remove CO_2 Remove H_2O Remove NH_3

Example:

$$\underset{\text{Pyruvic acid}}{CH_3-\overset{\overset{\displaystyle O}{\|}}{C}-\boxed{\overset{\overset{\displaystyle O}{\|}}{C}-O}-H} \xrightarrow{\text{Pyruvate decarboxylase}} \underset{\text{Acetaldehyde}}{CH_3-\overset{\overset{\displaystyle O}{\|}}{C}-H} + \boxed{CO_2}$$

Class	General reaction	Examples	Reactions catalyzed
Isomerases	Rearrange the atoms of a substrate	Isomerases Epimerases	Convert cis to trans; ketose to aldose Convert D- to L-isomer

Example:

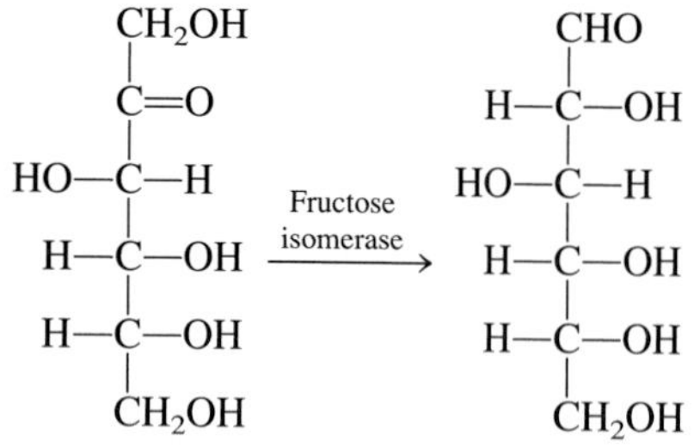

Fructose (CH_2OH—C=O—HO—C—H—H—C—OH—H—C—OH—CH_2OH) $\xrightarrow{\text{Fructose isomerase}}$ Glucose (CHO—H—C—OH—HO—C—H—H—C—OH—H—C—OH—CH_2OH)

Fructose Glucose

Class	General reaction	Examples	Reactions catalyzed
Ligases	Synthesis of larger molecules	Synthetases Carboxylases	Combine two molecules Add CO_2 to a substrate

Example:

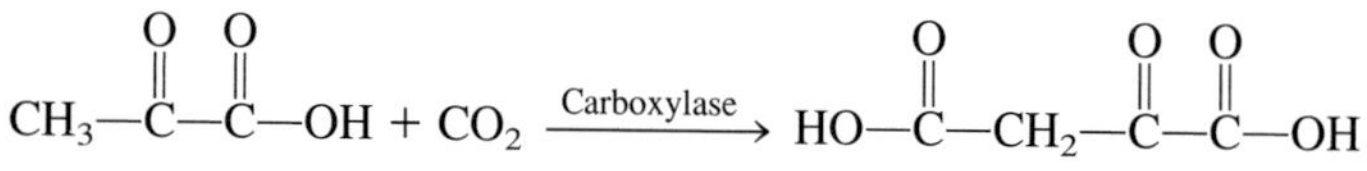

$$CH_3—\overset{O}{\overset{\|}{C}}—\overset{O}{\overset{\|}{C}}—OH + CO_2 \xrightarrow{\text{Carboxylase}} HO—\overset{O}{\overset{\|}{C}}—CH_2—\overset{O}{\overset{\|}{C}}—\overset{O}{\overset{\|}{C}}—OH$$

Pyruvic acid Oxaloacetic acid

Sample Problem 17.4
Classification of Enzymes

Classify the following enzymes according to the type of reaction they catalyze:

a. Glucose oxidase, an enzyme that oxidizes glucose to gluconic acid.
b. An enzyme that converts a cis fatty acid to a trans fatty acid.

Solution:

a. An oxidase would be classified as an oxidoreductase.
b. A conversion between a cis and trans isomer classifies this enzyme as an isomerase.

Study Check:
An acetyl CoA carboxylase adds CO_2 to acetyl CoA. How is this enzyme classified?

Answer: ligase

17.3 Factors That Affect Enzyme Activity

Learning Goal **Describe the effects of substrate and enzyme concentration, temperature, and pH on enzyme activity.**

Because enzymes are proteins, several factors affect the rate at which they catalyze reactions:

1. Substrate concentration
2. Enzyme concentration
3. Temperature
4. pH

Substrate Concentration

For a given amount of enzyme, a reaction will go faster as the amount of substrate increases. However, when all of the enzyme molecules are combined with substrate, a maximum reaction rate is reached. At this point, the enzyme molecules are saturated, and there can be no further increase in rate even if more substrate is added. (See Figure 17.4.)

Enzyme Concentration

In general, the concentration of substrate in a reaction is much higher than that of the enzyme. As the enzyme concentration is increased, the reaction rate increases, because more substrate molecules can undergo reaction. (See Figure 17.5.)

Temperature

Most enzymes show little activity at low temperatures because few substrate molecules have sufficient energy to undergo reaction. As the temperature increases, the rate of reaction also increases. Enzymes reach their greatest activity at an **optimum temperature,** usually around 37°C. Higher temperatures begin to denature an enzyme's protein, which leads to a loss of biological activity. (See Figure 17.6.)

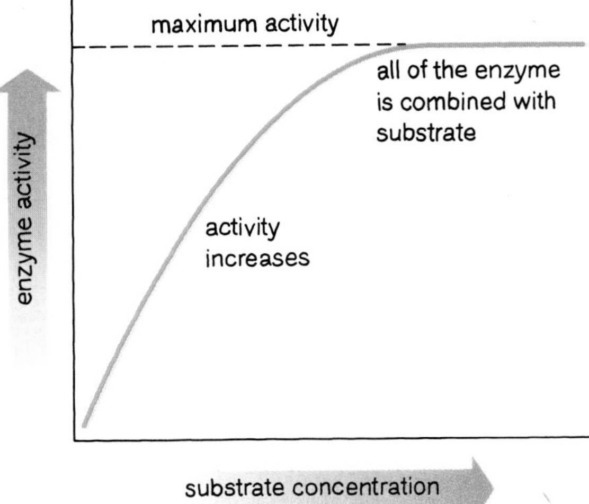

Figure 17.4
An increase in substrate concentration increases the rate of an enzyme-catalyzed reaction until all of the available enzyme has combined with substrate.

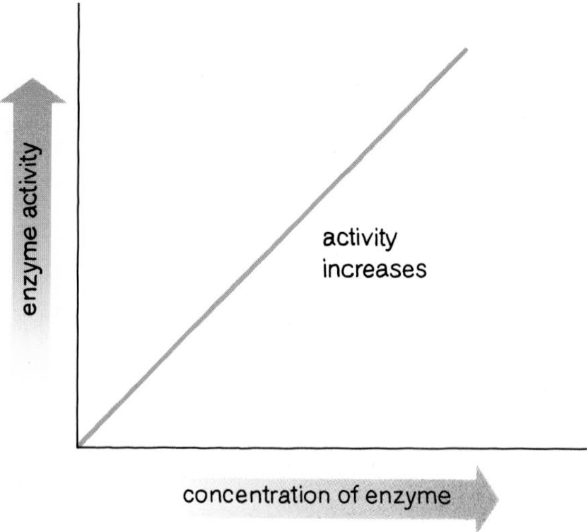

Figure 17.5
The rate of an enzymatic reaction increases when the concentration of enzyme is increased.

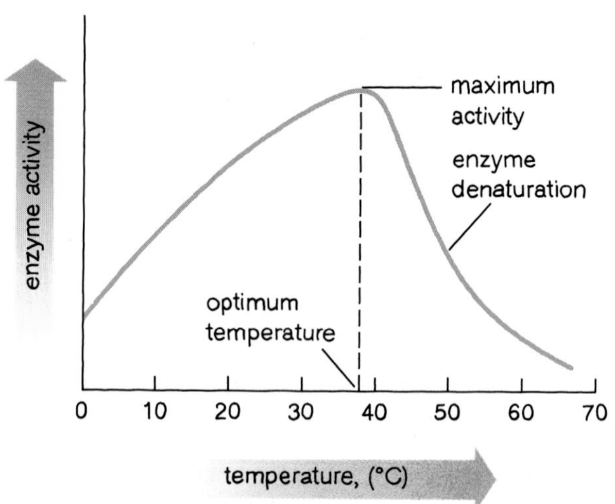

Figure 17.6
An increase in temperature increases enzyme activity up to its optimum temperature. Higher temperatures cause denaturation and loss of catalytic activity.

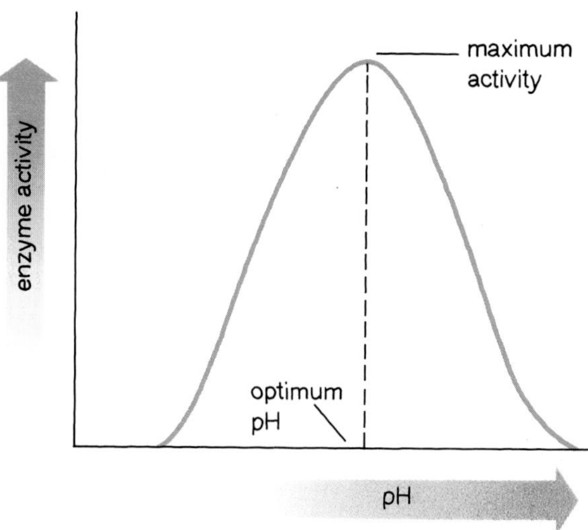

Figure 17.7
Enzymes are most active at their optimum pH. Values above or below optimum pH cause denaturation of the enzyme and loss of catalytic activity.

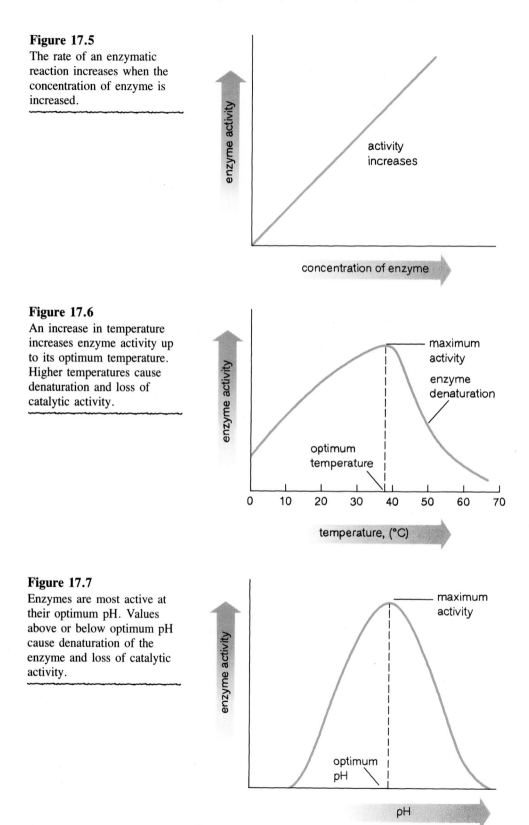

pH

Enzymes are most active at their **optimum pH** as shown in Figure 17.7. At this pH the protein maintains its proper tertiary structure. At higher or lower pH, changes in the ionic side chains of the enzyme interfere with its ability to bind properly to the substrate and prevent catalysis at the active site. If the pH change is great, the enzyme may be so denatured that it becomes completely inactive. For example, the digestive enzymes in the stomach, such as pepsin, have an optimum pH of 2. If the pH in the stomach rises to pH 4 or 5, pepsin shows little or no digestive activity. Another protein enzyme, trypsin, is most active in the small intestine at a pH of 8. At pH values of 6 or 10, it loses most or all of its catalytic activity. Table 17.4 lists the optimum pH values for some enzymes.

Table 17.4 Optimum pH for Some Enzymes

Enzyme	Location	Substrate	Optimum pH
Pepsin	Stomach	Protein	2.0
Pyruvate carboxylase	Liver	Pyruvate	4.8
Sucrase	Small intestine	Sucrose	6.2
Pancreatic amylase	Pancreas	Amylose	7.0
Trypsin	Small intestine	Polypeptides	8.0

Sample Problem 17.5
Factors Affecting Enzymatic Activity

Describe the effects of the following on the rate of the reaction catalyzed by urease.

$$H_2N\overset{\overset{\displaystyle O}{\|}}{C}NH_2 + H_2O \xrightarrow{\text{Urease}} 2NH_3 + CO_2$$

Urea

a. Increasing the urea concentration
b. Increasing the concentration of urease
c. Lowering the temperature to 10°C

Solution:

a. The rate of the enzyme-catalyzed reaction will increase with an increase in the urea concentration until all of the enzyme has combined with substrate. Then no further increase in rate will occur.
b. When more enzyme is added, more urea can react. Therefore, the rate of reaction increases.
c. Because 10°C is lower than the optimum temperature of 37°C, it will decrease the rate of the reaction.

Study Check:
Urease has an optimum pH of 5.0. What is the effect on the rate of reaction when the pH is lowered to 3.0?

Answer: The rate decreases.

17.4 Enzyme Inhibition

Learning Goal **Describe the effect of an inhibitor upon enzyme activity.**

Certain medicines stop bacterial infections because they inactivate an enzyme essential to the growth process of bacteria. A chemical compound that combines with an enzyme to make it inactive is called an **inhibitor.**

Competitive Inhibitors

In **competitive inhibition,** a substrate and an inhibitor compete for the active site on the enzyme. They are so similar in structure that the enzyme binds to the inhibitor by mistake. As long as the competitive inhibitor occupies the active site, no reaction of the substrate can take place. However, we can reverse competitive inhibition by adding more substrate that competes with the inhibitor for the active site. The addition of large amounts of substrate can completely reverse the inhibition. (See Figure 17.8.)

Malonic acid is an example of a competitive inhibitor of the enzyme succinic dehydrogenase. Because the inhibitor has a structure that is similar to the substrate, it and the substrate, succinic acid, compete for the active site. As long as malonic acid blocks the active site on the enzyme, no reaction occurs. Adding more succinic acid displaces the inhibitor and reverses the inhibition.

$$\mathrm{HO(O{=})C{-}CH_2{-}CH_2{-}C({=}O)OH} \xrightarrow{\text{Succinic dehydrogenase}} \mathrm{HO(O{=})C{-}CH{=}CH{-}C({=}O)OH}$$

Succinic acid (substrate) — Fumaric acid

$$\mathrm{HO(O{=})C{-}CH_2{-}C({=}O)OH} \xrightarrow{\text{Succinic dehydrogenase}} \text{NO REACTION}$$

Malonic acid, (competitive inhibitor)

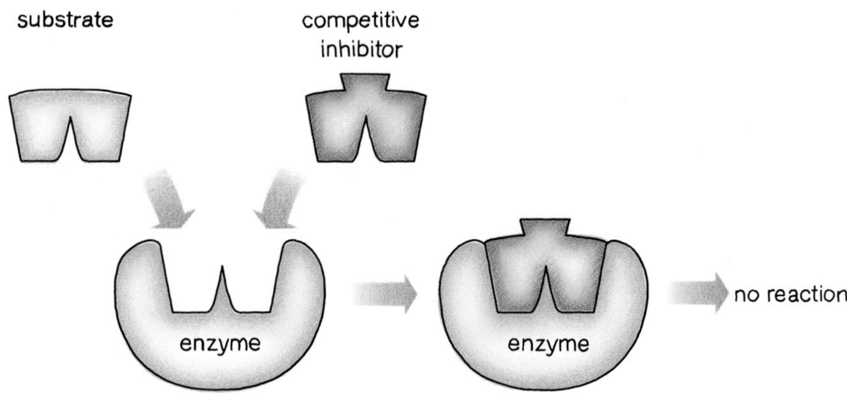

Figure 17.8
A competitive inhibitor competes for the active site because it has a structure very similar to the substrate.

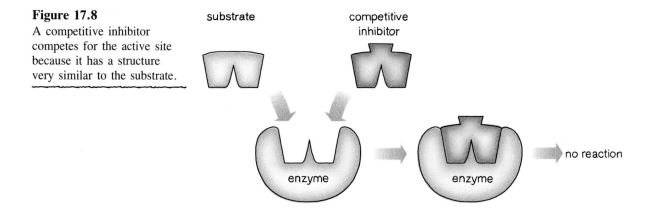

HEALTH NOTE

Competitive Inhibitors Used in Medicine

Illness is often caused by an invading microorganism such as a bacterium. The medical treatment may involve the use of an **antimetabolite,** which is a competitive inhibitor that inhibits a bacterial enzyme and disrupts its action. A classic example is sulfanilamide, one of the early sulfa drugs used to fight infection. Sulfanilamide competes with an essential substrate (metabolite), PABA, which bacteria use in the growth cycle. Sulfanilamide inhibits bacterial growth because it is chemically similar to PABA.

An **antibiotic** produced by bacteria or mold is used to inhibit bacterial growth in mammals. For example, penicillin inhibits an enzyme needed for the formation of bacterial cell walls, but not human cell membranes. With an incomplete cell wall, bacteria cannot survive, and the infection is stopped.

However, some bacteria have become resistant to penicillin because they form penicillinase, an enzyme that breaks down penicillin. New kinds of penicillin to which bacteria have not yet become resistant have been produced.

Substrate of Bacterial Growth — PABA (p-aminobenzoic acid)

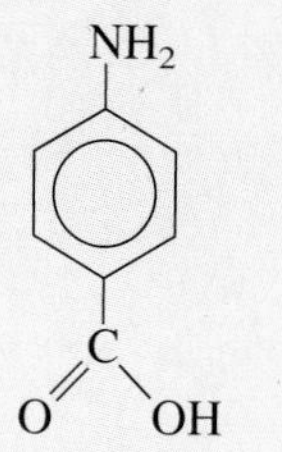

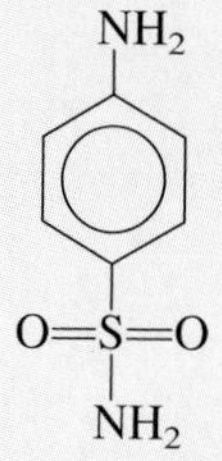

Inhibitor — Sulfanilamide

Penicillin

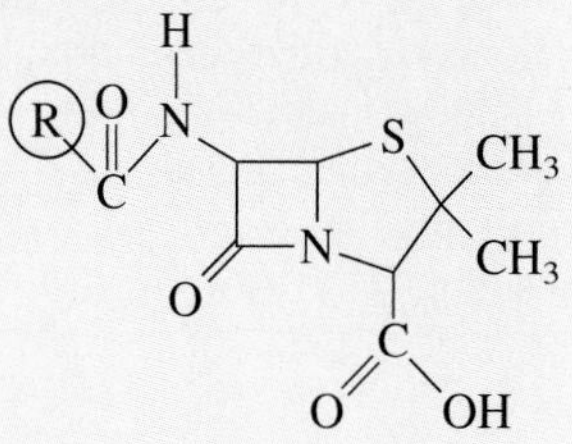

R ***Groups for Penicillin Derivatives***

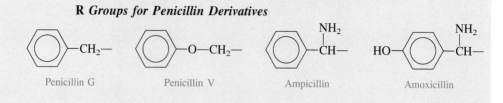

C_6H_5—CH_2— Penicillin G

C_6H_5—O—CH_2— Penicillin V

C_6H_5—CH(NH_2)— Ampicillin

HO—C_6H_4—CH(NH_2)— Amoxicillin

Another group of antibiotics—tetracycline, streptomycin, chloramphenicol, and aureomycin—stop bacterial infections by inhibiting the synthesis of protein in bacteria. They are also used to fight bacterial infections in people allergic to the penicillins.

Tetracycline

Noncompetitive Inhibition

A **noncompetitive inhibitor** binds to an enzyme, at a site other than the active site. Its structure does not resemble that of the substrate. When a noncompetitive inhibitor attaches to an enzyme, it alters the three-dimensional structure of the enzyme as well as the shape of the active site. As a result, the substrate cannot bind properly with the active site to form the ES complex. Because a noncompetitive inhibitor does not compete for the active site, adding more substrate will not reverse inhibition. (See Figure 17.9.)

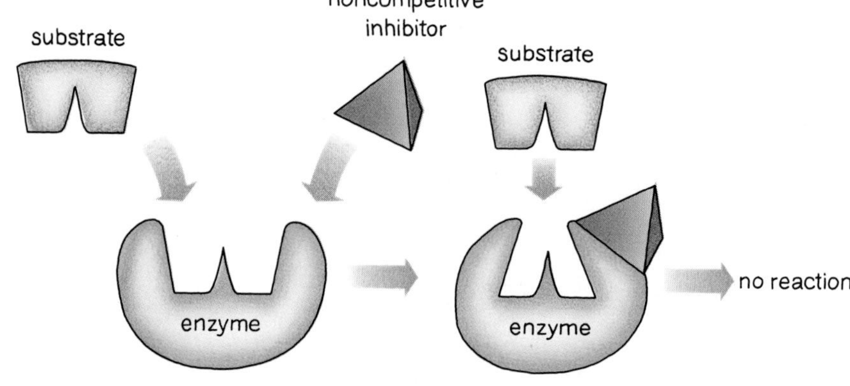

Figure 17.9
In noncompetitive inhibition, the inhibitor alters the shape of the enzyme, preventing the proper binding of the substrate.

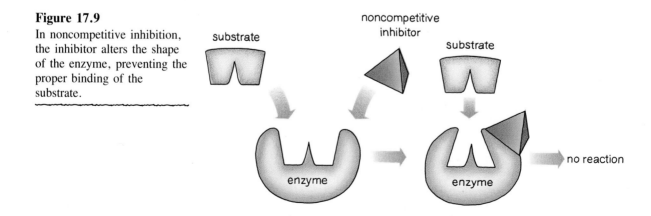

Examples of noncompetitive inhibitors are the heavy-metal ions Pb^{2+}, Ag^{+}, and Hg^{2+}. They form ionic bonds with the polar or ionic side chains such as —SH, $—COO^{-}$, or —OH which alters the three-dimensional shape of the enzyme. Some noncompetitive inhibitors can be removed by chemical reagents, which restores the catalytic ability of the protein.

Irreversible Inhibition

In an irreversible inhibition, an inhibitor forms a covalent bond with a side chain in the active site. This type of inhibition is also noncompetitive because the inhibitor does not have a structure similar to the substrate. However, an irreversible inhibitor cannot be removed from the protein without destroying it. Nerve gases act as irreversible inhibitors by forming a covalent bond with the side chain of serine ($—CH_2OH$), which is a part of the active site in acetylcholinesterase. When this enzyme is inhibited, the transmission of nerve impulses is blocked.

Sample Problem 17.6
Enzyme Inhibitors

State the type of inhibition in the following:

a. The inhibitor has a structure that is similar to the substrate.
b. This inhibitor binds to the surface of the enzyme, changing its structure in such a way that substrate cannot be converted to product.

Solution:

a. Competitive inhibition.
b. Noncompetitive inhibition.

Study Check:
Hydrogen cyanide (HCN) forms strong bonds with catalase, an enzyme that contains iron (Fe^{3+}). What type of inhibitor is HCN?

Answer: Irreversible

17.5 Enzymes in Digestion

Learning Goal **Describe the enzymes that hydrolyze carbohydrates, lipids, and proteins during digestion.**

Before we can use the carbohydrates, fats, and proteins in our food, these large polymers must be broken down to smaller molecules that can be absorbed through the intestinal walls. Mechanical digestion includes the chewing (mastication) of food and the contractions of muscles that push food along the digestive tract (peristalsis). Chemical digestion is carried out by enzymes that catalyze the chemical reactions of hydrolysis to break polysaccharides down to monosaccharides, lipids to fatty acids, and glycerol and proteins to amino acids.

Digestion of Carbohydrates

When a person chews food, the enzyme **α-amylase** in the saliva begins to break apart some of the α-1,4-glycosidic bonds in starch. Fragments called dextrins with three to eight glucose units and maltose units are produced.

After it is swallowed, the chewed food (bolus) and the partially digested starches encounter the highly acidic environment of the stomach, where little carbohydrate digestion occurs. In the small intestine, which has a pH of about 8, an α-amylase secreted by the pancreas continues the hydrolysis process to give maltose and glucose. The disaccharides, maltose, sucrose, and lactose, are hydrolyzed by the enzymes maltase, sucrase, and lactase. Their monosaccharide products, glucose, fructose, and galactose, are absorbed into the bloodstream and carried to the liver to be stored as glycogen, or to body tissues to be used for energy. The fiber (cellulose) in the diet is not digested because there are no enzymes in the human digestive tract that can hydrolyze the β-1,4-glycosidic bonds of cellulose.

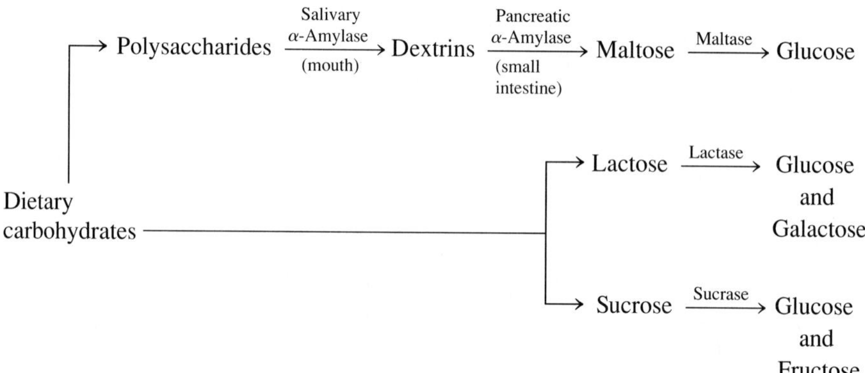

Sample Problem 17.7
Enzymes in Carbohydrate Digestion

Amylose is a polymer of glucose molecules bonded by α-1,4-glycosidic bonds. Describe its hydrolysis during digestion.

Solution:
Amylose consists of glucose molecules held together by α-1,4-glycosidic bonds. Salivary amylase begins to hydrolyze those bonds in the mouth. In the small intestine, pancreatic amylase hydrolyzes more glycosidic bonds of the remaining saccharides to give maltose. Finally, maltose is hydrolyzed to glucose.

Digestion of Fats

The water-insoluble fats and oils in our diets do not undergo digestion until they enter the small intestine. There, the triglycerides are mixed with the bile salts released from the gallbladder. The bile salts break apart the large globules of fat, forming an emulsion that provides a larger surface area for pancreatic lipases. The ester bonds of the triglycerides in fats and oils are split apart by the action of lipases and water to give fatty acids, monoglycerides, and glycerol. The shorter-chain fatty acids and glycerol diffuse directly into the bloodstream. However, in the intestinal mucosa the monoglycerides longer-chain fatty acids, and glycerol form triglycerides. These triglycerides combine with phospholipids, cholesterol, and protein to give more soluble lipoproteins called chylomicrons that can be transported through the lymphatic system to the bloodstream.

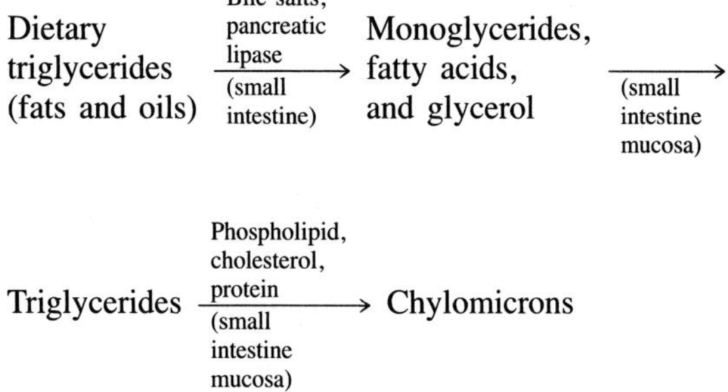

Sample Problem 17.8
Digestion of Fats

Describe the role of bile salts and pancreatic lipase in fat digestion.

Solution:
In the small intestine, fats are emulsified by the bile salts released from the gallbladder. Then the lipase enzymes from the pancreas split the ester bonds of the triglycerides to produce glycerol and fatty acids.

Study Check:
If long-chain fatty acids are insoluble in aqueous solutions, how are they transported in the body?

Answer: They are formed into triglycerides and combined with phospholipids and proteins that make them more soluble for transport.

Digestion of Proteins

Protein digestion begins in the stomach. When food is eaten, an increase in hydrochloric acid (pH 1–2) in the stomach converts pepsinogen into pepsin, an active protein-splitting enzyme. Pepsin hydrolyzes some peptide bonds to give a mixture of polypeptides.

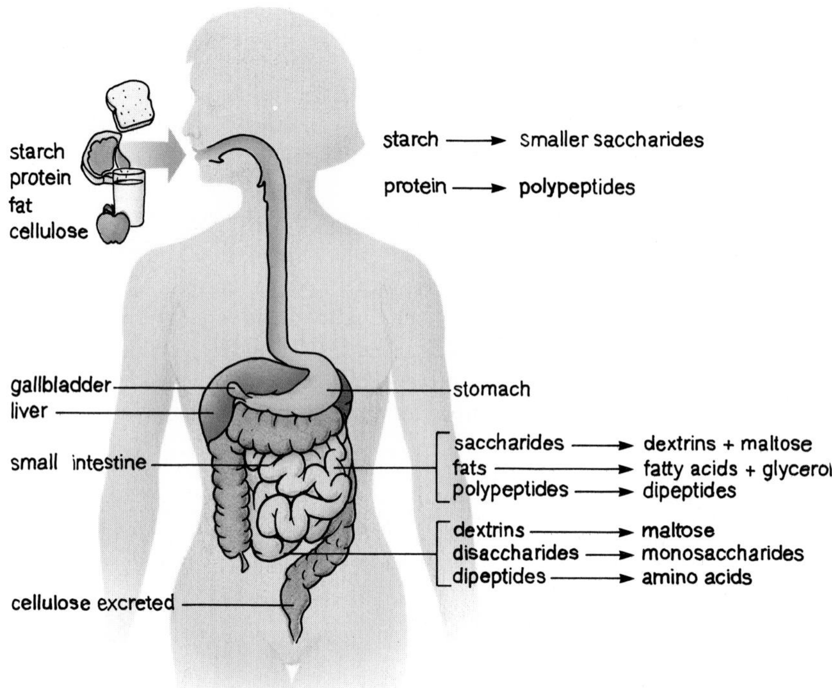

Figure 17.10
Digestion of foods along the gastrointestinal tract.

However, most of the peptide bonds are hydrolyzed in the basic environment (pH 8) of the small intestine by proteases such as trypsin and chymotrypsin. Finally, amino acids are hydrolyzed from ends of the peptides by peptidase. The amino acid end products of protein digestion are absorbed through the intestinal walls into the bloodstream for transport to tissues in the body.

$$\text{Dietary proteins} \xrightarrow[\text{(stomach)}]{\text{Pepsin}} \text{Polypeptides} \xrightarrow[\text{(small intestine)}]{\text{Trypsin Chymotrypsin}} \text{Peptides and amino acids} \xrightarrow[\text{(small intestine)}]{\text{Peptidases}} \text{Amino acids}$$

Figure 17.10 summarizes the digestion of foods in the gastrointestinal tract.

17.6 Vitamins

Learning Goal **Describe a vitamin in terms of water or fat solubility and biological function.**

Vitamins are organic substances required in small amounts by the cells in the body. Because most vitamins are not synthesized in the body, they must be obtained from the diet. If not, diseases such as scurvy, rickets, beriberi, and pellegra may develop. Foods that provide a daily source of vitamins include liver, whole grains, green, red, and yellow vegetables, yeast, eggs, and milk.

Vitamins are grouped as **water-soluble** and **fat-soluble.** The water-soluble vitamins of the B complex and C have polar groups that make them soluble in water. As a result, they are not stored but remain in body fluids. Any excess is excreted each day in the urine. The major role of many water-soluble vitamins is to supply the coenzymes that are needed by many enzymes for catalytic activity.

The fat-soluble vitamins A, D, E, and K are chemically related to cholesterol or the terpenes. They are nonpolar, which makes them soluble in body fat, where they are stored. The accumulation of too much of a fat-soluble vitamin can result in hypervitaminosis and damage to the body. The fat-soluble vitamins are important in blood-clotting, in the regulation of calcium and phosphorus in bone and body fluids, and in the formation of visual pigments for night vision. Table 17.5 lists the water-soluble vitamins, and Table 17.6 lists the fat-soluble vitamins.

Table 17.5 Water-Soluble Vitamins

Structure of vitamin	Coenzyme function	Deficiency symptoms
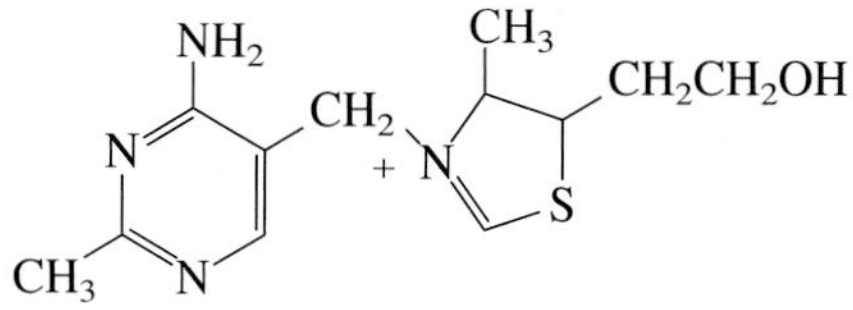 NH_2, CH_3, CH_2, CH_2CH_2OH, N, +, N, S, CH_3, N Thiamine (vitamin B_1)	Decarboxylation of α-Keto acids	Beriberi (fatigue, anorexia, nerve degeneration, paralysis, heart failure)
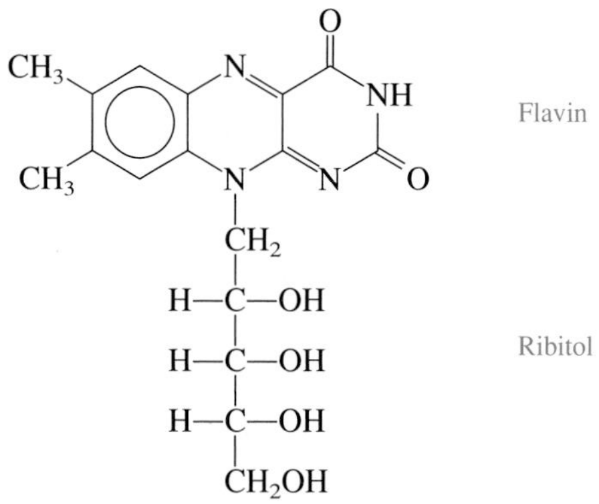 O, CH_3, N, NH, Flavin, CH_3, N, N, O, CH_2, H—C—OH, H—C—OH, Ribitol, H—C—OH, CH_2OH Riboflavin (vitamin B_2)	FAD (flavin adenine dinucleotide), FMN (flavin mononucleotide), used in biological oxidations	Dermatitis, glossitis (tongue inflammation), cataracts
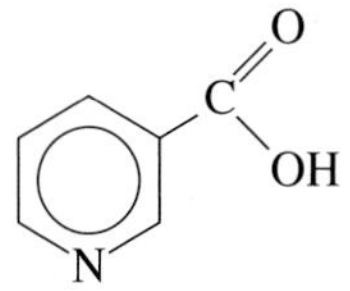 O, C, OH, N Niacin	NAD^+ (nicotinamide adenine dinucleotide) used in biological oxidations with respiratory chain	Pellegra (scaly skin, muscle fatigue, diarrhea, mouth sores, mental disorders)

Structure of vitamin	Coenzyme function	Deficiency symptoms
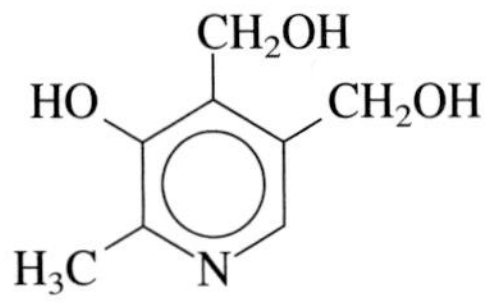 Pyridoxine (vitamin B_6)	Transamination reactions	Dermatitis, fatigue, anemia, irritability, convulsions in infants
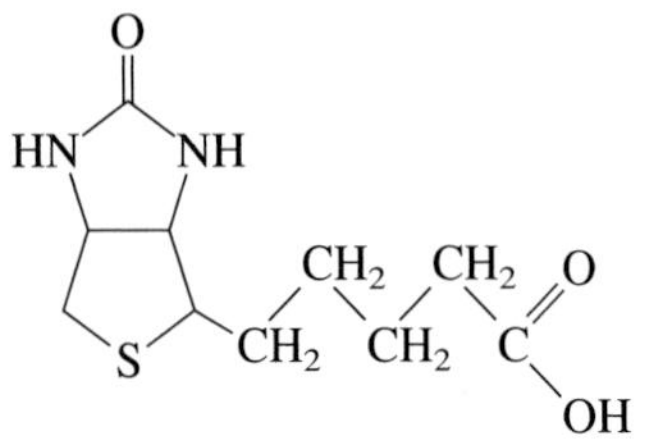 Biotin	Carboxylation of pyruvic acid to oxaloacetic acid, fatty acids, purines	Dermatitis, fatigue, anemia, nausea, mental depression
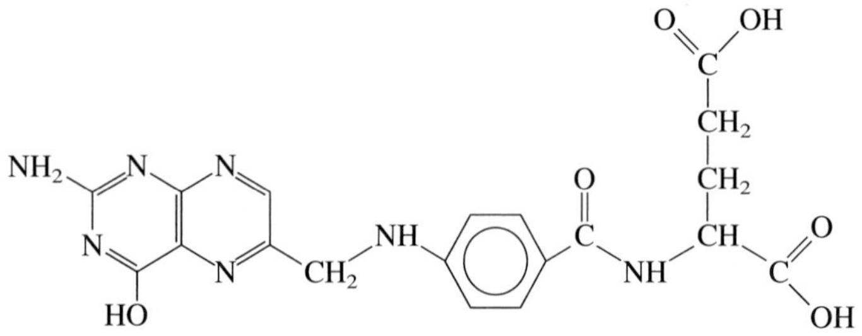 Folic acid	Transfer of methyl groups in biosynthesis reactions	Abnormal red and white blood cells, GI disturbances
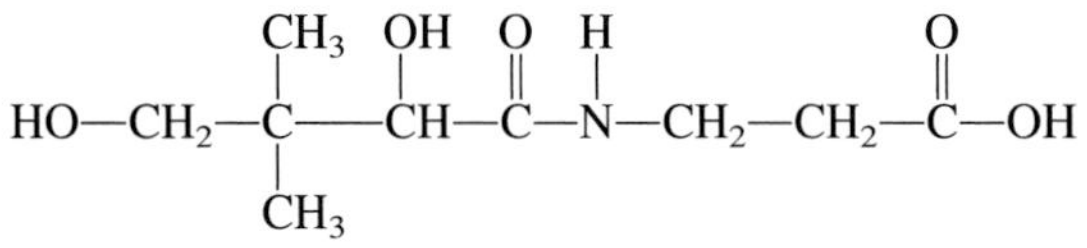 Pantothenic acid	As coenzyme A in the transfer of acetyl groups	No deficiency known, may cause fatigue, anemia

Structure of vitamin	Coenzyme function	Deficiency symptoms
Cobalamin (vitamin B_{12})	Transfer of methyl groups in biosynthesis of red blood cells, methionine, choline, purines	Pernicious anemia, malformed red blood cells, neurological disorders
Ascorbic acid (vitamin C)	Needed for collagen formation, protein metabolism, iron absorption, healing of wounds	Scurvy (bleeding gums, slow-healing wounds, muscle pain, anemia)

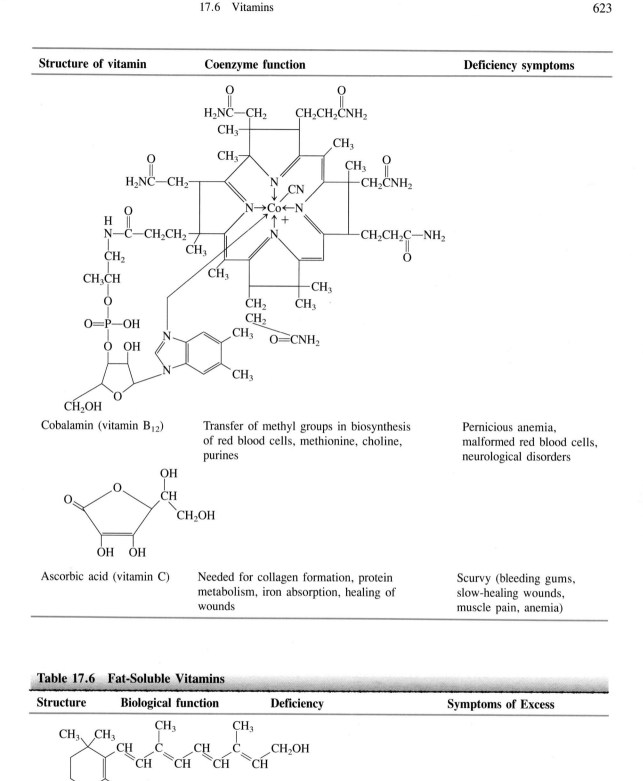

Table 17.6 Fat-Soluble Vitamins

Structure	Biological function	Deficiency	Symptoms of Excess
Retinol (vitamin A)	Formation of visual pigment; development of epithelial cells	Night blindness; dry, scaly skin; bacterial infections	Hypervitaminosis, irritability, sloughing of skin, joint pain, weight loss, liver enlargement

Structure	Biological function	Deficiency	Symptoms of Excess
Calciferol (vitamin D)	Absorption of calcium and phosphate; deposition of calcium and phosphate in bone	Rickets; bone decalcification; bone deformities	Hypervitaminosis D; muscle weakness, nausea, diarrhea, anorexia; calcification in tissues, kidney damage
α-Tocopherol (vitamin E)	May prevent oxidation of unsaturated fatty acids in cell membranes	Hemolysis of red blood cells; infertility in rats	
Phylloquinone (vitamin K)	Synthesis of prothrombin for blood clotting	Bruising, hemorrhaging from minor cuts, longer clotting time	

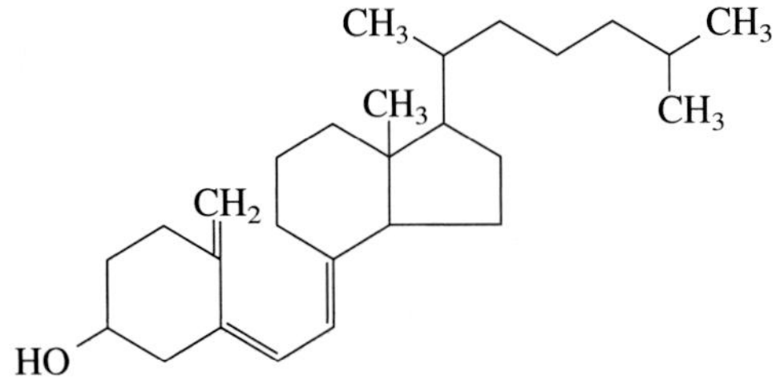

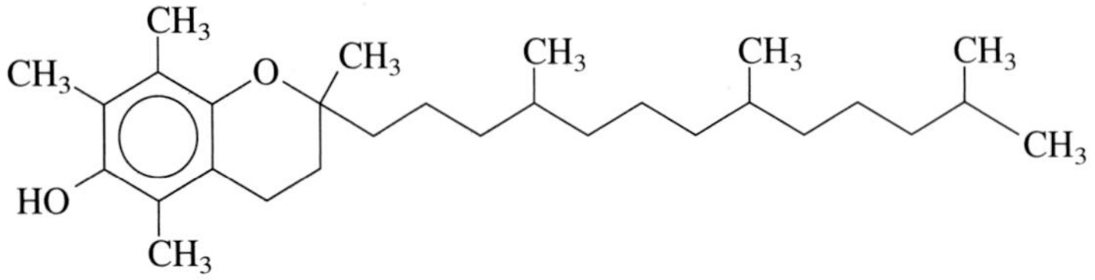

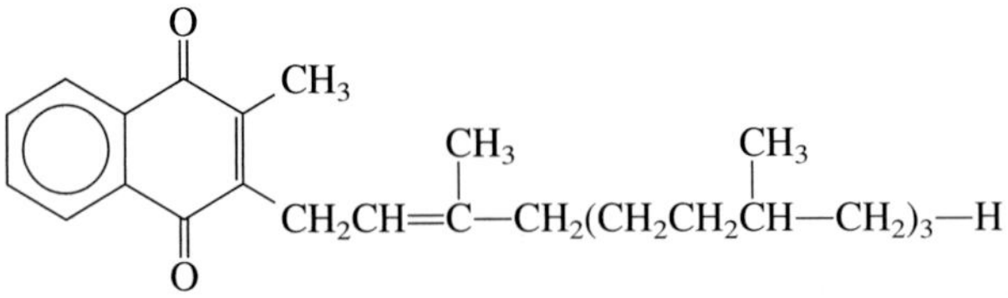

$CH_2CH{=}C(CH_3){-}CH_2(CH_2CH_2CH(CH_3){-}CH_2)_3{-}H$

Sample Problem 17.9
Functions of Vitamins

Indicate the vitamin involved in the following:

a. Blood clotting
b. Absorption of calcium
c. Formation of coenzyme NAD^+ (used in electron transport)

Solution:

a. Phylloquinone (vitamin K)
b. Calciferol (vitamin D)
c. Niacin

Study Check:

Why is hypervitaminosis a problem with fat-soluble vitamins, but not with water-soluble vitamins?

Answer: Fat-soluble vitamins can be stored in the body-fat for long periods of time and reach toxic levels. Water-soluble vitamins remain in body fluids, and excesses are excreted in the urine.

17.7 Hormones

Learning Goal **Describe some steroid and peptide hormones.**

The word **hormone** comes from the Greek "to arouse" or "to excite." Hormonal action is the stimulation of a cellular process somewhere in the body other than where the hormone is produced. Chemically, hormones are steroids, amino acids, or peptides that travel from the gland where they are produced through the bloodstream to their **target organs.** There, hormones may combine with receptor molecules to form a hormone–receptor complex that releases a second messenger within the cell. For many hormones, the cellular messenger is cyclic adenosine monophosphate or cyclic AMP.

Adenosine

Phosphate

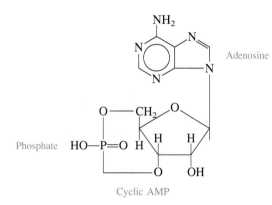

Cyclic AMP

Inside the target cell, the cyclic AMP messenger stimulates the synthesis of certain proteins, usually enzymes. The increase in enzyme concentration accelerates the reaction catalyzed by the enzyme. (See Figure 17.11.) Along with the nervous system, hormones serve as a kind of communication system from one part of the body to another. However, the nervous system acts rapidly, whereas the hormonal system sends very slow messages with long-lasting effects.

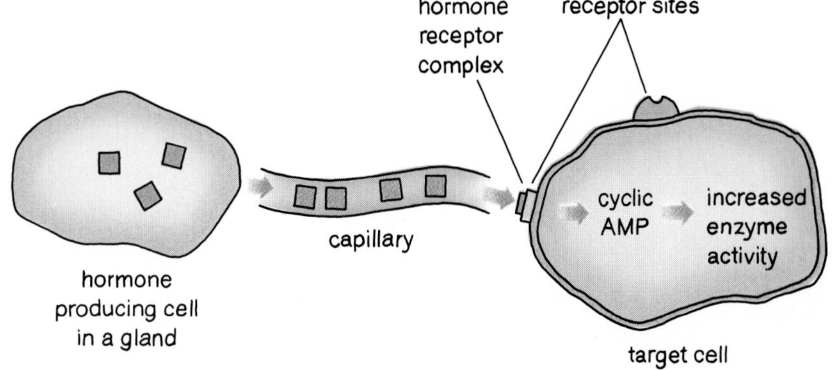

Figure 17.11
A hormone binds to a receptor on the surface of the target cell, triggering the release of cyclic AMP, which stimulates the synthesis of certain enzymes within the cell.

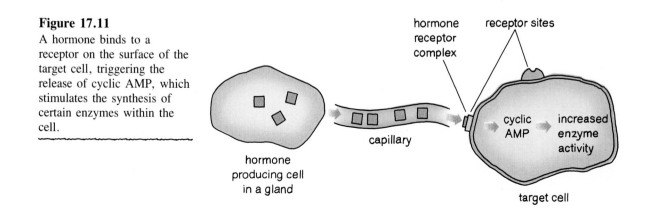

Steroid Hormones

The **steroid hormones** include the sex hormones and the adrenal steroids. All are closely related in structure to cholesterol and depend on cholesterol for their synthesis. In males, the testes produce the male sex hormones testosterone and androsterone. They are important in the development of the male sexual characteristics, muscle growth, facial hair, and the maturation of the male sex organs and of sperm.

In females, the female sex hormones such as progesterone and the estrogens are produced in the ovaries and uterus. The estrogens direct the development of female sexual characteristics: the uterus increases in size, fat is deposited in the breasts, the pelvis broadens, and body growth ceases. The most important function of progesterone is to prepare the uterus for the implantation of a fertilized egg. If an egg is not fertilized, the levels of progesterone and estrogen drop sharply, and menstruation follows. Synthetic forms of these hormones such as norethindrone and mestranol are used in birth-control pills. As with other kinds of steroids, side effects may include weight gain and edema, and with birth-control pills, there is a greater risk of blood-clot formation.

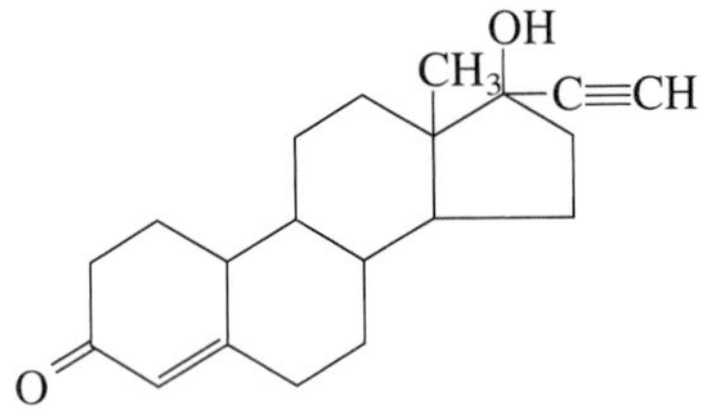

OH CH$_3$ C≡CH O

Norethindrone

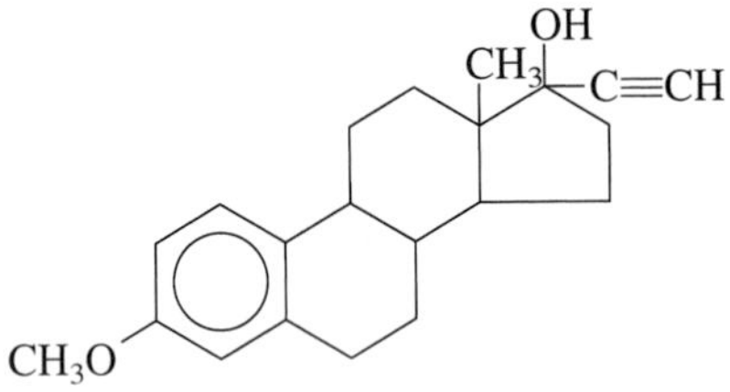

OH CH$_3$ C≡CH CH_3O

Mestranol

The adrenal glands, located on the top of each kidney, produce the adrenal steroids including mineralcorticoids and glucocorticoids. Aldosterone is a **mineralcorticoid** that increases the reabsorption of sodium ion by the kidneys and promotes water retention. Cortisol, a **glucocorticoid,** increases the blood glucose levels and stimulates the synthesis of glycogen in the liver from amino acids. Synthetic corticoids such as prednisone derived from cortisone are used medically for reducing inflammation and treating asthma and rheumatoid arthritis, although precautions are given for long-term use. Table 17.7 lists some of the important steroid hormones and their physiological effects.

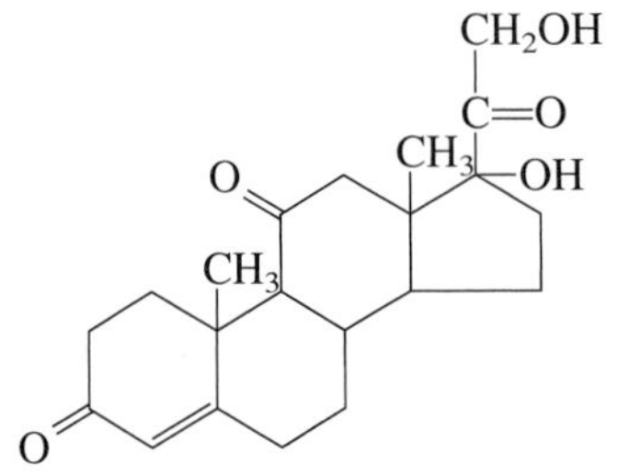

CH_2OH C=O CH$_3$ OH O CH$_3$ O

Cortisone

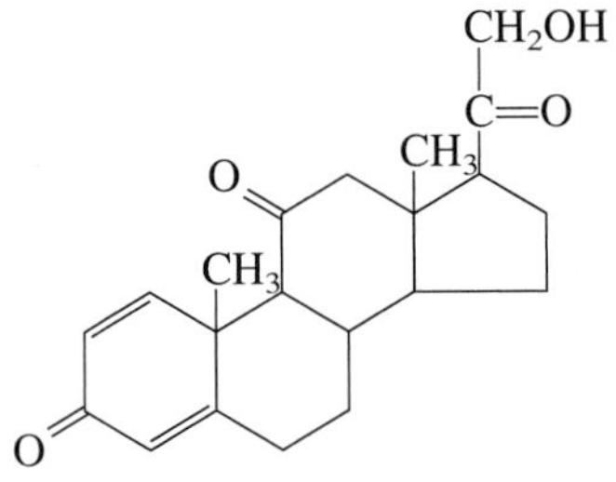

CH_2OH C=O CH$_3$ O CH$_3$ O

Prednisone

Table 17.7 Examples of Some Steroid Hormones Derived from Cholesterol

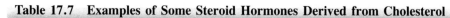

Hormone	Organ produced in	Biological effects
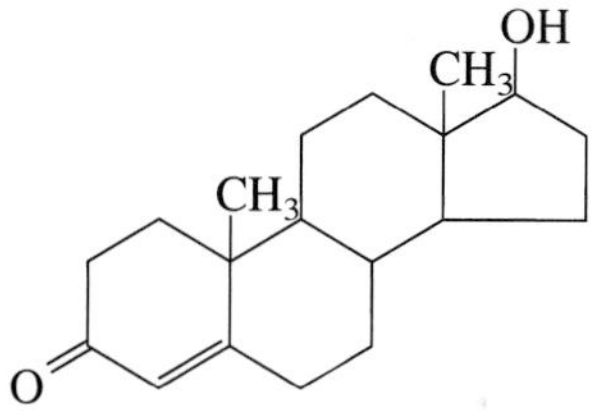 Testosterone (androgen)	Testes	Development of male organs; male sexual characteristics including muscles and facial hair; sperm formation
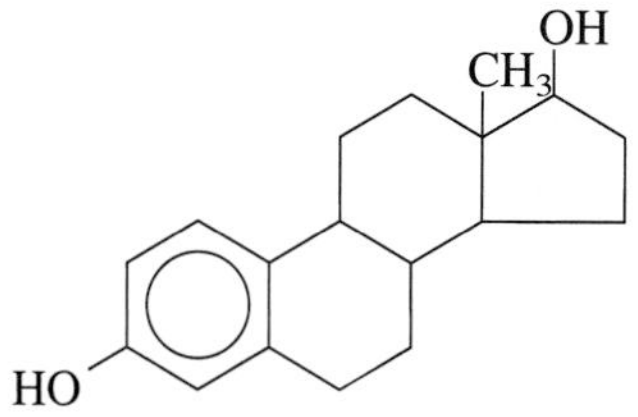 Estradiol (estrogen)	Ovaries	Development of female sexual characteristics; ovulation
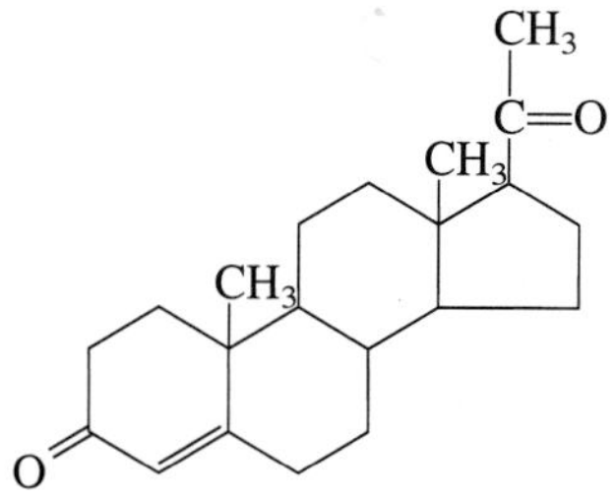 Progesterone	Ovaries	Prepare uterus for fertilized egg
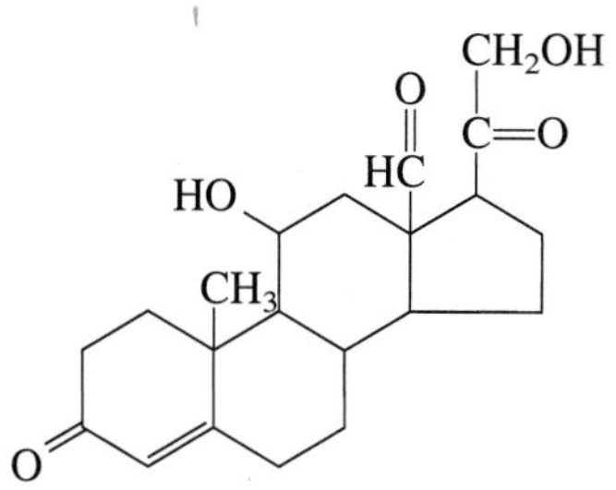 Aldosterone (mineralcorticoid)	Adrenal gland	Increase the reabsorption of Na^+ in kidneys; retention of water

Hormone	Organ produced in	Biological effects
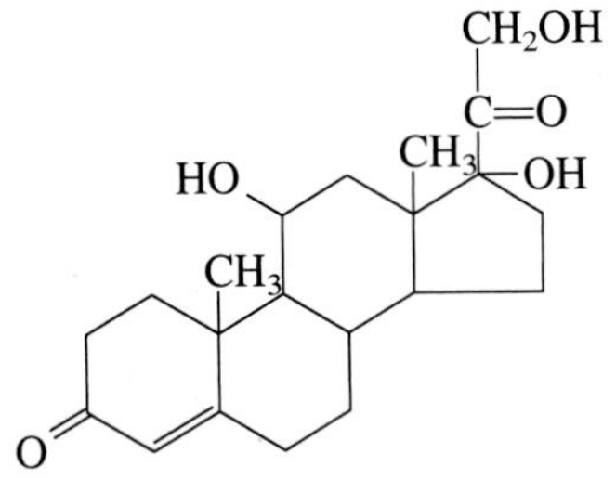 Cortisol (glucocorticoid)	Adrenal gland	Increase glucose blood levels; increase glycogen by forming glucose from amino acids

HEALTH NOTE

Anabolic Steroids

Some of the physiological effects of testosterone are to increase muscle mass and decrease body fat. Derivatives of testosterone called **anabolic steroids** that enhance these effects have been synthesized. Although they have some medical uses, anabolic steroids have been used in rather high dosages by some athletes in an effort to increase muscle mass. Such use is illegal.

Use of anabolic steroids in attempting to improve athletic strength can cause several side effects including hypertension, fluid retention, increased hair growth, sleep disturbances, and acne. Over a long period of time, their use can be devastating and may cause irreversible liver damage and decreased sperm production.

Some Anabolic Steroids

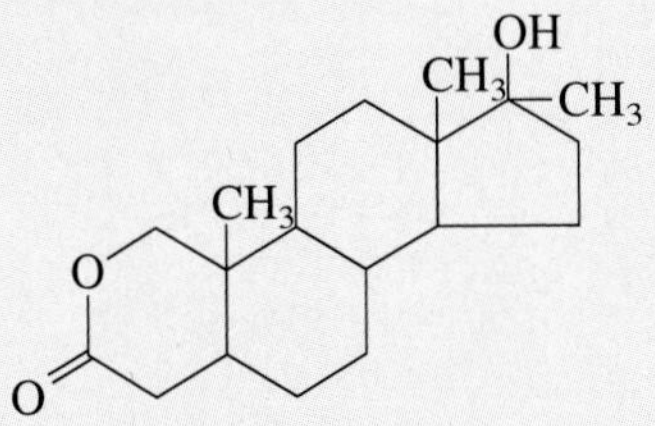

Anavar

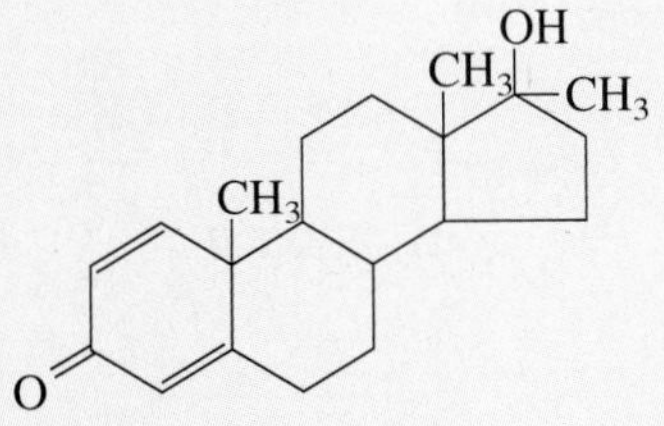

Dianabol

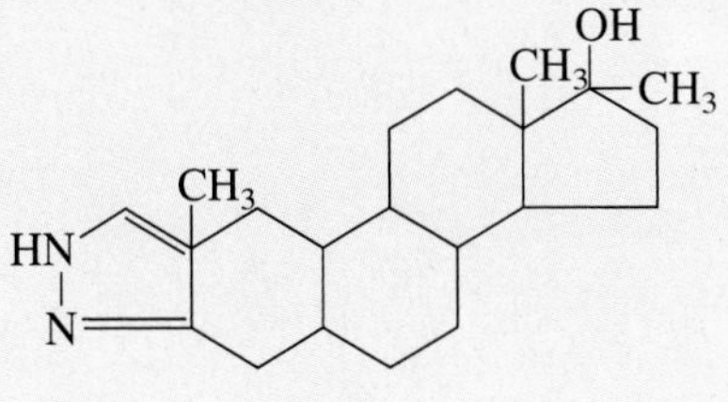

Winstrol

Sample Problem 17.10
Steroid Hormones

Name an important steroid hormone produced by males, one by females, and one by the adrenal gland.

Solution:
Testosterone is the most important male hormone. Progesterone or estrogen is an important female hormone. Important steroid hormones produced in the adrenal cortex are cortisol and aldosterone.

Peptide Hormones

Peptide hormones are released by the pituitary gland as well as by the pancreas, adrenal gland, and thyroid. Two of the hormones produced by the pituitary gland are vasopressin and oxytocin, peptides with nine amino acids. Vasopressin, or antidiuretic hormone (ADH), increases the reabsorption of water by the kidneys, which decreases urine formation. Oxytocin's structure differs from vasopressin by only two amino acids, but it has a much different function. One of its target organs is the uterus, where it causes strong labor contractions and is sometimes used to initiate labor. In the mammary glands, it stimulates the production of milk.

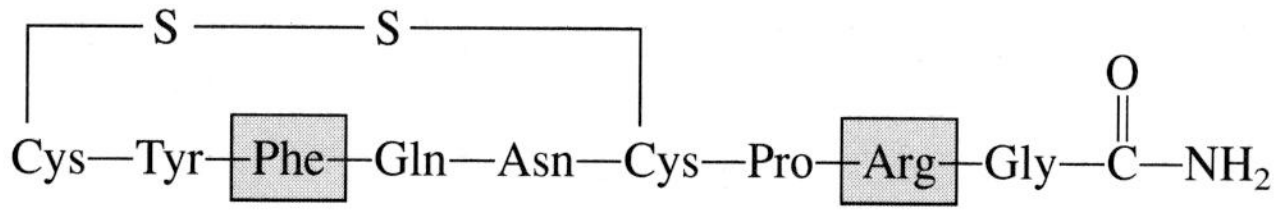

S——S (disulfide bridge between Cys 1 and Cys 6)

Cys—Tyr—Phe—Gln—Asn—Cys—Pro—Arg—Gly—C(=O)—NH_2

Vasopressin
(ADH antidiuretic hormone)

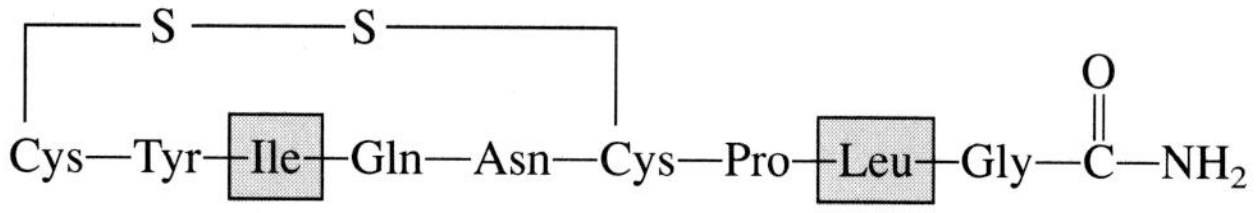

S——S (disulfide bridge between Cys 1 and Cys 6)

Cys—Tyr—Ile—Gln—Asn—Cys—Pro—Leu—Gly—C(=O)—NH_2

Oxytocin

Growth hormone (GH), also produced in the pituitary, is a protein with 188 amino acids. It promotes bone and body growth by increasing the rate of protein synthesis in all cells of the body. If a low amount of growth hormone is produced during childhood, growth rate is slowed to such an extent that as an adult the person's stature is that of a dwarf. If too much growth hormone is produced, tissue and bone growth accelerate, and the person can reach a height of 8 or 9 feet, a condition known as gigantism. When an adult produces too much growth hormone, the joints and cartilage in the face become enlarged, a condition called acromegaly. Figure 17.12 depicts some of the hormones secreted from the pituitary and their target organs.

The pancreas produces two important polypeptide hormones, insulin and glucagon, from a cluster of cells called the islets of Langerhans. Insulin is a small protein that increases the rate of glucose metabolism in the cells and promotes the storage of glycogen in the liver and muscles. If the pancreas does not produce sufficient insulin, glucose is not utilized, as indicated by high blood glucose levels, a condition called hyperglycemia.

Figure 17.12
Some hormones secreted from the pituitary gland and their target organs.

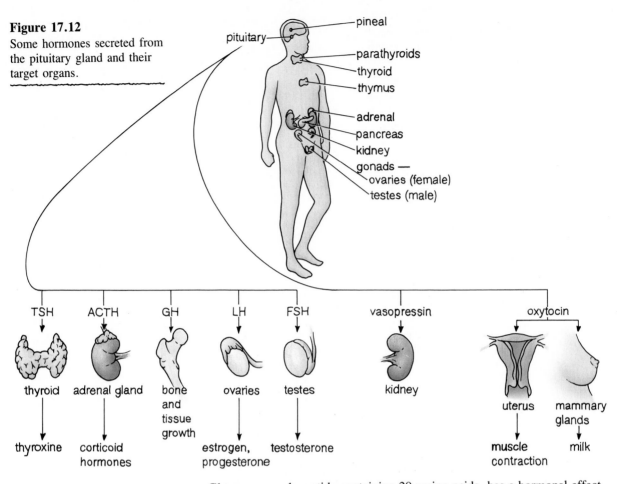

Glucagon, a polypeptide containing 29 amino acids, has a hormonal effect that opposes that of insulin. Glucagon increases the rate at which glycogen is broken down which increases the level of glucose in the bloodstream. It also promotes the synthesis of glucose from amino acids and lipids. Glucagon protects the body by preventing a major drop in the blood glucose level, a condition that can decrease the energy available to the brain. It is also used to treat an overdose of insulin by a diabetic. (See Figure 17.13.)

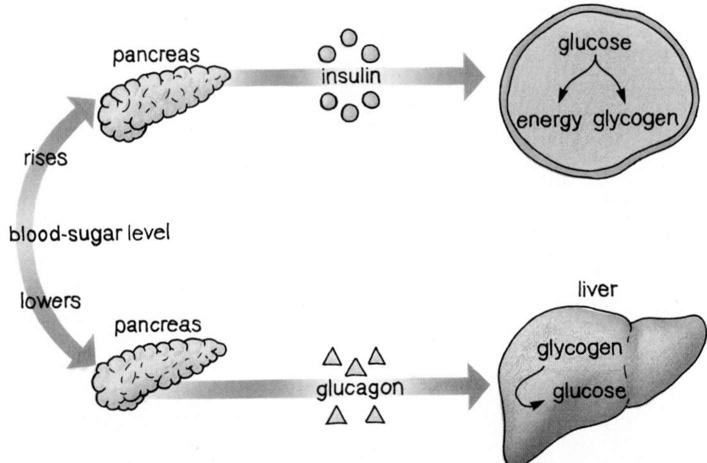

Figure 17.13
Regulation of blood glucose levels by polypeptide hormones of the pancreas.

A summary of some peptide hormones and their cellular functions is seen in Table 17.8.

Table 17.8 Some Peptide Hormones and Their Functions

Hormone	Composition	Target organ	Biological effect
From the pituitary gland			
Oxytocin	9 amino acids	Uterus, mammary glands	Stimulates contractions of uterus, and production of milk
Vasopressin	9 amino acids	Kidneys	Increases reabsorption of water by kidneys; decreases urine formation
Adrenocorticotrophin (ACTH)	39 amino acids	Adrenal cortex	Stimulates production of corticosteroids
Growth hormone (GH)	188 amino acids	Body tissues	Increases rate of protein synthesis; promotes growth of cells
Follicle-stimulating hormone (FSH)	Glycoprotein	Ovaries (female)	Stimulates follicle growth
Luteinizing hormone (LH)	Glycoprotein	Testes (male) Ovaries (female)	Stimulates production of testosterone, estrogens, and progesterone
From the pancreas			
Insulin	51 amino acids	Body tissues	Stimulates glucose metabolism; lowers blood glucose; increases glycogen storage
Glucagon	29 amino acids	Liver	Increases blood glucose by converting glycogen to glucose, and stimulates glucose synthesis from amino acids
From the parathyroid			
Parathyroid	84 amino acids	Body tissues	Regulates the calcium and phosphate levels in the body fluids
From the thyroid			
Thyroxine	Iodine-containing derivative of tyrosine	Body tissues	Increases metabolic rate, utilization of food, protein synthesis, and growth

Sample Problem 17.11
Peptide Hormones

Identify the hormone described in the following statements:

a. Increases use of glucose in the cells.
b. Increases rate of protein synthesis in body tissues.

Solution:

a. Insulin
b. Growth hormone

Study Check:

When blood glucose levels get low, there is an increase in the rate of converting glycogen into glucose. What hormone is responsible?

Answer: Glucagon

Glossary of Key Terms

active site The portion of the enzyme that catalyzes a reaction of the substrate.

amylase An enzyme in saliva and pancreatic juice that hydrolyzes α-1,4 glycosidic bonds in amylose and amylopectin.

anabolic steroid Synthetic derivatives of testosterone used to build muscle mass and decrease body fat.

antibiotic An inhibitor produced by a bacteria, mold, or yeast that is toxic to another bacteria.

antimetabolite A compound that inhibits an enzyme-catalyzed reaction.

bile The secretion from the pancreas that emulsifies fats in the small intestine.

catalyst A compound or enzyme that lowers the energy required for a reaction.

coenzyme An organic cofactor in an enzyme.

cofactor The nonprotein portion of an enzyme, such as a metal ion or a vitamin, that is necessary for enzyme activity.

competitive inhibition A type of inhibition in which an inhibitor with a structure similar to the substrate inhibits enzyme action by competing for the active site.

enzyme A protein that catalyzes biological reactions.

enzyme–substrate complex An intermediate in an enzyme-catalyzed reaction in which substrate combines with the enzyme.

fat-soluble vitamins Vitamins A, D, E, and K that are soluble in nonpolar solvents.

glucocorticoids Steroid hormones secreted by the adrenal glands that regulate glucose and protein metabolism levels; used to reduce inflammation, and to treat asthma and rheumatoid arthritis.

hormones Compounds, synthesized in the body, that are needed in small amounts to stimulate metabolic processes in other organs or cells of the body.

inhibitor A substance that interferes with the ability of an enzyme to react with a substrate.

isoenzymes Forms of an enzyme that vary slightly in structure depending on the tissue where the enzyme is active.

lock-and-key model A model of enzyme action that represents enzyme action as a key (substrate) fitting the correct lock (enzyme).

mineralcorticoids Steroid hormones secreted by the adrenal glands that regulate sodium, potassium and water levels.

noncompetitive inhibitor A substance that is not similar to the substrate that alters the shape of the enzyme to prevent the binding of the substrate at the active site.

optimum pH The pH at which an enzyme is most active.

optimum temperature A temperature at which an enzyme is most active.

steroid hormones Hormones related to cholesterol produced by adrenal glands, testes, ovaries, and the uterus.

substrate The reactant in an enzyme-catalyzed reaction.

substrate specificity The tendency of an enzyme to react with only specific substrates.

target organ The specific organ affected by a hormone.

vitamin An organic compound that must be obtained from the diet that is needed in small amounts to maintain proper metabolic activity in the cells of the body.

water-soluble vitamins Vitamins that are soluble in water and body fluids including vitamins B and C, pantothenic acid, biotin, niacin, and folic acid.

Problems

Enzyme Action *(Goal 17.1)*

17.1 Match these terms with the following phrases. (1) enzyme-substrate complex, (2) enzyme, (3) substrate, (4) active site, (5) lock-and-key theory

a. has a tertiary structure that recognizes the substrate
b. the combination of an enzyme with the substrate
c. the portion of an enzyme where catalytic activity occurs
d. the theory that accounts for the specificity of an enzyme
e. has a structure that complements the structure of the enzyme

17.2 Write an equation that represents an enzyme-catalyzed reaction.

a. Identify each of the symbols used in the equation.
b. How is the active site different from the whole enzyme structure?
c. After the products have formed, what happens to the enzyme?

17.3 The enzyme β-Galactosidase converts lactose to glucose and galactose.

a. What are the reactants and products of the reaction?
b. Draw an energy diagram for the reaction with and without the enzyme. Label the substrate and products and the energy requirement with and without enzyme.
c. Why would the enzyme make the reaction go faster?

Names and Classification of Enzymes *(Goal 17.2)*

17.4 What are cofactors in conjugated proteins?

17.5 Indicate whether the following statements describe a simple or a conjugated protein.

a. an enzyme that contains a sugar portion
b. an enzyme requiring Zn^{2+} for activity
c. an enzyme that contains only protein
d. an enzyme that needs a vitamin coenzyme
e. an enzyme composed of 155 amino acids

17.6 Name the substrate for the following enzymes:

a. lipase b. peptidase c. maltase d. esterase

17.7 What are the six classifications of enzymes?

17.8 Give the class of enzyme that would catalyze the reaction for each of the following reactions:

a. hydrolysis of sucrose
b. hydrolysis of galactose
c. hydrolysis of ester bonds in triglycerides
d. addition of oxygen
e. removal of hydrogen
f. isomerization of glucose to fructose
g. $CH_3—\overset{\overset{O}{\|}}{C}—\overset{\overset{O}{\|}}{C}—OH \longrightarrow CH_3—\overset{\overset{O}{\|}}{C}—OH + CO_2$

Pyruvic acid → Acetic acid

h. addition of water to a double bond (hydration)
i. transfer of an amino group to an α-keto acid
j. adding one molecule to another to form a larger compound

Factors That Affect Enzyme Activity *(Goal 17.3)*

17.9 Trypsin, a peptidase that hydrolyzes polypeptides, functions in the small intestine at an optimum pH of 8. How is the rate of a trypsin-catalyzed reaction affected by the following conditions?

a. lowering the concentration of polypeptides
b. increasing the trypsin concentration
c. decreasing the pH to 5
d. running the reaction at 15°C
e. running the reaction at 50°C

17.10 On the following graph are the curves for pepsin, urease, and an oxidase. Estimate the optimum pH for each.

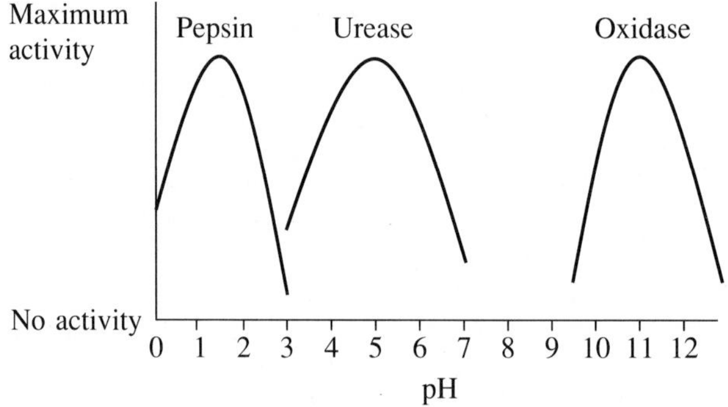

Enzyme Inhibition *(Goal 17.4)*

17.11 Indicate whether the following describe a competitive or a noncompetitive enzyme inhibitor:

a. has a structure similar to the substrate
b. its effect cannot be reversed by adding more substrate

c. competes with the substrate for the active site
d. can bind to the enzyme at the same time as the substrate
e. does not have a structure similar to the substrate
f. more substrate reverses the effect of the inhibitor

17.12 Oxaloacetic acid is an inhibitor of succinic dehydrogenase:

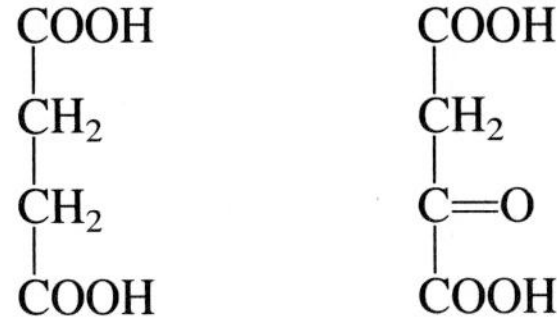

Succinic acid Oxaloacetic acid

a. Would you expect oxaloacetic acid to be a competitive or a noncompetitive inhibitor? Why?
b. How would you expect the inhibition of the enzyme to occur?
c. How could you reverse the effect of the inhibitor?

Enzymes in Digestion *(Goal 17.5)*

17.13 What is the major chemical process in the digestion of foods?

17.14 Describe the digestion of amylose and amylopectin.

17.15 What are some of the processes in the digestion of fats?

17.16 For proteins, consider the following:
a. Where does protein digestion begin?
b. When food is ingested, acid enters the stomach. Why?
c. What portion of protein digestion occurs in the small intestine? What enzymes are present?

Vitamins *(Goal 17.6)*

17.17 State whether each of the following statements is true or false:
a. Vitamins are needed in small amounts by the body.
b. Vitamins are not usually produced by the body.
c. All vitamins are water-soluble.
d. Vitamins often act as coenzymes in metabolic reactions.
e. A vitamin deficiency can cause illness.
f. Vitamins are proteins.
g. High doses of fat-soluble vitamins can be dangerous.

17.18 Give the chemical name for the following vitamins:
a. vitamin A
b. vitamin B_2
c. vitamin C
d. vitamin D
e. vitamin E

17.19 Classify the following as water- or fat-soluble.

a. vitamin A
b. vitamin B_6
c. vitamin D
d. vitamin C
e. folic acid
f. pantothenic acid
g. vitamin K

17.20 Which type of vitamin is lost more readily in the urine? Which type is stored in the body fat?

17.21 Identify the vitamin associated with each of the following functions or conditions:

a. night blindness
b. pellagra
c. beriberi
d. rickets
e. scurvy
f. flavin nucleotide (FAD)
g. blood clotting

Hormones *(Goal 17.7)*

17.22 The following synthetic estrogen is used in some oral contraceptive pills.

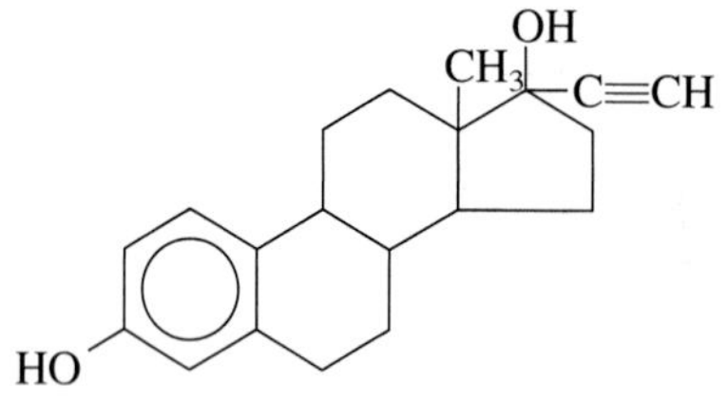

OH
CH_3 —C≡CH
HO

Ethinylestradiol

a. Why is it classified as a steroid?
b. Why would it cause physiological effects similar to the natural hormone, estradiol?

17.23 Supply the names of the hormones responsible for the following:

a. body growth
b. stimulating the contraction of the uterus
c. stimulating the production of corticosteroids in the adrenal gland
d. gigantism
e. increasing reabsorption of water by kidneys
f. stimulating production of testosterone
g. stimulating the breakdown of glycogen to glucose

17.24 Indicate the hormone responsible for the following:

a. increasing the rate of metabolic reactions
b. increasing the rate of carbohydrate metabolism
c. increasing glucose blood level
d. increasing the calcium level in the body fluids

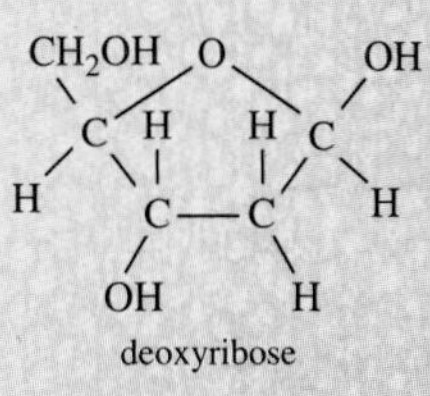
deoxyribose

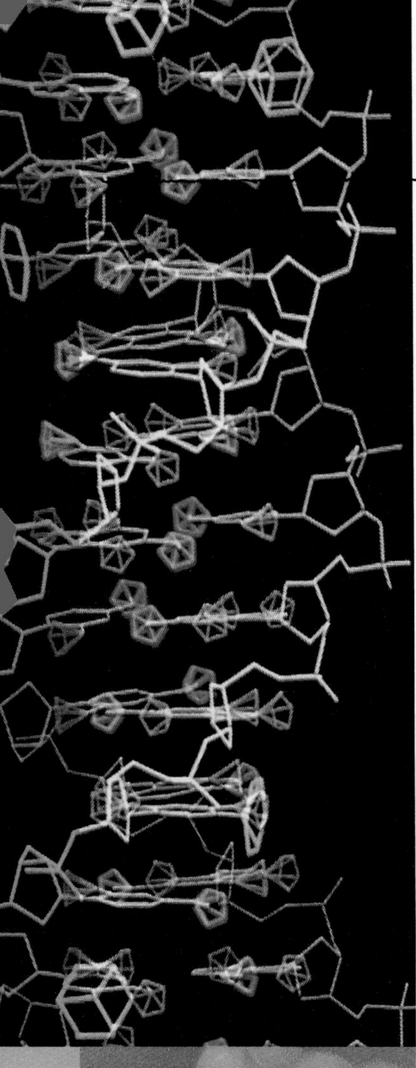

The double helix is the characteristic shape of the DNA molecules that transmit genetic information from one generation to the next.

18

Nucleic Acids and Heredity

Learning Goals

18.1 Describe the structural components of the nucleic acids.

18.2 Write the complementary base sequence for a segment of DNA.

18.3 Describe the replication process by which exact copies of DNA are made.

18.4 Describe the structure of RNA and its synthesis in the cell.

18.5 Indicate how triplets of nucleotides code for the amino acids of a protein.

18.6 Show how a DNA base sequence is converted into the sequence of amino acids in a protein.

18.7 Describe methods by which a cell controls the synthesis of protein.

18.8 Describe some ways in which DNA is altered to cause mutations and genetic disease.

SCOPE

In the preceding chapters we discussed the digestion and metabolism of carbohydrates, lipids, and proteins essential to the survival of living organisms. One more group of biomolecules must be included in our discussion of life—the nucleic acids.

Nucleic acids play a critical role in the control and direction of cellular growth and reproduction. The nucleic acid DNA—the genetic material in the nucleus of the cell—contains all the information needed for the development of a complete living system. The way you grow, your hair, your eyes, your physical appearance, and the activities of the cells in your body are all determined by a set of directions held in the nucleus of each of your cells. From the nucleus, the nucleic acid called DNA controls the functioning of the cells by regulating the synthesis of proteins. Another group of nucleic acids, RNA, transmits information from the DNA to the ribosomes, the cellular factories for the synthesis of protein. When the genetic blueprint is altered, incorrect information is transmitted to the ribosomes, leading to the formation of defective structural proteins and malfunctioning enzymes.

18.1 Nucleic Acids: DNA and RNA

Learning Goal **Describe the structural components of the nucleic acids.**

The **nucleic acids,** large molecules first discovered in the nuclei of cells, contain all the information needed to direct the activities of a cell and its reproduction. The two types, **deoxyribonucleic acid (DNA),** and **ribonucleic acid (RNA),** are composed of long chains built of **nucleotide** units. Thousands of nucleotides linked together form a nucleic acid much the way amino acids form proteins. Each nucleotide consists of three parts: a cyclic nitrogen base, a pentose sugar, and a phosphate.

Nitrogen base
+
Pentose sugar ⟶ Nucleotides ⟶ Nucleic acid (DNA, RNA)
+
Phosphate

Nitrogen Bases in Nucleotides of DNA and RNA

The nucleotides of DNA and RNA are made from four kinds of nitrogen bases derived from **pyrimidine** and **purine.** The pyrimidine bases in DNA are thymine (T) and cytosine (C), and the purine bases are adenine (A) and guanine (G). In RNA the nitrogen bases are much the same, except that uracil, another pyrimidine, replaces thymine. (See Figure 18.1.)

Nitrogen Bases

In DNA	*In RNA*
A Adenine	A Adenine
C Cytosine	C Cytosine
G Guanine	G Guanine
T Thymine	U Uracil

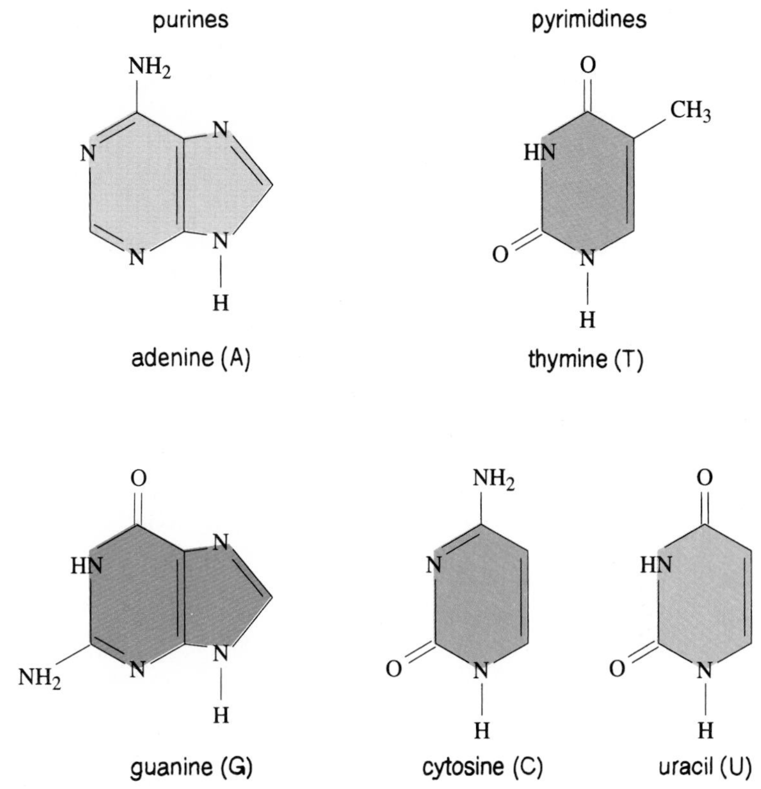

Figure 18.1
Nitrogen bases found in nucleic acids.

Pentose Sugars in Nucleotides

The pentose found in nucleotides of RNA is D-ribose, a five-carbon sugar. In DNA, the sugar is deoxyribose, a pentose similar to ribose except that there is no hydroxyl (−OH) group on carbon 2, thus the prefix *deoxy*. (See Figure 18.2.)

Formation of a Nucleotide

A nucleotide is formed when one of the nitrogen bases and a phosphate group attach to a sugar (ribose or deoxyribose). A bond forms between the nitrogen base and carbon 1 of the sugar, while carbon 5 of the sugar forms an ester bond

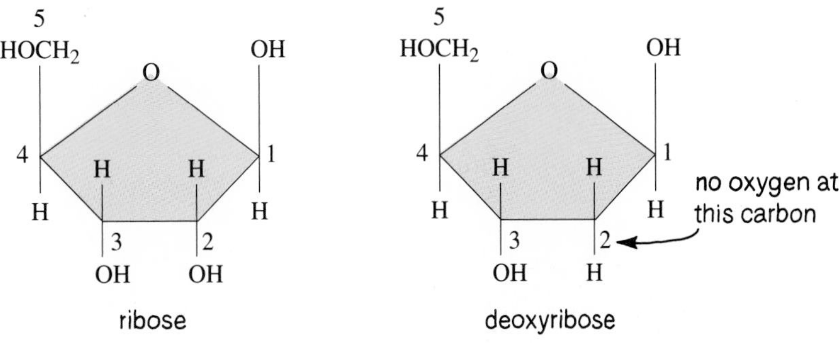

Figure 18.2
Pentose sugars in nucleic acids.

Figure 18.3
Formation of a nucleotide, cytidine monophosphate (CMP), found in RNA.

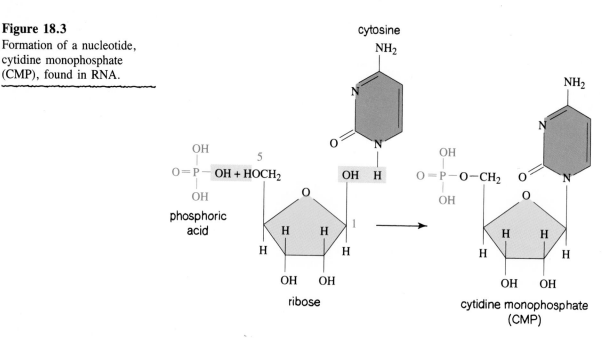

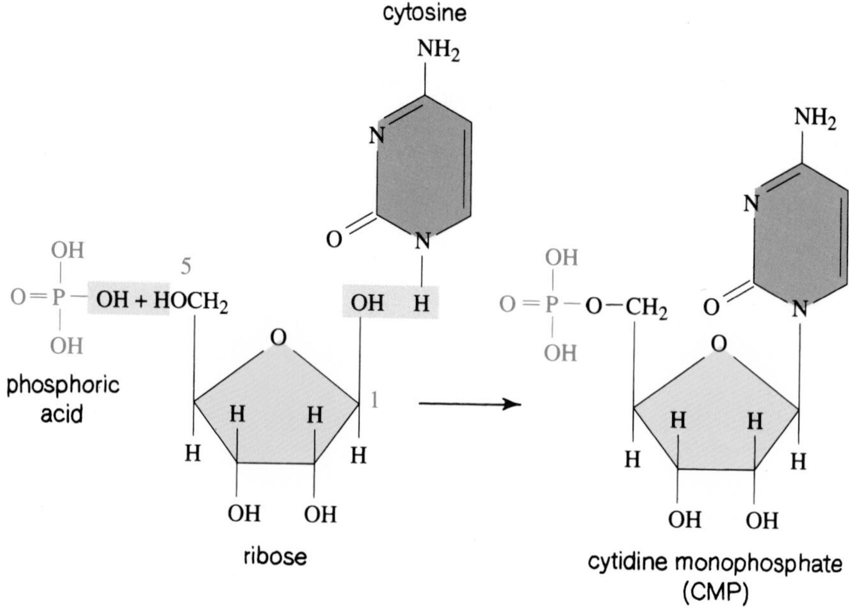

with a phosphate group. For example, ribose, cytosine, and phosphoric acid form a nucleotide of RNA named cytidine monophosphate (CMP). The ending of pyrimidine names (for example, cytosine) is changed to *idine* (cytidine); purine names end with *osine* (adenosine). (See Figure 18.3.)

The structures of the four nucleotides of DNA are shown in Figure 18.4. In the name, the prefix *deoxy* or its abbreviation d- refers to the deoxyribose sugar. If the sugar in dAMP, dGMP, and dCMP is replaced with ribose, the structures represent three nucleotides of RNA. The other nucleotide of RNA contains uracil instead of thymine.

Sample Problem 18.1
Nucleotides

Identify the nucleic acid where each of the following nucleotides would be found; give components of each nucleotide:

a. deoxyguanidine monophosphate (dGMP)
b. cytidine monophosphate (CMP)

Solution:

a. This is a nucleotide that would be found in DNA. It consists of deoxyribose (sugar), guanine (nitrogen base), and a phosphate.
b. This nucleotide would be found in RNA. It contains ribose, cytosine, and a phosphate.

Study Check:
What components are present in dAMP?

Answer: deoxyribose, adenine, and a phosphate

Figure 18.4
Nucleotides of DNA. Replacing the deoxyribose of dAMP, dGMP, and dCMP with ribose gives three nucleotides of RNA. In RNA, uracil replaces thymine.

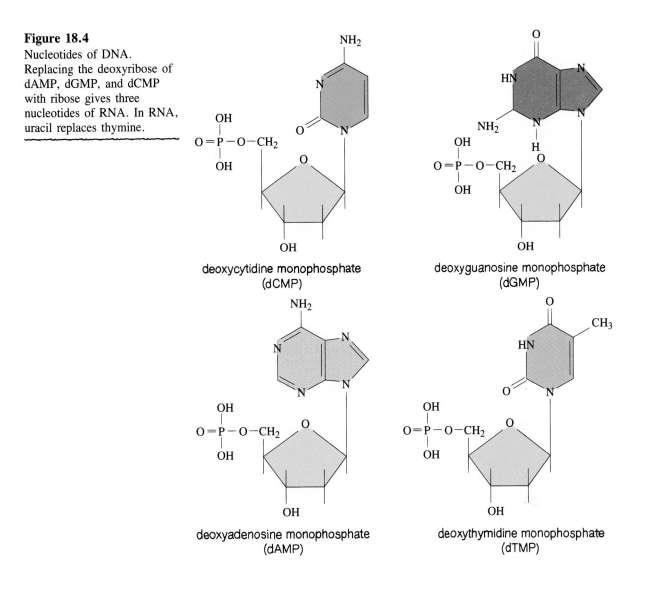

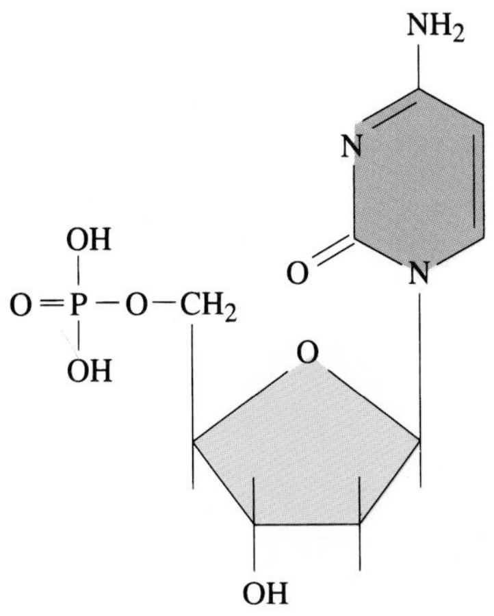

deoxycytidine monophosphate
(dCMP)

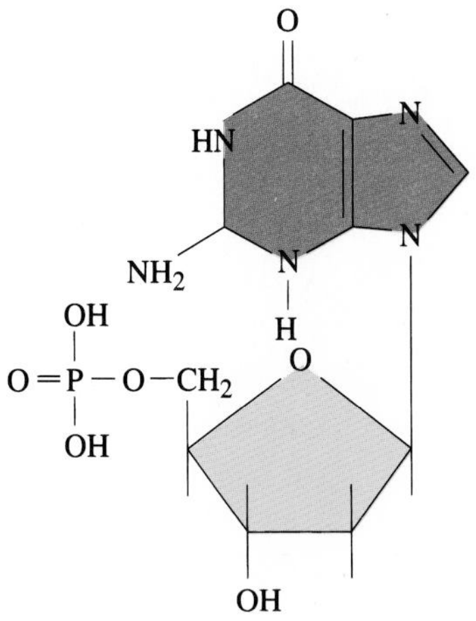

deoxyguanosine monophosphate
(dGMP)

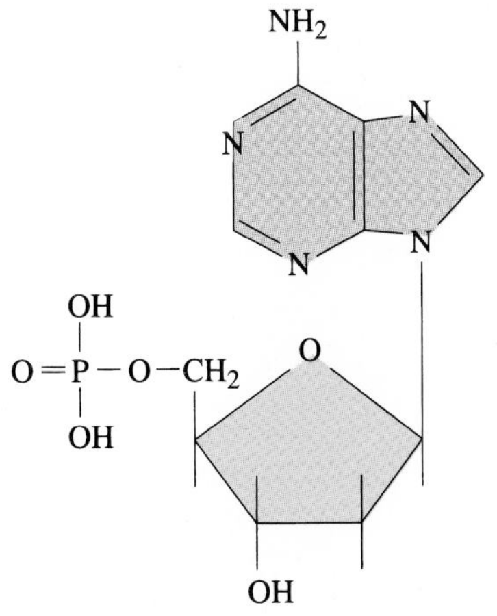

deoxyadenosine monophosphate
(dAMP)

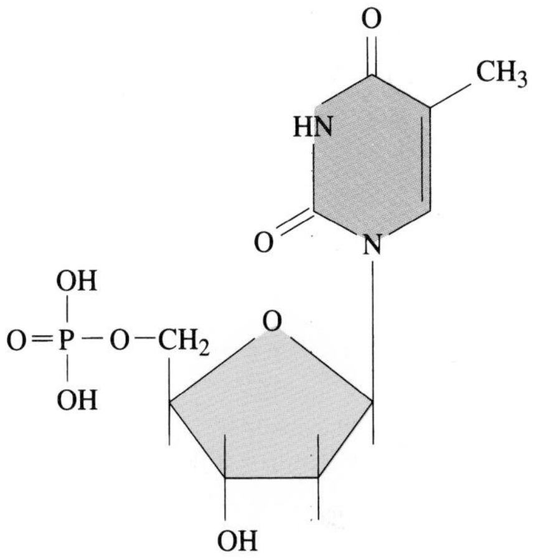

deoxythymidine monophosphate
(dTMP)

Nucleotides Form Polymers

The nucleic acids are polymers made up of an assortment of four different kinds of nucleotides in much the same way that 20 different amino acids form a protein, or glucose molecules form starch. In a nucleic acid chain as seen in Figure 18.5, the sugar and phosphate unit of each nucleotide join to form a backbone of repeating sugar units connected by phosphate groups. The bases attached to each sugar extend out from the backbone.

In the formation of the backbone of DNA, each phosphate group attaches by an ester bond to carbon 3 of the next sugar. There is a free (unbonded) phosphate group on the 5 carbon at one end of the chain (5 end). At the other end of the chain, there is a free —OH group on carbon 3 of the sugar (3 end).

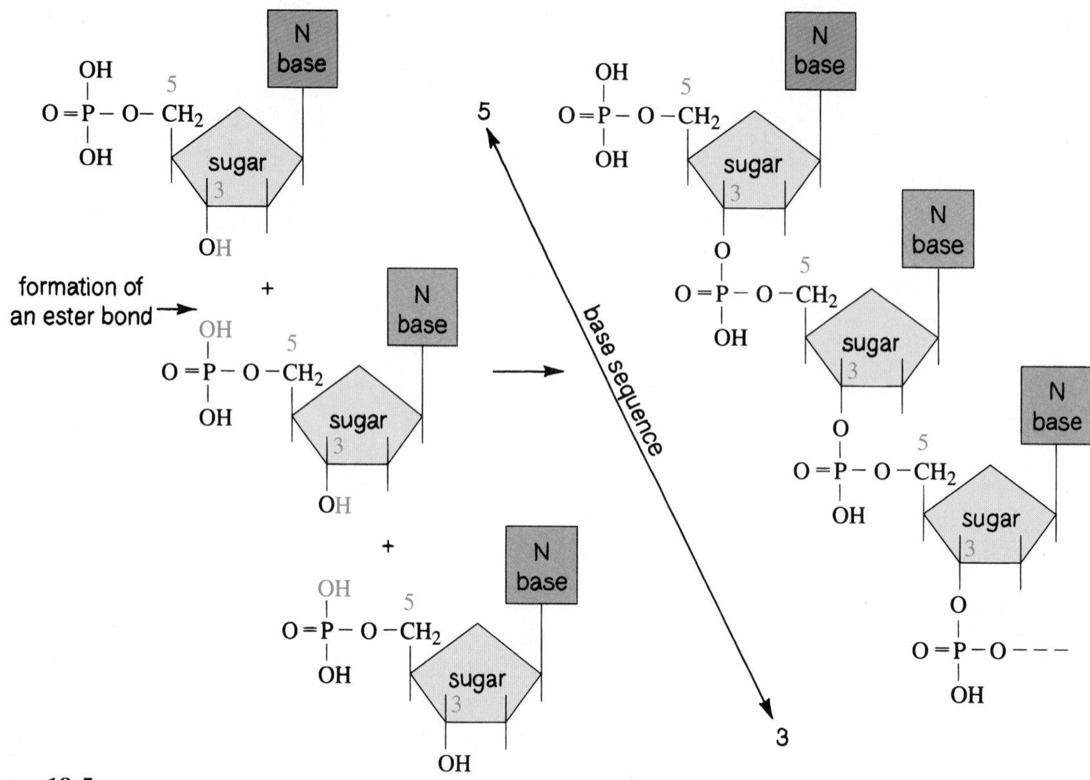

Figure 18.5
A portion of a nucleic acid in which alternating sugar-phosphate groups form a backbone with a nitrogen base attached on each sugar group.

The sequence of three nucleotides in a DNA chain is shown in Figure 18.6. The sequence of nucleotides, deoxyadenosine, deoxyguanosine, and deoxythymidine, is abbreviated AGT using the first letters of the different bases. If we used ribose as the sugar and substituted uracil for thymine, we would have a portion of an RNA chain, AGU.

Sample Problem 18.2
Nucleic Acid Chain

Write the structure for the dinucleotide CG(5 → 3) that would be found in a DNA chain.

Solution:

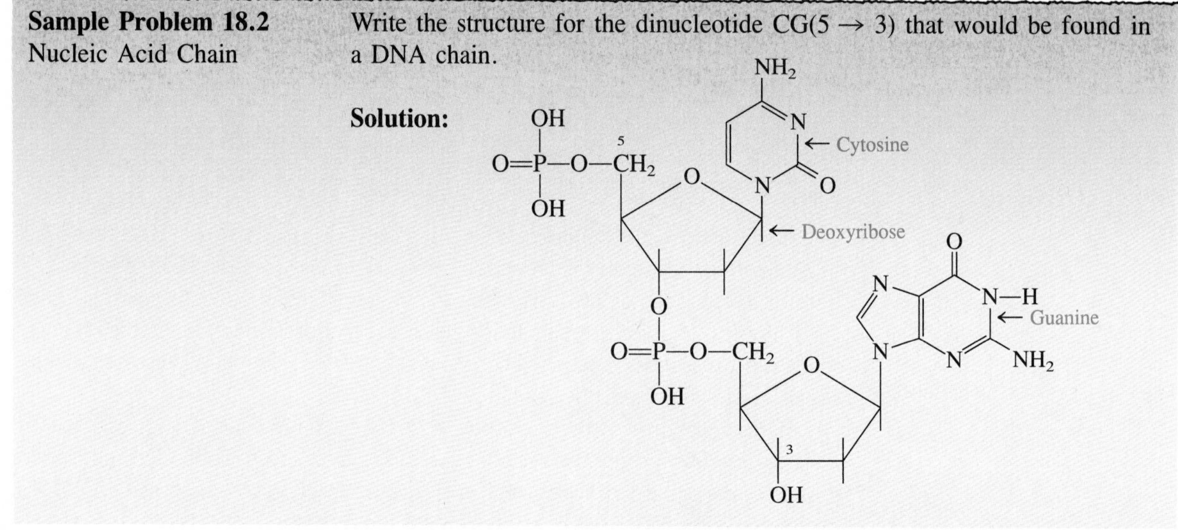

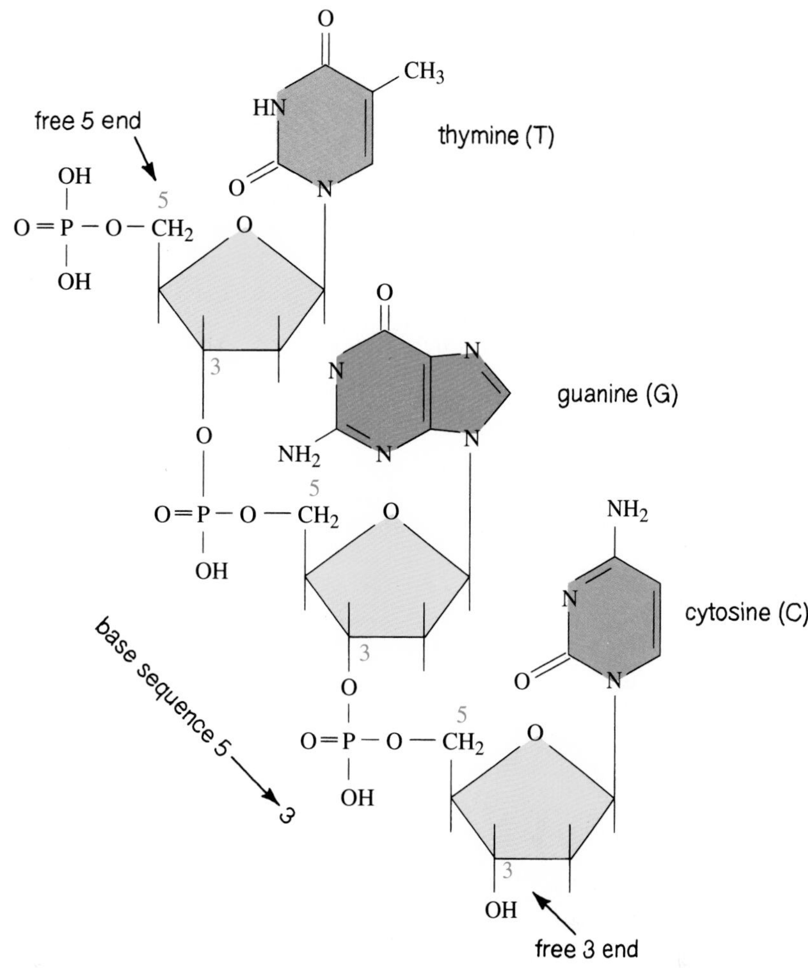

Figure 18.6
Structure of a trinucleotide from DNA.

18.2 Complementary Base Pairing in DNA

Learning Goal **Write the complementary base sequence for a segment of DNA.**

Figure 18.7
A set of 23 chromosomes stained in metaphase.

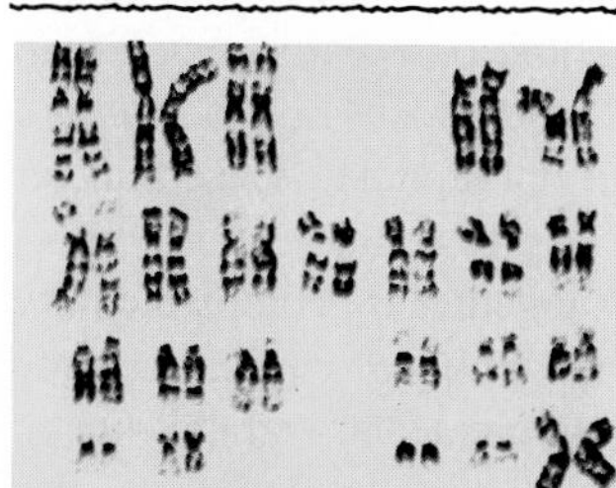

Deoxyribonucleic acid (DNA) exists as very large molecules, many times longer than proteins, in the nuclei of all our cells. Yet all the cells in our bodies have identical strands of DNA, although egg and sperm cells contain just half as much DNA as body cells. DNA, along with some proteins, makes up the material of the **chromosomes,** the threadlike structures observed in the nucleus during cellular division. In humans, every cell contains 23 pairs of chromosomes, or 46 chromosomes in all (except for egg and sperm cells, each of which contains 23). All of our characteristics develop from instructions given by the DNA. (See Figure 18.7.) Along the strands of DNA in each chromosome are sections called **genes** that contain the directions for every type of protein produced in our cells.

Figure 18.8
Complementary base pairing in a segment of DNA.

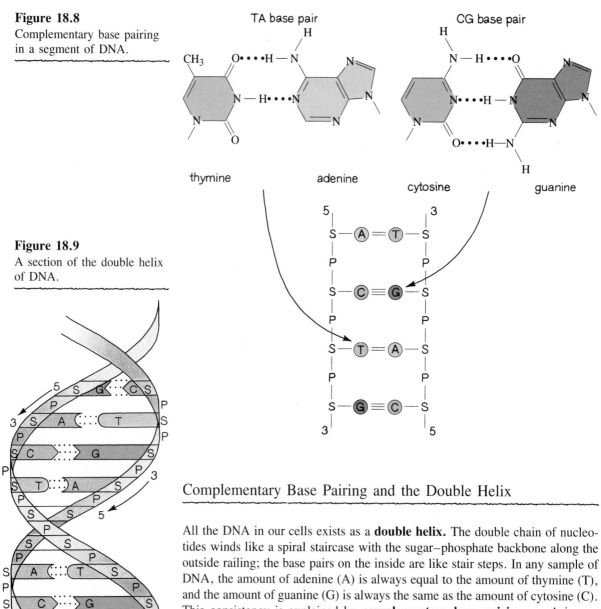

TA base pair

CG base pair

thymine

adenine

cytosine

guanine

5 S—A=T—S 3
P P
S—C≡G—S
P P
S—T=A—S
P P
S—G≡C—S
3 5

Figure 18.9
A section of the double helix of DNA.

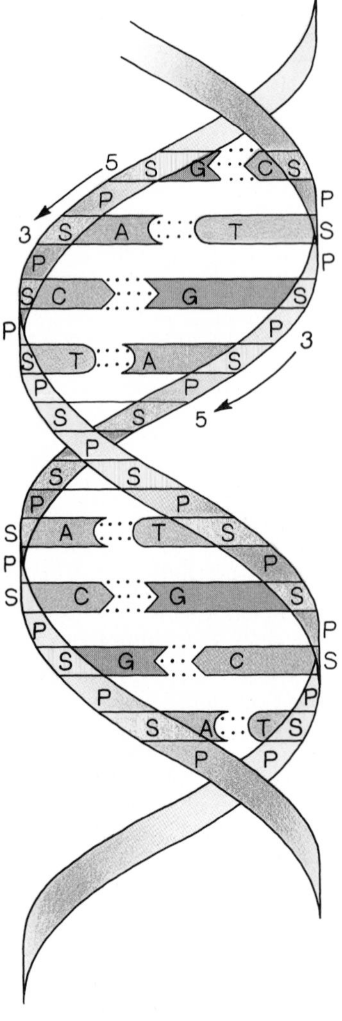

Complementary Base Pairing and the Double Helix

All the DNA in our cells exists as a **double helix.** The double chain of nucleotides winds like a spiral staircase with the sugar–phosphate backbone along the outside railing; the base pairs on the inside are like stair steps. In any sample of DNA, the amount of adenine (A) is always equal to the amount of thymine (T), and the amount of guanine (G) is always the same as the amount of cytosine (C). This consistency is explained by **complementary base pairing:** an A in one chain is always paired with a T in the other chain, while a G is always paired with a C. These pairs, A-T and G-C, are purine–pyrimidine pairs that best fit the space between the sugar–phosphate backbone. The pair G-C is held together by three hydrogen bonds, while the pair A-T has two hydrogen bonds. (See Figure 18.8.) Therefore, in DNA, there are only two possible base pairs: A::T, and G⋮⋮C, where the dots represent 2 and 3 hydrogen bonds, respectively. We will see the importance of complementary base pairing in cell replication, the transfer of hereditary information, and the survival of a living system.

In the double helix, the two DNA chains held together by complementary base pairs run in opposite directions. One side begins with free carbon 5 on the sugar and ends with a free carbon 3 ($5 \rightarrow 3$). The complementary strand runs in the opposite, $3 \rightarrow 5$, direction. (See Figure 18.9.)

Sample Problem 18.3
Complementary Base Pairing

One strand of DNA contains a base sequence of ACGAT. Write the base sequence for the complementary portion of the second strand in the double helix.

Solution:
The second strand of DNA contains bases that are complementary to the sequence given. The base adenine (A) always pairs with thymine (T); guanine (G) always pairs with cytosine (C). The backbone of the DNA strands is made of alternating sugar-phosphate groups.

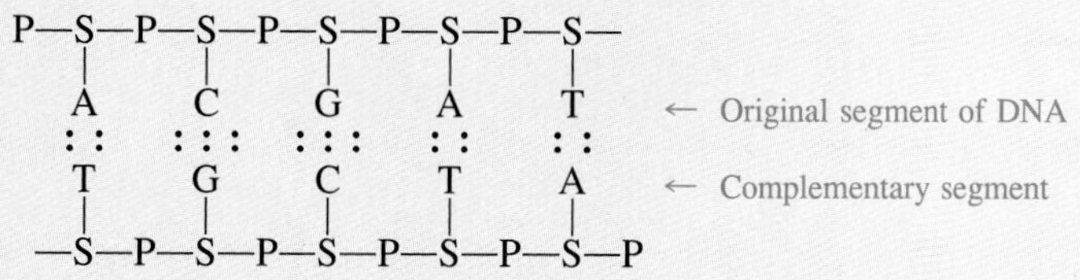

```
P—S—P—S—P—S—P—S—P—S—
  |     |     |     |     |
  A     C     G     A     T    ← Original segment of DNA
  ::    :::   :::   ::    ::
  T     G     C     T     A    ← Complementary segment
  |     |     |     |     |
 —S—P—S—P—S—P—S—P—S—P
```

We can abbreviate the writing of the complementary strands as

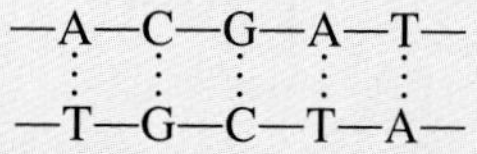

```
—A—C—G—A—T—
 :  :  :  :  :
—T—G—C—T—A—
```

Study Check:
What is the sequence of bases that is complementary to a portion of DNA with bases GGTTAACC?

Answer: CCAATTGG

18.3 DNA Replication

Learning Goal **Describe the replication process by which exact copies of DNA are made.**

One criterion of a living system is the ability of its cells to reproduce. Each time a cell divides, an exact copy of its DNA goes into each new (daughter) cell. The DNA in the parent cell provides the pattern for the synthesis of each new DNA. In **DNA replication,** the parent DNA is duplicated exactly every time the cell divides.

Central to the process of DNA replication is the concept of complementary base pairing. To begin replication, a portion of the DNA unwinds as an enzyme breaks the hydrogen bonds between the paired bases apart, much like opening a zipper. (See Figure 18.10.) DNA replication occurs along the separated DNA strands. Each of the nucleotides in the open strands form hydrogen bonds with their complementary nucleotide from the assortment available in the nucleus. The new DNA strand grows as DNA polymerase joins one nucleotide to the next by catalyzing the formation of sugar-phosphate bonds. Eventually the entire

Figure 18.10
Two exact copies of DNA are formed as complementary base pairs form between each DNA strand and appropriate nucleotides in the nucleus.

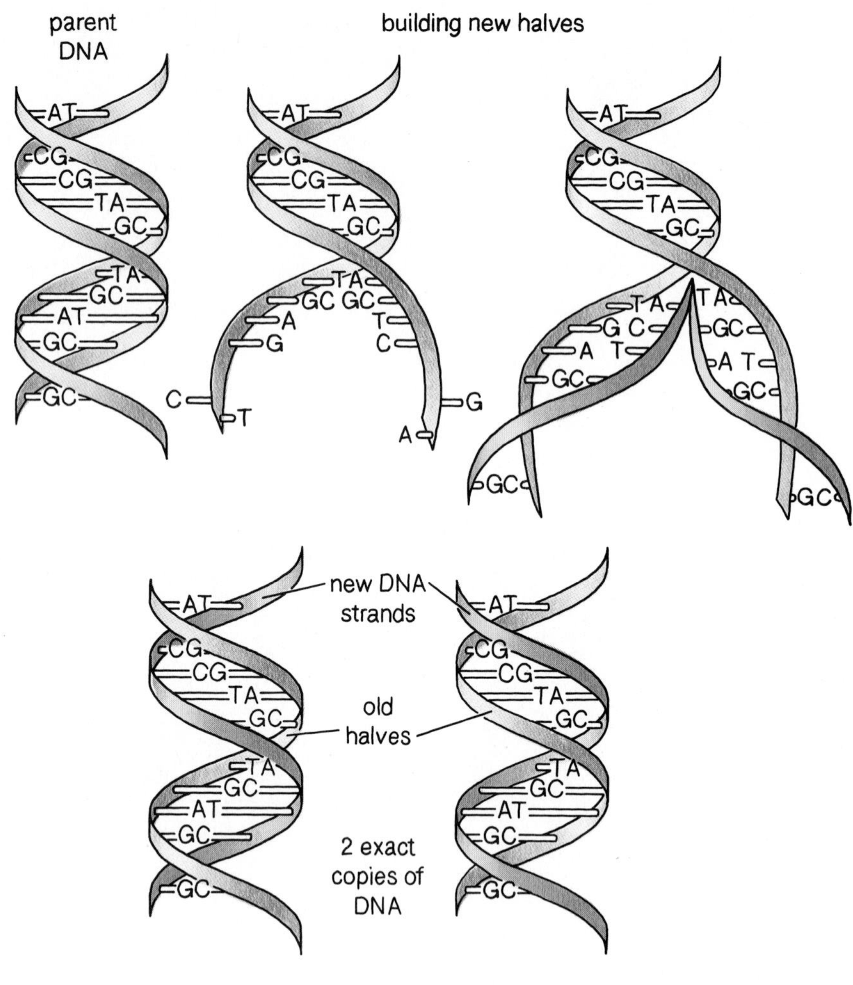

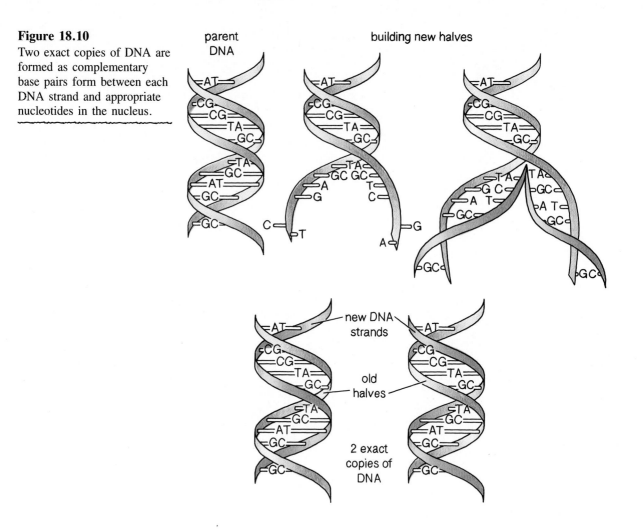

double helix unwinds and new strands of complementary DNA are synthesized for each of the parent DNA strands. In the process of DNA replication, complementary base pairing ensures the correct placement of bases for the new DNA strands. There are now two exact copies of the DNA for each of the new daughter cells.

Sample Problem 18.4
DNA Replication

How is a DNA molecule copied during replication?

Solution:
The double-stranded DNA separates and each half picks up new nucleotides using complementary base pairing. DNA polymerase forms sugar–phosphate bonds between the aligned nucleotides to complete the backbone, and hydrogen bonds hold the base pairs together. When the process is completed, two identical molecules of DNA have formed.

Recombinant DNA

Over the past 20 years, geneticists have been cutting, splicing, and rejoining DNA from different genes to form a new synthetic DNA, called **recombinant DNA.** Most of this experimentation has been done with a bacterium called *Escherichia coli (E. coli),* which contains a simple chromosome and several small cyclic DNA particles called **plasmids.**

When the *E. coli* cells are soaked in a detergent solution, they break open, releasing the plasmids, which are collected. Enzymes called restriction enzymes are used to remove a specific section of DNA from the plasmid. Then, a piece of DNA from another organism, which might be the gene that produces insulin or growth hormone, is placed in the cut region of the plasmid. The ends of the inserted DNA piece and the plasmid are joined by a DNA ligase. This process of gene splicing forms a hybrid DNA or recombinant DNA that can now be reabsorbed by the bacterium. The altered *E. coli* cells are then cloned. As they reproduce, the recombinant DNA in the plasmids is also replicated. The new *E. coli* cells will produce a protein they did not previously produce, for use in medicine, agriculture, or industry. (See Figure 18.11.)

Recombinant DNA has caused much excitement because of its potential medical uses. Already, recombinant DNA methods have produced enzymes, hormones, antibodies, clotting factors, and other compounds that some people do not produce in their cells as a result of genetic diseases. Recently, the DNA that synthesizes human insulin has been successfully incorporated into *E. coli* plasmids. The prospect is for a plentiful, inexpensive source of insulin and other

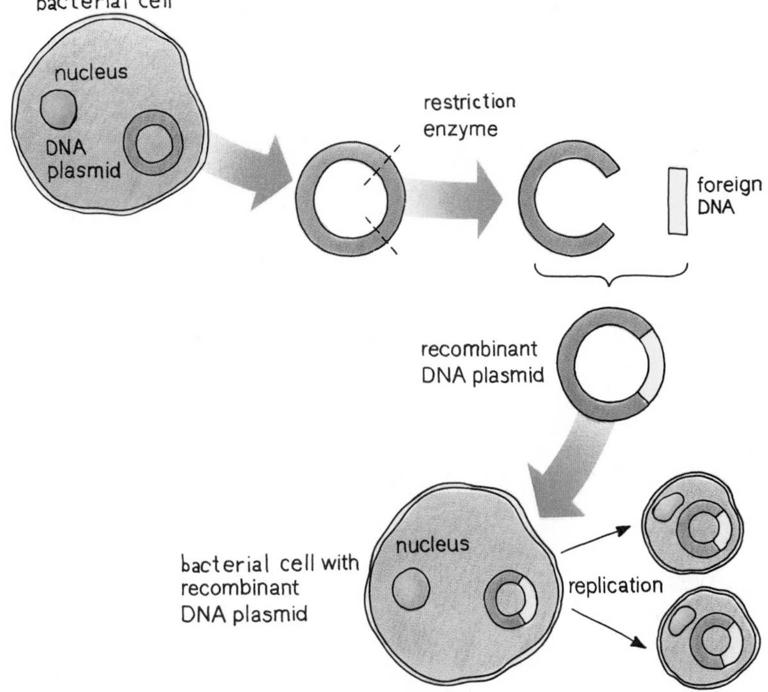

Figure 18.11
Formation of recombinant DNA by placing new DNA in plasmid DNA of bacterium. The bacterium can now produce a nonbacterium protein such as insulin or growth hormone.

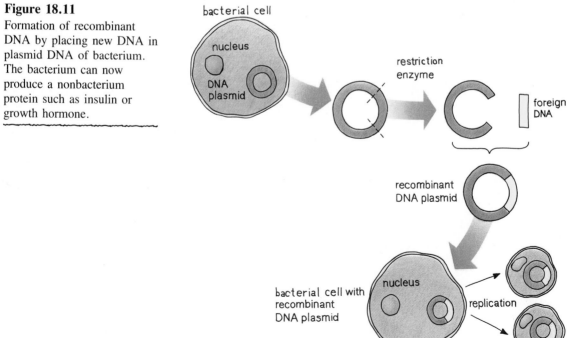

needed cellular components. The protein interferon, which helps fight viral infections and possibly cancer, has been produced by recombinant DNA technology in greater amounts than was previously available. Perhaps in the future defective genes that cause genetic disease will be corrected by recombinant DNA methods.

There has been concern over recombinant DNA since mistakes could produce toxic strains of bacteria for which there is no biological defense. Guidelines to control such possibilities have been established by the National Institutes of Health. As a control measure, the *E. coli* strain used for research is made deficient in one amino acid, so that it is dependent on laboratory conditions for survival.

18.4 RNA: Structure and Synthesis

Learning Goal **Describe the structure of RNA and its synthesis in the cell.**

Ribonucleic acid, RNA, makes up most of the nucleic acid found in the cell. RNA is continually produced by DNA to provide the information for the arrangement of amino acids in proteins. Although the structure of RNA is similar to DNA, there are some important differences. First, the sugar in RNA is ribose, not deoxyribose. Second, the nitrogen base uracil is used, not thymine. Third, RNA molecules are single strands of nucleotides, not double. Sometimes, however, a single RNA chain folds upon itself to form a double-stranded region that is held together by complementary base pairing.

Types of RNA

There are three different kinds of ribonucleic acids in the cells. (See Table 18.1.) **Messenger RNA (mRNA)** provides the pattern for the synthesis of a protein. (See Figure 18.12.) Every gene in the DNA produces its own mRNA. **Transfer RNA (tRNA),** a smaller RNA molecule, carries amino acids to the ribosomes to be incorporated into proteins. Each of the 20 amino acids has its own tRNA. **Ribosomal RNA (rRNA),** the most abundant RNA, makes up 60% of the structural material of the ribosomes. Its role in protein synthesis is not yet well understood.

Table 18.1 RNA Molecules

Type	Abbreviation	Percentage of total RNA	Function in the cell
Messenger RNA	mRNA	5–10	Carries information for the synthesis of a protein
Ribosomal RNA	rRNA	75	Major component of the ribosomes
Transfer RNA	tRNA	10–15	Carries amino acids to ribosome for protein synthesis

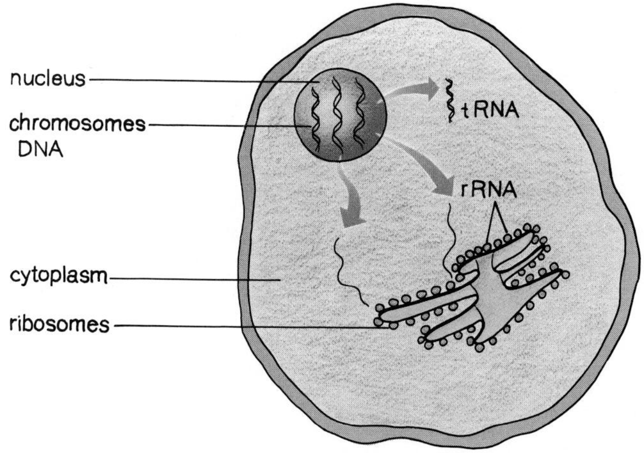

Figure 18.12
Nucleic acids in a cell.

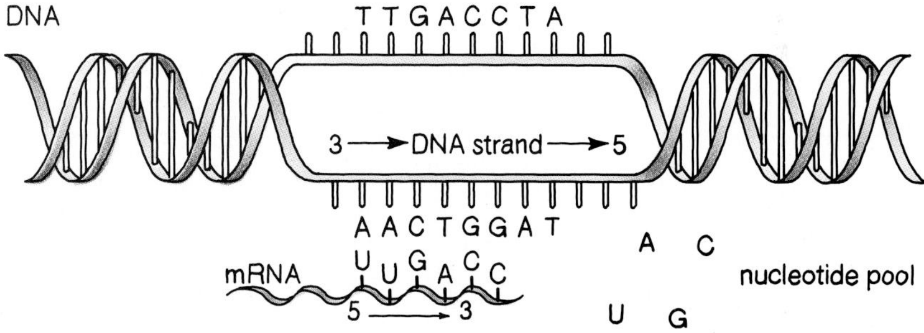

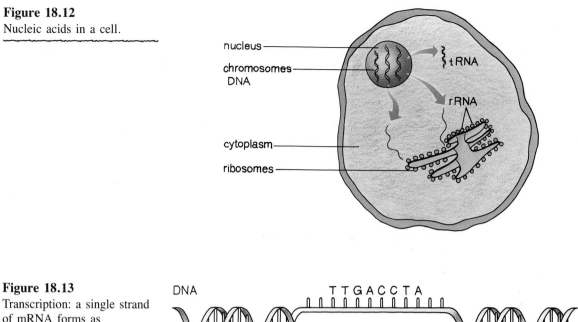

Figure 18.13
Transcription: a single strand of mRNA forms as nucleotides find their complementary bases on the DNA template.

Synthesis of Messenger RNA

In a process called **transcription,** the sequence of nucleotides on one strand of DNA is copied (transcribed) into a messenger RNA molecule. As in DNA replication, the DNA unwinds to give separate DNA strands. The appropriate RNA nucleotides pair up with their complementary bases along the DNA strand that is used as a template for RNA synthesis. Uracil bonds with adenine; cytosine bonds with guanine. The nucleotides of mRNA are joined by RNA polymerase that moves from the start point at the 3 end of the DNA template stand to a stop point at the 5 end. (See Figure 18.13.)

Sample Problem 18.5
Messenger RNA

The sequence of bases in a part of the DNA template for mRNA is CGATCA. What is the corresponding section of the mRNA produced?

Solution:
The corresponding portion of mRNA consists of the complementary bases of the DNA section.

DNA template	-C-G-A-T-C-A-
	: : : : : :
mRNA portion	-G-C-U-A-G-U-

Study Check:
Write the mRNA base sequence that is complementary to a DNA template section of CCCAAATTT.

Answer: GGGUUUAAA

18.5 The Genetic Code

Learning Goal **Indicate how triplets of nucleotides code for the amino acids of a protein.**

In the synthesis of a protein, the information transferred from the DNA to the mRNA must be translated into an amino acid sequence. The mRNA produced in the nucleus moves to a ribosome in the cytoplasm where tRNA is available to convert the nucleotide sequence into a sequence of amino acids in a protein. (See Figure 18.14.) To understand protein synthesis, we need to look at some of the different processes of translation beginning with the genetic code.

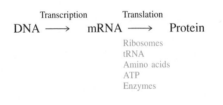

$$\text{DNA} \xrightarrow{\text{Transcription}} \text{mRNA} \xrightarrow{\text{Translation}} \text{Protein}$$

Ribosomes
tRNA
Amino acids
ATP
Enzymes

Genetic Code

The **genetic code** is a series of base triplets in mRNA called codons that code for a particular amino acid. Early work on protein synthesis found that an mRNA consisting of repeating triplets of uracil (UUU) produced a polypeptide that contained only phenylalanine.

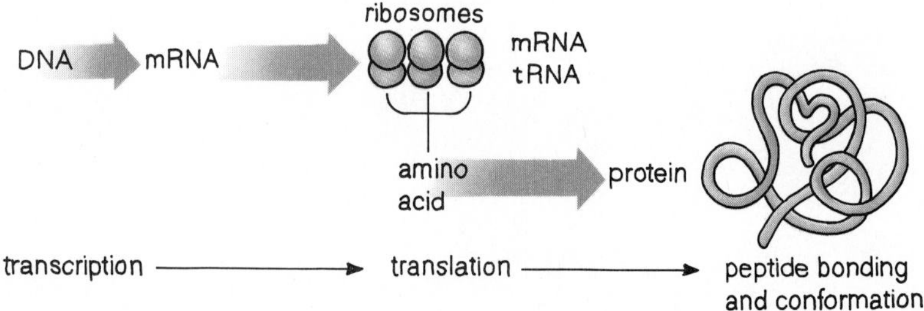

Figure 18.14
Protein synthesis (simplified).

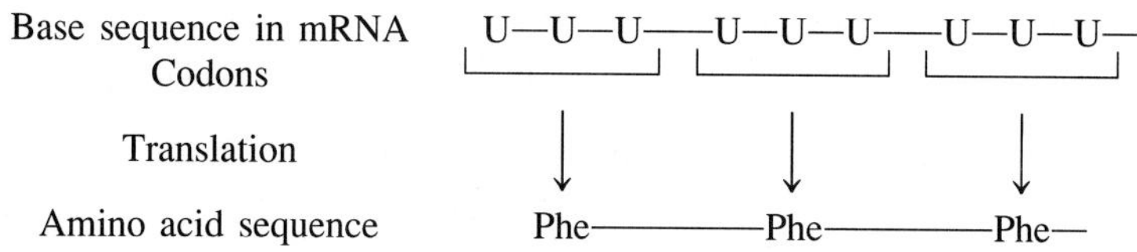

The mRNA codons have now all been determined as shown in Table 18.2. For example, AAA codes for lysine, and CCC codes for proline. In most cases, several codons code for the same amino acid. Two codons, AUG and GUG, also signal the start of a peptide chain. Three of the codons, UAG, UAA, and UGA, do not code for any amino acids: rather they signal the end of protein synthesis.

Table 18.2 mRNA Codons: The Genetic Code for Amino Acids

first letter ↓	second letter: U		C		A		G		third letter ↓
U	UUU	Phe	UCU	Ser	UAU	Tyr	UGU	Cys	U
	UUC		UCC		UAC		UGC		C
	UUA	Leu	UCA		UAA	STOP	UGA	STOP	A
	UUG		UCG		UAG	STOP	UGG	Trp	G
C	CUU	Leu	CCU	Pro	CAU	His	CGU	Arg	U
	CUC		CCC		CAC		CGC		C
	CUA		CCA		CAA	Gln	CGA		A
	CUG		CCG		CAG		CGG		G
A	AUU	Ile	ACU	Thr	AAU	Asn	AGU	Ser	U
	AUC		ACC		AAC		AGC		C
	AUA		ACA		AAA	Lys	AGA	Arg	A
	*AUG	Met	ACG		AAG		AGG		G
G	GUU	Val	GCU	Ala	GAU	Asp	GGU	Gly	U
	GUC		GCC		GAC		GGC		C
	GUA		GCA		GAA	Glu	GGA		A
	*GUG		GCG		GAG		GGG		G

*Codon that signals the start of a peptide chain.
STOP Codons signal the end of a peptide chain.

Sample Problem 18.6
Codons

State the amino acid coded for by each of the following:

a. GUC **b.** AGC **c.** UAA

Solution:

a. Valine (Val) **b.** Serine (Ser) **c.** none; (signals STOP)

Study Check:
Using Table 18.2, write the codons for proline and arginine.

Answer: proline: CCU, CCC, CCA, and CCG
arginine: CGU, CGC, CGA, CGG, AGA, AGG

Transfer RNA

The amino acids needed to produce a protein cannot directly recognize the sequence of bases in the mRNA. It is the job of the tRNA to pick up the correct amino acid and carry it to the ribosomes where the amino acid is placed in a peptide chain. While there are different tRNAs for the 20 kinds of amino acids, all tRNAs have similar structures.

Each tRNA exists as a single chain of 70-90 nucleotides. About half the nucleotides in the chain form hydrogen bonds with complementary nitrogen bases in another part of the same chain, which gives the tRNA an overall shape like a cloverleaf. At one end of the tRNA, there is an attachment site of bases (CCA) that bonds to the amino acid for that particular tRNA. In the center loop called the **anticodon loop,** there are three unpaired bases that are complementary to a codon on a mRNA. (See Figure 18.15.)

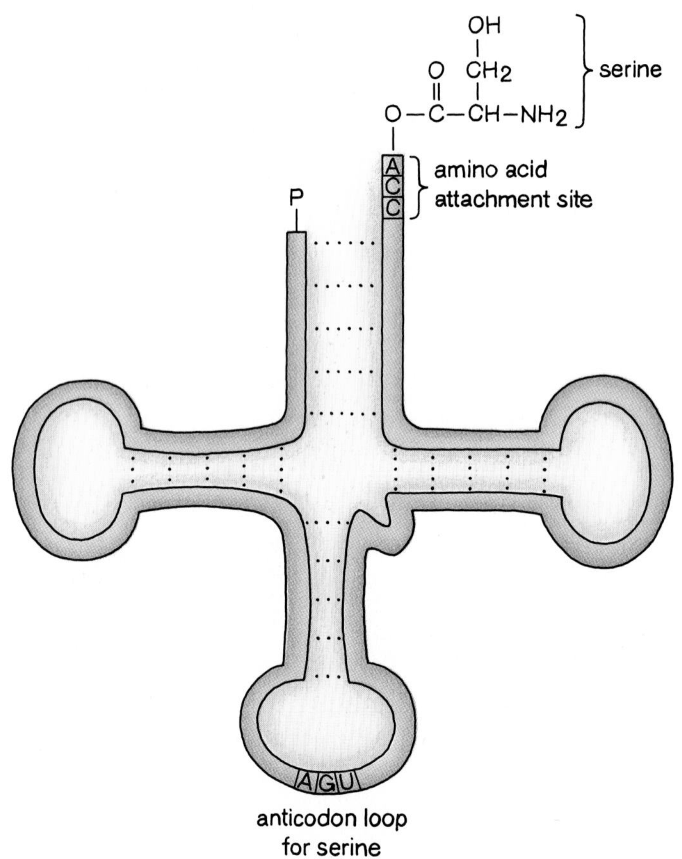

Figure 18.15
Diagram of a transfer RNA with an amino acid carried at the amino acid attachment site while the anticodon loop complements a codon on an mRNA.

18.6 Protein Synthesis

Learning Goal **Show how a DNA base sequence is converted into the sequence of amino acids in a protein.**

After an mRNA is synthesized, it migrates out of the nucleus into the cytoplasm, where it attaches to the ribosomes, the factories of the cell that produce protein. A typical ribosome contains proteins, ribosomal RNA and binding sites for the mRNA.

The **translation** of mRNA for the synthesis of a protein begins when a ribosome picks up an mRNA. The first codon in mRNA, usually AUG, one of the codons that starts protein, is placed in binding site 1. A tRNA carrying methionine bonds to the codon using the complementary bases on the anticodon loop.

Now the next codon on the mRNA must be translated. Another tRNA brings the second amino acid to the mRNA on the ribosome. Suppose that the second codon is UCC which codes for serine. A tRNA with an anticodon of AGG brings serine to the ribosome and attaches it at binding site 2. With the two amino acids held in the binding sites, an enzyme catalyzes the formation of a peptide bond. The first tRNA then leaves the ribosome and the second tRNA with the peptide moves over to binding site 1. Another tRNA can now bring an amino acid for the new codon now in binding site 2. (See Figure 18.16.)

The process of translation continues as the ribosome or several ribosomes (a polysome) moves along (translocation) the mRNA, one codon at a time. At each codon, a tRNA with the correct amino acid attaches to the mRNA. A peptide bond forms between the new amino acid and the last amino acid in the growing peptide chain. The polypeptide continues to grow as the ribosome moves along the mRNA codons.

Protein synthesis ends when the ribosome arrives at a stop codon, UGA, UAA, or UAG. Then the completed protein is released, and the ribosome is available to begin the synthesis of another protein.

As the peptide chain of the protein builds on the ribosome, it also forms secondary and tertiary cross-links. By the time the protein is released, it has assumed the secondary and tertiary structures that make it biologically active in the cell. In the case of quaternary structures, different mRNAs simultaneously produce subunits that combine as soon as they are released from the ribosome.

Sample Problem 18.7
Codons of the Genetic Code

What order of amino acids would you expect from the following mRNA segment: UCA-AAA-GCC-CUU?

Solution:

Dividing the mRNA into triplets of bases and referring to the codons in Table 18.2, we can write the following peptide:

```
UCA—AAA—GCC—CUU
└┬┘ └┬┘ └┬┘ └┬┘
Ser Lys Ala Leu
```

Study Check:

Where would protein synthesis stop in the following series of bases: GGG-AGC-AGU-UAG-GUU?

Answer: Protein synthesis would stop at UAG, a stop codon.

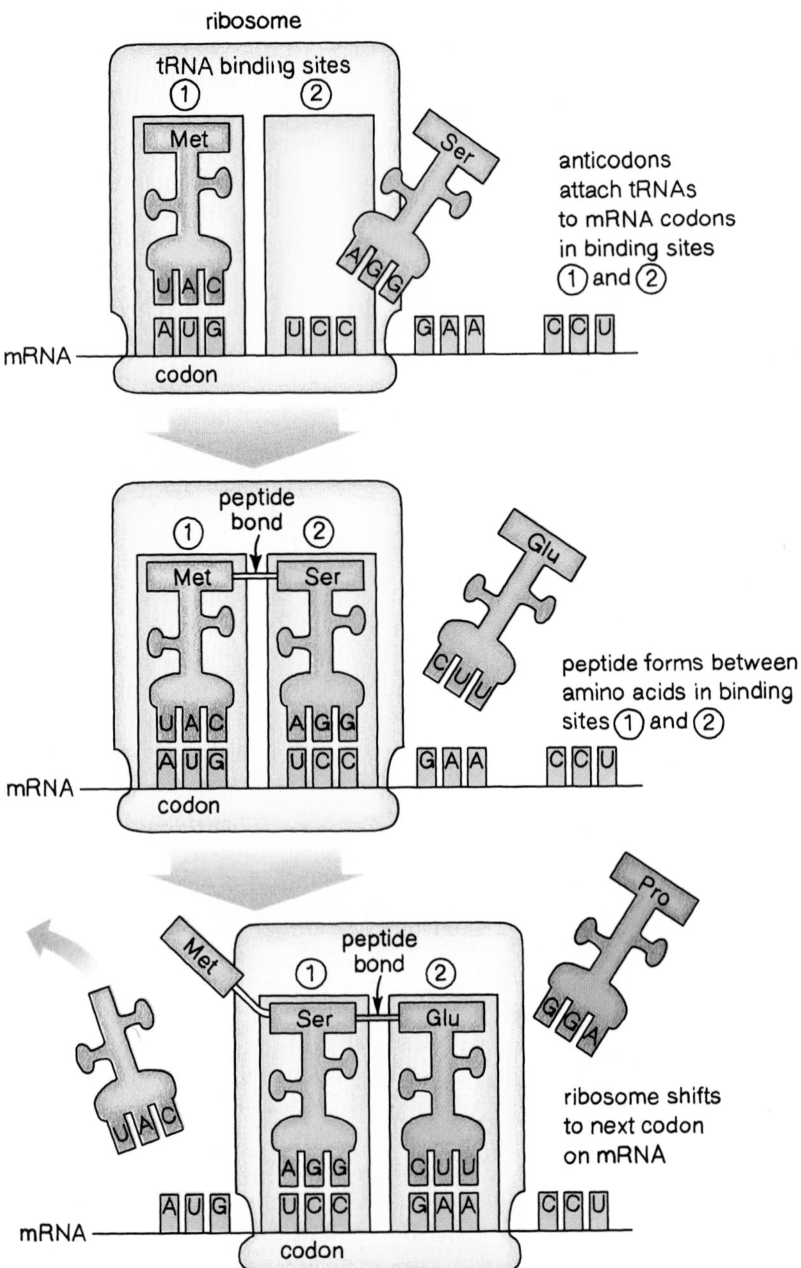

Figure 18.16
Translation: the RNA codons of the mRNA are paired with the anticodons of tRNA to bring the appropriate amino acid into the protein chain.

18.7 Cellular Control of Protein Synthesis

Learning Goal **Describe methods by which a cell controls the synthesis of protein.**

If a cell is to maximize available energy, it must operate efficiently. Therefore, materials are produced in a cell only as they are needed and are degraded when they are not needed. This means that when a substance enters a cell, the enzyme required to catalyze its reactions is produced via protein synthesis. We also need different proteins as we grow. Different organs and tissues in the body have different functions, and therefore require the synthesis of different proteins. In other words, mRNA is not randomly synthesized by DNA; rather, it is synthesized in response to cellular needs for certain materials.

Operon Model

In 1961, the French biologists François Jacob and Jacques Monod proposed models of cellular control to explain how the genes in *E. coli* bacteria respond to different nutrients. They proposed that some of the nucleotides in the DNA act as regulatory genes and operator genes to control the synthesis of mRNA. The structural genes contain the nucleotide sequences for the synthesis of different mRNA. The structural genes and the operator gene form a unit called an operon.

	Operon	
Regulator gene	Operator gene	Structural gene(s)

Enzyme Induction

One way a cell controls the synthesis of proteins is by **enzyme induction** in which a substrate induces the synthesis of its enzymes. The operator gene acts as a switch to turn on or turn off the production of mRNA by the structural genes. When the operator is functioning, the structural genes produce mRNA for protein synthesis. When it is turned off, the structural genes are repressed (not active) and do not produce mRNA. In many cases, one operator controls the protein synthesis of several structural genes. For example, a group of structural genes may produce the mRNAs for a series of enzymes in a particular pathway. When those enzymes are not needed by the cell, their structural genes would be turned off by the operator gene.

In enzyme induction, the function of an operator gene is controlled by a **repressor,** a protein produced by a regulator gene. (See Figure 18.17). As long as the repressor is bound to the operator gene, the structural genes cannot produce mRNA. The process of transcription is turned off. In fact, at any given time, most of the structural genes in a DNA are not operating, they are repressed.

Suppose now that a substrate enters the cell that requires enzymes. The subtrate binds to the repressor which makes its inactive and unable to bind to the operator gene. Without the repressor, the operator gene turns on the structural genes and mRNA is produced. The substrate acts as an inducer because it initiated the production of its own enzymes. As the substrate undergoes reaction, its concentration is lowered until the inducer is removed from the repressor. The repressor returns to the operator gene and stops protein synthesis.

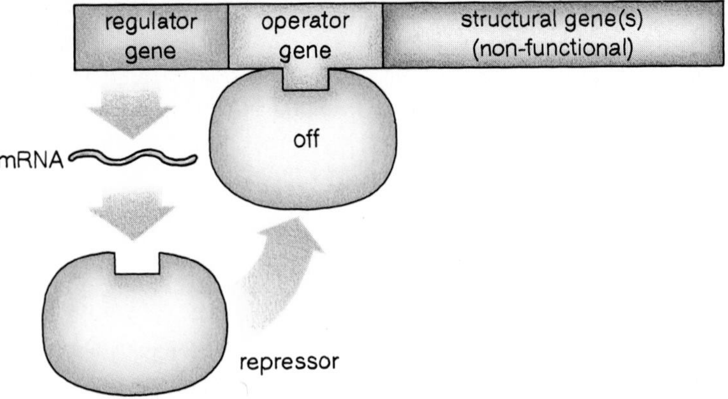

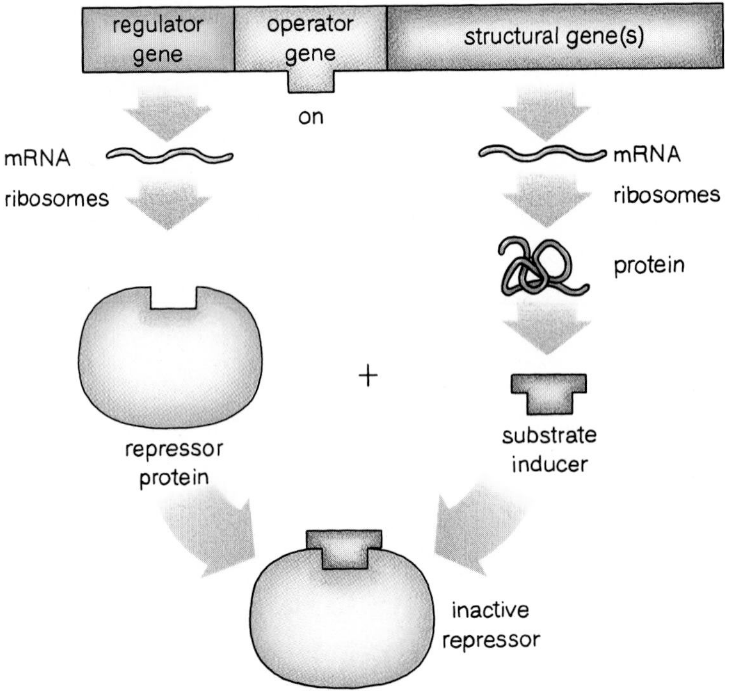

Figure 18.17
Enzyme induction: a repressor turns off the operator gene and prevents the synthesis of mRNA and its protein. If an inducer makes the repressor inactive, the operator gene is turned on, and mRNA and protein synthesis proceed.

Enzyme Repression

Another way to control protein synthesis is by **enzyme repression.** In this model, the operon normally produces mRNA for protein synthesis. (See Figure 18.18.) In repression, it is the end-product of an enzyme that regulates the synthesis of the protein. As long as the concentration of end-product is low, the structural gene produces mRNA. However, when the amount of end-product

HEALTH NOTE

Many Antibiotics Inhibit Protein Synthesis

Several antibiotics stop bacterial infections by interfering with the synthesis of proteins needed by the bacteria. Some antibiotics act only on bacterial cells by binding to the ribosomes in bacteria, but do not act on human cells. A description of some of these antibiotics is given in Table 18.3.

Table 18.3 Antibiotics That Inhibit Protein Synthesis in Bacterial Cells

Antibiotic	Effect on ribosomes to inhibit protein synthesis
Chloramphenicol	Inhibits peptide bond formation and prevents the binding of tRNAs
Erythromycin	Inhibits peptide chain growth by preventing the translocation of the ribosome along the mRNA
Puromycin	Causes release of an incomplete protein by ending the growth of the polypeptide early
Streptomycin	Prevents the proper attachment of tRNAs
Tetracycline	Prevents the binding of tRNAs ■

exceeds the needs of the cell, it combines with an inactive repressor produced by the regulator gene. The combination becomes an active repressor, binds to the operator gene, and represses protein synthesis. There is no further synthesis of the mRNA for the enzymes in that reaction.

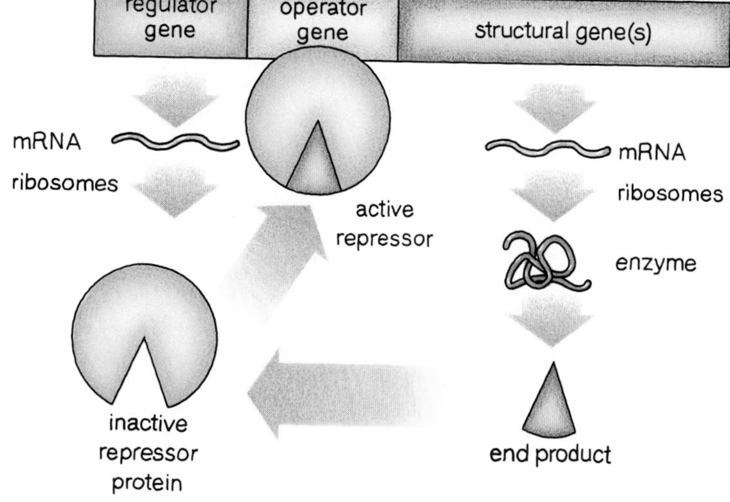

Figure 18.18
Enzyme repression: when the concentration of end products from an enzyme is low, protein synthesis proceeds. At high concentrations, the end product reacts with an inactive repressor, making an active repressor to block the synthesis of protein.

Sample Problem 18.8
Regulation of Protein Synthesis

Indicate whether the following statements are true or false:

a. In enzyme induction, the operon is normally repressed and must be turned on by the substrate.

b. In enzyme repression, the regulator protein is inactive when there is no end product of the enzyme present in the cell.

Solution:

a. True. A substrate binds with the repressor and prevents it from attaching to the operator gene. Then the operator gene turns on the structural gene to synthesize protein.

b. True. As long as the amount of end product is low, the operator gene directs the production of mRNA by the structural gene.

HEALTH NOTE

Viruses

Viruses are composed of a nucleic acid (DNA or RNA) contained within a protein-lipid coating. They invade living cells and cause viral infections including the common cold, influenza, measles, mumps, meningitis, hepatitis, herpes simplex, warts, and AIDS. (See Figure 18.19.)

When a virus attaches to the surface of a cell, enzymes contained in its protein coat dissolve some of the cellular membrane. Different viruses contain enzymes that are specific for different kinds of membranes. As a result, some viruses attack certain organs in the body but not others; some viruses attack plant cells but not animal cells. Once the virus has produced a hole in the cell membrane, it injects its DNA or RNA particles.

Viruses are considered a form of life, but they cannot reproduce without a host cell. However, by using the machinery available within the host cell, viruses make copies of their own nucleic acids along with the protein coat. In DNA viruses, the RNA polymerase of the host cell is used to produce copies of the viral RNA. Most viruses, however, contain viral RNA. Some called retroviruses direct the synthesis of their DNA by using the viral RNA as a template in a process called reverse transcription. Once the viral DNA is formed, it produces more copies of the viral RNA. Other kinds of RNA viruses copy their own RNA directly by using an enzyme called RNA replicase.

Eventually, in all types of viruses, so many new copies of the virus are produced that the host cell bursts. The viral particles released quickly infect more cells in a process called a viral infection. Because the virus uses the host cells for its synthesis, the biochemical pathways of infected and noninfected cells are very similar. As a result, agents that will destroy a viral-infected cell will also destroy normal cells. We do not now have any way to prevent or cure viral diseases, including the common cold. However, some viral infections can be avoided by receiving vaccinations. A vaccination using a dead virus causes host cells to produce antibodies that then recognize and destroy the virus in subsequent viral invasions.

We also know that viruses can produce tumors and cancers in laboratory animals. Some oncogenic viruses induce cancer such as lymphomas and leukemias in humans. A cell that is transformed by an oncogenic virus grows and multiplies without control, invading neighbor cells. Resistance of animal cells to a virus is increased by interferon, a protein produced and secreted by cells after some viral infections. The interferon inhibits the initiation step in protein synthesis. ■

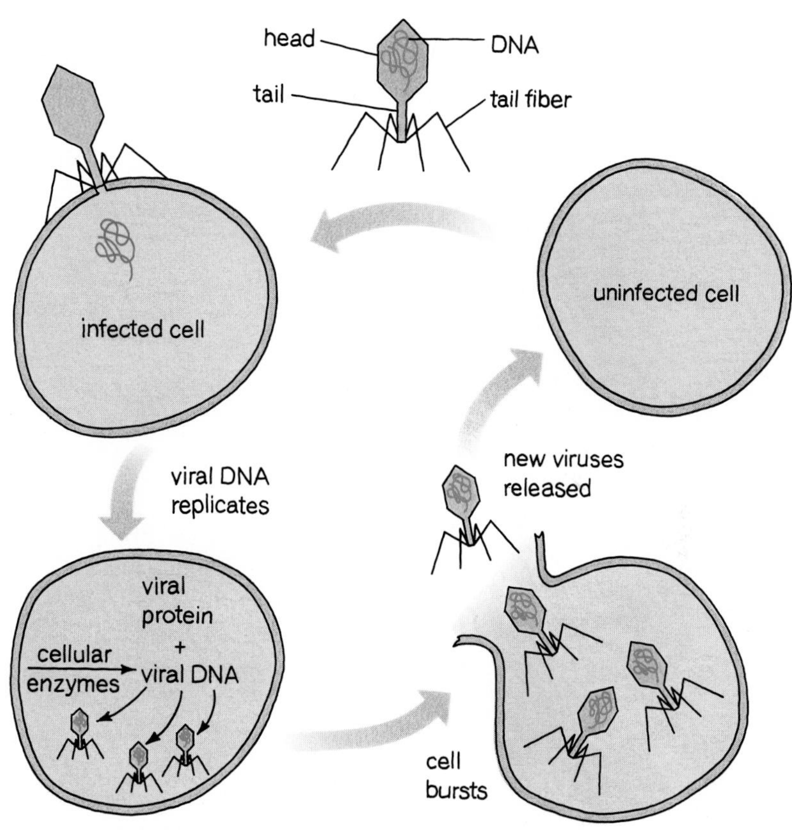

Figure 18.19
Action of a Virus.

18.8 Mutations

Learning Goal **Describe some ways in which DNA is altered to cause mutations and genetic disease.**

A **mutation** is an alteration in the DNA base sequence that changes the structure and function of a protein in the cell. Some mutations are known to result from x-rays, overexposure to sun(UV light), chemicals called mutagens, and possibly some viruses. If a change in DNA occurs in a cell other than a reproductive cell, that altered DNA will be limited to that cell and its daughter cells and will not cause much harm. However, if the mutation occurs in the DNA of eggs or sperm, then all the DNA produced in a new individual will contain the same genetic error. If the mutation greatly affects the catalysis of metabolic reactions or the formation of important structural proteins, the new cells may not survive.

Consider a triplet of bases such as CCC, which produces mRNA with the codon, GGG. Normally, a tRNA would place glycine in the peptide chain. Now, suppose the middle cytosine is replaced with adenine to give CAC. The resulting codon, GUG, will no longer code for glycine, but for valine. Substitution of one base in the DNA genetic code with another base called a point mutation is the most common way in which mutations occur. (See Table 18.4.)

Table 18.4 Effect of Point Mutation on Amino Acid Sequence

	Normal	Effect of Mutation
DNA	ACA—CCC—AGG—TTT	Change in base ↓ ACA—CAC—AGG—TTT
mRNA	UGU—GGG—UCC—AAA	Change in codon ↓ UGU—GUG—UCC—AAA
Amino acids	Cys—Gly—Ser—Lys	Change in amino acid in protein ↓ Cys—Val—Ser—Lys

In an insertion or deletion mutation, a base is added to or deleted from the DNA sequence. Then a frame shift occurs which leads to a misreading of all the codons following the base change. Table 18.5 illustrates insertion and deletion mutations.

Table 18.5 Insertion and Deletion Mutations

	Normal	Effect of Mutation	
DNA	ACA—CCC—AGG—TTT	A inserted ↓ ACA—CCC—AAG—GTT—T	A deleted ↓ ACA—CCC—GGT—TT
mRNA	UGU—GGG—UCC—AAA	Frame shift → UGU—GGG—UUC—CAA—A	Frame shift → UGU—GGG—CCA—AA
Amino acids	Cys—Gly—Ser—Lys	Cys—Gly—Phe—Gln— Incorrect	Cys—Gly—Pro— Incorrect

When a mutation causes a change in the amino acid sequence, the structure of the protein can be severely altered. The polypeptide that results from an insertion or deletion mutation may contain such a large section of incorrect amino acid sequence that it has no biological activity. If the protein is an enzyme, it may not properly bind to its substrate or the reaction of the substrate may be prevented at the active site. The inability of an enzyme to catalyze a reaction can cause certain substances to accumulate until they act as poisons in the cell. A defective enzyme in a major metabolic pathway or in the building of a cell membrane can be lethal. If the defect is in a structural protein, it can lead to detrimental anatomical changes. When a deficiency of a protein is genetic, the condition is called a **genetic disease.**

$$\text{DNA} \xrightarrow{\text{x-rays, sunlight; mutagens; viruses}} \text{Alteration of DNA base sequence} \longrightarrow \text{Defective protein} \longrightarrow \text{Genetic disease}$$

HEALTH NOTE

Some Genetic Diseases

Today, over 300 metabolic diseases are known to be inherited. These diseases result when defective enzymes and proteins disrupt metabolic processes of digestion, absorption, and metabolism. Phenylketonuria, PKU, occurs when DNA cannot produce the enzyme, phenylalanine hydroxylase, required for the conversion of phenylalanine to tyrosine. When phenylalanine cannot be metabolized, it accumulates in the cells of the body. In an attempt to break down the phenylalanine, the cells convert it to phenylpyruvic acid.

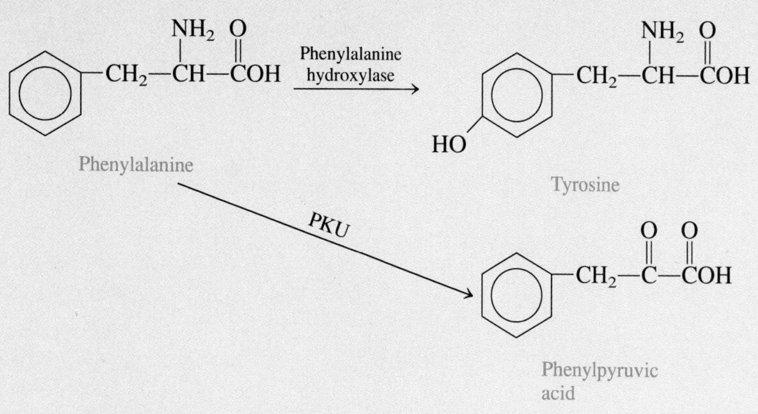

$$C_6H_5-CH_2-CH(NH_2)-COOH \xrightarrow{\text{Phenylalanine hydroxylase}} HO-C_6H_4-CH_2-CH(NH_2)-COOH$$

Phenylalanine → Tyrosine

$$C_6H_5-CH_2-CH(NH_2)-COOH \xrightarrow{\text{PKU}} C_6H_5-CH_2-C(=O)-COOH$$

Phenylalanine → Phenylpyruvic acid

The formation of phenylpyruvic acid in an infant will lead to neurological damage. However, severe brain damage can be averted through a test for PKU routinely performed at birth in most hospitals. If phenylpyruvic acid is detected, a diet is prescribed that eliminates all the foods that contain phenylalanine. Preventing the buildup of phenylalanine and phenylpyruvic acid ensures normal growth and development.

Galactosemia is a hereditary disorder that prevents the proper metabolism of galactose, a monosaccharide that is part of lactose, the milk sugar. Galactosemia is the result of a defective transferase required for the metabolism of galactose 1-phosphate. If lactose is included in the diet of an afflicted infant, the accumulation of galactose 1-phosphate in the cells causes the formation of cataracts and mental retardation. The effects of galactosemia can be avoided if the infant is placed on a galactose-free diet at birth. Table 18.6 lists some other common genetic diseases and the type of metabolism or area affected.

Table 18.6 Some Genetic Diseases

Genetic disorder	Result of defective protein
Gout	A deficiency of hypoxanthine-guanine phosphoribosyltransferase causes the formation of uric acid crystals that cause pain in joints and kidney damage
Cystic fibrosis	The most common inherited disease, thick mucus secretions make breathing difficult, and block pancreatic function
Diabetes mellitus, type I	A defective insulin leads to an accumulation of glucose in the blood and urine, hyperglycemia
Tay-Sachs disease	A defective hexosaminidase A causes an accumulation of gangliosides resulting in mental retardation, and loss of motor control
Sickle cell	A defective hemoglobin has a decreased ability to carry oxygen; sickled red blood cells aggregate, causing anemia, plugged capillaries, and low oxygen levels in affected tissues
Albinism	A defective tyrosinase prevents the formation of melanin pigment in the skin, hair, and retina of the eye
Hemophilia	One or more defective blood clotting factors leads to poor coagulation, excessive bleeding, and internal hemorrhages ■

Sample Problem 18.9
Mutations

In the mRNA sequence of codons CCC-UCA-AGA-GGG, the triplet AGA undergoes a point mutation to ACA. How does this affect the amino acid order in this section of mRNA?

Solution:
The normal mRNA codes for Pro-Ser-Arg-Gly. When the mutation occurs, the order of amino acids becomes Pro-Ser-Thr-Gly. The mutation places threonine in the protein chain in place of arginine.

Study Check:
What is the possible effect of the mutation described in this problem?

Answer: The change in amino acid order may modify the structure and activity of the protein to the extent that it is defective and unable to carry out the formation of structural protein or catalyze a metabolic reaction.

HEALTH NOTE

Cancer

In an adult, most cells in the body do not continue to reproduce. When cells in the body begin to grow and multiply without control, they invade neighboring cells and appear as a tumor or growth (neoplasm). When tumors interfere with normal functions of the body, they are cancerous. If they are limited, they are benign. Cancer can be caused by chemical and environmental substances, by radiation, or by oncogenic viruses.

Some reports estimate that 70–80% of all human cancers are initiated by chemical and environmental factors. A carcinogen is any substance that increases the probability of inducing a tumor. Known carcinogens include dyes, cigarette smoke, and asbestos. Over 90% of all persons with lung cancer are smokers. A carcinogen causes cancer by reacting with molecules in a cell, probably DNA, and altering the growth of that cell. Some known carcinogens are listed in Table 18.7.

Radiant energy from sunlight or medical radiation is another type of environmental factor. Skin cancer has become one of the most prevalent forms of cancer. It appears that DNA damage in the exposed areas of the skin causes mutations. The cells lose their ability to control protein synthesis, and uncontrolled cell division leads to cancer. The incidence of malignant melanoma, one of the most serious skin cancers, has been rapidly increasing. Some possible factors for this increase may be the popularity of suntanning as well as the reduction of the ozone layer, which absorbs much of the harmful radiation from sunlight.

Oncogenic viruses cause cancer when cells are infected. Several viruses associated with human cancers are listed in Table 18.8. Some cancers such as retinoblastoma and breast cancer appear to occur more frequently in families. There is some indication that a missing or defective gene may be responsible.

Table 18.7 Some Chemical and Environmental Carcinogens

Carcinogen	Tumor occurrence
Asbestos	Lung, respiratory tract
Arsenic	Skin, lung
Cadmium	Prostate, kidneys
Chromium	Lung
Nickel	Lung, sinuses
Aflatoxin	Liver
Nitrites	Stomach
Aniline dyes	Bladder
Vinyl chloride	Liver
Benzene	Leukemia

Table 18.8 Human Cancers Caused by Oncogenic Viruses

Virus	Disease
RNA viruses	
Human T-cell leukemia-lymphoma virus-I (HTLV-I)	Leukemia
Human immunodeficiency virus (HIV)	Acquired immune deficiency (AIDS)
DNA viruses	
Epstein-Barr virus (EBV)	Burkitt's lymphoma (cancer of white blood B cells), Nasopharyngeal carcinoma, Hodgkin's disease
Hepatitis B virus (HBV)	Liver cancer
Herpes simplex virus (type 2)	Cervical and uterine cancer
Papilloma virus	Cervical and colon cancer, genital warts ■

Glossary of Key Terms

anticodon loop The triplet of bases that occur in a loop of a tRNA that complement a codon on mRNA.

chromosome A series of genes made of DNA found in the nucleus of the cell.

codon A sequence of three bases in mRNA that specifies the incorporation of a certain amino acid into a protein chain. A few codons signal the start or end of transcription.

complementary base pairing The formation of specific combinations of base pairs. In DNA, adenine is always paired with thymine, and guanine is always paired with cytosine. In the formation of RNA, the adenine of DNA is paired with uracil in forming the single strand of RNA.

DNA Deoxyribonucleic acid—the genetic material of all cells found in the chromosomes of the nucleus containing deoxyribose sugar, phosphate, and four nitrogenous bases, adenine, thymine, guanine, and cytosine.

double helix The helical shape of the double chain of DNA that is like a spiral staircase with a sugar-phosphate backbone on the outside and the bases on the inside like stair steps.

enzyme induction A model in which protein synthesis is induced when a substrate inactivates a repressor.

enzyme repression A model in which protein synthesis is repressed when an end product forms a repressor.

gene A section of DNA that codes for a particular protein.

genetic code The order of codons in mRNA that specifies the amino acid order for the synthesis of protein.

genetic disease The manifestation of mutations in an individual by way of physical malformations or metabolic dysfunction.

mRNA Messenger RNA, produced in the nucleus by DNA, to carry the genetic information to the ribosomes for the construction of a protein.

mutation An alteration in the DNA base sequence that changes the structure and function of a protein in the cell.

nitrogen base Nitrogen-containing compounds, called purines and pyrimidines, found in DNA and RNA: adenine (A), thymine (T), cytosine (C), guanine (G), and uracil (U).

nucleic acids Large molecules, composed of nucleotides, found in the DNA of the chromosomes, or in the mRNA, tRNA, and rRNA.

nucleotide A building block of a nucleic acid consisting of a nitrogen base (purine or pyrimidine), a pentose sugar (ribose or deoxyribose), and a phosphate group.

operon A group of genes whose transcription is controlled by the same regulator and operator genes.

plasmid A small DNA found in *E. coli* bacteria that is used in recombinant DNA research.

purines The double-ringed nitrogen-containing compounds of adenine and guanine found in nucleic acids.

pyrimidines The single-ringed nitrogen-containing compounds of cytosine, uracil, and thymine found in nucleic acids.

rRNA Ribosomal RNA, the most prevalent type of RNA that is found in the ribosomes.

recombinant DNA The formation of a synthetic DNA by replacing a portion of a plasmid with a DNA from another organism. When placed back into bacterial cells the recombinant DNA undergoes replication and directs the synthesis of a protein previously not produced in that cell.

replication The process of duplicating DNA by pairing the bases on each parent strand with their complementary bases.

repressor A protein that interacts with the operator gene in an operon to prevent the transcription of mRNA.

RNA Ribonucleic acids that consist of a single chain of nucleotides that contains a ribose sugar, bases like DNA except uracil replaces thymine, and phosphate.

tRNA Transfer RNA, an RNA that places a specific amino acid into a peptide chain at the ribosome. There is at least one tRNA for each of the 20 different amino acids.

transcription The transfer of genetic information from DNA by the formation of an mRNA.

translation The interpretation of the codons in an mRNA as amino acids in a peptide.

virus A group of infectious nucleic acids that use the machinery of host cells to produce more viral particles and thereby destroy the host cells.

Problems

Nucleic Acids: DNA and RNA *(Goal 18.1)*

18.1 What are the components of the following?
a. nucleic acid
b. nucleotide

18.2 What is the difference in the sugar used in DNA and RNA?

18.3 What nitrogen bases are found in DNA? in RNA?

18.4 What are the four nucleotides of DNA? of RNA?

18.5 What are the components of the following:
a. deoxythymidine monophosphate
b. deoxyguanosine monophosphate
c. cytidine monophosphate
d. AMP
e. dGMP

18.6 How are nucleotides held together in a nucleic acid chain?

***18.7** Write the structure of the dinucleotide AC in DNA.

Complementary Base Pairing in DNA *(Goal 18.2)*

18.8 What is complementary base pairing in DNA?

18.9 Complete the double strand of DNA for each of the following DNA segments:
a. A-G-T-G-C-A
b. C-T-G-T-A-T-A-C
c. A-A-A-A-A-A
d. G-G-G-C-C-C-A-A-A

18.10 Describe the general structure of DNA.

DNA Replication *(Goal 18.3)*

18.11 How does the replication of DNA produce exact copies of DNA?

18.12 Diagram the replication of the following portion of a DNA molecule.

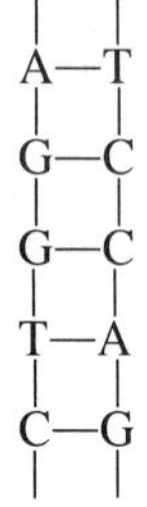

***18.13** How does recombinant DNA cause a bacterium to produce a protein from a non-bacterial organism?

RNA: Structure and Synthesis *(Goal 18.4)*

18.14 List the three types of RNA and their function.

18.15 What is the transcription step in protein synthesis?

18.16 Write the corresponding portion of the mRNA produced by the template in the following DNA section.

-C-C-G-A-A-G-G-T-T-C-A-C- ⟵ template

The Genetic Code *(Goal 18.5)*

18.17 State the amino acid coded for by each of the following codons:
a. CUU b. UCA c. GGU d. GCA

18.18 List the codons that can be used to code for each of the following:
a. glycine b. threonine c. alanine

18.19 What codons would be found at the beginning and end of an mRNA?

Protein Synthesis *(Goal 18.6)*

18.20 a. What is the structure of a tRNA?
b. How does a tRNA recognize a codon in the mRNA?
c. Why are there 20 different tRNAs?

18.21 What are the steps in the translation process of protein synthesis?

18.22 What amino acid sequence would you expect from the following mRNA segments:
a. -AAA-AAA-AAA-AAA-
b. -UUU-CCC-UUU-CCC-
c. -UAC-GGG-AGA-UGU-

***18.23** Write a possible base sequence for an mRNA that would code for the following amino acid sequence in a protein: Ala-Leu-Gly-Pro-Ser-His

Cellular Control of Protein Synthesis *(Goal 18.7)*

18.24 a. Why does the cell regulate protein synthesis?
b. Is protein synthesis controlled at transcription or translation? Why?

***18.25** In the operon model for enzyme induction, how does a substrate induce the synthesis of its reaction enzymes?

***18.26** In the model of enzyme repression, how is the repressor formed? What is its function?

Mutations *(Goal 18.8)*

18.27 In a mutation, the order of amino acids is altered.
a. What are the ways in which this happens?
b. What is the effect of an altered DNA on the protein it produces?
c. What is a genetic disease?

***18.28** A DNA contains a template sequence of CAA-GGA-TTT.
a. What is the normal order of amino acids in the peptide?
b. Suppose that a G is substituted for the C. What happens to the amino acid sequence of the peptide?

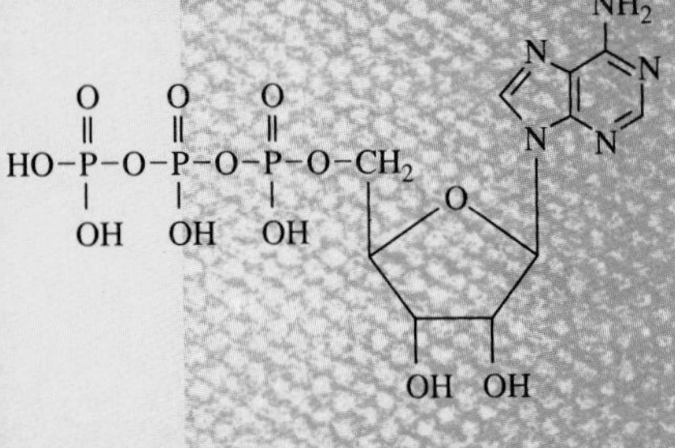

19

Metabolic Pathways and Energy Production

Learning Goals

19.1 Describe the role of ATP in metabolism.

19.2 Describe the production of ATP from NADH and $FADH_2$ by the electron transport chain.

19.3 Write an equation for the overall reaction of glucose in glycolysis.

19.4 Give the end products for reactions of pyruvic acid under aerobic and anaerobic conditions.

19.5 Describe the role of oxidation in the citric acid cycle. Account for the ATP formed in one turn of the citric acid cycle.

19.6 Account for the ATP produced by the complete oxidation of glucose.

19.7 Describe the beta(β)-oxidation of fatty acids and give the total ATP produced. Describe how acetyl CoA is used in the synthesis of fatty acids.

19.8 Describe reactions that prepare amino acids for energy production or synthesize nonessential amino acids from carbon compounds.

ATP energy from the metabolism of glucose stored in muscle cells is used to contract the muscles.

SCOPE

Metabolism includes all the reactions that take place in our cells. Several reactions, each one catalyzed by a specific enzyme, link together to form a metabolic pathway. Some of the pathways are involved in producing energy for the cell; others are using energy.

We saw earlier that the digestion of the foods we eat produced compounds such as glucose, fatty acids, and amino acids. After they are absorbed, they are carried through the bloodstream or lymphatic system to the cells of the body where they undergo oxidation reactions that break down their chemical bonds and release energy. That energy is stored in the cells as high-energy phosphate bonds of adenosine triphosphate (ATP). Then it is used to do work such as pumping your heart, synthesizing glycogen, proteins and nucleic acids, sending nerve impulses, and moving substances across cell membranes.

19.1 ATP: The Energy Storehouse

Learning Goal **Describe the role of ATP in metabolism.**

Although several thousand metabolic reactions occur in our cells, they can all be classified as energy-producing (**catabolic**) or energy-requiring (**anabolic**) reactions. Reactions produce energy by breaking down the chemical bonds of compounds in the cell. Reactions that require energy are building new and larger molecules for the cells and carrying out the work of the cell.

The link between energy-producing reactions and the energy-requiring reactions is a high-energy compound called adenosine triphosphate (**ATP**). ATP is the storage form of energy within the cell. It is composed of adenine, ribose, a pentose, and three phosphate groups. (See Figure 19.1.)

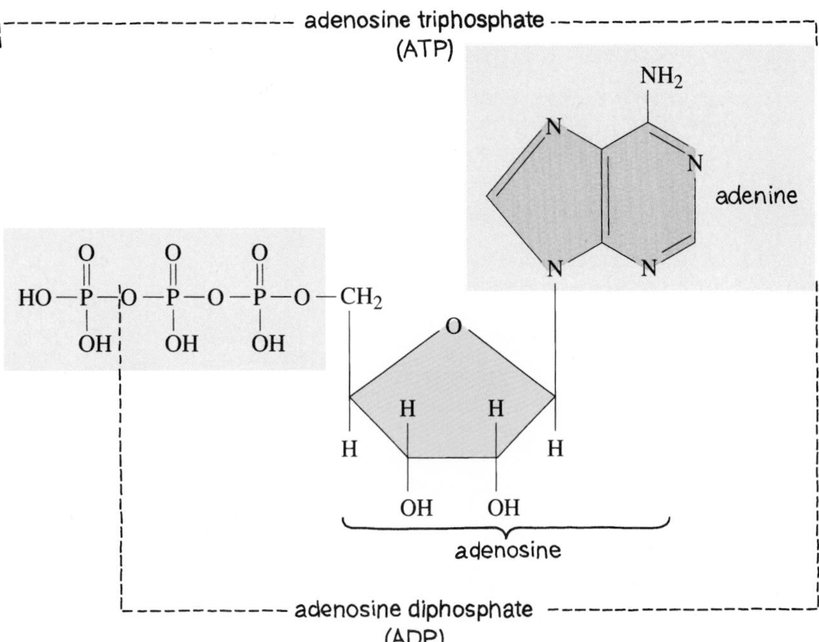

Figure 19.1
Structure of adenosine triphosphate, ATP, a compound that stores energy in the cell.

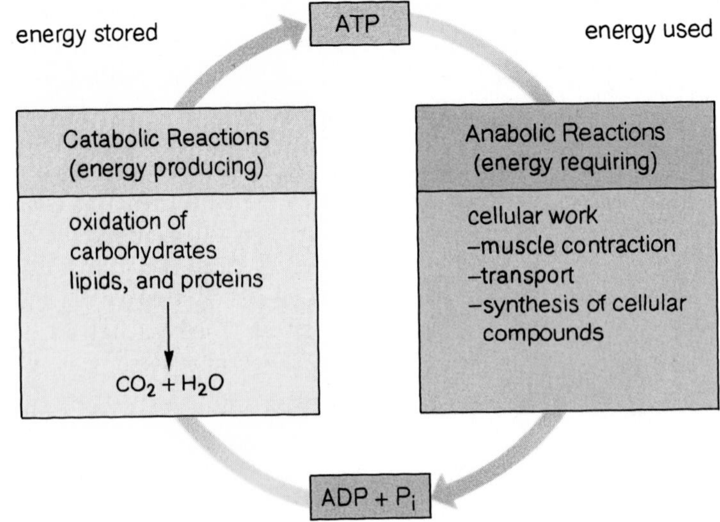

Figure 19.2
ATP, the energy-storage molecule, connects the energy-producing reactions with the energy-requiring reactions that do work in the cell.

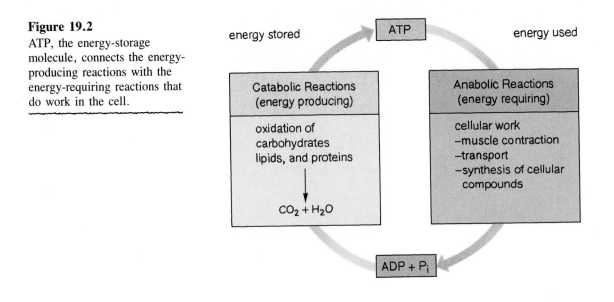

Hydrolysis of ATP Yields Energy

The hydrolysis of ATP by water yields a phosphate ion (P_i), adenosine diphosphate (**ADP**), and energy, 7.3 kcal per mole of ATP.

$$\text{ATP} + H_2O \longrightarrow \text{ADP} + P_i + \text{Energy}$$

Our cells utilize the energy from ATP hydrolysis to contract muscles and to move substances across cellular membranes. ATP is also hydrolyzed for the synthesis of larger molecules such as proteins, lipids, polysaccharides, and nucleic acids. When work is occurring in a cell, as many as 1–2 million ATPs may be hydrolyzed in one second. (See Figure 19.2.)

Sample Problem 19.1
Hydrolysis of ATP

Write an equation for the hydrolysis of ATP.

Solution:
The hydrolysis of ATP produces ATP, P_i, and energy.

$$\text{ATP} + H_2O \longrightarrow \text{ADP} + P_i + \text{Energy}$$

HEALTH NOTE

Use of ATP Energy for Muscle Contraction

A muscle consists of thousands of muscle fibers that lie parallel to one another. Within these muscle fibers are fibrils composed of two kinds of filaments. Thin filaments contain the protein actin, and thick filaments contain the protein myosin. They are arranged in alternating sections along the muscle fiber, with the thick myosin filaments overlapping the thin filaments. (See Figure 19.3.)

In a relaxed muscle, the Ca^{2+} concentration is low, ATP concentration is high, and myosin and actin filaments are stretched out. Contraction of a muscle begins when a nerve impulse releases Ca^{2+} in the muscle fiber. The Ca^{2+} activates the hydrolysis of ATP.

$$\text{Relaxed muscle} + \text{ATP} \xrightarrow{Ca^{2+}} \text{Contracted muscle} + \text{ADP} + P_i$$

The energy released by ATP hydrolysis is used to connect the myosin filaments to the actin filaments. The actin filaments are pulled inward along the myosin toward the center of the myosin, causing a shortening of the muscle fiber. Another nerve impulse releases more Ca^{2+}, hydrolyzes more ATP, and contracts the muscle again. When the nerve impulse ends, Ca^{2+} levels drop, and ATP concentration rises. The myosin and actin filaments return to their resting state as the muscle relaxes until the next impulse. ■

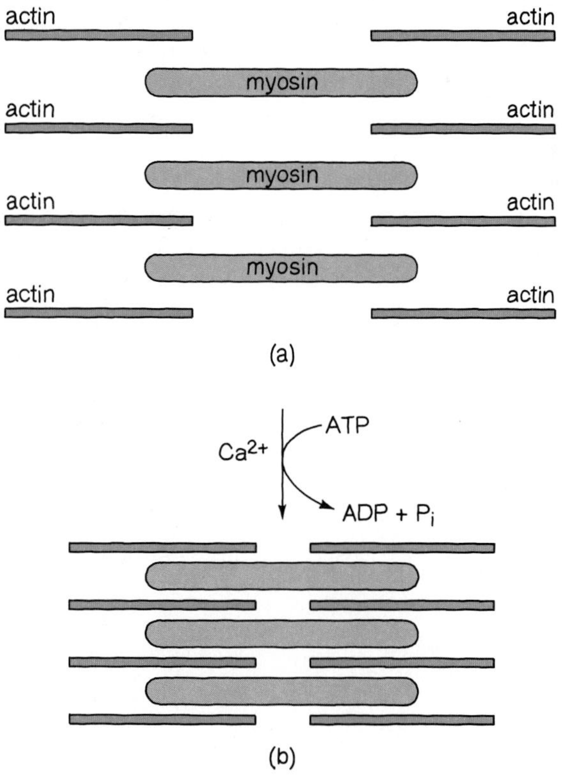

Figure 19.3
Muscle contraction occurs as actin filaments are pulled inward along myosin filaments. (a) relaxed muscle; (b) contracted muscle.

19.2 The Electron Transport Chain

Learning Goal **Describe the production of ATP from NADH and $FADH_2$ by the electron transport chain.**

Almost all of the ATP in the cell is produced by a series of oxidation-reduction reactions called the **electron transport chain** (or respiratory chain). Several steps along the electron transport chain release enough energy to synthesize molecules of ATP in a process called oxidative phosphorylation:

$$\text{ADP} + \text{P}_i + \text{Energy} \longrightarrow \text{ATP}$$

The electron transport chain is contained within the mitochondria, subcellular structures in a cell that contain an outer and inner membrane. (See Figure 19.4.) Along the highly convoluted inner membrane, there are many ridgelike folds containing a dense network of enzymes that catalyze energy-producing reactions. This is also the site for the oxidation of glucose, amino acids, and fatty acids, and for the formation of ATP. The mitochondria are considered the powerhouses of the cell because they contain all the machinery for producing energy.

Oxidation of Metabolites by NAD^+ and FAD

In metabolic oxidation, a compound called a **metabolite** (M) is oxidized by enzymes called dehydrogenases. Their coenzymes, NAD^+ or FAD, accept hydrogen and electrons which they carry to the electron transport chain. The reactions of the electron transport chain eventually combine the hydrogen and

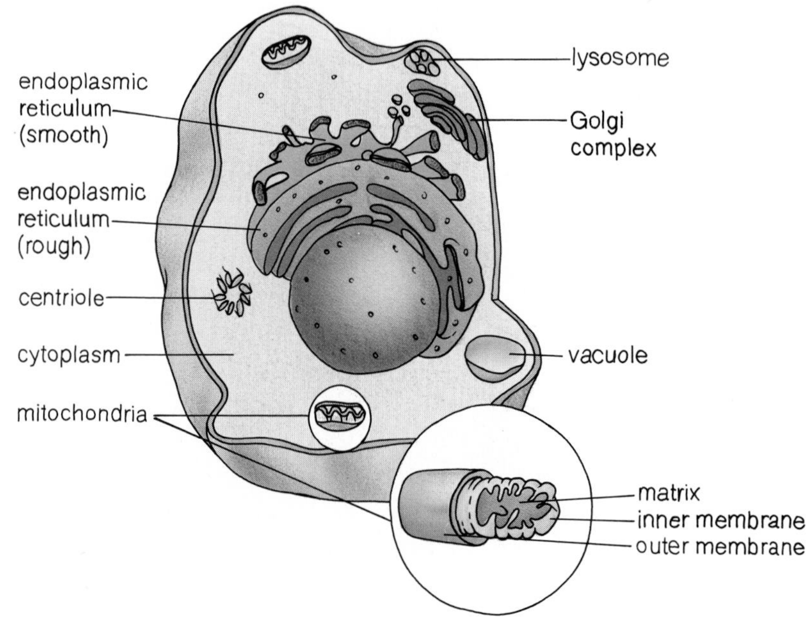

Figure 19.4
A simplified drawing of a cell: some degradative pathways occur in the cytoplasm, but most of the oxidation and energy production is done by the enzyme systems in the mitochondria, the powerhouse of the cell.

Figure 19.5
Structure of coenzyme NAD^+ and the formation of $NADH^+$.

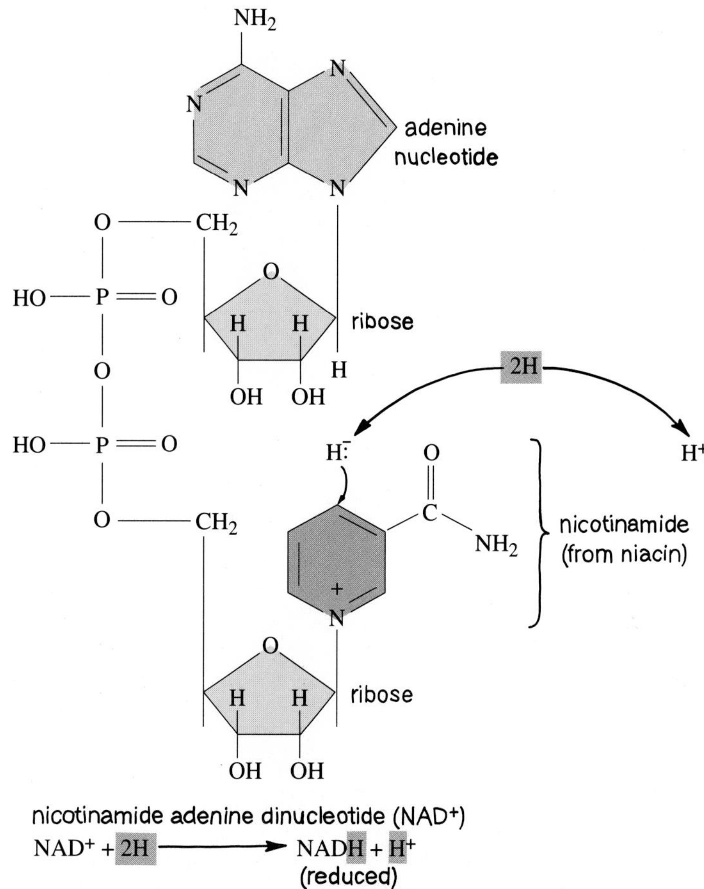

electrons with molecular oxygen obtained from respiration. The overall process is energy-producing because the end product, water, has a lower energy than the initial reactants. As energy is released, it is used to form ATP:

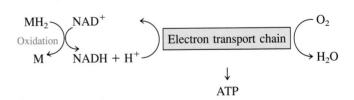

$$MH_2 \xrightarrow{\text{Oxidation}} M \quad NAD^+ \rightarrow NADH + H^+ \quad \boxed{\text{Electron transport chain}} \quad O_2 \rightarrow H_2O$$

$$\downarrow$$

$$\text{ATP}$$

The coenzyme, **NAD⁺** (nicotinamide adenine dinucleotide), is derived from niacin and an adenine nucleotide as seen in Figure 19.5. It is NAD^+ that accepts hydrogen from a metabolite that forms a carbon-oxygen double bond. For example, NAD^+ is the coenzyme for a dehydrogenase that oxidizes lactic acid to pyruvic acid. The reduced form of NAD^+ is NADH.

$$\underset{\text{Lactic acid}}{CH_3-\overset{OH}{\overset{|}{CH}}-COOH} + NAD^+ \xrightarrow{\text{Dehydrogenase}} \underset{\text{Pyruvic acid}}{\overset{\text{Oxidized}}{CH_3-\overset{O}{\overset{\|}{C}}-COOH}} + \underset{\text{Reduced}}{NADH + H^+}$$

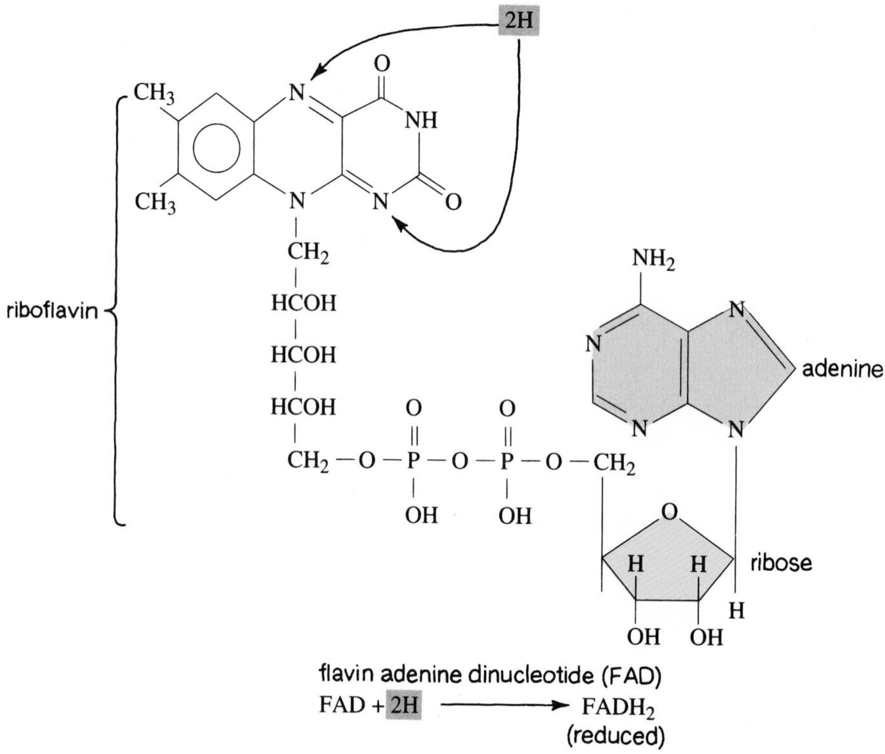

Figure 19.6
Structure of the FAD coenzyme and formation of its reduced form $FADH_2$.

Another coenzyme, **FAD** (flavin adenine dinucleotide), is derived from riboflavin (vitamin B_2). It is a hydrogen acceptor when two carbon atoms in a metabolite are oxided to give a carbon-carbon double bond. (See Figure 19.6.)

$$\underset{\text{Succinic acid}}{\text{COOH}-\boxed{\text{CH}_2-\text{CH}_2}-\text{COOH}} + \text{FAD} \xrightarrow{\text{Succinic acid dehydrogenase}} \underset{\text{Fumaric acid}}{\overset{\text{Oxidized}}{\text{COOH}-\boxed{\text{HC}=\text{CH}}-\text{COOH}}} + \underset{\text{Reduced}}{\text{FADH}_2}$$

Sample Problem 19.2
Coenzyme as Hydrogen Acceptors

When a metabolite is oxidized, what coenzymes might be used and what are their reduced forms?

Solution:
The coenzymes NAD^+ and FAD accept hydrogen when a metabolite is oxidized. NAD^+ is the coenzyme for an oxidation that involves carbon-oxygen bonds; FAD is used when an oxidation forms a carbon-carbon double bond. The reduced forms of the coenzymes are NADH and $FADH_2$.

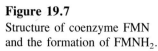

Figure 19.7
Structure of coenzyme FMN and the formation of $FMNH_2$.

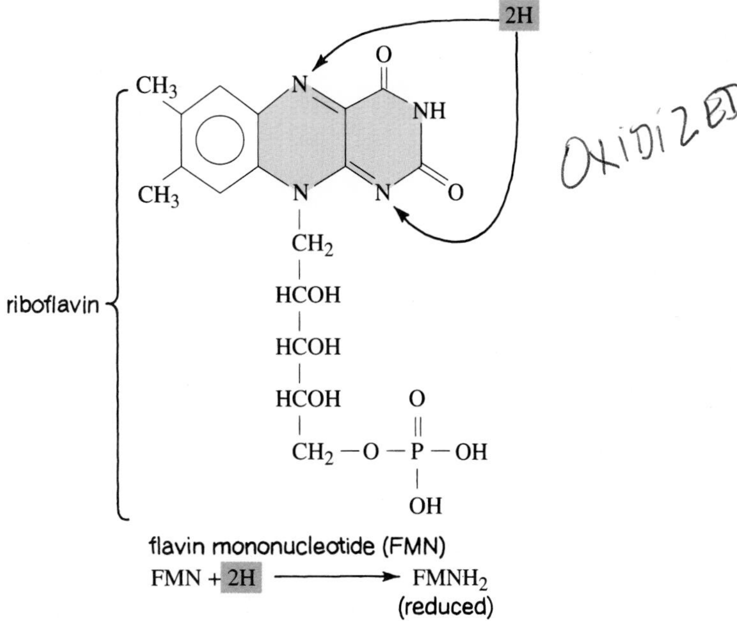

Starting the Electron Transport Chain

In the first step of the electron transport chain, NADH transfers hydrogen to **FMN** (flavin mononucleotide). (See Figure 19.7.) The FMN is reduced to $FMNH_2$ and the original carrier is reoxidized to NAD^+. This process of reduction and oxidation occurs all along the electron transport chain regenerating each component so it can be used again:

$$\underset{\text{Oxidized}}{\text{FMN}} + \underset{\text{Reduced}}{\text{NADH}} + H^+ \longrightarrow \underset{\text{Reduced}}{\text{FMNH}_2} + \underset{\text{Oxidized}}{\text{NAD}^+}$$

The next carrier in the electron transport chain is **coenzyme Q (Q),** a compound derived from quinone as seen in Figure 19.8. Coenzyme Q has a unique role in the chain because it accepts hydrogen from both $FMNH_2$ and $FADH_2$.

$$\underset{\text{Oxidized}}{\text{Q}} + \underset{\text{Reduced}}{\text{FMNH}_2} \longrightarrow \underset{\text{Reduced}}{\text{QH}_2} + \underset{\text{Oxidized}}{\text{FMN}}$$

and

$$\underset{\text{Oxidized}}{\text{Q}} + \underset{\text{Reduced}}{\text{FADH}_2} \longrightarrow \underset{\text{Reduced}}{\text{QH}_2} + \underset{\text{Oxidized}}{\text{FAD}}$$

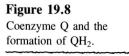

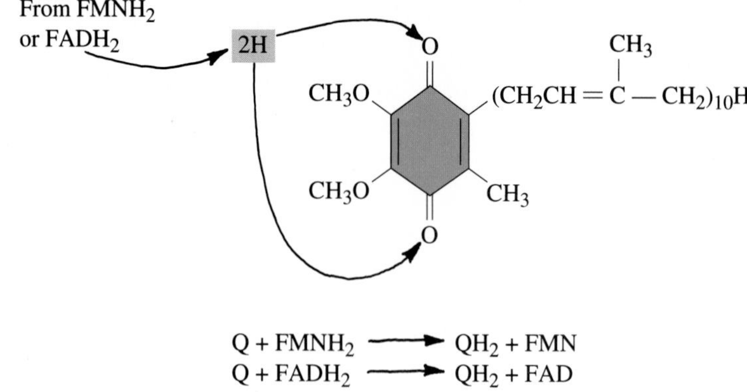

Figure 19.8
Coenzyme Q and the formation of QH_2.

Cytochromes and Oxygen Accept Electrons

From QH_2, protons ($2H^+$) are released inside the mitochondria, and the electrons ($2e^-$) are transferred to a group of electron acceptors called cytochromes, a group of four globular proteins. Each cytochrome contains an iron ion that alternates between Fe^{2+} and Fe^{3+}. When a cytochrome accepts an electron, the iron is reduced to Fe^{2+}; when the electron is transferred to the next cytochrome, the iron is oxidized back to Fe^{3+}.

$$\underset{\text{Reduced}}{QH_2} + \underset{\text{Oxidized}}{2\text{ Cyt b}(Fe^{3+})} \longrightarrow \underset{\text{Oxidized}}{Q} + \underset{\text{Reduced}}{2\text{ Cyt b }(Fe^{2+})} + 2\,H^+$$

The series of cytochromes are identified as cytochromes b, c_1, c, and a. As electrons move from one cytochrome to the next—much like buckets of water in a fire brigade—Fe^{3+} is reduced as electrons are accepted and oxidized as electrons are lost.

$$\underset{\text{Oxidized}}{2\text{ Cyt c }(Fe^{3+})} + 2e^- \longrightarrow \underset{\text{Reduced}}{2\text{ Cyt c }(Fe^{2+})}$$

In the last step of the electron transport chain, electrons are transferred from cytochrome a to a molecular oxygen (O_2). Protons are picked up, and a molecule of water is produced.

$$\underset{\text{Reduced}}{2\text{ Cyt a }(Fe^{2+})} + \underset{\text{Oxidized}}{\tfrac{1}{2}\,O_2} + 2\,H^+ \longrightarrow \underset{\text{Oxidized}}{2\text{ Cyt a }(Fe^{3+})} + \underset{\text{Reduced}}{H_2O}$$

Sample Problem 19.3
Electron Transport Chain

Identify the following steps in the electron transport chain as oxidation or reduction:

a. $FMN + 2H \longrightarrow FMNH_2$

b. $2\ Cyt\ c\ (Fe^{2+}) \longrightarrow 2\ Cyt\ c\ (Fe^{3+}) + 2e^-$

Solution:

a. The gain of hydrogen is reduction.

b. The loss of electrons is oxidation.

Study Check:

What are the final electron acceptor and product of the electron transport chain?

Answer: O_2, H_2O

Oxidative Phosphorylation by the Electron Transport Chain

The electron transport chain may seem to have little importance until we point out that three reactions in the chain yield a sufficient amount of energy to form ATP. The combination of the oxidation in the electron transport chain with the synthesis of ATP is called oxidative **phosphorylation:**

$$ADP + P_i + Energy \longrightarrow ATP$$

When NADH enters the electron transport chain, enough energy is released at three steps to generate three molecules of ATP:

$$NADH + H^+ + \tfrac{1}{2}\ O_2 + 3\ ADP \longrightarrow NAD^+ + H_2O + 3\ ATP$$

If $FADH_2$ is the carrier, it enters the chain at Q. Because Q is at a lower energy level in the chain, the energy released is sufficient to generate two molecules of ATP:

$$FADH_2 + \tfrac{1}{2}\ O_2 + 2\ ADP \longrightarrow FAD + H_2O + 2\ ATP$$

The production of ATP by the coenzymes FAD and NAD^+ are summarized below:

Initial Coenzyme	*Reduced Form*	*ATP Produced by Electron Transport Chain*
NAD^+	NADH	3 ATP
FAD	$FADH_2$	2 ATP

Figure 19.9 illustrates the electron transport chain and the energy changes that produce ATP.

Figure 19.9
Production of ATP by the electron transport chain.

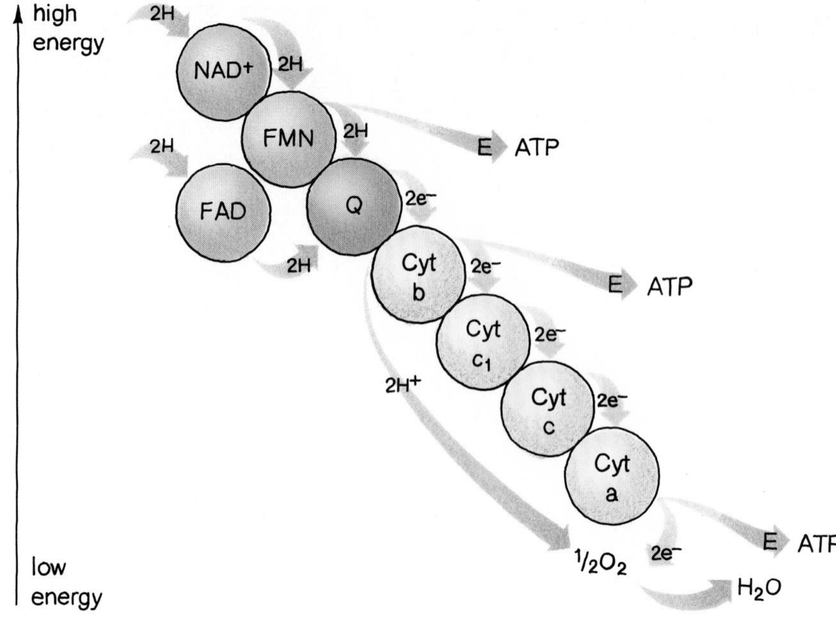

Sample Problem 19.4
Oxidation Phosphorylation

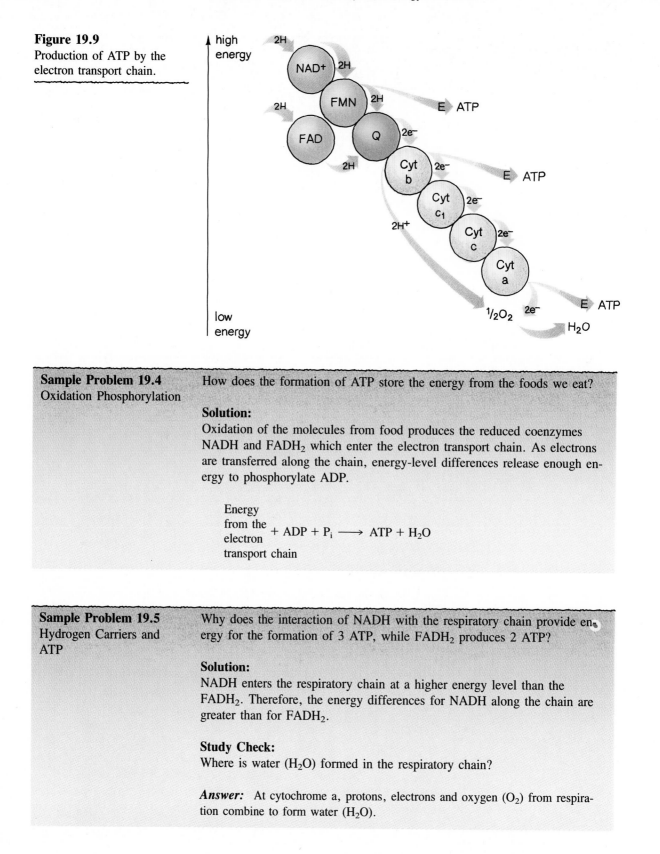

How does the formation of ATP store the energy from the foods we eat?

Solution:
Oxidation of the molecules from food produces the reduced coenzymes NADH and $FADH_2$ which enter the electron transport chain. As electrons are transferred along the chain, energy-level differences release enough energy to phosphorylate ADP.

$$\text{Energy from the electron transport chain} + ADP + P_i \longrightarrow ATP + H_2O$$

Sample Problem 19.5
Hydrogen Carriers and ATP

Why does the interaction of NADH with the respiratory chain provide energy for the formation of 3 ATP, while $FADH_2$ produces 2 ATP?

Solution:
NADH enters the respiratory chain at a higher energy level than the $FADH_2$. Therefore, the energy differences for NADH along the chain are greater than for $FADH_2$.

Study Check:
Where is water (H_2O) formed in the respiratory chain?

Answer: At cytochrome a, protons, electrons and oxygen (O_2) from respiration combine to form water (H_2O).

19.3 Glycolysis: Oxidation of Glucose

Learning Goal **Write an equation for the overall reaction of glucose in glycolysis.**

The glucose we obtain from the digestion of the starches and disaccharides in our diet is our major source of energy. When glucose is not immediately used by the cells for energy, it can be stored to a limited extent in the liver and muscles as glycogen, a polymer of glucose. The formation of glycogen from glucose is called **glycogenesis.** When the levels of glucose in the blood become low, a process called **glycogenolysis** breaks down the glycogen reserves, releasing glucose.

The process of extracting energy from glucose begins with glycolysis, a metabolic pathway that breaks down glucose to smaller molecules of pyruvic acid.

$$\text{Glycogen} \underset{\text{Glycogenesis}}{\overset{\text{Glycogenolysis}}{\longleftrightarrow}} \text{Glucose} \xrightarrow{\text{Glycolysis}} \text{2 Pyruvic acid}$$

Glycolysis

In the metabolic pathway called **glycolysis** (glyco = sugar, lysis = a breaking down), a glucose molecule (six carbon atoms) is oxidized to two molecules of pyruvic acid (three carbon atoms). Several steps are required for the process because the energy must be released in small amounts. If a lot of energy were released at one time, it would raise the temperature in the cells so much that the proteins in the cell would be destroyed. While many details are given for the pathway, it should not be necessary for you to memorize all the steps. However, it is important that you gain an understanding of the overall process. As we describe the steps in glycolysis, you may want to refer to Figure 19.10.

Step 1 First phosphorylation

When glucose enters a cell, it is converted to glucose 6-phosphate, a phosphorylation step that requires ATP. When the glucose level is high in the cell, insulin, a hormone secreted by the pancreas, accelerates this step.

$$\text{Glucose} + \boxed{\text{ATP}} \xrightarrow{\text{Hexokinase}} \text{Glucose 6-phosphate} + \text{ADP}$$

Step 2 Isomerization

In the next step, glucose 6-phosphate undergoes isomerization to give fructose 6-phosphate.

$$\text{Glucose 6-phosphate} \xrightarrow{\text{Isomerase}} \text{Fructose 6-phosphate}$$

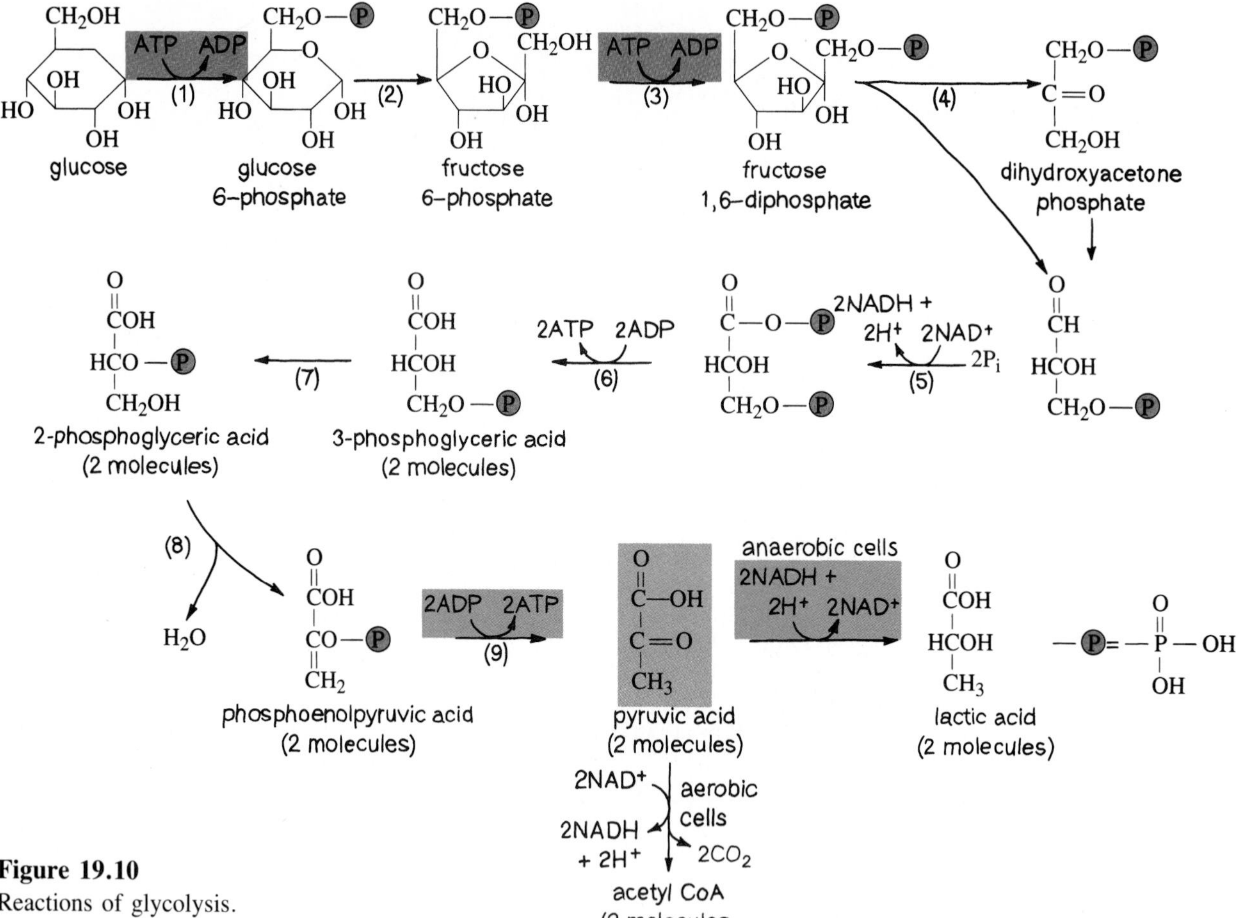

Figure 19.10
Reactions of glycolysis.

Step 3 Second phosphorylation

A second ATP is now used to add a second phosphate to fructose 6-phosphate to form fructose 1,6-diphosphate.

$$\text{Fructose 6-phosphate} + \boxed{\text{ATP}} \xrightarrow{\text{Phosphokinase}} \text{Fructose 1,6-diphosphate} + \text{ADP}$$

Step 4 The six-carbon molecule splits

At this point in glycolysis, the six-carbon molecule, fructose 1,6-diphosphate is split into two three-carbon molecules, dihydroxyacetone phosphate and glyceraldehyde 3-phosphate. Because dihydroxyacetone phosphate cannot oxidize further, it isomerizes to give a second molecule of glyceraldehyde 3-phosphate.

$$\text{Fructose 1,6-diphosphate} \xrightarrow{\text{Aldolase}}$$
$$\text{Dihydroxyacetone phosphate} + \text{Glyceraldehyde 3-phosphate}$$

$$\text{Dihydroxyacetone phosphate} \xrightarrow{\text{Isomerase}} \text{Glyceraldehyde 3-phosphate}$$

At this point in glycolysis, one glucose molecule has been converted to two triose phosphate molecules. The rest of the reactions in glycolysis occur with two molecules of substrate from the initial glucose molecule.

Step 5 Oxidation by NAD^+

An oxidative phosphorylation of glyceraldehyde 3-phosphate by the coenzyme NAD^+ yields NADH and a high-energy compound.

$$\text{Glyceraldehyde 3-phosphate} + NAD^+ + P_i \xrightarrow{\text{Dehydrogenase}} \text{1,3-Diphosphoglyceric acid} + NADH + H^+$$

Step 6 Formation of first ATP

When a phosphate group from the high-energy 1,3-diphosphoglyceric acid is transferred to ADP, ATP is produced. This process is known as a direct substrate-phosphate transfer. Recall that there are two substrate molecules from glucose in this reaction which would produce two ATP.

$$\text{1,3-Diphosphoglyceric acid} + \text{ADP} \xrightarrow{\text{Phosphokinase}} \text{3-Phosphoglyceric acid} + \boxed{\text{ATP}}$$

Step 7 Isomerization

To use the 3-phosphoglyceric acid further in glycolysis, the phosphate group is moved from carbon 3 to carbon 2 to give 2-phosphoglyceric acid.

$$\text{3-Phosphoglyceric acid} \xrightarrow{\text{Isomerase}} \text{2-Phosphoglyceric acid}$$

Step 8 Dehydration

When a molecule of water is removed from 2-phosphoglyceric acid, another high-energy phosphate compound, phosphoenolpyruvic acid (PEP) is produced.

$$\text{2-Phosphoglyceric acid} \xrightarrow{\text{Enolase}} \text{Phosphoenolpyruvic acid} + H_2O$$

Step 9 Formation of a second ATP

In the last step of glycolysis, pyruvic acid is formed by a transfer of the phosphate group from PEP to ADP to form another molecule of ATP. Again, there is a yield of two ATP.

$$\text{Phosphenolpyruvic acid} + \text{ADP} \xrightarrow{\text{Kinase}} \text{Pyruvic acid} + \boxed{\text{ATP}}$$

ATP Production from Glycolysis Under Aerobic Conditions

In the steps of glycolysis, we used two ATP for the initial phosphorylation steps (1 and 3). Later reactions (steps 6 and 9) involving direct phosphate transfer to ADP yielded a total of four molecules of ATP. This gives a net gain of two ATP for glycolysis. In addition, two NADH molecules were produced in the oxidation of glyceraldehyde 3-phosphate (step 5).

Summary of Glycolysis:

$$\text{Glucose} \longrightarrow \text{2 Pyruvic acid}$$

$$\text{2 ADP} \longrightarrow \text{2 ATP}$$

$$\text{2 NAD}^+ \longrightarrow \text{2 NADH}$$

When glycolysis operates in oxygen (aerobic conditions), additional ATP can be produced. Earlier we learned that when NADH enters the electron transport, it produces three molecules of ATP. However, glycolysis takes place in the cytoplasm, which is separated by a membrane from the mitochondria. The two NADH from glycolysis must cross that membrane, a process that uses the energy of two ATP. The ATP production from glycolysis under aerobic conditions is summarized in Table 19.1.

Table 19.1 Total ATP from Glycolysis (Aerobic Conditions)

Produced by glycolysis		ATP produced
Phosphorylation (1,3)		−2 ATP
Oxidation of glyceraldehyde 3-phosphate (5) ⟶ 2 NADH		6 ATP
Transport of 2 NADH across membrane		−2 ATP
Phosphorylation of 2 ADP (6)		2 ATP
Phosphorylation of 2 ADP (9)		2 ATP
	Total	6 ATP

The overall reaction for ATP produced by aerobic glycolysis can be written:

Glycolysis (aerobic):

$$\text{Glucose} \longrightarrow \text{2 Pyruvic acid} + \text{6 ATP}$$

19.4 Pathways for Pyruvic Acid

Learning Goal **Give the end products for reactions of pyruvic acid under aerobic and anaerobic conditions.**

All cells produce pyruvic acid from glucose in glycolysis. However, pyruvic acid can enter several pathways depending on oxygen in the cell and the type of cell. When oxygen is present, pyruvic acid is converted to acetyl CoA for further oxidation. If oxygen is not present, pyruvic acid is converted to lactic acid. In yeast cells, pyruvic acid is converted to ethanol under anaerobic conditions.

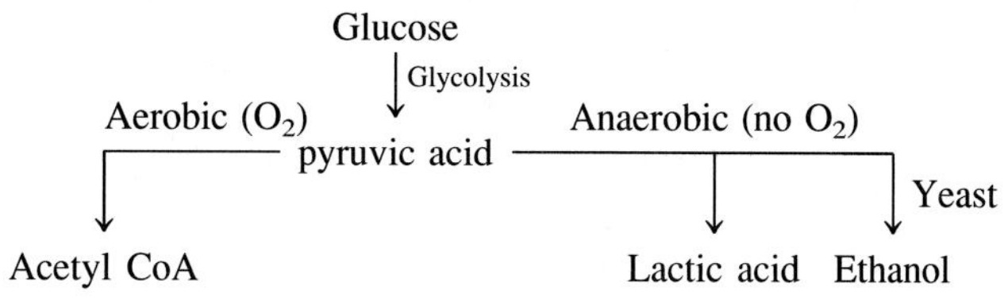

Glucose
↓ Glycolysis
pyruvic acid
Aerobic (O_2) → Acetyl CoA
Anaerobic (no O_2) → Lactic acid; Yeast → Ethanol

Formation of Acetyl CoA (Aerobic)

The further oxidation of glucose to CO_2 and water and much more energy occurs when there is oxygen in the cell. Under aerobic conditions, the pyruvic acid from glycolysis moves from the cytoplasm into the mitochondria. The oxidation is catalyzed by a multienzyme called pyruvic acid dehydrogenase complex that includes coenzyme A (pantothenic acid). The coenzyme A is composed of cysteine, an amino acid, pantothenic acid, a vitamin, and a molecule of ATP. (See Figure 19.11.) When pyruvic acid undergoes oxidation, it is also decarboxyl-

Figure 19.11
Structure of acetyl coenzyme A.

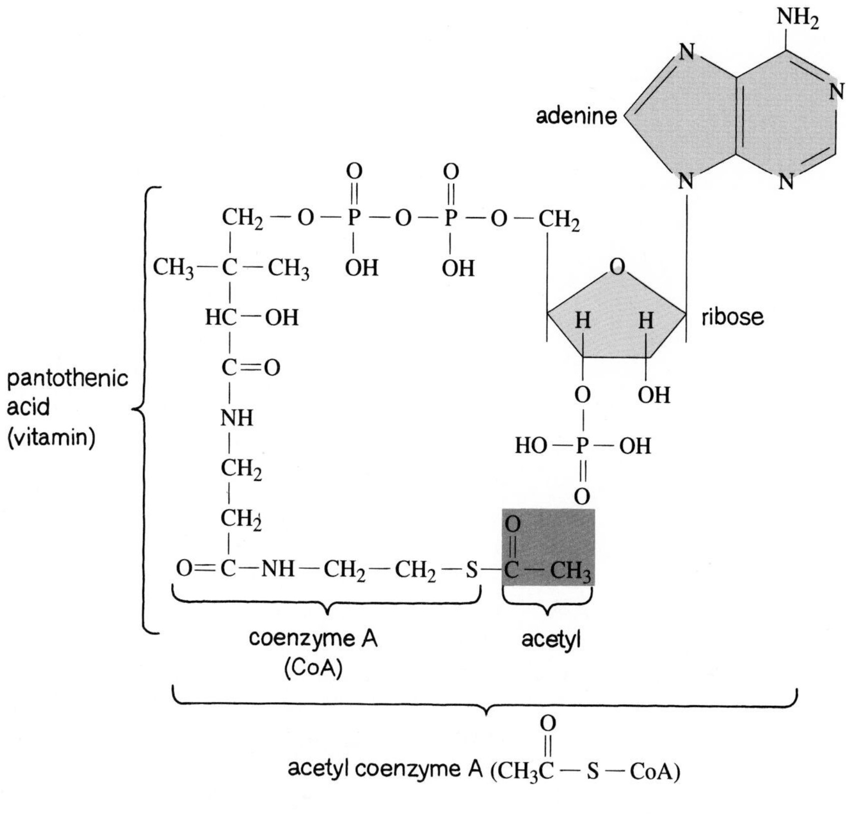

ated. A carbon atom is removed as CO_2 leaving a two-carbon acetyl unit attached to coenzyme A, **acetyl CoA,** an important intermediate in metabolic reactions.

$$\underset{\text{Pyruvic acid}}{CH_3-\overset{\overset{\large O}{\|}}{C}-\boxed{COOH}} + CoA + NAD^+ \xrightarrow[\text{dehydrogenase}]{\text{Pyruvic acid}}$$

$$\underset{\text{Acetyl CoA}}{CH_3-\overset{\overset{\large O}{\|}}{C}-CoA} + \boxed{CO_2} + NADH + H^+$$

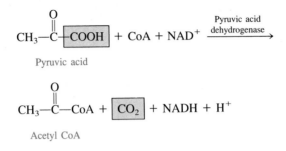

The formation of acetyl CoA provides NADH that can be converted to three ATP molecules for each acetyl CoA produced.

$$\text{Pyruvic acid} \longrightarrow \text{Acetyl CoA} + 3\ \text{ATP}$$

Sample Problem 19.6
Production of ATP

Indicate the ATP produced by each of the following:

a. The oxidation of pyruvic acid to acetyl CoA.
b. The aerobic conversion of glucose to acetyl CoA.

Solution:

a. The oxidation of pyruvic acid to give acetyl produces one NADH which yield three ATP.
b. Six ATP are produced from glycolysis. Since two pyruvic acids result form glycolysis, another six ATP result by the oxidation to acetyl CoA.

Glucose ⟶ 2 Pyruvic acid	6 ATP
2 Pyruvic acid ⟶ 2 Acetyl CoA	6 ATP
Total	12 ATP

Formation of Lactic Acid (Anaerobic)

While muscle cells normally have sufficient amounts of oxygen present (**aerobic**), it is used up quickly during strenuous exercise. The muscle cells must then meet energy demands without oxygen (**anaerobic**). Without oxygen, the electron transport chain in the mitochondria does not operate and cannot be used to reoxidize NADH or to produce ATP. In the absence of oxygen, NADH produced by glycolysis is used to reduce pyruvic acid to lactic acid, a process that regenerates the NAD^+ in the cytoplasm. (See Figure 19.12.) In this way NAD^+ is regenerated to oxidize glyceraldehyde 3-phosphate and produce ATP.

Glycolysis is the only pathway producing ATP while the cell is working under anaerobic conditions. However, large amounts of glucose are expended, seven times more than in aerobic conditions, so the process cannot go on long

Figure 19.12
Under anaerobic conditions, pyruvic acid from glycolysis is converted to lactic acid to regenerate NAD^+; under aerobic conditions pyruvic acid can enter the mitochondria for complete oxidation.

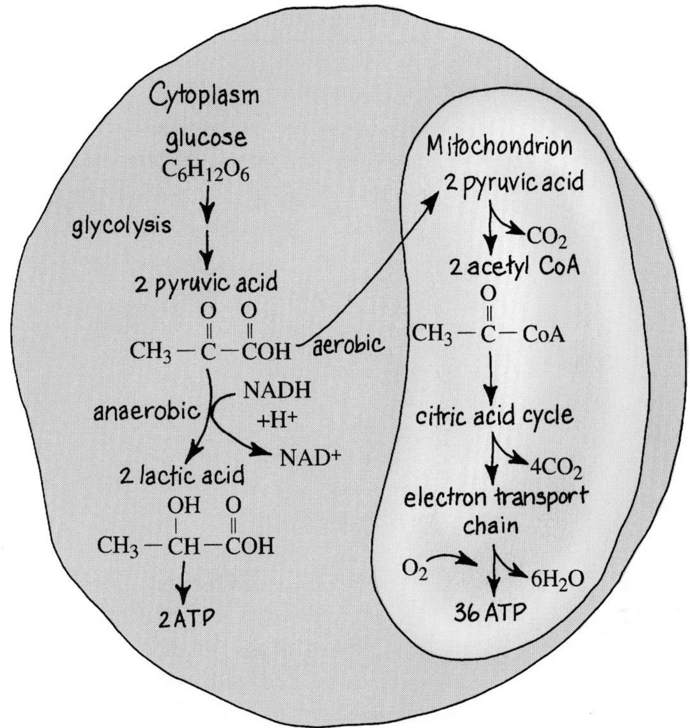

before exhaustion occurs. The rapid accumulation of lactic acid causes the muscles to tire rapidly and become sore. Most of the lactic acid is carried from the muscle to the liver, where it is converted to CO_2 and H_2O. Since oxygen is needed for this oxidation process, a person breathes rapidly after strenuous exercise in order to repay the oxygen debt incurred during the activity.

$$\underset{\text{Pyruvic acid}}{CH_3-\overset{O}{\overset{\|}{C}}-\overset{O}{\overset{\|}{C}}OH} + NADH + H^+ \xrightarrow[\text{(anaerobic)}]{\text{Lactic acid dehydrogenase}} \underset{\text{Lactic acid}}{CH_3-\overset{OH}{\overset{|}{CH}}-\overset{O}{\overset{\|}{C}}-OH} + NAD^+$$

The only ATP production in glycolysis under anaerobic condition occurs during the steps that directly phosphorylate ADP, giving a net gain of two ATP molecules.

Glycolysis (anaerobic):

$$\text{Glucose} \longrightarrow 2 \text{ Lactic acid} + 2 \text{ ATP}$$

Several bacteria convert pyruvic acid to lactic acid under anaerobic conditions. In the preparation of kimchee, cabbage is covered with a salt brine. Bacteria convert starches in the cabbage to lactic acid. The resulting acid environment acts as a preservative for the pickled cabbage because it prevents the growth of other bacteria. The pickling of olives and cucumbers gives similar products. When cultures of bacteria that produce lactic acid are added to milk, the change in pH denatures the milk protein to give products such as sour cream and yogurt.

Fermentation by Yeast Produces Ethanol

Most cells typically produce pyruvic acid from glucose. However, several microorganisms, particularly yeast, convert pyruvic acid to ethanol, an alcohol, under anaerobic conditions. In **fermentation,** a carbon atom is removed as CO_2 from pyruvic acid to give acetaldehyde, which is then reduced to ethanol. In this way, yeast cells regenerate NAD^+ needed for glycolysis under anaerobic conditions.

$$\underset{\text{Pyruvic acid}}{CH_3-\overset{\overset{O}{\|}}{C}-\boxed{\overset{\overset{O}{\|}}{C}OH}} \xrightarrow[\text{decarboxylase}]{\text{Pyruvic acid}} \underset{\text{Acetaldehyde}}{CH_3-\overset{\overset{O}{\|}}{C}H} + \boxed{CO_2}$$

$$\underset{\text{Acetaldehyde}}{CH_3-\overset{\overset{O}{\|}}{C}H} + NADH + H^+ \xrightarrow[\text{dehydrogenase}]{\text{Alcohol}} \underset{\text{Ethanol}}{CH_3CH_2-OH} + NAD^+$$

The process of fermentation by yeast is one of the oldest chemical reactions. Enzymes in yeast are used to convert the sugars in a variety of carbohydrate sources to glucose and then to ethanol. The type of carbohydrate determines the taste associated with a particular alcoholic beverage. Beer is made from the fermentation of barley malt, wine from the sugars in grapes, vodka from potatoes, sake from rice, and whiskeys from corn or rye. Fermentation produces ethanol solutions up to about 15% alcohol by volume. At this concentration, the alcohol denatures the yeast and prevents further activity. Higher concentrations of alcohol are obtained by distilling and concentrating the alcohol.

19.5 Citric Acid Cycle

Learning Goal **Describe the role of oxidation in the citric acid cycle. State the number of ATP molecules formed in one turn of the citric acid cycle.**

The **citric acid cycle** (or Krebs cycle) is a metabolic pathway in which oxidation reactions provide NADH and $FADH_2$ for the electron transport chain. The enzymes of the citric acid cycle are located in the mitochondria so that NADH and $FADH_2$ can enter the electron transport system to be oxidized.

Reactions of the Citric Acid Cycle

As we follow the reactions in the citric acid cycle, we will observe two important types. One is oxidation which produces NADH or $FADH_2$ for ATP production by the electron transport chain. The other is decarboxylation in which a carbon chain is shorted by the removal of a carbon atom as CO_2. As we discuss the steps of the citric acid cycle, you may wish to refer to the diagram in Figure 19.13 for the structure of the compounds that undergo reaction.

Figure 19.13
Reactions in the citric acid cycle.

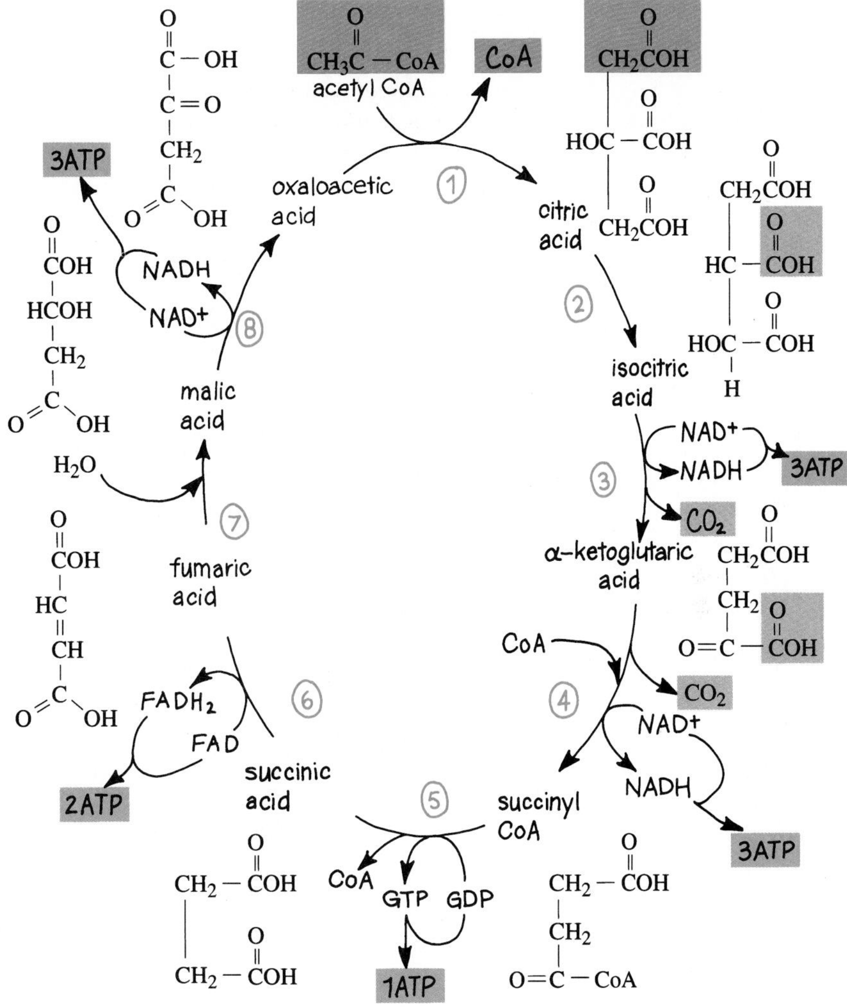

Step 1 Formation of citric acid

We can think of the first step in the citric acid cycle as the combination of the two-carbon acetyl CoA with a four-carbon molecule, oxaloacetic acid, to produce a six-carbon compound, citric acid:

$$\text{Acetyl CoA} + \text{Oxaloacetic acid} \xrightarrow[\text{synthetase}]{\text{Citric acid}} \text{Citric acid} + \text{CoA}$$

Step 2 Isomerization to isocitric acid

In this step, citric acid undergoes a rearrangement to isocitric acid, an isomer. The process involves a dehydration followed by hydration to change the position of the hydroxyl (—OH) group.

$$\text{Citric acid} \xrightarrow{\text{Aconitase}} \text{Isocitric acid}$$

Step 3 Oxidation and decarboxylation

At this point in the citric acid cycle, an enzyme called isocitric acid dehydrogenase catalyzes the first oxidative decarboxylation. The hydroxyl (—OH) group of isocitric acid is oxidized to an α-keto group, using NAD^+. The chain is shortened by the removal of the carboxylic acid group as CO_2, producing a five-carbon compound, α-ketoglutaric acid.

$$\text{Isocitric acid} + NAD^+ \xrightarrow[\text{dehydrogenase}]{\text{Isocitric acid}} \alpha\text{-Ketoglutaric acid} + CO_2 + NADH + H^+$$

Step 4 Second oxidation and decarboxylation

Another oxidative decarboxylation occurs as a carboxyl group is removed as CO_2. When the α-ketoglutaric acid undergoes oxidation in the presence of coenzyme A and NAD^+, it produces succinyl CoA, a high-energy compound.

$$\alpha\text{-Ketoglutaric acid} + \text{CoA} + NAD^+ \xrightarrow[\text{dehydrogenase}]{\alpha\text{-Ketoglutaric acid}} \text{Succinyl CoA} + NADH + H^+ + CO_2$$

Step 5 Hydrolysis

When succinyl CoA is hydrolyzed, sufficient energy is released to add a phosphate to guanosine diphosphate (GDP), forming guanosine triphosphate (GTP), a high-energy compound.

$$\text{Succinyl CoA} + \text{GDP} + P_i \xrightarrow[\text{synthetase}]{\text{Succinyl CoA}} \text{Succinic acid} + \text{GTP} + \text{CoA}$$

ATP is formed when a phosphate is transferred directly from GTP to ADP. This is the only place in the citric acid cycle where ATP is formed by a direct substrate phosphorylation:

$$\text{GTP} + \text{ADP} \xrightarrow{\text{Phosphokinase}} \text{GDP} + \text{ATP}$$

Step 6 Third oxidation

In this step, succinic acid is oxidized to give fumaric acid. The formation of a carbon–carbon double bond reduces the coenzyme FAD to $FADH_2$.

$$\text{Succinic acid} + \text{FAD} \xrightarrow[\text{dehydrogenase}]{\text{Succinic acid}} \text{Fumaric acid} + FADH_2$$

Step 7 Hydration

In a hydration step, water is added to the double bond of fumaric acid to yield malic acid:

$$\text{Fumaric acid} + H_2O \xrightarrow{\text{Fumarase}} \text{Malic acid}$$

Step 8 Fourth oxidation

The final oxidation of the citric acid occurs when the hydroxyl (—OH) group of malic acid is oxidized using NAD^+ to give the four-carbon compound oxaloacetic acid and NADH.

$$\text{Malic acid} + NAD^+ \xrightarrow[\text{dehydrogenase}]{\text{Malic acid}} \text{Oxaloacetic acid} + NADH + H^+$$

This completes one full turn of the citric acid cycle by regenerating the starting compound. The cycle starts again with two more carbon atoms in the form of acetyl CoA combining with the new oxaloacetic acid. The cycle provides three NADH and one $FADH_2$ for the electron transport chain and one ATP by direct phosphorylation:

$$CH_3{-}\overset{\overset{\large O}{\|}}{C}{-}CoA + 3\ NAD^+ + FAD + ADP \xrightarrow{\text{Citric acid cycle}} 2\ CO_2 + 3\ NADH + 3H^+ + FADH_2 + ATP$$

Sample Problem 19.7
NADH, $FADH_2$, and ATP in the Citric Acid Cycle

When acetyl CoA combines with oxaloacetic acid, it enters a pathway called the citric acid cycle. How many NADH and $FADH_2$ are produced in one turn of the cycle?

Solution:
There are three oxidation steps that oxidize alcohol groups to α-keto groups and yield three NADH. Another oxidation forms a C=C bond and yields one $FADH_2$.

Study Check:
How much ATP is produced by direct phosphorylation during one turn of the citric acid cycle?

Answer: one ATP

ATP Energy from the Citric Acid Cycle

The NADH and $FADH_2$ produced in one turn of the citric acid cycle can enter the electron transport chain where each NADH yields three ATP, and each $FADH_2$ yields two ATP. Table 19.2 describes the calculation of the total ATP energy provided by the citric acid cycle.

Table 19.2 ATP Energy from One Turn of the Citric Acid Cycle

Step in citric acid cycle	ATP
NADH from oxidation of isocitric acid (3)	3 ATP
NADH from oxidation of α-ketoglutaric acid (4)	3 ATP
ATP from direct phosphorylation of ADP by GTP (5)	1 ATP
$FADH_2$ from oxidation of succinic acid (6)	2 ATP
NADH from oxidation of malic acid (8)	3 ATP
total ATP	12 ATP

The total ATP produced when one acetyl CoA goes through the citric acid cycle is 12 molecules of ATP. The overall reaction can be written as follows:

Citric Acid Cycle:

$$\text{Acetyl CoA} \longrightarrow 2\ CO_2 + 12\ \text{ATP}$$

Sample Problem 19.8
ATP from the Citric Acid Cycle

How is the oxidative decarboxylation of isocitric acid related to the production of ATP in the cell?

Solution:
The hydroxyl (—OH) group of isocitric acid is oxidized to give α-ketoglutaric acid. The NADH produced by the oxidation is carried to the electron transport chain, where three ATP are produced:

Study Check:
Write a general equation for the ATP produced when one acetyl CoA passes through the citric acid cycle.

Answer:

$$\text{Acetyl CoA} \longrightarrow 2\ CO_2 + 12\ \text{ATP}$$

Sample Problem 19.9
Citric Acid Cycle

In the citric acid cycle, succinic acid is oxidized by FAD to oxaloacetic acid. How many ATP molecules are generated from the reaction?

Solution:
When an oxidation produces a carbon–carbon double bond, the coenzyme FAD is reduced to $FADH_2$ which produces two ATP molecules in the electron transport chain.

19.6 Energy Production from Glucose

Learning Goal **Account for the ATP produced by the complete oxidation of glucose.**

We can now determine the total ATP production for glucose when it undergoes complete combustion to carbon dioxide and water. The overall relationship of the metabolic pathways leading to the production of ATP is shown in Figure 19.14. For our calculations, recall that glucose produces two pyruvic acid in glycolysis. When oxidized, they provide two acetyl CoA that will go through two turns of the citric acid cycle. In Table 19.3, the ATP from aerobic glycolysis, acetyl CoA formation, citric acid cycle and the electron chain is calculated.

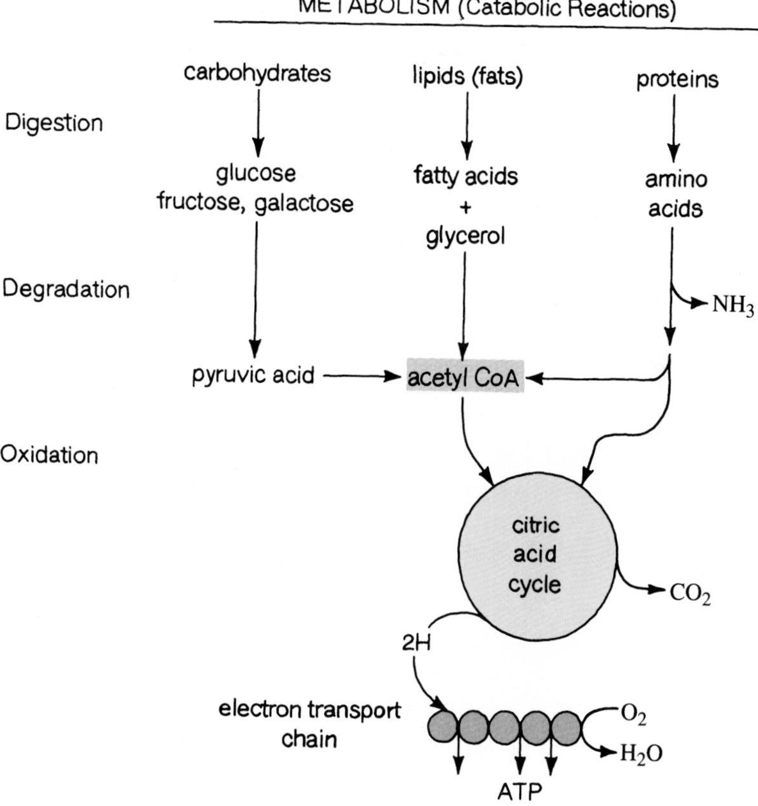

Figure 19.14
Overall relationship of degradative pathways that provide acetyl CoA for the citric acid cycle and the production of ATP.

Table 19.3 ATP Produced by the Complete Combustion of Glucose

Reaction Pathway		ATP for one glucose
Glycolysis (aerobic):		
Phosphorylation (two)		−2 ATP
Oxidation of glyceraldehyde 3-phosphate (2 NADH)		6 ATP
Transport of 2 NADH across membrane		−2 ATP
Direct phosphorylation (four)		4 ATP
Summary:		
Glucose → 2 pyruvic acid		6 ATP
Pyruvic acid to Acetyl CoA:		
2 Pyruvic acid → 2 Acetyl CoA (2 NADH)		6 ATP
Citric acid cycle:		
Oxidation of 2 isocitric acid (2 NADH)		6 ATP
Oxidation of 2 α-ketoglutaric acid (2 NADH)		6 ATP
2 direct phosphorylations		2 ATP
Oxidation of 2 succinic acid (2 $FADH_2$)		4 ATP
Oxidation of 2 malic acid (2 NADH)		6 ATP
Summary:		
2 Acetyl CoA $\longrightarrow$ 2 CO_2		24 ATP
Summary:		
Glucose $\xrightarrow{\text{Glycolysis, Acetyl CoA formation, Citric acid, Electron transport system}}$ 6 CO_2	Total ATP	36 ATP

The overall production of ATP from the complete oxidation of glucose can be written as follows:

$$C_6H_{12}O_6 + 6O_2 \longrightarrow 6CO_2 + 6H_2O + 36\ \text{ATP}$$

We have seen that glucose can be degraded to provide energy for the cell, or it can be stored as glycogen. In addition, glucose can be obtained from glycogen or synthesized from noncarbohydrate sources (**gluconeogenesis**) if glycogen stores are depleted. It is the balance of all these reactions that maintains the necessary blood glucose level in our cells and provides the necessary amount of ATP for our continued energy needs. (See Figure 19.15.)

Figure 19.15
The metabolic pathways of glucose in the cell.

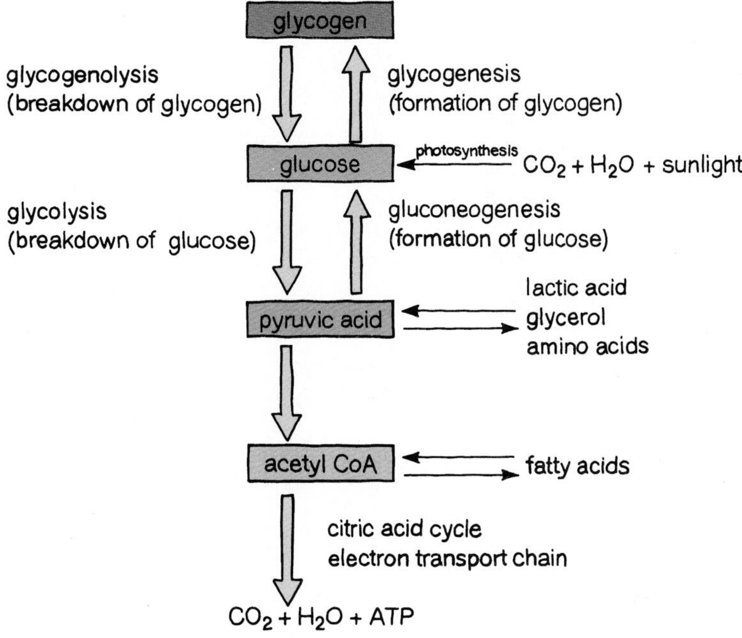

Sample Problem 19.10
Production of ATP

a. How much ATP can be produced from one turn of the citric acid cycle and the electron transport chain?
b. What metabolic pathways are involved in the complete combustion of glucose?
c. How much ATP is produced?

Solution:
a. One turn of the citric acid cycle produces a total of 12 ATP.
b. The metabolic pathways of glycolysis, the conversion of pyruvic acid to acetyl CoA, the citric acid cycle, and the electron transport chain.
c. A total of 36 ATP are produced as a result of direct phosphorylation and oxidation reactions that provide NADH and $FADH_2$ for ATP production in the respiratory chain.

19.7 Metabolic Pathways for Fatty Acids

Learning Goal **Describe the beta (β)-oxidation of fatty acids and give the total ATP produced. Describe how acetyl CoA is used in the synthesis of fatty acids.**

Most of our energy is stored as triglycerides in the fat cells or adipose tissues. (See Figure 19.16.) If we use up all our reserves of glucose and glycogen, we obtain energy by breaking down our body fats.

Figure 19.16
Photomicrograph of a fat cell filled with fat.

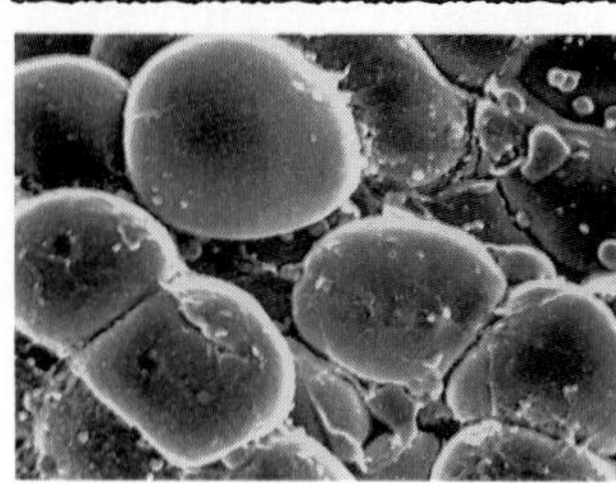

$$
\begin{array}{l}
CH_2{-}O{-}\overset{\overset{\displaystyle O}{\|}}{C}{-}(CH_2)_{16}CH_3 \\
| \\
CH{-}O{-}\overset{\overset{\displaystyle O}{\|}}{C}{-}(CH_2)_{16}CH_3 \\
| \\
CH_2{-}O{-}\overset{\overset{\displaystyle O}{\|}}{C}{-}(CH_2)_{16}CH_3
\end{array}
+ 3\,H_2O \xrightarrow{\text{Lipase}}
\begin{array}{l}
CH_2OH \\ | \\ CHOH \\ | \\ CH_2OH
\end{array}
+ 3\,CH_3(CH_2)_{16}\overset{\overset{\displaystyle O}{\|}}{C}OH
$$

Fat (tristearin) — Glycerol — Fatty acids (stearic acid)

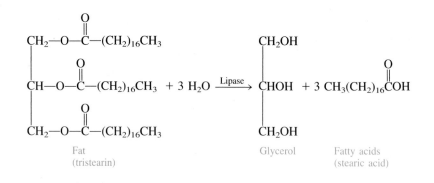

Beta (β)-Oxidation of Fatty Acids

Fatty acids resulting from the hydrolysis of triglycerides undergo oxidation in a series of reactions called beta (β)-oxidation. During β-oxidation, the bond between the alpha (α) and beta (β) carbons of the fatty acid chain is broken, yielding a shortened fatty-acid chain and a two-carbon unit of acetyl CoA.

$$
R{-}CH_2{-}\overset{\beta}{C}H_2{-}\overset{\alpha}{C}H_2{-}\overset{\overset{\displaystyle O}{\|}}{C}{-}OH \xrightarrow{\text{CoA}} R{-}CH_2{-}\overset{\overset{\displaystyle O}{\|}}{C}{-}OH + CH_3{-}\overset{\overset{\displaystyle O}{\|}}{C}{-}CoA
$$

bond broken (between β and α carbons)

Fatty acid — Fatty acid (shorter by two carbon atoms) — Acetyl CoA

Activation of the Fatty Acid

Before fatty acid oxidation, a fatty acid must be activated by combining with coenzyme A. The energy for this activation step is provided by the hydrolysis of ATP to give adenine monophosphate (AMP). Because two phosphate groups are hydrolyzed, the energy cost is equivalent to using two ATP. The activated fatty acid can now be carried into the mitochondria to undergo β-oxidation.

$$
R{-}CH_2{-}CH_2{-}\overset{\overset{\displaystyle O}{\|}}{C}{-}OH + ATP + CoA \xrightarrow[\text{synthetase}]{\text{Acetyl CoA}}
$$

Fatty acid

$$
R{-}CH_2{-}CH_2{-}\overset{\overset{\displaystyle O}{\|}}{C}{-}CoA + AMP + PP_i
$$

Activated fatty acid

Beta (β)-Oxidation

Beta (β)-oxidation is a series of reactions that removes a two-carbon acetyl unit from a fatty acid to give a shorter carbon chain. The cycle is repeated until the original fatty acid is degraded into two-carbon units. The number of carbon

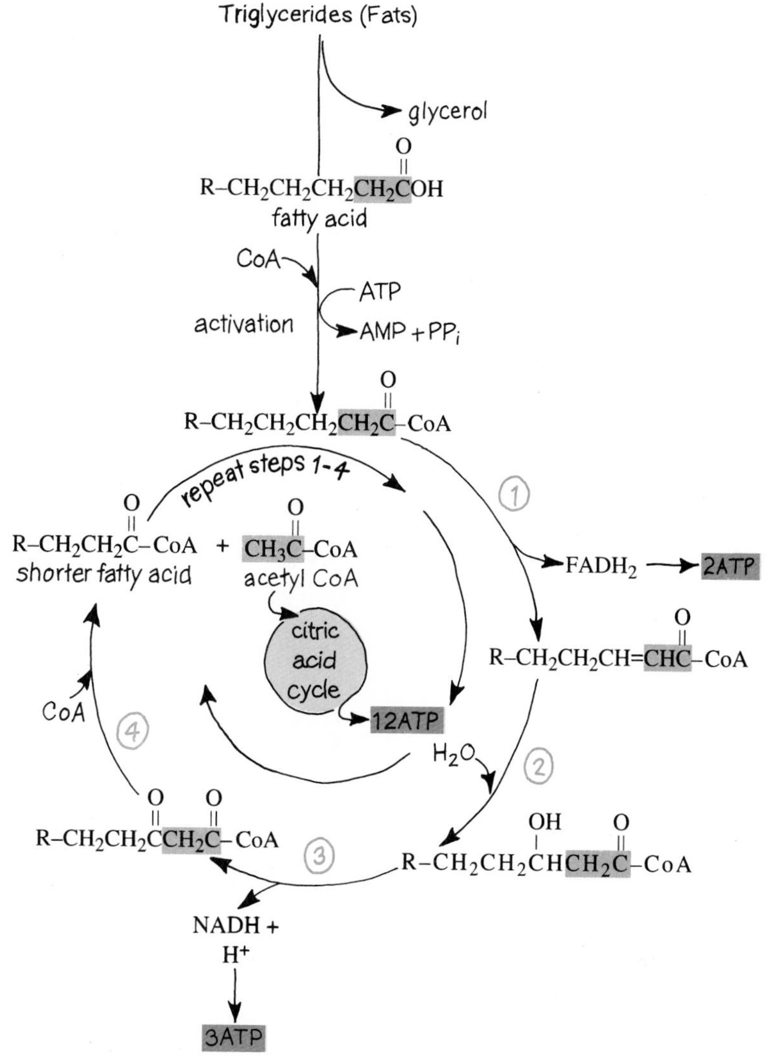

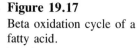

Figure 19.17
Beta oxidation cycle of a fatty acid.

atoms in the fatty acid determines the number of times the cycle is repeated and the number of acetyl units produced. Each cycle of β-oxidation also produces one NADH and one $FADH_2$. (See Figure 19.17.)

Step 1 Oxidation

In the first step, an oxidation by FAD of the alpha (α) and beta (β) carbons forms a carbon–carbon double bond and $FADH_2$.

Step 2 Hydration

In the second step, water is added to the double bond to give a beta (β)-hydroxy compound.

Step 3 Oxidation

In the third step, the alcohol group is oxidized by a dehydrogenase enzyme and NAD^+ to form a β-keto group and NADH.

Step 4 Splitting the chain

In this step, the chain is split between the alpha and beta carbons to give acetyl CoA and a shorter fatty acid that repeats the cycle again.

Reaction Steps in the β-Oxidation of a Fatty Acid

Step 1 Oxidation (FAD)

$$R{-}CH_2{-}\overset{\beta}{C}H_2{-}\overset{\alpha}{C}H_2{-}\overset{O}{\overset{\|}{C}}{-}CoA$$

$\downarrow$ FAD → $FADH_2$ ⟶ 2 ATP by electron transport chain

$$R{-}CH_2{-}\overset{\beta}{C}H{=}\overset{\alpha}{C}H{-}\overset{O}{\overset{\|}{C}}{-}CoA$$

Step 2 Hydration

$\downarrow$ H_2O

$$R{-}CH_2{-}\underset{\beta}{\overset{OH}{\overset{|}{C}H}}{-}\underset{\alpha}{C}H_2{-}\overset{O}{\overset{\|}{C}}{-}CoA$$

Step 3 Oxidation (NAD^+)

$\downarrow$ NAD^+ → $NADH + H^+$ ⟶ 3 ATP by electron transport chain

$$R{-}CH_2{-}\underset{\beta}{\overset{O}{\overset{\|}{C}}}{-}\underset{\alpha}{C}H_2{-}\overset{O}{\overset{\|}{C}}{-}CoA$$

Step 4 Splitting the Chain

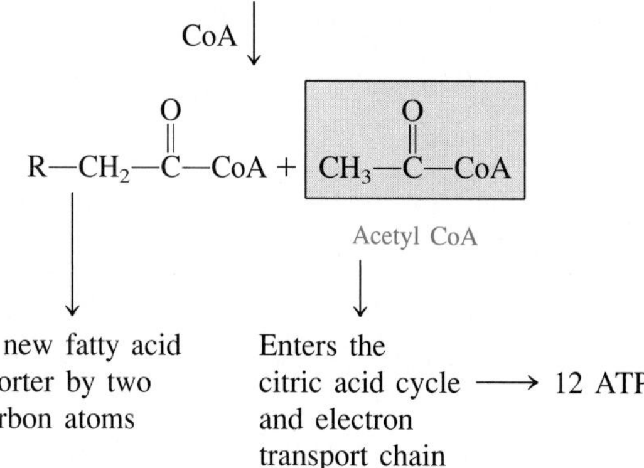

For any fatty acid, the number of acetyl CoA can be determined by dividing the number of carbon atoms by 2. The number of times that the fatty acid repeats the cycle is one less than the number of acetyl CoA produced. For example, caprylic acid contains 8 carbon atoms. It will be split into 4 acetyl CoA by repeating β-oxidation three times. (See Table 19.4.)

Table 19.4 Units of Acetyl CoA, NADH, and $FADH_2$ Produced by One Caprylic Acid Molecule

		third cycle ↓ second cycle ↓ first cycle ↓
Number of carbon atoms	8	{C—C—}—{—C—C—}—{—C—C—}—{—C—C}
Number of acetyl CoA Units produced (8/2)	4	
Number of cycles (4-1)	3	
NADH produced	3	
$FADH_2$ produced	3	

ATP Production from β-Oxidation

The amount of ATP energy produced by the complete oxidation of a fatty acid depends upon the length of the carbon chain. In each cycle, one NADH, one $FADH_2$, and one acetyl CoA are produced. In the electron transport chain, NADH yields 3 ATP, and $FADH_2$ yields 2 ATP. In the citric acid cycle, the oxidation of acetyl CoA produces an additional 12 ATP. Because two ATP were required to activate the fatty acid, two ATP must be subtracted from the total ATP for the oxidation of the fatty acid. Table 19.5 summarizes the calculation of ATP for the eight-carbon caprylic acid.

Table 19.5 Total ATP Produced by Caprylic Acid (8 Carbon Atoms)

Number of carbon atoms	8
Number of β-oxidation cycles	3
3 oxidations (3 $FADH_2$)	6 ATP
3 oxidations (3 NADH)	9 ATP
4 acetyl CoA (x 12 ATP/citric acid cycle)	48 ATP
Activation of fatty acid	− 2 ATP
Summary:	
Total ATP from caprylic acid =	61 ATP

HEALTH NOTE

Ketone Bodies

When carbohydrates are not available to meet energy needs, the body breaks down large quantities of body fat. The oxidation of fatty acids produces so many units of acetyl CoA that the citric acid cycle is overwhelmed. At the same time, less oxaloacetic acid is available for the citric acid cycle because it is used to synthesize glucose. Without oxaloacetic acid, acetyl CoA from β-oxidation cannot be completely utilized by the citric acid cycle. Instead, it enters a pathway called ketogenesis in which four-carbon ketone bodies, acetoacetic acid and β-hydroxybutyric acid, are formed from the two acetyl CoA units. Acetoacetic acid is then converted to acetone, another ketone body.

Dietary fats

↓

Fatty acids

$$2\ CH_3{-}\overset{\overset{\large O}{\|}}{C}{-}CoA \longrightarrow \text{Citric acid cycle}$$

Acetyl CoA

↓

$$\underset{\text{Acetone}}{CO_2 + CH_3{-}\overset{\overset{\large O}{\|}}{C}{-}CH_3} \xleftarrow{\text{Decarboxylation}} \underset{\text{Acetoacetic acid}}{CH_3{-}\overset{\overset{\large O}{\|}}{C}{-}CH_2{-}\overset{\overset{\large O}{\|}}{C}{-}OH} \xrightarrow{\text{Reduction}} \underset{\beta\text{-Hydroxybutyric acid}}{CH_3{-}\overset{\overset{\large OH}{|}}{CH}{-}CH_2{-}\overset{\overset{\large O}{\|}}{C}{-}OH}$$

←——— Ketone bodies ———→

In a condition called ketosis, the level of ketone bodies becomes so high that they cannot be metabolized by the body. The fruity odor of acetone can be detected on the breath of a person with ketosis. The excess ketone bodies in the blood are excreted in the urine, taking large amounts of sodium with them. The body's sodium supply is depleted, causing a large output of urine and a strong sensation of thirst.

Since two of the ketone bodies are acids, the pH of the blood is lowered leading to acidosis which could lead to a coma. The symptoms of ketosis and acidosis are seen in cases of severe diabetes and also from diets that are high in fat and low in carbohydrate and starvation. ■

The Synthesis of Fatty Acids

When the body has met all its energy needs and glycogen stores are full, excess glucose is converted to fatty acids and stored in the fat cells. In the cytoplasm, a process called lipogenesis links together two-carbon units obtained from acetyl CoA to form a fatty acid with an even number of carbon atoms. Several of the steps reverse the reactions we saw in β-oxidation. The typical fatty acid product of lipogenesis is palmitic acid from which other fatty acids are derived. (See Figure 19.18.)

Figure 19.18
Lipogenesis produces fatty acids from units of acetyl CoA.

$$CH_3-\overset{O}{\overset{\|}{C}}-CoA + CO_2 + ATP \xrightarrow{\text{acetyl CoA carboxylase}} HO-\overset{O}{\overset{\|}{C}}-CH_2-\overset{O}{\overset{\|}{C}}-CoA + ADP + P_i$$

malonyl CoA

$CH_3-\overset{O}{\overset{\|}{C}}-CoA$ acetyl CoA

$CH_3-CH_2-CH_2-\overset{O}{\overset{\|}{C}}-CoA$ four-carbon fatty acid; CO_2; malonyl CoA

$CH_3-CH_2-CH_2-CH_2-CH_2-\overset{O}{\overset{\|}{C}}-CoA$ six-carbon fatty acid; CO_2; malonyl CoA

cycle repeats; CO_2; malonyl CoA

$CH_3(CH_2)_{14}-\overset{O}{\overset{\|}{C}}OH$ palmitic acid; CO_2

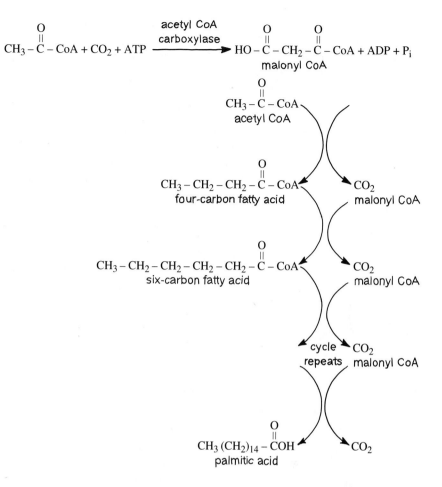

Sample Problem 19.11
ATP Production from β-Oxidation

How much ATP will be produced from the β-oxidation of myristic acid, a 14-carbon fatty acid?

Solution:

A 14-carbon fatty acid will produce seven acetyl CoA units and go through six β-oxidation cycles.

Number of carbon atoms	14	
Number of cycles	6	
6 NADH × 3 ATP/NADH		= 18 ATP
6 $FADH_2$ × 2 ATP/$FADH_2$		= 12 ATP
7 acetyl CoA × 12 ATP/citric acid cycle		= 84 ATP
Activation of fatty acid		= −2 ATP
Total ATP from β-oxidation of myristic acid		112 ATP

19.8 Metabolic Pathways for Amino Acids

Learning Goal **Describe reactions that prepare amino acids for energy production or synthesize nonessential amino acids from carbon compounds.**

Amino acids obtained from the digestion of proteins are absorbed into the bloodstream and carried to the cells in the body, where they are used to build proteins. Normally, amino acids are not used for energy. However, in conditions such as starvation, the body must extract energy from amino acids after the fat reserves have been used up. Over a period of time, the use of amino acids for energy can deplete body protein and destroy body tissues.

Transamination

To utilize amino acids for energy, the amino group must be removed. The resulting compounds are carboxylic acids that become energy sources as intermediates of the citric acid cycle. Many of the three-carbon amino acids are converted to pyruvic acid; four-carbon amino acids provide oxaloacetic acid; five-carbon chains are converted to α-ketoglutaric acid. (See Figure 19.19.) By entering the citric acid cycle, these derivatives of the amino acids undergo oxidation and the formation of ATP.

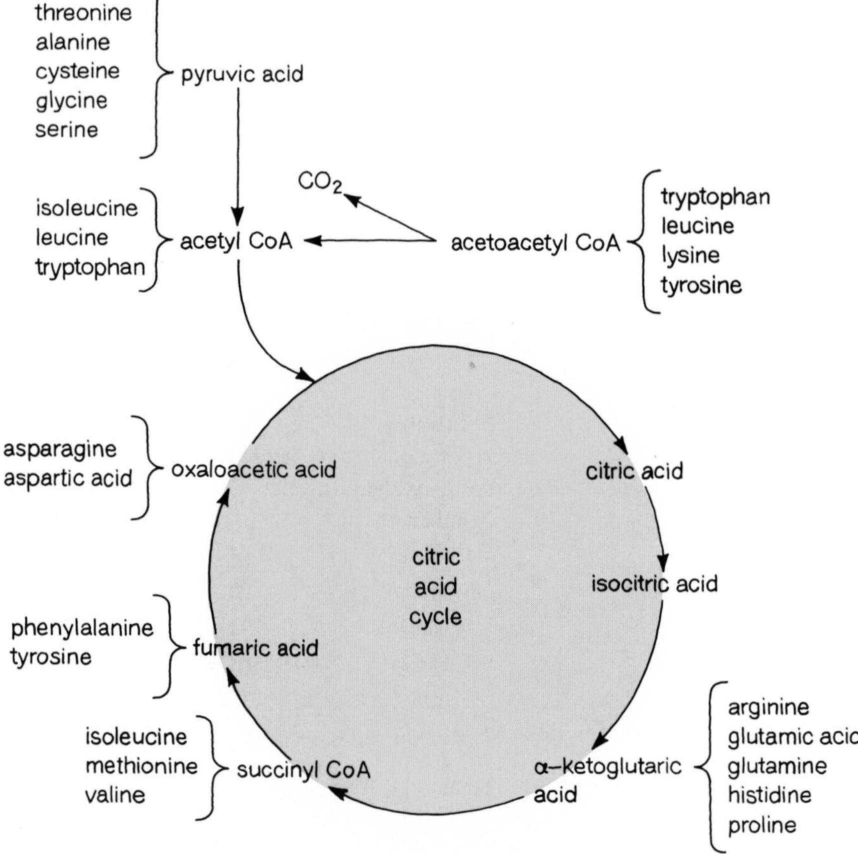

Figure 19.19
The loss of the amino group converts amino acids to intermediates of the citric acid cycle.

One way to remove an amino group is to transfer it to another acid in a reaction called **transamination.** These reactions are catalyzed by transaminases, enzymes that require pyridoxal phosphate, a coenzyme derived from pyridoxine (vitamin B). The following equation illustrates the transfer of the amino group of alanine to α-ketoglutaric acid to form glutamic acid, an amino acid. In this example of transamination, the transfer of the amino group converts alanine to pyruvic acid. After pyruvic acid is oxidized to acetyl CoA, it can enter the citric acid cycle.

$$\underset{\text{Alanine}}{CH_3-\overset{\boxed{NH_2}}{\overset{|}{C}H}-\overset{O}{\overset{\|}{C}}-OH} + \underset{\alpha\text{-Ketoglutaric acid}}{HO-\overset{O}{\overset{\|}{C}}-CH_2-CH_2-\overset{O}{\overset{\|}{C}}-\overset{O}{\overset{\|}{C}}-OH} \xrightarrow{\text{transaminase}}$$

$$\underset{\text{Pyruvic acid}}{CH_3-\overset{O}{\overset{\|}{C}}-\overset{O}{\overset{\|}{C}}-OH} + \underset{\text{Glutamic acid}}{HO-\overset{O}{\overset{\|}{C}}-CH_2-CH_2-\overset{\boxed{NH_2}}{\overset{|}{C}H}-\overset{O}{\overset{\|}{C}}-OH}$$

Amino group transferred

Oxidative Deamination

Many of the transamination reactions produce glutamic acid which can be converted to α-ketoglutarate, an intermediate of the citric acid cycle. In a reaction called oxidative deamination, the amino group of glutamic acid is removed as NH_4^+. The NH_4^+ is converted to urea and excreted in the urine.

$$\underset{\text{Glutamic acid}}{HO-\overset{O}{\overset{\|}{C}}-CH_2-\overset{NH_2}{\overset{|}{C}H}-\overset{O}{\overset{\|}{C}}-OH} + NAD^+ \xrightarrow{\text{Glutamic acid dehydrogenase}}$$

$$\underset{\alpha\text{-Ketoglutaric acid}}{HO-\overset{O}{\overset{\|}{C}}-CH_2-\overset{O}{\overset{\|}{C}}-\overset{O}{\overset{\|}{C}}-OH} + NH_4^+ + NADH$$

Sample Problem 19.12
Using Amino Acid for Energy

Serine can be converted to pyruvic acid by a series of reactions including transamination. How does this reaction represent a way in which amino acids can be utilized as an energy source?

Solution:
Amino acids can be converted to compounds that are a part of the energy-producing citric acid cycle. After the removal of the amino group of serine (deamination), it is converted to pyruvic acid which is converted to acetyl CoA. The acetyl CoA enters the citric acid cycle, where it produces 12 ATP.

HEALTH NOTE

What Makes You Feel Hungry?

When you say "I'm hungry," you probably go looking for food. You are responding to a stimulation of the neurons in the hunger center located in the hypothalamus of the brain. When another group of nearby neurons called the satiety center is stimulated, you feel full, and stop eating. Although the hunger center is operational all the time, it is inhibited by the satiety center.

In the glucostatic theory, a low blood glucose level decreases the activity of the satiety center so it no longer inhibits the feeding center. We respond by eating a meal. Our blood glucose levels rise, increasing the activity of the satiety center, which inhibits the feeding center.

An increase in the level of fatty acids in the blood also depresses the hunger feeling. The lipostatic theory suggests that the higher levels of fatty acids found in the blood as the amount of body fat increases may play a role in the regulation of food intake. Higher levels of amino acids also depress hunger. Therefore, the levels of fatty acids and amino acids in the blood act in a similar fashion to glucose, and cause an inhibition of the feeding center.

Our body temperature also affects our food intake. In a cold environment, food intake increases, which accelerates metabolic reactions, increases body temperature, and provides more body fat for insulation. When we are hot, less body fat is needed and we tend to undereat.

In summary, our desire for food depends on the level of nutrients stored in our bodies. When the stores of nutrients are low, we feel hungry and eat. When the levels of nutrients in the blood are high, the activity of the satiety center inhibits our desire for food. ■

Formation of Amino Acids

All of the nonessential amino acids can be synthesized in the body by transferring an amino group from glutamic acid to several of the α-keto acids obtained from the citric acid cycle or glycolysis. Some examples of amino acids formed by transamination are shown in Table 19.6. The carbon skeletons for essential amino acids are not available; those amino acids must obtained from the diet.

Table 19.6 Some Nonessential Amino Acids Formed by Transamination with Glutamic Acid

α-Ketoacid	Nonessential amino acids
Pyruvic acid ⟶	alanine, valine, leucine, serine, cysteine, glycine
Oxaloacetic acid ⟶	aspartic acid, threonine, lysine, isoleucine
α-Ketoglutaric acid ⟶	glutamine, proline

Overview of Metabolism

Figure 19.20 summarizes the energy-producing and energy-requiring reactions we have discussed. The oxidative pathways degrade large molecules to small molecules that can be used for energy production via the citric acid cycle and the electron transport chain. These same molecules can also enter pathways that lead to the synthesis of larger molecules in the cell. Such synthesis is energy consuming and depends on a supply of ATP.

The pathways in metabolism have several branch points from which compounds may be degraded for energy or used to synthesize larger molecules. For example, glucose can be converted to glycogen for storage, degraded to acetyl CoA for the citric acid cycle to produce energy, or used to synthesize fatty acids when energy stores are full. When glycogen stores are depleted, fatty acids can be oxidized for energy. Amino acids are normally used to synthesize proteins for the cells, but they can be converted to intermediates of the citric acid cycle in an energy crisis. The α-keto acids of the citric acid may be used in the synthesis of several nonessential amino acids by transamination reactions.

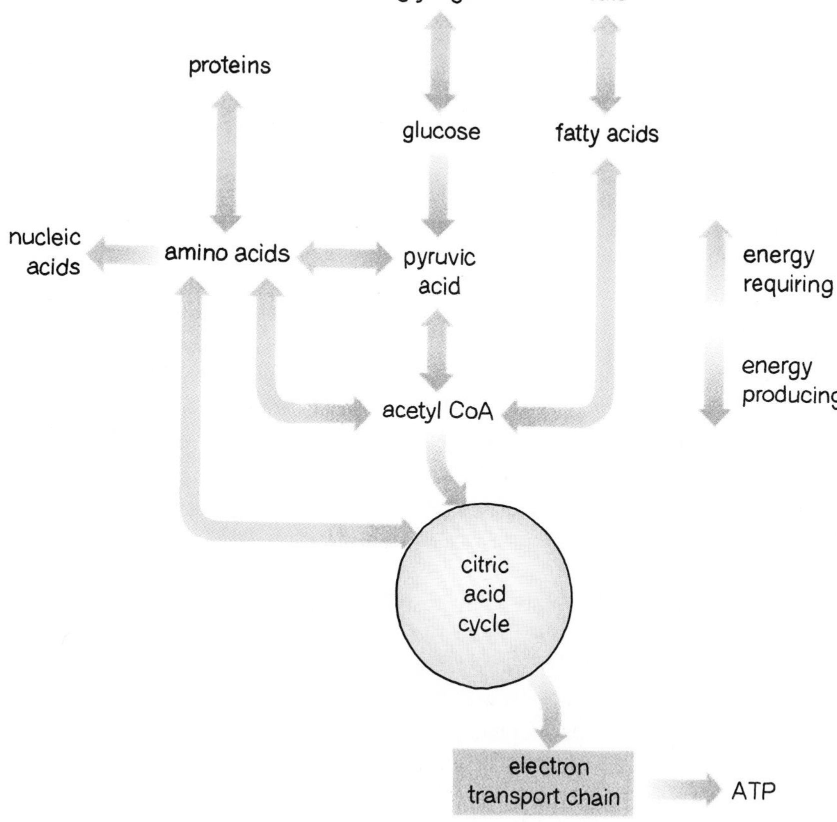

Figure 19.20
Overall relationship of energy-producing and energy-requiring pathways in metabolism.

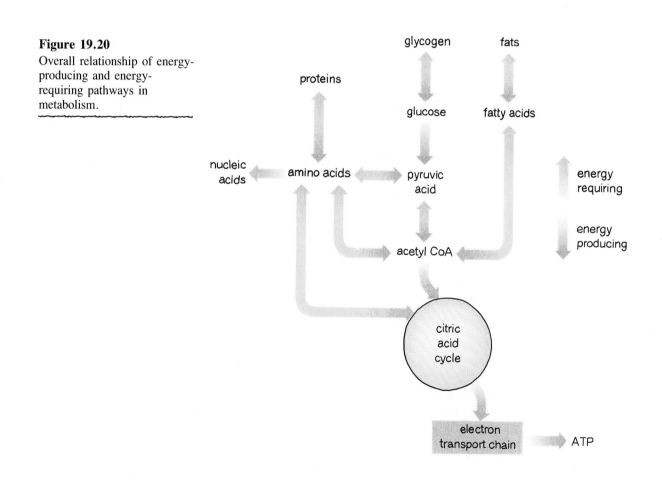

Glossary of Key Terms

acetyl CoA A 2-carbon acetyl unit attached to coenzyme A. As an end product of glucose, fatty acid, and protein metabolism, it enters the citric acid cycle for oxidation that leads to the production of ATP or it is used in the biosynthesis of larger molecules such as fatty acids or glucose.

ADP Adenosine diphosphate is a compound consisting of adenine, a ribose sugar, and two phosphate groups that forms when a high-energy phosphate is cleaved from ATP to release energy.

aerobic An oxygen-containing environment in the cells.

anabolic A metabolic reaction that is energy-requiring.

anaerobic A cellular environment in which oxygen has been depleted.

ATP Adenosine triphosphate is a compound consisting of adenine, a ribose sugar, and three phosphate groups formed by the respiratory chain of the mitochondria to store and transport energy in the cells.

beta (β)-oxidation A series of reactions that separate fatty acids into acetyl CoA units that can enter the citric acid cycle or synthesize other cellular molecules.

catabolic A metabolic reaction that produces energy for the cell by the degradation of digested food products such as glucose, fatty acids and amino acids.

citric acid cycle A series of oxidation reactions in the mitochondria that convert acetyl CoA to CO_2 and provide NADH and $FADH_2$ for ATP energy production by the electron transport chain.

coenzyme Q (Q) A quinone in the respiratory chain that accepts electrons from $FMNH_2$ or $FADH_2$ in the respiratory chain and releases two protons (H^+) into the inner membrane space of the mitochondria.

cytochromes Electron acceptors in the respiratory chain that use the oxidation states of iron (Fe^{3+}, Fe^{2+}) to transfer electrons from NADH and $FADH_2$ to oxygen to form water. The accompanying energy changes are coupled to the production of ATP.

electron transport chain A series of reactions occurring in the mitochondria in which electrons are transferred among coenzymes and cytochromes to oxygen to produce water and energy in the form of ATP molecules.

FAD A coenzyme (flavin adenine dinucleotide) for dehydrogenase enzymes that remove hydrogen to form carbon–carbon double bonds. Its reduced form, $FADH_2$, transfers electrons to Q in the electron transport chain.

fermentation The production of ethanol + CO_2 from pyruvic acid by yeast under anaerobic conditions.

FMN (flavin mononucleotide) The initial electron acceptor in the electron transport chain.

gluconeogenesis The series of reactions that synthesizes glucose from noncarbohydrate sources such as pyruvic acid.

glycogenesis The anabolic process of converting glucose to glycogen for storage.

glycogenolysis The hydrolysis of glycogen to glucose.

glycolysis The oxidation reactions of glucose to yield two pyruvic acid molecules under aerobic conditions and two lactic acid molecules under anaerobic conditions.

metabolite A compound that is part of a series of metabolic reactions.

NAD^+ A coenzyme that removes hydrogen in the formation of carbon–oxygen double bonds for transport to the electron transport chain to yield three ATP molecules.

oxidation The loss of hydrogen atoms or electrons by a metabolite in a chemical reaction. In metabolism, it is generally associated with the degradation of large molecules and the production of energy.

phosphorylation The transfer of a phosphate group to a substrate; phosphorylation of ADP produces ATP.

reduction The gain of hydrogen or electrons by a metabolite in a chemical reaction.

transamination The transfer of an amino group from an α-amino acid to an α-keto acid to produce intermediates in the citric acid cycle that can be oxidized for the production of energy in the cell.

Problems

ATP: The Energy Storehouse *(Goal 19.1)*

19.1 What is the difference between a catabolic and an anabolic reaction in metabolism?

19.2 Consider the formation of ATP:
a. Write an equation for the formation of ATP from ADP.
b. Write an equation to represent the hydrolysis of ATP.

19.3 What is the role of ATP in the cells?

The Electron Transport Chain *(Goal 19.2)*

19.4 a. What coenzymes accept hydrogen directly from metabolites undergoing oxidation?
*b. Write an equation to illustrate the transfer of hydrogen atoms from a metabolite MH_2 to FAD; to NAD^+.

19.5 Write an equation to illustrate the following:
a. Oxidation of NADH
b. Reduction of Q
c. Oxidation of $FMNH_2$

19.6 What is the function of the cytochromes in the electron transport chain?

19.7 Indicate whether the following represent oxidation or reduction reactions:
a. 2 Cyt c $(Fe^{3+}) + 2e^- \longrightarrow$ 2 Cyt c (Fe^{2+})
b. 2 Cyt a $(Fe^{2+}) \longrightarrow$ 2 Cyt a $(Fe^{3+}) + 2e^-$

19.8 Complete the blanks in the following series of reactions in the electron transport chain:
a. $FMNH_2$ + ______ $\longrightarrow$ QH_2 + ______
*b. 2 Cyt b (Fe^{2+}) + ______ $\longrightarrow$ 2 Cyt c_1 (Fe^{2+}) + ______
c. $FADH_2$ + Q $\longrightarrow$ ______ + ______

19.9 a. What is the end product of the electron transport chain?
b. How many ATP are produced when NADH is oxidized?
c. How many ATP are produced when $FADH_2$ is oxidized?

Glycolysis: Oxidation of Glucose *(Goal 19.3)*

19.10 a. What is the starting reactant for glycolysis?
b. What is the end product of glycolysis under aerobic conditions?

19.11 The first three steps of glycolysis involve the phosphorylation of glucose.
a. How is ATP used in these initial steps?
b. How many ATP molecules are used?

19.12 a. What is the coenzyme used in glycolysis?
b. How does direct phosphorylation account for the production of ATP in glycolysis?

***19.13** The NADH + H^+ in glycolysis is produced in the cytoplasm and must be transported across the mitochondrial membranes to be oxidized. How does this condition affect the ATP produced by NADH + H^+ in glycolysis?

Pathways for Pyruvic Acid *(Goal 19.4)*

19.14 a. Under aerobic conditions, how is pyruvic acid used to produce energy?
b. Write the overall equation for the oxidative decarboxylation of pyruvic acid.
c. What is the energy yield in ATP molecules when pyruvic acid undergoes oxidative decarboxylation?

Citric Acid Cycle *(Goal 19.5)*

19.15 a. What is the function of acetyl CoA in the citric acid cycle?
b. What are the 6-carbon compounds in the citric acid cycle?
c. What are the 5-carbon compounds?
d. What are the 4-carbon compounds?
e. How is the number of carbon atoms of the compounds in the citric acid cycle decreased?
f. What are the end products of the citric acid cycle?

***19.16** Why is NAD^+ used for the oxidation of isocitric acid and malic acid, while FAD is the coenzyme used for the oxidation of succinic acid?

19.17 How is the citric acid cycle linked to the production of ATP by the electron transport chain?

19.18 State the acceptor for hydrogen or phosphate in each of the following reactions; state the number of ATP produced.

	Acceptor	ATP
a. Isocitric acid $\longrightarrow$ α-ketoglutaric acid	______	______
b. α-ketoglutaric acid $\longrightarrow$ succinyl CoA	______	______
c. Succinyl CoA $\longrightarrow$ succinic acid	______	______
d. succinic acid $\longrightarrow$ fumaric acid	______	______
e. Malic acid $\longrightarrow$ oxaloacetic acid	______	______

19.19 Write an overall equation for the citric acid cycle showing the number of ATP molecules produced in one turn of the cycle.

19.20 How much ATP is produced when the following numbers of acetyl CoA molecules enter the citric acid cycle?
a. five molecules acetyl CoA
b. two moles acetyl CoA
c. 25 molecules acetyl CoA
*d. 10 moles acetyl CoA

Energy Production from Glucose *(Goal 19.6)*

19.21 Indicate the amount of ATP associated with each of the following conversions:

a. Glucose $\xrightarrow{\text{Aerobic}}$ 2 pyruvic acid

b. 2 Pyruvic acid $\xrightarrow{\text{Aerobic}}$ 2 acetyl CO_2 + 2 CO_2

c. 2 Acetyl CoA $\xrightarrow{\text{Citric acid cycle}}$ 4 CO_2

d. Glucose + 6 O_2 $\xrightarrow{\text{Aerobic}}$ 6 CO_2 + 6 H_2O

e. Glucose $\xrightarrow{\text{Anaerobic}}$ 2 lactic acid

*f. Maltose + 12 O_2 $\longrightarrow$ 12 CO_2 + 12 H_2O

Metabolic Pathways for Fatty Acids *(Goal 19.7)*

19.22 What is the activation step that prepares a fatty acid for oxidation?

19.23 a. What are the four major reactions in β-oxidation?
b. In which steps are hydrogen atoms removed?
c. Why are different coenzymes used?
d. What is the yield of ATP from one cycle of β-oxidation?

19.24 Palmitic acid, $CH_3(CH_2)_{14}COOH$, is a common fatty acid in butter, tallow, and human fat. The activated form of palmitic acid can be written:

$$CH_3(CH_2)_{12}—CH_2—CH_2—\overset{\overset{\displaystyle O}{\|}}{C}—SCcA$$

a. Indicate the α- and β-carbon atoms in the compound.
*b. Write equations for palmitic acid going through one cycle of β-oxidation.

19.25 For palmitic acid (Problem 19.24), give the following information for complete oxidation:
a. What are the total number of carbon atoms in palmitic acid?
b. How many acetyl CoA units will be produced when β-oxidation is complete?
c. How many cycles of β-oxidation are needed to completely oxidize palmitic acid?
d. Account for the ATP yield by completing the following table:

Units Produced from Palmitic Acid

________ acetyl CoA	________ ATP
________ $FADH_2$	________ ATP
________ NADH	________ ATP
Activation step	________ ATP
Total	________ ATP

***19.26** a. How does the excessive breakdown of fats lead to the undesired formation of ketone bodies?
b. What are the dangers of high levels of ketone bodies?

19.27 What is lipogenesis?

Metabolic Pathways for Amino Acids *(Goal 19.8)*

***19.28** Write a transamination reaction for the following amino acids and α-ketoglutaric acid:

a. Alanine $CH_3-\underset{}{\overset{NH_2}{\overset{|}{C}}}H-COOH$

b. Aspartic acid $HOOC-CH_2-\overset{NH_2}{\overset{|}{C}}H-COOH$

19.29 How are amino acids converted to α-ketoacids by oxidative deamination?

19.30 How are amino acids used as fuel in the cells?

19.31 A person has eaten an overabundance of carbohydrate. What metabolic pathways are most likely to occur?

19.32 If a nonessential amino acid is low in a cell, how might it be prepared from intermediates of the citric acid cycle?

Appendix A
Some Useful Equalities for Conversion Factors

Length: *metric (SI):* meter (m)

1 kilometer (km) = 1000 meters

1 meter = 1000 millimeters (mm)

= 100 centimeters (cm)

metric-American:

1 inch = 2.54 centimeters

1 mile = 1.61 kilometers

Volume: *metric:* liter (L); *SI:* cubic meter (m^3)

1 m^3 = 1000 liters

1 liter = 1000 milliliters (mL)

= 1000 cubic centimeters (cm^3)

= 10 deciliters (dL)

1 milliliter = 1 cubic centimeter

metric-American:

1 liter = 1.06 quarts

1 quart = 0.946 liter = 946 milliliters

Mass: *metric:* gram (g); *SI:* kilogram (kg)

1 kilogram = 1000 grams

1 gram = 1000 milligrams (mg)

metric-American:

1 pound = 454 grams

1 kilogram = 2.20 pounds

Density: mass/volume (g/mL or g/L)

density of water (4°C) = 1.00 g/mL

Temperature: *metric:* Celsius (°C); *SI:* Kelvin (K)

$°F = 1.8°C + 32°$

$$°C = \frac{°F - 32}{1.8}$$

$K = °C + 273°C$

Heat and energy: *metric:* calorie (cal); *SI:* joule (J)

1 calorie = 4.18 joules

1 kilocalorie (kcal) = 1000 calories

$$\text{specific heat of water} = \frac{1 \text{ calorie}}{(1 \text{ g})(1°C)}$$

heat of fusion for water = 80 cal/g

heat of vaporization for water = 540 cal/g

Pressure: *metric:* atmosphere (atm); *SI:* pascal (Pa)

1 atmosphere = 760 mm Hg (torr)

1 mole gas (STP) = 22.4 liters

metric-American:

1 atmosphere = 14.7 lb/in^2

Acids and bases:

$K_w = [H^+][OH^-] = 1.0 \times 10^{-14}$

$pH = -\log[H^+]$

Appendix B
Using Percent in Calculations

The term percent means parts per hundred. It represents a comparison of each item to the total items in the group. For example, suppose you have a basket of fruit containing 10 apples and 30 oranges. There is a total number of 40 items (apples and oranges) in this group of items called the whole. To find the percent of apples in the basket, we compare the number of apples to the whole by dividing the number of apples by the whole and multiplying by 100. The percentage of oranges is found in the same way.

Percent apples

$$\frac{\text{parts}}{\text{whole}} = \frac{\text{number of apples}}{\text{total number of fruit}} = \frac{10\ \text{apples}}{40\ \text{fruit}} \times 100 = 25\%\ \text{apples}$$

Percent oranges

$$\frac{\text{parts}}{\text{whole}} = \frac{\text{number of oranges}}{\text{total number of fruit}} = \frac{30\ \text{oranges}}{40\ \text{fruit}} \times 100 = 75\%\ \text{oranges}$$

Note that the total of the percents of all the items is equal to 100% (25% + 75%).

Sample Problem B.1
Calculating a Percentage

In one evening at a restaurant, 42 people ordered pasta, 54 people ordered salmon, and 24 people ordered steak. What was the percent of salmon dinners ordered?

Solution:
The total number of orders was 120 (42 + 54 + 24). The part of the whole that ordered salmon was 54. The percentage is calculated as

$$\frac{\text{parts}}{\text{whole}} = \frac{\text{number of salmon dinners}}{\text{total dinners ordered}} = \frac{54}{120} \times 100 = 45\%$$

That night, 45% of all the dinners ordered were salmon dinners.

Study Check:
In the above problem, what were the percent of pasta dinners and the percent of steak dinners?

Answer: pasta dinners: 35%; steak dinners: 20%

Using Percents as Conversion Factors

Sometimes, percentages are given in a problem. To work with a percent, it is convenient to write it as a conversion factor. To do this, we choose units from the problem to express the numerical relationship of the part to 100 parts of the whole. For example, an athlete might have 18% body fat by weight. We can use a choice of units for weight or mass to write the following conversion factors:

$$\frac{18 \text{ kg body fat}}{100 \text{ kg body mass}} \qquad \frac{18 \text{ lb body fat}}{100 \text{ lb body weight}}$$

If a question is asked about the number of kilograms of body fat in a marathon runner with a total body mass of 48 kg, we would state the units in the factor in kg body fat. Note that the units in the factor for the percentage cancel the given.

$$\underset{\text{given}}{48 \cancel{\text{ kg body mass}}} \times \underset{\text{percent factor}}{\frac{18 \text{ kg body fat}}{100 \cancel{\text{ kg body mass}}}} = 8.6 \text{ kg body fat}$$

If the problem asked about the pounds of body fat in a dancer weighing 126 lb, we would state the factor in lb body fat.

$$126 \cancel{\text{ lb body weight}} \times \frac{18 \text{ lb body fat}}{100 \cancel{\text{ lb body weight}}} = 23 \text{ lb body fat}$$

Sample Problem B.2
Using Percent Factors

A wine contains 13% alcohol by volume. How many milliliters of alcohol are contained in a glass containing 125 mL of the wine?

Solution:
The factor for 13% alcohol can be written using mL units as

$$\frac{13 \text{ mL alcohol}}{100 \text{ mL wine}}$$

The factor is then used with the given to set up the problem.

$$125 \cancel{\text{ mL}} \times \frac{13 \text{ mL alcohol}}{100 \cancel{\text{ mL}} \text{ wine}} = 16 \text{ mL alcohol}$$

Study Check:
There are 85 students in a chemistry class. On the last test, 20% received grades of A. How many A's were given on the test?

Answer: 17 A grades

Appendix C Scientific (Exponential) Notation

Powers of 10

When numbers are very large or very small, it is more convenient to use scientific notation to represent those numbers in **powers of 10.** For example, 1000 is the same as 10 multiplied three times. The three is written as a power of 10 (exponent). Some examples of powers of 10 are listed in Table C.1.

$$1000 = 10 \times 10 \times 10 = 10^{3} \quad \leftarrow \text{Exponent}$$

Table C.1 Some Examples of Positive Powers of 10

$10^5 = 10 \times 10 \times 10 \times 10 \times 10 = 100{,}000$
$10^4 = 10 \times 10 \times 10 \times 10 = 10{,}000$
$10^3 = 10 \times 10 \times 10 = 1000$
$10^2 = 10 \times 10 = 100$
$10^1 = 10$
$10^0 = 1$

We can represent 2400 in scientific notation by converting the number to 2.4 (a value between 1 and 10) and multiplying by 10 three times.

$$2400 = 2.4 \times 1000 = \underbrace{2.4}_{\substack{\text{number}\\\text{between}\\\text{1 and 10}}} \times \underbrace{10^3}_{\substack{\text{power of}\\10}}$$

To convert a large number into scientific notation, the decimal point is moved to the left until it is located after the first digit. The number of places the decimal point was moved becomes the power of 10 (exponent). Suppose we want to write the number 18,000 in scientific notation. We move the decimal point to give 1.8. (Only the significant numbers are retained.)

1 8 0 0 0.

4 3 2 1

Since we had to move the decimal point four places to the left, the power of 10 will be 4.

$$18\ 000 = 1.8 \times 10^4$$

Sample Problem C.1
Writing Large Numbers in Scientific Notation

Express the following numbers in scientific notation:

a. 35,200
b. 5,000,000
c. 8100

Solution:
For numbers larger than 10, we need to move the decimal point from the end of the number to the left until it is located behind the first number.

decimal point moves four places to the left

a. 3 5 2 0 0. $\quad 3.52 \times 10^4$
b. 5, 0 0 0, 0 0 0. $\quad 5 \times 10^6$
c. 8 1 0 0. $\quad 8.1 \times 10^3$

Study Check:
Convert the following numbers to their decimal form:

a. 5.8×10^3
b. 2×10^5

Answer:

a. $5.8 \times 1000 = 5800$
b. $2 \times 100{,}000 = 200{,}000$

Small numbers can also be written using scientific notation. For example, 0.01 or 1/100 is the same as 10^{-2} and 0.001 or 1/1000 is the same as 10^{-3}.

$$0.01 = \frac{1}{100} = \frac{1}{(10)(10)} = \frac{1}{10^2} = 10^{-2}$$

$$0.001 = \frac{1}{1000} = \frac{1}{(10)(10)(10)} = \frac{1}{10^3} = 10^{-3}$$

Some examples of negative powers of 10 are given in Table C.2.

Table C.2 Some Examples of Negative Powers of 10

$$10^{-1} = \frac{1}{10} = \frac{1}{10^1} = 0.1$$

$$10^{-2} = \frac{1}{100} = \frac{1}{(10)(10)} = \frac{1}{10^2} = 0.01$$

$$10^{-3} = \frac{1}{1000} = \frac{1}{(10)(10)(10)} = \frac{1}{10^3} = 0.001$$

$$10^{-4} = \frac{1}{10{,}000} = \frac{1}{(10)(10)(10)(10)} = \frac{1}{10^4} = 0.0001$$

$$10^{-5} = \frac{1}{100{,}000} = \frac{1}{(10)(10)(10)(10)(10)} = \frac{1}{10^5} = 0.00001$$

To convert a small number into scientific notation, the decimal point must be moved to the right until it is located after the first nonzero digit. For example, to write 0.00086 in scientific notation, the decimal point must be moved to the right until it is between the 8 and the 6, 8.6.

0.0 0 0 8 6 — decimal point moves four places to the right.
1 2 3 4

When the decimal point moves four places to the right, it is the same as dividing by 10 four times. This is expressed as a negative exponent in the power of 10.

$$0.00086 = 8.6 \times \frac{1}{10^4} = 8.6 \times 10^{-4}$$

Sample Problem C.2
Writing Small Numbers in Scientific Notation

Write the following numbers in scientific notation:

a. 0.0042
b. 0.00000255
c. 0.000108

Solution:
Moving the decimal to the right converts each number to scientific notation form with negative powers of 10:

a. 0.0 0 4 2 — 4.2×10^{-3}
b. 0.0 0 0 0 0 2 5 5 — 2.55×10^{-6}
c. 0.0 0 0 1 0 8 — 1.08×10^{-4}

Study Check:
Express the following numbers in decimal form:

a. 8.5×10^{-2}
b. 3×10^{-5}

Answer:

a. $8.5 \times 1/100 = 0.085$
b. $3 \times 1/100{,}000 = 0.000\ 03$

Appendix D
Using Your Calculator

A calculator makes it possible to carry out mathematical operations quickly. While there are different kinds of calculators, most have similar procedures. Sometimes, the steps are a little different from what you remember from basic math. Learning to use your calculator correctly can help you be efficient while avoiding mistakes.

Adding and Subtracting

When you need to add or subtract, you may follow these steps:

1. Use the number keys to enter the first number.
2. Press the function key for the correct operation, + to add, − to subtract.
3. Enter the second number.
4. Press the equals key =.
5. Read your answer in the display area.

If there are additional numbers to add or subtract, repeat steps 2, 3, and 4. To give the correct answer, be sure to round off the display answer to give the correct number of significant figures. For addition and subtraction, round off to the last digit where all the numbers have a significant figure.

Sample Problem D.1
Adding and Subtracting on a Calculator

Give the correct answer for the following calculations:

a. 4.82 + 25.3
b. 5.18 − 0.926

Solution:

a.

Press key(s)	Display reads	
4.82	4.82	
+	4.82	
25.3	25.3	
=	30.12	final answer 30.1

Since one of the numbers has significant figures only to the tenths (0.1) place, the correct answer, 30.1, is obtained by rounding off the display answer.

b.

Press key(s)	Display reads	
5.18	5.18	
−	5.18	
0.926	0.926	
=	4.254	final answer 4.25

The final answer, 4.25, is obtained by rounding off the display answer.

Study Check:
Carry out the following calculations:
a. 5.104 + 25.2 − 14.38 **b.** 37 − 85.42

Answer:
a. 15.9 **b.** −48

Multiplying and Dividing

When you need to multiply or divide, you may follow these steps:

1. Use the number keys to enter the first number.
2. Press the function key for the correct operation, × to multiply, ÷ to divide.
3. Enter the second number.
4. Press the equals key =.
5. Read your answer in the display area.

To give the correct answer, be sure to round off the display answer to give the correct number of significant figures. For multiplication and division, round off to give as many digits as the number with the fewest significant figures.

Sample Problem D.2
Multiplying and Dividing on a Calculator

Carry out the following operations:

a. 15.3×0.45 **b.** $\frac{85.1}{44.65}$

Solution:

a.

Press key(s)	Display reads	
1.35	1.35	
×	1.35	
0.87	0.87	
=	1.1745	final answer 1.2

The final answer, 1.2, is obtained by rounding off the display answer to two significant figures.
b. The number above the line is called the numerator; the number below the line is the denominator. The line means to divide the numerator by the denominator.

Press key(s)	Display reads	
85.1	85.1	
÷	85.1	
44.65	44.65	
=	1.9059351	final answer 1.91

The final answer, 1.91, is obtained by rounding off the display answer to three significant figures. (The input number 85.1 has only three, whereas 44.65 has four significant figures.)

Study Check:
Carry out the following calculations:

a. $\dfrac{185 - 22}{34}$

b. $(2 \times 18.0) + (4 \times 30.1)$; the 2 and 4 are exact.

Answer:
a. 4.8 **b.** 156

Positive and Negative Numbers

A negative number has the same magnitude as a positive number but has a negative sign in front. Most positive numbers are written without the positive sign.

positive twenty-five	+25 or 25
negative four point eight	−4.8

To multiply numbers, the rule of signs is used.

$(+1) \times (+1) = +1$
$(+1) \times (-1) = -1$
$(-1) \times (+1) = -1$
$(-1) \times (-1) = +1$

Suppose we want to solve 1.8×-14. To enter a number with a negative sign, use the change sign key, +/−. DO NOT USE THE SUBTRACT OPERATION KEY. This is a multiplication of a negative number, not a subtraction.

Press key(s)	Display reads	
1.8	1.8	
×	1.8	
14	14	
+/−	−14	
=	−25.2	final answer −25

The display answer with its negative sign is rounded to two significant figures to give −25, the final answer.

Chain Calculations

The advantage of the calculator is that you can carry out several mathematical operations in a sequence. For example, suppose you want to solve the following calculation

$$\frac{(18.6)(2.1)}{4.6}$$

The multiplication and division can all be done without writing down any intermediate answer, only the final answer. After each operation press = to review the current progress.

Press key(s)	Display reads	
18.6	18.6	
×	18.6	
2.1	2.1	
=	39.06	
÷	39.06	
4.6	4.6	
=	8.4913043	final answer 8.5

If there are two or more numbers in the denominator, be sure to divide by each, one at a time. Consider the following

$$\frac{(1.45)(26.8)}{(16.8)(4.56)}$$

The problem should be read as 1.45 multiplied by 26.8 divided by 16.8 divided by 4.56. The steps are as follows:

Press key(s)	Display reads	
1.45	1.45	
×	1.45	
26.8	26.8	
=	38.86	
÷	38.86	
16.8	16.8	
=	2.3130952	
÷	2.3130952	
4.56	4.56	
=	0.5072577	final answer 0.507

Sample Problem D.3
Chain Calculations with a Calculator

Perform the following calculations using the calculator:

a. $\frac{(1.6)(85.2)}{125}$

b. $\frac{(68)(1.52)}{(24)(5.16)}$

Solution:

a.

Press key(s)	Display reads	
1.6	1.6	
×	1.6	
85.2	85.2	
=	136.32	
÷	136.32	
125	125	
=	1.09056	final answer = 1.1

b.

Press key(s)	Display reads	
68	68	
×	68	
1.52	1.52	
=	103.36	
÷	103.36	
24	24	
=	4.3066667	
÷	4.3066667	
5.16	5.16	
=	0.8346253	final answer = 0.83

Study Check:

Solve $\frac{(0.485)(-22)}{(0.028)(5.15)}$

Answer: −74

Numbers in Scientific Notation

Some calculators will convert an answer that is a large number or small number into scientific notation. For example, in the display area you may see the following numbers separated by a space or raised.

7.5 04 and 8^{-03}

These numbers have been given in scientific notation. The second part of the number is the power of 10. To write them in proper exponential form, the term × 10 must be inserted.

Display reads		Scientific Notation
7.5 04	=	7.5×10^4
$8\ ^{-03}$	=	8×10^{-3}

To enter numbers in scientific notation, use the EE key or the exp key. For example, the steps to enter the number 2.5×10^4 are:

Press key(s)	Display reads
2.5	2.5
EE (exp)	2.5 00
4	2.5 04

If the number has a negative exponent, use the +/− key to change its sign. For example, 1.8×10^{-6} is entered by

Press key(s)	Display reads
1.8	1.8
EE (exp)	1.8 00
6	1.8 06
+/−	1.8 −06

Sample Problem D.4
Scientific Notation in Calculations

Solve the following using the calculator:

a. $(2.7 \times 10^5) \times (3.6 \times 10^2)$
b. $(3.1 \times 10^{-2}) \times (1.8 \times 10^4)$

Solution:

a. Press key(s)	Display reads	
2.7	2.7	
EE (exp)	2.7 00	
5	2.7 05	
×	2.7 05	
3.6	3.6	
EE (exp)	3.6 00	
2	3.6 02	
=	9.72 07	final answer 9.7×10^7

b. Press key(s)	Display reads	
3.1	3.1	
EE (exp)	3.1 00	
2	3.1 02	
+/−	3.1 −02	
×	3.1 −02	
1.8	1.8	
EE (exp)	1.8 00	
4	1.8 04	
=	5.58 02	final answer 5.6×10^2

Study Check:

Solve $\dfrac{5.2 \times 10^{-6}}{3.8 \times 10^3}$

Answer: Display reads: 1.3684211 −09 final answer 1.4×10^{-9}

Appendix E
Answers to Problems

Chapter 1

1.1 The metric system uses one unit, the meter (m), to measure length; the American system uses several different units including the inch (in), foot (ft), and mile (mi). For volume, the metric system uses the liter (L), while the American system uses many volume units including teaspoon (tsp), pint (pt), quart (qt), and gallon (gal). Mass is measured in grams (g) in the metric system; the American system uses ounces (oz) and pounds (lb).

1.2 **a.** meter; length **b.** gram; mass **c.** milliliter; volume **d.** meter; length

1.3 In the metric system, prefixes are used to indicate the decimal multiples or fractions of the base unit.

1.4 **a.** mg **b.** dL **c.** km **d.** kg

1.5 **a.** centimeter **b.** millimeter **c.** deciliter **d.** kilogram

1.6 **a.** 0.01 (1/100) **b.** 1000 **c.** 0.001 (1/1000) **d.** 0.1 (1/10)

1.7 **a.** deci- **b.** deka- **c.** kilo- **d.** centi-

1.8 **a.** milli-, centi-, kilo-
b. micro-, milli-, centi-
c. milli-, deci-, deka-, mega-
d. mg, cg, dg, g, kg
e. mm, dm, m, hm, km
f. mL, cL, L, kL

1.9 **a.** 100 cm **b.** 1000 m **c.** 0.001 m **d.** 1000 mL **e.** 10 dL **f.** 0.1 L **g.** 0.001 kg **h.** 1000 mg

1.10 It is important to count significant figures when you are working with measured numbers or quantities that are derived from measured numbers. Counting numbers and defined relationships are exact and their significant figures are not considered.

1.11 **a.** measured **b.** exact **c.** exact **d.** exact **e.** measured

1.12 Zeros are significant when they are part of a measured number. If they precede any recorded number or occur in large numbers, they are not significant.

1.13 **a.** 5 **b.** 2 **c.** 2 **d.** 5 **e.** 4

1.14 **a.** 1.9 **b.** 184.2 **c.** 0.00474 **d.** 88 **e.** 1.83

1.15 **a.** 1.6 **b.** 27.6 **c.** 3.5 **d.** 0.005 **e.** 0.0055

1.16 **a.** 53.54 cm **b.** 127.6 g **c.** 121.5 mL **d.** 0.50 L

1.17 Either factor in the equality can be in the numerator and the other in the denominator.

1.18 **a.** 3 ft = 1 yd: 1 yd/3 ft and 3 ft/1 yd
b. 1 min = 60 sec: 1 min/60 sec and 60 sec/1 min
c. \$1 = 4 quarters: \$1/4 quarters and 4 quarters/\$1
d. 1 gal = 4 qt: 1 gal/4 qt and 4 qt/1 gal
e. 1 mile = 5280 ft: 1 mile/5280 ft and 5280 ft/1 mile
f. 7 days = 1 week: 7 days/1 week. 1 week/7 days

1.19 **a.** 100 cm = 1 m: 1 m/100 cm and 100 cm/1 m
b. 1000 mg = 1 g: 1 g/1000 mg and 1000 mg/1 g
c. 1 cm = 10 mm: 1 cm/10 mm and 10 mm/1 cm
d. 1 L = 1000 mL: 1 L/1000 mL and 1000 mL/1 L
e. 1 dL = 100 mL: 1 dL/100 mL and 100 mL/1 dL
f. 1000 g = 1 kg: 1 kg/1000 g and 1000 g/1 kg

1.20 **a.** 2.54 cm = 1 inch: 1 in/2.54 cm and 2.54 cm/1 in
b. 2.20 lb = 1 kg: 1 kg/2.20 lb and 2.20 lb/1 kg
c. 946 mL = 1 qt: 1 qt/946 mL and 946 mL/1 qt
d. 454 g = 1 lb: 1 lb/454 g and 454 g/1 lb
e. 1.06 qt = 1 L: 1 L/1.06 qt and 1.06 qt/1 L

1.21 You could tell your friend to use conversion factors and be sure that the units cancel out to leave the desired unit in the answer.

1.22 **a.** 8.0 yd **b.** 900 or 9.0×10^2 sec **c.** 0.88 gal **d.** 14 games **e.** 13,000 ft

1.23 **a.** 1.75 m **b.** 5.5 L **c.** 55 g **d.** 0.8 g **e.** 85 mL **f.** 2.84 g

1.24 **a.** 710 mL **b.** 75.0 kg **c.** 495 mm **d.** 20 gal

1.25 **a.** 110 g **b.** 2400 mL **c.** 1.8 ft **d.** 9.5 kg **e.** 200 mm

1.26 **a.** 66. gal **b.** 3 tablets **c.** 1800 mg **d.** 187 lb **e.** 8 tablets **f.** 0.65 mL

1.27 Dear Cousin:
If you want the density of that diamond, you'll need to buy a balance and a graduated cylinder. You can measure the mass in grams of that diamond on the balance. Since its shape is probably irregular, you first place some water in the cylinder and record its level. Then you place the diamond under the water and record the new water level. Then subtract the first volume measurement from the second to find the volume of the diamond. Don't forget the units! Density is calculated by dividing the mass you recorded by the volume of the diamond. Your answer will have units of g/cm^3 or g/mL. If you get 3.51 g/cm^3, you probably have a real diamond and should take it to a jewelry store. If the density doesn't match, your diamond may have just been a piece of glass. Sorry. Well, good luck!
Sincerely, Cousin Emma

1.28 **a.** 1.2 g/mL **b.** 4.4 g/mL **c.** 3.10 g/mL **d.** 1.28 g/mL **e.** 1.01 g/mL

1.29 **a.** 210 g **b.** 575 g **c.** 62 oz **d.** 250 lb **e.** 159 mL

1.30 **a.** 1.030 **b.** 1.13 **c.** 0.85 g/mL **d.** 220 g **e.** 382 mL **f.** 2.0 qt

Chapter 2

2.1 Tell her that in the United States, we still use the Fahrenheit temperature scale. In Fahrenheit, normal body temperature is 98.6°F. On a Celsius thermometer, she would have a temperature of 37.7°C. Therefore, on either temperature scale, she has only a mild fever.

2.2 **a.** 98.6°F **b.** 77°F **c.** 311°F **d.** 18.5°C **e.** 43°C **f.** −32°C

2.3 **a.** 335 K **b.** 246 K **c.** 295 K **d.** 272°C **e.** −49°C **f.** 1116°F

2.4 **a.** 41°C
b. Yes. The temperature is 101.7°F
c. 63°C (62.8°C)

2.5 **a.** 69.1°F **b.** 294 K

2.6 As a roller coaster climbs a hill, it slows as kinetic energy is changed into potential energy. At the top of the hill, it has all potential energy. As the coaster goes down the other side, potential energy is changed to kinetic energy and the coaster speeds up.

2.7 **a.** potential **b.** kinetic **c.** potential **d.** potential **e.** potential **f.** potential **g.** kinetic **h.** kinetic **i.** kinetic

2.8 **a.** electrical energy changes to heat energy
b. radiant energy changes to electrical energy
c. chemical energy changes to mechanical energy
d. radiant energy changes to heat energy
e. electrical energy changes to radiant energy

2.9 The pans are made from metals (copper, iron, aluminum) that conduct heat and heat up quickly. The handles are made of wood, so they do not conduct heat.

2.10 The mass, the specific heat, and the change in temperature.

2.11 **a.** 550 cal **b.** 4500 J **c.** 130 cal

2.12 30 kcal

2.13 **a.** 6300 cal **b.** 110,000 J **c.** 0.81 kcal

2.14 **a.** gas **b.** gas **c.** solid **d.** gas **e.** gas **f.** liquid

2.15 **a.** evaporation **b.** freezing **c.** boiling **d.** sublimation **e.** condensation **f.** melting

2.16 **a.** As sweat (water) evaporates from the body, heat is lost, and the body is cooled.
b. At higher temperatures, water evaporates faster from the wet clothes.
c. As the substance evaporates, so much heat is lost that the skin becomes numb from the cold.

2.17

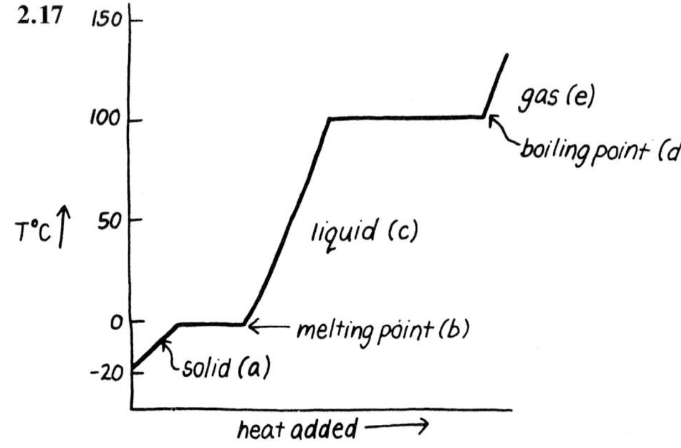

2.18

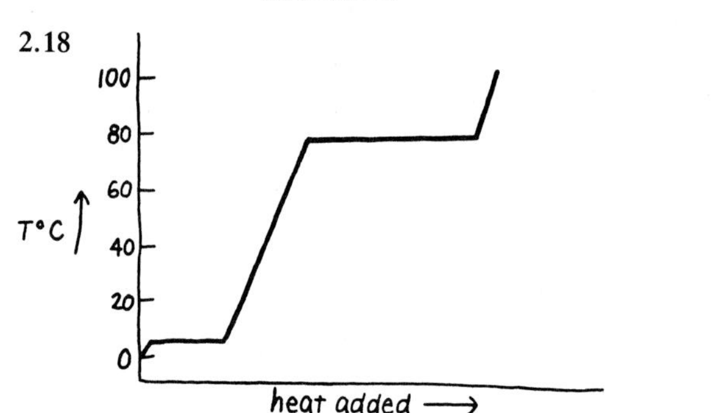

2.19

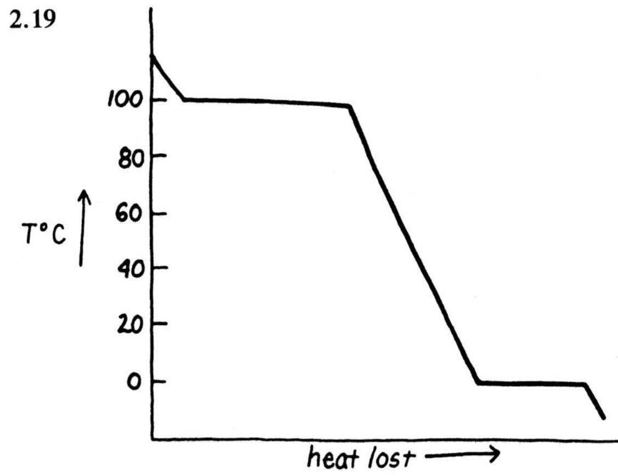

2.20 **a.** 9200 cal **b.** 25,000 J **c.** 7300 cal

2.21 22 kcal

2.22 **a.** 27.0 kcal **b.** 12,300 cal **c.** 58 kcal

2.23 99,000 cal

2.24 82.8 kcal

2.25 **a.** As the water freezes, heat is released into the air surrounding the trees. The temperature will not go below zero until all the water has frozen. **b.** 480 kcal

2.26 **a.** 5 kcal **b.** 210 kcal **c.** 25 kcal

2.27 **a.** 68 kcal **b.** 18 g carbohydrate **c.** 275 kcal **d.** 37 g fat

2.28 208 kcal

2.29 1800 kcal

2.30 925 kcal

2.31 **a.** 38 g protein, 113 g carbohydrate, 44 g fat **b.** 68 g protein, 203 g carbohydrate, 80 g fat **c.** 98 g protein, 293 g carbohydrate, 116 g fat

2.32 2.0 hr

Chapter 3

3.1 **a.** Cu **b.** Si **c.** K **d.** Co **e.** Fe **f.** Ba **g.** Pb **h.** Ne **i.** O **j.** Li **k.** S **l.** Al **m.** Unq **n.** Unh **o.** H

3.2 **a.** carbon **b.** chlorine **c.** iodine **d.** phosphorus **e.** silver **f.** fluorine **g.** argon **h.** zinc **i.** magnesium **j.** sodium **k.** helium **l.** nickel **m.** mercury **n.** calcium **o.** unnilpentium

3.3 **a.** electron **b.** proton **c.** electron **d.** neutron **e.** electron **f.** neutron **g.** proton and neutron **h.** electron

3.4 **a.** atomic number **b.** mass number—atomic number **c.** mass number **d.** atomic number

3.5 **a.** atomic number 15, mass number 31, P
b. atomic number 35, mass number 80, Br
c. atomic number 11, mass number 23, Na
d. atomic number 26, mass number 56, Fe
e. atomic number 8, mass number 18, O

3.6 **a.** 20 neutrons **b.** 20 neutrons **c.** 24 neutrons **d.** 36 neutrons

3.7

	Atomic Number	Mass Number	Protons	Neutrons
a.	13	27	13	14
b.	12	24	12	12
c.	6	13	6	7
d.	16	31	16	15
e.	16	34	16	18
f.	20	42	20	22

	Electrons	Name	Symbol
a.	13	Aluminum	Al
b.	12	Magnesium	Mg
c.	6	Carbon	C
d.	16	Sulfur	S
e.	16	Sulfur	S
f.	20	Calcium	Ca

3.8 **a.** 13 protons, 14 neutrons, and 13 electrons
b. 24 protons, 28 neutrons, and 24 electrons
c. 16 protons, 18 neutrons, and 16 electrons
d. 26 protons, 30 neutrons, and 26 electrons
e. 1 proton, 1 neutron, and 1 electron
f. 30 protons, 40 neutrons, and 30 electrons

3.9 **a.** ${}^{44}_{20}Ca$ **b.** ${}^{59}_{28}Ni$ **c.** ${}^{24}_{11}Na$ **d.** ${}^{80}_{35}Br$ **e.** ${}^{35}_{17}Cl$ **f.** ${}^{109}_{47}Ag$

3.10 **a.** ${}^{32}_{16}S$ ${}^{33}_{16}S$ ${}^{34}_{16}S$ ${}^{36}_{16}S$
b. They all have the same number of protons and electrons. (They all have the same atomic number.)
c. They have different numbers of neutrons—different mass numbers.
d. The atomic weight of sulfur is the weighted average mass of all its isotopes.

3.11 **a.** 16.0 **b.** 14.0 **c.** 55.8 **d.** 1.0 **e.** 24.3 **f.** 35.5 **g.** 23.0 **h.** 31.0

3.12 The elements in a group have similar chemical and physical properties. Each horizontal row is a period of elements that differ in physical and chemical properties.

3.13 **a.** period **b.** group **c.** group **d.** period **e.** group **f.** period

3.14 **a.** Li **b.** H **c.** S **d.** Ca **e.** Kr **f.** I **g.** Ba

3.15 **a.** group **b.** period **c.** group **d.** period **e.** group **f.** group

3.16 **a.** Metals are shiny, ductile, and malleable. They are usually good conductors of heat and electricity, and melt at high temperatures. Nonmetals are usually not shiny, ductile, or malleable. They are poor conductors of heat and electricity; they usually have low melting points and low densities.
b. Metals are located to the left of the zig-zag line; nonmetals are on the right.

3.17 **a.** nonmetal **b.** metal **c.** metal **d.** nonmetal **e.** metal **f.** nonmetal **g.** nonmetal **h.** nonmetal **i.** metal **j.** metal

3.18 **a.** metal **b.** metal **c.** metal **d.** nonmetal **e.** metal **f.** nonmetal **g.** metal **h.** metal

3.19 **a.** 2, 4 **b.** 2, 8, 8 **c.** 2, 8, 3 **d.** 2, 8, 6 **e.** 2, 8, 8, 1 **f.** 2, 8, 5 **g.** 2, 5 **h.** 2, 8

3.20 **a.** 2, 8, 3 **b.** 2, 8, 8 **c.** 2, 6

3.21 **a.** Li **b.** Mg **c.** H **d.** Cl **e.** O

3.22 **a.** absorb **b.** emit

3.23 **a.** B 2, 3 Al 2, 8, 3 **b.** 3 **c.** 3A

3.24 **a.** $2e^-$, 2A **b.** $7e^-$, 7A **c.** $6e^-$, 6A **d.** $5e^-$, 5A **e.** $1e^-$, 1A **f.** $8e^-$, 8A **g.** $4e^-$, 4A **h.** $8e^-$, 8A

3.25 Mg, Ca, and Sr are all in group 2A, and all have 2 electrons in their outer energy levels. Thus, they would be expected to have similar chemical and physical properties.

3.26 **a.** 2 **b.** 6 **c.** 18 **d.** 2 **e.** 2

3.27 **a.** $1s^2\ 2s^2\ 2p^6\ 3s^2$
b. $1s^2\ 2s^2\ 2p^6\ 3s^2\ 3p^3$
c. $1s^2\ 2s^2\ 2p^6\ 3s^2\ 3p^6$
d. $1s^2\ 2s^2\ 2p^6\ 3s^2\ 3p^4$
e. $1s^2\ 2s^2\ 2p^6\ 3s^2\ 3p^5$
f. $1s^2\ 2s^2\ 2p^6\ 3s^2\ 3p^6\ 4s^1$

3.28 **a.** H **b.** N **c.** Na **d.** F **e.** S **f.** Ca

Chapter 4

4.1 **a.** 2, 5; 5 valence electrons
b. 2, 6; 6 valence electrons
c. 2, 8, 8; 8 valence electrons
d. 2, 8, 8, 1; 1 valence electron
e. 2, 8, 6; 6 valence electrons
f. 2, 8, 1; 1 valence electron
g. 2, 8, 3; 3 valence electrons
h. 2, 8, 7; 7 valence electrons

4.2 **a.** ·S̈·, 6A **b.** ·Ṅ·, 5A **c.** Ca·, 2A **d.** Na·, 1A **e.** K·, 1A **f.** ·Ċ·, 4A **g.** ·Ö·, 6A **h.** :F̈·, 7A **i.** Li· 1A **j.** :C̈l·, 7A

4.3 **a.** 2A **b.** 5A **c.** 7A **d.** 1A **e.** 3A

4.4 **a.** 2, 8; stable **b.** 2, 6; gain electrons **c.** 2, 1; lose electrons **d.** 2, 8, 8; stable **e.** 2, 8, 5; gain electrons

4.5 **a.** lose 2 **b.** gain 3 **c.** gain 1 **d.** lose 1 **e.** lose 3 **f.** gain 2 **g.** lose 2 **h.** gain 1 **i.** lose 1 **j.** gain 3

4.6 **a.** F^- **b.** Mg^{2+} **c.** P^{3-} **d.** Ne **e.** K^+ **f.** Ca^{2+}

4.7 **a.** 2, 8 **b.** 2, 8, 8 **c.** 2, 8, 8 **d.** 2, 8, 8 **e.** 2, 8

4.8 **a.** Cl^- **b.** Mg^{2+} **c.** K^+ **d.** O^{2-} **e.** Al^{3+} **f.** F^- **g.** Ca^{2+} **h.** S^{2-} **i.** Na^+ **j.** Li^+

4.9 **a.** sodium **b.** nitride **c.** fluoride **d.** chloride **e.** oxide **f.** aluminum **g.** calcium **h.** sulfide

4.10 **a.** iron(II); or ferrous
b. copper(II); or cupric
c. iron(III); or ferric
d. lead(IV); or plumbic
e. silver
f. copper(I); or cuprous
g. zinc
h. tin(II); or stannous

4.11 **a.** K· + ·C̈l: ⟶ K^+ :C̈l:$^-$ ⟶ KCl

b. Mg· + ·C̈l: + ·C̈l: ⟶ Mg^{2+} 2[:C̈l:$^-$] ⟶ $MgCl_2$

c. Na· + Na· + Na· + ·N̈· ⟶ $3Na^+$:N̈:$^{3-}$ ⟶ Na_3N

d. Mg· + ·S̈: ⟶ Mg^{2+} :S̈:$^{2-}$ ⟶ MgS

e. ·Al· + ·C̈l: + ·C̈l: + ·C̈l: ⟶ Al^{3+} 3[:C̈l:$^-$] ⟶ $AlCl_3$

4.12 **a.** Na_2O **b.** $AlCl_3$ **c.** $BaCl_2$ **d.** MgO **e.** $ZnCl_2$ **f.** Al_2S_3 **g.** Li_2S **h.** KI **i.** $FeCl_3$ **j.** CuS **k.** Cu_2O

4.13 **a.** Na_2S **b.** Al_2O_3 **c.** $CaCl_2$ **d.** $BaBr_2$ **e.** LiCl **f.** K_3N

4.14 **a.** aluminum oxide **b.** calcium chloride **c.** sodium oxide **d.** magnesium nitride **e.** sodium sulfide **f.** potassium phosphide **g.** magnesium oxide **h.** lithium bromide **i.** iron(II) chloride **j.** tin(IV) chloride

4.15 **a.** K^+ **b.** Cu^{2+} **c.** Fe^{3+} **d.** Fe^{2+} **e.** Ag^+ **f.** Zn^{2+}

4.16

	Cl^-	O^{2-}	N^{3-}
K^+	KCl potassium chloride	K_2O potassium oxide	K_3N potassium nitride
Cu^{2+}	$CuCl_2$ copper(II) chloride	CuO copper(II) oxide	Cu_3N_2 copper(II) nitride
Mg^{2+}	$MgCl_2$ magnesium chloride	MgO magnesium oxide	Mg_3N_2 magnesium nitride
Fe^{3+}	$FeCl_3$ iron(III) chloride	Fe_2O_3 iron(III) oxide	FeN iron(III) nitride

4.17 **a.** Sn^{2+} **b.** Fe^{2+} **c.** Cu^+ **d.** Ag^+ **e.** Fe^{3+}

4.18 **a.** iron(II) chloride **b.** copper(II) oxide **c.** iron(III) sulfide **d.** copper(I) chloride **e.** silver phosphide **f.** sodium sulfide **g.** zinc fluoride **h.** aluminum phosphide

4.19 **a.** $MgCl_2$ **b.** Na_2S **c.** Cu_2O **d.** Zn_3P_2 **e.** Ba_3N_2 **f.** Fe_2O_3 **g.** BaF_2 **h.** $AlCl_3$ **i.** Ag_2S **j.** $CuCl_2$

4.20 **a.** :B̈r:B̈r: **b.** H:S̈: with H below S **c.** H:F̈: **d.** :F̈:Ö: with :F: below O

e. :C̈l:N̈:C̈l: with :C̈l: below N **f.** :C̈l:C:C̈l: with :C̈l: above and below C

g. H:Ö: with H below O **h.** :F̈:Si:F̈: with :F̈: above and below Si

4.21 H_2, N_2, O_2, F_2, Cl_2, Br_2, I_2

4.22 **a.** :Ö::S:Ö: **b.** H:C:H with :O: double-bonded (::) above C **c.** :N:::N: **d.** :C̈l:N::N:C̈l:

4.23 **a.** $\overset{\delta+}{H}—\overset{\delta-}{F}$ **b.** $\overset{\delta+}{C}—\overset{\delta-}{Cl}$ **c.** $\overset{\delta+}{N}—\overset{\delta-}{O}$ **d.** $\overset{\delta-}{O}—\overset{\delta+}{H}$ **e.** $\overset{\delta-}{O}—\overset{\delta+}{S}$

4.24 **a.** covalent **b.** none **c.** covalent **d.** ionic **e.** polar **f.** ionic **g.** polar **h.** ionic **i.** covalent **j.** polar

4.25 **a.** hydrogen **b.** carbon tetrabromide **c.** sulfur dichloride **d.** hydrogen monofluoride **e.** nitrogen triiodide **f.** carbon disulfide **g.** diphosphorus pentoxide **h.** dichlorine monoxide

4.26 **a.** CCl_4 **b.** CO **c.** PCl_3 **d.** N_2O_4 **e.** OF_2 **f.** HCl **g.** N_2O **h.** Cl_2

4.27 **a.** sulfate **b.** carbonate **c.** phosphate **d.** nitrate **e.** hydroxide **f.** sulfite

4.28 **a.** HCO_3^- **b.** NH_4^+ **c.** PO_4^{3-} **d.** HSO_4^- **e.** NO_2^- **f.** SO_3^{2-} **g.** OH^- **h.** PO_3^{3-}

4.29

a.	$Na_2(CO_3)$	sodium carbonate
b.	$(NH_4)Cl$	ammonium chloride
c.	$Li_3(PO_4)$	lithium phosphate
d.	$Cu((NO_2))_2$	copper(II) nitrite; or cupric nitrite
e.	$Fe(SO_3)$	iron(II) sulfite; or ferrous sulfite
f.	$K(OH)$	potassium hydroxide
g.	$Na(NO_3)$	sodium nitrate
h.	$Cu(CO_3)$	copper(II) carbonate; or cupric carbonate
i.	$Na(HCO_3)$	sodium bicarbonate; or sodium hydrogen carbonate
j.	$Ba(SO_4)$	barium sulfate

4.30

	OH^-	NO_2^-	CO_3^{2-}	HSO_4^-	PO_4^{3-}
Li^+	$LiOH$ lithium hyroxide	$LiNO_2$ lithium nitrite	Li_2CO_3 lithium carbonate	$LiHSO_4$ lithium hydrogen sulfate	Li_3PO_4 lithium phosphate
Cu^{2+}	$Cu(OH)_2$ copper(II) hydroxide	$Cu(NO_2)_2$ copper(II) nitrite	$CuCO_3$ copper(II) carbonate	$Cu(HSO_4)_2$ copper(II) hydrogen sulfate	$Cu_3(PO_4)_2$ copper(II) phosphate
Ba^{2+}	$Ba(OH)_2$ barium hydroxide	$Ba(NO_2)_2$ barium nitrite	$BaCO_3$ barium carbonate	$Ba(HSO_4)_2$ barium hydrogen sulfate	$Ba_3(PO_4)_2$ barium phosphate
NH_4^+	NH_4OH ammonium hydroxide	NH_4NO_2 ammonium nitrite	$(NH_4)_2CO_3$ ammonium carbonate	NH_4HSO_4 ammonium hydrogen sulfate	$(NH_4)_3PO_4$ ammonium phosphate
Al^{3+}	$Al(OH)_3$ aluminum hydroxide	$Al(NO_2)_3$ aluminum nitrite	$Al_2(CO_3)_3$ aluminum carbonate	$Al(HSO_4)_3$ aluminum hydrogen sulfate	$AlPO_4$ aluminum phosphate
Pb^{4+}	$Pb(OH)_4$ lead(IV) hydroxide	$Pb(NO_2)_4$ lead(IV) nitrite	$Pb(CO_3)_2$ lead(IV) carbonate	$Pb(HSO_4)_4$ lead(IV) hydrogen sulfate	$Pb_3(PO_4)_4$ lead(IV) phosphate

4.31 **a.** $Ba(OH)_2$ **b.** Na_2SO_4 **c.** $Fe(NO_3)_2$ **d.** $Zn_3(PO_4)_2$ **e.** $SiCl_4$ **f.** $AlCl_3$ **g.** $(NH_4)_2O$ **h.** $Mg(HCO_3)_2$ **i.** $NaNO_2$ **j.** Cu_2SO_4

4.32 **a.** aluminum oxide **b.** potassium carbonate **c.** oxygen difluoride **d.** sodium phosphate **e.** ammonium sulfate **f.** nitrogen **g.** nitrogen tribromide **h.** iron(III) nitrate **i.** magnesium oxide **j.** copper(II) chloride

Chapter 5

5.1 **a.** two aluminum; three oxygen
b. one aluminum; three oxygen; three hydrogen
c. two aluminum; three sulfur; twelve oxygen
d. two nitrogen; eight hydrogen; one carbon; three oxygen
e. one magnesium; two hydrogen; two oxygen
f. 14 carbon; 29 hydrogen, 1 nitrogen, 4 oxygen, 1 sulfur

5.2 **a.** 74.6 amu **b.** 73.8 amu **c.** 183.1 amu **d.** 132.1 amu **e.** 46.0 amu **f.** 444.0 amu

5.3 **a.** 23.0 g/mole **b.** 35.5 g/mole **c.** 207.2 g/mole **d.** 63.5 g/mole **e.** 55.8 g/mole **f.** 24.3 g/mole **g.** 1.0 g/mole **h.** 12.0 g/mole

5.4 **a.** 42.0 g/mole **b.** 322.0 g/mole **c.** 80.0 g/mole **d.** 78.0 g/mole **e.** 84.0 g/mole **f.** 430.0 g/mole

5.5 **a.** 80 g **b.** 3.9 g **c.** 270 g **d.** 54 g **e.** 59 g **f.** 0.54 g **g.** 39 g **h.** 78 g

5.6 **a.** 8.5 g **b.** 120 g **c.** 66 g **d.** 180 g **e.** 350 g **f.** 15 g

5.7 600 g

5.8 11 g

5.9 11 g

5.10 7.1 g

5.11 **a.** 0.463 mole **b.** 0.630 mole **c.** 0.833 mole **d.** 2.7 mole **e.** 12 mole **f.** 0.0534 mole

5.12 0.0869 mole

5.13 0.178 mole

5.14 336 mole

5.15 **a.** 0.500 mole **b.** 10 mole **c.** 2.00 mole **d.** 1.6 mole **e.** 2.00 mole **f.** 0.00107 mole

5.16 12 mole

5.17 2330 mole

5.18 0.0087 mole

5.19 **a.** physical **b.** chemical **c.** physical **d.** chemical **e.** physical **f.** chemical **g.** chemical

5.20 **a.** two molecules of nitrogen monoxide react with one molecule of oxygen to produce two molecules of nitrogen dioxide
b. two molecules of dihydrogen sulfide react with three molecules of oxygen to produce two molecules of sulfur dioxide and two molecules of water
c. one molecule of ethanol reacts with three molecules of oxygen to produce two molecules of carbon dioxide and three molecules of water
d. 1 formula unit of potassium sulfate reacts with two formula units of silver nitrate to produce 1 formula unit of silver sulfate and 2 formula units of potassium nitrate

5.21 **a.** $4NH_3 + 3O_2 \longrightarrow 2N_2 + 6H_2O$
b. $4Fe + 3O_2 \longrightarrow 2Fe_2O_3$
c. $2C_3H_8O + 9O_2 \longrightarrow 6CO_2 + 8H_2O$

5.22 **a.** 2Na, 2Cl **c.** 4P, 16O, 12H
b. 2N, 8H, 4O **d.** 5C, 16O, 12H

5.23 **a.** $N_2 + O_2 \longrightarrow 2NO$
b. $2HgO \longrightarrow 2Hg + O_2$
c. $4Fe + 3O_2 \longrightarrow 2Fe_2O_3$
d. $2Na + Cl_2 \longrightarrow 2NaCl$
e. $2Cu_2O + O_2 \longrightarrow 4CuO$

5.24 **a.** $2Al + 3Cl_2 \longrightarrow 2AlCl_3$
b. $P_4 + 5O_2 \longrightarrow P_4O_{10}$
c. $C_3H_8 + 5O_2 \longrightarrow 3CO_2 + 4H_2O$
d. $Sb_2S_3 + 6HCl \longrightarrow 2SbCl_3 + 3H_2S$
e. $Fe_2O_3 + 3C \longrightarrow 2Fe + 3CO$

5.25 **a.** $Mg + 2AgNO_3 \longrightarrow Mg(NO_3)_2 + 2Ag$
b. $CuCO_3 \longrightarrow CuO + CO_2$
c. $2Al + 3CuSO_4 \longrightarrow 3Cu + Al_2(SO_4)_3$
d. $Pb(NO_3)_2 + 2NaCl \longrightarrow PbCl_2 + 2NaNO_3$
e. $Zn + H_2SO_4 \longrightarrow ZnSO_4 + H_2$
f. $Al_2(SO_4)_3 + 6KOH \longrightarrow 2Al(OH)_3 + 3K_2SO_4$
g. $K_2SO_4 + BaCl_2 \longrightarrow BaSO_4 + 2KCl$
h. $CaCO_3 \longrightarrow CaO + CO_2$

5.26 **a.** exothermic **b.** endothermic **c.** exothermic

5.27 **a.** One mole NaCl reacts with one mole $AgNO_3$ to produce one mole AgCl and one mole $NaNO_3$.
b. Four mole Al react with three mole oxygen to produce two mole Al_2O_3

5.28 **a.** $N_2 + O_2 = 60.0$ g; $2 \times NO = 60.0$ g
b. $CaCO_3 = 100.1$ g; $CaO + CO_2 = 100.1$ g
c. $2 \times SO_2 = 128.2$ g $2 \times SO_3 = 160.2$ g
$O_2 =$ 32.0 g
160.2 g
d. $4 \times Al = 108$ g $2 \times Al_2O_3 = 204$ g
$3 \times O_2 =$ 96 g
204 g

5.29 **a.** 1 mole NaCl/1 mole $NaNO_3$;
1 mole $NaNO_3$/1 mole NaCl
1 mole NaCl/1 mole $AgNO_3$;
1 mole $AgNO_3$/1 mole NaCl
1 mole NaCl/1 mole AgCl;
1 mole AgCl/1 mole NaCl
1 mole $AgNO_3$/1 mole AgCl;
1 mole AgCl/1 mole $AgNO_3$
1 mole $AgNO_3$/1 mole $NaNO_3$;
1 mole $NaNO_3$/1 mole $AgNO_3$
1 mole AgCl/1 mole $NaNO_3$;
1 mole $NaNO_3$/1 mole AgCl
b. 4 mole Al/3 mole O_2;
3 mole O_2/4 mole Al
3 mole O_2/2 mole Al_2O_3;
2 mole Al_2O_3/3 mole O_2
4 mole Al/2 mole Al_2O_3;
2 mole Al_2O_4/4 mole Al

5.30 **a.** 1.0 mole S **b.** 5.0 mole Cu_2S
c. 200 g Cu_2S

5.31 **a.** 3.0 mole H_2 **b.** 15 g H_2 **c.** 68 g NH_3

5.32 **a.** 6.0 mole O_2 **b.** 140 g N_2 **c.** 80.0 g O_2
d. 0.12 lb H_2O

5.33 **a.** 5.56 mole CO_2 **b.** 123 g ethanol

5.34 **a.** $C_2H_6O + O_2 \longrightarrow CO_2 + H_2O$
b. $C_2H_6O + 3O_2 \longrightarrow 2CO_2 + 3H_2O$
c. 12 mole O_2 **d.** 28.2 g H_2O **e.** 96 g O_2

Chapter 6

6.1 All are isotopes of K, with $19p^+$ and $19e^-$. Each has a different number of neutrons, and the radioisotopes, ^{40}K and ^{41}K, do not have stable nuclei and will undergo radioactive decay. The nonradioactive isotope, ^{39}K, is stable; it will not decay.

6.2

Medical use	Isotope symbol	Mass number
Spleen imaging	$^{51}_{24}Cr$	51
Malignancies	$^{60}_{27}Co$	60
Blood volume	$^{59}_{26}Fe$	59
Hyperthyroidism	$^{131}_{53}I$	131
Leukemia treatment	$^{32}_{15}P$	32

Number of protons	Number of neutrons
24	27
27	33
26	33
53	78
15	17

6.3 **a.** α or $^{4}_{2}He$ **b.** $^{1}_{0}n$ **c.** β or $^{0}_{-1}e$ **d.** $^{1}_{1}H$ **e.** γ

6.4 **a.** e **b.** He **c.** n **d.** γ **e.** H

6.5 Since β particles are smaller with a higher energy than α particles, they penetrate farther into solid before they are stopped.

6.6 Ionizing radiation produces ions or radicals within the cells that disrupt activity so much that the cell may not continue to function.

6.7 Because cancer cells are immature cells that divide rapidly, they are more susceptible to the effects of ionizing radiation.

6.8 Certain substances placed between the radioactive source and the body will absorb some or all the radiation, thereby providing shielding. Clothing and skin shield a person from alpha particles. Heavy clothing provides shielding from beta particles while lead or concrete is needed to provide protection from gamma rays. Exposure is also minimized by spending as little time as possible near the source of radioactivity. Increasing the distance that a person stands from a radioactive source decreases the intensity of the radiation received.

6.9 A lead apron protects the rest of the patient's body from radiation while x-rays of the teeth are taken.

6.10 Make sure that both the total of mass numbers and the total of atomic numbers of the reactants and products are equal.

6.11 **a.** $^{208}_{84}Po \longrightarrow {}^{4}_{2}He + {}^{204}_{82}Pb$

b. $^{232}_{90}Th \longrightarrow {}^{4}_{2}He + {}^{228}_{88}Ra$

c. $^{251}_{102}No \longrightarrow {}^{4}_{2}He + {}^{247}_{100}Fm$

d. $^{220}_{86}Rn \longrightarrow {}^{4}_{2}He + {}^{216}_{84}Po$

6.12 **a.** $^{243}_{96}Cm \longrightarrow {}^{4}_{2}He + {}^{239}_{94}Pu$

b. $^{252}_{99}Es \longrightarrow {}^{4}_{2}He + {}^{248}_{97}Bk$

c. $^{251}_{98}Cf \longrightarrow {}^{4}_{2}He + {}^{247}_{96}Cm$

6.13 **a.** $^{25}_{11}Na \longrightarrow {}^{0}_{-1}e + {}^{25}_{12}Mg$

b. $^{20}_{8}O \longrightarrow {}^{0}_{-1}e + {}^{20}_{9}F$

c. $^{92}_{38}Sr \longrightarrow {}^{0}_{-1}e + {}^{92}_{39}Y$

d. $^{42}_{19}K \longrightarrow {}^{0}_{-1}e + {}^{42}_{20}Ca$

6.14 $^{59}_{26}Fe \longrightarrow {}^{0}_{-1}e + {}^{59}_{27}Co$

$^{60}_{26}Fe \longrightarrow {}^{0}_{-1}e + {}^{60}_{27}Co$

6.15 **a.** $^{28}_{14}Si$ **b.** $^{87}_{36}Kr$ **c.** $^{0}_{-1}e$ **d.** $^{238}_{92}U$ **e.** $^{4}_{2}He$

f. $^{207}_{81}Tl$ **g.** $^{35}_{17}Cl$

6.16 Radiation from a radioactive source strikes a detection tube of the Geiger counter. The radiation interacts with gas molecules inside the tube, causing the formation of ions. The charged particles cause an electric current that is detected and amplified as clicks that are counted.

6.17 **a.** Curie (Ci)
b. Radiation absorbed dose (rad)
c. Radiation equivalent in humans (rem)

6.18 2.2×10^{12} disintegrations in 20 seconds

6.19 294 μCi

6.20 1 mCi

6.21 50 mrem for gamma particles; 1000 mrem for alpha particles. The radiation effect is 20 times greater for alpha particles.

6.22 When pilots are flying at high altitudes, there is less atmosphere to protect them from radiation.

6.23 A patient might experience nausea, vomiting, fatigue, a reduction in white cells, diarrhea, loss of hair, and infections.

6.24 Since the elements Ca and P are part of bone, their radioactive isotopes will also become part of the bony structures of the body. The radioactive Sr also seeks out bone since Sr acts much like Ca. Once the radioactive isotopes are in the bone, their radiation can be used to diagnose and reduce a bone lesion or bone tumor.

6.25 ^{32}P seeks out bone where its radiation will decrease some of the bone marrow that produces red cells.

6.26 ^{131}I locates in the thyroid, where its radiation will destroy some of the thyroid cells that produce thyroid hormone.

6.27 **a.** When radioisotopes are given to a patient, the radiation emitted is picked up by a detection tube and converted to an image.
b. A CT scan is obtained by measuring the amount of absorption of 30,000 x-ray beams directed at successive layers of a tissue. The differences in densities produce a series of images called a CT scan.
c. In MRI, a nonradioactive technique, imaging is achieved from the energy emitted by changes in atomic nuclei (hydrogen) that occur when placed in strong magnetic fields.
d. In a PET scan, very short lived radioisotopes are used that emit a positron, a positively charged particle, that quickly combines with an electron with a burst of gamma energy that is detected and used to produce a scan.

6.28 Radioisotopes with short half-lives release much of their radiation soon after they are given to a patient. Therefore, smaller amounts can be used to provide sufficient radiation for detection. Since the radioisotopes decay rapidly, they leave only small amounts of active isotope in the body.

6.29 **a.** 40.0 mg **b.** 20.0 mg **c.** 10.0 mg **d.** 5.0 mg

6.30 0.75 mCi

6.31 128 days; 192 days

6.32 16 μCi/g

6.33 12.5 mg are still active (330 min is 3 half-lives)

6.34 11,460 years old (two half-lives)

6.35 **a.** ${}^{10}_{4}Be$ **b.** ${}^{0}_{-1}e$ **c.** ${}^{27}_{13}Al$ **d.** ${}^{4}_{2}He$ **e.** ${}^{239}_{92}U$ **f.** ${}^{14}_{7}N$

6.36 ${}^{249}_{98}Cf + {}^{18}_{8}O \longrightarrow {}^{263}_{106}Unh + 4\,{}^{1}_{0}n$

6.37 After a neutron strikes a uranium atom, the fission releases three neutrons. Each of these neutrons strikes uranium atoms, and more neutrons are released. Therefore, the neutron products from one fission become the bombarding neutrons for the next, causing a chain of events.

6.38 ${}^{103}_{42}Mo$

6.39 **a.** fission **b.** fusion **c.** fission **d.** fusion **e.** fusion **f.** fusion **g.** fusion **h.** both **i.** fission

Chapter 7

7.1 **a.** Gases are composed of very small particles (atoms or molecules).
b. Particles of a gas move rapidly in all directions.
c. The particles of a gas are far apart.
d. There are no attractive forces between gas particles.
e. Gases move faster at high temperatures.
f. Gas particles move at high speeds.

7.2 Pressure, volume, temperature, and quantity.

7.3 **a.** temperature **b.** pressure **c.** amount **d.** temperature **e.** volume **f.** volume

7.4 Items a, d, and e describe gas pressure.

7.5 **a.** 1520 torr **b.** 29.4 lbs/in^2 **c.** 1520 mmHg

7.6 **a.** 0.967 atm **b.** 735 torr **c.** 28.9 inHg

7.7 **a.** boiling point **b.** vapor pressure **c.** atmospheric pressure **d.** boiling point

7.8 **a.** Boiling would not occur.
b. Boiling would occur.
c. Boiling would occur.
d. Boiling would not occur.

7.9 **a.** The atmospheric pressure at the top of Mt. Whitney is less than 760 torr, so boiling occurs at a temperature lower than 100°C.
b. The pressure inside a pressure cooker is greater than 760 torr, so boiling occurs above 100°C, and the food cooks more quickly.
c. Water does not boil until 120°C, when the pressure in an autoclave is 2.0 atm.

7.10 **a.** increases **b.** increases **c.** decreases

7.11 **a.** 328 torr **b.** 1210 torr **c.** 2.97 atm

7.12 **a.** 25 L **b.** 622 mL **c.** 300 L

7.13 **a.** inspiration **b.** expiration **c.** expiration **d.** inspiration

7.14 **a.** decreases **b.** increases **c.** decreases

7.15 **a.** 2970 mL **b.** 6170 mL **c.** 2240 mL

7.16 **a.** 410°C **b.** −191°C **c.** −102°C

7.17 **a.** increases **b.** decreases **c.** decreases

7.18 **a.** 770 torr **b.** 1.51 atm **c.** 11.9 atm

7.19 **a.** 1400°C **b.** −223°C

7.20 **a.** 1400 torr **b.** 0.829 atm **c.** 1500 L **d.** 150 mL

7.21 **a.** 5.00 L **b.** 1320 mL **c.** 3.0 moles

7.22 **a.** 2.00 mole O_2 **b.** 0.179 mole CO_2 **c.** 10.1 g Ne **d.** 0.071 mole H_2

7.23 **a.** 56 L **b.** 9410 mL **c.** 4.48 L **d.** 55,400 mL

7.24 740 torr

7.25 1.2 atm

7.26 1805 torr

7.27 **a.** 713 mmHg **b.** 210 mL

7.28 Partial pressure He = 600 torr
Partial pressure O_2 = 1800 torr

7.29 4.0 atm

7.30 56 L

7.31 29 mole O_2

7.32 44 L O_2

7.33 0.0309 mole

7.34 0.226 g NH_3

Chapter 8

8.1 **a.** NaCl, solute; water, solvent **b.** water, solute; ethanol, solvent **c.** O_2, solute; N_2, solvent **d.** mercury, solute; silver, solvent **e.** sugar, solute; water, solvent

8.2 **a.** surface tension **b.** hydrogen bonding **c.** polar solvent

8.3 **a.** oxygen atom **b.** hydration **c.** hydrogen atom

8.4 **a.** water **b.** water **c.** hexane **d.** hexane **e.** hexane **f.** water

8.5 **a.** unsaturated **b.** saturated **c.** unsaturated **d.** unsaturated **e.** unsaturated

8.6 **a.** saturated **b.** unsaturated **c.** unsaturated **d.** saturated

8.7 Heating the water will make the $Cl_2(g)$ less soluble. Bubbles of $Cl_2(g)$ form and leave the solution.

8.8 **a.** unsaturated **b.** saturated **c.** 60°C

8.9 **a.** increase **b.** decrease **c.** decrease **d.** increase **e.** increase

8.10 A 5% (w/w) glucose contains 5 g of glucose in 100 g of solution; a 5% (w/v) glucose contains 5 g of glucose in 100 mL of solution.

8.11 **a.** 2.0% **b.** 13.% **c.** 15% **d.** 7.5% **e.** 6.00% **f.** 16%

8.12 **a.** $\frac{10\text{ g NaOH}}{100\text{ g solution}}$ and $\frac{100\text{ g solution}}{10\text{ g NaOH}}$

b. $\frac{2.5\text{ g NaCl}}{100\text{ mL solution}}$ and $\frac{100\text{ mL solution}}{2.5\text{ g NaCl}}$

c. $\frac{15\text{ mL CH}_3\text{OH}}{100\text{ mL solution}}$ and $\frac{100\text{ mL solution}}{15\text{ mL CH}_3\text{OH}}$

d. $\frac{5.0\text{ g glucose}}{100\text{ mL solution}}$ and $\frac{100\text{ mL solution}}{5.0\text{ g glucose}}$

e. $\frac{20\text{ g K}_2\text{CO}_3}{100\text{ g solution}}$ and $\frac{100\text{ g solution}}{20\text{ g K}_2\text{CO}_3}$

8.13 **a.** 2.5 g KCl **b.** 22.5 g NaCl **c.** 50 g NH_4Cl **d.** 30 g NaOH

8.14 **a.** 25 mL acetic acid **b.** 60 mL isopropyl alcohol **c.** 110 mL methyl alcohol

8.15 **a.** 20 g mannitol **b.** 480 g mannitol

8.16 **a.** 60 g amino acids, 375 g dextrose, 100 g lipids **b.** 240 kcal from amino acids, 1500 kcal from dextrose, and 900 from lipids; 2640 kcal total

8.17 **a.** 1000 g **b.** 20 g **c.** 400 g

8.18 **a.** 3000 mL **b.** 4000 mL **c.** 2500 mL **d.** 25 mL

8.19 2.0 L

8.20 300 mL alcohol

8.21 **a.** 2.0 M KOH **b.** 0.50 M glucose **c.** 10 M NaOH **d.** 2.5 M NaCl

8.22 **a.** 1.0 M HCl **b.** 1.0 M NaOH **c.** 3.6 M glucose

8.23 **a.** $\frac{6.00\text{ mole HCl}}{1\text{ L solution}}$ and $\frac{1\text{ L solution}}{6.00\text{ mole HCl}}$

b. $\frac{0.250\text{ mole NaHCO}_3}{1\text{ L solution}}$ and $\frac{1\text{ L solution}}{0.250\text{ mole NaHCO}_3}$

c. $\frac{1.0\text{ mole H}_2\text{SO}_4}{1\text{ L solution}}$ and $\frac{1\text{ L solution}}{1.0\text{ mole H}_2\text{SO}_4}$

d. $\frac{0.50\text{ mole KBr}}{1\text{ L solution}}$ and $\frac{1\text{ L solution}}{0.50\text{ mole KBr}}$

8.24 **a.** 3.0 mole NaCl **b.** 10 mole $CaCl_2$ **c.** 0.80 mole glucose **d.** 0.50 mole sucrose

8.25 **a.** 40 g NaOH **b.** 600 g KCl **c.** 29 g NaCl **d.** 48 g NaOH **e.** 135 g glucose

8.26 **a.** 1.0 L **b.** 10 L **c.** 2 L **d.** 0.50 L

8.27 **a.** 40 mL **b.** 150 mL **c.** 900 mL **d.** 80 mL **e.** 30 mL

8.28 **a.** 3.0% NaCl **b.** 5% KOH **c.** 1.0 M $CaCl_2$ **d.** 1.00 M sucrose

8.29 **a.** solution **b.** colloid **c.** colloid **d.** suspension **e.** suspension **f.** solution **g.** suspension

8.30 **a.** into the starch solution **b.** the starch solution **c.** the starch solution

8.31 **a.** from the 0.1% albumin solution into the 2% albumin solution **b.** the 2% albumin solution **c.** the 2% albumin solution

8.32 **a.** 5% glucose, 0.9% NaCl **b.** 10% NaCl **c.** H_2O, 1% glucose **d.** 10% NaCl **e.** H_2O, 1% glucose **f.** 5% glucose, 0.9% NaCl

8.33 **a.** NaCl, amino acids **b.** KCl and glucose **c.** urea and NaCl

8.34 The dialysate is prepared with a high potassium level so the potassium will move from the dialysate into the patient's blood. The low levels of sodium and urea in the dialysate will cause these compounds to move out of the patient's blood into the dialysate solution.

Chapter 9

9.1 **a.** $KCl(s) \xrightarrow{H_2O} K^+(aq) + Cl^-(aq)$

b. $Na_2SO_4(s) \xrightarrow{H_2O} 2Na^+(aq) + SO_4^{2-}(aq)$

c. $CaCl_2(s) \xrightarrow{H_2O} Ca^{2+}(aq) + 2Cl^-(aq)$

d. $LiNO_3(s) \xrightarrow{H_2O} Li^+(aq) + NO_3^-(aq)$

e. $K_3PO_4(s) \xrightarrow{H_2O} 3K^+(aq) + PO_4^{3-}(aq)$

f. $Ba(NO_3)_2(s) \xrightarrow{H_2O} Ba^{2+}(aq) + 2NO_3^-(aq)$

9.2 **a.** ions only **b.** molecules only **c.** molecules and some ions **d.** ions only **e.** molecules only

9.3 **a.** strong **b.** weak **c.** non **d.** non **e.** weak

9.4 **a.** $KOH + HI \longrightarrow KI + H_2O$
b. $Mg(OH)_2 + 2HCl \longrightarrow MgCl_2 + 2H_2O$
c. $2KOH + H_2SO_4 \longrightarrow K_2SO_4 + 2H_2O$
d. $LiOH + HNO_3 \longrightarrow LiNO_3 + H_2O$
e. $3Ca(OH)_2 + 2H_3PO_4 \longrightarrow Ca_3(PO_4)_2 + 6H_2O$

9.5 **a.** yes **b.** no **c.** no **d.** yes **e.** yes **f.** no **g.** yes **h.** yes **i.** no **j.** no **k.** yes

9.6 **a.** yes: $2NaCl(aq) + Pb(NO_3)_2(aq) \longrightarrow PbCl_2(s) + 2NaNO_3(aq)$
b. no
c. yes: $Na_2S(aq) + CuCl_2(aq) \longrightarrow CuS(s) + 2NaCl(aq)$
d. yes: $BaCl_2(aq) + Na_2SO_4(aq) \longrightarrow BaSO_4(s) + 2NaCl(aq)$
e. no
f. yes: $2Na_3PO_4(aq) + 3MgCl_2(aq) \longrightarrow Mg_3(PO_4)_2(s) + 6NaCl(aq)$

9.7 **a.** strong **b.** strong **c.** weak **d.** weak

9.8 **a.** $HCl \xrightarrow{H_2O} H^+(aq) + Cl^-(aq)$
$HCl + H_2O \longrightarrow H_3O^+(aq) + Cl^-(aq)$
b. $HNO_3 \xrightarrow{H_2O} H^+(aq) + NO_3^-(aq)$
$HNO_3 + H_2O \longrightarrow H_3O^+(aq) + NO_3^-(aq)$
c. $HNO_2 \overset{H_2O}{\rightleftharpoons} H^+(aq) + NO_2^-(aq)$
$HNO_2 + H_2O \rightleftharpoons H_3O^+(aq) + NO_2^-(aq)$
d. $H_2CO_3 \overset{H_2O}{\rightleftharpoons} H^+(aq) + HCO_3^-(aq)$
$H_2CO_3 + H_2O \rightleftharpoons H_3O^+(aq) + HCO_3^-(aq)$

9.9 **a.** $LiOH \xrightarrow{H_2O} Li^+(aq) + OH^-(aq)$
b. $Mg(OH)_2 \xrightarrow{H_2O} Mg^{2+}(aq) + 2OH^-(aq)$
c. $KOH \xrightarrow{H_2O} K^+(aq) + OH^-(aq)$

9.10 $NH_3 + H_2O \rightleftharpoons NH_4^+(aq) + OH^-(aq)$

9.11 **a.** acid **b.** base **c.** base **d.** base **e.** acid

9.12 **a.** $[H^+] = 1 \times 10^{-7}$ M; $[OH^-] = 1 \times 10^{-7}$ M
b. $[H^+][OH^-] = 1 \times 10^{-14}$

9.13 **a.** 1×10^{-3} **b.** 3.3×10^{-10} **c.** 1×10^{-13} **d.** 1.2×10^{-3}

9.14 5.0×10^{-13}

9.15 **a.** 1×10^{-3} M **b.** 1×10^{-12} M **c.** 5.0×10^{-11} M **d.** 1.7×10^{-7}

9.16 2.50×10^{-13} M

9.17 **a.** basic **b.** acidic **c.** basic **d.** acidic **e.** acidic **f.** neutral **g.** acidic **h.** basic **i.** basic **j.** basic

9.18 **a.** 0.4, 4.5, 6.8, 13.0
b. 1.6, 2.3, 7.1, 8.5, 11.7
c. 2.9, 3.3, 4.4, 9.8, 14.0
d. 7.4, 8.8, 9.7, 11.4, 13.4

9.19 **a.** acidic **b.** neutral **c.** basic **d.** acidic **e.** acidic **f.** acidic **g.** basic **h.** neutral **i.** acidic **j.** basic

9.20 **a.** 4.0 **b.** 7.0 **c.** 11.3 **d.** 4.0 **e.** 1.3 **f.** 2.0 **g.** 10.0 **h.** 7.0 **i.** 4.0 **j.** 11.0

9.21

$[H^+]$	$[OH^-]$	pH	Acidic, Basic, Neutral
1×10^{-8} M	1×10^{-6} M	8	basic
1×10^{-2} M	1×10^{-12} M	2	acidic
1×10^{-5} M	1×10^{-9} M	5	acidic
1×10^{-7} M	1×10^{-7} M	7	neutral
1×10^{-10} M	1×10^{-4} M	10	basic

9.22 **a.** $HCl + KOH \longrightarrow KCl + H_2O$
b. $2HNO_3 + Ca(OH)_2 \longrightarrow Ca(NO_3)_2 + 2H_2O$
c. $H_3PO_4 + 3LiOH \longrightarrow Li_3PO_4 + 3H_2O$
d. $3H_2SO_4 + 2Al(OH)_3 \longrightarrow Al_2(SO_4)_3 + 6H_2O$

9.23 **a.** $2NaOH + H_2SO_4 \longrightarrow Na_2SO_4 + 2H_2O$
b. $KOH + HNO_3 \longrightarrow KNO_3 + H_2O$
c. $H_3PO_4 + 3NaOH \longrightarrow Na_3PO_4 + 3H_2O$
d. $H_2CO_3 + Ca(OH)_2 \longrightarrow CaCO_3 + 2H_2O$

9.24 **a.** 2.2 M HCl **b.** 0.76 M H_2SO_4 **c.** 0.53 M H_3PO_4

9.25 Items c and e are buffers systems. Item c contains a weak base (NH_3) and its salt, NH_4Cl. Item e contains the weak acid (H_2CO_3) and its sodium salt, $NaHCO_3$.

9.26 **a.** Maintain the pH of a system
b. Provide $C_2H_3O_2^-$, which reacts with H^+
c. $C_2H_3O_2^- + H^+ \longrightarrow HC_2H_3O_2$
d. $HC_2H_3O_2 + OH^- \longrightarrow H_2O + C_2H_3O_2^-$

9.27 **a.** 36.5 g/equiv **b.** 29.2 g/equiv **c.** 32.7 g/equiv **d.** 55.6 g/equiv

9.28 **a.** 3.3 N **b.** 4.46 N **c.** 0.724 N

9.29 5.25 N

9.30 50 mL

9.31 3.54 g Na^+, 5.47 g Cl^-

9.32 155 meq/L

Chapter 10

10.1 **a.** inorganic **b.** organic **c.** organic **d.** organic **e.** inorganic **f.** organic

10.2 **a.**

```
     H  H  H  H  H
     |  |  |  |  |
  H—C—C—C—C—C—H
     |  |  |  |  |
     H  H  H  H  H
```

b.

```
        H
        |
      H—C—H
     H  |  H  H
     |  |  |  |
  H—C—C—C—C—H
     |  |  |  |
     H  H  H  H
```

c.

```
        H
        |
      H—C—H
     H  |  H
     |  |  |
  H—C—C—C—H
     |  |  |
     H  |  H
      H—C—H
        |
        H
```

d.

```
        H        H
        |        |
      H—C—H  H—C—H
     H  |        |  H  H
     |  |        |  |  |
  H—C—C————C—C—C—H
     |  |        |  |  |
     H  |        H  H  H
      H—C—H
        |
        H
```

10.3 **a.** $CH_3—CH_2—CH_2—CH_2—CH_3$
b. $CH_3—CH_2—CH_3$

c.

```
        CH3 CH3
         |   |
   CH3—CH—CH—CH2—CH3
```

d.

```
        CH3
         |
   CH3—CH—CH2—CH2—CH2—CH3
```

10.4 **a.**

```
     H  H
     |  |
  H—C—C—H
     |  |
     H  H
```

b.

```
     H  H  H  H  H
     |  |  |  |  |
  H—C—C—C—C—C—H
     |  |  |  |  |
     H  H  H  H  H
```

c.

```
        H
        |
      H—C—H
     H  |  H  H
     |  |  |  |
  H—C—C—C—C—H
     |  |  |  |
     H  H  |  H
         H—C—H
           |
           H
```

d.

```
        H
        |
      H—C—H
     H  |  H  H  H  H
     |  |  |  |  |  |
  H—C—C—C—C—C—C—C—H
     |  |  |  |  |  |  |
     H  H  H  |  H  H  H
            H—C—H
              |
            H—C—H
              |
              H
```

10.5 **a.** butane **b.** ethane **c.** heptane **d.** methane

10.6 **a.** A straight chain means that the carbon atoms are directly connected; there are no branches.
b. No, the carbon atoms are not linear; they form a zigzag pattern.

10.7 **a.** $CH_3—CH_2—CH_3$
b. $CH_3—CH_2—CH_2—CH_2—CH_2—CH_2—CH_3$
c. $CH_3—CH_2—CH_2—CH_2—CH_3$
d. $CH_3—CH_2—CH_2—CH_2—CH_2—CH_2—CH_2—CH_3$

10.8 **a.** A straight-chain alkane has all of its carbons connected in a line; a branched-chain alkane has carbon branches.
b. 1. straight chain 4. branched chain
2. branched chain 5. straight chain
3. straight chain

10.9 **a.** 2-methylpropane
b. 2,2-dimethylbutane
c. 2,3-dimethylpentane
d. 3-ethyl-3,5-dimethylheptane
e. 5-ethyl-3,3-dimethyloctane
f. 5-ethyl-2,8-dimethyldecane

10.10 **a.**

```
        CH3
         |
   CH3—CH—CH2—CH2—CH3
```

b.

```
              CH3
               |
   CH3—CH2—C—CH2—CH2—CH3
               |
              CH3
```

c.
$$\begin{array}{c} \quad CH_3 \\ CH_3-CH-CH-CH_3 \\ \qquad\quad CH_3 \end{array}$$

d.
$$\begin{array}{l} \quad\ CH_3 \\ CH_3-CH-CH-CH_2-CH_2-CH_3 \\ \qquad\qquad CH_2 \\ \qquad\qquad CH_3 \end{array}$$

e.
$$\begin{array}{l} \qquad\qquad\qquad CH_3 \\ \quad\ CH_3 \qquad\quad CHCH_3 \\ CH_3-CH-CH_2-CH-CH_2-CH_2-CH_3 \end{array}$$

f.
$$\begin{array}{l} \quad\ CH_3 \qquad\quad CH_3 \\ CH_3-C-CH_2-C-CH_3 \\ \quad\ CH_3 \qquad\quad CH_3 \end{array}$$

10.11 $CH_3-CH_2-CH_2-CH_2-CH_3$

Pentane

$$\begin{array}{l} \quad\ CH_3 \\ CH_3-CH-CH_2-CH_3 \end{array}$$

2-Methylbutane

$$\begin{array}{l} \qquad\ CH_3 \\ CH_3-C-CH_3 \\ \qquad\ CH_3 \end{array}$$

2,2-Dimethylpropane

10.12
$$\begin{array}{l} \quad\ CH_3 \\ CH_3-C-CH_2-CH_2-CH_3 \\ \quad\ CH_3 \end{array}$$

2,2-Dimethylpentane

$$\begin{array}{l} \quad\ CH_3 \\ CH_3-CH-CH-CH_2-CH_3 \\ \qquad\qquad CH_3 \end{array}$$

2,3-Dimethylpentane

$$\begin{array}{l} \qquad\qquad\quad CH_3 \\ CH_3-CH_2-C-CH_2-CH_3 \\ \qquad\qquad\quad CH_3 \end{array}$$

3,3-Dimethylpentane

$$\begin{array}{l} \quad\ CH_3 \qquad\quad CH_3 \\ CH_3-CH-CH_2-CH-CH_3 \end{array}$$

2,4-Dimethylpentane

$$\begin{array}{l} \qquad\qquad\quad CH_2-CH_3 \\ CH_3-CH_2-CH-CH_2-CH_3 \end{array}$$

3-Ethylpentane

10.13 **a.** isomers **b.** isomers **c.** identical **d.** identical

10.14 **a.** cyclobutane
b. 1-ethyl-2-methylcyclobutane
c. cyclopentane
d. 1,4-dimethylcyclohexane
e. 1,1-dimethylcyclopentane

10.15 **a.**

d.

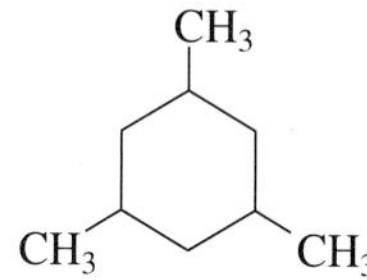

b. CH_3

e.

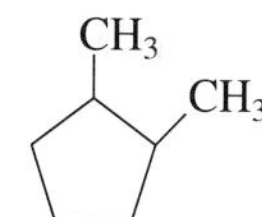

c. CH_2CH_3

10.16 CH_3 CH_3

1,1-dimethylcylopentane

CH_3 CH_3

1,2-dimethylcyclopentane

CH_3 CH_3

1,3-dimethylcyclopentane

10.17

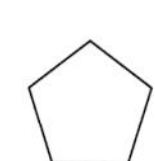

Cyclopentane

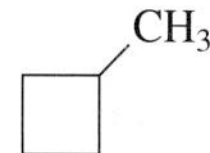

Methylcyclobutane

CH_3 CH_3

1,1-dimethylcyclopropane

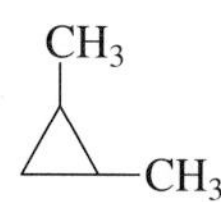

1,2-dimethylcyclopropane

10.18 **a.** 1-bromopropane
b. 2-bromo-4-chloropentane
c. 1,1-dichloro-2,2-difluoroethane
d. 2,4-dichloro-2,4-difluorohexane

10.19 chloromethane CH_3Cl
chloroethane CH_3-CH_2-Cl

10.20 **a.**
$$\begin{array}{l} \qquad\quad Cl \\ CH_3-CH-CH_3 \end{array}$$

b.
$$\begin{array}{l} \qquad\qquad\quad Cl \\ Br-CH_2-CH-CH_3 \end{array}$$

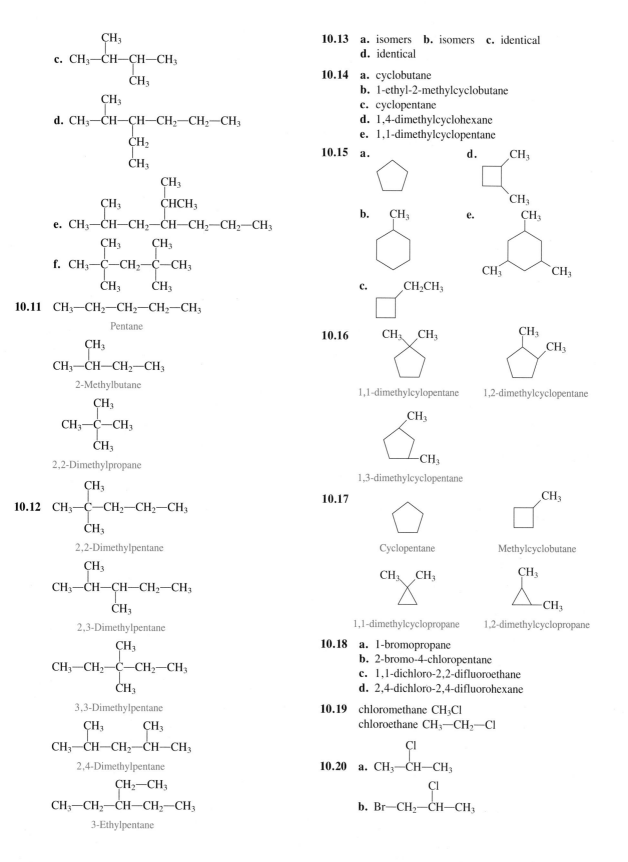

c. F Cl (on cyclopropane ring)

d.
$\quad\quad Cl \quad CH_3 \quad Cl$
$CH_3-CH-CH-CH-CH_2-CH_3$

e.
$\quad\quad CH_3 \quad Br$
$CH_3-CH-C-CH_2-CH_3$
$\quad\quad\quad\quad Br$

f. Cl Cl (on cyclohexane ring)
Cl Cl

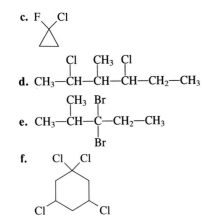

10.21 a. $CH_4 + 2O_2 \longrightarrow CO_2 + 2H_2O$
b. $C_3H_8 + 5O_2 \longrightarrow 3CO_2 + 4H_2O$
c. $2C_4H_{10} + 13O_2 \longrightarrow 8CO_2 + 10H_2O$
d. $C_4H_8 + 6O_2 \longrightarrow 4CO_2 + 4H_2O$
e. $2C_8H_{18} + 25O_2 \longrightarrow 16CO_2 + 18H_2O$

10.22 a. $CH_3-CH_2-Br + HBr$
b. $CH_3-CH_2-CH_2-Cl$ and
$\quad\quad Cl$
$CH_3-CH-CH_3 + 2HCl$
c. no reaction
d. $CH_3Br + HBr$

10.23 a. Br (on cyclopropane ring) b. Cl (on cyclobutane ring)
c. no reaction d. Br (on cyclohexane ring)

Chapter 11

11.1 Cyclohexane is a six-carbon cycloalkane. It is a saturated alkane that is not very reactive, although it will undergo substitution reactions and combustion. Cyclohexene is a six-carbon cycloalkene. It is unsaturated and much more reactive, undergoing addition reactions at the site of the double bond.

11.2 a. ethene; ethylene
b. 5-bromo-2-pentene
c. 3-methyl-2-pentene
d. 2-ethyl-1-pentene
e. 1-methylcyclopentene
f. 2,3-dichlorocyclobutene

11.3 a. $CH_3-CH=CH_2$ b.
$\quad\quad CH_3$
$CH_3-C=CH_2$

c. (cyclopentene) d. Cl (on cyclopentene ring)

e. CH_3 CH_3 (on cyclohexene ring) f.
$\quad\quad\quad\quad\quad\quad CH_3$
$Cl-CH_2-C=C-CH_3$
$\quad\quad\quad\quad Cl$

11.4 Cl Cl (on cyclobutene ring) Cl Cl (on cyclobutene ring)

1,2-dichlorocyclobutene 1,3-dichlorocyclobutene

2,3-dichlorocyclobutene 3,3-dichlorocyclobutene

Cl Cl (on cyclobutene ring)

3,4-dichlorocyclobutene

11.5 a. trans-2,3-dichloro-2-butene
b. cis-3-hexene
c. trans-3,4-dibromo-3-heptene

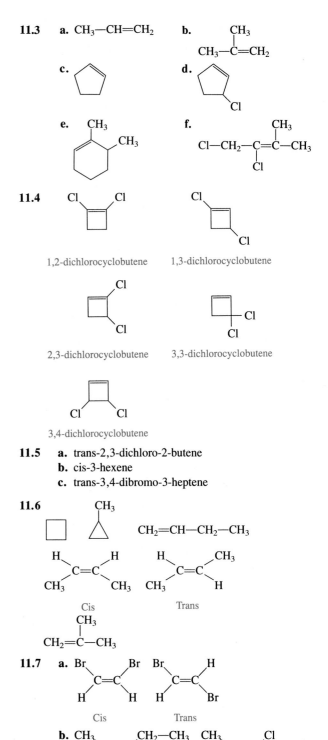

11.6 (cyclobutane) CH_3 (on cyclopropane ring) $CH_2=CH-CH_2-CH_3$

H H / C=C / CH_3 CH_3 H CH_3 / C=C / CH_3 H

Cis Trans

$\quad CH_3$
$CH_2=C-CH_3$

11.7 a. Br Br / C=C / H H Br H / C=C / H Br

Cis Trans

b. CH_3 CH_2-CH_3 / C=C / Cl Cl CH_3 Cl / C=C / Cl CH_2-CH_3

Cis Trans

c. Br(H)C=C(Br)(CH$_2$—CH$_3$) Cis; Br(H)C=C(CH$_2$—CH$_3$)(Br) Trans

d. CH$_3$(H)C=C(CH$_3$)(H) Cis; CH$_3$(H)C=C(H)(CH$_3$) Trans

e. CH$_3$—CH$_2$(H)C=C(CH$_2$—CH$_3$)(H) Cis

CH$_3$—CH$_2$(H)C=C(H)(CH$_2$—CH$_3$) Trans

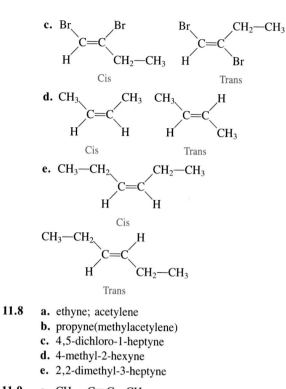

11.8 **a.** ethyne; acetylene
b. propyne(methylacetylene)
c. 4,5-dichloro-1-heptyne
d. 4-methyl-2-hexyne
e. 2,2-dimethyl-3-heptyne

11.9 **a.** $CH_3—C{\equiv}C—CH_3$
b. $HC{\equiv}CH$
c. $CH_3—C{\equiv}C—CCl_2—CH_3$
d. $CH_3—CH_2—C{\equiv}CH$
e. $HC{\equiv}C—CH(CH_3)—CH_3$

11.10 Since alkenes and alkynes have multiple bonds, they may add, while alkanes are saturated and must undergo substitution.

11.11 **a.** $CH_3—CH_2—CH_2—CH_3$
b. $CH_3—CH(Br)—CH(Br)—CH(CH_3)—CH_3$
c. 1,2-dichlorocyclobutane (Cl, Cl on adjacent ring carbons)
d. bromocyclohexane (Br on ring)
e. $CH_3—CH_2—C(CH_3)(Cl)—CH_2—CH_3$
f. $CH_3—CH(OH)—CH_3$

11.12 $CH_3—CH_2—CH_2—CH_3$

11.13 **a.** $CH_3—C(CH_3){=}CH_2 + H_2 \xrightarrow{Pt} CH_3—CH(CH_3)—CH_3$
b. $+ Cl_2 \longrightarrow$

cyclopentene $+ Cl_2 \longrightarrow$ 1,2-dichlorocyclopentane (Cl, Cl)

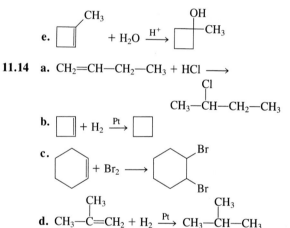

c. $CH_3—CH{=}CH—CH_2—CH_3 + Br_2 \longrightarrow CH_3—CH(Br)—CH(Br)—CH_2—CH_3$

d. $CH_3—C(CH_3){=}CH—CH_3 + HBr \longrightarrow CH_3—C(CH_3)(Br)—CH_2—CH_3$

e. 1-methylcyclobutene (CH$_3$) $+ H_2O \xrightarrow{H^+}$ 1-methylcyclobutanol (OH, CH$_3$ on same carbon)

11.14 **a.** $CH_2{=}CH—CH_2—CH_3 + HCl \longrightarrow CH_3—CH(Cl)—CH_2—CH_3$
b. cyclobutene $+ H_2 \xrightarrow{Pt}$ cyclobutane
c. cyclohexene $+ Br_2 \longrightarrow$ 1,2-dibromocyclohexane (Br, Br)
d. $CH_3—C(CH_3){=}CH_2 + H_2 \xrightarrow{Pt} CH_3—CH(CH_3)—CH_3$

11.15 Benzene is an aromatic compound that is made stable by the unique arrangement of electrons within the six-carbon ring. Cyclohexene is a reactive compound because of the unsaturated site (double bond) within its six-carbon ring. Cyclohexane is a saturated compound that contains only single bonds between the carbon atoms in its ring.

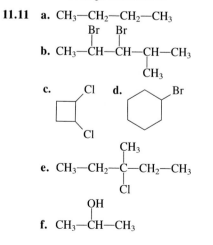

11.16 **a.** benzene
b. toluene
c. 1,2-dichlorobenzene; *o*-dichlorobenzene
d. *m*-xylene
e. ethylbenzene
f. 3-chlorotoluene; *m*-chlorotoluene
g. *o*-xylene
h. 3,4-dichlorotoluene

11.17 **a.**

b. CH_3, CH_3

c. CH_3, Br

d. Br, Br

e. Cl, Cl

f. Br, Br, Br

g. $CH_2—CH_3$

h. $CH_2—CH_3$, CH_3

11.18 **a.** $+ Cl_2 \xrightarrow{Fe}$ Cl

b. $+ HNO_3 \xrightarrow{H_2SO_4}$ NO_2

c. $+ SO_3 \xrightarrow{H_2SO_4}$ SO_3H

d. $+ CH_3—CH_2—Cl \xrightarrow{AlCl_3}$ $CH_2—CH_3$

11.19 **a.** alkene, 1-butene
b. alkane; 2,2,5-trimethylhexane
c. cycloalkene; 4-methylcyclopentene
d. haloalkane; 2-chloro-4-methylpentane
e. aromatic; 1,3-dichlorobenzene, *m*-dichlorobenzene
f. alkyne; 2-pentyne

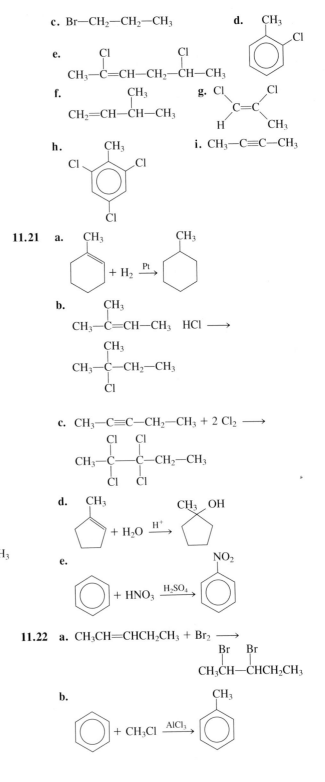

11.20 **a.**

b. Cl, Cl

c. $Br—CH_2—CH_2—CH_3$

d. CH_3, Cl

e. $CH_3—\underset{}{C}(Cl)=CH—CH_2—CH(Cl)—CH_3$

f. $CH_2=CH—CH(CH_3)—CH_3$

g. Cl, Cl, C=C, H, CH_3

h. CH_3, Cl, Cl, Cl

i. $CH_3—C≡C—CH_3$

11.21 **a.** CH_3 $+ H_2 \xrightarrow{Pt}$ CH_3

b. $CH_3—C(CH_3)=CH—CH_3$ HCl ⟶

$CH_3—C(CH_3)(Cl)—CH_2—CH_3$

c. $CH_3—C≡C—CH_2—CH_3 + 2\ Cl_2 \longrightarrow$

$CH_3—CCl_2—CCl_2—CH_2—CH_3$

d. CH_3 $+ H_2O \xrightarrow{H^+}$ CH_3 OH

e. $+ HNO_3 \xrightarrow{H_2SO_4}$ NO_2

11.22 **a.** $CH_3CH=CHCH_2CH_3 + Br_2 \longrightarrow$

$CH_3CH(Br)—CH(Br)CH_2CH_3$

b. $+ CH_3Cl \xrightarrow{AlCl_3}$ CH_3

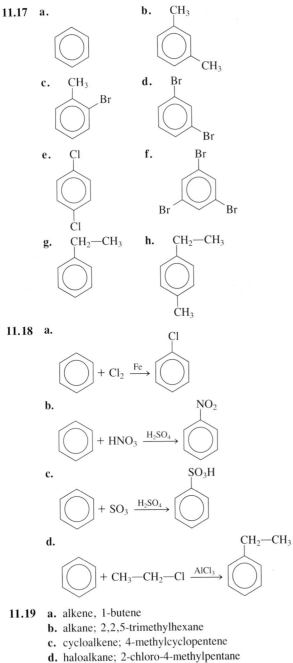

c. cyclohexene + HCl ⟶ chlorocyclohexane

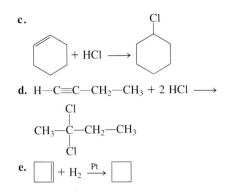

d. $H-C\equiv C-CH_2-CH_3 + 2\ HCl \longrightarrow CH_3-CCl_2-CH_2-CH_3$

e. cyclobutene + H_2 $\xrightarrow{Pt}$ cyclobutane

Chapter 12

12.1 **a.** ethanol; ethyl alcohol
b. 2-butanol
c. cyclohexanol
d. 1-chloro-2-methyl-2-propanol
e. 2-methylcyclopentanol
f. 1-methylcyclobutanol
g. 2,3-dimethyl-1-pentanol
h. 3-methyl-3-pentanol

12.2 **a.** $CH_3CH_2CH_2OH$
b. CH_3CH_2OH
c. $CH_3C(OH)(CH_3)CH_2CH(CH_3)CH_3$ **d.** $HOCH_2CH(Br)-CH(CH_3)CH_3$
e. cyclopentane with OH and CH_2CH_3 on adjacent carbons **f.** cyclohexane with OH, Cl (adjacent carbon), and Cl (para to OH)
g. $CH_3CH(OH)CH_3$

12.3 **a.** ethanol **b.** propanol **c.** propanol

12.4 **a.** Ethanol can form hydrogen bonds to water; ethane cannot.
b. The 3 carbons in propanol give it a smaller nonpolar portion than the 4 carbons of butanol.
c. Hexanol, with 6 carbons, has a large nonpolar section.

12.5 Phenols are found in a, b, d, and e.

12.6 **a.** 2,4-dinitrophenol
b. 3-chlorophenol; *m*-chlorophenol
c. 4-ethylphenol, *p*-ethylphenol

12.7 **a.** phenol with Br (meta) and CH_3 (para) **b.** phenol with CH_2CH_3 (meta)
c. phenol with para phenyl (biphenyl-4-ol) **d.** phenol with Cl, Cl (ortho) and Cl (para)
e. phenol with CH_3 (ortho), Cl (meta), Cl (para)

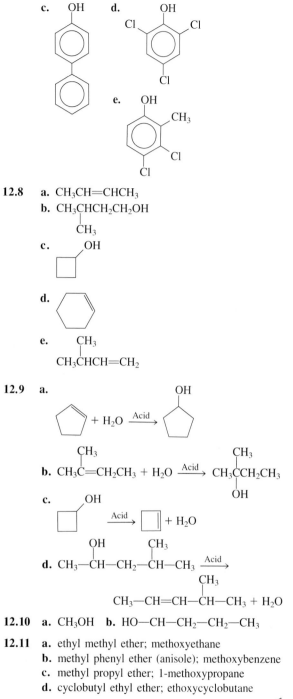

12.8 **a.** $CH_3CH{=}CHCH_3$
b. $CH_3CH(CH_3)CH_2CH_2OH$
c. cyclobutanol
d. cyclohexene
e. $CH_3CH(CH_3)CH{=}CH_2$

12.9 **a.** cyclopentene + H_2O $\xrightarrow{Acid}$ cyclopentanol
b. $CH_3C(CH_3){=}CH_2CH_3 + H_2O \xrightarrow{Acid} CH_3C(CH_3)(OH)CH_2CH_3$
c. cyclobutanol $\xrightarrow{Acid}$ cyclobutene + H_2O
d. $CH_3-CH(OH)-CH_2-CH(CH_3)-CH_3 \xrightarrow{Acid} CH_3-CH{=}CH-CH(CH_3)-CH_3 + H_2O$

12.10 **a.** CH_3OH **b.** $HO-CH-CH_2-CH_2-CH_3$

12.11 **a.** ethyl methyl ether; methoxyethane
b. methyl phenyl ether (anisole); methoxybenzene
c. methyl propyl ether; 1-methoxypropane
d. cyclobutyl ethyl ether; ethoxycyclobutane

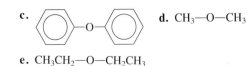

12.12 **a.** $CH_3-O-CH_2CH_2CH_3$ **b.** CH_3CH_2-O-cyclopropyl
c. diphenyl ether (phenyl–O–phenyl) **d.** CH_3-O-CH_3
e. $CH_3CH_2-O-CH_2CH_3$

12.13 **a.** CH_3OCH_3 **b.** 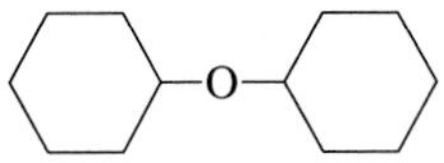 (cyclohexyl)—O—(cyclohexyl)

12.14 butane, ethyl methyl ether, 1-propanol

12.15 **a.** sulfide **b.** thiol **c.** thiol **d.** disulfide

12.16 **a.** $CH_3CH_2CH_2SH$
b. $CH_3CH_2—S—S—CH_2CH_3$
c. $CH_3—\overset{CH_3}{\overset{|}{C}H}—CH_2—CH_2—SH$
d. $CH_3—S—CH_3$

12.17 **a.** ethanal; acetaldehyde
b. methanal; formaldehyde
c. 4-chloropentanal
d. pentanal
e. 3-methylbutanal; β-methyl butyraldehyde

12.18 **a.** $CH_3—CH_2—\overset{O}{\overset{\|}{C}}H$ **b.** $H\overset{O}{\overset{\|}{C}}H$ **c.** $CH_3—\overset{O}{\overset{\|}{C}}H$
d. $CH_3—\overset{Cl}{\overset{|}{C}H}—CH_2—\overset{O}{\overset{\|}{C}}H$
e. $CH_3—CH_2—\overset{CH_3}{\overset{|}{C}H}—CH_2—\overset{O}{\overset{\|}{C}}H$
f. 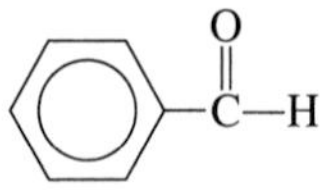 (benzene ring)$—\overset{O}{\overset{\|}{C}}—H$

12.19 **a.** 2-butanone; ethyl methyl ketone
b. 3-hexanone; ethyl propyl ketone
c. cyclopentanone
d. 4-bromo-2-pentanone
e. 2-bromocyclohexanone

12.20 **a.** $CH_3CH_2\overset{O}{\overset{\|}{C}}CH_2CH_3$ **b.** $CH_3\overset{O}{\overset{\|}{C}}CH_3$
c. 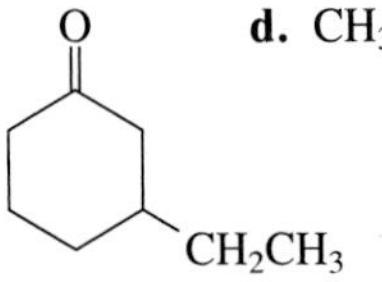 (cyclohexanone ring: O; CH_2CH_3) **d.** $CH_3—\overset{O}{\overset{\|}{C}}—\overset{CH_3}{\overset{|}{C}H}CH_3$

12.21 **a.** aldehyde; propanal, propionaldehyde
b. alcohol; 2-propanol, isopropyl alcohol
c. alcohol; 1-propanol, propyl alcohol
d. ether; methyl propyl ether
e. ketone; 2-propanone, dimethyl ketone, acetone
f. ketone; cyclohexanone

12.22 **a.** 1° **b.** 2° **c.** 2° **d.** 3° **e.** 2° **f.** 3° **g.** 1°
h. 3°

12.23 **a.** $CH_3CH_2CH_2\overset{O}{\overset{\|}{C}}—H$
b. $CH_3CH_2\overset{O}{\overset{\|}{C}}CH_2CH_3$
c. $CH_3\overset{CH_3}{\overset{|}{C}H}—\overset{O}{\overset{\|}{C}}—H$
d. 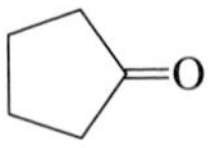 (cyclopentane ring)=O
e. no reaction

12.24 **a.** 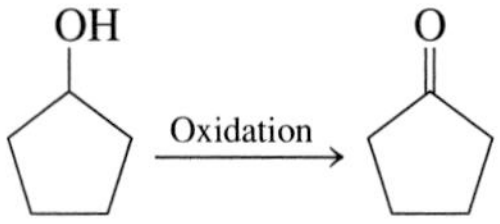 (cyclopentanol: OH) $\xrightarrow{\text{Oxidation}}$ (cyclopentanone: O)
b. 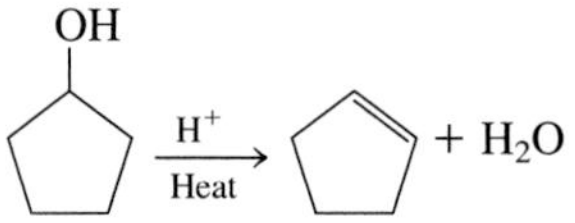 (cyclopentanol: OH) $\xrightarrow[\text{Heat}]{H^+}$ (cyclopentene) $+ H_2O$
c. $CH_3CH_2CH_2OH \xrightarrow{\text{Oxidation}} CH_3CH_2\overset{O}{\overset{\|}{C}}H + H_2O$
d. $CH_3CH_2CH_2OH \xrightarrow[\text{Heat}]{H^+} CH_3CH{=}CH_2 + H_2O$

12.25 **a.** CH_3OH **b.** 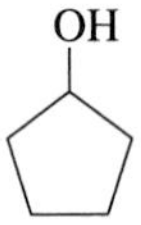 (cyclopentane ring: OH) **c.** $CH_3\overset{OH}{\overset{|}{C}H}CH_2CH_3$
d. 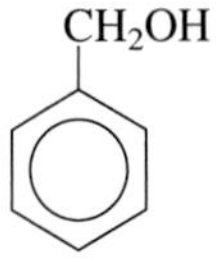 (benzene ring: CH_2OH) **e.** 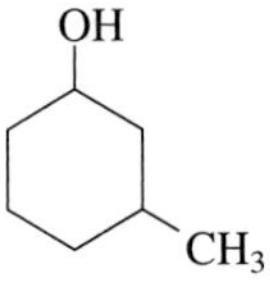 (cyclohexane ring: OH; CH_3)

12.26 **a.** pentanal **b.** propionaldehyde
c. formaldehyde

12.27 **a.** $CH_3\overset{OH}{\overset{|}{C}H}CH_3$ **b.** $CH_3CH_2CH_2CH_2CH_2OH$
c. $CH_3\overset{OH}{\overset{|}{C}H}CH_2\overset{CH_3}{\overset{|}{C}H}CH_3$ **d.** 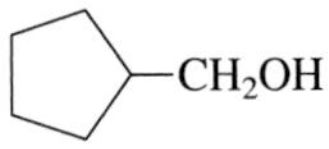 (cyclopentane ring)$—CH_2OH$
e. 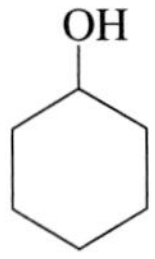 (cyclohexane ring: OH)

12.28 **a.** alkane **b.** haloalkane **c.** alcohol **d.** alkene
e. haloalkane **f.** thiol **g.** alkyne **h.** alcohol
i. ether **j.** sulfide **k.** alcohol **l.** ether
m. ketone

12.29 **a.** 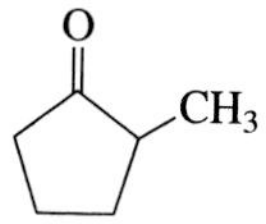 O, CH_3 (cyclopentanone ring) **b.** 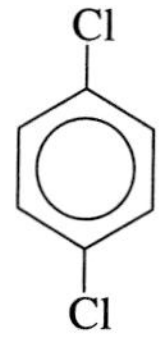 Cl, Cl (benzene ring) **c.** $CH_3\overset{Cl}{\overset{|}{C}}H-\overset{O}{\overset{\|}{C}}H$

d. 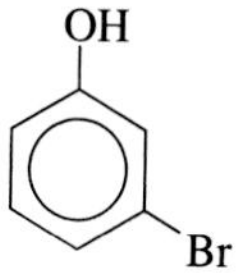 OH, Br (benzene ring) **e.** $CH_3CH_2OCH_2CH_3$ **f.** CH_3SH

g. $CH_3\overset{Cl}{\overset{|}{C}}HCH_2\overset{O}{\overset{\|}{C}}H$ **h.** $HOCH_2CH_2\overset{CH_3}{\overset{|}{C}}HCH_3$

12.30 **a.** $HOCH_2\overset{CH_3}{\overset{|}{C}}HCH_3 \xrightarrow{H_2SO_4} CH_2{=}\overset{CH_3}{\overset{|}{C}}-CH_3$

$$+ H_2O \xrightarrow{H^+} CH_3-\overset{OH}{\overset{|}{\underset{CH_3}{\underset{|}{C}}}}-CH_3$$

b. $CH_2{=}CHCH_2CH_3 + H_2O \xrightarrow{H^+}$

$$CH_3\overset{OH}{\overset{|}{C}}HCH_2CH_3 \xrightarrow{\text{Oxidation}} CH_3-\overset{O}{\overset{\|}{C}}-CH_2CH_3$$

c. 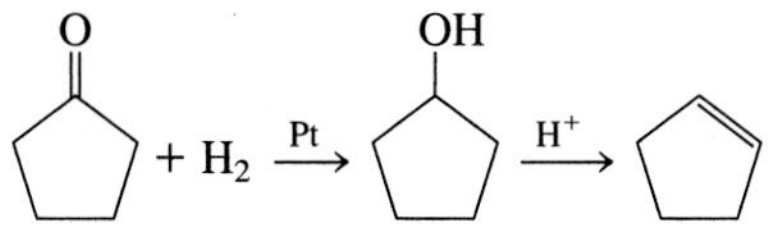 O (cyclopentanone) $+ H_2 \xrightarrow{Pt}$ OH (cyclopentanol) $\xrightarrow{H^+}$ (cyclopentene)

12.31 **a.** alcohol, phenol
b. alcohol, ketone
c. alkene, ketone, alcohol
d. aldehyde, ether, phenol

Chapter 13

13.1 **a.** ethanoic acid, acetic acid
b. butanoic acid, butyric acid
c. methanoic acid, formic acid
d. benzoic acid
e. propanoic acid, propionic acid
f. 3-chloropropanoic acid, β-chloropropionic acid
g. 2-methylbutanoic acid, α-methyl butyric acid
h. 2-hydroxypropanoic acid; α-hydroxypropionic acid

13.2 **a.** $CH_3\overset{O}{\overset{\|}{C}}OH$ **b.** 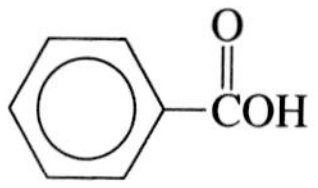 (benzene ring)$-\overset{O}{\overset{\|}{C}}OH$

c. $CH_3\overset{CH_3}{\overset{|}{C}}HCH_2\overset{O}{\overset{\|}{C}}OH$ **d.** $CH_3CH_2\overset{O}{\overset{\|}{C}}OH$

e. $H\overset{O}{\overset{\|}{C}}OH$ **f.** 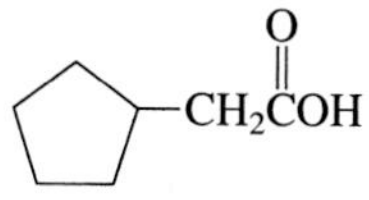 (cyclopentane ring)$-CH_2\overset{O}{\overset{\|}{C}}OH$

13.3 **a.** $CH_3\overset{CH_3}{\overset{|}{C}}HCH_2\overset{O}{\overset{\|}{C}}OH$ **b.** $CH_3CH_2\overset{Cl}{\overset{|}{C}}HCH_2\overset{O}{\overset{\|}{C}}OH$

c. $BrCH_2CH_2CH_2\overset{O}{\overset{\|}{C}}OH$ **d.** $CH_3\overset{OH}{\overset{|}{C}}H-\overset{O}{\overset{\|}{C}}OH$

e. 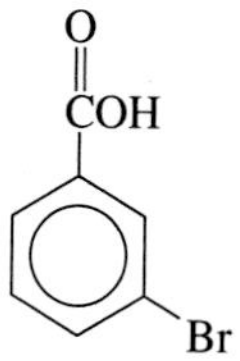 $\overset{O}{\overset{\|}{C}}OH$, Br (benzene ring) **f.** 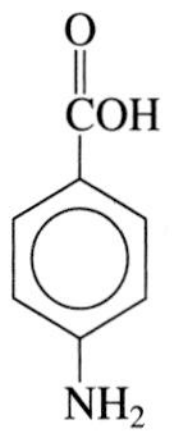 $\overset{O}{\overset{\|}{C}}OH$, NH_2 (benzene ring)

13.4 **a.** The hydrogens in the carboxylic acid groups of the acetic acid form hydrogen bonds with carboxyl oxygen atom in other acetic acid molecules.
b. When hydrogen bonds form between molecules of a substance, more energy is required to break the bonds which results in higher melting and boiling points.

13.5 **a.** $CH_3CH_2\overset{O}{\overset{\|}{C}}-OH$, hydrogen bonds give a higher boiling point

b. $CH_3\overset{O}{\overset{\|}{C}}-OH$, it can form more hydrogen bonds

13.6 **a.** $CH_3\overset{O}{\overset{\|}{C}}-OH$, can hydrogen bond to water and dissolve

b. $CH_3CH_2\overset{O}{\overset{\|}{C}}-OH$, can form more hydrogen bonds and dissolve

13.7 **a.** $CH_3\overset{O}{\overset{\|}{C}}OH \underset{}{\overset{H_2O}{\rightleftharpoons}} CH_3\overset{O}{\overset{\|}{C}}-O^- + H^+$

b. 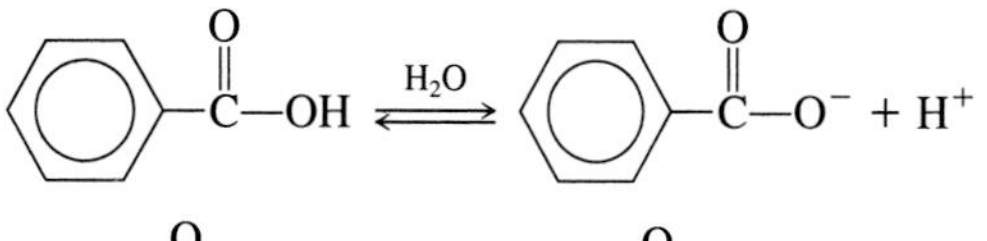 (benzene ring)$-\overset{O}{\overset{\|}{C}}-OH \overset{H_2O}{\rightleftharpoons}$ (benzene ring)$-\overset{O}{\overset{\|}{C}}-O^- + H^+$

c. $CH_3CH_2\overset{O}{\overset{\|}{C}}-OH \overset{H_2O}{\rightleftharpoons} CH_3CH_2\overset{O}{\overset{\|}{C}}-O^- + H^+$

13.8 **a.** $CH_3\overset{O}{\overset{\|}{C}}-OH + KOH \longrightarrow CH_3\overset{O}{\overset{\|}{C}}-O^-\ K^+ + H_2O$

b. 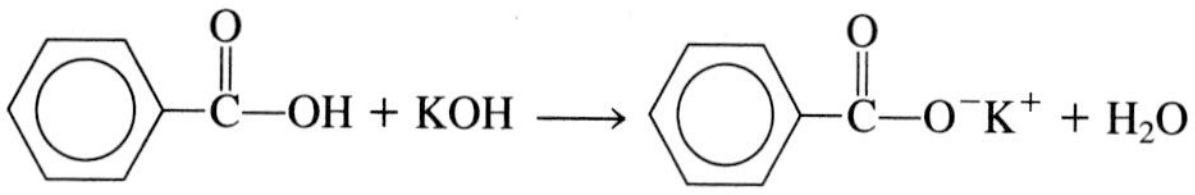 (benzene ring)$-\overset{O}{\overset{\|}{C}}-OH + KOH \longrightarrow$ (benzene ring)$-\overset{O}{\overset{\|}{C}}-O^-K^+ + H_2O$

c. $CH_3CH_2\overset{O}{\overset{\|}{C}}-OH + KOH \longrightarrow CH_3CH_2\overset{O}{\overset{\|}{C}}-O^-K^+ + H_2O$

13.9 Aspirin: benzene ring bearing $COOH$ and $OCOCH_3$ (ester group circled); $CH_3(CH_2)_{14}COO(CH_2)_{29}CH_3$ (ester group circled); $CH_3COOCH_2CH_2CH_2CH_2CH_3$ (ester group circled)

13.10 **a.** $C_6H_5-COOCH_3$ **b.** $CH_3CH(CH_3)-C(=O)-OCH_3$

c. $Cl-CH_2C(=O)-OCH_3$

13.11 **a.** $CH_3CH_2C(=O)-OCH(CH_3)CH_3$

b. $C_6H_5-C(=O)OCH_2CH_2CH_3$

c. $CH_3CH_2CH_2C(=O)-O-C_6H_5$

d. $CH_3C(=O)O-CH_2CH_2CH_2-OC(=O)CH_3$

e. benzene ring bearing two adjacent $-C(=O)-OCH_3$ groups

13.12 **a.** carboxylic acid group and phenol(alcohol)

b. benzene ring bearing $-C(=O)-OH$ and adjacent $-OC(=O)CH_3$

c. benzene ring bearing $-C(=O)-OCH_3$ and adjacent $-OH$

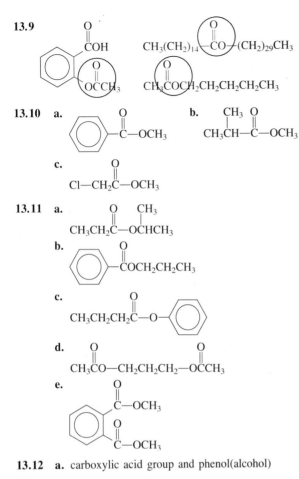

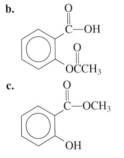

13.13 **a.** methyl ethanoate; methyl acetate
b. ethyl ethanoate, ethyl acetate
c. propyl methanoate, propyl formate
d. methyl benzoate
e. ethyl pentanoate
f. phenyl ethanoate; phenyl acetate
g. propyl benzoate

13.14 **a.** $CH_3CH_2CH_2C(=O)-OCH_2CH_3$
b. $HC(=O)-OCH_2CH_3$
c. $CH_3CH_2CH_2C(=O)-OCH_2CH_2CH_2CH_2CH_3$
d. $CH_3C(=O)-OCH_2CH_2CH_2CH_2CH_2CH_2CH_2CH_3$

13.15 **a.** $HC(=O)-OH + CH_3CH_2CH_2-OH$
b. $CH_3CH_2C(=O)-OH + HO-C_6H_5$
c. $CH_3C(=O)-OH + CH_3CH_2CH_2CH_2-OH$
d. $C_6H_5-C(=O)-OH + CH_3CH_2OH$

13.16 **a.** $CH_3CH_2C(=O)-O^-\ Na^+ + CH_3OH$
b. $C_6H_5-C(=O)-O^-\ K^+ + CH_3CH(CH_3)CH_2OH$
c. $CH_3CH_2C(=O)-O^-Na^+ + C_6H_5-CH_2OH$
d. $K^+{-}O^-C(=O)-CH_2CH_2CH_2-C(=O)-O^-K^+ + 2\ CH_3OH$
e. $3\ CH_3(CH_2)_{12}C(=O)-O^-Na^+ +$ $HO-CH_2CH(OH)-CH_2-OH$

13.17 **a.** $CH_3CH_2CH_2-NH_2$ Propylamine (1°) $CH_3CH_2-NH-CH_3$ Ethylmethylamine (2°)

$CH_3-N(CH_3)-CH_3$ Trimethylamine (3°)

13.18 **a.** primary **b.** secondary **c.** primary
d. tertiary **e.** secondary

13.19 **a.** ethylamine **b.** butyl amine **c.** aniline
d. diethylamine **e.** ethylmethylpropylamine
f. *N*-ethyl-*N*-methyl aniline **g.** *N*-methyl aniline

13.20 **a.** $CH_3CH_2NH_2$ **b.** C_6H_5—NH—CH_3 (N-methylaniline: H—N—CH_3 on benzene ring) **c.** $C_6H_5NH_2$ (benzene ring with NH_2)

d. CH_3—NH—CH_3 (H on N) **e.** CH_3—NH—$CH(CH_3)CH_3$ (H on N, CH_3 on C) **f.** $C_6H_5NH_2$ (benzene ring with NH_2)

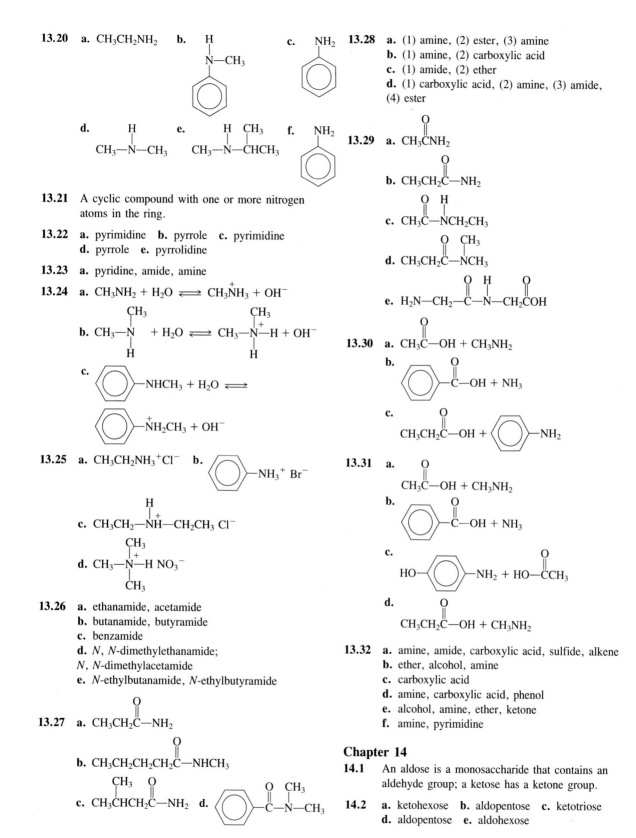

13.21 A cyclic compound with one or more nitrogen atoms in the ring.

13.22 **a.** pyrimidine **b.** pyrrole **c.** pyrimidine **d.** pyrrole **e.** pyrrolidine

13.23 **a.** pyridine, amide, amine

13.24 **a.** $CH_3NH_2 + H_2O \rightleftharpoons CH_3\overset{+}{N}H_3 + OH^-$

b. CH_3—N(H)(CH_3) $+ H_2O \rightleftharpoons CH_3$—$\overset{+}{N}$(H)($CH_3$)—H $+ OH^-$

c. C_6H_5—$NHCH_3 + H_2O \rightleftharpoons$

C_6H_5—$\overset{+}{N}H_2CH_3 + OH^-$

13.25 **a.** $CH_3CH_2NH_3{}^+Cl^-$ **b.** C_6H_5—$NH_3{}^+$ Br^-

c. CH_3CH_2—$\overset{+}{N}H$(H)—CH_2CH_3 Cl^-

d. CH_3—$\overset{+}{N}$(CH_3)(CH_3)—H $NO_3{}^-$

13.26 **a.** ethanamide, acetamide
b. butanamide, butyramide
c. benzamide
d. *N*, *N*-dimethylethanamide; *N*, *N*-dimethylacetamide
e. *N*-ethylbutanamide, *N*-ethylbutyramide

13.27 **a.** $CH_3CH_2\overset{O}{\overset{\|}{C}}$—$NH_2$

b. $CH_3CH_2CH_2CH_2\overset{O}{\overset{\|}{C}}$—$NHCH_3$

c. $CH_3\overset{CH_3}{\overset{|}{C}}HCH_2\overset{O}{\overset{\|}{C}}$—$NH_2$ **d.** C_6H_5—$\overset{O}{\overset{\|}{C}}$—$\overset{CH_3}{\overset{|}{N}}$—$CH_3$

13.28 **a.** (1) amine, (2) ester, (3) amine
b. (1) amine, (2) carboxylic acid
c. (1) amide, (2) ether
d. (1) carboxylic acid, (2) amine, (3) amide, (4) ester

13.29 **a.** $CH_3\overset{O}{\overset{\|}{C}}NH_2$

b. $CH_3CH_2\overset{O}{\overset{\|}{C}}$—$NH_2$

c. $CH_3\overset{O}{\overset{\|}{C}}$—$\overset{H}{\overset{|}{N}}CH_2CH_3$

d. $CH_3CH_2\overset{O}{\overset{\|}{C}}$—$\overset{CH_3}{\overset{|}{N}}CH_3$

e. H_2N—CH_2—$\overset{O}{\overset{\|}{C}}$—$\overset{H}{\overset{|}{N}}$—$CH_2\overset{O}{\overset{\|}{C}}OH$

13.30 **a.** $CH_3\overset{O}{\overset{\|}{C}}$—$OH + CH_3NH_2$

b. C_6H_5—$\overset{O}{\overset{\|}{C}}$—$OH + NH_3$

c. $CH_3CH_2\overset{O}{\overset{\|}{C}}$—$OH$ + C_6H_5—NH_2

13.31 **a.** $CH_3\overset{O}{\overset{\|}{C}}$—$OH + CH_3NH_2$

b. C_6H_5—$\overset{O}{\overset{\|}{C}}$—$OH + NH_3$

c. HO—C_6H_4—NH_2 + HO—$\overset{O}{\overset{\|}{C}}CH_3$

d. $CH_3CH_2\overset{O}{\overset{\|}{C}}$—$OH + CH_3NH_2$

13.32 **a.** amine, amide, carboxylic acid, sulfide, alkene
b. ether, alcohol, amine
c. carboxylic acid
d. amine, carboxylic acid, phenol
e. alcohol, amine, ether, ketone
f. amine, pyrimidine

Chapter 14

14.1 An aldose is a monosaccharide that contains an aldehyde group; a ketose has a ketone group.

14.2 **a.** ketohexose **b.** aldopentose **c.** ketotriose **d.** aldopentose **e.** aldohexose

14.3

CH_2OH
$C{=}O$
$H{-}C{-}OH$
CH_2OH

14.4 **a.** achiral **b.** chiral **c.** achiral **d.** achiral **e.** chiral **f.** chiral **g.** chiral **h.** achiral

14.5 Chiral compounds are c, d, f, g.

14.6 When polarized light is passed through a sample of one of the chiral forms of lactic acid, the plane of light is rotated by the asymmetrical atoms and emerges at a different angle.

14.7 **a.** D-threose **b.** D-xylulose **c.** L-mannose **d.** D-allose **e.** L-sorbose **f.** L-ribose

mirror images

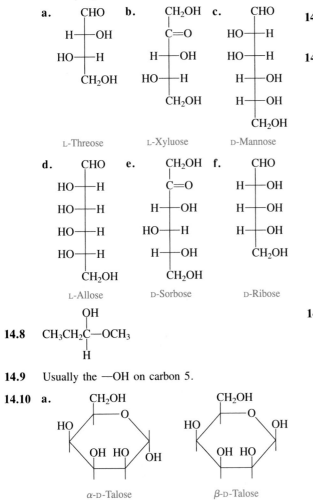

a. CHO; H—OH; HO—H; CH_2OH — L-Threose

b. CH_2OH; C=O; H—OH; HO—H; CH_2OH — L-Xylulose

c. CHO; HO—H; HO—H; H—OH; H—OH; CH_2OH — D-Mannose

d. CHO; HO—H; HO—H; HO—H; HO—H; CH_2OH — L-Allose

e. CH_2OH; C=O; H—OH; HO—H; H—OH; CH_2OH — D-Sorbose

f. CHO; H—OH; H—OH; H—OH; CH_2OH — D-Ribose

14.8 $CH_3CH_2C(OH)(H){-}OCH_3$

14.9 Usually the —OH on carbon 5.

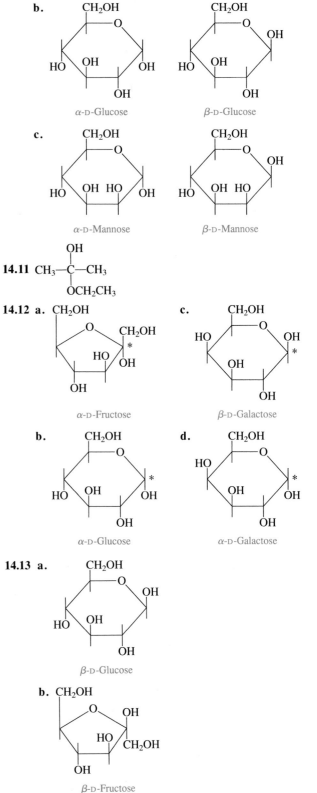

14.10 **a.** CH_2OH, O, HO, OH HO, OH

α-D-Talose

CH_2OH, O, HO, OH, OH HO

β-D-Talose

b. CH_2OH, O, HO, OH, OH, OH

α-D-Glucose

CH_2OH, O, OH, HO, OH, OH

β-D-Glucose

c. CH_2OH, O, HO, OH HO, OH

α-D-Mannose

CH_2OH, O, OH, HO, OH HO

β-D-Mannose

14.11 $CH_3{-}C(OH)(OCH_2CH_3){-}CH_3$

14.12 **a.**

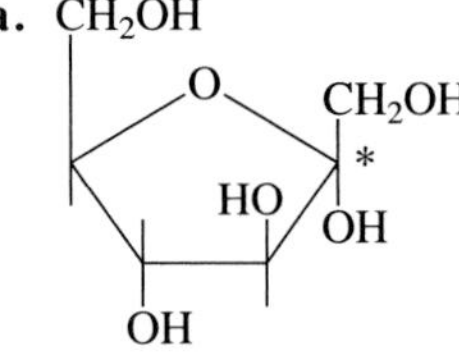

α-D-Fructose

b. CH_2OH, O, *, HO, OH, OH, OH

α-D-Glucose

c. CH_2OH, O, HO, OH, *, OH, OH

β-D-Galactose

d. CH_2OH, O, HO, *, OH, OH, OH

α-D-Galactose

14.13 **a.** CH_2OH, O, OH, HO, OH, OH

β-D-Glucose

b. CH_2OH, O, OH, HO, CH_2OH, OH

β-D-Fructose

14.14

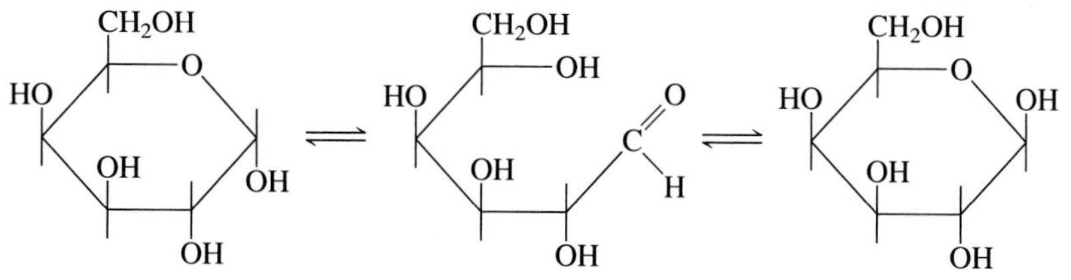

CH₂OH, O, HO, OH, OH, OH ⇌ CH₂OH, OH, HO, O, C, H, OH, OH ⇌ CH₂OH, O, HO, OH, OH, OH

α-Galactose Open chain β-Galactose

14.15 **a.** fruits, vegetables, corn syrup, honey
b. fruit juices, honey
c. lactose (milk sugar)

14.16 A reducing sugar reduces Cu^{2+} to Cu^+ in Benedict's reagent, or Ag^+ to Ag in Tollen's reagent, when the sugar is able to undergo mutarotation to give an aldehyde group that can be oxidized to an acid.

14.17 The carboxylic acid of D-galactose and $Cu_2O(s)$.

14.18 The keto group undergoes an isomerization to give a free aldehyde group.

14.19

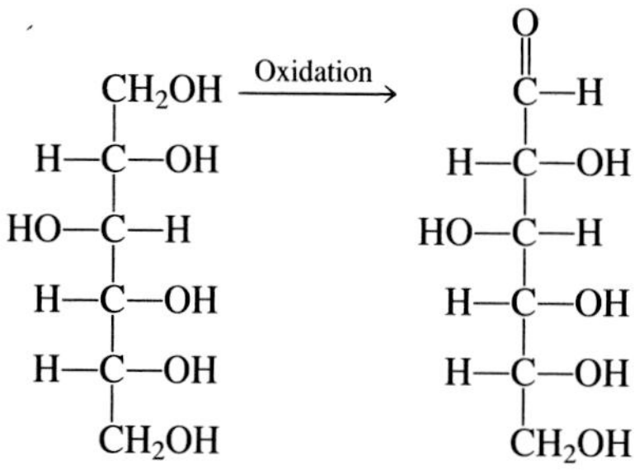

D-Sorbitol	Oxidation →	D-Glucose
CH₂OH		O=C—H
H—C—OH		H—C—OH
HO—C—H		HO—C—H
H—C—OH		H—C—OH
H—C—OH		H—C—OH
CH₂OH		CH₂OH

14.20 a.

CH₂OH, O, HO, OH, O—CH₃, OH CH₂OH, O, O—CH₃, HO, OH, OH

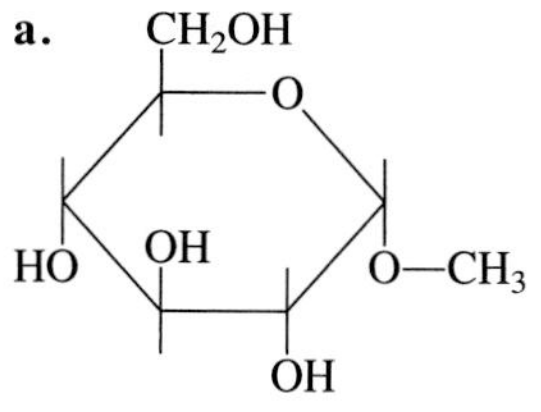

α-Methyl-D-glucoside 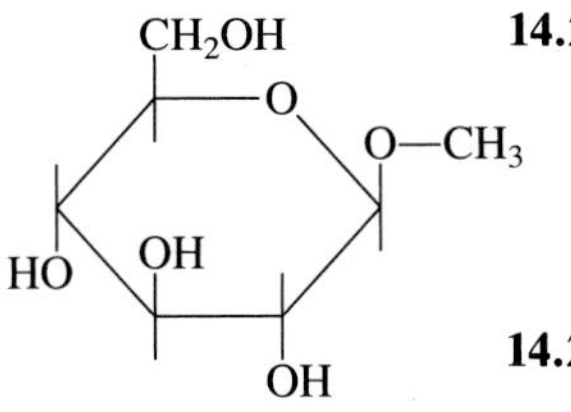 β-Methyl-D-glucoside

b.

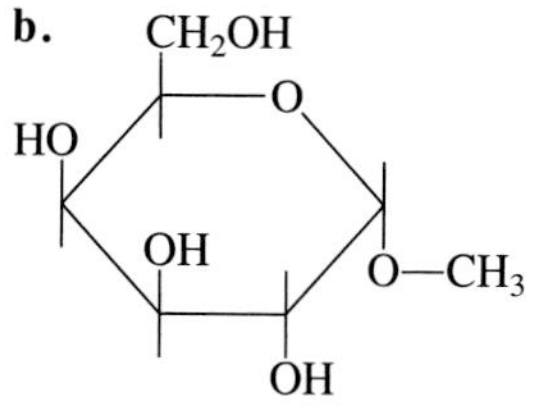

CH₂OH, O, HO, OH, O—CH₃, OH CH₂OH, O, HO, O—CH₃, OH, OH

α-Methyl-D-galactoside 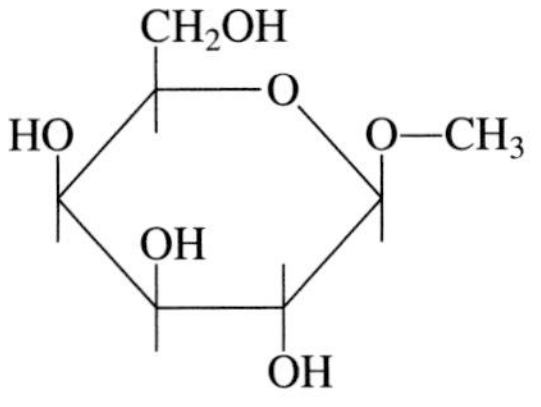 β-Methyl-D-galactoside

14.21 Galactose is bonded to glucose, and glucose is bonded to fructose.

14.22 **a.** β-galactose and β-glucose; β-1,4; β-lactose; a reducing sugar
b. α-glucose and β-glucose; α-1,4; β-maltose; a reducing sugar
c. α-glucose and α-glucose; α-1,4, α-maltose; a reducing sugar
d. α-glucose and β-fructose; α-1, β-2; sucrose; a nonreducing sugar.

14.23 β-cellobiose would consist of two glucose molecules linked by a β-1,4 glycosidic bond.

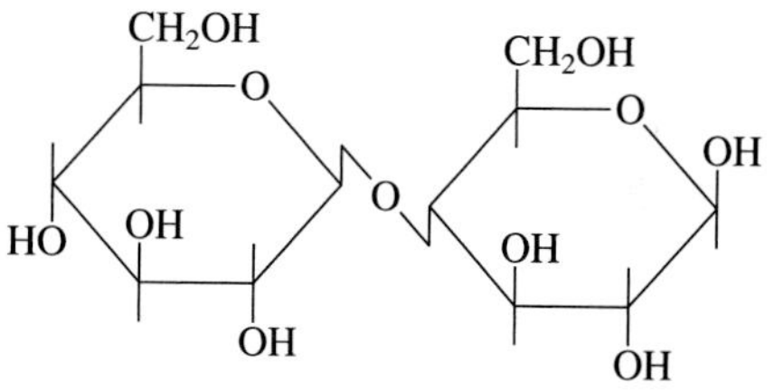

CH₂OH, O, HO, OH, OH, O, CH₂OH, O, OH, OH, OH

14.24 Amylose; α-glucose; straight chain; α-1,4 bonds Amylopectin, α-glucose; branched; α-1,4 and α-1,6 bonds Glycogen; α-glucose, branched; α-1,4 and α-1,6 bonds Cellulose; β-glucose; straight chain; β-1,4

14.25 **a.** cellulose **b.** amylopectin or glycogen
c. amylose

14.26 Humans have enzymes that hydrolyze the α-1,4 bonds of the starches, amylose and amylopectin, but we do not have enzymes that can break down the β-1,4 glycosidic bonds of cellulose. Therefore, we cannot digest cellulose.

14.27 **a.** $Cu(OH)_2$ is added to sugar solution. A free aldehyde group on a reducing sugar reduces the Cu^{2+} to Cu^+ which forms a reddish-orange color. Positive tests occur with reducing sugars, which include all monosaccharides and disaccharides except sucrose.
b. When yeast is added to certain sugars, they ferment to ethanol and CO_2. Glucose, fructose, maltose, and sucrose undergo fermentation.
c. Iodine reacts with polysaccharides to give a blue-black color with amylose and reddish-purple and brown colors with other polysaccharides.

14.28

	Iodine	Fermentation	Benedict's
Galactose	None	None	Positive
Maltose	None	Positive	Positive
Sucrose	None	Positive	None
Starch	Positive	None	None
Lactose	None	None	Positive

14.29 **a.** sucrose **b.** maltose **c.** sucrose **d.** amylose **e.** glycogen **f.** cellulose **g.** galactose **h.** galactose **i.** maltose **j.** cellulose

Chapter 15

15.1 All fatty acids contain a long chain of carbon atoms with a carboxylic acid group. Saturated fats contain only single carbon-to-carbon bonds, whereas unsaturated fats contain one or more double bonds. More saturated fats are found in animal fats, whereas vegetable oils contain more unsaturated fats.

15.2 Saturated fatty acids have regular shapes that fit close together with other saturated fatty acids. To break the attractive forces between them requires higher melting points. Unsaturated fatty acids have one or more cis double bonds that form a bend in the chain. The bent shapes cannot align themselves in a regular pattern and therefore do not have the strong molecular attractions found in the saturated fatty acids. Thus, less energy and a lower melting point will separate (melt) unsaturated fatty acids.

15.3 **a.** $CH_3(CH_2)_{12}\overset{O}{\overset{\|}{C}}OH$ **b.** $CH_3(CH_2)_{14}\overset{O}{\overset{\|}{C}}OH$

c. $CH_3(CH_2)_{16}\overset{O}{\overset{\|}{C}}OH$

15.4 linoleic acid **a.** unsaturated **b.** vegetables **c.** liquid
palmitic acid **a.** saturated **b.** animal fats **c.** solid

15.5 **a.** $CH_3(CH_2)_{14}\overset{O}{\overset{\|}{C}}O(CH_2)_{29}CH_3$ beeswax

b. $CH_3(CH_2)_{18}\overset{O}{\overset{\|}{C}}OCH_2(CH_2)_{20}CH_3$ jojoba wax

15.6 **a.**
$H_2C—OH$
$HC—OH$
$H_2C—OH$

b.
$H_2C—O\overset{O}{\overset{\|}{C}}(CH_2)_{16}CH_3$
$HC—O\overset{O}{\overset{\|}{C}}(CH_2)_{16}CH_3$
$H_2C—O\overset{O}{\overset{\|}{C}}(CH_2)_{16}CH_3$

15.7 palmitic; linoleic; oleic

15.8
$H_2C—O\overset{O}{\overset{\|}{C}}(CH_2)_{14}CH_3$
$HC—O\overset{O}{\overset{\|}{C}}(CH_2)_{14}CH_3$
$H_2C—O\overset{O}{\overset{\|}{C}}(CH_2)_7CH{=}CH(CH_2)_7CH_3$

$H_2C—O\overset{O}{\overset{\|}{C}}(CH_2)_{14}CH_3$
$HC—O\overset{O}{\overset{\|}{C}}(CH_2)_7CH{=}CH(CH_2)_7CH_3$
$H_2C—O\overset{O}{\overset{\|}{C}}(CH_2)_{14}CH_3$

15.9 Safflower oil contains fatty acids with two or three double bonds; olive oil contains a high amount of oleic acid, which has a single (monounsaturated) double bond.

15.10 Olive oil contains more fatty acids that are unsaturated.

15.11 Although coconut oil comes from a vegetable, it has high amounts of saturated fatty acids, and very small amounts of unsaturated fatty acids.

15.12 According to Table 15.3, coconut oil contains 9% unsaturated fats, while olive oil contains 90% unsaturated fats.

15.13 **a.** Triolean contains three unsaturated fatty acids. Tristearin contains only saturated stearic acid. **b.** A triglyceride containing two molecules of linoleic acid and one molecule of oleic acid has more double bonds.

15.14 **a.** 410 kcal; 49%
b. 260 kcal; 50%
c. 160 kcal; 29%
d. 190 kcal; 41%
e. 250 kcal; 52%
The fats are mostly saturated since they are from animal sources.

15.15
$$\begin{array}{l} CH_2-OC(=O)(CH_2)_{16}CH_3 \\ | \\ CH-OC(=O)(CH_2)_{14}CH_3 \\ | \\ CH_2-OC(=O)(CH_2)_{16}CH_3 \end{array}$$

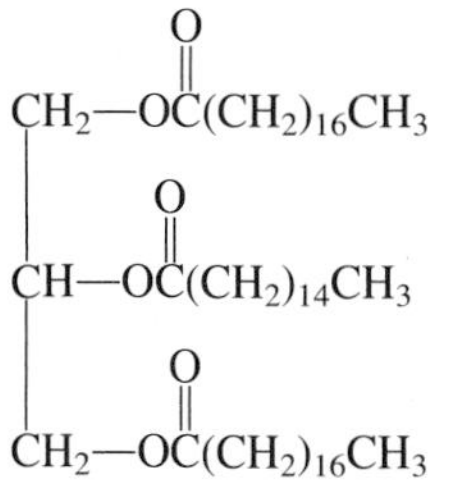

15.16 **a.** Some of the double bonds in the unsaturated fatty acids have been converted to single saturated bonds.
b. It is mostly saturated fatty acids.

15.17 **a.** Safflower oil has 7% saturated fats; corn oil has 15% saturated fats.
b. Yes.

15.18 **a.** 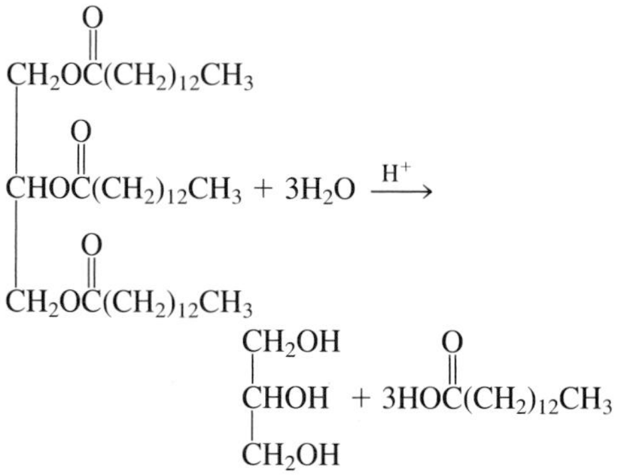
$$\begin{array}{l} CH_2OC(=O)(CH_2)_{12}CH_3 \\ | \\ CHOC(=O)(CH_2)_{12}CH_3 + 3H_2O \xrightarrow{H^+} \\ | \\ CH_2OC(=O)(CH_2)_{12}CH_3 \end{array}$$
$$\begin{array}{l} CH_2OH \\ | \\ CHOH \\ | \\ CH_2OH \end{array} + 3HOC(=O)(CH_2)_{12}CH_3$$
Glycerol Myristic acid

b. 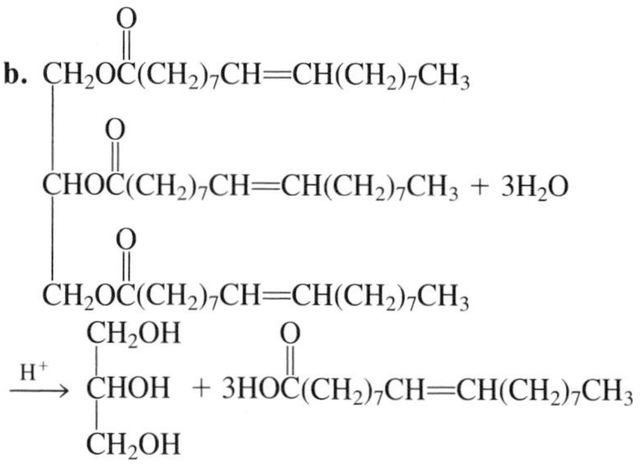
$$\begin{array}{l} CH_2OC(=O)(CH_2)_7CH{=}CH(CH_2)_7CH_3 \\ | \\ CHOC(=O)(CH_2)_7CH{=}CH(CH_2)_7CH_3 + 3H_2O \\ | \\ CH_2OC(=O)(CH_2)_7CH{=}CH(CH_2)_7CH_3 \end{array}$$
$$\xrightarrow{H^+} \begin{array}{l} CH_2OH \\ | \\ CHOH \\ | \\ CH_2OH \end{array} + 3HOC(=O)(CH_2)_7CH{=}CH(CH_2)_7CH_3$$
Glycerol Oleic acid

15.19 **a.** 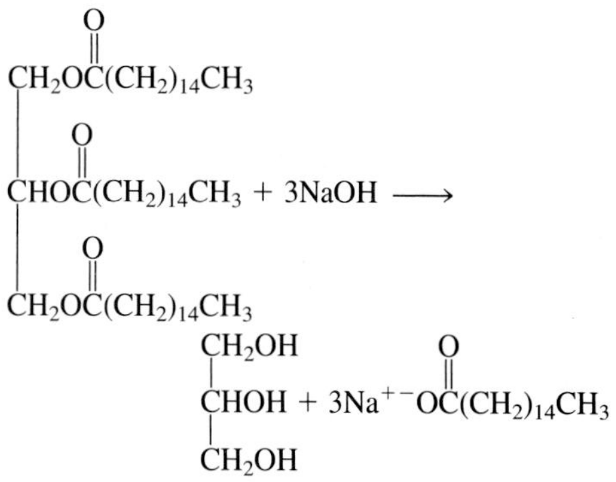
$$\begin{array}{l} CH_2OC(=O)(CH_2)_{14}CH_3 \\ | \\ CHOC(=O)(CH_2)_{14}CH_3 + 3NaOH \longrightarrow \\ | \\ CH_2OC(=O)(CH_2)_{14}CH_3 \end{array}$$
$$\begin{array}{l} CH_2OH \\ | \\ CHOH \\ | \\ CH_2OH \end{array} + 3Na^{+}\,{}^{-}OC(=O)(CH_2)_{14}CH_3$$

b. 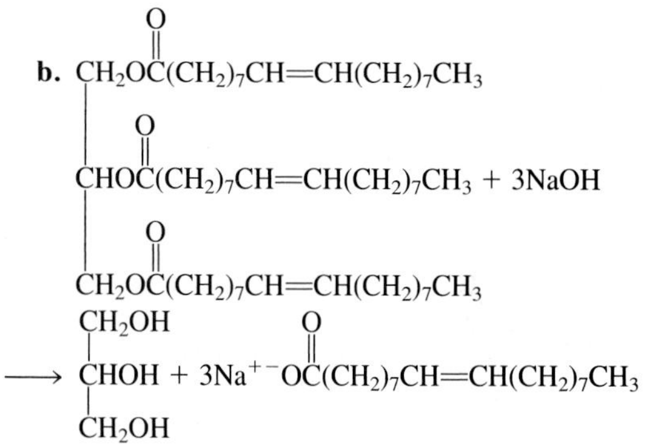
$$\begin{array}{l} CH_2OC(=O)(CH_2)_7CH{=}CH(CH_2)_7CH_3 \\ | \\ CHOC(=O)(CH_2)_7CH{=}CH(CH_2)_7CH_3 + 3NaOH \\ | \\ CH_2OC(=O)(CH_2)_7CH{=}CH(CH_2)_7CH_3 \end{array}$$
$$\longrightarrow \begin{array}{l} CH_2OH \\ | \\ CHOH \\ | \\ CH_2OH \end{array} + 3Na^{+}\,{}^{-}OC(=O)(CH_2)_7CH{=}CH(CH_2)_7CH_3$$

15.20 The unsaturated sites can react with oxygen faster when it is warm to give smaller chain fatty acids with unpleasant rancid odors.

15.21 **a.** a saturated triglyceride
b. glycerol and fatty acids
c. glycerol and the salts of the fatty acids

15.22 One of the fatty acids has been replaced by a phosphate group that is attached to an amino alcohol.

15.23 The phosphate group and amino alcohol make the molecule polar, which makes it more soluble in water.

15.24 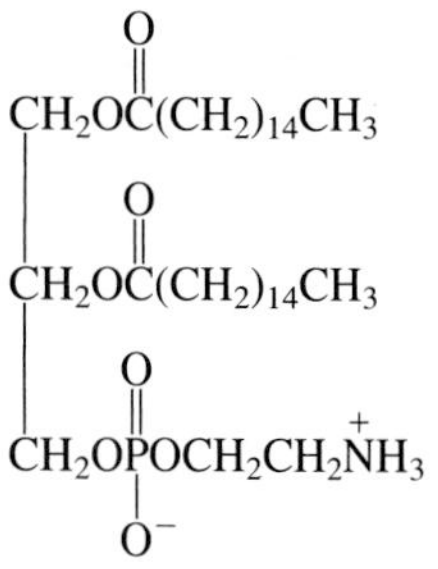
$$\begin{array}{l} CH_2OC(=O)(CH_2)_{14}CH_3 \\ | \\ CH_2OC(=O)(CH_2)_{14}CH_3 \\ | \\ CH_2OP(=O)(O^-)OCH_2CH_2\overset{+}{N}H_3 \end{array}$$
Cephalin

15.25 Lecithin contains the amino alcohol choline whereas a cephalin contains serine or ethanolamine.

15.26 Sphingosines use the alcohol sphinogosine in place of glycerol.

15.27 One end of a phospholipid is nonpolar and soluble in nonpolar environment; the polar end is soluble in water.

15.28 In a cell membrane, the phospholipids are arranged in a double row with the polar parts which are hydrophilic on the outsides and the nonpolar hydrophobic fatty acids in the middle.

15.29 A lecithin contains glycerol as the alcohol; sphingomyelin uses sphingosine. Yes, both are phospholipids.

15.30 **a.** steroid nucleus **b.** isoprene

15.31 **a.** terpene **b.** steroid **c.** terpene **d.** steroid **e.** steroid

15.32 **a.** wax **b.** steroid **c.** phospholipid **d.** terpene **e.** sphingolipid **f.** triglyceride

15.33 **a.** long chain alcohol and fatty acid
b. steroid nucleus
c. two fatty acids, glycerol, phosphate and amino alcohol
d. glycerol and three fatty acids
e. isoprene units
f. salts of fatty acids
g. sphingosine, fatty acids, and a monosaccharide
h. steroid nucleus
i. isoprene units

Chapter 16

16.1 All amino acids contain a carboxylic acid group and an amino group on the alpha carbon. The distinguishing feature is a side group(R) that may be one hydrogen atom or several carbon atoms. Sometimes an oxygen atom or a sulfur atom is present in the side group.

16.2 **a.** $H_2H-CH(CH_2SH)-C(=O)OH$ Cysteine **b.** $H_2H-CH(CH_2OH)-C(=O)OH$ Serine

c. $H_2N-CH(CH_2CH_2CH_2CH_2NH_2)-C(=O)OH$ Lysine **d.** $H_2N-CH(CH_2C(=O)OH)-C(=O)OH$ Aspartic acid

e. $H_2N-CH(CH_2-C_6H_4-OH)-C(=O)OH$ Tyrosine **f.** $H_2N-CH(CH_3)-C(=O)OH$ Alanine

16.3 **a.** phenylalanine(Phe) **c.** valine(Val)
b. glutamic acid(Glu) **d.** isoleucine(Ile)

16.4 **a.** hydrophobic **b.** hydrophilic **c.** hydrophobic **d.** hydrophobic

16.5 An essential amino acid is not synthesized in the body and must be provided by the diet.

16.6 **a.** yes **b.** yes **c.** no **d.** yes **e.** no **f.** yes

16.7 Kwashiorkor

16.8 **a.** A zwitterion has an ionized amino group (NH_3^+) and carboxyl group (COO^-).
b. A hydrogen lost by the ionization of the carboxylic acid group is gained by the lone pair of electrons on the amino group.
c. The pH at which the zwitterion of an amino acid is electrically neutral.

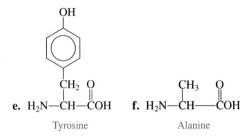

16.9 **a.** $\overset{+}{N}H_3-CH_2-C(=O)O^-$ **b.** $\overset{+}{N}H_3-CH(CH_2C_6H_5)-C(=O)O^-$

c. $\overset{+}{N}H_3-CH(CH_2OH)-C(=O)O^-$ **d.** $\overset{+}{N}H_3-CH(CH_2-C_6H_4-OH)-C(=O)O^-$

e. $\overset{+}{N}H_3-CH(CH_2CH_2-S-CH_3)-C(=O)O^-$ **f.** $\overset{+}{N}H_3-CH(CH_2CH(CH_3)_2)-C(=O)O^-$

16.10 **a.** and **b.** $\overset{+}{N}H_3-CH(CH_2CH(CH_3)_2)-C(=O)OH$ pH = 2.0, 4.0

$CH_3\ \ CH_3$
CH
$CH_2\quad O$
c. $\overset{+}{N}H_3-CH-CO^-$ pH = 6.0
(At isoelectric point)

$CH_3\ \ CH_3$
CH
$CH_2\quad O$
d. and **e.** $NH_2-CH-CO^-$ pH = 8.0, 10.0

O
COH
$CH_2\quad O$
16.11 **a.** $H_3\overset{+}{N}-CH-COH$ pH = 2.0

O
CO^-
$CH_2\quad O$
b, c, d, e $H_2N-CH-CO^-$ pH = 4.0, 6.0, 8.0, 10.0

SH
$CH_3\ O\ \ H\ \ CH_2\ O$
16.12 **a.** $H_2N-CH-C-N-CH-COH$

CH_3
$OH\qquad CH_2$
$CH_2\ O\ \ H\ \ CH-CH_3$
b. $H_2N-CH-C-N-CH-COOH$

$CH_3\qquad NH_2$
$S\qquad C{=}O$
$CH_2\qquad CH_2$
$CH_2\ O\ \ H\ \ CH_2\ O$
c. $H_2N-CH-C-N-CH-COH$

$CH_3\ \ CH_3$
CH
$O\ \ H\ \ CH_2\ O$
16.13 **a.** $H_2N-CH_2-C-N-CH-COH$ Glycylleucine Gly-Leu

$CH_3\ \ CH_3$
CH_2
$CH_2\ O\ \ H\qquad O$
$H_2N-CH-C-N-CH_2-COH$ Leucylglycine Leu-Gly

OH
(benzene ring)
$CH_3\ O\ \ H\qquad O$
b. $H_2N-CH-C-N-CH-COH$ alanyltyrosine Ala-Tyr

OH
(benzene ring)
$O\ \ H\ \ CH_3\ O$
$H_2N-CH-C-N-CH-COH$ tyrosylalanine Tyr-Ala

$O\ \ H\ \ CH_3\ O\ \ H\ \ CH_3\ O$
16.14 **a.** $H_2N-CH_2-C-N-CH-C-N-CH-COH$
Gly Ala Ala

$CH_3\qquad OH$
S (benzene ring) (benzene ring)
CH_2
$CH_2\ O\ \ H\ \ CH_2\ O\ \ H\ \ CH_2$
b. $H_2N-CH-C-N-CH-C-N-CH-COOH$
Met Tyr Phe

O
$COH\qquad\qquad CH_3$
$CH_2\quad CH_3\ \ CH_3\quad CH_2$
$CH_2\ O\ \ H\ \ CH\ \ O\ \ H\ \ CH-CH_3$
c. $H_2N-CH-C-N-CH-C-N-CH-COOH$
Glu Val Ile

NH_2
$C{=}O$
$CH_3\ \ CH_3\quad OH\qquad SH\qquad CH_2$
$CH\ \ O\ \ H\ \ CH_2\ O\ \ H\ \ CH_2\ O\ \ H\ \ CH_2\ O$
16.15 **a.** $H_2N-CH-C-N-CH-C-N-CH-C-N-CH-COH$
Val Ser Cys Gln

OH
O
$COH\quad CH_3\ \ CH_3$ (benzene ring)
$CH_2\ O\ \ H\ \ CH\ \ O\ \ H\qquad O\ \ H\qquad O$
b. $H_2N-CH-C-N-CH-C-N-CH_2-C-N-CH-COH$
Asp Val Gly Tyr

16.16 **a.** Tyr-Phe-Glu
b. Cys-Val-Ala-His
c. Thr-Gly-Arg-Leu-Glu
d. Lys-Met-Cys-Lys-Val-Gly

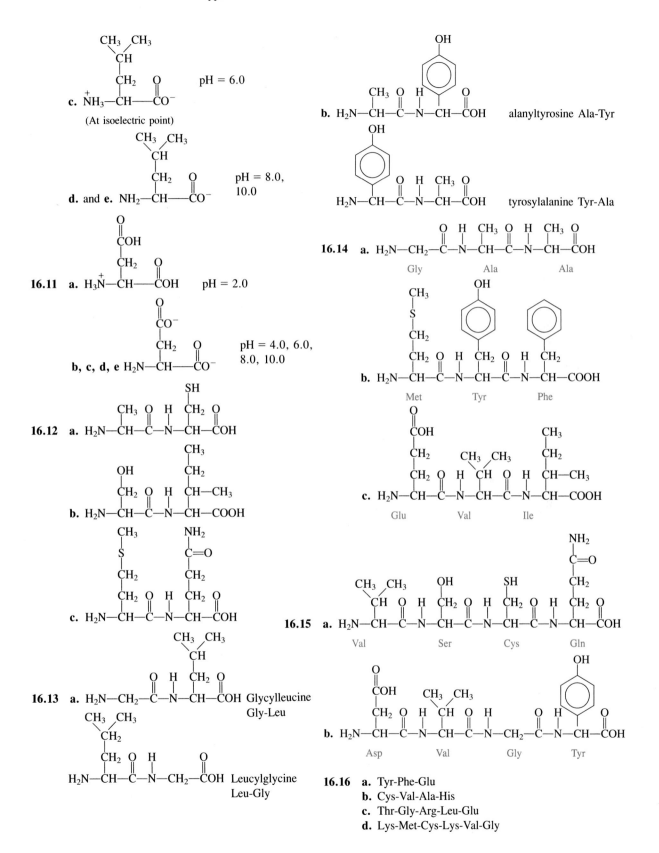

16.17 In an α-helix, hydrogen bonding occurring between the amide groups and carbonyl groups (—N—H---O=C—) makes the polypeptide chain take on a coiled structure.

16.18 In a pleated sheet, several polypeptide chains are held close together by forming hydrogen bonds between the adjacent chains.

16.19 **a.** In collagen, three polypeptide chains containing high amounts of proline and glycine are woven together like a braid called a triple helix.
b. Fibers of the triple helixes of collagen make up 30% of the protein in the human body including connective tissue. α-keratins consist of several α-helixes wrapped together to form fibrous proteins of hair, wool, skin, and nails.

16.20 **a.** On the inside of the protein structure.
b. On the outside.
c. Myoglobin contains polar R groups on the surface of the protein, which helps make it more soluble in water. Silk has small nonpolar R groups, so it would not be expected to be soluble in water.

16.21 **a.** disulfide **b.** salt bridge **c.** hydrogen bonds **d.** hydrophobic attractions

16.22 **a.** cysteine **b.** leucine and valine; they are hydrophobic
c. aspartic acid; soluble in water
d. The R group on the amino acids in the chain will determine where cross-links occur and how the chain folds to place the nonpolar R groups in the center and the polar R groups on the surface.

16.23 **a.** A secondary structure involves hydrogen bonds to form an α-helix or pleated sheet; a tertiary structure involves side group attractions such as disulfide bonds, salt bridges, hydrophobic, and hydrogen bonds to cause a folding of the chain into a three-dimensional shape.
b. An essential amino acid must be supplied by the diet; a nonessential amino acid is synthesized by the body.
c. Polar amino acids have R groups with oxygen or ionic groups; they are hydrophilic. Nonpolar amino acids have R groups with only carbon atoms that make the amino acids hydrophobic.
d. A dipeptide contains two amino acids with one peptide bond; a tripeptide has three amino acids attached by two peptide bonds.
e. A salt bridge is a tertiary feature that results from the ionic attraction between acidic and basic side groups. A disulfide bond is a tertiary feature that occurs when two cysteine side groups bond.
f. A fibrous protein uses β-pleated sheet or a collagen braid to form structural components in the body such as muscle, hair, nails, and tendons. Globular proteins such as antibodies and enzymes have a compact, three-dimensional shape.
g. In an α-helix, the hydrogen bonding occurs between the amino acids in a single polypeptide chain. In the β-pleated sheet, the hydrogen bonding occurs between aligned polypeptide chains.
h. An α-helix is the secondary structure that forms the spiral chain in a polypeptide. A triple helix consists of 3 α-helical polypeptide chains.
i. The tertiary structure is a three-dimensional shape of a single protein caused by side group attractions. The quaternary structure combines two or more of the tertiary structures.

16.24 **a.** tertiary **b.** quarternary **c.** primary
d. secondary **e.** secondary **f.** tertiary
g. secondary

16.25 **a.** globular **b.** fibrous **c.** globular **d.** fibrous
e. fibrous **f.** globular

16.26 **a.** simple **b.** conjugated **c.** conjugated
d. simple **e.** simple

16.27 **a.** transport(3) **b.** structural(2) **c.** structural(2)
d. catalytic(1) **e.** hormonal(6) **f.** protective(5)
g. storage(4) **h.** catalytic(1)

16.28 **a.** Heat breaks apart the hydrogen bonds, causing a loss of the tertiary and secondary structures.
b. The protein will retain the primary structure, the sequence of the amino acids.

16.29 Acid causes the disruption of ionic bonds and hydrogen bonds; tertiary and secondary structures break down.

16.30 Alcohol disrupts hydrogen bonds, causing a loss of the tertiary and secondary structures of the proteins in the bacteria on the skin.

16.31 **a.** Destroys bacteria by denaturing their proteins.
b. Coagulates proteins, forming a cover over the burn area.
c. Denatures milk protein; coagulation occurs.
d. Denatures protein in bacteria.
e. Coagulates protein of mold and bacteria.

16.32 **a.** $H_2N—CH(CH_3)—\overset{O}{\overset{\|}{C}}OH + H_2N—CH(CH_2OH)—\overset{O}{\overset{\|}{C}}OH$

b. $H_2N—CH_2—\overset{O}{\overset{\|}{C}}OH + H_2N—CH(CH(OH)CH_3)—\overset{O}{\overset{\|}{C}}OH + H_2N—CH(CH_2C_6H_5)—\overset{O}{\overset{\|}{C}}OH$

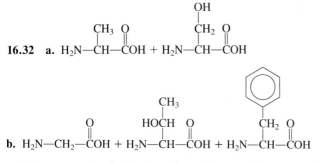

16.33 Denaturation is the disruption of the secondary and tertiary structural features of the protein that make it biologically active; hydrolysis is the breaking apart of the amino acids in the primary structure.

16.34 **a.** The albumin, which has a net negative charge.
b. The lysozyme, which will have a net positive charge.
c. The hemoglobin, which will be electrically neutral since the buffer is the same pH as the isoelectric point of hemoglobin.

16.35 **a.** biuret test; a tripeptide or polypeptide forms a violet color with the cupric ion.
b. xanthoproteic test; amino acids with aromatic R groups give a yellow color when treated with HNO_3.
c. sulfur test; cysteine with a sulfur group reacts with lead acetate to form a black precipitate of PbS.
d. biuret test
e. ninhydrin; the free amino group produces a blue to purple color when heated with ninhydrin.

Chapter 17

17.1 **a.** 2 **b.** 1 **c.** 4 **d.** 5 **e.** 3

17.2 **a.** $\underset{\text{Enzyme}}{E} + \underset{\text{Substrate}}{S} \rightleftharpoons \underset{\text{Enzyme-substrate}}{ES} \longrightarrow \underset{\text{Enzyme}}{E} + \underset{\text{Product}}{P}$

b. The active site, the small pocket where the substrate fits, is composed of a group of amino acids that catalyze the breaking or forming of bonds on the substrate.
c. The products leave the enzyme and the enzyme is again available to combine with substrate.

17.3 **a.** reactant: lactose; products: glucose and galactose

b.

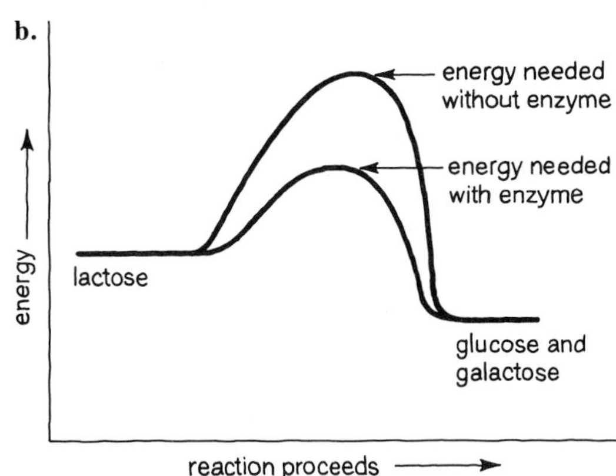

c. The enzyme catalyzes the reaction by lowering the energy required. Therefore, more substrate molecules have the energy to undergo reaction.

17.4 A cofactor is a nonprotein component of an enzyme such as a metal ion, vitamin, or hormone required for activity by some enzymes.

17.5 **a.** conjugated **b.** conjugated **c.** simple **d.** conjugated **e.** simple

17.6 **a.** lipid **b.** peptide **c.** maltose **d.** ester

17.7 oxidoreductase; transferase, hydrolase, lyase, ligase, and isomerase

17.8 **a.** hydrolase **b.** hydrolase **c.** hydrolase **d.** oxidoreductase **e.** oxidoreductase **f.** isomerase **g.** lyase **h.** lyase **i.** transferase **j.** ligase

17.9 **a.** slower **b.** faster **c.** slower **d.** slower (body temperature is 37°C) **e.** slower or stops

17.10 pepsin, pH 1.5; urease, pH 5; oxidase, pH 11

17.11 **a.** competitive **b.** noncompetitive **c.** competitive **d.** noncompetitive **e.** noncompetitive **f.** competitive

17.12 **a.** competitive; structure is similar
b. by competing for the active site
c. increase the concentration of succinic acid

17.13 hydrolysis

17.14 Digestion of amylose and amylopectin begins in the mouth where salivary amylase hydrolyzes α-1,4 bonds. Maltose, glucose, smaller polysaccharides, and dextrins result. Further hydrolysis continues in the small intestine with pancreatic amylase. The disaccharides sucrose (from diet), maltose, and lactose are hydrolyzed in the small intestine by the enzymes sucrase, maltase, and lactase to give glucose, fructose, and galactose.

17.15 In the small intestine, lipases from the pancreas hydrolyze emulsified fats into fatty acids, monoglycerides, and glycerol. Triglycerides re-formed in the mucosal cells join with proteins, phospholipids and cholesterol to be carried in a more soluble form through the lymph and bloodstream.

17.16 **a.** In the stomach.
b. An increase in the acidity of the stomach activates the inactive enzyme (pepsinogin) to give pepsin, which hydrolyzes proteins. If acid were always present in high amounts in the stomach, the formation of the active form of pepsin will cause the destruction of the proteins in the stomach lining and result in ulcers.
c. The smaller polypeptides resulting from hydrolysis of proteins in the stomach continue their degradation in the small intestine by enzymes such as trypsin and chymotrypsin. The dipeptides are hydrolyzed in the mucosa by peptidases to give amino acids, end products of protein digestion, which can be absorbed through the intestinal walls into the bloodstream.

17.17 **a.** T **b.** T **c.** F **d.** T **e.** T **f.** F **g.** T

17.18 **a.** retinol **b.** riboflavin **c.** ascorbic acid **d.** calciferol **e.** tocopherol

17.19 **a.** fat **b.** water **c.** fat **d.** water **e.** water **f.** water **g.** fat

17.20 Water-soluble vitamins; fat-soluble vitamins

17.21 **a.** A, retinol **b.** niacin **c.** B_1, thiamine, **d.** D, calciferol **e.** C, ascorbic acid **f.** B_2, riboflavin **g.** K, phylloquinone

17.22 **a.** It contains the steroid nucleus.
b. It is structurally similar to estrogen.

17.23 **a.** growth hormone **b.** oxytocin **c.** adrenocorticotrophin **d.** growth hormone **e.** vasopressin **f.** luteinizing hormone **g.** glucagon

17.24 **a.** thyroxine **b.** insulin **c.** glucagon **d.** parathyroid

Chapter 18

18.1 **a.** The nucleic acids, DNA and RNA, are composed of four different kinds of nucleotides.
b. Nucleotides consist of a nitrogen base, a sugar, and a phosphate.

18.2 In DNA, the sugar is deoxyribose; RNA nucleotides contain ribose.

18.3 In DNA, the nitrogen bases are adenine, thymine, guanine, and cytosine. In RNA, uracil replaces thymine.

18.4 DNA: deoxyadenosine monophosphate(dAMP), deoxycytidine monophosphate(dCMP), deoxyguanosine monophosphate (dGMP), and deoxythymidine monophosphate(dTMP)
RNA: adenosine monophosphate(AMP), cytidine monophosphate (CMP), guanosine monophosphate(GMP), and uridine monophosphate(UMP)

18.5 **a.** deoxyribose, thymine, and phosphate
b. deoxyribose, guanine, and phosphate
c. ribose, cytosine, and phosphate
d. ribose, adenine, and phosphate
e. deoxyribose, guanine, and phosphate

18.6 Bonds form between the —OH group on carbon 3 of the sugar and the phosphate on carbon 5 of the next sugar to give an alternating sugar-phosphate-sugar-phosphate backbone. The nitrogen bases extend out from the sugars.

18.7

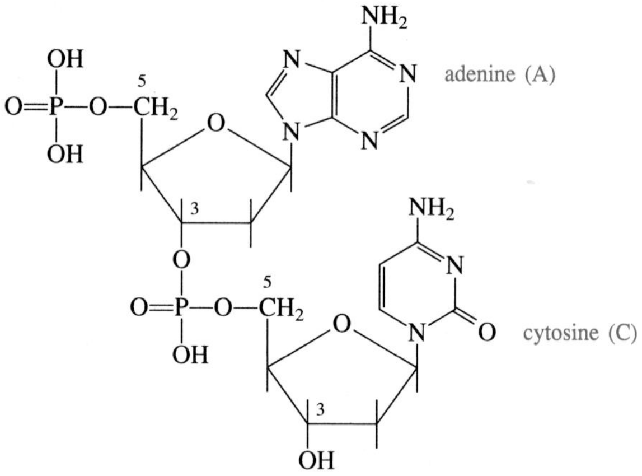

18.8 Complementary base pairing is the bonding requirement between DNA strands of adenine with thymine, and cytosine with guanine.

18.9 **a.** T-C-A-C-G-T **b.** G-A-C-A-T-A-T-G **c.** T-T-T-T-T-T **d.** C-C-C-G-G-G-T-T-T

18.10 The DNA polymer is a coiled shape called an α helix with the sugar-phosphate backbones on the outside like the railing of a spiral staircase, and the base pairs, A-T and G-C, as the steps.

18.11 An enzyme separates the base pairs along the DNA chain. Each separated chain acts as a template, picking up the complementary base for each nucleotide, which exactly duplicates each half and produces two identical DNA copies.

18.12

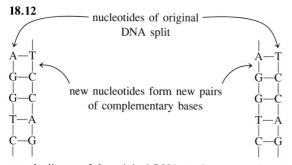

nucleotides of original DNA split

```
A—T        A—T
G—C        G—C
G—C        G—C
T—A        T—A
C—G        C—G
```

new nucleotides form new pairs of complementary bases

two duplicates of the original DNA portion are produced

18.13 A section of a plasmid from a bacterium is replaced by a gene(DNA) from a nonbacterial organism. When the altered plasmid enters bacteria, it is cloned and able to produce a new protein.

18.14 Messenger RNA(mRNA) carries information from DNA for the construction of a protein.
Ribosomal RNA(rRNA), making up 60% of the ribosomes, is involved in the synthesis of proteins.
Transfer RNA(tRNA) consists of 20 or more small RNA that recognize and transport amino acids to the mRNA.

18.15 Transcription is the transfer of the instructions for protein synthesis from the DNA to an mRNA that is synthesized from the template DNA strand using complementary base pairing and using uracil to complement adenine(A-U).

18.16 -G-G-C-U-U-C-C-A-A-G-U-G-

18.17 **a.** leucine **b.** serine **c.** glycine **d.** alanine

18.18 **a.** GGU, GGC, GGA, or GGG
b. ACU, ACC, ACA, and ACG
c. GCU, GCC, GCA, and GCG

18.19 The beginning codon would be AUG or GUG; the ending would be UGA, UAA, or UAG.

18.20 **a.** Several sections of the tRNA have doubled over to form complementary pairs giving an overall appearance of a cloverleaf.
b. At one of the loops, there is a triplet called an anticodon which is complementary to the codons for the amino acids on the mRNA.
c. Each tRNA picks up one of the twenty kinds of amino acids to place in the polypeptide chain.

18.21 In translation, the tRNA with the correct amino acid finds its complementary codon on the mRNA at binding site 2. The amino acid fits close to the one in binding site 1 and a peptide bond forms. The tRNA with the peptide moves to binding site 1, and a new codon moves into binding site 2. When a stop codon is reached, the peptide is released from the ribosome.

18.22 **a.** Lys-Lys-Lys-Lys
b. Phe-Pro-Phe-Pro
c. Tyr-Gly-Arg-Cys

18.23 G-C-U-C-U-U-G-G-U-C-C-U-U-C-U-C-A-U

18.24 **a.** To make efficient use of available amino acids.
b. At transcription, in which mRNA is synthesized by the gene. When enzymes are not needed, their mRNA is not synthesized.

18.25 When a substrate requires an enzyme, it combines with the repressor to make it inactive. Then the operator gene can turn on the synthesis of mRNA by its structural genes.

18.26 The repressor is a protein formed by the regulator gene. It is active when end products of a reaction combine with the repressor and turn off the operator gene.

18.27 **a.** A substitution can occur in one of the nucleotides, or there could be a deletion or addition of a nucleotide in the DNA sequence.
b. The altered DNA transfers incorrect information to its mRNA, which in turn contains codons for incorrect amino acids. For a substitution, there may be only one incorrect amino acid. For deletions or additions, there can be many, since there is a shift in reading the nucleotide sequence after the mutation.
c. A genetic disease results from a mutation that produces an error in the primary structure of a protein that may cause it not to function or to function poorly.

18.28 **a.** Val-Pro-Lys
b. In DNA, CAA changes to GAA, which produces an mRNA that now codes for an amino acid order of Leu-Pro-Lys.

Chapter 19

19.1 In a catabolic reaction such as oxidation, energy is released; energy is required in anabolic reactions such as muscle contraction, transport, and synthesis of cellular compounds.

19.2 **a.** $ADP + P_i + \text{energy} \longrightarrow ATP$
b. $ATP + H_2O \longrightarrow ADP + P_i + \text{energy}$

19.3 ATP is a storage form of energy in cells. Its hydrolysis provides energy for energy-requiring processes.

19.4. **a.** NAD^+, FAD
b. $MH_2 + FAD \longrightarrow M + FADH_2$
$MH_2 + NAD^+ \longrightarrow M + NADH + H^+$

19.5 **a.** $NADH + H^+ \longrightarrow NAD^+ + 2H$
b. $Q + 2H \longrightarrow QH_2$
c. $FMNH_2 \longrightarrow FMN + 2H$

19.6 The cytochromes transfer electrons from coenzyme Q to oxygen(O_2) to form H_2O.

19.7 **a.** reduction **b.** oxidation

19.8 **a.** Q: FMN
b. 2 Cyt c_1 (Fe^{3+}); 2 Cyt b(Fe^{3+})
c. FAD; QH_2

19.9 **a.** H_2O and ATP **b.** 3 **c.** 2

19.10 **a.** glucose **b.** pyruvic acid

19.11 **a.** ATP is used first to activate glucose.
b. 2

19.12 **a.** NAD^+
b. In two of the steps, phosphate is transferred directly to ADP; ADP + $P_i \longrightarrow$ ATP

19.13 Two ATP are used up in the transport of NADH + H^+ into the mitochondria. The net production of ATP in glycolysis is 4 ATP (6 ATP − 2 ATP).

19.14 **a.** Pyruvic acid is oxidized as it undergoes decarboxylation to give acetyl CoA. The hydrogen acceptor NAD^+ provides 3 ATP.
b. pyruvic acid + $NAD^+ \longrightarrow$ acetyl CoA + CO_2 + NADH + H^+
c. 3 ATP.

19.15 **a.** Acetyl CoA is the end product of several metabolic pathways that oxidize food stuffs. It serves as the fuel for the citric acid cycle in the production of energy.
b. citric acid, isocitric acid
c. α-ketoglutaric acid
d. succinic acid, fumaric acid, malic acid, oxaloacetic acid
e. decarboxylation of isocitric acid and α-ketoglutaric acid
f. CO_2 and ATP

19.16 The hydrogen acceptor NAD^+ is used when hydrogen atoms are removed from atoms in a carbon-oxygen bond. FAD is used when hydrogen atoms are removed from the atoms in a carbon-carbon bond.

19.17 Oxidation reactions of compounds in the citric acid cycle provide the hydrogen atoms for the electron transport system where ATP is produced.

19.18 **a.** NAD^+; 3 ATP **b.** NAD^+; 3 ATP **c.** GDP; 1 ATP **d.** FAD; 2 ATP **e.** NAD^+; 3 ATP

19.19 Acetyl CoA $\longrightarrow$ $2CO_2$ + 12 ATP

19.20 **a.** 60 molecules ATP **b.** 24 moles ATP
c. 300 molecules ATP **d.** 120 moles ATP

19.21 **a.** 6 **b.** 2(3) = 6 **c.** 2(12) = 24 **d.** 36 **e.** 2 **f.** 72

19.22 Activation is necessary before a fatty acid can enter the mitochondria. The activation combines a fatty acid with coenzyme A using the energy of 2 ATP.

19.23 **a.** (1) Oxidation forms a double bond between the α and β carbons. (2) Hydration adds water to give a beta (β)-alcohol. (3) Oxidation gives a beta (β)-keto group. (4) A splitting of the chain gives the two-carbon unit, acetyl CoA, and a shorter fatty acid.
b. First and third oxidation steps
c. FAD is used in first oxidation because hydrogen atoms are removed to form a carbon-carbon double bond. In the second oxidation, NAD^+ is used because hydrogen atoms are removed to form a carbon-oxygen double bond.
d. 5 ATP.

19.24 **a.** $CH_3(CH_2)_{12}\text{—}\underset{\beta}{CH_2}\text{—}\underset{\alpha}{CH_2}\text{—}\overset{O}{\overset{\|}{C}}\text{—CoA}$

b. oxidation

$$CH_3(CH_2)_{12}\text{—}CH_2\text{—}CH_2\text{—}\overset{O}{\overset{\|}{C}}\text{—CoA} + FAD \longrightarrow CH_3(CH_2)_{12}\text{—}CH\text{=}CH\text{—}\overset{O}{\overset{\|}{C}}\text{—CoA} + FADH_2$$

hydration

$$CH_3(CH_2)_{12}\text{—}CH\text{=}CH\text{—}\overset{O}{\overset{\|}{C}}\text{—CoA} + H_2O \longrightarrow CH_3(CH_2)_{12}\text{—}\overset{OH}{\overset{|}{C}H}\text{—}CH_2\text{—}\overset{O}{\overset{\|}{C}}\text{—CoA}$$

oxidation

$$CH_3(CH_2)_{12}\text{—}\overset{OH}{\overset{|}{C}H}\text{—}CH_2\text{—}\overset{O}{\overset{\|}{C}}\text{—CoA} + NAD^+ \longrightarrow CH_3(CH_2)_{12}\text{—}\overset{O}{\overset{\|}{C}}\text{—}CH_2\text{—}\overset{O}{\overset{\|}{C}}\text{—CoA} + NADH + H^+$$

splitting the chain

$$CH_3(CH_2)_{12}\text{—}\overset{O}{\overset{\|}{C}}\text{—}CH_2\text{—}\overset{O}{\overset{\|}{C}}\text{—CoA} + CoA \longrightarrow CH_3(CH_2)_{12}\text{—}\overset{O}{\overset{\|}{C}}\text{—CoA} + CH_3\text{—}\overset{O}{\overset{\|}{C}}\text{—CoA}$$

19.25 **a.** 16 **b.** 16/2 = 8 acetyl CoA units
c. 8 − 1 = 7 turns **d.** 8 acetyl CoA (96 ATP); 7 $FADH_2$ (14 ATP); 7 NADH (21 ATP); Activation step (−2 ATP); Total = 129 ATP

19.26 **a.** When too many acetyl CoA result, they cannot all enter the citric acid cycle. Instead, they form acetoacetyl CoA, a four-carbon compound that gives β-hydroxybutyric acid or acetone, both ketone bodies.
b. Since several of the ketone bodies are acids, the pH of the blood is lowered causing acidosis and dehydration.

19.27 Lipogenesis is the formation of fatty acids from two-carbon acetyl CoA units that result from the break down of glucose but are not needed for energy.

19.28 **a.**

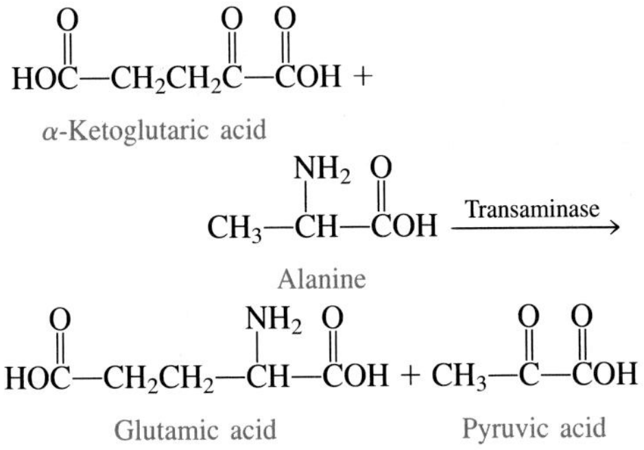

$$HO\overset{\overset{O}{\|}}{C}-CH_2CH_2\overset{\overset{O}{\|}}{C}-\overset{\overset{O}{\|}}{C}OH + \underset{\text{Alanine}}{CH_3-\overset{\overset{NH_2}{|}}{C}H-\overset{\overset{O}{\|}}{C}OH} \xrightarrow{\text{Transaminase}}$$

α-Ketoglutaric acid

$$\underset{\text{Glutamic acid}}{HO\overset{\overset{O}{\|}}{C}-CH_2CH_2-\overset{\overset{NH_2}{|}}{C}H-\overset{\overset{O}{\|}}{C}OH} + \underset{\text{Pyruvic acid}}{CH_3-\overset{\overset{O}{\|}}{C}-\overset{\overset{O}{\|}}{C}OH}$$

b.

$$HO\overset{\overset{O}{\|}}{C}-CH_2CH_2\overset{\overset{O}{\|}}{C}-\overset{\overset{O}{\|}}{C}OH + \underset{\text{Aspartic acid}}{HO\overset{\overset{O}{\|}}{C}CH_2\overset{\overset{NH_2}{|}}{C}H\overset{\overset{O}{\|}}{C}OH} \xrightarrow{\text{Transaminase}}$$

α-Ketoglutaric acid

$$\underset{\text{Glutamic acid}}{HO\overset{\overset{O}{\|}}{C}-CH_2CH_2\overset{\overset{NH_2}{|}}{C}H-\overset{\overset{O}{\|}}{C}OH} + \underset{\text{Oxaloacetic acid}}{HO\overset{\overset{O}{\|}}{C}CH_2\overset{\overset{O}{\|}}{C}-\overset{\overset{O}{\|}}{C}OH}$$

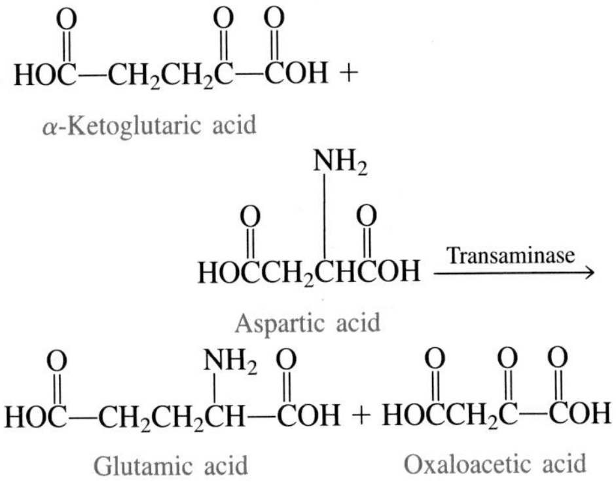

19.29 In oxidative deamination, the amine portion of an amino acid is oxidized to ammonia, leaving an acid with a keto group.

19.30 The α-ketoacids resulting from transamination and deamination can enter the citric acid cycle as three, four, or five-carbon intermediates for energy production.

19.31 Usually, we take in only enough carbohydrate to meet our energy needs. If we have more, the surplus glucose can be used to replenish glycogen stores. If those are full, glucose is converted to acetyl CoA which is used to build fatty acids and is stored as body fat.

19.32 An α-keto intermediate of the citric acid with the same number of carbon atoms can be transaminated to make it into an amino acid.

Acknowledgments

Illustration Credits

All artwork has been prepared as a collaborative effort by Precision Graphics and Rolin Graphics.

Photo Credits

Unless otherwise acknowledged, all photographs are the property of ScottForesman. Page abbreviations are as follows: (C) center, (B) bottom, (L) left, (R) right.

viii: Tom Pantages
xii: John D. Cunningham/Visuals Unlimited
xv: Dr. Earl Rees, Portland State Univ.
1: Jay Freis/The Image Bank
30: Yoav Levy/Phototake
31: Davis/Courtesy University of California
38: E. R. Degginger
51: Judi Buie/Bruce Coleman, Inc.
74: David Sutherland/Tony Stone Worldwide
79: Lawrence Berkeley Laboratory/ Courtesy University of California
90 (TL): E. R. Degginger
90 (TC): E. R. Degginger
90 (TR): E. R. Degginger
90 (B): J. Pasachoff/Visuals Unlimited
97: The National Museum of American History/Smithsonian Institution
98: The University of Texas, M. D. Anderson Cancer Center
109: Dan McCoy/Rainbow
122: E. R. Degginger
155: Jim McGarth/Rainbow
160: E. R. Degginger
170 (all): Tom Pantages
171 (all): Tom Pantages
179: Leonard Lee Rue IV/Bruce Coleman, Inc.
194: Dan McCoy/Rainbow
200: Yoav Levy/Phototake
212 (T): Dan McCoy/Rainbow
212 (BL, BR): SIU/Visuals Unlimited
214 (L): Hank Morgan/Rainbow
214 (R): Courtesy of Dr. Michael E. Phelps and Dr. John Mazziotta, UCLA
215: Yoav Levy/Phototake
221: Courtesy of Dupont Radiopharmaceuticals
224: Jessica Ehlers/Bruce Coleman, Inc.
231: Pictor/Uniphoto
245: Patrick H. Marrow/First Light/ Phototake
258: SIU/Visuals Unlimited
262: NASA
267: Roger Tully/Tony Stone Worldwide
268: Tom Pantages
275: E. R. Degginger
293: E. R. Degginger/Bruce Coleman, Inc.
301: SIU/Bruce Coleman, Inc.
307: Raymond G. Barnes/Tony Stone Worldwide
311: E. R. Degginger
312: Herbert Wagner/Phototake
313 (L): Tom Pantages
313 (R): E. R. Degginger
332 (all): Tom Pantages
349: Pictor/Uniphoto
382: Charles Thatcher/Tony Stone Worldwide
395 (all): Tom Pantages
415: Pictor/Uniphoto
427: Stephen Kline/Bruce Coleman, Inc.
450: Barry Seidman/The Stock Market
472: Hans Reinhard/Bruce Coleman, Inc.
492: Frank Siteman/Uniphoto
518: John D. Cunningham/Visuals Unlimited
534: Philip A. Harrington/The Image Bank
563: Rainbow
566: Pictor/Uniphoto
588: Stanley Flegler/Visuals Unlimited
601: E. R. Degginger
637: R. Langride/Rainbow
643: Visuals Unlimited
668: Focus On Sports
694: Fred Hossler/Visuals Unlimited

Index

Formulas and Names of Some Common Ions

Group number	Formula of ion	Name of ion	Group number	Formula of ion	Name of ion
1A	H^+ (aq)	hydrogen ion	7A	F^-	fluoride ion
	Li^+	lithium ion		Cl^-	chloride ion
	Na^+	sodium ion		Br^-	bromide ion
				I^-	iodide ion
2A	Mg^{2+}	magnesium ion			
	Ca^{2+}	calcium ion	Transition metals	Fe^{2+}	iron(II) ion
	Ba^{2+}	barium ion		Fe^{3+}	iron(III) ion
				Cu^+	copper(I) ion
3A	Al^{3+}	aluminum ion		Cu^{2+}	copper(II) ion
				Ag^+	silver ion
5A	N^{3-}	nitride ion		Zn^{2+}	zinc ion
	P^{3-}	phosphide ion			
6A	O^{2-}	oxide ion			
	S^{2-}	sulfide ion			

Formulas and Names of Some Common Polyatomic Ions

Charge	Formula of ion	Name of ion
1+	NH_4^+	ammonium ion
1−	OH^-	hydroxide ion
	NO_2^-	nitrite ion
	NO_3^-	nitrate ion
	HCO_3^-	hydrogen carbonate (bicarbonate) ion
	HSO_4^-	hydrogen sulfate (bisulfate) ion
	$C_2H_3O_2^-$	acetate ion
2−	CO_3^{2-}	carbonate ion
	SO_3^{2-}	sulfite ion
	SO_4^{2-}	sulfate ion
3−	PO_4^{3-}	phosphate ion